D1369528

Simulation
in
Social
and
Administrative
Science

SIMULATION IN SOCIAL AND ADMINISTRATIVE SCIENCE

Overviews and Case-Examples

HAROLD GUETZKOW

PHILIP KOTLER

RANDALL L. SCHULTZ

PRENTICE-HALL, INC., *Englewood Cliffs, New Jersey*

ISBN: 0-13-810382-8

Library of Congress Catalog Card No.: 74-162352

10   9   8   7   6   5   4   3   2   1

Prentice-Hall International, Inc., *London*
Prentice-Hall of Australia, Pty. Ltd., *Sydney*
Prentice-Hall of Canada, Ltd., *Toronto*
Prentice-Hall of India Private Limited, *New Delhi*
Prentice-Hall of Japan, Inc., *Tokyo*

Contributors

Irma Adelman, *Northwestern University*
Wayne L. Francis, *University of Washington*
Jeanne E. Gullahorn, *Michigan State University*
John T. Gullahorn, *Michigan State University*
Philip Kotler, *Northwestern University*
Gilbert K. Krulee, *Northwestern University*
John C. Loehlin, *The University of Texas at Austin*
Duane F. Marble, *Northwestern University*
Robert C. Meier, *University of Washington*
Thomas H. Naylor, *Duke University*
William T. Newell, *University of Washington*
Gustave J. Rath, *Northwestern University*
Randall L. Schultz, *University of Pittsburgh*
Paul Smoker, *University of Lancaster and University of British Columbia*
Edward M. Sullivan, *Northwestern University*

Contents

Foreword

One can use computers without doing simulation, and do simulation without using computers. In fact, many examples in this volume illustrate the latter possibility. Nevertheless, the computer has brought with it a great burgeoning of interest in the use of simulation as a technique for studying and understanding the behavior of complex systems.

Although each of us is acquainted with these developments in our own area of activity, we tend to be only dimly aware of the parallel developments going on in other fields within the social sciences, and of advances in conception and technique that may have value for our own endeavors. The editors of this volume have done us all a good deed by bringing together this survey of the present state of the art of simulation across a wide range of social science areas.

Like most innovations, simulation was born of both opportunity and necessity. The computer, with its unprecedented power for consuming (and disgorging) symbols provided the opportunity. The poverty of our armory of analytic techniques in the face of the complex social and psychological systems we wish to understand presented the necessity.

We must be careful about regarding simulation as merely a second-best alternative, a substitute to be brought into the game when analytic techniques fail us. As simulation matures, it acquires its own sophistication of technique—for modeling systems and for evaluating outcomes. Sensitivity analysis is just one example of a growing body of method and theory that enable us to extract more fundamental and more general information from our simulations.

The progress of simulation, however, is still largely imbedded in specific applications. It is for this reason that, if we wish to use the tool effectively, we cannot limit ourselves to knowledge of its uses in our field of research, but must search out ideas and techniques through a much broader literature. This volume, with its wealth of examples and references, gives us good access to that literature. I expect it to cause many acts of cross-pollination, and to have some sturdy hybrids among its progeny.

Herbert A. Simon

Preface

During the last decade simulation has assumed an important role as an aid in developing social and administrative science. The purpose of this book is to examine critically the progress that has been made and the work that remains to be done. In a sense, this book is a sequel to an earlier volume, *Simulation in Social Science: Readings* (Guetzkow, 1962), in that it brings together case-examples of simulations in the various diciplines. In a broader sense, though, the present volume represents a new addition to the social science, administrative science, and simulation literature by providing a handbook-like review of past accomplishments and guide to future research.

This book is comprised of a set of overviews, case-examples, and readings. Two of the overviews—the introductory chapter on developments in simulation and the chapter on methodological considerations—are designed to provide a general framework for evaluating simulation in social and administrative science. The remaining overviews present specific critical examinations of the use of simulation in different fields or disciplines. Each of these field overviews is accompanied by a case-example showing a relevant application of simulation in that field. The readings supplement the other chapters by providing a discipline-based but generalizable discussion of theory-building and comparison with simulation.

This book was written for a wide audience in social and administrative science and for decision-makers in organizations. The utility of simulation for modeling decision processes and in assisting decision-making is an important theme running through the various chapters. Although our readers will vary in their knowledge of simulation and its application in particular social and adminis-trative science areas, we hope that the multidisciplinary character of the volume will open new perspectives into research, teaching, and decision-making. Thus, the book may be used in classes where the focus is on simulation in one field or many or by individuals in organizations who have an interest in research or practical decision making.

The eclectic approach we have taken seems suited for evaluating past accomplishments and stimulating future work and thought.

The editors are grateful for the advice and support of several individuals and organizations. The original support afforded to us by the Gordon Scott Fulcher Chair in Decision-Making at Northwestern University was important in launching the project and bringing it together. Our contributors, besides writing individual overviews, shared in the development process and helped us sharpen our multidisciplinary perspective. Professor Herbert Simon, as the general editor of the series in which this book appears, counseled us in the later stages of writing, as did Mr. Chester Lucido at Prentice-Hall. We also wish to thank the individuals who allowed us to reprint their significant research articles and essays as case-examples. At each stage we hope that we have profited by this good advice and generous support.

There are always others too numerable to thank individually and so we thank them all collectively, with two exceptions. The late Edna Goss, Fulcher Chair Assistant at Northwestern, provided us with such high quality services that we are especially grateful to her. Our families deserve the highest gratitude and one of the wives, Bobbi Schultz, served beyond this supportive role as a conscientious proofreader. Although the chapters represent for the most part the views of their authors, the book itself represents our view of simulation and we alone are responsible for these macro errors of omission or commission.

Harold Guetzkow
Philip Kotler
Randall L. Schultz

Simulation
in
Social
and
Administrative
Science

PART ONE

Introduction

1

Developments in Simulation
in Social and Administrative Science

RANDALL L. SCHULTZ
EDWARD M. SULLIVAN

Simulation has brought together a new community of interest among social and administrative scientists. This interest centers on the use of simulation as a research methodology, as a technique for theory building and comparison, as a teaching and training device, and as an aid to practical decision-making in organizations. This community has looked to simulation as a method for dealing with complex processes in situations where experimentation or analytical techniques are not feasible. Efforts during the past decade have shown that simulation is a powerful tool in many fields of social and administrative science. The purpose of this book is to present the achievements and promise of simulation in a number of these disciplines.

While the advocates and users of simulation have grown rapidly, there is still much uncertainty in the larger academic and professional communities as to the nature of simulation and its place in scientific research. Simulation is not classical experimentation, nor is it classical mathematical analysis. Rather, simulation combines features of experimentation and analysis in a form that provides the user with great flexibility in modeling social and administrative systems. The purpose of this chapter is to examine the nature and use of simulation models. It is hoped that the reader will gain an overview of the advantages and disadvantages of the technique that will be beneficial in examining the simulations discussed in the rest of the book. Methodological considerations are discussed in the final part of the book.

The book covers individual behavior simulations of cognitive processes and personality, social and cultural system simulations, economic system simulations, political process simulations, international relations simulations, human geography simulations, business system simulations, marketing system simulations, and public system simulations. Although simulations have appeared in a few other social and administrative science fields, these are not explicitly discussed here. Equally exciting developments have occurred in the biological and physical sciences and in engineering but are outside the scope of this volume.

MEANINGS OF SIMULATION

The word *simulation* has various meanings. A common denotation is that a simulation is an *operating model of a real system:* It is "operating" in the sense that the simulation model can replicate, to some degree, the behavior of the real system over time. Usage, however, does not always agree with this, ranging from more general to more specific connotations. Some researchers, for example, consider almost any kind of model-building to be simulation. Others consider only models with special characteristics, such as complex, interactive, stochastic, or behavioral elements, to be simulation models. More specific connotations include the use of *simulation* to refer to so-called Monte Carlo analysis, computer models, operational gaming, or heuristic programming.

Some other viewpoints were recently collected in a survey of leading users of simulation by one of the authors for this chapter. A psychologist said that simulation (in his field) was "usually the modeling of some sample of cognitive or perceptual behavior, sometimes based on human protocols." A sociologist defined simulation as the modeling of social processes "through establishing rules of play and a goal structure for players that mirrors that found in some area of social life." In economics, it was reported that simulation often refers to "large-scale computer models used for experimentation." And a geographer described simulation as "Monte Carlo techniques for solving geographic problems."

The connotations within the same field were sometimes close. In political science, for example, simulation was defined as "an operating model of a complex system or experimentation on or in a model of a complex system; an attempt to mimic a decision-making or developmental process under laboratory conditions; and laboratory and mathematical representation of complex social processes." In the field of business, the replies were slightly more divergent. Simulation was seen as "a model in which the implications are determined by actually running it on the computer rather than solving it analytically—primarily because the latter can't be done; detailed process modeling as opposed to analytic solution; artificial duplication of some process; and (a method) for assistance in and obtaining insights about decision-making."

The results of this survey seem to indicate that if there is a convergence in meanings, it centers on the concepts of *modeling* and the application of modeling to *processes*. This commonality is supported by the literature; most simulation studies can be described as *the use of a process to model a process*. While this description omits those computer models that bear little if any internal structural resemblance to referent processes (some examples of which are reported in this volume), it can serve as a frame of reference for discussing most simulations in social and administrative science.

According to this formulation, a simulation is a special kind of model—one that is itself a process. This, of course, does not preclude that the process we call simulation may be the *operation* of some (otherwise static) model. Thus, the concept of simulation as an operating model of a real system or process is consistent with our broader perspective. There are many alternative discussions of the meaning of simulation. A classic account is the paper by Churchman (1963). Other basic references include Dawson (1962), Emshoff and Sisson (1970), Naylor *et al.* (1966), Meier, Newell, and Pazer (1969), and Barton (1970).

REASONS FOR USING SIMULATION

If simulation is taken to be a special way of modeling, its merits should be evaluated by looking first at the advantages and disadvantages of modeling in general and then at the advantages and disadvantages of simulation as opposed to other forms of modeling. Indeed, many of the advantages often claimed for simulation are in reality advantages of modeling itself rather than of simulation as a particular form of modeling. This section examines the nature, types, and usefulness of models in general, the specific differences between simulation and other forms of modeling, and the advantages and disadvantages of simulation as a special form of modeling.

The Usefulness of Models

Models, in this context, are taken to be analogues of existing or conceivable systems, resembling their referent systems in form but not necessarily in content (Kaplan, 1965). In some way they exhibit, display, or demonstrate structural relationships among elements found in the referent system. At the same time, they are abstractions and idealizations, omitting some aspects of the referent systems and duplicating only those that are of interest for the purposes at hand. There is an isomorphic (one-to-one) mapping between relevant aspects of the referent system and the model, but a homomorphic (many-to-one) mapping between the total referent system and the model.

A useful distinction can be drawn between representation and explanation and hence between models and theory. Models need only represent the referent system; explanation is the role of theory. In this scheme, theory can be considered to be a set of conceptual causal relationships developed to provide a logically acceptable chain of reasoning starting from well-defined assumptions and proceeding to deductions or conclusions which conform to observation of the referent system. While many theories may, in fact, be models, not all are, for as Kaplan suggests (1965) theories need not actually exhibit the structure they assert the referent system to possess. On the other hand, models that do not purport to explain are not theories, though usually models presuppose or embody theory, even if only in an implicit way. Models can also be used to predict phenomena without necessarily explaining them.

The use of models in scientific research is well established; they have long been considered the "central necessity of scientific procedure" (Rosenblueth and Wiener, 1945). Models can increase understanding of the referent systems, aid in development of theory, and serve as a framework for experimentation. They can aid in understanding by simplification (ignoring irrelevant aspects of the referent system), idealization (attributing "pure" or ideal properties to the referent system), organizing data, structuring experience, objectifying theory, sensitizing perception, giving quantitative and qualitative predictions, and in other ways (Morris, 1967). Models can help theory development by raising questions, demonstrating gaps, helping discriminate between the important and the unimportant, generating testable hypotheses, and serving as vehicles for the communication or comparison of theories. Finally, they are especially useful for analyzing consequences of changes in the referent system where controlled manipulation of the system itself is impossible, or at least impractical or undesirable due to time, cost, inaccessibility, political or moral considerations, or other reasons. Wherever scientists or engineers have been unable to study a real system directly they have resorted to a model of it, whether symbolic or physical. For many of the same reasons, models are also useful as teaching or training devices, as will be discussed in more detail later in this chapter.

The disadvantages of modeling in general, including simulation, stem from the model's artificiality, abstraction or simplification, and idealization, and the consequent difficulties and dangers in making inferential leaps from a model to the real world. Questions of validity and inference in the model-building process are very important on both theoretical and pragmatic grounds. As a representation of a real system, the model's "representativeness" of that system is a crucial issue. The usefulness of models is similarly linked with the confidence with which the model-builder can make inferences about the referent system. A discussion of these issues with respect to simulation models follows.

Classifying Models

There are innumerable ways of classifying models. On a dimension indicating the relation between the model and the referent system, we may distinguish between models existing at the conceptual level as a set of symbolic relationships and models existing physically either as a set of recorded symbols or in other physical form. Nonsymbolic models can represent the properties of the referent system by the same properties with a change in scale (iconic models, such as a model airplane) or by other properties (analogue models) (Ackoff, 1962). An example of the last would be the use of electric currents or hydraulic flows to represent money flows, as in an analogue economic simulation.

Symbolic models are of special interest in social and administrative science and may take various forms, such as verbal, analytic (in the mathematical sense), or numerical. In addition, flow chart and similar pictorial models are becoming increasingly popular. Verbal models, of course, are the most time honored and need no further explanation; along with pictorial models they are essentially qualitative. Analytic and numerical models are essentially quantitative. While analytic models deal with mathematical *functions* or relationships between whole classes of quantities or numbers, numerical models deal with relationships between quite specific *numbers*—that is, with specific cases rather than general cases. However, numerical models are often used for inferring or checking general solutions obtained from analytic models. As an example of the distinction between the two, whereas economic theory would propose a general analytic model of the behavior of a firm, the set of books maintained by the firm's accountants would constitute a numerical model of the firm.

How Simulation Differs from Other Modeling

As has already been indicated, there is no unique meaning currently attached to the word *simulation.* Testing a model of an airplane or missile in a wind tunnel is simulation, as is testing a laboratory model of an industrial chemical process. Within social and administrative science the term might refer, at one extreme, to the computation of a numerical solution from a deterministic mathematical model such as a linear programming model. In this case the actual computations are usually done by computer simply as a means of quickly completing a tedious task, but the use of the computer is not an essential element. At the other extreme, simulation can refer to interactions among people playing specific roles, with no quantification or computerization at all.

The deterministic-nondeterministic distinction refers to the degree to which the exact outcome of a process is not completely determined by the initial specification of the model but depends on "chance" events. A formula or specified sequence of steps that always gives exactly the same result each time it is used with the same input data is deterministic—an example of what is called an algorithm. On the other hand, if the final result is in part dependent on chance considerations, such as tossing a coin or rolling a die, then the model is nondeterministic. We prefer to exclude from the meaning of simulation those deterministic mathematical models like linear programming that bear no internal structural resemblance to the referent systems—that is, models that do not resemble in form the changes that take place in the referent systems. There is really nothing that substantially distinguishes such models from other forms of analytic or numerical modeling, except that in some fields it has become popular to apply the term *simulation* to them. For this reason a few are reported in various chapters of this volume. Rather, we restrict the term simulation to *processes* that in some way resemble in *form* the processes that they model. In other words, simulation is both a model and a process: as noted above, we would define it as *the modeling of a process by a process.*

The very nature of process implies extension in time and a succession of states of the system of interest. Each state in the referent system can be represented by a (static) model, and a more inclusive model can be constructed by linking together these static models of the successive states to form a state history model. At the simplest, one can have two such state models—one of the initial and one of the final state of the system—the second produced from the first (or vice versa), but with the linkup between the two in no way resembling in *form* the actual processes involved in the referent system. We would still be reluctant to qualify this as a simulation.

If, however, there is a model of at least one *intermediate* state, and if each state model is produced in turn from the preceding one, then the production of such models is a process that begins to resemble in form the processes involved in the referent system, and the *process* of producing the successive state models may be regarded as a simulation.

We may go one step further. Two states may be regarded as being connected by a black box labeled "change." Simulation, in our view, is an attempt to model what goes on inside of that black box. It is an attempt to model not only the states but also the steps that lead from one state to the next. To paraphrase Ackoff (1962, p. 346), simulation not only represents a process but imitates it. Thus, although simulation is a model itself, it contains within it three kinds of submodels. The first is comprised of the static models of the initial and final states. The second is a model of the changes leading from one to the other, which at its simplest may merely consist of discrete models of intermediate states (like a motion picture) or at some complex level may consist of continuous changes, as in some all-man simulations. The third is an iconic model of time itself in the sense that duration of activities or sequence of events or both are directly modeled in real time by duration or sequence in the simulation. With these three kinds of submodels, simulation is inherently a complex model.

It is precisely the greater degree of fidelity in modeling changes themselves, reproducing intermediate states within an iconic model of time rather than modeling only an end state, or beginning and end states, that distinguishes simulation from other models. Modeling in real time has led to the popular definition of a simulation as an *operating* model. But when that definition is interpreted as the operation *of* a model or the doing of something *to* a model (Evans, Wallace, and Sutherland, 1967, p. 6), it suffers because attention is directed to one submodel of the simulation, a static one that is usually the initial state model, and away from the idea that the simulation itself is a model. It is rather like defining an orchestral performance as the playing of a score. Or like saying that the score itself is music but the performance is not; that the playwright's script is drama but the enactment of it is not.

The essence of simulation, then, is that it represents real system changes by changes of the same kind (in some sense) in the model. Where control is exercised over changes in order to learn about the relationship between the changes and the resulting state of the system undergoing change, one has an experimental situation. In verbal or symbolic models change is represented only symbolically, not by real changes in the models themselves. Although one can "experiment" by actually making changes to such models, this type of change represents changes *to* the system and does not directly represent or imitate changes taking place *in* the referent system, and such a manipulation is certainly not regarded as an experiment in the usual scientific sense. With simulation the manipulations can produce changes of the same kind that take place in the real system. Experimentation can take place in a vicarious way. Thus, simulation can serve as an experiment; it is not a model to be experimented *on* but actually *is* the experiment, albeit a vicarious one using a surrogate submodel of the real system.

Advantages of Simulation

The unique advantages and disadvantages of simulation over other forms of modeling stem from its characteristic attribute of permitting greater fidelity in modeling processes than that given by other forms of modeling. The major advantages are discussed below, and then the major disadvantages are presented in the following section.

First, the unique advantage of simulation in providing a vehicle for experimentation has already been indicated. In simulation as experimentation, pure-machine or all-computer simulations have several further advantages over other forms of simulation or experimentation: The modeler has complete control over all variables, and he can replicate conditions for various runs to a degree impossible in any other form of experimentation; large quantities of artificial data (output) can quickly and efficiently be generated, and this output can be presented in the form of a time series which can immediately be tested (Fishman and Kiviat, 1967).

Second, by including not only system states but also processes linking them, simulation allows for a more complete model of the referent system (by definition simulation models include things omitted or less adequately handled by other forms of modeling). In principle, a greater degree of complexity can be accommodated in a simulation model, and simulation is often thought of in terms of modeling of complex systems. The introduction of high-speed computers, with their capability for replicating processes by manipulating at incredible speeds the symbols representing the different elements or properties of the system states, has made simulation a powerful tool for modeling complex processes. Weiss has defined a complex situation as one in which many interrelated phenomena are studied simultaneously and the situation as a whole is studied rather than a particular element within the situation (Weiss, 1966). Simulation's capabilities for handling complexities make possible precisely this sort of holistic approach.

In contrast to computer simulations, simulations involving human participants can transcend mathematical, verbal, and other symbolic (including computer) models in their ability to include many more variables—environmental, physical, cognitive, emotional, social, etc. Their nonprogrammed nature (i.e., using real "chance" events) allows them to go beyond previous assumptions in the evolution of the system process and model form, as well as in the questions and possibilities they raise. Such simulations can be designed to be understood by a wide audience, as Smoker (1969) suggests, and possibly even help identify and explore alternatives for such an audience. To the participants they can furnish an experience and insight that will be hard to gain in any other way short of direct participation in the referent system, largely by virtue of the continuous and direct feedback regarding model response to their own actions and interventions.

Even if the processes under study are not complex, simulation can give a better picture of them because of its greater likeness to them than other models, and this is useful not only for instructional purposes but also for research. Where the causal relationships are not well understood (that is, where theory is not well developed), sometimes the best one can do is attempt to imitate the change process itself in the hope of learning more about such relationships. Thus, the model becomes an aid to theory development. Guetzkow (1966), for example, has given this as a chief reason for using simulation in the study of international relations. With such a model one can also try out a theory through manipulation of the model processes to see if results conform to real-world observations.

With the construction of more complete models the simulation approach can be eclectic, permitting combinations of appropriate theories and methodologies. The high degree of explication required in the model can often help to ferret out logical gaps and contradictions, ambiguities, vagueness, and redundancy in the theory on which it may be based. Competing theories may be compared and evaluated within the structure of a simulation, whereas with most other forms of modeling this will be very difficult if not impossible. Simulation also allows different methodologies to be combined where each is relevant to a different part of the system process.

Third, although simulation is not the only modeling technique that explicitly models time, it is the only one that permits manipulation of an *iconic* model of time. This is especially advantageous where models are desired to represent dynamic processes without abstraction of the time element or mere symbolic representation of it. In the special discrete simulation languages for computers, some sort of timing or "clock" routine always plays a central role. Lagged effects, nonlinearities, and other refinements that may be extremely difficult or impossible to handle with other forms of modeling can directly be incorporated into simulations. Interactive processes over time, event sequences, and transitional properties can explicitly be studied. Furthermore, time can be expanded or contracted to any degree desired. One reservation must be made when speaking of digital computer simulations, however: Simultaneity in the real system can be modeled only by a pseudosimultaneity in the computer, which must proceed in a step-by-step fashion. Analogue computers do not have this limitation.

Finally, real random or pseudorandom variables can directly be introduced into the model, because the model is a succession of changes over time. Nondeterministic simulations permit the study of stochastic or random processes that cannot otherwise be studied, either because of the lack of development of appropriate analytic tools for such complex situations as can be modeled by a simulation or because of the lack of mathematical sophistication in the use of such tools on the part of the modeler. Probabilistic techniques are useful where statistical predictability can be presumed to be present but individual unpredictability is encountered, for example in the computer modeling of human decision processes. Chance can also be introduced by using human decision-makers as components of the model. When either technique is used, however, there remains the problem of determining what will constitute an adequate sample in a particular case in order to make valid inferences. Nevertheless, the potential for experimentation is apparent.

Chance is commonly introduced in nondeterministic computerized simulations by picking numbers from a random number table or by using computerized pseudorandom number generators. The pseudorandom numbers, in turn, are used to produce pseudorandom variates according to any of a wide variety of probability distributions that may be appropriate to the case at hand. The prefix "pseudo" is used because the numbers are, in fact, generated by an algorithm (i.e., deterministically), the essential requirement being that when large quantities of such pseudorandom numbers are generated they will meet the requirements of statistical tests for randomness. Similarly, the random variates produced by, say, a stochastic generator for normally distributed numbers will produce a set of numbers that will pass the statistical tests identifying the set as having come from a normal distribution, even though the computer routine generating the numbers will always generate the same sequence of numbers given the same starting point. This would be an example of a nondeterministic (i.e., stochastic) programmed operation. It can be made nonprogrammed (Guetzkow, chap. 13) simply by using an unpredictable starting point, in which case the algorithm producing the pseudorandom numbers will generate each time a different set of numbers, all still meeting the same statistical tests. Where a "seed number" is used as the starting point, all one has to do is make this seed number dependent on some truly chance variable, such as the exact second of the hour when the program happens to be run on the computer. From a practical standpoint, it is convenient to control the probabilistic variables by always using the same seed number in test cases until the program is debugged and ready for use. Another example of a nonprogrammed operation would be the use of human decision-makers incorporated into a simulation using their own decision rules, not completely specified in advance by the designers of the simulation.

The use of pseudorandom variates is sometimes erroneously called the Monte Carlo technique (Dawson, 1962, p. 11), although the term *Monte Carlo* refers to a very specific use of such variates that has nothing to do with simulation. The Monte Carlo technique is a numerical method of solving deterministic analytic problems by taking advantage of the fact that an integral can be regarded as representing a probability distribution. The region to be integrated can be approximated by enclosing it in a larger region of known size and generating a large number of random points falling within this larger region. The proportion of these random points that also fall within the smaller enclosed region representing the desired integral gives the proportionate size of the smaller region relative to the larger one, at least approximately. Since the size of the larger region is known, the approximate value of the desired integral can directly be computed.

The straightforward use of random variables in a simulation simply as a direct representation of random processes in the real world is thus quite different from the Monte Carlo technique. Still, the misuse of the term *Monte Carlo* to refer to such an inclusion of stochastic variables in a simulation seems sufficiently widespread to warrant its acceptance. In fact, some even equate computer simulation with the use of such stochastic variables, although simulation is a much broader concept.

One difficulty in the algorithmic generation of pseudorandom numbers should be mentioned. In many generators each new pseudorandom number is generated in some way from the preceding one in a deterministic process. This means that whenever a number previously used is again generated a cycle has developed which will keep repeating itself. Considerable attention has been devoted to developing generators with long cycles that will not interfere with simulation results (Emshoff and Sisson, 1970, p. 176). Pseudorandom number generators are discussed in detail by Naylor *et al.* (1966, chap. 4).

Disadvantages of Simulation

There are at least three major disadvantages resulting from simulation's greater fidelity in modeling processes: decreased generality of results, greater effort required, and dangers of forgetting the limitations of the model.

First, the greater the fidelity the less the abstraction, and hence the less the generality of results of a particular simulation run. As a process, a simulation exists only during a particular period of time: It is a case-example of one particular replication or set of replications of the modeled process. Simulations involving people are dependent on the unique personal characteristics of the people involved as well as on certain environmental influences operating at the time that cannot be controlled by the modeler. Computer simulations are numerical rather than analytic models, and the process being modeled must ultimately be reduced, at least inside the machine, to a set of very specific numbers. Thus simulation gives only individual case results, although a number of such case results can be collected for analysis by running the simulation a number of times. Accordingly, Gordon (1969) suggests that "the use of simulation should be planned as a series of experiments." Naylor has long argued that simulations are experiments and hence experimental design is a fundamental issue (see chap. 12). The need for replications in order to develop more general results is, however, a disadvantage of simulation in contrast to other forms of modeling.

A corollary of simulation's inability to give general solutions is its general inability to yield theoretically optimal solutions in normative models. But there are ways of compensating for this deficiency. One way is to couple to the simulation an optimum-seeking module, such as linear programming, which utilizes the findings of the simulation to produce an optimal solution. Another way is to use a series of simulations iteratively (Geisler, 1962, p. 244), building on previous results by changing the variables so as to produce increasingly better solutions until an optimal one is found. Emshoff and Sisson (1970, chap. 9) discuss direct search processes at some length. The IBM 360 Scientific Sub-routine Package contains two such search subroutines. Various heuristic devices can also be built into a simulation, such as a model of human search and decision-making, to develop solutions better than merely "acceptable" ones. Several examples of such simulations are given in this book.

A second disadvantage resulting from simulation's greater fidelity to the process being modeled is that because simulation models tend to be more complex they usually require greater effort in construction. McKenney (1967, p. 114) reports that whereas computer costs in running a business game at the Harvard Business School were quite small, game development was another story: "the setup costs of programming and developing an appropriate simulation was on the order of tens of thousands of dollars and man-years of faculty time." Abelson (1968) points out that in the field of social psychology the simulator has to specify much more detail than originally planned. In most cases, simulations —especially computer simulations—require that elements be specified in fine detail and with great precision. Validation likewise requires close attention to operationalization of concepts and measurement of variables in the referent systems. Although this applies to other forms of modeling as well, it is true to a greater degree in simulations where explicit attention must be paid to the mechanisms of change linking successive system states.

Whether the greater attention required is an advantage or a disadvantage, however, depends on one's purposes and point of view. Often the stimulus for careful analysis and the discipline imposed on the modeler's activity are considered as alone worth the entire effort, in addition to anything that may be learned from the simulation itself. Some (e.g., Brody, 1963) have regarded this as the greatest potential of simulation. As already noted, part of the usefulness of models lies in the fact that the construction process itself can be a powerful aid to understanding and a spur to theory development, and some feel that simulation's most significant contributions in particular research fields have been made in this very way. (See "Achievements and Potentials of Simulation" later in this chapter.)

The problems and costs of data collection and parameter estimation, large enough in most classical models, may grow exponentially in complex simulations. Yet simulation in many cases will justify its cost, expecially where it is the only or major promising methodological avenue to scientific research.

In terms of both effort and money, then, simulation is relatively expensive compared with other forms of modeling. Mainly for this reason it has often been thought of as a "last resort" technique, although it can be a "first resort" technique in modeling many real-world complexities. The relevant question becomes how much complexity must be modeled for the purpose at hand. On this point, Cyert (1966) has cautioned that the degree of complexity allowed by simulation can serve to encourage the construction of models so complex that the purpose of modeling is defeated. This is the trap of "pictorial realism," or regarding the model as more a picture of reality than a device for learning more about reality (Kaplan, 1965).

In fact, utility of a simulation is sometimes confused with validity. The one refers to its usefulness for some purpose, whereas the other refers to its degree of correspondence with the real world. Since utility usually requires some degree of validity, some authors speak of a model as having been "validated" by some use to which it has been put. Validity of a model, however, is not an end in itself but merely a means of enhancing the utility of the model—and usually only up to a point. Both validity and utility are commonly matters of degree. A model valid enough and hence useful for one purpose may not be valid enough and hence not useful for another; on the other hand, inordinate preoccupation with validity may even detract from the usefulness of the model, as Cyert suggests. Another sense of the term "validity" is put forth by Raser, Campbell, and Chadwick (n.d.), who would apply it to the processes by which inferences are drawn from a model rather than to the model itself. Yet utility and validity remain distinct concepts. While validity is the ultimate test of a theory, the ultimate test of a model is its utility. Decisions about model development should be made on the basis of a utility-cost tradeoff.

The validation of a simulation as an accurate representation of the referent system is one of the thorniest problems in the field (Hermann, 1967; Naylor and Finger, 1967) and is further discussed later in this chapter under "Problems and Prospects" and in chapters 12 and 13. Although ability to predict accurately is usually considered the ultimate test of a model, Cyert (1966) has pointed out that in simulation many parameters often have to be set arbitrarily, and even with a large number of runs to vary them and obtain model results fitting real-world data, there is still no assurance that one has an accurate model rather than merely a fortuitously happy choice of parameter values. In general, however, the problems of validation are more a result of the methodological state of the art than an inherent disadvantage of simulation.

A third disadvantage due to simulation's greater fidelity in modeling processes is that there is greater danger of forgetting the limitations of the model, especially in the case of decision-makers, who tend to be operationally oriented (Koopman and Berger, 1967). A variation of this danger is what Kaplan (1965) terms *map-reading,* or attaching too literal an interpretation to all the details of a model, regarding inessential features as essential. While these dangers exist with any model, they can be somewhat greater in the case of simulation models because of the more apparent realism of the models.

Specific kinds of simulations suffer from several further disadvantages. Computer models, for example, require special knowledge of computer capabilities, languages (to be discussed shortly), and programming talents, not to mention the availability of the machines themselves. Once crystallized, they may require large-scale revision when slight conceptual changes are made in the model. Where a number of estimated probabilities are used in the absence of known actual probabilities, there can be a large buildup of errors giving very erroneous results misleading by their apparent precision (Koopman and Berger, 1967). One remedy lies in sensitivity analysis of the probabilistic variables, but again there is a proliferation of the number of runs needed. Furthermore, the modeler cannot observe a computer simulation directly, but only a state history of it, a fact that has led some to define computer simulation as the construction of a state history of a model (Evans, Wallace, and Sutherland, 1967). Lastly, the computer's ability to generate data has tended to run ahead of capabilities for adequate statistical analysis of the type of output data produced (Teichroew, 1965), but this is another state-of-the-art deficiency rather than an inherent disadvantage.

Summary of Advantages
and Disadvantages of Simulation

Simulation shares in the advantages and disadvantages of modeling in general. In the preceding sections we have discussed some particular advantages and disadvantages of simulation as a special kind of model. To summarize, the advantages of simulation are:

1. It provides a method for vicarious experimentation.
2. It allows more complete models with greater degrees of complexity—a holistic approach, and can be eclectic, permitting combinations of theories and methodologies.
3. It permits manipulation of an iconic model of time.
4. It allows random or pseudorandom variables to be introduced directly into the model.

The disadvantages of simulation are:

1. It provides specific case results, requiring replications to produce more general results.
2. It usually requires more effort in constructing the model.
3. It can lead to more apparent realism and consequently greater danger of forgetting the limitations of the model.

CLASSIFICATION OF SIMULATIONS
AND COMPUTER SIMULATION LANGUAGES

This section deals with two important topics: the classification of simulations and computer simulation languages. First, a metaclassification system for simulations is presented. Second, general computer simulation languages and special purpose simulation languages are discussed.

The Classification of Simulations

There are several good reasons for using classification schemes for simulations: They are useful for exposition and technical communication; they help bring out differences between simulations and give better understanding of the range and trend of simulation studies; and they are useful for describing the features of particular simulations by relating them to specific types.

Many contributing authors to this volume have adopted some sort of classification scheme in presenting the simulations covered in their reviews. Adelman (chap. 5), for example, broadly distinguishes between simulation studies used for prediction and those used for policy formulation. Loehlin (chap. 3), Francis (chap. 6), and Naylor (chap. 12) explicitly concentrate only on digital computer simulations as a subclass of simulations in general. Kotler and Schultz (chap. 10) class the simulations they describe into those dealing with marketing systems, those dealing with specific decisions, and marketing games, with subclassifications into different kinds of systems in the first class and into different decision areas in the second. The Gullahorns (chap. 4) dichotomize the simulations they describe into those dealing with total social systems and micro-behavioral simulations. Others use still different classification systems.

Beyond these examples, there are many other ways of classifying simulations, since they can be classed according to any of their attributes. Table 1-1 suggests a number of possible classification dimensions within the framework of a metaclassification system based on the genesis of the model, the system modeled, the model itself, and the relationship between the model and the world. It makes no pretense at comprehensiveness but merely suggests a number of quite different dimensions along which classification can be made.

While the table is largely self-explanatory, we can make a few footnotes. Under *genesis of the model,* classification schemes could be based on the discipline involved (the scheme by which this book is divided into chapters); when, where, and by whom the model was developed; the method employed in model construction (synthetic or analytic, cf. Morgenthaler, 1961); and a variety of purposes.

In the second category, *the system modeled,* we introduce a system-level classification that could be useful for some purposes: macroanalytic, microanalytic, and microbehavioral. The distinction is whether the simulation focuses on the entities or components of the system in aggregate as "typical" units or as separate units. In economics, for example, a simulation of a particular national industry or of the nation as a whole would be macroanalytic; one that breaks the economy down into sectors of firms and/or individuals where each is taken as "typical" of its class would be microanalytic; and one that starts with a set of separate, behaviorally unique entities, concentrating on social-psychological laws of behavior, and then derives aggregate behavior from their interaction would be microbehavioral.

In the third category, criteria based on *the model itself*, we include three subclasses first distinguished by Tocher (1966b): discrete-event networks, continuous-variable models, and recurrence-relation models, which in itself is a

TABLE 1-1 Some Possible Bases for Classifying Simulations

Genesis of the Model	*The Model Itself*
Discipline involved	Elements used in constructing the model
Psychology	Gaming
Sociology	All-man simulations
Others	Man-machine simulations
When the model was developed	All-machine simulations
Where developed	Analogue computer
By whom developed	Digital computer
Country	Type of language used
Institution	General purpose (procedure oriented)
Method of development	Special purpose simulation language
Formal or informal	Discrete-event language
Synthetic or analytic	Continuous-variable language
Purpose of the model	List-processing language
Research, pure or applied	Degree to which randomness enters
Theory construction or development	Deterministic
Theory comparison	Nondeterministic
Theory verification or falsification	Programmed
Experimentation	Nonprogrammed
Decision-making or design	Handling of time
Develop alternatives	Faster or slower than life
Evaluate alternatives	Continuous time
Optimizing	Discrete-interval time
Nonoptimizing	Fixed-interval time
Verification or control	Variable-interval time
Pedagogy	Structure of the model
Teaching	Discrete-event network
Training	Continuous-variable model
	Recurrence-relation model
The System Modeled	Relationship to other models
	Compatibility
System type	Modularity
Physical	Reliability
Behavioral	Replicability
Financial	
Others	*Relationship between the Model and the World*
Problem area	
System context	Degree of abstraction involved
System form	Iconic
Type of entities involved	Analogue
System level	Symbolic
Aggregate or detailed	Degree of analysis of model results
Macroanalytic, microanalytic, micro-	Validity
behavioral	Of the model
	Of the inferential processes
	Utility of the model
	Use actually made of the model

Note: Although the term *model* generally refers to the simulation model itself, it can also in some cases refer to the submodels of the simulation.

partial metaclassification scheme for various types of languages employed in simulations.

Finally, classification criteria based on *the relationship between the model and the world* could include degree of abstraction involved, as in Ackoff's (1962) classification of models into iconic, analogue, and symbolic; degree to which the model faithfully represents the modeled system (validity); and utility of the model and uses to which the model has actually been put or degree to which it has actually been used.

Most of the individual classification criteria indicated above serve as the bases for particular classification schemes adopted either explicitly or implicitly at different points in this book, including the present chapter, which employs different schemes in different contexts. There is at present no agreed-upon typology, nor is it clear that this is any loss, so long as communication purposes are served. The utility of any particular scheme for purposes at hand seems sufficient to justify its use.

One of the most useful distinctions, simply because it occurs so often, remains that between simulations that employ human participants and those that do not. By "those that do not" is almost invariably meant simulations done entirely on a digital computer when speaking of the social or administrative sciences, though analogue computers and various other machines are commonly used in other fields.

Computer Simulation Languages

A comparison of the contents of the present volume with those of Guetzkow's earlier reader on simulation in social science (1962) reveals an increasing ascendancy of computerized simulations over all-man or man-computer simulations. The use of computer simulations requires the use of computer simulation languages. These languages fall into three general classes: the general purpose algorithmic or procedure languages such as FORTRAN and ALGOL; the special simulation languages such as GPSS, SIMSCRIPT, and the continuous simulation languages; and the list-processing languages such as IPL-V and LISP.

Many introductory volumes on computer simulation discuss computer languages, including Barton (1970), Chorafas (1965), Emshoff and Sisson (1970), Evans, Wallace, and Sutherland (1967), Gordon (1969), Martin (1968), Meier, Newell, and Pazer (1969), Mize and Cox (1968), McLeod (1968), McMillan and Gonzales (1965), Naylor et al. (1966), Schmidt and Taylor (1970), Tocher (1963), and Wyman (1970). Most of these books tend to concentrate mainly on simulations involving the first two classes of languages, with a somewhat greater emphasis on the second class. There are also a number of books dealing strictly with engineering types of simulation (e.g., McLeod, 1968; Shigley, 1967), which are beyond the scope of the present volume. We shall briefly discuss each of the three classes of computer languages in turn. For further details and descriptions of some of the languages the reader is referred to the references indicated, especially to Emshoff and Sisson (1970), Gordon (1969), Naylor *et al.* (1966), and Wyman (1970).

General Purpose Algorithmic Languages

Languages in this first class are the most common and are useful for many purposes. They include FORTRAN, ALGOL, COBOL, and PL/1. FORTRAN is the most popular of the languages. ALGOL is internationally developed and machine independent. COBOL is a widely used business language. PL/1 is a multipurpose language implemented on the IBM System/360. While these languages lack a specific structure designed for simulation, they are sufficiently flexible and available to allow simulations that may not otherwise be possible and to encourage replication and adaptation of simulations by different researchers at different installations. In addition, depending on the type of problem and how the various languages are implemented for particular computers, the general purpose languages may sometimes provide greater program efficiency in terms of actual machine running time than the more specialized languages.

General purpose languages may be supplemented by special local packages of routines or procedures useful for simulation. An example is Northwestern University's SPURT (A *S*imulation *P*ackage for *U*niversity *R*esearch and *T*eaching), a set of FORTRAN subroutines providing eleven psuedorandom variate generators for various probability distributions, seven statistical routines, three output routines for matrix and graphical output, a clock routine for event scheduling and sequencing, eight list-processing routines, and analogue simulation routines. SPURT is modeled after the popular special purpose simulation languages and was originally designed as an interim device before such languages became available to Northwestern users.

Of the sixty-three corporate models reported in a recent survey by Gershefski (1970), 65 percent were developed using FORTRAN, 20 percent using COBOL, and 4 percent using PL/1.

Special Purpose Simulation Languages

Two basic divisions of this class of languages are the discrete-event languages and the continuous-variable languages. Discrete-event languages are oriented toward physical systems and operations research applications. They emphasize entities, activities, attributes, events or transactions, and some sort of "clock" or timing device for scheduling events or activities and advancing time between them. Various pseudorandom variate generators are included for stochastic simulations, plus statistical routines. These languages advance time in discrete intervals between events, which may be exogenously scheduled by the programmer or endogenously determined and dynamically scheduled during program execution. Such languages still leave a choice between variable-interval scheduling (advancing time from event to event) and unit-time or fixed-interval scheduling (advancing time in fixed intervals and checking to see if events are scheduled to occur, with all events scheduled to occur in a time interval being lumped together at an end point of the interval). Tocher (1966b) reports that there are about a dozen languages of the first kind enjoying international use in addition to a number of local ones. He notes that discrete-event languages were the first and probably the most successful of the problem-oriented languages, since there were originally developed by the users themselves for their own purposes. These languages greatly simplify programming and debugging of programs for the types of problems for which they were designed. Programming and debugging time can be reduced to a tenth of that required for the general purposes languages, and the language itself can be a help in problem formulation (Emshoff and Sisson, 1970, p. 117).

GPSS is a language that has become popular in the United States because it is simply the automation of a flow chart, and thus much of the programming is already done when a flow chart has been constructed. SIMSCRIPT (Markowitz, Hansner, and Karr, 1964; Wyman, 1970) is also a popular American language; it is more flexible but harder to use. A new version, SIMSCRIPT II (Kiviat, Villanueva, and Markowitz, 1968), has also become available. GASP is not a complete language but rather a set of FORTRAN subroutines which can be used on any computer that accepts FORTRAN. Other American discrete-event languages include MILITRAN, SIMPAC, and SIMULA, the last being an extension of ALGOL. SIMPAC, not widely used, is a fixed-interval language.

European discrete-event languages include ESP, CSL, and MONTECODE. While the American languages tend, in general, to emphasize entities, with events that occur to them, the European languages tend to emphasize activities, which commence and terminate with events. The former are referred to as material-based or particle-oriented languages and the latter as machine-based or activity-oriented languages (Tocher, 1965, 1966a; Gordon, 1969; Emshoff and Sisson, 1970). ESP is based on ALGOL, and CSL has been implemented on some IBM and Honeywell computers.

Reviews and comparisons of simulation languages are given by Krasnow and Merikallio (1964), Teichroew and Lubin (1966), and Tocher (1965). The first deals particularly with SIMSCRIPT, CSL, GPSS, SIMPAC, and DYNAMO; the second with SIMSCRIPT, CSL, GASP, GPSS, and a number of others (providing a good bibliography), while the last leans more heavily toward the European languages. Teichroew and Lubin also give references for special languages for simulating continuous-change models on hybrid analogue-digital computers. Buxton (1968) and Krasnow (1967) also treat simulation languages.

The second subclass of special simulation languages are the continuous-variable languages. These are designed to focus on continuous-valued variables and the rates at which such variables change, the underlying mathematical concept being differential equations, or difference equations, their discrete analogue, since digital computers must use numerical rather than analytic methods. One of the most notable of such languages is DYNAMO (Forrester, 1961), which applies a difference equation model to the flow of men, money, material, information, and capital goods in an industrial firm. While it represented a pioneering effort, it has been criticized as using crude methods of solution, ignoring much work of numerical analysts in obtaining more accurate solutions and avoiding error accumulation (Tocher, 1966b). Naylor and his associates (1966, p. 300) suggest the possibilities of DYNAMO in modeling nonindustrial systems, such as physical, social, or biological systems. Other continuous simulation languages include CSMP, CSML, MIMIC, SIMIC, CSSL, DSL/ 90, and MIDAS (Chu, 1969; Clancy and Fineberg, 1965; Emshoff and Sisson, 1970; Gordon, 1969), MIDAS and CSML apparently being the more popular ones. In general, these languages permit a digital computer to simulate an analogue computer, so that analogue computer techniques can be used in turn in simulating continuous-change systems. They are designed for models formulated as nonlinear integral-differential equations and use more advanced integration rules than DYNAMO. Analogue computers themselves are not widely used for simulation in social and administrative science. The journal *Simulation* is largely devoted to analogue computer applications in other fields.

Finally, SIMULATE is a program designed for econometric applications. It "begins by analyzing an econometric model in terms of recursive blocks of equations and then proceeds to solve any linear systems contained in the model by matrix inversion and the nonlinear systems by iterative methods" (Naylor *et al.,* 1966, p. 305).

Despite the widespread use of special purpose simulation languages in areas such as operations research and management science, they seem to have found little acceptance in some areas of social science, due probably in part to their more or less specific "world views" and in part perhaps to ignorance. Abelson, in his article on simulation in the *Handbook of Social Psychology* (1968), dismisses them with the following comment: "It is conceivable that some of these languages have something to offer social scientists, but to the writer's knowledge this possibility has not been adequately explored." One of the goals of the present volume is to acquaint researchers with work that is being done in other areas in the hope that fruitful explorations of this type will result. The use of continuous languages in particular has many interesting possibilities.

List-Processing Languages

Focusing on research in social psychology, Abelson thinks rather in terms of using general purpose languages (our first class, described above) or list-processing languages, the third of the three main classes of languages used for simulation in social and administrative science. Languages of this third type are characterized by list structures and are especially useful for symbol manipulation when it is desirable to construct and modify (add to, delete from, or rearrange) lists of symbols and to scan selectively such lists in particular patterns. Since they are not oriented toward flow or scheduling-type problems such as the discrete simulation languages tend to be, they do not have the timing routines, statistical routines, and similar features of such languages. Instead, list-processing languages are used for constructing tree structures and chains of symbols (words) for the study of problem solving, natural language processing, and cognitive processes. The use of such languages for psychological research is described by Abelson (1968) and by Krulee (chap. 2).

Examples of list-processing languages are IPL-V, LISP, SLIP, and COMIT, which are discussed and compared by Bobrow and Raphael (1967). Foster (1967) discusses list processing in general, and Newell (1964) covers IPL-V. The string-processing language SNOBOL is described by DeSautels and Smith (1967) and Griswold, Poage, and Polonsky (1969). PL/1 also contains list-processing capabilities.

Finally, in addition to the three main classes of languages, there is always the possibility of using an assembly language for the particular computer involved. The assembly language is the next closest thing to "machine language," one major difference being the replacement of numerical codes of the machine language by mnemonic symbols. Such languages are machine-specific and their utilization requires a good familiarity with the particular computer.

Regardless of the language in which a program is originally written, it is ultimately compiled into assembly language at some point in the process before it can be run, although this is usually effected by a "compiler program" that performs the actual translation. Because compilers are designed to handle general classes of problems, they may not always produce the most efficient program (in terms of machine running time) to accomplish the desired purpose in a particular case, which can result in a lot of wasted machine time if the program is to be run a great number of times. For such programs an alternative would be to repro- gram inefficient segments directly in assembly language (although this requires a considerable amount of specialized knowledge, especially for debugging). The major purpose of the three classes of languages discussed above is that they permit the user to handle specific kinds of problems without having to concern himself unduly with the technical problems of how the machine will actually carry out his instructions. He can concentrate on the development of a good model rather than on the mechanics of implementation.

USES FOR SIMULATION

Simulation techniques in social and administrative science can be divided into three general areas of application: research, decision-making, and instruc- tional, or pedagogical, situations. This section reviews the use of simulation for these major purposes and provides some guidance to specific applications discussed in the following chapters.

A rapidly growing literature is indicative of the interest in and use of simulation. Recent bibliographies on simulation and its applications include Duke and Schmidt (1965), Hartman (1966), Holmen (1969), IBM (1966), International Relations Program (1968), Johnson (1969), Naylor (1968), Pitts (1968), Roeckelein (1967), and Werner (1968). Some earlier bibliographies include Kraft and Wensrick (1961), Malcolm (1960), and Shubik (1960). The Naylor bibliography is probably the most recent comprehensive one of direct interest to social and administrative scientists. Reviews dealing with simulation in particular social or administrative science fields have been written by Abelson (1968), Cohen and Cyert (1965), Coleman (1964), Coleman, Boocock, and Schild (1966), Holland (1965), Sisson (1969), Tocher (1966b), and others, and Guetzkow has edited two earlier volumes dealing with applications in the social sciences (Guetzkow, 1962; Guetzkow *et al.,* 1963). Books dealing with applications in specific fields include Balderston and Hoggatt (1962), Bonini (1963), Clarkson (1962), Mattessich (1964), and Tomkins and Messick (1963). More recent additions include Amstutz (1967), Coplin (1968), Hamilton *et al.,* (1969), Hollingdale (1967), Loehlin (1968), Raser (1969), Schreiber (1970), Scott, Lucas, and Lucas (1968), and Siegel and Wolf (1969), in addition to a number of volumes cited later in this section. Gullahorn and Gullahorn are also preparing a book on this subject. There are two journals devoted to simulation: *Simulation,* published by Simulation Councils, and *Simulation and Games,* which began publication in March 1970. The wide range of other journals containing articles on simulation can be found by perusing the references at the end of each chapter.

Research Uses

Simulation serves as a tool for research in at least two ways: as a special kind of modeling and as a vehicle for experimentation.

Simulation as modeling can aid in research and theory development, testing, comparison, and communication in all the ways that modeling is useful. Some of these ways we·have noted include aiding in the organization of data and structuring of experience, objectifying theory, predicting, raising questions, generating testable hypotheses, differentiating important aspects of rival theories, and so forth. Not only the simulation itself but also the very process of constructing the simulation model can aid in theory development. Simulation's particular advantage, already discussed at some length, is its greater fidelity in modeling processes, making possible both more complex models and models of more complex systems, the iconic modeling of time, the incorporation of chance events and nonprogrammed events, and the possibilities of vicarious experimentation.

Raser, Campbell, and Chadwick (1970), in a thought-provoking essay exploring how gaming and simulation can contribute to the development of social theory, suggest five primary roles for it. Two of these relate to simulation as experimentation, to be discussed below. The other three relating to simulation as modeling are:

1. Increase degrees of freedom through replications not possible in the real . . . system.

2. Provide initial constructs to bring order to partial theories and isolated bodies of data.
3. Provide a goad to stimulate creative scholarship.

First, the "degrees of freedom" issue is this: With a number of explanatory variables, one needs at least one more than that number of observations of a system or process to make a valid inference about the relationship between any variable and observed results. Modeling in general and simulation modeling in particular make possible the generation of more observation points (runs) than the number of variables under consideration. An example would be the simulation of international relations, where although there are many variables of interest, there is only one real-world international system (Guetzkow, 1966). In this case, simulation can provide alternate systems (as many as needed) to isolate the effects of particular variables across differently developed systems.

Second, building a simulation can be a useful heuristic in generating subtheories and integrating data. The generation of subtheories is aided by the tentative design of an overall model. These theories receive a kind of validation, or increase in credibility, from the consistency of the internal logic of the whole in which they are embedded. "Gaming and simulation, more than any other scientific technique we know, explicitly utilize this 'way of learning'" (Raser, Campbell, and Chadwick, 1970). Here, we may observe, the holistic approach afforded by simulation can be of particular importance.

Third, they note, the very process of constructing the simulation can be a spur to theory development in several ways:

1. Confrontation—vague generalizations crumble when put to the test of modeling.
2. Explication—theory must be made explicit, logical, and precise in order to build a simulation model.
3. Expansion—the tendency to a holistic approach in simulation forces a broadening of one's horizon, a looking into other relevant fields for ideas.
4. Communication—problem-oriented simulations lead to jumping of disciplinary boundaries, less parochialism.
5. Involvement—it can be fun, the construction process motivates the modeler to attempt to fill in the knowledge gaps.

In addition, the simulation itself, especially in its developmental forms, often leads to unexpected findings. Defining gaming as a kind of presimulation model with less formal correspondences between the real-world variables and the model variables, these authors cite a number of cases to support their thesis that this serendipity "occurs with such regularity in gaming that it is an almost certain outcome of the endeavor." (Raser, Campbell, and Chadwick, 1970, p. 43).

All-computer simulations tend to be most akin to analytical models, differing from them mainly in that the model is a process extended in time, represents a particular case history, and requires less stringent assumptions.

Analytical models have appeared in all of the disciplines covered here, but a growing number of researchers have realized that if progress is to be made toward more realistic and hence more useful theories of behavior, the complexities of the real world, including relationships that are dynamic, nonlinear, stochastic, lagged, and interactive, cannot be omitted from the analysis. Simulation has consequently found increasing use as a method for capturing the substance of a set of behavior processes in a model. It is a complementary alternative to the use of analytic techniques when those techniques provide a good fit with the real world. Where they prove inadequate, it becomes a supplement for creating and preserving a high degree of correspondence between the model and the system modeled. On the other hand, a criticism that has been voiced is that simulation is sometimes used too readily in place of analysis and has also been used to produce models too complex to allow discrimination between the important and the unimportant (Morgenthaler, 1961; Adelman, chap. 5).

As opposed to the all-computer models, simulations employing human participants are more closely akin to verbal models, commonly differing from them by including the iconic modeling of processes, with human decision-makers being used to represent human decision-makers. One of the more notable early large-scale manned exercises was the RAND Corporation's series of simulations over the years 1952 to 1954 of an air defense direction center, which was used to study the process of organizational learning and adaptation to increasing task pressures (Chapman and Kennedy, 1955; Chapman *et al.,* 1962). Not only were the actual human processes iconically modeled but also the environment of such a center was simulated in its physical, organizational, cultural, and task dimensions. Obviously, some degree of simulation of the environment is a necessary condition for the effective simulation of processes taking place within that environment, but a distinction should be maintained between these two kinds of simulations, the one being at the service of the other. The accounts of the RAND experience provide early documentation of the ability of such a non-programmed simulation to produce serendipitous results. Further examples of manned simulations are discussed later in this book, particularly in chapters 7 and 13, since a significant part of the research applications appear in international relations.

Simulation as experimentation differs from traditional experimentation in that a model rather than the real system is manipulated. Experimentation has long been the hallmark of research in the biological and physical sciences. The ability of natural scientists to exercise some control over their variables and research environments without significantly distorting the quality of the phenomena under study has resulted in great advances in knowledge. The social scientist, however, faces not only great difficulties in controlling variables for experimentation but also difficulties resulting from the impossibility or impracticability of reproducing social environments. Experiments on administrative systems are sometimes equally difficult. Some systems are either too large or too costly for experimentation, such as the economy or a business firm. Yet simulation models of both have permitted vicarious experimentation to explore consequences of various national and management policies. For systems that are too complex or otherwise unsuitable for experimentation, simulation provides a method for controlling variables in a model, if not in the real world. (Hunter and Naylor, 1970; Naylor, Burdick, and Sasser, 1967; Naylor, 1969).

Raser, Campbell, and Chadwick (1970) see an important role of simulation as a complex laboratory in which "the technique of representative sampling of subjects is extended to representative sampling of situations." They criticize the "naive faith" of small group experimenters, arguing that the inclusion of complexity in an experiment is not only as valid as simplification but also less arbitrary. The more closely the model matches the real world in its complexity, the more validly one can make inferences from the model to the world. Manned simulation, they point out, manipulates no more variables than small group experiments, but it holds a great many other variables constant that are ignored by other experimenters.

As the philosophy of simulation becomes better developed, simulation will possibly overthrow the small group experiment paradigm (Kuhn, 1962). Although not so labeled, many laboratory experiments involving artificially formed groups are, in fact, simulations insofar as the artificial groups are iconic models of other groups encountered in the real world rather than genuine representatives of such groups. One can argue that a more realistic view would treat them as simulations.

In the RAND simulation, although the first participants employed were surrogate representatives, in later runs the team was an actual crew for an air defense direction center, although the environment and task were still simulated. One of the main purposes of this simulation was the development of techniques for controlling large-scale experiments. The Inter-Nation Simulation has also been tested using participants from the State Department and diplomatic representatives from the United Nations, with results not notably different from those obtained using lay participants (Robinson and Wyner, 1965).

Raser, Campbell, and Chadwick (1970), noting that no science can conclusively prove causality, suggest that simulation as experimentation can aid in sorting out cause and effect by the falsification of theory, the ruling out of alternative explanations. A somewhat similar use is the postulational approach (Gordon, 1969), in which simulation can aid in the parsimonious refinement of theory by producing results conforming to real world observations without including variables previously considered significant. The credibility of hypotheses built into the model can be increased by their power to explain total system behavior under a wide range of circumstances. Examples of models employing this concept in the study of legislative behavior are given in chapter 6.

Research uses of simulation are reported in virtually all of the chapters of this volume. Some contributors (such as Francis, Krulee, Loehlin, the Gullahorns) are concerned almost exclusively with applications in pure research, while others (such as Newell and Meier, Kotler and Schultz, and Rath) are more concerned with its uses in applied research in particular areas of interest. Francis reports its use to model community voter behavior and legislative behavior, a couple of models in the latter area offering interesting examples of how simulation can suggest theory or modifications to theory through demonstrating the explanatory power of certain key variables in the model. Krulee considers the two companion fields of cognitive process simulation, especially problem solving, and artificial intelligence studies. Although artificial intelligence studies are designed to develop computer capabilities for symbol manipulation rather than to contribute to psychological theory, they do have interesting implications for the latter and, along with the cognitive simulations, have contributed to both psychological research and theory development.

The Gullahorns consider applications of simulation to kinship marriage rules, inheritance rules, population growth, information diffusion, urban environment and political socialization processes, reference group processes, and individual processing of social communication. Loehlin reports several personality models all built around an attitude structure that is in some way altered by experience and controls behavior, such experience in some cases including interaction with a similar "model personality"! Not only do these models demonstrate the descriptive resources of computers, according to Loehlin, but the modelers claim that such simulations will be useful for evaluating personality theories by generating predictions that can be subjected to empirical testing.

Smoker's chapter discusses the use of simulation in international relations. Other uses of simulation for research purposes are reported in the chapters on economics, business, marketing, and public systems, where simulation has been used for the study of such diverse activities as interactions of various segments of the economy, economic cycles, vulnerability of the economy to depression, demographic prediction, inventory systems, risk analysis of capital investments, waiting lines, vertical marketing structures in the shoe and lumber industries, the importance of social and chance factors in consumer behavior toward a new product; the merits of competitive marketing strategies, the probation process, and the Supreme Court.

Simulation as an Aid to Decision-Making

The major appeal of simulation as an aid to practical decision-making is basically the same as its appeal to researchers, and much of what has been said in favor of research uses of simulation relates with equal force to the applied research setting as well. An additional advantage in some may be that particular simulations are more intuitively understandable to decision-makers than are other forms of modeling.

As March, Simon, and Guetzkow (1958) and others (e.g., Meyer, 1970) have pointed out, some sort of model is always used in decision-making, namely, the decision-maker's "definition of the situation." Recent developments of normative approaches to decision-making emphasize the importance of making formal and explicit models. The use of such models is the hallmark of systems analysis and of operations research (Morris, 1967; Quade, 1969). Of the avalanche of books in the past decade on quantitative techniques in decision-making and management, it is very rare to find one that does not contain an extended discussion of simulation as a formal model-building technique, and often a complete chapter is devoted to this subject.

There are four broad ways that modeling in general and simulation in particular can help in decision-making: as a tool for increasing general understanding of the system for purposes of future decision-making, as an aid in the development of alternatives for specific decisions, as a device for evaluation of alternatives, and as an instrument for verification or control. The simulations that are employed for such purposes are quite commonly simulations of nonbehavioral or physical processes, in contrast to other uses of simulation discussed in this book, although simulations of behavioral processes are used too (e.g., in marketing simulations, chap. 10). It is in the modeling of these nonbehavioral processes that the special computer simulation languages, especially the discrete languages, receive their widest use, since most of the simulations in aid of decision-making are of the all-machine variety.

First, simulation can be used *to increase general understanding of the system of interest as an aid in making future decisions.* This usage is perhaps most similar to the uses of simulation in research. The major difference from strictly research uses is that the simulation is oriented toward a specific organization but not necessarily toward a particular problem. Simulation in this context can serve as a self-education device and is usually an instrument developed for future use. An example of the former purposes was the RAND Corporation's simulation of the air defense direction center, which focused on a particular kind of organizational entity in which the Air Force was then vitally interested (Chapman and Kennedy, 1955). An example of the latter purpose is given by corporate simulation models, a tool developed in the 1950s, which began receiving significant application only in the last several years (Schrieber, 1970).

While a number of early operations research groups attempted to build global models of their firms, attention subsequently shifted away from such efforts (Radnor, Rubenstein, and Bean, 1968). In the last half of the 1960s, however, a significant increase in the extent of corporate model-building took place, possibly as a result of the increase of operations research and management science activities in corporate planning departments and at other organizational levels close to top management. Gershefski (1970), reporting on a survey of 323 companies, found sixty-three such models in use or under development, with 95 percent of them being simulation models. Another 39 companies planned to begin development of a model within a year. These models were found to be in use or under development in almost every type of industry. On the average, development of the models lagged initiation of formal planning by four and one-half years and required three and one-half man-years of effort. Most models were finance oriented, and roughly two-thirds were begun as overall views of the company with little detail, whereas the others were constructed on a function-by-function basis. Eighty-eight percent were deterministic, the computer merely playing the role of a fast adding machine.

In the same study, 91 percent of the respondents were satisfied that the results from their models were worth the effort. On the basis of many discussions with persons working on corporate model building, Gershefski lists a number of advantages they offer (Gershefski, 1970, p. 33):

> They can give low-cost answers quickly, and allow management to experiment cheaply. They permit the study of ramified effects of changes. They demand unambiguous specification of calculations. They help specify decision-makers' information needs, the methodology of model-building being similar to that of designing an information system. All parts of the organization become equally visible. Long-term impacts of short-term decisions can be appraised.

Second, simulation can aid in *the development of alternative plans or courses of action* to be considered by decision-makers. It can do this either by generating the alternatives or by assisting analysts or decision-makers heuristically in generating and synthesizing alternatives. An example of the first way is an advertising media selection model (reported in chap. 10) which contains a heuristic program that develops an advertising media schedule superior to those developed by an advertising agency. As an example of the second, twenty of the sixty-four corporate models reported to Gershefski (1970) were used to help determine corporate goals. Smoker (1969) urges the use of simulation of international processes for suggesting new possibilities and generating alternatives for public policy. The generation of alternatives is usually the most potentially creative step in the decision-making process, and the serendipitous effects of gaming in particular can help in going beyond prior assumptions and frames of reference traditional in organizations. As an experimental technique, simulation can help identify key variables through sensitivity analysis, suggesting things that should receive maximum attention.

As a heuristic guide for the generation of alternatives, simulation can be used to "meliorize" (our word), meaning to find preferred solutions, change "satisficing" decisions into "meliorizing" ones, and in some cases even find "approximately optimal" solutions. Dalkey (1966) has urged that one of the greatest challenges to operations researchers is the introduction of more "analycity" into simulations, meaning the computation of "preferred" or "optimal" solutions. Direct search techniques, previously discussed, can be used in conjunction with simulation experiments to discover optimizing alternatives (Emshoff and Sisson, 1970, chap. 9). In some cases simulation models are supplemented with optimizing modules, such as linear programming, and significantly more activity has been predicted in this area in the near future (Dickson, Mauriel, and Anderson, 1970).

Third, simulation is useful for *evaluating the merits of alternative courses of action* submitted to the model or generated by the model. (This may be combined with the use of simulation to aid in developing alternatives.) Most of the simulation models described in the administrative science chapters of this volume (Part Three) were designed for this purpose. They are oriented toward quite specific problem areas, and in at least one case (the Indus River study, chaps. 5 and 11) toward evaluation of one specific solution. This type of application is a kind of experimental use of simulation, in which the effects of the proposed action or of alternative actions can be assessed without having to learn the hard way by experimenting with the real system. In investment decisions and in research and development planning, "risk analysis" is receiving increasing use (Hertz, 1964, 1968). Simulations can also automate "programmed decisions" (Simon, 1965, p. 52 ff.) either by using a set of decision rules developed expressly for the situation (not truly a simulation but tested by a simulation) or by simply reproducing the decision rules that human decision-makers employ in evaluating and selecting from among alternatives (i.e., simulating human decision-making). Cyert's model of the pricing behavior of a department store buyer (chap. 10) is an example of a model of the latter type.

Fourth, simulation can be used for *verification or control.* It can be used to test analytic results or to provide norms or standards against which performance can be measured. Eight of Gershefski's respondents (1970) reported that their models were used to validate manually prepared projections and existing procedures. In particular, stochastic simulation models can be used as a check against results obtained by analysis. The use of simulation in providing norms is suggested by Cyert's pricing decision model. A validated computer model embodying policy could provide standards against which performance could be measured in specific cases, thus alerting decision-makers to departures from standards so that appropriate problem-solving action could be taken.

In general, simulation is receiving rather wide use today as a decision-making aid. While in 1964 King and Smith reported at an international conference that the technique was not generally accepted for many industrial applications, being used in only one or two areas in any single industry (King and Smith, 1966), at another conference the next year Farmer and Collcutt predicted that it would certainly become one of the most widely used techniques of operations research "even if it has not reached that stage already" (Farmer and Collcutt, 1967). At yet another conference the following year, Tocher of U.S. Steel in England asserted that in operations research "simulation has become probably the most widely used technique and its use has been extended into more fields than any other" (Tocher, 1966b). At the same conference Dalkey (1966) noted that in the preceding three years in the United States alone there had been in support of military decisions "something of the order of 200 to 300 major simulation projects."

Most of the applications of simulation as an aid to practical decision-making found in this book are contained in the chapters on economics (Adelman), business (Newell and Meier), marketing (Kotler and Schultz), public systems (Rath), and international processes (Smoker), the remaining chapters being primarily or exclusively concerned with research applications. While there were a number of studies reported that were designed for or could be useful as an aid to decision-making, considerably fewer were explicitly reported as having been used for this purpose. Undoubtedly some of these studies were that are not so reported.

In economics, Adelman indicates the use of simulation as an aid to policy formulation in both free enterprise and socialist countries, noting a shift in emphasis from simulation as a tool for pure research to its use in forecasting and controlling real economies. She gives examples of a model developed to aid in production management in Russia, and another for the study of government expenditures in Venezuela. In business management, Newell and Meier describe several "total firm" simulations which have important policy implications. Forrester's model of the firm, for example, focuses on the dynamic, inter-dependent behavior of various functional areas (marketing, production, finance) and thus provides insight into decisions regarding the flow of orders, manu-facture, capital, and so forth. Another model, built by Kuehn and Hamburger, provides specific "heuristics" for locating warehouses in a distribution system. In the field of marketing, Kotler and Schultz report on a number of studies either specifically designed as aids to decision-making or in their opinion suitable for such a purpose. Areas of application include estimating costs and customer service times for a distribution system; planning marketing strategies; evaluating competitive strategies; making pricing, dealing, and product line decisions; selecting advertising media; planning salesman calls on customers; evaluating new product strategies; and predicting costs of field interviewing. In one case a model designed for market planning is criticized as being too complex and costly (in terms of data required) for such a purpose, though perhaps useful for research. Unfortunately, the extent to which the studies described have actually been used as an aid in decision-making is rarely indicated.

As might be expected, in the somewhat vague area that this book labels "public systems," the great majority of the simulation studies Rath reports were specifically designed as an aid to decision-making. Some of the diverse problem areas are land-use forecasting for regional development planning, freight car scheduling for a railroad network, evaluating an ocean-going maritime fleet, forecasting overall distribution of transportation demand, improving hospital admission policies, improving doctor-patient scheduling at a clinic, determining whether to install an autoanalyzer in a clinical pathology lab, evaluating alterna-tive resource allocation policies for a school district, predicting departmental budget and staff requirements for a university, evaluating alternative water reservoir plans, assisting in policy function by the Canadian government regard-ing uranium mines, investigating a specific irrigation plan, improving police squad car response to calls, and aiding city planners in predicting the impact of various city plans. Again, it is not always clear to what extent the simulations have actually directly been used in decision-making, although a number have— notably the Canadian uranium mines study, the irrigation study for the Indus River project, the maritime fleet evaluation, and the CAMPUS model, which "is becoming a key tool at the University of Toronto for planning, data collection and analysis."

Pedagogical Uses

The use of simulation in both teaching and training situations is also quite extensive. Simulation has been used to teach principles of business management and executive decision-making (Graham and Gray, 1969; Dale and Klasson, 1964), international relations and political decision-making (Guetzkow *et al.,* 1963), as well as a host of other subjects at all educational levels including economics at the elementary school level, history at the junior high school level, evolution for senior high school students, and urban affairs for college students (Boocock and Schild, 1968). We do not include here a number of traditional activities that qualify as simulations, although they are not usually labeled as such—moot courts in law schools or any form of organized role playing such as takes place in classes at all educational levels. Simulations have also been used to train airplane flight crews (Adams, 1962) and future plant managers (Plattner and Harron, 1962), as well as in many other job-specific training situations. Such devices as the Link Trainer and the Dilbert Dunker, used for the training of military pilots, received much publicity in earlier years. Recently, the use of simulators for the training of astronauts has been well publicized. Military gaming (Thomas, 1961) has a venerable tradition.

Simulation used for instructional purposes in social and administrative science almost always uses human participants and is commonly called gaming, although the word can also refer to simulations or presimulations (Raser, Campbell, and Chadwick, 1970, p. 19) involving human participants that are used for research purposes (chap. 13). In the latter context it is sometimes called operational gaming. In gaming the focus is on participants who assume specific roles in the model and are concerned with structure and activity as much as with game outcomes (Geisler and Ginsberg, 1966), although in traditional games as such the focus is usually more on the outcome of the game.

The concept of role is an important one in gaming just as it is in social and organizational life in the real world. The difference in gaming is that the roles are prescribed and shaped by the "rules of the game," which are commonly of five types (Coleman, 1968):

1. Rules relating to participant goals: these establish the internal game goals to which participants are to orient their actions, and the means by which such goals can be achieved (e.g., the rules for "winning the game").
2. Behavior constraint rules: these set bounds on what is considered acceptable role behavior in the game or simulation.
3. Procedural rules: these prescribe the general order of activity, a subtype being mediation rules, which indicate how deadlocks are to be resolved.
4. Police rules: these establish what happens to a participant if he violates one of the other rules.
5. Environmental response rules: these are designed to simulate the ways in which the environment would actually respond to the participants' behavior, and are the most important of the rules, being what distinguishes simulation gaming from games as such.

Once the rules for the game or simulation are prescribed, the roles are defined and hence the interdependencies among players are prescribed (Schild,

1968). The social environment of each participant is thus specified, although it is of course a simulated social environment modeling that of the real world. The core of the game as a simulation is not so much the simulation of the social environment, much stressed by some writers, as it is the process of role enactment by the participants, which imitates role enactment in the real-life environment that is simulated in the model. But since social role behavior is an interactive process, simulation of the social environment is a necessary part of a social simulation. In some cases the actual physical environment may also be simulated, as in the RAND experiments. The physical environment also gives role behavior cues in real life.

Gaming is distinguished from what is commonly termed role play in psychological applications (even though the latter does meet our definition of simulation) by its different purpose. Whereas gaming for pedagogical purposes is generally designed to help the participant learn about the system—either how it operates or how to be an effective part of it—role play as such is oriented toward producing greater empathy on the part of the role player with the real-life occupant of the role (Schild, 1968). A different pedagogical use of simulation that is oriented toward nonparticipants is suggested by Smoker's (1969) discussion of the possibility of using televised simulations of international processes for the instruction of the public to help them take on a more significant role in foreign policy formulation. Such a simulation has appeared on educational television in the Chicago area.

Boocock and Schild (1968) briefly trace the history of simulation gaming, pointing out that up to 1962 or 1963 there was great enthusiasm for it, but little real basis for that feeling other than faith in the efficacy of gaming as an instructional device. The years 1963-1965 saw some controlled experiments on the effectiveness of games as learning devices, and subsequent years have been marked by "realistic optimism, based on accumulated experience with a large number of games." These later years have seen field testing of games in many settings, the amassing of data on the teaching effectiveness of particular games and some evaluations by persons different from the games' designers, and clarification of ideas about the mechanisms by which learning from games takes place (Boocock and Schild, 1968, pp. 17-18). Their volume contains several readings discussing the effectiveness of gaming as a learning device, in an attempt "to demonstrate by specific instances the *potential* of the method."

A primary advantage of simulation as an instructional tool would seem to be its focus on the *operation* of a system, with immediate feedback to the participant trainees or students of system response to their own actions. They become part of the model itself, getting an "inside view." Where computers are used their function is usually one of providing logistic support to "implement" decisions of the participants, generate consequences, and feed back results. As a teaching or training device gaming also has the attendant advantages of more complete participant involvement, especially emotional. Snyder has noted that

Gaming—in its several varieties—offers an opportunity to play out a strategy over a period of time and to observe concrete consequences of decisions. Moreover, the importance of theory is easier to demonstrate when a system is actually in operation. The degree of transferability of experience will depend heavily on the particular domain of social behavior, on the state of research and theory, and on the sophistication with which games are designed. But the explicitness required by calculations, decisions, and actions that characterize business and political gaming is usually missing in reading a case study or listening to a lecture. (Snyder, 1963, p. 12).

Alger has summarized the benefits of simulation gaming (in political science) as (1) a heightening of student interest and motivation, (2) an opportunity for applying and testing knowledge, (3) a greater understanding of the world as seen and experienced by the decision-maker, and (4) a simplification of the world that is easier to comprehend (Alger, 1963, pp. 152-159). In business education, Cyert (1966) sees the benefit of management games in terms of the experience gained in organizing a large mass of data, exercising and applying quantitative decision-making techniques, and working effectively with both the student's team and with outside groups. Thorelli and Graves (1964) point to the increase in the student's understanding of different business functions; interrelationships between different functions and different parts of an enterprise, and between an enterprise and its environment; and problems of organization, policy, and decision-making in general.

Business management games are in widespread use. Newell and Meier (chap. 9) remark that as early as 1961 it was estimated that there were over one hundred games in existence, which had been played by over thirty thousand executives. Dale and Klasson (1964) reported a survey of the American Association of Collegiate Schools of Business in which sixty-four of ninety reporting schools used gaming, forty reporting that it had been firmly integrated into their curricula. This particular survey found twenty-nine different types of noncomputerized and twenty-eight computerized games in use, although over eighty business games had been cataloged by this time. The noncomputerized games tended to be oriented toward different functions of the business, and the computerized games tended to be oriented toward general management. Games were used in all business areas except accounting and finance, but not widely used at the undergraduate level. In 1967 Graham and Gray surveyed 125 business school deans, finding that of the 92 respondents all but 8 were using at least one game, the majority using more than one (Graham and Gray, 1969). They also surveyed the five hundred largest United States corporations to develop an informative handbook describing over two hundred business games in current use, a number of them industry-specific. A wide range of industries is involved.

Some examples of business games discussed at book length by those involved in their development or use are the Carnegie Tech Management Game (Cohen *et al.*, 1960; Cohen *et al.*, 1964), the International Operations Simulation (Thorelli and Graves, 1964), the Harvard Business School Game (McKenney, 1967), and the Management Game (McFarlan *et al.,* 1970). A general discussion of business simulation in industrial and university education appears in Greenlaw, Herron, and Rawdon (1962).

A major disadvantage of the more formalized simulation games as instructional devices is expense: the cost of building and testing the model, the cost of implementing the system (training of operators, computer programming, etc.), and the cost of running the system. Amstutz (1967, p. 431) regards the demands on instructors, students, and support personnel as "substantial."

Game development is a major expense. McKenney's remarks about the setup costs for the Harvard Business School Game have already been noted (McKenney, 1967). Nearly 40 percent of the respondents to the Dale and Klasson survey reported that their own faculty members had participated in developing the games in use or in modifying existing ones. To avoid such costs Shubik (1966) entered a plea for multipurpose games with, for example, parameters that could be adjusted for different industries. On the other hand, the educational benefits of constructing a game are not to be overlooked. Rath (chap. 11) has, in private conversation, suggested the idea of designing a game for designing a game, precisely to reap these benefits.

For games in educational institutions, faculty involvement during the operation of the game, although an absolute cost, is simply an alternative to other forms of faculty involvement with students. On the basis of the Harvard experience, McKenney considers such involvement in the gaming exercise and intelligent counseling as important determinants of a successful game. At Harvard, faculty members acted as "boards of directors" of the student "firms" in the game, setting objectives and policies, and even giving students specific assignments. The game lasted over a period of eight weeks, covering twelve quarters of simulated activity for the four- or five-man "firms," requiring a maximum of four hours per week of student preparation time per "move" and about one-half to one hour of faculty time per move for each firm.

Once a game has been developed, computer running costs need not be high. McKenney stresses the low cost of the Harvard game, reporting that for the version used in 1963 it amounted to less than twelve hundred dollars for a student body of 660 students, with about a third of that being for paper output.

One disadvantage noted by the same author is that the competitive aspects of gaming lead some students to emphasize "winning the game" at the expense of using it most effectively to learn, such as by experimenting with less-conservative strategies.

In this volume, examples of instructional uses of simulation are discussed in chapters 7 (international relations), 9 (business systems), and 11 (public systems). The Boocock and Schild volume (1968) contains in its appendixes "A Selective Bibliography on Simulation Games as Learning Devices" as well as a list of major centers involved in research and development of games with simulated environments.

ACHIEVEMENTS
AND POTENTIALS OF SIMULATION

The researchers polled in the survey mentioned at the beginning of this chapter held differing views about the achievements of simulation, whether those achievements were general or specific. In the field of business administration Richard Cyert flatly asserted that there had been no significant achievements to date, and Roman Tuason said that, in marketing, achievements were only superficial, that simulations promised a lot but delivered little. Charles Bonini, on the other hand, thought that simulations had made contributions to the improvement of real systems, understanding of decision processes, and specification of previously assumed relationships. Philip Kotler, too, felt they had resulted in richer specification of variables and functions, as well as directing attention to the system level by pointing out feedbacks that tend to be neglected in partial analyses.

In psychology, simulation was credited with reawakening an interest in "higher" mental processes and providing a means of studying symbolic behavior, and in sociology James Coleman considered it as a half-way point between verbal speculative theory and formal theory, aiding the development of such theory through concretizing the functioning of social processes. In political science, William Coplin cited its use at numerous colleges and high schools for teaching about international relations, its aid in forming concepts for analyzing international politics, and its lesser use as an aid to policy makers. Philip Burgess also thought it had helped more explicit theory-building, creating greater awareness of the need for and requirements of theory, and had increased sensitivity to possibilities of applying experimental methods to issues in political science research, as well as giving more options for teaching. Others, too, thought it had helped make theory more explicit.

There were equally differing views on the potentials of this tool. Cyert does not find the outlook encouraging in economics but anticipates increasing computerization of business decisions. Kotler sees a high potential not only for aiding decision-making but also for theory construction in marketing. In psychology and in communication-diffusion studies, simulation is also seen to have a high potential for theory-building activity if used in conjunction with empirical and experimental studies. Abelson (1968) sees it as useful as a bridge builder in social psychology—both in synthesis of propositions and in connecting "islands of theoretical analysis," such as the psychology of individuals and of groups. A somewhat different kind of integrative bridge building is also suggested by Krulee's analysis in this book (chap. 2) of the relationships between psychological theory, cognitive process simulations, and artificial intelligence studies.

In political science, the major potentials are seen in all three applications of research, practical decision-making, and teaching—though here, also, there were disagreements. Burgess, for example, confesses that he is "not excited" about heuristic uses or potential of simulation but foresees its usefulness in testing hypotheses about human behavior, especially when applied in tandem with other research strategies. Wayne Francis forecasts only isolated successes rather than theoretical innovation for the next five to ten years, until new mathematical procedures are developed for exhausting the predictive and postdictive power of existing data banks. Another researcher active in the study of international relations sees a wider use of simulation for teaching at all levels, including elementary schools, and sees simulation pacing the field in theory development and in creating data requirements. He believes that simulation has special advantages in international affairs, where experimentation is hazardous, and as a policy tool for generating alternatives. In this latter area, Kenneth Janda expects the consequences of simulation for government decision-making "to be tremendous" and the decision-making process in foreign policy to undergo radical changes with better information available about the state of the world and better predictions of the consequences of policy decisions.

Problems and Prospects

Dalkey (1966) has suggested that one challenge to the field is the development of a richer theoretical structure favoring the study of general characteristics of simulation models, and the suggestion seems well taken. The remark of the Gullahorns at the end of their chapter (chap. 4) that simulation has been very much a do-your-own-thing enterprise may be extended to the general areas of application covered by this book. There is at present little in the way of an established theoretical framework or recognized canons of behavior. Ill-planned studies generating large masses of print-out data that the researcher doesn't know what to do with, reported by more than one author as a common fault, may be in part due to this lack of structure. The problem of validating simulations remains one of the most difficult areas, both in concept and in development of appropriate means, although there is a growing literature on the subject (see chap. 12). Many simulation studies either sidestep the issue or rest content with the most superficial forms of validity checks. Abelson (1968) suggests four reasons why statistical validation treatments are undeveloped in social behavior simulations: stress on getting a running model, impracticability of repeated runs due to slow computing facilities in some instances, innocence of the need, and ignorance of techniques. A third lacuna lies in the reporting of simulation studies, in which key questions are often glossed over (accounting for some of the lack of specification of detail for many of the simulations reviewed herein) and useful facts omitted. Tocher (1966b, p. 700) complains that in operations research:

> The literature on simulation case studies nearly always dismisses the details of the actual model-building and concentrates on the results of the study. . . . failure to report on the details of the modeling techniques is a barrier to the dissemination of knowledge and holds back the advancement of expertise in simulation modeling.

Better communication between researchers in the open literature can lead to more integrated efforts, such as more general multipurpose business games as Shubik (1966) suggests to decrease costs, or modular or compatible models, or refinement or adaptation of models for other research purposes. More attention, in general, could be given to the relationships between simulations—for example, the design of simulations at the macro level that could be decomposed and implemented with more fine structure at the microbehavioral level, or micro-level simulations whose outputs could be used in macro-models. Dalkey (1967) has suggested the use of "families of models" at various levels of aggregation as a useful heuristic in zeroing in on details for further study at considerable savings of computer time, and he describes the use of such a family by the RAND Corporation. In the field of social psychology, Abelson (1968) considers it "obvious" that future simulations will focus on the dynamic coupling of opinion or attitude variables (discussed in several chapters of this book, especially Loehlin's) and social interaction (discussed particularly by the Gullahorns). Indeed, the very classification system the Gullahorns use in their review chapter (system level and microbehavioral simulations) underscores the dichotomy that must be bridged over.

If simulation in the social sciences has reached the consolidation phase, as the Gullahorns suggest it has in their area of interest, the above will be fruitful areas to study.

One likely prospect suggested by Bonini in response to the questionnaire is the development of more special purpose computer languages for simulation. It is indeed likely that the future will see the development of more special purpose simulation languages as the usefulness of particular conceptual frameworks in special areas of study becomes more apparent. For example, a general "proposition network" language (Sullivan, 1969) conceptualizing behavioral variables as functions of time and propositions relating such variables as mathematical functions (not necessarily in the analytically tractable sense but in the wide sense of a rule of correspondence between two variables), with provisions for feedback loops utilizing control theory concepts, probabilistic application of propositions based on confidence levels, and handling of nonlinear and lagged effects, would seem a useful tool in computer modeling of organizational processes as well as in other fields in which operationally quantifiable variables and propositions relating them were of interest. It would readily lend itself to simulation of organizational process models such as described by March, Simon, and Guetzkow (1958) and by many writers since.

The development of more specialized languages will have advantages in giving new and better tools for specific purposes, possibly resulting in better simulation investigations. Additional "world views" may be suggestive of different kinds of models, and the general acceptance of particular languages within specialized areas of research may promote greater replicability of models and perhaps syntheses of modularly constructed models. On the other hand, a proliferation in the number of languages will make it increasingly difficult to keep abreast of developments and will result in increased fragmentation of the field with more intensive specialization.

Future prospects thus seem to point toward further development in both theory and methodology, but not without attendant problems. The set of critical overviews and case-examples that follows gives some indication of the state of the art (or science) today. In the future, simulation will undoubtedly play a significant role in the further development of social and administrative science.

REFERENCES

Abelson, Robert P., "Simulation of Social Behavior," in Gardner Lindzey and Elliot Aronson (eds.), *The Handbook of Social Psychology, Vol. II, Research Methods* (2nd ed.), pp. 274-356. Reading, Mass.: Addison-Wesley Publishers, 1968.

Ackoff, Russell L. (with S. K. Gupta and J. S. Minas), *Scientific Method: Optimizing Applied Research Decisions.* New York: John Wiley & Sons, Inc., 1962, 464 pp.

Adams, Jack A., "Some Considerations in the Design and Use of Dynamic Flight Simulators," in Harold Guetzkow (ed.), *Simulation in Social Science: Readings,* pp. 29-47. Englewood Cliffs, N.J.: Prentice-Hall, Inc., 1962.

Alger, Chadwick F., "Use of Inter-Nation Simulation in Undergraduate Teaching," in H. Guetzkow, *et al., Simulation in International Relations: Developments for Research and Teaching,* pp. 150-89. Englewood Cliffs, N.J.: Prentice-Hall, Inc., 1963.

Amstutz, Arnold E., *Computer Simulation of Competitive Market Response.* Cambridge, Mass.: M.I.T. Press, 1967.

Balderston, Frederick E., and Austin C. Hoggatt, *Simulation of Market Processes.* Berkeley: University of California, Institute of Business and Economic Research, 1962.

Barton, Richard F., *A Primer on Simulation and Gaming.* Englewood Cliffs, N.J.: Prentice-Hall, Inc., 1970.

Bobrow, D. G., and B. Raphael, "A Comparison of List-Processing Computer Languages," in Saul Rosen (ed.), *Programming Systems and Languages,* pp. 490-511. New York: McGraw-Hill Book Company, 1967.

Bonini, Charles, *Simulation of Information and Decision Systems in the Firm.* Englewood Cliffs, N.J.: Prentice-Hall, Inc., 1963.

Boocock, Sarane S., and E. O. Schild, *Simulation Games in Learning.* Beverly Hills, Calif.: Sage Publications, 1968.

Brody, Richard A., "Some Systematic Effects of the Spread of Nuclear Weapons Technology: A Study Through Simulation of a Multi-nuclear Future," *The Journal of Conflict Resolution,* 7 (December 1963), 663-753.

Buxton, J. N. (ed.), *Simulation Programming Languages.* Amsterdam: North Holland Publishing Co., 1968. (Proceedings, IFIP Working Conference, Oslo, 1967.)

Chapman, Robert L., and John L. Kennedy, "The Background and Implications of the Rand Corporation Systems Research Laboratory Studies," in Albert H. Rubenstein and Chadwick J. Haberstroh (eds.), *Some Theories of Organization* (rev. ed.), pp. 149-56. Homewood, Ill.: Richard D. Irwin, Inc., and The Dorsey Press, 1966. (Reproduced from RAND Paper P-740, September 21, 1955.)

Chapman, Robert L., John L. Kennedy, Allen Newell, and William C. Biel, "The Systems Research Laboratory's Air-Defense Experiments," in Harold Guetzkow (ed.), *Simulation in Social Science: Readings,* pp. 172-88. Englewood Cliffs, N.J.: Prentice-Hall, Inc., 1962.

Chorafas, Dimitris, N.,*Systems and Simulation.* New York: Academic Press, Inc., 1965, 487 pp.

Chu, Yaohan (with Frederick J. Sanson and Harry E. Peterson), *Digital Simulation of Continuous Systems.* New York: McGraw-Hill Book Company, 1969, 423 pp.

Churchman, C. West, "An Analysis of the Concept of Simulation," in Austin C. Hoggatt and Frederick E. Balderston (eds.), *Symposium on Simulation Models,* pp. 1-12. Cincinnati, Ohio: South-Western Publishing Co., 1963.

Clancy, J., and M. Fineberg, "Digital Simulation Languages: A Critique and a Guide," in *Proceedings of the IFIP Congress,* pp. 22-36, Baltimore: Spartan Books, 1965.

Clarkson, Geoffrey P.E., *Portfolio Selection: A Simulation of Trust Investment.* Englewood Cliffs, N.J.: Prentice-Hall, Inc., 1962.

Cohen, Kalman J., and Richard M. Cyert, "Simulation of Organizational Behavior," in James G. March (ed.), *Handbook of Organizations,* pp. 304-34. Chicago: Rand McNally & Co., 1965.

Cohen, Kalman J., *et al.* "The Carnegie Tech Management Game," *The Journal of Business,* 33 (October 1960), 303-9.

Cohen, Kalman J., William R. Dill, Alfred A. Kuehn, and Peter R. Winters, *The Carnegie Tech Management Game: An Experiment in Business Education.* Homewood, Ill.: Richard D. Irwin, Inc., 1964, 389 pp.

Coleman, James S., "Mathematical Models and Computer Simulation," in Robert E. L. Faris (ed.), *Handbook of Modern Sociology,* pp. 1027-62. Chicago: Rand McNally and Company, 1964.

———, "Social Processes and Social Simulation Games," in Sarane S. Boocock and E. O. Schild (eds.), *Simulation Games in Learning,* pp. 29-51. Beverly Hills, Calif.: Sage Publications, 1968.

Coleman, James S., Sarane S. Boocock, and E. O. Schild (eds.), "Simulation Games and Learning Behavior," *The American Behavioral Scientist,* 10 (October and November 1966), the entire issues.

Coplin, William D., *Simulation Models of the Decision Maker's Environment.* Chicago: Markham Publishing Company, 1968.

Cyert, Richard M., "A Description and Evaluation of Some Firm Simulations," *Proceedings IBM Scientific Computing Symposium Simulation Models and Gaming,* pp. 3-22. White Plains, N.Y.: IBM Data Processing Division, 1966.

Dale, A. G., and C. R. Klasson, *Business Gaming: A Survey of American Collegiate Schools of Business.* Austin, Tex.: The University of Texas, Bureau of Business Research, 1964, 64 pp.

Dalkey, Norman C., "Families of Models," in S. H. Hollingdale (ed.), *Digital Simulation in Operational Research,* pp. 38-46. New York: American Elsevier Publishing Co., Inc. 1967.

———, "Simulation of Military Conflict," in David B. Hertz and J. Melese (eds.), *Proceedings of the Fourth International Conference on Operational Research,* pp. 717-19. New York: John Wiley & Sons, Inc., 1966.

Dawson, Richard E., "Simulation in the Social Sciences," in Harold Guetzkow (ed.), *Simulation in Social Science: Readings,* pp. 1-15. Englewood Cliffs, N.J.: Prentice-Hall, Inc., 1962.

DeSautels, E. J., and D. K. Smith, "An Introduction to the String Manipulation Language SNOBOL," in Saul Rosen (ed.), *Programming Systems and Languages,* pp. 419-54. New York: Mc-Graw Hill Book Company, 1967.

Dickson, Gary W., John J. Mauriel, and John C. Anderson, "Computer Assisted Planning Models: A Functional Analysis," in Albert N. Schrieber (ed.), *Corporate Simulation Models,* pp. 43-70. Seattle, Wash.: University of Washington, Graduate School of Business Administration, 1970.

Duke, Richard D., and Allen H. Schmidt, *Operational Gaming and Simulation in Urban Research: An Annotated Bibliography.* East Lansing: Michigan State University, Institute for Community Development, Continuing Education Services, January 1965.

Emshoff, James R., and Roger L. Sisson, *Design and Use of Computer Simulation Models.* New York: The Macmillan Company, 1970, 302 pp.

Evans, G. W., II, G. F. Wallace, and G. L. Sutherland, *Simulation Using Digital Computers.* Englewood Cliffs, N.J.: Prentice-Hall, Inc., 1967, 198 pp.

Farmer, J. A., and R. H. Collcutt, "Experience of Digital Simulation in a Large O.R. Group," in S. H. Hollingdale (ed.), *Digital Simulation in Operational Research,* pp. 166-75. New York: American Elsevier Publishing Co., Inc., 1967.

Fishman, George J., and Philip J. Kiviat, "The Analysis of Simulation-Generated Time-Series," *Management Science,* 13, No. 7 (March 1967), 525-57.

Forrester, J. W., *Industrial Dynamics.* Cambridge, Mass.: M.I.T. Press, 1961, 464 pp.

Foster, J. M., *List Processing.* New York: American Elsevier Publishing Co., Inc., 1967.

Geisler, Murray A., "Appraisal of Laboratory Simulation Experience," *Management Science,* 8, No. 3 (April 1962), 239-45.

Geisler, Murray A., and A. L. Ginsberg, "Man Machine Simulation Experience," *Proceedings IBM Scientific Computing Symposium Simulation Models and Gaming.* pp. 225-42. White Plains, N.Y.: IBM Data Processing Division, 1966.

Gershefski, George N., "Corporate Models—The State of the Art," in Albert N. Schrieber, (ed.), *Corporate Simulation Models,* pp. 26-42. Seattle: University of Washington, Graduate School of Business Administration, 1970; also in *Management Science,* 16, No. 6 (February 1970), B303-21.

Gordon, G., *System Simulation.* Englewood Cliffs, N.J.: Prentice-Hall, Inc., 1969, 303 pp.

Graham, Robert G., and Clifford E. Gray, *Business Games Handbook.* New York: American Management Association, 1969, 480 pp.

Greenlaw, Paul S., Lowell W. Herron and Richard H. Rawdon, *Business Simulation in Industrial and University Education.* Englewood Cliffs, N.J.: Prentice-Hall, Inc., 1962.

Griswold, R. E., J. F. Poage, and I. P. Polonsky, *The SNOBOL 4 Programming Language.* Englewood Cliffs, N.J.: Prentice-Hall, Inc., 1969.

Guetzkow, Harold, "Simulation in International Relations," in *Proceedings of the IBM Symposium on Simulation Models and Gaming*, pp. 249-78. York, Pa.: Maple Press, 1966.

Guetzkow, Harold (ed.), *Simulation in Social Science: Readings.* Englewood Cliffs, N.J.: Prentice-Hall, Inc., 1962, 199 pp.

Guetzkow, Harold, Chadwick Alger, Richard Brody, Robert Noel, and Richard Snyder, *Simulation in International Relations: Developments for Research and Teaching.* Englewood Cliffs, N.J.: Prentice-Hall, Inc., 1963, 248 pp.

Gullahorn, John T., and Jeanne E. Gullahorn, *Explorations in Computer Simulation,* New York: The Free Press, forthcoming.

Hamilton, H. R., S. E. Goldstone, and J. W. Milliman, A. L. Pugh III, E. B. Roberts, and A. Zellner, *Systems Simulation for Regional Analysis.* Cambridge, Mass.: M.I.T. Press, 1969.

Hartman, John J., *Annotated Bibliography on Simulation in the Social Sciences.* Ames: Iowa State University, Department of Sociology and Anthropology, Rural Sociology Report No. 53, 1966.

Hermann, Charles F., "Validation Problems in Games and Simulations with Special Reference to Models of International Politics," *Behavioral Science,* 12, No. 3 (May 1967), 216-31.

Hertz, David B., "Investment Policies that Pay Off," *Harvard Business Review,* 46 (January-February 1968), pp. 96-108.

———, "Risk Analysis in Capital Investment," *Harvard Business Review,* 42 (January-February 1964) pp. 95-106.

Holland, Edward P., "Principles of Simulation," *American Behavioral Scientist,* 9, No. 1 (September 1965), 6-9.

Hollingdale, S. H. (ed.), *Digital Simulation in Operational Research.* New York: American Elsevier Publishing Co., Inc., 1967.

Holmen, Milton G., *Bibliography of Games and Simulations,* preliminary edition. Los Angeles: University of Southern California, Graduate School of Business Administration, February 18, 1969.

Hunter, J. S., and Thomas H. Naylor, "Experimental Designs for Computer Simulation Experiments," *Management Science,* 16, No. 7 (March 1970), 422-34.

IBM (International Business Machines Corporation), *Bibliography on Simulation,* White Plains, New York, 1966.

Inbar, Michael, and C. S. Stoll (eds.), *Social Science Simulations.* New York: The Free Press, 1970.

International Relations Program, Northwestern University, *Another Partial Bibliography on Simulation in International Relations.* Evanston, Ill.: Northwestern University, August 1968.

Johnson, Edward R., *Simulation and Gaming in Business and Economics in the 1960s: A Bibliography.* Iowa City: University of Iowa, College of Business Administration, 1969, 48 pp.

Kaplan, A., *The Conduct of Inquiry: Methodology for Behavioral Science.* San Francisco, Calif.: Chandler Publishing Co., 1965, 428 pp.

King, E. P. and R. N. Smith, "Simulation in an Industrial Environment," in *Proceedings IBM Scientific Computing Symposium Simulation Models and Gaming,* pp. 43-58. White Plains, N.Y.: IBM Data Processing Division, 1966.

Kiviat, P. J., R. Villanueva, and H. M. Markowitz, *The SIMSCRIPT II Programming Language.* Englewood Cliffs, N.J.: Prentice-Hall, Inc., 1968.

Koopman, Bernard O., and Howard M. Berger, "Use and Misuse of Simulations," in S. H. Hollingdale (ed.), *Digital Simulation in Operational Research,* pp. 19-25. New York: American Elsevier Publishing Co., Inc., 1967.

Kraft, R., and C.J. Wensrick, *Monte Carlo Methods: A Bibliography Covering the Period 1947-1961.* Washington, D.C.: Office of Technical Services, Department of Commerce, 1961.

Krasnow, Howard S., "Computer Languages for System Simulation," in Melvin Klerer and Granino Korn (eds.), *Digital Computer User's Handbook,* pp. I-258-77. New York: McGraw-Hill Book Company, 1967.

Krasnow, Howard S., and Reino O. Merikallio, "The Past, Present, and Future of General Simulation Languages," *Management Science,* 11, No. 2 (November 1964), 236-67.

Kuhn, Thomas S., *The Structure of Scientific Revolutions.* Chicago: University of Chicago Press, 1962, 172 pp.

Linebarger, R.N., and R.D. Brennan, "A Survey of Digital Simulation: Digital-Analog Programs," *Simulation,* 3, No. 6 (December 1964), 22-26.

Loehlin, John C., *Computer Models of Personality.* New York: Random House, Inc., 1968, 177 pp.

McFarlan, F. Warren, James L. McKenney, and John A. Seiler, *The Management Game: Simulated Decision Making.* New York: The Macmillan Company, 1970, 153 pp.

McKenney, James L., *Simulation Gaming for Management Development.* Boston, Mass.: Harvard University, Division of Research, Graduate School of Business Administration, 1967, 189 pp.

McLeod, John, *Simulation: the Modeling of Ideas and Systems with Computers.* New York: McGraw-Hill Book Company, 1968.

McMillan Claude, and Richard F. Gonzalez, *Systems Analysis: A Computer Approach to Decision Models.* Homewood, Ill.: Richard D. Irwin, Inc., 1965.

Malcolm, D.G., "Bibliography on the Use of Simulation in Management Analysis," *Operations Research,* 8 (March-April 1960), 169-77.

March, James G., and Herbert A. Simon (with Harold Guetzkow), *Organizations.* New York: John Wiley & Sons, Inc., 1958, 262 pp.

Markowitz, H., B. Hansner, and H. Karr, *SIMSCRIPT.* Englewood Cliffs, N. J.: Prentice-Hall, Inc., 1964.

Martin, Francis F., *Computer Modeling and Simulation.* New York: John Wiley & Sons, Inc., 1968.

Mattessich, Richard, *Simulation of the Firm Through a Budget Computer Program.* Homewood, Ill.: Richard D. Irwin, Inc., 1964.

Meier, Robert C., William T. Newell, and Harold L. Pazer, *Simulation in Business and Economics.* Englewood Cliffs, N.J.: Prentice-Hall, Inc., 1969.

Meyer, Robert J., "Development and Research: The Curious Inversion," in Albert N. Schrieber (ed.), *Corporate Simulation Models,* pp. 466-579. Seattle: University of Washington, Graduate School of Business Administration, 1970.

Mize, Joe H., and J. Grady Cox, *Essentials of Simulation.* Englewood Cliffs, N.J.: Prentice-Hall, Inc., 1968, 234 pp.

Morgenthaler, George W., "The Theory and Application of Simulation in Operations Research," in Russell L. Ackoff (ed.), *Progress in Operations Research,* I, 361-419. New York: John Wiley & Sons, Inc., 1961.

Morris, W. T., "On the Art of Modeling," *Management Science,* 13 (August 1967), B707-17.

Naylor, Thomas H., *Bibliography on Simulation and Gaming.* Durham, N.C.: Duke University Press, June 1, 1968, (TIMS College on Simulation and Gaming.)

———(ed.), *The Design of Computer Simulation Experiments.* Durham, N. C.: Duke University Press, 1969, 417 pp.

Naylor, Thomas H., Joseph L. Balintfy, Donald S. Burdick, and Kong Chu, *Computer Simulation Techniques.* New York: John Wiley & Sons, Inc., 1966, 352 pp.

Naylor, Thomas H., Donald S. Burdick, and W. Earl Sasser, "Computer Simulation Experiments with Economic Systems: The Problem of Experimental Design," *Journal of the American Statistical Association,* 62 (December 1967, 1315-37.

Naylor, Thomas H., and J. M. Finger, "Verification of Computer Simulation Models," *Management Science,* 14, No. 2 (October 1967), B92-101.

Naylor, Thomas H., *et al.,* "Methods for Analyzing Data from Computer Simulation Experiments," *Communications of the Association for Computing Machinery,* 10, No. 11 (November 1967), 703-10.

Newell, A. (ed.), *Information Processing Language–V* (2nd ed.). Englewood Cliffs, N.J.: Prentice-Hall, Inc., 1964.

Olson, C.A., E. E. Sorenson, and W. J. Sullivan, "Medium-Range Scheduling for a Freighter Fleet," *Operations Research,* 17, No. 4 (July-August 1969), 565-82.

Pitts, Forrest R., *Simulation Bibliography.* Honolulu: University of Hawaii, Social Science Research Institute, ca. 1968.

Plattner, John W., and Lowell W. Harron, "Simulation: Its Use in Employee Selection and Training," American Management Association, Personnel Division, Management Bulletin No. 20, 1962.

Quade, E.S., "Systems Analysis Techniques for Planning-Programming-Budgeting," in David I. Cleland and William R. King (eds.), *Systems, Organizations, Analysis, Management: A Book of Readings,* pp. 193-205. New York: McGraw-Hill Book Company, 1969.

Radner, Michael, Albert H. Rubenstein, and Alden S. Bean, "Integration and Utilization of Management Science Activities in Organizations," *Operational Research Quarterly,* Pergamon Press, 19 (June 1968), 117-41.

Raser, John R., *Simulation and Society.* Boston, Mass.: Allyn & Bacon, Inc., 1969.

Raser, John R., Donald T. Campbell, and Richard W. Chadwick, "Gaming and Simulation for Developing Theory Relevant to International Relations," manuscript of a paper to appear in *General Systems,* Vol. XV, 1970.

Robinson, James A., and A. Wyner, "Information Storage and Search in an Inter-Nation Simulation," Columbus: Ohio State University, May 1965.

Roeckelein, John E., *Simulation of Organizations: An Annotated Bibliography.* Washington, D.C.: George Washington University Human Relations Research Office (HumRRO) Technical Report 67-14, December 1967.

Rosenblueth, Arturo, and Norbert Wiener, "The Role of Models in Science," *Philosophy of Science,* 12 (October 1945), 316-21.

Schild, E. O., "Interaction in Games," in Sarane S. Boocock and E. O. Schild (eds.), *Simulation Games in Learning,* pp. 93-104. Beverly Hills, Calif.: Sage Publications, Inc., 1968.

Schmidt, J. W., and R. E. Taylor, *Simulation and Analysis of Industrial Systems.* Homewood, Ill.: Irwin-Dorsey, 1970.

Schreiber, Albert N. (ed.), *Corporate Simulation Models.* Seattle: University of Washington, Graduate School of Business Administration, 1970, 614 pp.

Scott, Andrew M. (with William A. Lucas and Trudi M. Lucas), *Simulation and National Development.* New York: John Wiley & Sons, Inc., 1968.

Shigley, J. E., *Simulation of Mechanical Systems.* New York: McGraw-Hill Book Company, 1967, 352 pp.

Shubik, Martin, "Bibliography on Simulation, Gaming, Artificial Intelligence and Allied Topics," *Journal of the American Statistical Association,* 55 (December 1960), 736-51.

———, "Some Comments on Gaming for Teaching and Research Purposes," *Proceedings IBM Scientific Computing Symposium Simulation and Gaming,* pp. 243-48. White Plains, N.Y.: IBM Data Processing Division, 1966.

Siegel, Arthur I., and Jay Wolf, *Man Machine Simulation Models: Psychological and Performance Interaction.* New York: John Wiley & Sons, Inc., 1969.

Simon, Herbert A., *The Shape of Automation: for Men and Management.* New York: Harper & Row, Publishers, 1965, 111 pp.

Sisson, R. L., "Simulation: Uses," in Julius Aronofsky (ed.), *Progress in Operations Research,* III, 71-113. New York: John Wiley & Sons, Inc., 1969.

Smoker, Paul, "Social Research for Social Anticipation," *The American Behavioral Scientist,* 12, No. 6 (July-August 1969), 7-13.

Snyder, Richard C., "Some Perspectives on the Use of Experimental Techniques in the Study of International Relations," in Harold Guetzkow, *et al. Simulation in International Relations: Developments for Research and Teaching,* pp. 1-23. Englewood Cliffs, N.J.: Prentice-Hall, Inc., 1963.

Starbuck William A., and J. M. Dutton (eds.), *Computer Simulation in Human Behavior.* New York: John Wiley & Sons, Inc., 1970.

Sullivan, Edward M., "Simulation as a Tool for Organizational Research and Design," Northwestern University, Department of Industrial Engineering and Management Sciences, Evanston, Ill., January 1969.

Teichroew, Daniel, "A History of Distribution Sampling Prior to the Era of the Computer and its Relevance to Simulation," *Journal of the American Statistical Association,* 60 (March 1965), 27-49.

Teichroew, Daniel, and J. Lubin, "Computer Simulation—Discussion of the Technique and Comparison of Languages," *Communications of the Association for Computing Machinery,* 9, No. 10 (October 1966), 723-41.

Thomas, Clayton J., "Military Gaming," in Russell L. Ackoff (ed.), *Progress in Operations Research,* I, 361-419, New York: John Wiley & Sons, Inc., 1961.

Thorelli, H. B., and R. L. Graves, *International Operations Simulation.* New York: The Free Press of Glencoe, 1964, 405 pp.

Tocher, Karl D., *The Art of Simulation.* London: English Universities Press, 1963 (also published as *The Art of Simulation.* Princeton, N.J.: D. Van Nostrand Co., Inc., 1964).

——, "Review of Simulation Languages," *Operational Research Quarterly,* 16 (June 1965), 189-218.

——, "Some Techniques of Model Building," *Proceedings IBM Scientific Computing Symposium Simulation and Gaming,* pp. 119-56. White Plains, N.Y.: IBM Data Processing Division, 1966a.

——, "The State of the Art of Simulation—A Survey," in D. B. Hertz and J. Melese (eds.), *Proceedings of the Fourth International Conference in Operational Research* pp. 693-701. New York: John Wiley & Sons, Inc., 1966b.

Tomkins, S. S., and S. Messick (eds.), *Computer Simulation of Personality.* New York: John Wiley & Sons, Inc., 1963, 325 pp.

Weiss, R. S., "Alternative Approaches in the Study of Complex Situations," *Human Organization,* 25 (Fall 1966), 198-206.

Werner, Roland, *A Bibliography of Social System Simulations.* San Diego, Calif.: United States International University, Department of Sociology, August 30, 1968.

Wyman, Forrest Paul, *Simulation Modeling: A Guide to Using SIMSCRIPT.* New York: John Wiley & Sons, Inc., 1963, 325 pp.

PART TWO

Simulation in Social Science

2

Individual Behavior Simulations
– Cognitive Processes

GILBERT K. KRULEE

INTRODUCTION

In 1956, there appeared in print a paper entitled "The Logic Theory Machine," written by Newell and Simon (1956); a related paper describing some empirical explorations appeared a year later (Newell, Shaw, and Simon, 1957). For all practical purposes, these papers mark the beginning of a new field of activity concerned with computer simulations of cognitive processes. In what follows, I want to review briefly some characteristic features of this field and its accomplishments to date. In addition, I want to discuss more intensively what is happening at the present time. My purpose is twofold: to describe how the field has been changing and to identify certain problems and issues that are likely to be of continuing interest.

As to the scope of this review, I have included selected material from the companion fields of cognitive simulation and artificial intelligence. In these two fields, similar methods are used, although there are usually differences in objectives. In either field, one attempts to write a computer program that resembles some form of "intelligent" behavior. When simulating human cognition, one's purpose is to contribute to the development of psychological theory. Assuming that psychologists study relationships among stimulus-organism-response, then a cognitive simulation can be interpreted as a model or a representation of an organism accepting stimuli and generating responses. Moreover, this research is often anchored to a sample of human behavior or a set of protocols as an initial statement of the problem.

In the field of artificial intelligence, one's objectives are rather different. The purpose of these studies is often to expand the capabilities of a digital computer in handling problems that are essentially nonnumerical and that depend on a computer's ability to manipulate symbols. The interest in psychological theory is either absent or peripheral. However, one may wish to design a system that will augment human intelligence or that may replace it as, for example, when one designs a robot that will behave intelligently (Feigenbaum, 1968).

In what follows, I propose to review the implications of these two fields of research for the development of psychological theory. Given this objective, why should one even include any studies from the field of artificial intelligence? There are two main reasons for their inclusion. First of all, surprisingly few studies have been completed that actually simulate human behavior. More importantly, many studies of artificial intelligence have interesting implications for the development of psychological theory. At the very least, an artificial intelligence program can be reinterpreted as a model of a cognitive process; one can attempt to infer those psychological assumptions that are inherent in the program. Although these assumptions may be evaluated as either bad—or perhaps naive—psychology, they may still be suggestive or may focus explicitly on some important psychological issues.

As to the amount of detail to be included, there already exist some excellent reviews and I will not attempt to duplicate them. *Computers and Thought* by Feigenbaum and Feldman (1963) is the best available summary of research completed prior to 1963. Abelson (1968) has summarized more recent accomplishments in the field of computer simulations, whereas the reviews by Hunt (1968) and Feigenbaum (1968) give greater emphasis to progress in the field of artificial intelligence. There are also available summaries of specialized fields of interest, such as the collections of papers on pattern recognition edited by Uhr (1966) and by Kolers and Eden (1968), on information retrieval and natural language edited by Minsky (1968), or the book by Hunt (1962) on concept formation. Material on the simulation of personality will not be included because it is covered elsewhere in a chapter by Loehlin. However, it would not be unreasonable to assume that cognitive processes are essential to the study of personality. Also, no attempt has been made to review material on the simulation of learning. This is an area that is both difficult and important, and a thorough discussion of it is beyond the scope of this review.

COMPUTERS THAT THINK: AN OVERVIEW

As one observes the field of cognitive simulations, one is certainly impressed by its high rate of activity and by its vitality. In many respects, research in this area has been influential, although it has also been controversial (Taube, 1961; Dreyfus, 1965, 1967). But how does one explain the nature of this influence or of the controversy? In order to consider these questions, I want to refer to Kuhn's discussion of scientific revolutions (1962) and to explore its implications for the field of cognitive simulation. Kuhn makes a distinction between "normal" science and "revolutionary" science—normal science takes place within a shared paradigm, whereas revolutionary science has the effect of stating a new or significantly revised paradigm. It may seem an exaggeration to characterize such scientific developments as revolutionary. Admittedly, most changes in paradigm are more nearly a matter of evolution, whereas only a few major developments accomplish a revolution. What does Kuhn mean by a paradigm? I infer that a paradigm is in part a shared value system about what problems are important, what will constitute an adequate theory or explanation, and how one *ought* to proceed in order to make further scientific contributions. To the extent that a paradigm establishes criteria for the development of theory or for evaluating theoretical proposals, it may be said to function as a theory about theory, that is, as a metatheory. By analogy, when one works within the framework of normal science, one plays a game according to a particular set of rules; when one advances a new paradigm one in effect proposes a new game and a new set of rules that should govern the behavior of scientists.

Let us suppose that a new paradigm has been advanced. What would one expect to happen in the immediate future? One would first expect an initial period of transition characterized by conflict between competing paradigms. Eventually, disagreements would be resolved and scientists would settle down to another period of normal science but now governed by a newly accepted or revised paradigm. Along these lines, Kuhn is most persuasive in suggesting that disagreements between competing paradigms are not really settled by logical argument or even by experiment. To a significant extent, belief in a paradigm is an act of faith. After all, a paradigm is essentially metatheory rather than theory, and metatheoretical propositions are not necessarily verifiable in the usual experimental sense.

There is another way in which Kuhn's argument can be restated. Consider any piece of research. It can be evaluated as a contribution either to theory or to metatheory, and it may be influential in either sense. At the level of theory, a piece of research makes a substantive contribution, as, for example, by identifying certain relationships among a set of variables. At the level of metatheory, it may help to modify or change the rules that guide the activities of scientists and the development of theory.

Under what circumstances will a new paradigm be developed? I would infer from Kuhn's discussion that one can usually identify some predisposing intellectual developments to prepare the way, and these predispositions may well be unintentional. There must also be some precipitating influences that stimulate one or more individuals to advance a new paradigm as a synthesis of all of these influences.

In what sense do these views about developments in science help us to understand what has been happening in the field of cognitive simulations? In what follows, I want to propose that the computer simulations of cognitive processes represent the development of a new paradigm or perhaps of two complementary ones for the study of perceptual and cognitive processes. One paradigm is most closely associated with developments in the field of pattern recognition and concept formation, whereas a second is more relevant for studies in the fields of problem solving and symbol manipulation. I do not mean to imply that these are necessarily better paradigms but merely that these developments can best be understood as proposing a revised set of metatheoretical propositions which are of considerable interest to psychologists.

PATTERN RECOGNITION

Let us begin by trying to define the problem of pattern recognition. Consider an experimental situation in which letters and/or numbers are to be exposed visually one at a time and the viewer is to indicate by some appropriate response which symbol has been displayed. (A similar experiment could be conducted using an auditory presentation.) How we recognize these patterns is essentially the problem of pattern recognition.

Contemporary approaches to the study of how pattern recognition works depend substantially upon the neural network ideas of McCulloch and Pitts (1943) and the quantitative theory of information transmission developed by Shannon (1948) and by Shannon and Weaver (1949). What McCulloch and Pitts proposed was that there existed a formal correspondence between the logical functions of the propositional calculus and a neural network synthesized out of some rather simple neural elements. In the present-day terms of a computer programming language, any branching series of "If . . . . Then" statements can be represented by a neural network. Later, they discussed in a rather general way how these ideas about neural networks could be applied to the perception of visual and auditory forms (Pitts and McCulloch, 1947). Hebb (1948) was influenced by these models of neural networks. In his volume on *The Organization of Behavior,* he managed to acquaint psychologists with these ideas and to suggest that they were psychologically respectable. Ashby's *Design for a Brain* (1952) had a somewhat similar influence.

Shannon's influence is a rather complementary one. From his quantitative theory of information transmission, one can abstract the following qualitative propositions. First of all, it is misleading to talk about a system for the recognition of a single letter or word. Recognition is a process in which one first becomes aware of a set of available alternatives, and recognition is equivalent to selecting from among them. Moreover, one may make a partial choice, such as that the letter is *o, c,* or *q* and definitely not *a, v,* or *w.* In psychological terms, there is first a perceptual set that limits the possibilities; the function of communication is to reduce uncertainty about which symbol has been selected.

Second, there is the question of how patterns are encoded. To quote from Neisser's interesting discussion of pattern recognition, "To [Hebb], and no doubt to most of those concerned with the problem today, it seems likely that patterns are identified in terms of their 'attributes' (1967, p. 51). What are attributes? They are features or dimensions that have values, such as round versus square shape, the presence or absence of a horizontal line, or a form that is open or closed at the top. By analogy, each feature corresponds to a position in a code and the value assigned to the feature is equivalent to the value assigned to that position. In information theory, one talks about binary codes, bits, and a logarithmic relationship between the length of the code and the number of alternatives, but these characteristics of the quantitative theory are not essential for the development of a theory of pattern recognition.

Is there any other way in which the problem of pattern recognition may be formulated? Neisser refers to a theory of template-matching or of canonical forms which would imply that each stimulus input would be compared to a series of templates in order to obtain a match (1967, p. 50). Unfortunately, this view of pattern recognition appears to be unacceptable.

To review Shannon's influence on the study of pattern recognition, it is important to note that it has primarily been at the level of metatheory rather than of substantive theory; that is, he has helped to establish a new paradigm for the study of pattern recognition. One infers from the subject of information theory that one *should* try to abstract a set of features for describing a set of patterns, and one is left with the substantive problem of identifying an acceptable set of features, as, for example, the set that has been proposed for letters by E. J. Gibson (1969, p. 88).

To continue, let us consider a possible problem in pattern recognition. Let the set of alternatives include the letters *a, c, o, q, v,* and *w.* A possible recognition procedure is described in figure 2-1. The complete procedure will be called a decision tree, although it may also be called a discrimination net. The nodes, in the tree, are the "states" of uncertainty about what has been displayed, and a form can be identified when there is no longer any uncertainty. Each F_{ij} refers to the *j* the value of the *i*th feature. For example, F_{11} identifies those forms made up of straight lines *(a, v, or w)*, and F_{12} identifies the curved letters *(c, o, or q)*. We can now identify three theoretical problems that are pertinent to any theory of pattern recognition. First, one must identify the features and a pro-

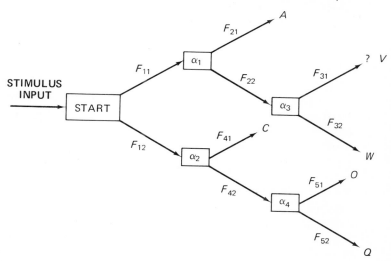

FIGURE 2-1   A decision tree for the identification of forms.

cedure for obtaining the alternative values. Second, there is the overall problem of performance, of how features are organized into a decision tree. Finally, there is the problem of learning, of how one acquires the ability to identify features and to organize them into a decision tree.

Visual Pattern Recognition

There are three major approaches to the study of visual pattern recognition, each of which can be simulated by a computer program. The "perceptron" as developed originally by Rosenblatt (1958, 1962) is one very general approach to the study of pattern recognition. Although it was originally proposed as a model of brain functioning in the form of a general purpose neural network, the applications of the perceptrons that have been worked out in detail are primarily for problems in pattern recognition. A simple perceptron consists of sensory elements, association units, and response units. The association units provide for a network of connections between the sensory elements and the responses.

It is customary to think of the sensory elements as a simulated retina and, more specifically, as a matrix array of elements that will "fire" or exhibit neural responses when stimulated. Presumably, different input patterns on this retina lead to a different pattern of activity at the sensory level. Initially, the association units are connected in random fashion to the response units, but these connections have variable thresholds and are trainable. The problem for a given perceptron is to follow a training sequence in such a way that the pattern of association becomes organized; the end result of a training sequence is that stimulus patterns should be appropriately connected to the response units.

Basically, the perceptron learns by reinforcement and, as Block has written, ". . . the device changes its internal functional properties. The resultant behavior exhibits, as is shown in the test, interesting aspects of learning, discrimination, generalization, and memory" (1962, p. 123). A first important question is whether or not such a randomly organized network is actually trainable by a reasonable sequence of reinforcements. Theoretical results have been obtained by Rosenblatt (1958), Block (1962), and Block, Knight, and Rosenblatt (1962). They have shown that the behavior of a perceptron can be represented as a stochastic process; given a long enough sequence of reinforcements, the network ought to learn to make appropriate discriminations among the stimuli. However, computer simulations have been most useful for studying the rate at which learning will take place and the effect of experimental variations on performance.

Results with these computer simulations indicate that the learning exhibited by the simple perceptrons is rather unsatisfactory. Their performance as self-organizing systems can be improved in a variety of ways. The "multiple-layer" perceptrons have several association layers connected in series between stimulus and response elements. Connections between successive pairs of association layers are trainable, as are the connections between the last association layer in the series and the response units. For example, Block, Knight, and Rosenblatt (1962) have described a four-layer series coupled perceptron with two association layers. By comparison with the simple perceptrons, these more complex systems do show an improved ability to learn by reinforcement. Another method of approach is to assume that the initial pattern of connections is not random and to introduce somewhat more complex methods of reward. Results along these lines obtained by Roberts (1960) are, in general, encouraging.

We began this discussion with the definition of a decision tree, and in certain respects the perceptrons do not fit this model of pattern recognition. In the theory of perceptrons, the emphasis is on learning and on modifying the properties of a neural network. One does not talk explicitly about feature detection or decision trees. On the other hand, the ability to identify features and to distinguish among their possible values is implicit in the functioning of these perceptrons, particularly the multiple-layer systems. As the total network becomes more organized, certain stimulus elements begin to function together as do certain association elements within a given association layer. Although the identification of features is never explicit, it is implicit as one aspect of the self-organization that is taking place, and the result of this process of organization can be interpreted as leading to the creation of a multiple-level decision tree.

There is a second approach to the problem of pattern recognition that follows rather closely the point of view inherent in the perceptron research. In the field of electrical engineering, considerable effort has been devoted to the development of efficient character recognition devices. The problem is enormously simplified as long as one makes use of a standard alphabet; under these circumstances, template-matching schemes are quite satisfactory. However, for the reading of handwriting or text with unstandardized alphabets, more subtle schemes must be developed. In particular, a variety of complex statistical schemes have been proposed as theories of character recognition. These include the use of correlational techniques (Horwitz and Shelton, 1961), and particularly discriminant analysis. It is this latter approach that I want to describe in some detail. With the conventional uses of discriminant analysis, one begins with a finite number of known groups or populations, such as students who succeed or fail. There is also a set of tests such that measures on these tests are available for the subjects from each of these populations. It is then possible to combine the tests into one or more linear functions (the discriminant functions) so as to give the "best possible" separation between the known populations. Subsequently, these functions can be used as the basis for a classification scheme which will be used in the following way. Given a new subject, let us assume that his set of test scores is known but that his membership in a population is unknown. Then the discriminant function values can be computed and used to assign the subject to one of the groups, and the classification procedure will be optimal in some defined statistical sense. See, for example, the discussions by Cooley and Lohnes (1962) on the use of discriminant functions in the behavioral sciences or by Nilsson (1965) on applications in the field of pattern recognition.

When applied to problems of character recognition, these discriminant functions can be reinterpreted as follows. Let each population be associated with a single character to be recognized and there will be as many populations as there are elementary characters. Instead of a set of tests, there will be an array of sensory elements to be stimulated, and a test score is equivalent to the intensity with which a particular sensory element is being stimulated at any given time. Then the discriminant functions will be defined on the readings associated with the sensory elements; the problem of classification is to assign a particular stimulus to one of the populations, which means effectively to recognize it as an example of a particular character.

As a typical example of this approach, let us review a pattern recognition procedure that was designed by Highleyman (1962). He made use of ten alternative patterns in the form of handwritten characters. The patterns were written on a 12 times 12 matrix; by analogy there were 144 tests used to describe an example of a pattern. A matrix element could be either filled or unfilled so that the only possible test scores were 0 or 1. In designing the pattern recognition procedure, one must first define the particular set of linear discriminant functions that are to be employed with numerical values assigned to all relevant parameters. Highleyman obtained samples of each character from each of fifty subjects (500 samples in all) and used these data for the design of the pattern recognition procedure. Subsequently, he tested the adequacy of the procedure with 120 additional examples, 12 samples per class. The results were that 61.6 percent of these characters were correctly identified; 19.2 percent could not be classified.

In this particular example, the machine did not actually follow a training sequence, although it is certainly possible to design an iterative procedure for the development of these discriminant functions; under these circumstances, a computer could effectively be trained to recognize patterns as was done in the research on perceptrons. (See, for example, the discussion by Nilsson (1965) on machines that learn to recognize patterns.)

In much of this research, designs are based on the use of linear discriminant functions. More general nonlinear discriminant functions may also be used, as in the research of Specht (1967), Cooper (1965), and Sebestyen (1962). In summary, these statistical approaches to the design of pattern recognition procedures have much in common with the perceptron research. The design procedures make use of known statistical results, and the implications for psychological theory are, at best, unclear. Certainly, with the use of discriminant functions there is no claim or intention that the results have any implications for psychology. In addition, there is no emphasis in either of these approaches on the design of a decision tree or on the identification of features that will be part of the sequence of decisions. However, the identification of features can be inferred (perhaps with difficulty) as implicit in the functioning of the decision procedures.

The third approach to the study of pattern recognition is based on the explicit identification of features and their organization into a decision tree. Using this approach, one must explicitly identify a set of features for describing the alternative patterns. Second, one must organize the recognition of features into a decision tree. How does one obtain a suitable set of features? These might be constructed on a priori or logical grounds or one might write a program that could construct its own features as it learned from a series of experiences.

Much of the early work on pattern recognition was stimulated by the writings of Selfridge (1955) and Selfridge and Dineen (1955), who suggested the emphasis on feature identification and helped to clarify some of the complexities inherent in the problem of pattern recognition. Selfridge also suggested the possibility of some preprocessing as an aid in the detection of features. Subsequently, Selfridge proposed a system called Pandemonium (1959) which would permit detection of features in parallel and would have some ability to select new features based on experience. Further work on the analysis of characters into features was undertaken by Selfridge and Neisser (1960). Doyle (1960) explored the possibilities of a Pandemoniumlike system that learns for use with handprinted characters. Other related schemes for the analysis of handwritten characters have been developed by Eden (1962), Eden and Halle (1961), and by Grimsdale, *et al.* (1959). Eden has also been concerned with a related problem, that of form generation, and particularly with a system for the generation of handprinted characters (Mermelstein and Eden, 1964; Eden, 1968). This system uses features or line segments as the basis for its character generation.

Undoubtedly, one of the most interesting efforts to write programs that will create and evaluate their own features is that of Uhr (1960) and Uhr and Vossler (1961), and their latter paper is appropriately entitled "A Pattern-Recognition Program That Generates, Evaluates, and Adjusts Its Own Operators." Basically, their program is able to make guesses about features (operators) that may be included in a recognition procedure. The adequacy of a set of operators can then be evaluated against experience; operators can be saved or discarded depending on their apparent contribution to successful discrimination; similarly, new operators can be devised and added to the procedure. What specifically are operators? They are generated within a 5 times 5 matrix of elements and may be either preprogrammed or generated as part of a training sequence. Examples taken from Uhr and Vossler (1963) of both types of operators are given in figure 2-2. Elements in an array may take on the value of zero, one, or blank. When used to scan a pattern, a "0" implies that there should be an absence of stimulus in the corresponding input location, a "1" implies that there should be a stimulus, and a blank implies indifference. In a sense, these matrices are measuring local characteristics of a pattern. With the preprogrammed operators, it is relatively easy to describe in words the feature that one hopes to identify, but such descriptions for the derived operators are more difficult to obtain. Of course, it is not essential that one be able to interpret verbally the operators that are being constructed.

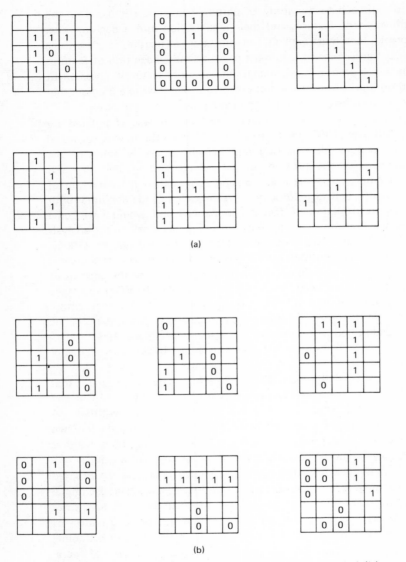

(a)

(b)

FIGURE 2-2   (a) Examples of preprogrammed operators and (b) examples of
operators generated by a program, from Uhr and Vossler, 1963.

These operators are used in the following way. First of all, unknown input
patterns are presented to the computer in the form of a 20 times 20 array whose
elements have the value of zero, one, or blank. The existing list of operators is
used in order to scan the pattern and to record for each operator—"(1) the
number of matches, (2) the average horizontal position of the matches within
the rectangular mask, (3) the average vertical position of the matches, and (4)

the average value of the square of the radial distance from the center of the mask" (1963, p. 253-54). This information is then used to characterize the input pattern. In addition, the computer remembers lists of characteristics, one for each pattern previously encountered. The information obtained from the scanning process is then compared with the information recorded on these category lists. The program then selects for its response the name of that pattern whose stored characteristics most closely resemble the input information.

We have described three rather different approaches to the simulation of pattern recognition. In what ways are they similar or different? In all three, there is an emphasis on training sequences, on how we learn to distinguish patterns. In the first two approaches, the emphasis is on rather sophisticated procedures for learning based upon methods of mathematical statistics. In the third approach, learning is more nearly heuristic. With the approach of Uhr and Vossler, there is also an explicit emphasis on the identification of features to be recognized; with the other approaches, the features remain implicit and to some extent buried in the mathematics of the pattern classifications. Moreover, the programs of Uhr and Vossler resemble in many ways "problem-solving" programs, such as those that will be discussed in a later section of this paper. This problem-solving or symbol manipulation characteristic is particularly noticeable when operators are being generated, evaluated, and modified. There is one distinction that can be made between the two more mathematical approaches and that proposed by Uhr and Vossler. With the former, there are few presuppositions about the selection of features for inclusion in the process of recognition, and the only presuppositions are mathematical rather than psychological. To be more precise, the programs are able to group together sensory elements in an enormous variety of ways, although certain groupings of elements may be impossible because of the presuppositions inherent in the underlying mathematics. On the other hand, the approach of Uhr and Vossler favors the selection of simple shapes, such as oriented lines, curvature, corners. This difference in the way in which operators are constructed is undoubtedly responsible for the relative lack of success in learning obtained by the approach based on random nets (Minsky and Selfridge, 1961). For example, let us consider the problem of selecting an effective set of features as a kind of search among a set of alternatives, the space of all conceivable sets of operators. With the statistical approaches, the space of all possible operators is extremely large, and the search for a reasonable set is slow and inefficient. With the approach of Uhr and Vossler, one focuses on a more limited family of possibilities, and, hopefully, the search procedure will be more efficient. We should add that this more focused search for a recognition procedure is appropriate only if a possible "answer" in the form of an efficient set of operators is actually included within the space of alternatives that are being considered. Otherwise, the search is doomed to failure. In the future, one might anticipate a continuing emphasis on the selection of features to be included in a recognition procedure and in the logic for their selection and evaluation.

Research on the simulation of visual pattern recognition has been continuing for at least fifteen years, and it is interesting to attempt to evaluate the results. In many respects, the achievements to date are impressive, although it is important that we do not overestimate their magnitude. Let us take the ability of humans to recognize forms as a standard of comparison. What is relatively easy for people proves to be surprisingly hard to simulate on a computer. While we have no difficulty in reading letters that vary in size and orientation, in reading our own handwriting or the handwriting of others, or in reading handwriting that has been carelessly executed, these variations pose major difficulties for a computer program. It is not inaccurate to say that the computer simulations that now exist perform rather well only as long as the range of permissible variation is limited. There are two additional sources of variations that people handle easily and that cannot yet be mastered by the existing simulations. In the research undertaken by Eden and his colleagues on the generation and recognition of handwriting (Eden, 1968), one encounters the following difficulty. When people read handwritten script (i.e., words or sentences as distinguished from single handwritten characters), they are somehow able to segment sentences into words and words into letters: to recognize the boundaries between successive units. Recognizing the boundaries between words is not a major problem, but recognizing the boundaries between letters in a word is much more difficult. Indeed, Mermelstein and Eden (1964) have paid a good deal of attention to this problem of letter segmentation. Eden has also discussed a related issue, namely, that the recognition of words and of letter boundaries can be simplified when one can take advantage of the context within which the word appears. He has quite persuasively suggested that when reading we use a knowledge of syntax as well as semantics in order to predict what is likely to occur and to help resolve ambiguities in the reading of particular words. This type of research is of considerable interest. We can infer from it that perception should not necessarily be studied by an isolated system. Indeed, the successful simulation of perceptual phenomena may be impossible unless we imbed perceptual capabilities in a more inclusive cognitive and problem-solving system.

It is also worth noting that research on the simulation of visual processes has focused primarily on the problem of pattern recognition, and this is perhaps the simplest of all visual phenomena. For example, it is illuminating to read the review article by Kolers (1968) entitled "Some Psychological Aspects of Pattern Recognition" in which he discusses a wide variety of visual processes that may legitimately be included under the heading of pattern recognition. Research on the simulation of these more complex processes is essentially nonexistent.

Speech Recognition

Just as there is research on the simulation of visual pattern recognition, there has also been an extensive effort devoted to developing programs that would analyze as well as synthesize speech.[1] Surprisingly enough, the problem

---

[1] I have not attempted to include material on the synthesis of speech. See for example the chapter on speech synthesis included in Flanagan (1965), the collection of papers edited by Wathen-Dunn (1967), and the chapter on speech analysis and synthesis by Fant (1968).

of speech recognition has proved to be rather more difficult. Lindgren (1965) reviewed much of this research on automatic speech recognition and concluded that the effort had been frustratingly unsuccessful. Similar views have been expressed by Hill (1967) and Flanagan (1965). Why should this be so? Why should automatic recognition of speech be more difficult than the automatic recognition of visual forms?

To begin, let us describe a possible approach based upon the use of decision trees (fig. 2-1) as they were applied to the recognition of visual forms. In this case, the input will be the acoustic information. The desired outputs would be a series of phonemic units out of which words might be constituted. The decision tree would attempt to analyze the signal into a set of distinctive features, such as have been proposed by Jakobsen, Fant, and Halle (1952), or a set of phonetic parameters, to use the more neutral term introduced by Halle and Stevens (1964). This approach would seem to be a straightforward one and the difficulties encountered are illuminating. For example, Davis, Biddulph, and Balashek (1952) devised a procedure called Audrey for recognizing the ten spoken digits based upon an evaluation of the first and second formants. It works reasonably well as long as inputs are limited to a single speaker. The system does not have provisions for automatically adjusting for differences among speakers. However, accuracy can be improved by permitting some manual adjustments which compensate for each additional speaker. Some extensions of this approach have been undertaken by Dudley and Balashek (1958).

Wirren and Stubbs (1956) have adopted a rather different approach. They have constructed a scheme which can quite easily be described as a decision tree which uses the distinctive features of Jakobsen, Fant and Halle (1952) to analyze the acoustic signal and produce as output a series of phonemes. A variety of other approaches are summarized in Lindgren (1965). The earlier approaches have several characteristics in common. Their analysis focuses primarily on the acoustic signal. They are reasonably successful with very limited vocabularies and are usually limited to the analysis of a single word or syllable at a time, that is, they do not attempt to analyze phrases or sentences. Although reasonably accurate for a single speaker, they are unable to adjust to differences among speakers.

How might improvements be introduced? Along these lines, there have been various attempts to improve recognition procedures without introducing any major reorientation in point of view. One possibility is to make use of the statistics of sound patterns in the form of the probabilities with which one sound will follow another. A machine based upon this form of prediction was devised by Denes (1959) with some success. Shearne and Leach (1968) have been concerned with a related complication, that of taking into account the characteristic variations among speakers. Their system attempts to recognize thirty-two words with a population of ten different speakers. Instead of storing a single representation of a word, they store several representations of each word based on inputs from several speakers. Essentially, the system can remember how a word sounded when uttered by different speakers. Then inputs are matched against a family of templates, and the closest match is selected for response. With nine templates per word, accuracy approaches 90 percent. Unfortunately, it is hard to believe that this approach is either psychologically reasonable or generally useful for a realistic vocabulary and for a larger number of speakers.

One other method for taking into account individual differences is of theoretical as well as psychological interest. The principle is essentially as follows. Consider a decision schema or a family of decision trees with some unspecified parameters which would be used to characterize the differences among speakers; a given speaker could be represented (approximately) by a particular set of values for these parameters. Furthermore, let us assume that our recognition procedure be self-adjusting in that it could choose the values of these parameters. Upon listening to a new voice, the system would first set the parameters so that it could be tuned for the particular speaker whose speech it was to recognize. Such a self-adjusting scheme called MAUDE was first devised by Gold (1959) and applied to the problem of the machine recognition of hand-sent Morse code. In this case, there are characteristic variations among telegraph operators; individuals tend to define rather differently the dots, dashes, and spaces that make up the code. Gold's machine-recognition procedure has a self-adjusting capability that significantly improves its performance. MAUDE is also discussed in Selfridge and Neisser (1960). More recently, Cannon (1968) has described a method of analysis for spoken vowels which can automatically compensate for individual differences among speakers by a method of parametric adjustment. This method is promising and should receive continuing attention.

Parenthetically, this problem of individual differences also arises with the machine recognition of handwriting. It seems intuitively clear that our ability to read a given individual's handwriting improves with experience. We appear to be able to adjust in order to take into account systematic differences between individuals. To the best knowledge of this reviewer, little research if any has been devoted to this problem, and I am unaware of any machine-recognition procedure for handwriting analysis that compensates for individual differences.

What limits the success of this research on speech recognition is that our understanding of the auditory signal is still surprisingly limited and that our understanding of the relationship between auditory signals and phonemic analysis is inadequate. As a consequence, there has been an extraordinary amount of effort devoted to the study of the underlying sensory processes in an attempt to determine how information is conveyed by acoustic inputs. Liberman *et al.* have written an excellent review of much of this research (1967a). A significant source of difficulty is that the phonetic parameters or distinctive features are not independent of each other. The form of a particular feature will be influenced by other features with which it is associated. As stated by Liberman *et al.*: "a finding which turns up over and over again in a variety of different yet related forms is that there is a complex relation between perceived language and the acoustic signal which conveys it. More specifically, the acoustic signal is quite commonly not invariant with respect to the phoneme as perceived." (1967b, p. 74).

One radically different approach proposed by Halle and Stevens (1964) is likely to be of increasing importance in the future. Their point of view is substantially motivated by their pessimistic evaluation of much of the research that has already been undertaken. "The analysis procedure that has enjoyed the widest acceptance postulates that the listener first segments the utterance and then identifies the individual segments with particular phonemes. No analysis scheme based on this principle has ever been successfully implemented. This failure is understandable in the light of the preceding account of speech production, where it was observed that segments of an utterance do not in general stand in a one-to-one relation with the phonemes. The problem, therefore, is to devise a procedure which will transform the continuously changing speech signal into a discrete output without depending crucially on segmentation." (Halle and Stevens, 1964, p. 606.)

We can make use of an analogy as a way of describing the shift in point of view that has been proposed by Halle and Stevens. In the field of computer science, most procedures for the syntactic analysis of programming languages (such as FORTRAN) have been characterized as either "bottom-up" or "top-down" methods of analysis. (See Feldman and Gries (1968) for a review of these alternatives.) With bottom-up analysis, one proceeds from part to whole. One emphasizes the identification of the smallest units making up a sentence and implicitly assumes that, having identified the parts, the identification of the whole will follow. With top-down analysis, one proceeds from whole to part: that is, one keeps in mind a prediction about the larger unit that is being identified and uses this prediction as an aid for the identification of the elements as they are being processed.

The procedure being implemented by Halle and Stevens is called "analysis by synthesis." In many respects, it can be described as "top-down," whereas the more customary approaches can be described as "bottom-up." With analysis by synthesis, the system must store a set of generative rules for transforming phonemes into their related phonetic parameters or distinctive features, that is, for generating speech. On the basis of a preliminary analysis of the speech signal, certain decisions are reached that limit very sharply the set of phonemes that are to be considered as possibilities. Using the generative rules, the system then synthesizes in sequence the sound pattern associated with each phoneme and compares the synthesized pattern and the received signal. A conclusion is reached to select the phoneme that leads to the closest match between the pair of signals.

At the present time, the main effort to implement this proposal has been limited to a system that accepts an uncorrelated series of inputs, akin to nonsense syllables or words. For this purpose, the only generative rules that need be included are for the generation of the sounds associated with a given phoneme. However, other forms of generative analysis could also be included for the processing of sentences or continuous text. To quote Halle and Stevens, "If the speech material to be recognized consists of words, phrases, or continuous text, then the output of the present analysis scheme would have to be processed further to take account of the constraints imposed by the morphological and syntactic structure of the language" (1964, p. 610). Presumably, one could at least include a second level of generative rules that could aid in the synthesis of well-formed sentences as well as sounds.

Interestingly enough, Eden, Halle, and Mermelstein have adapted these ideas about analysis by synthesis for use in their research on the machine recognition of handwritten script (Eden and Halle, 1961; Mermelstein and Eden, 1964; Eden, 1968). They define a set of generative rules for the synthesis of handwritten words based on a small number of elementary forms. Using these generative rules, the system synthesizes letters and words which are used in the analysis of the material that is being read. As in the case of speech recognition, additional generative rules at the syntactic level could be added to extend the capabilities of the system.

Recently, Neisser (1967) has suggested that analysis by synthesis is characteristic of a variety of cognitive processes. From a psychological point of view, analysis by synthesis has at least the following implications. It is potentially a very powerful strategy for making predictions about patterns and facilitates predictive processes at more than one level of analysis. Second, it avoids any sharp separations between sensory, perceptual, and cognitive phenomena. It suggests a simulation of a cognitive process which may include in an integrated way provisions for the accompanying perceptual levels of analysis. Finally these separate approaches of Stevens, Halle, and Eden to the machine recognition of speech and handwriting are quite consistent with a variety of experimental data. We know that meaningful sentences are easier to process than are unrelated strings of words. Provisions for the inclusion of generative rules at both the syntactic and the phonetic levels suggest at least one possible explanation for the observed experimental data with human subjects.

PROBLEM SOLVING

Current approaches to the simulation of problem solving have a variety of intellectual antecedents, including those already discussed under the heading of pattern recognition. In addition, there is a strong emphasis on the purposive nature of human behavior and on mechanisms for error correction, monitoring, and control. Thus, this research can be viewed as an extension of the early work of Cannon (1939) on homeostatic mechanisms and of the cybernetic revolution associated with the writings of Wiener (1948) and even earlier by Rosenblueth, Wiener, and Bigelow (1943). Experiences with general purpose digital computers have also had an important influence on the development of this point of view. First of all, the simulations of problem solving would be impossible if computers did not exist. More importantly, one borrows theoretical concepts particularly from computer programming for the representation of psychological processes. The idea of a stored program itself is of theoretical interest, since it can be interpreted as a representation of a plan, a hierarchy of plans, or a purposive system with provisions for error correction. See, for example, the stimulating discussion by Miller, Galanter, and Pribram (1960) which considers these analogies.

In the previous discussion, I suggested that Shannon's work on a mathematical theory of information had been influential in the establishment of a metatheory about pattern recognition but that information theory as a substantive body of knowledge had only a minor influence. This same observation can be made about the influence of Wiener as well as the developments in the design of feedback systems. The book *Cybernetics* (Wiener, 1948) is a stimulus to research but is certainly not a substantive blueprint for it. Moreover, although there exists an extensive body of theory about the design of control systems, this body of theory has had little impact if any on the simulations of problem solving that have been undertaken. One borrows concepts, but the mathematical results are not directly applicable.

Although "problem solving" can be defined in a variety of ways, we need to consider how the term is used in the field of computer simulations. In psychology, it is often equated with "thinking," so that one ambiguous phrase is replaced by another that is equally ambiguous. Problem solving is also used as a residual category to include whatever cannot be assigned elsewhere. In the influential text by Woodworth and Schlosberg, a chapter entitled "Problem Solving: Thinking" includes the following statement: "A problem, we may say, exists when 0's activity has a goal but no clear or well-learned route to the goal. He has to explore and find a route." (Woodworth and Schlosberg, 1965, p. 814.) Thus, problem solving becomes essentially a search for means to achieve an end. Along similar lines, Johnson, in his influential book on thought and judgment, states: "A person may be said to have a problem if he is motivated toward a goal and his first goal-directed response is unrewarding. Typically, he varies his activity, and perhaps by some subsequent response he does attain his goal or progress toward it. Problem solving, then, which is one meaning of thinking, is the activity . . . of an organism trying to attain a not easily attained goal." (Johnson, 1955, p. 63.)

By way of comparison, the computer simulations of problem solving are often referred to as "information-processing theories" in which the analogy between computer programs and thought processes is emphasized. Indeed Newell and Simon (1965) have entitled their recent review of information-processing theories of human behavior "Programs as Theories of Higher Mental Processes." Elsewhere, Ernst and Newell have written, "The problem is initially expressed in some external representation, which is converted by a translator into an internal representation inside the computer. The internal representation is processed by a set of problem-solving techniques, and the result of this processing is (hopefully) the solution" (1969, p. 7). The point of view implied by the computer simulations of problem solving appears to have the following characteristics. Most importantly, the simulating program can be interpreted as a psychological theory for explaining how people respond to a series of sensory inputs. It is clearly a dynamic model in the sense that it has "moving parts" and operates in time. Moreover, the problem-solving process can be broken down into three major subsystems. One of these must simulate the ways people accept sensory inputs and translate them into an internal representation (the problem of perception). A second must simulate the manipulation of an internal problem representation and the internal processes of inference, deduction, induction, and recall. A third subsystem must simulate the way people finally produce a response, that is, how they translate from an *internal* representation of a response into a sequence of actions (the problem of response control). If one is willing to accept this formulation, then perception becomes a part of problem solving: It is the first subsystem that has just been identified. Moreover, our abilities to understand natural language, whether written or spoken, fall within this definition of problem solving. After all, problem statements are usually expressed in language, and one cannot begin to solve the problem until it has been translated into an internal representation. In practice, most research focuses on only one of these subsystems at a time. For example, in Newell and Simon's Logic Theorist (1956), all the emphasis was on the internal problem-solving processes and the complications of input and output were effectively bypassed.

As one reviews this research on the simulation of problem solving, it becomes apparent that problem solving is being given a broad interpretation. Clearly, proving a theorem, playing chess, writing an essay, would all be accepted as examples of problem solving. Yet within this information-processing point of view, remembering what my wife asked me to buy at the grocery store or giving the correct answer to "How much is two plus two?" are in some sense also examples of problem solving. Apparently, problems fall on a continuum in which the internal processing required before responding will vary qualitatively as well as quantitatively. And this point of view is not seriously inconsistent with the definitions given by Woodworth and Schlosberg or Johnson in spite of the difference in emphasis.

I would now like to give an overview of what has been accomplished during the approximate period of 1956-69. Some of these accomplishments are methodological while others are substantive, having to do with the variety of problem-solving behaviors that have been simulated and the implications for a general theory of problem solving. The substantive accomplishments can be described at two levels of abstraction. On the one hand, there exist programs that will simulate the ability to solve certain classes of problems, such as theorem proving, game playing, or verbal learning. On the other hand, there is research on the general nature of problem solving and its representation to deal with a variety of problems in a unified way. We shall first of all discuss this general model of problem solving and use it as the basis for further discussion of a variety of examples.

A General Problem Solver

Over a period of years, Newell and others have been engaged in the design of a general problem solver, known as GPS. An early version of GPS was described in Newell, Shaw, and Simon (1957b). Some results with GPS were discussed in a paper given by Newell and Simon (1961b), and a detailed description of GPS appears in Newell (1963). More recently, Newell and Ernst have written about GPS as illustrating "The Search for Generality" (Newell and Ernst, 1965). The results of this twelve-year program of research have now been reported in a book by Ernst and Newell (1969), appropriately entitled *GPS: A Case Study in Generality and Problem Solving.*

What are the characteristic features of GPS? GPS has a way of representing a given task environment as a set of objects, operators, differences among objects, and a set of relationships between differences and operators. The objects are state descriptions such as an expression to be integrated, a position in a game of chess, or a set of simultaneous equations to be solved. The operators are the rules for introducing changes into a state description, such as a possible move in chess or a substitution to be made within a set of equations. Differences permit one to compare objects and, presumably, are essential to the evaluation of progress.

In addition, GPS makes use of a problem-solving executive which coordinates a series of moves in an effort to solve a problem. In GPS, the executive program relies on a very general strategy for solving problems, known as heuristic search. "A heuristic-search problem is:

"Given: An initial situation represented as an object. A desired situation represented as an object. A set of operators.

"Find: A sequence of operators that will transform the initial situation into the desired situation.

"The first operator of the solution sequence is applied to the initial situation, the other operators are applied to the result of the application of the preceding operator, and the result of the application of the last operator in the sequence is the desired situation." (Ernst and Newell, 1969, p. 17.)

We can make this discussion more concrete by considering a specific example and how it may be represented in GPS. The problem is to simulate how people prove trigonometric identities. In many respects this is a typical example of mathematical problem solving. It resembles proving theorems in symbolic logic or geometry and has been discussed by Newell, Shaw, and Simon (1960). Here are four possible identities to be established:

1. $$\cos \frac{\theta}{2} \, \tan \frac{\theta}{2} = \sin \frac{\theta}{2}$$

2. $$\sec \theta + \tan \theta = \frac{\cos \theta}{1 - \sin \theta}$$

3. $$\frac{\sec \theta + 1}{\tan \theta} = \frac{\tan \theta}{\sec \theta - 1}$$

4. $$\frac{\csc^2 \theta - \cot^2 \theta}{\sec^2 \theta} = \cos^2 \theta$$

A set of protocols obtained from a subject in solving the fourth of these problems is given in figure 2-3, and the protocols summarize the series of expressions obtained by the subject as well as his justification for the operations he performs.

Now let us consider a possible representation of this task in GPS:

STEP	EXPRESSION	JUSTIFICATION
1)	$$\dfrac{\csc^2\theta - \cot^2\theta}{\sec^2\theta}$$	Left-hand-side
2)	$$\dfrac{\dfrac{1}{\sin^2\theta} - \cot^2\theta}{\sec^2\theta}$$	Substitution: $\csc^2\theta = \dfrac{1}{\sin^2\theta}$
3)	$$\dfrac{\dfrac{1}{\sin^2\theta} - \dfrac{\cos^2\theta}{\sin^2\theta}}{\sec^2\theta}$$	Substitution: $\cot^2\theta = \dfrac{\cos^2\theta}{\sin^2\theta}$
4)	$$\dfrac{\dfrac{1 - \cos^2\theta}{\sin^2\theta}}{\sec^2\theta}$$	Factoring
5)	$$\dfrac{\dfrac{\sin^2\theta}{\sin^2\theta}}{\sec^2\theta}$$	Substitution: $\sin^2\theta = 1 - \cos^2\theta$
6)	$$\dfrac{1}{\sec^2\theta}$$	Cancellation
7)	$\cos^2\theta = \cos^2\theta$ Q. E. D.	Substitution: $\cot^2\theta = \dfrac{1}{\sec^2\theta}$

FIGURE 2-3   Sample protocols for proving a trigonometric identity.

1. The initial objects are the left-hand and right-hand side of each identity that is to be established. From the protocols it is apparent that the subject selects $\cos^2\theta$ as the desired object and

$$\frac{\csc^2\theta - \cot^2\theta}{\sec^2\theta}$$

as the initial object; his goal is to transform the latter into the former. Note that this particular choice is arbitrary and probably depends upon the subject's style for dealing with problems as well as past experience.

2. The set of operators used in transforming objects includes some that are trigonometric and some that are algebraic. For example, in steps 1 and 3 he introduces the trigonometric definitions

$$\csc\theta = \frac{1}{\sin\theta} \quad \text{and} \quad \cot\theta = \frac{\cos\theta}{\sin\theta}$$

In step 5 he uses the algebraic identity that $A/A = 1$ where A is any well-formed expression. One can also raise some interesting questions about how operators are remembered. It is quite likely that he constructs the operator $\cot \theta = \cos \theta / \sin \theta$ out of some more fundamental ones: i.e., since

$$\tan \theta = \frac{\sin \theta}{\cos \theta} \qquad \text{and} \qquad \cot \theta = \frac{1}{\tan \theta};$$

therefore,
$$\cot \theta = \frac{\cos \theta}{\sin \theta}.$$

Moreover, after proving one identity, it may then be reclassified as an operator for use with subsequent problems.

3. We can infer from the protocols that the heuristic search makes use of objectives at two levels of abstraction. The primary goal is to obtain two identical objects as the outcome of a legal series of moves. But this goal is not immediately operable. As a subgoal, one can attempt to eliminate differences between objects in order to make progress toward the primary objective.

4. This goal-seeking behavior implies the ability to compare objects and to identify in what respects they differ. For example, in the first three steps, $\csc \theta$ and $\cot \theta$ are eliminated from the initial object, apparently because they are absent from the desired object. In step 4, for some unknown reason the subject chooses to simplify the obtained expression rather than to eliminate $\sec^2 \theta$, perhaps because the obtained object has both numerator and denominator, whereas the desired object is not expressed in fractional form.

This goal-seeking behavior also implies some ability to select operators that are appropriate to a particular subgoal.

5. Finally, we can give a graphical interpretation to this heuristic search procedure, as is shown with the tree description given in figure 2-4. "Each node of the tree represents an object, and each branch from a node represents the application of an operator to the object represented by the node." (Ernst and Newell, p. 17.) Let A_I represent the initial object and A_D the desired object, and the operators employed can arbitrarily be designated as $X_1 X_2, \ldots X_6$. Then $(X_1, X_2, X_3, X_4, X_5, X_6)$ is the solution to the problem. In figure 2-4, the path from A_I to A_8 using X_7 might be interpreted as an example of a path that was explored and subsequently abandoned.

Problem Representations in GPS.

Clearly the problem of proving trigonometric identities can be given an adequate representation in GPS. We now want to consider the generality of this formulation and its empirical adequacy for the representation of other simulations of problem-solving behavior. First of all, it will be necessary to generalize the previous definition of heuristic search in the following way: "The initial

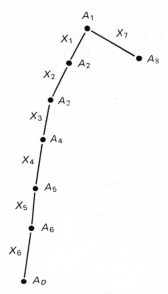

FIGURE 2-4   An example of heuristic search as represented by a tree of objects.

situation may be more than one object that can be represented by an object schema or a list of objects and object schemas." (Ernst and Newell, 1969, p. 19.) An object schema is an expression with one or more variables such that it specifies an explicit object upon substitution of constants for variables. Then, one class of problems to which GPS readily applies are those problems for which the initial and final situations as well as the list of operators can *explicitly* be defined. Trigonometric identities are an example of such a class of problems. The proofs of identities in any formal system are classes of problems for which the initial and desired situation are well defined.

Ernst and Newell (1969, p. 3) also discuss the set of problems known as the Missionaries and Cannibals Task. They give an example of a typical problem "in which there are three missionaries and three cannibals who want to cross a river. The only means of conveyance is a small boat with a capacity of two people, which all six know how to operate. If at any time there are more cannibals than missionaries on either side of the river, those missionaries will be eaten by the cannibals. How can all six get across the river without any missionaries being eaten?" For this set of problems, the initial situation is that there are a certain number of missionaries and cannibals on one side of the river. The desired situation is that they have all been transported safely to the other side. The various ways in which the boat can be loaded are the operators for modifying one object into another.

In the statement calculus, there is a class of problems in which one is given a set of premises and a possible conclusion. The problem is to verify whether or not the conclusion is a valid consequence of the initial premises. Here is a typical problem taken from Stoll (1961, p. 90), first given in words and then in symbols:

Wages will increase only if there is inflation. If there is inflation, then the cost of living will increase. Wages will increase. Therefore, the cost of living will increase.

In symbolic form, we will obtain

where *I*, *W*, and *C* are the three prime sentences appearing in the premises and *C* (the cost of living will increase) is asserted as a valid consequence. For this class of problems, the initial situation is defined as a set of objects (the premises), the desired situation is a single object (the conclusion), and the operators are the axioms and rules of inference that are to be used within the statement calculus.

There is a related class of tasks in which both the initial and the desired situations can be said to be defined. Programs have been written to prove theorems in geometry (Gelernter, 1959), in the statement calculus (Newell, Shaw, and Simon, 1957a), and the first-order predicate calculus (Ernst and Newell, 1969). In each case, a theorem is stated to be proved and this is the desired object; some initial premises are given and these are the initial objects. Certain operators permit the system to make substitutions and inferences. Moreover, in proving a theorem, it is always admissible (and often necessary) to add one or more axioms of the system as additional premises or to introduce a theorem that has already been proved as an additional premise. We can define as an operator that procedure that permits the introduction of additional premises into the system.

We now want to describe some additional tasks which have the following characteristics. The initial situation is explicitly defined, whereas the desired situation is given an implicit definition. By an implicit definition, we imply that the desired situation cannot be described in precise form, but one can introduce a set of criteria or a set of tests to be applied to a series of objects. The desired object is an object that satisfies these criteria. Consider, for example, the task of differentiation which has been successfully simulated by Moses (1967) as part of his program for the simulation of integration. A sample problem is given in figure 2-5. The initial object is explicitly defined, and the differentiation operator, *Dx*, has been introduced to indicate that the expression is to be differentiated with respect to *x*. The desired object is simply a new expression from which all references to the differentiation operator have been eliminated. The operators are a set of elementary differentiation formulas which can be applied

INITIAL OBJECT

$$D_x \ (X^2 \ \tan X)$$
$$= \quad X^2 \ D_x \ \tan X + \tan X \ D_x \ X^2$$
$$= \quad X^2 \ \sec^2 X \ D_x \ (X) + \tan X \ (2X \ D_x \ (X))$$

DESIRED OBJECT $= \quad X^2 \ \sec^2 X + 2X \ \tan X$

FIGURE 2-5   An example of differentiation when formulated as a problem
in GPS.

to the series of expressions. For this task, a solution algorithm exists and a suitable program for the simulation of differentiation can easily be written.

There are a variety of other tasks that can be characterized in a similar way. Slagle (1963) has described a program that can solve problems in symbolic integration, and a more powerful version has been developed by Moses (1967). As in differentiation, the initial object is a symbolic expression, and the operator ∫ is attached to signify that the expression is to be integrated. The criterion to be met by the desired object is that it be an expression from which the integration operator has been eliminated.

There also exist programs for obtaining solutions to a variety of word problems. A word problem is simply a problem stated in words (English) that must first be translated into a suitable formal representation before a solution can be obtained. The problem of translation into a formal representation will be discussed subsequently. For the moment, we shall limit our attention to the manipulation of the formal representation.

Bobrow (1964) has designed a system for obtaining solutions to algebraic word problems. A typical problem is the following: "Bill's father's uncle is twice as old as Bill's father. Two years from now Bill's father will be three times as old as Bill. The sum of their ages is ninety-two. Find Bill's age." Along related lines, Kuck and Krulee have devised a system for obtaining solutions to simple problems in physics as well as the algebraic problems dealt with by Bobrow. A typical physics problem might be the following: "A current of 15 amperes flows through an electrical appliance of 30 ohms resistance and a lamp of 190 ohms resistance. What is the voltage drop across the lamp?" (See Kuck, 1963, and Kuck and Krulee, 1964.) Finally, Charniak (1968) has written a program called CARPS which solves word problems in calculus. Here is a typical problem that his system may solve:

Water is flowing into a conical filter at the rate of 15.0 cubic inches per second. If the radius of the base of the filter is 5.0 inches and the altitude is 10.0 inches, find the rate at which the water level is rising when the volume is 100.0 cubic inches.

In what sense are these problems that may be represented in GPS? In each case, the word problem is translated into a set of simultaneous equations or logical relations, and one or more variables are designed as the desired unknown. This set of relations is the initial situation. The desired situation is that a certain variable be assigned a numerical value. The operators are the rules of substitution and inference for manipulating the initial situation, and the problem-solving process will continue until the desired object has been attained. Alternatively, we may say that the desired situation is given as an object schema, and the problem solving will continue until the variable in the schema has been assigned a value so that a particular instance of the schema can be obtained. In the algebraic word problem taken from Bobrow (1964), the desired situation might be represented initially as EQ (Age, x) where we have been asked to find the value of Bill's age that makes this a true statement. The desired object will be obtained when it can be shown that the above relation is true for a value of $x = 22$.

Games, such as chess or checkers, are also examples of tasks for which the initial object is explicitly given and the desired object is implicitly given. Let us consider first the game of checkers for which a series of interesting simulation programs have been written by Samuels (1959, 1963). In writing a simulation of checker playing, the initial situation is explicitly defined by the arrangement of pieces on the board. Assuming that one's objective is to win, then the desired object is a winning configuration that is defined by the criterion for determining who wins the game. The operators define legal moves and how one situation is transformed into another. In principle, the simulation program could think through a series of moves and opponent responses until a winning sequence had been determined. In practice, the attainment of this ultimate objective is not possible and must be replaced with some alternative and approximate procedure for evaluating a sequence of moves. Samuels' program looks ahead a fixed number of moves and uses an evaluation polynomial to score the resulting board positions. Then it makes use of a minimax procedure to select a particular move. The evaluation polynomial uses certain criteria, such as maintaining a relative piece advantage, to evaluate each of the board positions. Finally, the choice of each move can be viewed as a new problem to be solved. The board position at that time is taken as the initial situation, and the desired situation is defined implicitly by the characteristics of the evaluation polynomial.

From the point of view of GPS, chess resembles checkers, although it is a significantly more complex task. As with checkers, the desired situation is to checkmate one's opponent, although it is not in general possible to achieve this objective directly. Instead the simulation programs look for a good position that involves the evaluation of possible board positions according to certain desirable features. These may include relative material advantage, control of the center of the board, maintaining an attack on one's opponent, etc. There now exist several programs that play chess, including those of Bernstein *et al.* (1958), Newell *et al.* (1963), Newell and Simon (1965), Baylor and Simon (1966), Greenblatt, Eastlake, and Crocker (1967). Interestingly enough one of them (Greenblatt's) plays surprisingly good (although not expert) chess. More importantly, all of these programs use evaluation functions to evaluate possible positions and select a move, and all of them can be interpreted within the framework of GPS.

There is one additional example to be discussed within the framework of GPS, that of binary choice. "In the binary choice experiment, the subject is asked to predict which of two events, E_1 or E_2, will occur on each of a series of trials. After the subject makes a prediction, he is told which event actually occurred." (Feldman, Tonge, and Kantor, 1963, p. 55.) As the experiment continues, the subject makes additional predictions, each of which is either confirmed or disconfirmed. The original simulation of subject protocols was written by Feldman (1961), and a revised version was written within the framework of GPS by Feldman, Tonge, and Kantor (1963). In this task, the program makes a series of hypotheses about the sequence of events, and the purpose of these hypotheses is to predict subsequent events. In general, the initial situation is again a hypothesis that will satisfy a criterion, namely, the ability to predict subsequent events. The operators are the rules for modifying a hypothesis based upon experience with predicting a sequence of events, and the operators are used after every event when a current hypothesis has been either confirmed or disconfirmed. From this description, it is not surprising that this task behavior can also be represented within the GPS.

We conclude this section with a brief review of methodological accomplishments in the simulation of problem-solving behavior. These methodological contributions relate primarily to the special purpose list-processing and string manipulation languages which have been developed for use in this field. Among the more important languages are IPL (Newell, 1965), LISP (McCarthy *et al.*, 1960), COMIT (Yngve, 1962), and more recently SNOBOL (Farber, Griswold, and Polonsky, 1966; Forte, 1967).[2]

It is factually correct to state that programs to simulate problem solving have usually been written in one of these list-processing languages.[3] Why should this be so? Why were the programs not written in a more conventional language such as FORTRAN or ALGOL? To answer this question, let us refer to our discussion of GPS with its emphasis on objects, operators, differences, and a strategy for the manipulation of objects. In GPS, one represents both objects and operators as lists or as list structures (list of lists). One is continually examining lists to identify the presence or absence of certain properties and subsequently using the operators to modify these lists. These list-processing features are difficult to simulate within more conventional languages; list-processing languages have specifically been designed to facilitate the simulation of these problem-solving behaviors.

COMPUTER PROGRAMS
WITH NATURAL LANGUAGE INPUTS

Most problem statements are initially represented in natural language, and thinking about a problem cannot begin until the problem statement has been

---

[2] Any detailed discussion of these languages is beyond the scope of this review. The interested reader should refer to Bobrow and Raphael (1964) for some comparative comments or to Jean Sammet's review of list-processing languages (1969).

[3] This comment does not apply to the research on visual pattern recognition or pattern recognition in general. These programs are less complex and can be written in FORTRAN or machine language.

understood and translated into some suitable internal representation. Thus, it is not surprising that considerable effort has been devoted to the design of problem-solving systems that can accept the initial problem statement in natural form, that is, English. In what follows, we shall primarily discuss research on the process of translation from external to internal representation. The problem-solving process itself has already been discussed in the preceding section.

Question-Answering Systems

One of the earliest systems that could accept natural language inputs was a program known as BASEBALL, written by Green *et al.* (1961), and it has a number of interesting characteristics. BASEBALL answers questions about baseball and sample questions might be the following: "Whom did the Red Sox lose to on July 5, 1947?" "Did Babe Ruth hit sixty home runs in 1927?" "What was the score of the game played between the Yankees and the Red Sox on April 29, 1953?"[4] Questions in BASEBALL are translated into a formal representation known as a specification list which summarizes the known facts or conditions as well as the unknown that must be retrieved. Each specification list is made up of a series of attribute value pairs, some of which are complete and some of which may be incomplete. In any question-answering system, it is useful to recognize the existence of two classes of questions for which answers can readily be obtained. One class is the yes-no questions for which every attribute in the specification list has been assigned a value and the objective of search is to verify whether or not the specification list represents a valid statement. "Did Babe Ruth hit sixty home runs in 1927?" is an example of such a question. There are other questions which are associated with the interrogative pronouns when, where, who, what, and how, for which the specification list is necessarily incomplete. There is at least one attribute to which a value is unassigned, and the objective is to retrieve the necessary value or values that complete the statement and make it valid. Examples of specification lists for each type of question are given in figure 2-6.

BASEBALL also makes use of a file of records about games, and records about players can easily be added. A dictionary is also essential to the functioning of the system. The program is in two parts. In the first part, the question is subjected to a syntactic analysis in order to translate into a specification list. In a sense, this ability to translate correctly into a specification list is equivalent to being able to understand the question. In the second part, the program uses the specification list to control its search of the records, and the answer is contingent upon the successful completion of this search.

Other question-answering systems have subsequently been written which resemble BASEBALL, and these have generally been designed with a particular

---

[4]These questions are illustrative of reasonable questions about baseball, but these specific questions are not necessarily answerable in BASEBALL as it was actually programmed. For example, questions about games can be answered but not about players.

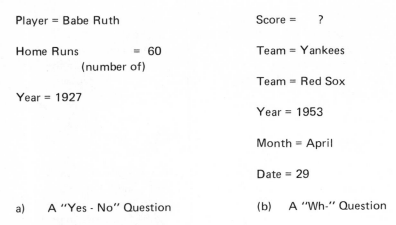

a)   A "Yes - No" Question   (b)   A "Wh-" Question

FIGURE 2-6   Specification lists for two classes of questions.

subject matter in mind. For example, Woods (1967) has designed a system that will answer questions about the *Official Airline Guide.* Vallée has programmed a system called ALTAIR that will answer questions of interest to astronomers about a star catalog (Vallée, 1967; Vallée, Krulee, and Grau, 1968). PROTO-SÝNTHEX, designed by Simmons, Klein, and McConlogue (1963) answers questions about the contents of an encyclopedia. The DEACON BREADBOARD system discussed by Craig (1963) and Thompson (1964) is another example of a question-answering system that accepts questions in English. Some general reviews of question-answering systems with natural language inputs are also available, and the reader may wish to consult them. See, for example, the rather optimistic discussions by Simmons (1965) or by Bross (1969) as well as the more skeptical review by Kasher (1967).

These systems differ substantially in the way in which they are implemented and in the subject matter for which they are designed. Yet there are certain general observations that one can make that are relevant to the simulation of cognitive processes. First of all, each of these systems translates from a question in natural language into some formal representation, not unlike the specification list used by BASEBALL, and the ability to obtain such a formal representation would appear to be essential to understanding the question. Moreover, these formal representations can be given a theoretical interpretation as statements or expressions in symbolic logic: that is, they are logically equivalent to a statement in the first-order predicate calculus. Theoretical analyses along these lines have been undertaken by Belnap (1963), Harrah (1963), Kuhns (1967), Levien and Maron (1967), and Howe (1969).

In addition, these question-answering systems can be given an interesting interpretation within the generalized problem-solving framework of GPS. Consider, first, the process of translation into a formal representation. We can think of the initial situation as defined by the question that is being asked, and this is always explicitly given. The desired situation is implicitly defined as an object schema, as, for example, when one attempts to translate into a set of attribute-value pairs or a logical statement in the predicate calculus. The operators are the rules for syntactic analysis of the question and for the additional semantic analysis that must be undertaken. Similarly, the search phase itself can be given representation within GPS. The initial object is the formal specification of the question. Again, the desired situation is implicitly defined. For yes-no questions, one must obtain a truth value for the statement; for all other questions, one must find values that will complete the statement and make it valid.

Finally, it is quite apparent that the ability to carry out a syntactic analysis of a question is central to the process of translation from natural language into a formal representation. At one extreme, there are systems such as ALTAIR that make no use of a procedure for syntactic analysis. Instead, the analysis is based upon a kind of Key-Word-In-Context (KWIC) form of analysis. Such analyses are reasonably successful in a restricted technical field in which the meaning of a question is communicated by a limited and very specialized vocabulary. At the other extreme, one ought to include provisions for a much more elaborate form of syntactic analysis in order to obtain the deep-structure interpretation of the question in the sense defined by Chomsky (1965). For an extended discussion of a transformational grammar for English see Jacobs and Rosenbaum (1968). Most existing systems fall somewhere in between; they carry out some syntactic analysis, but their capability is limited to rather restricted grammatical forms and to a very limited subset of the English language. In order to make substantial progress in this field, I predict that much more sophisticated provisions for syntactic analysis will be essential. Along these lines, it is important to note that progress in the field of computer simulations of natural language behavior is currently limited by the need for additional theoretical developments in linguistics, particularly as they relate to the syntactic analysis of English on a computer. Perhaps the best system currently available for the transformational analysis of English on a computer has been written by Petrick (1965, 1966). Additional efforts to develop methods for transformational analysis on a computer are being undertaken by Friedman (1968) and at the MITRE Laboratory (Zwicky *et al.,* 1965). Further efforts in improved transformational analysis of English will undoubtedly be forthcoming.

There are two important respects in which these systems for answering questions are deficient: they are unable to detect errors in presuppositions or to make inferences from what is directly known. Research designed to eliminate these deficiencies is currently underway. Questions in which presuppositions must be taken into account might be described as when-did-you-stop-beating-your-wife type questions. For example, consider the following pair of questions that may be given to a BASEBALL-like system. "How many home runs did Babe Ruth hit in 1947?" "What was Casey Stengel's batting average in 1955?" A knowledgeable person might respond to the former, "None. He was no longer an active player," and to the latter, "The question isn't meaningful because Stengel was the manager of the Yankees and not a player." In general, every question implies some presuppositions if it is to be accepted as meaningful or if a direct response to the question is to be accepted as appropriate. "A presupposition of a

question, A, is any statement which is logically implied by a direct answer of A."
(Howe, 1969, p. 115.) Theoretical work on the analysis of presuppositions has
been undertaken by Belnap (1963), Leonard (1957), and Howe (1969). As yet,
the results of this research have not been incorporated into a working simulation
of a question-answering system.

Interestingly enough, the analysis of presuppositions implies the ability to
make logical inferences from known facts, which is an additional capability that
should be added to these systems. For example, in answering the previous question about Babe Ruth, let us suppose that we were able to consult a set of
records about players. Such a record might state that Babe Ruth's career as a
player began in 1913 and ended in 1935. Thus in 1947 he was no longer an
active player. Only active players are officially recorded as hitting home runs.
Therefore, the original question is misleading because it begins with a presupposition that is incorrect, namely, that Babe Ruth was an active player in 1947.

How we make inferences from known facts is an extremely important
question; it is also a very complex one. Most of the theoretical developments
make use of the methods of formal logic in order to design a system of logical
inference. Cooper (1964) has described a possible system in which statements in
natural form would first be translated into some artificial logical language, that
is, a formal representation. Then the standard rules of inference would be executed on the formal representation in order to reach a conclusion. Along similar
lines, Black (1968) has designed a question-answering system with deductive
capabilities, and some related efforts on the problem of inference have been
undertaken by Raphael (1964).

How are we to assess this research on the simulation of inferential capabilities in a question-answering system? Clearly, this ability to make inferences is
characteristic of human thought, and it is an ability that we often take for
granted in everyday life. Thus, any model that simulates human thought must
include provisions for making inferences. On the other hand, this skill is a very
complex one and our present understanding of it is surprisingly limited. The
theoretical efforts directed toward an increased understanding of inferential
capabilities are important and ought to be continued. Perhaps one note of caution ought to be introduced. Current theoretical work has been strongly influenced by developments in the field of symbolic logic. By implication, we tend to
be assuming that human thought processes can be represented by a system of
formal inference. It is not unreasonable to assume that humans possess an
equivalent capability. However, it is not necessarily true that people use the
same formal representation of "facts" or even that they follow the same method
of inference in attempting to reach a conclusion.

Problem Solvers With Natural Language Inputs

There is an additional area in which computer simulations that will accept
national language inputs have been developed. This is the area of problem solving
in which the problems may initially be stated in natural form. We have already

given examples of typical word problems in the statement calculus, high school algebra, and freshman-level physics or calculus. Systems have been designed by Darlington (1965) and Manelski and Krulee (1965) for translating word problems in the statement calculus into a formal representation, by Bobrow (1964) and Kuck (1963) for translating word problems in algebra into a set of equations, and by Charniak (1968, 1969) for translating calculus word problems into a set of equations. In some important respects, these systems closely resemble the systems already described for translating questions in English into a specification list or an equivalent logical expression. In all these systems, one makes use of a formal representation of the problem as an intermediate result. The objective of a translation phase is to obtain this formal representation from the initial problem statement. The objective of a problem-solving phase is to use this formal representation as input and to generate an answer that will satisfy the problem requirements.

The final example we shall discuss is a system for the simulation of long-term memory designed by Tharp and Krulee (1969). It is designed to represent how an individual may remember connected discourse—such as a short story—and subsequently answer questions about what he can remember.[5] The stories to be remembered are about famous inventions, as taken from a standard encyclopedia for children. As with other simulations, the assumption is made that the natural language inputs are first transformed into a formal representation that can be stored internally. The formal representation consists of a series of relations and property lists. The property lists summarize descriptive information about a person or an object, with a single list for each. The relations are used as a representation for what happened, since a relation usually summarizes an event with one or more actors or objects involved. To some extent, the system functions like an information-retrieval system. Many questions relate directly to the stored relations and property lists; answers can be given as soon as the system recovers the appropriate relation or property list. The system can also answer questions about the temporal ordering of events by consulting a time-priority structure that has been constructed as part of the memory process.

In addition, the system attempts to answer two general questions about the story as a whole: to determine the most important person in the story and the central theme. For these purposes, the system compares relations in order to identify those with common arguments (i.e., actors). In a sense, the system is able to associate actors with events. The most important person takes part in many events, and the central theme is one of the key events with which the most important person is associated.

In designing this system, it has been necessary to introduce a number of key psychological assumptions about long-term memory. Several of these are very likely inadequate. Let us try to identify some of the more important assumptions and to consider how they may be replaced by others that will

---

[5]For other approaches to the simulation of long-term memory, see Lindsay (1963), Raphael (1964), and Quillian (1966).

psychologically be more reasonable. First of all, the system is unable to make logical inferences based upon its recollection of elementary events or its descriptions of people and objects. Deductive capabilities could be added that would be similar to those discussed previously for use with information-retrieval systems. Second, there is considerable evidence to suggest that memory is essentially a process of reconstruction and that it depends on more abstract forms of recoding than have been attempted in this system. This point of view about memory was initially proposed by Bartlett (1932) and later expanded by Allport and Postman (1947). An extensive discussion of such a constructive theory of memory has recently been completed by Neisser (1966), and similar views have been expressed by Miller, Galanter, and Pribram (1960). Consider, for example, how many of us remember a joke or an anecdote. If we tell the joke more than once, we are not likely to repeat ourselves exactly, that is, we probably do not memorize word for word. More likely, we remember (or store) some general outline of the joke, and we may remember word for word the punch line. In retelling it, we reconstruct the details from the general outline, or what Bartlett refers to as the *schema.*

In the Tharp-Krulee system, recall is not particularly constructive. The memory for a story is more nearly literal, except that sentences are translated into a formal representation. As an alternative, one might design a system that made use of some general ideas about "plots" as a basis for the organization of memory, and recall would be based upon the ability to use one of these plots to reconstruct the story. A plot might be in the form of a synopsis for identifying a central theme; a beginning, middle, and end, and so forth. My point in introducing these speculations about long-term memory is not to propose a new theory of memory. However, it is important to note that any simulation of long-term memory implies some assumptions about the organization of memory and the nature of recall. And the assumptions being made in this particular simulation are certainly not the only ones possible.

There is an additional characteristic to this particular simulation that needs to be made explicit. In this system, the computer "reads" each story but once and exhibits "one-trial perfect learning." Thus, it remembers at first the entire story, although it subsequently discards those details that appear to be peripheral or of low priority. Clearly, such a process is psychologically unreasonable. A human subject would perhaps need to read and reread the story, and his ability to remember would improve with each rereading. An alternative model might make use of the idea of a schema or plot as essential for remembering the story. It is conceivable that our reading of a story is guided by some general expectations about plots and their organization. On a first reading, the main outline of the story would be assimilated; on subsequent readings, additional information would be assimilated to expand the outline. Interestingly enough, such a schema could be viewed as another example of analysis by synthesis. I am suggesting that one's ability to remember a story makes use of a constructive process that attempts to resynthesize the telling of the story and incidentally to predict what is necessary in order to make it complete.

Finally, the method used for identifying the central theme and the central character deserves some critical examination. In this system, the central character and theme are derived from the formal representation of the story. The method borrows from the theory of connected graphs and makes use of a centrality index for determining the central person in the graph. If a person were to read these stories, he could very likely tell us what was the central theme and who was the main character, but his answers would not be likely to depend upon a knowledge of graph theory. How do people make such judgments? I do not know and will not attempt any speculations. Yet, this ability to identify a central theme and the main character is psychologically an important one and well deserving of further attention.

THE COMPUTER AS ARTIFICIAL SUBJECT

Over a period of several years, Feigenbaum and Simon have developed an information-processing theory of verbal learning in the form of a computer simulation known as EPAM. An article summarizing this effort has been included in this volume as an example of a cognitive simulation. I also want to add some general comments about the nature of their approach and the possible implications for psychology.

In experimental psychology, there has been a continuing interest in the study of verbal learning and specifically in the learning of lists of paired associates. Primarily, one is interested in the ease with which a given list will be learned and in studying the influence of certain independent variables on the ability of *S*s to learn. Feigenbaum and Simon have taken the experimental literature on paired-associate learning and have written a program that, in some sense, summarizes the results of a series of experiments. They referred to their program as an "artificial subject," since it could be interpreted as a model of a subject as he might take part in these experiments.

Why is their approach interesting? The program can be viewed as a theory that integrates the findings from a number of experiments. Their theory of necessity makes some explicit assumptions about the *process* of learning and about *what* is learned. Thus, they assume that the end product of a series of trials is the development of a "discrimination net." Using this net, a subject identifies a stimulus and retrieves the correct associated response. Although their assumptions may be incorrect, they are at least explicit and make clear how some assumptions about memory are essential to a theory of verbal learning. In addition, there is no limit to the number of variables that can be included within a single simulation model of an artificial subject. By way of contrast, experiments with "real" subjects are usually of limited complexity. As one attempts to include more and more variables, the experimental design begins to become more and more complex. In practice, there are limits to the number of subjects one can use in a single experiment and to the amount of time one can reasonably expect from a single subject. These limitations are avoided by the use of an artificial subject.

Once a simulation program has been obtained, it could also be used to suggest additional variations that might be undertaken in the form of further experiments. Along these lines, Paige and Simon (1966) have studied the processes that humans use in solving algebra word problems, and their research was stimulated by some previous work with computer programs for accepting natural language inputs. Similarly, some research undertaken by Kennedy, Eliot, and Krulee (1969) is an outgrowth of some computer simulations of problem solving with natural language inputs. They derived some hypotheses about differences between able and less able students in solving word problems that were used in an experiment with high school subjects.

The computer functions as artificial subject in other examples of research. Feldman, Tonge, and Kantor (1963), in their research on binary choice, have written a program that simulates the findings obtained from a number of related experiments. Simon and Kotovsky (1963) have also programmed a computer as artificial subject in their research on sequential pattern recognition.

CONCLUSIONS AND EVALUATIONS

In evaluating what has been accomplished by these computer simulations of cognitive processes, it is useful to identify an initial period of exploration which is now being succeeded by a second period of consolidation. In the initial period, emphasis was on the development of appropriate methods, such as the list-processing computer languages, on the possible advantages of computer simulations, and on demonstrations that they are indeed possible. Interestingly enough, in such an exploratory period, one uses a particular set of criteria for the evaluation of research. Viewed as a pioneering effort, any simulation of a cognitive process is important, and one tends not to worry about the cumulative effects of the research or about the relationship of one simulation to another. Each simulation is a kind of demonstration of feasibility. Moreover, in such a period, one tends to be relatively tolerant of imperfections. For example, the Logic Theorist is an important piece of research. We are impressed with its very existence and with what it accomplished and to be unconcerned with its imperfections as a completed theory of problem solving. Similarly, BASEBALL is an important piece of research and one that has stimulated additional research. Again, we are impressed with its accomplishments rather than with its limitations. After all, its ability to handle natural language is quite restricted, and it cannot begin to answer many important types of questions. Why have these researches been so influential? Undoubtedly, because they communicate a new point of view and provide the stimulus for some new and rather different research efforts. It is not my intention to downgrade the substantive accomplishments of this initial period but rather to emphasize that there were other accomplishments, particularly in the establishment of a new or revised paradigm.

We now appear to be entering a second period of activity in which increasing emphasis is on substantive contributions, on the cumulative effects of research, and on the assimilation of this new point of view into the mainstream of psychological research. One effect of this shift of emphasis is that our standards for the evaluation of research are beginning to change. One more simulation will no longer be acknowledged as a contribution; it should represent a significant contribution to the development of theory and should probably have some useful relationship to what has already been accomplished.

In earlier sections, I have discussed simulations of a variety of behaviors, ranging from pattern recognition to problem solving and natural language processing. In all these areas, there are significant obstacles to progress, although when one is being optimistic these may as well be described as challenges or opportunities. Interestingly enough, the significant obstacles are no longer methodological but substantive. The existing list-processing languages are not necessarily as efficient as they can be or as easy to use as one may desire; yet they are adequate for most purposes. The major obstacles are theoretical, and it is unmistakably clear that good computer simulations of behavior must be based upon a good theoretical understanding of that same behavior.

Let us review in more detail some of those theoretical developments that are most essential for progress with cognitive simulations. In the research on the analysis and snythesis of speech, difficulties in the analysis of the auditory signal into distinctive features or some equivalent units have been particularly formidable. The proposals of Stevens, Halle, and others about the importance of analysis by synthesis relate to a second major obstacle. It is reasonable to assume that analysis at the sensory level is steered or controlled by some higher-order predictive processes which make use of a knowledge of syntax as well as semantics. As yet, analysis by synthesis is essentially an intriguing speculation; additional research is needed to clarify how such a scheme may operate and how the relevant predictive skills may be acquired.

Difficulties in the simulation of visual phenomena parallel those encountered in the field of audition. In the field of pattern recognition, one can make some reasonable assumptions about discrimination at the level of the retina as involving the detection of elementary shapes which serve as features or attributes. These assumptions are essential to the design of any system for the discrimination of forms. However, with other visual phenomena, our lack of understanding of the underlying sensory processes continues to be an obstacle. Similarly, with visual phenomena, it is reasonable to assume that decoding at the sensory level is controlled by a higher-order predictive system, perhaps the same system that controls the recognition of speech. Any detailed understanding of how this system may operate is essentially nonexistent.

In the related areas of problem solving and natural language processing, rather significant progress has been made, perhaps because some reasonable theoretical developments have become available. Research on the simulation of problem solving has had a cumulative effect as evidenced by GPS and its interpretation as an information-processing theory of problem-solving behavior. GPS is of interest in at least three different ways. It is a practical guide for use in the simulation of behavior in some task environment. The recipe is as follows: For any given task, define, if possible, the objects, operators, differences, and relations between operators and differences. These become the main structural elements to be included in the problem-solving system. GPS can also be interpreted as a comparative theory of problem solving. The main structural concepts of GPS can be used to compare two or more tasks and to identify the similarities and differences among them. Finally, GPS is a partial or preliminary answer to the search for some generalized cognitive processes that apply in a variety of task environments.

In simulations that accept natural language inputs, the ability to analyze sentences syntactically is essential. In this respect, parallel developments in theoretical linguistics have made a most valuable contribution. I refer specifically to the theory of transformational grammars and its implications for syntactic analysis.

Interestingly enough, as progress is made by dealing effectively with one set of problems, others come even more sharply into focus and become the new obstacles to further progress. For example, in simulations of problem solving and natural language processing, one is clearly introducing some *ad hoc* assumptions about the organization of long-term memory, about processes for searching that memory in order to recall, and about an intermediate level of memory that functions essentially as a short-term storage of information. Indeed, there is a growing interest in more explicit assumptions about the organization and functioning of human memory, although relatively little progress has been made as yet.

I have suggested that the use of simulation methods as applied to the study of cognitive processes is beginning to be assimilated into psychology. Generalizing from past experiences, this process is likely to continue in at least two ways. Let us first consider the computer as artificial subject. In this type of research, one begins with an existing body of experimental knowledge and creates an information-processing model that summarizes the findings. If the model is successful, it will suggest additional possibilities for experimentation. As an information-processing model, it will also make explicit some critical assumptions about those internal processes that mediate between stimulus and response. One would hope that further research along these lines would continue. It is important to note that this type of research is of necessity interdisciplinary. One must know something about the methods of simulation. More importantly, one must have a reasonable mastery of the relevant efforts in psychology.

Computer simulations are also important for psychology as they expand the range of phenomena that is actively being investigated. It seems not unreasonable to conclude that computer simulations of problem solving and theoretical contributions of GPS are responsible for a resurgence of interest among psychologists in these and related phenomena. Similarly, research on natural language processing is one of the factors that is encouraging an increasing interest in the experimental study of human abilities with language. As questions about memory become more important in the field of computer simulations, one notes a parallel growth of interest in experimental studies of memory. Thus, in a variety of ways, one would anticipate that computer simulations would continue to influence the growth of knowledge in the field of cognitive processes.

REFERENCES

Abelson, R. P., "Simulation of Social Behavior," in G. Lindzey and E. Aronson (eds.), *Handbook of Social Psychology*, pp. 274-356. Reading, Mass.: Addison-Wesley, Publishers, 1968.

Adey, W. R., "Aspects of Cerebral Organization: Information Storage and Recall," in W. C. Corning and M. W. Balaban (eds.), *The Mind: Biological Approaches to Its Functioning*, pp. 69-100. New York: Interscience Press, 1968.

Allen, M., "A Concept Attainment Program That Simulates a Simultaneous Learning Strategy," *Behavioral Science*, 7 (1962), 247-50.

Allport, G. W. and L. Postman, *The Psychology of Rumor*. New York: Holt, 1947.

Amarel, S., "An Approach to Heuristic Problem Solving and Theorem Proving in the Propositional Calculus," in J. F. Hart and S. Takasu (eds.), *Systems and Computer Science*, Toronto: University of Toronto Press, 1967.

———, "On Representations of Problems of Reasoning about Actions," in D. Michie (ed.), *Machine Intelligence 3*, Edinburgh: Edinburgh University Press. 1968.

———, "On the Mechanization of Creative Processes," *IEEE Spectrum* (April 1966).

Ashby, W. R., *Design for a Brain*. New York: John Wiley & Sons, Inc. 1952.

Baker, F., "An IPL-V Program for Concept Attainment," *Educ. Psychol. Measmt.*, 24 (1965), 119-27.

———, "CASE: A Program for Simulation Concept Learning," *Proceedings Fall Joint Computer Conference*, AFIPS, 27 (1964), 979-84.

Balzer, R. A., "A Mathematical Model for Performing a Complex Task in a Card Game," *Behavioral Science*, 11 (1966), 219-27.

Bar-Hillel, Y., "The Present Status of Automatic Translation of Languages," in F. L. Alt (ed.), *Advances in Computers*, New York: Academic Press, Inc., 1960.

Bartlett, F. C., *Remembering*. Cambridge, Eng.: Cambridge Univ. Press, 1932.

Baylor, G., and H. Simon, "A Chess Mating Combinations Program," *Proceedings Spring Joint Computer Conference*, AFIPS, 28 (1966), 431-47.

Becker, Joseph D., "The Modeling of Simple Analogic and Inductive Processes in a Semantic Memory System," *Proceedings, IJCAI*, Washington, 1969, 655-68.

Bellman, R., "Adaptive Processes and Intelligent Machines," in L. Lecam and J. Neyman (eds.), *Proceedings of the Fifth Berkeley Symposium on Mathematical Statistics and Probability*. Berkeley: University of California Press, 1967. 11-14.

Belnap, N. D., Jr., "An Analysis of Questions: Preliminary Report," Technical Memo TM-1287, System Development Corp., Santa Monica, Calif., 1963.

Bernstein, A., T. Arbuckle, M. De V. Roberts, and M. A. Belsky, "A Chess Playing Program for the IBM 704," *Proceedings Western Joint Computer Conference,* 13 (1958), 157-59.

Black, F., "A Deductive Question-Answering System," in M. L. Minsky (ed.), *Semantic Information Processing,* pp. 354-402. Cambridge, Mass.: M.I.T. Press, 1968.

Bledsoe, W., and I. Browning, "Pattern Recognition and Reading by Machine," *Proceedings Eastern Joint Computer Conference,* 16 (1959), 225-32.

Block, H.D., "Analysis of Perceptrons," *Proceedings Western Joint Computer Conference,* 19 (1961), 281-89.

———, "The Perceptron: A Model for Brain Functioning I," *Review of Modern Physics,* 34 (1962), 123-35.

Block, H. D., B. W. Knight, and F. Rosenblatt, "Analysis of a Four-Layer Series Coupled Perceptron II," *Review of Modern Physics,* 34 (1962), 135-42.

Block, H. D., N. Nilsson, and W. Duda, "Determination and Detection of Features in Patterns," in J. Tou and R. Wilcox (eds.), *Computers and Information Sciences,* pp. 75-110. Baltimore, Md.: Spartan Press, 1964.

Bobrow, D., "A Question Answering System for High School Algebra Word Problems," *Proceedings Fall Joint Computer Conference,* AFIPS, 26 (1964), 591-614. (Reprinted in Minsky, M. L. (ed.), *Semantic Information Processing.* Cambridge, Mass.: M.I.T. Press. 1968.)

———, "Syntactic Analysis of English by Computer—a Survey," *Proceedings Fall Joint Computer Conference,* AFIPS, 24 (1963), 365-87.

Bobrow, D., and B. Raphael, "A Comparison of List Processing Languages," *Communications of the Association for Computing Machinery,* 7 (1964), 231-40.

Bobrow, D. G., and J. B. Fraser, "An Augmented State Transition Network Analysis Procedure," *Proceedings, IJCAI,* Washington, 1969, 557-67.

Bourne, L., "Learning and Utilization of Conceptual Rules," in B. Kleinmuntz (ed.), *Concepts and the Structure of Memory,* pp. 1-32. New York: John Wiley & Sons, Inc., 1967.

Broadbent, D. E., "Flow of Information within the Organism," *Journal of Verbal Learning and Verbal Behavior,* 2 (1963), 34-39.

———, *Perception and Communication.* London: Pergamon Press, 1958.

Bross, I.D.J., *et al.,* "Feasibility of Automated Information Systems in the Users' Natural Language," *American Scientist,* 57 (1969), 193-205.

Bruner, J. S., J. J. Goodnow, and G. A. Austin, *A Study of Thinking.* New York: John Wiley & Sons, Inc., 1956.

Cannon, M. W., "A Method of Analysis and Recognition for Voiced Vowels," *IEEE Transac. on Audio and Electroacoustics,* AU-16 (1968), 154-58.

Cannon, W. B., *The Wisdom of the Body.* New York: W. W. Norton & Company, Inc., 1939.

Charniak, E., *Carps, A Program Which Solves Calculus Word Problems.* Report MAC-TR-51, Project MAC, M.I.T., Cambridge, Mass., July 1968.

———, "Computer Solution of Calculus Word Problems," *Proceedings, IJCAL,* Washington, 1969, 303-16.

Chien, Y.T., and K.S. Fu, "A Modified Sequential Recognition Machine Using Time-Varying Stopping Boundaries," *IEEE Transactions on Information Theory,* IT-12 (1966), 206-14.

Chomsky, N., *Aspects of the Theory of Syntax.* Cambridge, Mass.: M.I.T. Press, 1965.

———, "Formal Properties of Grammars," in R. Luce, R. Bush, and E. Galanter (eds.), *Handbook of Mathematical Psychology,* II, 323-418. New York: John Wiley & Sons, Inc., 1963.

Chomsky, N., and G. Miller, "Introduction to the Formal Analysis of Natural Languages," in R. Luce, R. Bush, and E. Galanter (eds.), *Handbook of Mathematical Psychology,* II, 269-321. New York: John Wiley & Sons, Inc., 1963.

Chu, J. T., and J. C. Cheu, "Error Probability in Decision Functions for Character Recognition," *J. Assoc. Computing Machinery,* 14 (1967), 273-80.

Coles, L. S., "Syntax Directed Interpretation of Natural Language." Ph.D. dissertation, Carnegie Institute of Technology, 1967.

———, "Talking with a Robot in English," *Proceedings, IJCAI,* Washington, 1969, 587-96.

Collins, N. L., and D. Michie (eds.), *Machine Intelligence I.* Edinburgh: Oliver and Boyd, 1967.

Cooley, W. W., and P. R. Lohnes, *Multivariate Procedures for the Behavioral Sciences.* New York: John Wiley & Sons, Inc., 1962.

Cooper, P. W., "Quadratic Discriminant Functions in Pattern Recognition," *IEEE Trans. on Information Theory,* IT-11 (1965), 313-15.

Cooper, W. S., "Fact Retrieval and Deductive Question-Answering Information Retrieval Systems," *Journal of the Association for Computing Machinery,* 11 (1964), 117-37.

Corning, W. C., and M. W. Balaban (eds.), *The Mind: Biological Approaches to Its Functioning.* New York: Interscience Press, 1968.

Craig, J. A., "Grammatical Aspects of a System for Natural Man-Machine Communication," RM63TMP-31, Tempo, General Electric Co., Santa Barbara, Calif., 1963.

Culbertson, J. T., *The Minds of Robots.* Urbana: University of Illinois Press, 1963.

Daly, J. A., R. D. Joseph and D. M. Ramsey, "Perceptrons as Models of Neural Processes," in R. Stacy and B. Waxman (eds.), *Computers and Biomedical Research I,* pp. 525-45. New York: John Wiley & Sons, Inc., 1965.

Darlington, J. D., "Machine Methods for Proving Logical Arguments Expressed in English," *Mechanical Translation,* 8 (1965), 3-4.

Darlington, J. L., "Theorem Provers as Question Answerers," *Proceedings, IJCAI,* Washington, 1969, 317-18.

Davis, K. H., R. Biddulph, and S. Balashek, "Automatic Recognition of Spoken Digits," *Journal of the Acoustic Society of America,* 24 (1952), 637-43.

De Groot, A. D., "Perception and Memory versus Thought: Some Old Ideas and

Recent Findings," in B. Kleinmuntz (ed.), *Problem Solving*. New York: John Wiley & Sons, 1966.

———, *Thought and Choice in Chess*. The Hague, Netherlands: Mouton, 1965.

Denes, P., "The Design and Operation of the Mechanical Speech Recognizer at University College, London," *J. British Institute of Radio Engineers*, 19 (1959), 219-29.

Doyle, W., "Recognition of Sloppy, Hand-printed Characters," *Proceedings Western Joint Computer Conference*, 1960, pp. 133-142.

Dreyfus, H., *Alchemy and Artificial Intelligence*. Rand Corporation Technical Report. P-3244. Santa Monica, Calif.: Rand Corporation, 1965.

———, "Why Computers Must Have Bodies in Order to Be Intelligent," *Review of Metaphysics*, 21 (1967), 13-32.

Dudley, H., and S. Balashek, "Automatic Recognition of Spoken Digits," *Journal of the Accoustic Society of America*, 30 (1958), 721-32.

Eden, M., "Handwriting and Pattern Recognition," *IRE Transactions on Information Theory*, IT-8 (1962), 160-66.

———, "Handwriting Generation and Recognition," in P. Kolers and M. Eden (eds.), *Recognizing Patterns*, pp. 138-154. Cambridge, Mass.: M.I.T. Press, 1968.

Eden, M., and M. Halle, "The Characterization of Cursive Writing," in C. Cherry (ed.), *Information Theory, Fourth London Symposium*, pp. 287-99. Washington: Butterworth, 1961.

Elcock, E. W., and A. M. Murray, "Experiments with a Learning Component in a Go-Muku Playing Program," in N. L. Collins and D. Michie, *Machine Intelligence I*, pp. 87-104. Edinburgh: Oliver and Boyd, 1967.

Ernst, G. W. and A. Newell, "Some Issues of Representation in a General Problem Solver," *Proceedings Spring Joint Computer Conference*, AFIPS, 30 (1967), 583-600.

———, *GPS: A Case Study in Generality and Problem Solving*. New York: Academic Press, Inc., 1969.

Evans, T. G., "A Heuristic Program to Solve Analogies Problems," *Proceedings tory Cortex of the Un-anaesthetized and Unrestrained Cat," *Journal of Physiology*, 171 (1964), 476-93.

Evans, T. G. "A Heuristic Program to Solve Analogies Problems," *Proceedings Spring Joint Computer Conference*, AFIPS, 28 (1964), 327-38.

Fant, G., "Analysis and Synthesis of Speech Processes," in B. Malmberg (ed.), *Manual of Phonetics*, pp. 173-277. Amsterdam: North Holland, 1968.

Farber, D. J., R. E. Griswold, and I. P. Polonsky, "The SNOBOL 3 Programming Language," *Bell System Technical Journal*, 45 (1966), 895-944.

Farley, B., "A Neural Network Model," in R. Stacy and B. Waxman (eds.), *Computers and Biomedical Research*, I, 265-94. New York: John Wiley & Sons, Inc., 1965.

Feigenbaum, E. A., *An Information Processing Theory of Verbal Learning*. Rand Corporation Technical Report P-1817. Santa Monica, Calif.: Rand Corporation, 1959.

——— "Artificial Intelligence: Themes in the Second Decade," *Proceedings, IFIP, 68 Congress,* 1968.

———, "Information Processing and Memory," in L. Lecam and J. Neyman (eds.), *Proceedings, Fifth Berkeley Symposium on Mathematical Statistics and Probability,* pp. 37-51. Berkeley: University of California Press, 1967.

———, "The Simulation of Verbal Learning Behavior," *Proceedings Western Joint Computer Conference,* 19 (1961), 121-32.

Feigenbaum, E. A., and J. Feldman (eds.), *Computers and Thought.* New York: McGraw-Hill Book Company, 1963.

Feigenbaum, E. A., and J. Lederberg, "Mechanization of Inductive Inference in Organic Chemistry," in B. Kleinmuntz (ed.), *Formal Representation of Human Judgments,* pp. 187-218. New York: John Wiley & Sons, Inc., 1968.

Feigenbaum, E., and H. Simon, "A Theory of the Serial Position Effect," *British Journal of Psychology,* 53 (1962), 307-20.

Feldman, J., "An Analysis of Predictive Behavior in a Two-Choice Situation," Ph.D. dissertation, Carnegie Institute of Technology, 1959.

———, "Recognition of Patterns in Periodic Binary Sequences," in L. Lecam and J. Neyman (eds.), *Proceedings, Fifth Berkeley Symposium on Mathematical Statistics and Probability,* pp. 53-64. Berkeley: University of California Press, 1967.

———, "Simulation of Behavior in the Binary Choice Experiment," *Proceedings Western Joint Computer Conference,* 19 (1961), 133-44.

Feldman, J., and D. Gries, "Translator Writing Systems," *Communications of the ACM,* 11 (1968), 77-113.

Feldman, J., and J. Hanna, "The Structure of Responses to a Sequence of Binary Events," *Journal of Mathematical Psychology,* 2 (1966), 371-87.

Feldman, J., F. Tonge, and H. Kantor, "Empirical Explorations of a Hypothesis Testing Model of Binary Choice Behavior," in A. Hoggat and F. Balderston (eds.), *Symposium on Simulation Models.* Cincinnati: South-Western Publishing Co., 1963.

Findler, N. "Some Remarks on the Game 'Dama' Which Can Be Played on a Digital Computer," *Computer Journal,* 3 (1960), 40-44.

Flanagan, J. L., *Speech Analysis, Synthesis, and Perception.* New York: Academic Press, Inc., 1965.

Forte, A., *SNOBOL 3 Primer: An Introduction to the Computer Programming Language.* Cambridge, Mass.: M.I.T. Press, 1967.

Frank, L. K. *et al.,* "Teleological Mechanisms," *Annals, New York Academy of Science,* 50 (1948), 189.

Friedberg, R. M., "A Learning Machine, Part I," *IBM Journal of Research and Development,* 2 (1958), 3-13.

Friedberg, R. M., B. Dunham, and J. A. North, "A Learning Machine, Part II," *IBM Journal of Research and Development,* 3 (1959), 282-87.

Friedman, J., *A Computer System for Transformational Grammar,* Report CS-84, AF-21, Computer Science Department, Stanford University, 1968.

Frijda, N. H., "Problems of Computer Simulation," *Behavioral Science,* 12 (1967), 59-67.

Fu, K. S., *Sequential Methods in Pattern Recognition and Machine Learning.* New York: Academic Press, Inc., 1968.

Gelernter, H., "A Note of Syntactic Symmetry and the Manipulation of Formal Systems by Machine," *Information Control,* 2 (1959), 80-89.

———, "Realization of a Geometry Theorem-Proving Machine," *Proceedings of the International Conference on Information Processing,* 273-82. Paris: Unesco House, 1959.

Gelernter, H., J. Hansen, and D. Loveland, "Empirical Explorations of the Geometry Theorem Machine," *Proceedings Western Joint Computer Conference,* 17 (1960), 143-47.

Gelernter, H., and N. Rochester, "Intelligent Behavior in Problem Solving Machines," *IBM Journal of Research and Development,* 3 (1958), 336-45.

Gibson, E. J., *Principles of Perceptual Learning and Development.* New York: Appleton-Century-Crofts, 1969.

Gilmore, P., "A Program for the Production of Proofs for Theorems in the 1st Order Predicate Calculus from Axioms," *Proceedings of the International Conference on Information Processing,* 265-73. Paris: Unesco House, 1959.

Giulano, V. E., "Comments on Simon's Paper," *Communications of the Association for Computing Machinery,* 8 (1965), 69-70.

Glushkov, V., *Introduction to Cybernetics.* New York: Academic Press, Inc., 1966.

Gold, B., "Machine Recognition of Hand-Sent Morse Code," *IRE Transactions on Information Theory,* IT-5 (1959), 17-24.

Goldstine, H. H., and J. Von Neumann, "Planning and Coding of Problems for an Electronic Computing Instrument," Part II, Vols. I-III. Report prepared for U.S. Army Ordinance Department, 1947 and 1948. (Reprinted in J. Von Neumann, *Collected Works,* Vol. V. New York: The Macmillan Company, 1963.)

Golomb, S. W., and L. D. Baumert, "Backtrack Programming," *Journal of the ACM,* 12 (October 1965), 4.

Green, B. F., A. Wolf, C. Chomsky, and K. Laughrey, "Baseball: An Automatic Question Answerer," *Proceedings Western Joint Computer Conference,* 19 (1961), 219-24.

Green, C., "Application of Theorem Proving to Problem Solving," *Proceedings, IJCAI,* Washington, 1969, 219-239.

Green, C., and B. Raphael, "The Use of Theorem-Proving Techniques in Question-Answering Systems," *Proceedings 23rd National Conference, ACM.* Washington, D.C.: Thompson Book Co., 1968.

Greenblatt, G., D. Eastlake, and S. Crocker, "The Greenblatt Chess Program," *Proceedings of the Fall Joint Computer Conference,* Anaheim, Calif., 1967.

Gregg, L., "Internal Representation of Sequential Concepts," in B. Kleinmuntz (ed.), *Concepts and the Structure of Memory,* pp. 107-42. New York: John Wiley & Sons, Inc., 1967.

Gregg, L. W., and H. A. Simon, "An Information-Processing Explanation of One-Trial and Incremental Learning," *Journal of Verbal Learning and Verbal Behavior,* 6 (October 1967), 5.

Grimsdale, R. L., F. H. Sumner, C. J. Tunis, and T. A. Kilburn, "A System for the Automatic Recognition of Patterns," *Proceedings, Institute of Electrical Engineers,* 106 (1959), 210-21.

Halle, M., and K. W. Stevens, "Speech Recognition: A Model and a Program for Research," in J. A. Fodor and J. J. Katz, *The Structure of Language.* Englewood Cliffs, N. J.: Prentice-Hall, Inc., 1964.

Harrah, D., *Communication: A Logical Model.* Cambridge, Mass.: M.I.T. Press, 1963.

Harré, Rom, "The Formal Analysis of Concepts," in H. J. Klausmeier and C. W. Harris (eds.), *Analysis of Concept Learning,* pp. 3-17. New York: Academic Press, Inc., 1966.

Haygood, R., and L. Bourne, "Attribute and Rule Learning Aspects of Conceptual Behavior," *Psychological Review,* 72 (1965), 175-95.

Hebb, D. W., *The Organization of Behavior.* New York: John Wiley & Sons, Inc., 1948.

Highleyman, W. H., "Linear Decision Functions, with Application to Pattern Recognition," *Proceedings, Institute of Radio Engineers,* 50 (1962), 1501-14.

———, "The Design and Analysis of Pattern Recognition Experiments," *Bell System Technical Journal,* 41 (1962), 723-44.

Hill, D. R., "Automatic Speech Recognition: A Problem in Machine Intelligence," in N. L. Collins and D. Michie (eds.), *Machine Intelligence,* pp. 199-228. Edinburgh: Oliver and Boyd, 1967.

Horwitz, L. P., and G. L. Shelton, Jr., "Pattern Recognition Using Autocorrelation," *Proceedings, Institute of Radio Engineers,* 49 (1961), 175-85.

Hovland, C. E., "A 'Communication Analysis' of Concept Learning," *Psychological Review,* 59 (1962), 461-72.

Howe, W. G., "A Logic of English Questions with Emphasis on Automated Query Systems." Ph.D. dissertation, Northwestern University, 1969.

Hubel, D., and T. Wiesel, "Receptive Fields of Single Neurons in the Cat's Visual Cortex," *Journal of Physiology,* 148 (1959), 574-91.

Hunt, E. B., "Computer Simulation: Artificial Intelligence Studies and Their Relevance to Psychology," *Annual Review of Psychology,* 19 (1968), 135-68.

———, *Concept Learning: An Information Processing Problem.* New York: John Wiley & Sons, Inc., 1962.

———, "Simulation and Analytic Models of Memory," *Journal of Verbal Learning and Verbal Behavior,* 2 (1963), 49-59.

———, "Utilization of Memory in Concept Learning Systems," in B. Kleinmuntz (ed.), *Concepts and the Structure of Memory,* pp. 77-106. New York: John Wiley & Sons, Inc., 1967.

Hunt, E. B., and C. I. Hovland, "Order of Consideration of Different Types of Concepts," *Journal of Experimental Psychology,* 59 (1960), 220-25.

———, "Programming a Model of Human Concept Formulation," *Proceedings Western Joint Computer Conference,* 19 (1961), 145-55.

Hunt, E. B., J. Marin, and P. J. Stone, *Experiments in Induction.* New York: Academic Press, Inc., 1966.

Jacobs, R. A., and P. D. Rosenbaum, *English Transformational Grammar.* Waltham, Mass.: Blaisdell, 1968.

Jakobsen, R., G. M. Fant, and M. Halle, "Preliminaries to Speech Analysis: The Distinctive Features and Their Correlates," Cambridge, Acoustics Laboratory: M.I.T. Technical Report No. 13, 1952.

Johnson, D., and A. D. C. Holden, "Computer Learning in Theorem Proving," *IEEE Convention Record,* 14, Part 6 (1966), 51-60.

Johnson, D. M., *The Psychology of Thought and Judgment.* New York: Harper & Row, Publishers, 1955.

Johnson, E., "An Information Processing Model of One Kind of Problem Solving," *Psychological Monographs,* 78 (4) (1964), 31 pp.

———, *Outline of an Information Processing Model for Simulating One Kind of Problem Solving.* Tech. Rept. 10, Contract Nonr 609 (20) (NR 150-166). Yale University, Department of Psychology, 1966.

Kasher, A., "Data-Retrieval by Computer: A Critical Survey," in M. Kochen (ed.), *The Growth of Knowledge,* pp. 292-324. New York: John Wiley & Sons, Inc., 1967.

Kelly, J. L., and O. G. Selfridge, "Sophistication in Computers: A Disagreement," *IRE Transactions on Information Theory,* IT-8 (1962), 78-80.

Kennedy, G., J. Eliot, and G. K. Krulee, "Error Patterns in Problem Solving Formulations," *Psychology in the Schools,* 7 (1969), 93-99.

Kleinmuntz, B. (ed.), *Concepts and the Structure of Memory.* New York: John Wiley & Sons, Inc., 1967.

Kirsch, R. A., "Computer Interpretation of English Text and Picture Patterns," *IEEE Transaction on Electronic Computers,* Vol. EC-13 (August 1964).

Kister, J., P. Stein, S. Ulam, W. Walden, and W. Wells, "Experiments in Chess," *Journal of the Association of Computing Machinery,* 4 (1957), 174-77.

Klein, S., S. Lieman, and G. Lindstrom, *Diseminar: A Distributional-Semantics Inference Maker,* Carnegie Institute of Technology Technical Report, 1966.

Klein, S., and R. F. Simmons, "A Computational Approach to Grammatical Coding of English Words," *Journal of the Association of Computing Machinery,* 10 (1963), 334-47.

Kolers, P., "Some Psychological Aspects of Pattern Recognition," in P. Kolers and M. Eden (eds.), *Recognizing Patterns,* pp. 4-61. Cambridge, Mass.: M.I.T. Press, 1968.

Kolers, P., and M. Eden (eds.), *Recognizing Patterns.* Cambridge, Mass.: M.I.T. Press, 1968.

Krulee, G. K., and D. J. Kuck, "Finite State Models for Perception," *Journal of Mathematical Psychology,* 1 (1964), 316-35.

Kuck, D. J., "A Multicontext Problem Solving System with Natural Language Inputs." Ph.D. dissertation, Northwestern University, 1963.

Kuck, D. J., and G. K. Krulee, "A Problem Solver with Formal Descriptive

Inputs," in J. T. Tou and R. H. Wilcox (eds.), *Computer and Information Sciences,* pp. 344-74. Washington: Spartan Books, 1964.

Kuhn. T. S., *The Structure of Scientific Revolutions.* Chicago: University of Chicago Press, 1962.

Kuhns, J. L., "Answering Questions by Computer: A Logical Study," The Rand Corporation, Memo, RM-5428-PR, Santa Monica, Calif., 1967.

Kuno, S., "The Predictive Analyzer and a Path Elimination Technique," *Communications of the Association for Computing Machinery,* 8 (1959), 453-62.

Lashley, K. S., "The Problem of Serial Order in Behavior," L. A. Jeffress (ed.), *Cerebral Mechanisms in Behavior,* pp. 112-36. New York: John Wiley & Sons, Inc., 1951.

Laughrey, K., and L. Gregg, "Simulation of Human Problem Solving Behavior," *Psychometrika,* 27 (1962), 265-82.

Ledley, R. S., "High Speed Automatic Analysis of Biomedical Pictures," *Science,* 146 (1964), 216-23.

Ledley, R. S., *Programming and Utilization of Digital Computers.* New York: McGraw-Hill Book Company, 1962.

Leonard, H. S., *An Introduction to Principles of Right Reason.* New York: Henry Holt, 1957.

Lettvin, R., H. Maturana, W. McCulloch, and W. Pitts, "What the Frog's Eye Tells the Frog's Brain," *Proceedings, Institute of Radio Engineers,* 47 (1959), 1940-51.

Levien, A., and M.E. Maron, "A Computer System for Inference Execution and Data Retrieval," *Communications of the Association for Computing Machinery,* 10 (1967), 715-21.

Levine, M., "Hypothesis Behavior in Humans During Discrimination Learning," *Journal of Experimental Psychology,* 71 (1966), 331-38.

Liberman, A. M., *et al.*, "Perception of the Speech Code," *Psychological Review,* 74 (1967a), 431-61.

———, "Some Observations on a Model for Speech Perception," in W. Wathen-Dunn (ed.), *Models for the Perception of Speech and Visual Form,* pp. 68-87. Cambridge, Mass.: M. I. T. Press, 1967b.

Lindgren, N., "Machine Recognition of Human Language," I, II, III, *IEEE Spectrum,* 1965, 2, March, 114-36; April, 44-59; May, 104-116.

Lindsay, R. K., "Inferential Memory As the Basis of Machines Which Will Understand Natural Language," in E. Feigenbaum and J. Feldman (eds.), *Computers and Thought,* pp. 217-38. New York: McGraw-Hill Book Company, 1963.

McCarthy, J., *et al., LISP 2 Programmer's Manual.* Cambridge, Mass.: M.I.T. Press, 1960.

McConlogue, K., and R. Simmons, "Analyzing English Syntax with a Pattern Learning Parser," *Communications of the Association for Computing Machinery,* 11 (1965), 687-98.

McConnell, J. V., "The Modern Search for the Engram," in W. C. Corning and M. W. Balaban (eds.), *The Mind: Biological Approaches to Its Functioning,* pp. 49-68. New York: Interscience Press, 1968.

McCulloch, W. S., "Why the Mind Is in the Head," in L. A. Jeffries (ed.), *Cerebral Mechanisms in Behavior: The Hixon Symposium*, pp. 42-74. New York: John Wiley & Sons, Inc., 1951.

McCulloch, W. S., and W. Pitts, "A Logical Calculus of the Ideas Imminent in Nervous Activity," *Bulletin of Mathematical Biophysics,* 5 (1943), 115-37.

MacKay, D. M., "Ways of Looking at Perception," in W. Wathen-Dunn (ed.), *Models for Perception of Speech and Visual Form*, pp. 25-43. Cambridge, Mass.: M.I.T. Press, 1967,

Manelski, D. M., and G. K. Krulee, "A Heuristic Approach to Natural Language Processing," Preprint No. 18, 1965. International Conference on Computational Linguistics, New York, 1965.

Mason, S. J., and J. K. Clemems, "Character Recognition in an Experimental Reading Machine for the Blind," in P. Kolers and M. Eden (eds.), *Recognizing Patterns*, pp. 156-66. Cambridge, Mass.: M.I.T. Press, 1968.

Mermelstein, P., "Computer Simulation of Articulatory Activity in Speech Production," *Proceedings, IJCAI*, Washington, 1969, 447-54.

Mermelstein, P., and M. Eden, "Experiments on Computer Recognition of Connected Handwritten Words," *Information & Control,* 7 (1964), 155-270.

Miller, G. A., *Language and Communication.* New York: McGraw-Hill Book Company, 1951.

Miller, G. A., E. Galanter, and K. Pribram, *Plans and the Structure of Behavior.* New York: Holt, Rinehart & Winston, Inc., 1960.

Milner, P., "The Cell Assembly-Mark II," *Psychological Review,* 64 (1957), 245-52.

Minsky, M., "Artificial Intelligence," *Scientific American,* 215 (1966), 246-63.

Minsky, M., "Steps Toward Artificial Intelligence," in E. Feigenbaum and J. Feldman (eds.), *Computers and Thought*, pp. 406-50. New York: McGraw-Hill Book Company, 1963.

Minsky, M., and O. G. Selfridge, "Learning in Random Nets," in C. Cherry (ed.), *Information Theory, Fourth London Symposium,* pp. 335-47. Washington: Butterworth, 1961.

Moses, J., *Symbolic Integration.* Report MAC-TR-47, Project MAC, MIT, Cambridge, Mass., December 1967.

Neisser, U., *Cognitive Psychology.* New York: Appleton-Century-Crofts, 1967.

———, "The Imitation of Man by Machine," *Science,* 193 (1963).

Newell, A., "A Guide to the General Problem-Solving Program," GPS-2-2, RM 3337-PR, Rand Corporation, Santa Monica, Calif., 1963.

———, "Eye Movements and Problem Solving," *Computer Science Research Review,* Carnegie-Mellon University, Pittsburgh, December 1967.

———, *On the Analysis of Human Problem Solving Protocols.* Carnegie Institute of Technology Technical Report, Pittsburgh, 1966.

———, "Some Problems of Basic Organization in Problem- Solving Programs," *Self-Organizing Systems,* pp. 393-424. Washington, D.C.: Spartan Press, 1961.

———, *Information Processing Language V. Manual.* Englewood Cliffs, N.J.: Prentice-Hall, Inc., 1965.

Newell, A., and G. Ernst, "The Search for Generality," in W. A. Kalenich (ed.), *Information Processing 1965: Proceedings of IFIP Congress 65,* 17-24. Washington: Spartan Books, 1965.

Newell, A., and J. C. Shaw, "Computers in Psychology," in R. Luce, R. Bush, and E. Galanter (eds.), *Handbook of Mathematical Psychology,* 1, 261-428. New York: John Wiley & Sons, Inc., 1963.

Newell, A., J. C. Shaw, and H. Simon, "Chess Playing and the Problem of Complexity," in E. Feigenbaum and J. Feldman (eds.), *Computers and Thought,* pp. 39-70. New York: McGraw-Hill Book Company, 1963.

———, "Elements of a Theory of Human Problem Solving," *Psychological Review,* 65 (1958), 151-66.

———, " Empirical Explorations with the Logic Theory Machine," *Proceedings Western Joint Computer Conference,* 11 (1957a), 218-30.

———, "Preliminary Description of General Problem Solving Program I" (GPS-I), Working Paper No. 7, Carnegie Institute of Technology, Pittsburgh, 1957.

———, "Report on a General Problem Solving Program," *Proceedings, International Conference on Information Processing,* 256-64. Paris: Unesco House, 1960.

Newell, A., and H. Simon, "An Example of Human Chess Play in the Light of Chess Playing Programs," in N. Wiener and P. Schade (eds.), *Progress in Biocybernetics,* 3, 19-75. Amsterdam: Elsevier, 1965.

———, "Computer Simulation of Human Thinking," *Science,* 134 (1961a), 2011-17.

———, "GPS, a Program That Simulates Human Thought," in H. Billing (ed.), *Lernende Automation,* pp. 109-24. Munich, Germany: R. Oldenbourg. (Reprinted in Feigenbaum and Feldman, 1963.)

———, "Overview: Memory and Process in Concept Formation," in B. Kleinmuntz (ed.), *Concepts and the Structure of Memory,* pp. 241-62. New York: John Wiley & Sons, Inc., 1967.

———, "Programs as Theories of Higher Mental Processes," in R. Stacy and B. Waxman, *Computers in Biomedical Research,* II, 141-72. New York: John Wiley & Sons, Inc., 1965.

———, "The Logic Theory Machine," *IRE Transactions on Information Theory,* IT-2, No. 3 (1956), 61-79.

Newman, C., and L. Uhr, "BOGART: A Discovery and Induction Program for Games," in *Association for Computing Machinery Proceedings of the 20th National Conference,* New York, 1965, 176-86.

Nilsson, J., "A New Method for Searching Problem-Solving and Game Playing Trees," Working Paper (Artificial Intelligence Group, Stanford Research Institute, Menlo Park, Calif., November 1967). Paper presented at IFIP68 Congress, Edinburgh.

Nilsson, N. J., *Learning Machines.* New York: McGraw-Hill Book Company, 1965.

Norman, D. A., *Memory and Attention.* New York: John Wiley & Sons, Inc., 1969.

Paige, G., and H. Simon, "Cognitive Processes in Solving Algebra Word

Problems," in B. Kleinmuntz (ed.), *Problem Solving: Research, Method, and Theory*, pp. 51-119. New York: John Wiley & Sons, Inc., 1966.

Petrick, S. R., *A Program for Transformational Syntactic Analysis.* Government document AFCRL-66-698, October 1966.

———, "A Recognition Procedure for Transformational Grammars." Ph.D. dissertation, Massachusetts Institute of Technology, 1965.

Pingle, K. K., J. A. Singer, and W. M. Wichman, "Computer Control of a Mechanical Arm through Visual Input." Paper presented at IFIP68 Congress, Edinburgh.

Pitts, W., and W. S. McCulloch, "How We Know Universals: The Perception of Auditory and Visual Forms," *Bulletin of Mathematical Biophysics,* 9 (1947), 127-47.

Quillian, M. R., "Semantic Memory," Tech. Rept. AFCRL-66-189 (Bolt, Beranek, and Newman, Cambridge, Mass., 1966).

Quinlan, J. R., "A Task-Independent Experience-Gathering Scheme for a Problem Solver," *Proceedings, IJCAI,* Washington, 1969, 193-97.

Quinlan, J. R., and E. B. Hunt, *A Formalization of a General Problem Solver.* Technical Report, Seattle, University of Washington, Department of Psychology, 1967.

Raphael, B., "A Computer Program which 'Understands,'" *Proceedings Fall Joint Computer Conference,* AFIPS, 26 (1964), 577-90.

———, "Programming a Robot." Working paper (Artificial Intelligence Project, Stanford Research Institute, Menlo Park, Calif.). Paper presented at the IFIP68 Congress, Edinburgh.

Reitman, W., *Cognition and Thought.* New York: John Wiley & Sons, Inc., 1965.

———, "Modeling the Formation and Use of Concepts, Percepts, and Rules," in L. Lecam and J. Neyman (eds.), *Proceedings, Fifth Berkeley Symposium on Mathematical Statistics and Probability,* pp. 65-79. Berkeley: University of California Press, 1967.

Reitman, W., R. B. Grove, and R. G. Shoup, "Argus: An Information Processing Model of Thinking," *Behavioral Science,* 9 (1964), 270-80.

Roberts, L. G., "Pattern Recognition with an Adaptive Network," *Record 1960 Institute of Radio Engineers International Convention,* Part 2 (1960), 66-70.

Robinson, J., "A Machine Oriented Logic Based on the Resolution Principle," *Journal of the Association of Computing Machinery,* 12 (1965), 23-41.

Robinson, J. A., "Heuristic and Complete Processes in the Mechanization of Theorem Proving," in J. F. Hart and S. Takasu (eds.), *Systems and Computer Science.* Toronto: University of Toronto Press, 1967.

Rochester, N., J. Holland, L. Haibt, and W. Duda, "Test on a Cell Assembly Theory of the Action of the Brain, Using a Large Digital Computer," *IRE Transactions on Information Theory,* IT-2 (1965), 80-93.

Rosen, C., "Pattern Classification by Adaptive Machines," *Science,* 156 (1967), 38-44.

Rosenblatt, F., *Principles of Neurodynamics.* Washington: Spartan Press, 1962.

———, "The Perceptron: A Probabilistic Model for Information Storage and Organization in the Brain," *Psychological Review,* 65 (1958), 368-407.

Rosenblueth, A., N. Weiner, and J. Bigelow, "Behavior, Purpose and Teleology," *Philosophy of Science,* 10 (1943), 18-24.

Sammet, J., *Programming Languages: History and Fundamentals.* Englewood Cliffs, N. J.: Prentice-Hall, Inc., 1969.

———, "Survey of Formula Manipulation," *Communications of the Association for Computing Machinery,* 9 (1966), 555-69.

Samuels, A., "Programming Computers to Play Games," in F. Alt (ed.), *Advances in Computers,* I, 165-192. New York: Academic Press, Inc., 1960.

———, "Some Studies in Machine Learning Using the Game of Checkers," *IBM Journal of Research and Development,* 3 (1959), 210-29.

———, "Some Studies in Machine Learning Using the Game of Checkers," in E. Feigenbaum and J. Feldman (eds.), *Computers and Thought,* 71-108. New York: McGraw-Hill Book Company, 1963.

Sebestyen, G. S., *"Decision Making Processes in Pattern Recognition."* New York: The Macmillan Company, 1962.

Selfridge, O. G., "Pandemonium: A Paradigm for Learning," in D. V. Blake and A. M. Uttley (eds.), *Proceedings of the Symposium on the Mechanisation of Thought Processes,* 511-29. H. M. Stationary Office, London, 1959.

———, "Pattern Recognition and Modern Computers," in *Proceedings of the 1955 Western Joint Computer Conference,* March 1955, 91-93.

Selfridge, O. G., and G. P. Dineen, "Programming Pattern Recognition," *Proceedings of the Western Joint Computer Conference,* Los Angeles, 1955, 94-100.

Selfridge, O. G., and U. Neisser, "Pattern Recognition by Machine," *Scientific American,* 203 (1960), 60-68.

Shannon, C. E., "A Chess Playing Machine," in J. Newman (ed.), *The World of Mathematics,* IV, 2124-33. New York: Simon and Schuster, Inc., 1956.

———, "The Mathematical Theory of Communications," *Bell System Technological Journal,* 27 (1948), 379-423, 623-56.

Shannon, C. E., and W. Weaver, *The Mathematical Theory of Communications.* Urbana: University of Illinois Press, 1949.

Shearne, J. N., and P. F. Leach, "Some Experiments with a Simple Word Recognition System," *IEEE Transactions on Audio and Electroacoustics,* AU-16 (1968), 256-61.

Siklossy, L., "Natural Language Learning by Computer." Ph.D. dissertation, Carnegie Institute of Technology, 1968.

Simmons, R. F., "Answering English Questions by Computer: A Survey," *Communications of the Association for Computing Machinery,* 8 (1965), 53-69.

———, "Storage and Retrieval Aspects of Meaning in Directed Graph Structures," *Communications of the Association for Computing Machinery,* 9 (1966), 211-15.

———, "Synthetic Language Behavior," *Data Processing Management,* 5 (12), 11-18.

Simmons, R. F., S. Klein, and K. L. McConlogue, "Indexing and Dependency Logic for Answering English Questions," *American Documentation,* 15 (1964), 196-204.

Simon, H. A., "Experiments with a Heuristic Compiler," *Journal of the Association for Computing Machinery,* 10 (October 1963), 4.

———, "Motivational and Emotional Controls of Cognition," *Psychological Review,* 74 (1967), 29-39.

———, "Scientific Discovery and the Psychology of Problem Solving," in R. G. Colodny (ed.), *Mind and Cosmos: Essays in Contemporary Science and Philosophy.* Pittsburgh: University of Pittsburgh Press, 1966.

———, "Thinking by Computers," in R. G. Colodny (ed.), *Mind and Cosmos: Essays in Contemporary Science and Philosophy,* 3, 3-21. Pittsburgh: University of Pittsburgh Press, 1966.

Simon, H. A, and M. Barenfield, "Information-Processing Analysis of Perceptual Processes in Problem Solving," *Psychological Review,* 76 (1969), 473-83.

Simon, H. A., and E. Feigenbaum, "An Information Processing Theory of Some Effects of Similarity, Familiarization and Meaningfulness in Verbal Learning," *Journal of Verbal Learning and Verbal Behavior,* 3 (1964), 385-96.

Simon, H. A., and K. Kotovsky, "Human Acquisition of Concepts for Sequential Patterns," *Psychological Review,* 70 (1963), 534-46.

Simon, H. A., and P. Simon, "Trial and Error Search in Solving Difficult Problems: Evidence from the Game of Chess," *Behavioral Science,* 7 (1962), 425-29.

Simon, H. A., and R. K. Sumner, "Pattern in Music," in B. Kleinmuntz (ed.), *Formal Representation of Human Judgment,* pp. 219-50. New York: John Wiley & Sons, Inc., 1968.

Slagle, J. R., "A Heuristic Program That Solves Symbolic Integration Problems in Freshman Calculus," *Journal of the Association for Computing Machinery,* 10 (1963), 507-20.

———, "Experiments with a Deductive Question-Answering Program," *Communications of the Association for Computing Machinery,* 8 (1965), 792-98.

Slagle, J., and P. Bursky, *Experiments with a Multipurpose Theorem Proving Heuristic Program That Learns.* Tech. Rept. UCRI, 70051, University of California, Lawrence Radiation Laboratory, Berkeley, 1966.

Slukin, W., *Minds and Machines.* London: Penguin, 1960.

Specht, D. F., "Generation of Polynomial Discriminant Functions for Pattern Recognition," *IEE Transactions on Electronic Computers,* EC-16 (1967), 308-18.

Stefferud, E., *The Logic Theory Machine: A Model Heuristic Program.* RAND Corporation Technical Report RM-3731-CC. Santa Monica, Calif.: Rand Corporation, 1963.

Stoll, R. R., *Sets, Logic and Axiomatic Theories.* San Francisco: W. H. Freeman, 1961.

Taube, M., *Computers and Common Sense: The Myth of Thinking Machines.* New York: Columbia University Press, 1961.

Tharp, A. L., and G. K. Krulee, "Using Relational Operators to Structure Long-term Memory," *Proceedings, International Joint Conference on Artificial Intelligence,* Washington, 1969, 579-86.

Thompson, F. B., "English for the Computer," *Proceedings Fall Joint Computer Conference,* AFIPS, 29 (1966), 349-56.

Thompson, F. B., *et al.,* "Deacon Breadboard Summary," RM64TMP-9, TEMPO, General Electric Co., Santa Barbara, Calif., 1964.

Thorne, J. P., P. Bratley, and H. Dewar, "The Syntactic Analysis of English by Machine," in D. Michie (ed.), *Machine Intelligence 3,* pp. 281-310. Edinburgh: The University Press, 1968.

Travis, L., "Experiments with a Theorem Utilizing Program," *Proceedings Spring Joint Computer Conference,* AFIPS, 25 (1964), 339-58.

Turing, A., "Digital Computers Applied to Games," in B. Bowden (ed.), *Faster than Thought,* pp 286-310. London: Sir Isaac Pitman & Sons, Ltd. 1958.

Uhr, L., "Complex Dynamic Models of Living Organisms," in R. W. Stacy and B. Waxman (eds.), *Computers in Biomedical Research,* I, 15-31. New York: Academic Press Inc., 1965.

———, "Intelligence in Computers: The Psychology of Perception in People and in Machines," *Behavioral Science,* 5 (1960), 177-82.

———, "Pattern Recognition Computers as Models for Form Perception," *Psychological Bulletin,* 60 (1963), 40-73.

———, "Pattern-String Learning Programs," *Behavioral Science,* 3 (1964), 258-70.

———, "Recognition of Letters, Pictures, and Speech by a Discovery and Learning Program," *Wescon Convention Review,* 8 (1964), 1-5.

———, *Pattern Recognition.* New York: John Wiley & Sons, Inc., 1966.

Uhr, L., and Sara Jordan, "The Learning of Parameters for Generating Compound Characterizers for Pattern Recognition," *Proceedings, International Joint Conference on Artificial Intelligence,* Washington, 1969, 381-415.

Uhr, L., and C. Vossler, "A Pattern Recognition Program That Generates, Evaluates, and Adjusts Its Own Operators," in E. Feigenbaum and J. Feldman (eds.), *Computers and Thought,* pp. 251-68. New York: McGraw-Hill Book Company, 1963.

———, "A Pattern Recognizer That Generates, Evaluates, and Adjusts Its Own Operators," *Proceedings Western Joint Computer Conference,* 19 (1961), 555-69.

Uhr, L., C. Vossler, and J. Uleman, "Pattern Recognition over Distortion by Human Subjects and by a Computer Simulation Model for Visual Pattern Recognition," *Journal of Experimental Psychology,* 63 (1962), 227-34.

Vallée, J., "Search Strategies and Retrieval Languages." Ph.D. dissertation, Northwestern University, 1967.

Vallée, J., G. K. Krulee, and A. A. Grau, "Retrieval Formulae for Inquiry Systems," *Information Storage and Retrieval,* 4 (1968), 13-26.

Von Neumann, J., *The Computer and the Brain.* New Haven, Conn.: Yale University Press, 1958.

Wales, R. J., and J. C. Marshall, "The Organization of Linguistic Performance," in J. Lyons and R. J. Wales (eds.), *Psycholinguistic Papers,* pp. 19-95. Edinburgh: The University Press, 1967.

Wang, H., "Formalization and Automatic Theorem Proving," in W. A. Kalenich (ed.), *Information Processing 1965: Proceedings of IFIP Congress,* 65, 51-58. Washington: Spartan Books, 1965.

Wathen-Dunn, W. (ed.), *Models for the Perception of Speech and Visual Form.* Cambridge, Mass.: M.I.T. Press, 1967.

Weizenbaum, J., "Contextual Understanding by Computers," in P. A. Kolers and M. Eden (eds.), *Recognizing Patterns,* pp. 170-95. Cambridge, Mass.: M.I.T. Press, 1968.

Wiener, N., *Cybernetics.* New York: John Wiley & Sons, Inc., 1948.

Wickelgren, W., "A Simulation Program for Concept Attainment by Conservative Focusing," *Behavioral Science,* 7 (1962), 245-47.

Williams, P. G., "Some Studies in Game Playing with a Digital Computer." Ph.D. dissertation, Carnegie-Mellon University, 1965.

Wirren, J., and H. C. Stubbs, "Electronic Binary Selection System for Phoneme Classification," *Journal of the Acoustical Society of America,* 28 (1956), 1082-91.

Woods, W. A., *Semantics for a Question-Answering System.* Report No. NSF-19, Mathematical Linguistics and Automatic Translation. Cambridge, Mass.: Harvard University Press, 1967.

Woodworth, R. S., and H. Schlosberg, *Experimental Psychology* (rev. ed.). New York: Holt, Rinehart & Winston, 1965.

Wynn, H., "An Information Processing Model of Certain Aspects of Paired Associates Learning." Ph.D. dissertation, University of Texas, 1966.

Yngve, V., "Comit As an IR Language," *Communications of the Association for Computing Machinery,* 5 (1962), 19-28.

Young, J. Z., *A Model of the Brain.* London: Oxford University Press, 1964.

———, *The Memory System of the Brain.* Berkeley: University of California Press, 1966.

Zwicky, A., *et al.,* "The MITRE Syntactic Analysis Procedure for Transformational Grammars," *Proceedings, AFIPS Conference,* Fall 1965.

AN INFORMATION-PROCESSING THEORY OF SOME EFFECTS OF SIMILARITY, FAMILIARIZATION, AND MEANINGFULNESS IN VERBAL LEARNING

HERBERT A. SIMON

EDWARD A. FEIGENBAUM

Among the most commonly used paradigms in the study of verbal learning is the learning of nonsense syllables by the paired-associate or serial anticipation methods. The variables that have been shown to have important effects on the rate of learning include the levels of familiarity and meaningfulness of the syllables, the amount of similarity among them, and the rate of presentation. In addition, in the learning of lists, there are well-known serial position effects.

In previous papers (Feigenbaum, 1959, 1961; Feigenbaum and Simon, 1962b), a theory has been set forth that undertakes to explain the performance and learning processes underlying the behavior of *S*s in verbal learning experiments. The theory, in its original version, makes correct quantitative predictions of the shape of the serial position curve (Feigenbaum and Simon, 1962a) and the effect of rate of presentation on learning (Feigenbaum, 1959; Feigenbaum and Simon 1962a) as well as predictions of certain qualitative phenomena (for example, oscillation) (Feigenbaum and Simon, 1961).

In this paper, a simplified and improved version of the theory is reported that retains these properties of the earlier theory while providing correct quantitative predictions of the effects of the other important variables: familiarization, meaningfulness, and similarity. The tests of the theory discussed here are based on comparisons of the performance of human *S*s, as reported in published experiments on paired-associate learning (Bruce, 1933; Chenzoff, 1962; Underwood, 1953; Underwood and Schulz, 1960), with the performance predicted by the theory in the same experimental situations with the same, or equivalent, stimulus material.

The theory to be described is a theory of the information processes underlying verbal learning. The precise statement of such a theory is most readily made in the information-processing language of a digital computer, i.e., the language of computer programs.

The formal and rigorous statement of the theory is a program called the *Elementary Perceiver and Memorizer* (third version), or EPAM-III. This program is a closed model and is used as an "artificial subject" in standard verbal learning

Reprinted with permission from the *Journal of Verbal Learning and Verbal Behavior,* 3 (1964), 385-96. Copyright ©1964 by Academic Press, Inc.

experiments (the latter being also simulated within the computer by means of an Experimenter program). Imbedded in the theory are hypotheses about the several kinds of processes that are involved in the performance of verbal learning tasks. These hypotheses take the form of subroutines that are component parts of the total program. Thus, there are performance subroutines which allow the program to produce responses that have previously been associated with stimuli, subroutines for learning to discriminate among different stimuli, and subroutines for acquiring familiarity with stimuli. Top-level executive routines, which organize these subroutines into a program, represent hypotheses about the S's understanding of the experimental instructions and the learning strategy he employs. The computer simulation of verbal learning behavior using the EPAM-III theory is, in essence, generation (by the computer) of the remote consequences of the information-processing hypotheses of the theory in particular experimental situations.

In the first part of the paper a brief description of EPAM-III is provided. Since other descriptions of the program are available in the literature (Feigenbaum, 1959, 1961; Feigenbaum and Simon, 1962b), only so much of the detail will be presented as is essential to an understanding of the experiments and the interpretation of their results. In the second part of the paper, the results will be reported of comparisons of the behavior of EPAM-III with the behavior of human Ss in paired-associate learning where similarity is the independent variable. In the third part of the paper, the results will be reported of comparisons in which familiarization and meaningfulness are the independent variables.

A BRIEF DESCRIPTION OF EPAM-III

The computer language in which EPAM-III is written is known as IPL-V (Newell, Tonge, Feigenbaum, Green, and Mealy, 1964). A companion program simulates an experimental setting, more specifically, a memory drum capable of exposing stimulus materials to EPAM, in either the serial or paired-associate paradigm. The simulated drum-rotation rate can be altered as desired, as can the stimulus materials. An interrupt system is provided so that the simulated experimental environment and the simulated S can behave simultaneously, to all effects and purposes, and can interact, the S having access to the stimulus material presented in the memory-drum window.

The Performance System

EPAM-III incorporates one major performance system and two learning processes (Feigenbaum, 1959, 1961; Feigenbaum and Simon, 1962b). When a stimulus (a symbol structure) is presented, EPAM seeks to recognize it by sorting it through a *discrimination net*. At each node of the net, some characteristic of the stimulus is noticed, and the branch corresponding to that characteristic is followed to the next node. With each terminal node of the net is associated an *image* that can be compared with any stimulus sorted to that node. If the two are similar, in the characteristics used for comparisons, the stimulus has been successfully recognized. We call such a stimulus *familiar,* i.e., it has a recognizable image in the discrimination-net memory.

An image is the internal informational representation of an external stimulus configuration that the learner has stored in memory. An image, thus, is comprised of the information the learner knows about, and has associated with, a particular stimulus configuration. An image may be elementary or compound. A compound image has, as components, one or more elementary or compound images which may themselves be familiar and which may possess their own terminal nodes in the discrimination net. For simplicity, in the current representation, letters of the Roman alphabet are treated as elementary stimuli whose characteristics may be noticed but which are not decomposable into more elementary familiar stimuli. On the other hand, syllables are compound stimuli, their components being, of course, letters.

A compound stimulus image, viewed from the bottom up, may be regarded as an association among the component stimuli. Thus, the net may contain stimulus images that represent pairs of syllables, these compound images having as components other compound images, the individual syllables.

In performing the paired-associate task, the program uses the stimulus, present in the window of the memory drum, to construct a compound symbol representing the pair comprised of the stimulus and its associated responses. We may designate this compound symbol by S———, for the second response member is not then visible in the drum window. The compound symbol, S-———, is sorted through the net, and the image associated with the terminal is——— retrieved. We will designate this image by S'-R', for if the previous learning has been successful, it will be comprised of two components: an image of the stimulus syllable and an image of the associated response syllable. The response image, R', which has just been retrieved as the second component of the compound image, S'-R', identifies a net node where an image, say R', is stored. R' will have as its components symbols designating the constituent letters of the syllable, say X', Y', and Z'. Each of these, in turn, identifies a terminal node. Associated with the terminal for a letter is not only an image of the usual kind (an afferent image), but also the information required to produce the letter in question, i.e., to print it. This information, which we may call the efferent image, is used to produce the response. Thus, the final step in the sequence is for the program to respond, say, X', Y', Z'.

It is a fundamental characteristic of this program that elementary symbols and compound symbols of all levels are stored in the discrimination net in exactly the same way. Thus, a syllable is simply a list of letters, and an S-R is simply a list of syllables. A single interpretive process suffices to sort a letter, a syllable, an S-R pair, or any other symbol, elementary or compound. Moreover, the symbols discriminated by the net are not restricted to any specific sensory or effector mode. All modes can be accommodated by a single net and a single interpretive process. Afferent symbols belonging to different sensory modes will possess different attributes; phonemes will have attributes like "voicing," "tongue position," and so on; printed letters will have attributes like "possession of diagonal line," and so on. Because they possess entirely different attributes, they will be sorted to different parts of the net. Finally, symbols may be of mixed mode. In a symbol, S-R, for example, S may be in the visual mode, R in the oral.

The Learning System

EPAM-III uses just two learning processes, one to construct and elaborate images at terminal nodes of the net (*image building*),[1] and the other to elaborate the net by adding new branches (*discrimination learning*). The first learning process also serves to guide the second.

When a stimulus, S, is in view and is sorted to a terminal, the stimulus can be compared with the image, S', stored at the terminal. If there is no image at the terminal, the image-building process copies a part of S and stores the copy, S', as the initial image at the terminal. If there is already an image, S', at the terminal, one or more differences between S and S' are detected, and S' is corrected or augmented to agree more closely with S.

When a positive difference (not a mere lack of detail) is detected between a stimulus, S, and its image, S', the discrimination learning process can use this difference to construct a new test that will discriminate between S and S'. The terminal node with which S' was associated is then changed to a branch node, the test associated with the node S', is associated with a new terminal on one of the branches, and a new image of S is associated with a new terminal on another branch. Thus, the discrimination-learning process adds a new pair of branches to the discrimination net and attaches initial images to the branches.

Note that a stimulus, S, can be sorted to a terminal, T, only if S satisfies all the tests that point to the branches leading to T. But the image, S', stored at T must also satisfy these tests. Hence, there can be a positive difference between S' and S only if S' contains more information than is necessary to sort S to T. For instance, let S be the syllable KAW, and suppose that all the tests leading to the terminal T happen to be tests on the first letter, K. Then the image, S', stored at T must have K as its first symbol, but may differ from KAW in other characteristics. It might be, for example, the incomplete syllable K-B. The discrimination-learning process could detect the difference between the W and the B in the final letters of the respective syllables, construct a test for this difference, and append the test to a new net node. The redundancy of information in the image, in this case the letter B, permits the further elaboration of the net.

Thus, learning in EPAM-III involves cycles of the two learning processes. Through image building, the stimulus image is elaborated until it contains more information than the minimum required to sort to its terminal. Through discrimination, this information is used to distinguish between new stimuli and the stimulus that generated this terminal and grew its image. On the basis of such distinctions, the net is elaborated. The interaction of these two processes is fundamental to the whole working of EPAM. It is not easy to conjure up alternative schemes that will permit learning to proceed when the members of a pair of stimuli to be discriminated are not present simultaneously.

The stimuli that EPAM-III can make familiar and learn to discriminate are symbols of any kind, elementary or compound. Thus, the letters of the alphabet can be first made familiar, and the net elaborated to discriminate among them.

---

[1] We use the term "image building" rather than the less clumsy and more descriptive term "familiarization" to resolve a dilemma of nomenclature that will become obvious in a later section.

Then EPAM can make familiar and learn to discriminate among syllables, using the now-familiar letters as unitary building blocks. But now, paired-associate-learning can take place without the introduction of any additional mechanisms. Instead of postulating a new associational process, we suppose that an S-R pair is associated simply by making familiar and learning to discriminate the compound object SR.

The entire EPAM-III paired-associate learning scheme is completed by an executive routine that determines under what circumstances the several image-building and discrimination-learning processes will be activated. The executive routine makes use of a kind of knowledge of results. When the simulated *S* detects that he has made an incorrect response to a stimulus syllable, he engages in a rudimentary diagnostic activity: distinguishing between no response and a wrong response, and determining to what extent the response syllable, the stimulus syllable, and the S-R pair are familiar. Depending on the outcome of the diagnosis, various image-building and discrimination-learning processes are initiated.

There are many details of the EPAM-III program we have not described here, but this general sketch will give us a sufficient basis for discussing the behavior of the program in standard paired-associate learning situations.

EFFECTS OF INTRALIST AND INTERLIST SIMILARITY

The adequacy of EPAM-III as a theory of human rote verbal learning has been tested initially by replicating, with the program, experiments of Underwood (1953) on intralist similarity; of Bruce (1933) on interlist similarity; and a number of authors (Underwood and Schulz, 1960) on stimulus and response familiarization and meaningfulness. In this section the experiments employing similarity as the independent variable will be discussed; in the next section the experiments on familiarization and meaningfulness will be considered.

Underwood (1953) studied paired-associate learning of nonsense syllables under various conditions of intralist similarity of stimulus syllable and response syllables. If we use L. M, and H to designate low, medium, and high intralist similarity, respectively; and let, e.g., L-M stand for "low intralist similarity of stimuli, medium intralist similarity of responses," then Underwood's five experimental conditions are L-L, M-L, H-L, L-M, and L-H. Underwood also studied three different conditions of distribution of practice, but since he found no significant differences in his data, we shall not consider this variable further.

In summary, Underwood found (a) that intralist similarity of *responses* had virtually no effect on ease or difficulty of learning;[2] (b) that trials required for learning increased with degree of intralist similarity of *stimuli,* the difference being about 30% between the LL and HL conditions. The first row in Table 1

---

[2]In the Underwood experiment the effect of response similarity is inconsequential. In general, the evidence on the impeding or facilitating effects of response similarity is mixed. What does stand out, however, is this: The effects of response similarity, if any, are quantitatively small and insignificant when compared with the large effects of stimulus similarity.

summarizes Underwood's findings averaged over the three conditions of distribution of practice. The numbers are relative numbers of trials to criterion, with the number for the LL condition taken as 100.

TABLE 1   Comparison of EPAM with Underwood's (1953) Data
on Intralist Similarity[a]

	Condition of stimulus and response similarity				
Data	*L-L*	*L-M*	*M-L*	*L-H*	*H-L*
Underwood	100	96	109	104	131
EPAM-III ("visual only")	100	88	141	91	146
EPAM-III ("aural only")	100	100	100	100	114
EPAM-III ("visual" and "aural" mixed, 1:1)	100	94	121	96	130
EPAM-III ("visual" and "aural" mixed, 1:2)	100	96	114	97	125

[a]Relative number of trials to criterion, LL = 100

The syllables employed in the EPAM simulation were the same as those used by Underwood.[3] Row 2 in Table 1 summarizes the data from the EPAM tests. Response similarity facilitated learning very slightly, while stimulus similarity impeded learning by as much as 40%. Since relative learning times are reported in both cases, there is one free parameter available for matching the two series. (In the normal course of events, the compound images, S'-R', are discriminated from each other on the basis of stimulus information, not response information. High intralist stimulus similarity makes difficulties for EPAM in discriminating and retrieving these images, and hence impedes learning. Response similarity, of course, has no such effect.)

The qualitative fit of the EPAM predictions to the Underwood data is better than the quantitative fit, although, considering the (a priori) plausible range of impact of the similarity variable on difficulty of learning, even the quantitative fit is not bad. Nevertheless, we sought a much better quantitative prediction. This search led us into the following considerations. The prediction that is seriously out of line in Table 1 is the prediction for the M-L condition. The more carefully one scrutinizes the Underwood experiment and the Underwood materials, the more puzzling the experimental results become. Why do *S*s, as the results indicate, respond in the M-L condition so similarly to the way they respond in the L-L condition, while their responses in the H-L condition are so different from responses in either the M-L or L-L conditions? The answer is not to be found in the Underwood materials. We have analyzed the Underwood

[3]We are indebted to Professor Underwood for making these sets of syllables available to us.

definition of "medium similarity" in terms of the information necessary to discriminate the items on a list of given length (in EPAM-like fashion) and have found that Underwood's definition is quite careful and correct. By his definition, one should expect "medium similarity" lists to be midway in effect between his "low similarity" and his "high similarity" lists.

The answer, we believe, lies in the recoding, or "chunking," behavior of *S*s, which would make the "medium similarity" stimulus list formally identical with the "low similarity" stimulus list under Underwood's definition. Suppose that many *S*s were pronouncing the Underwood CVC's, i.e., recoding the items into the aural mode, instead of dealing with them directly in the visual-literal (presentation) mode. The recoded ("aural") syllables will be "chunked" into two parts: a consonant-vowel pair, and a consonant. In other words, the visual-literal stimulus objects of three parts (CVC) quite naturally recode into "aural" stimulus objects of two parts (C'C or CC''). In Underwood's "medium similarity" lists, none of the C' chunks are duplicated, nor are any of the C'' chunks. The recoding, therefore, has transformed the "medium similarity" list into a "low similarity" list, by Underwood's definition.

To test this hypothesis for sufficiency from the point of view of the theory, we reran the EPAM (simulated) experiments using "aural" recodings of the original syllables. The modified predictions are given in Row 3 of Table 1. As the analysis above indicates, the M-L condition is now no different from the L-L condition, but the prediction of difficulty for the H-L condition is too low. Assuming that some *S*s are processing in the visual-literal mode and some in the "aural" mode, we have computed the average (1:1) of the two sets of EPAM predictions. This is given in Row 4 of Table 1. If we weight the average 2:1 in favor of *S*s doing "aural" recoding, the result is as given in Row 5 of Table 1. Each of these averaging procedures gives a prediction which is much better than that for the Underwood lists non-recoded.

It is clear that we still have much to learn about this low-vs.-medium similarity problem. In this regard, we are currently attempting a direct experimental test of the "aural" recoding hypothesis.

Bruce's *S*s (1933) learned two successive lists of paired-associate nonsense syllables. On the second list, response syllables, or stimulus syllables, or neither, could be the same as the corresponding syllables on the first list. Thus, using current designations, Bruce's three conditions were (A-B,C-D), (A-B,A-C), and (A-B,C-B), respectively. In summary, he found that learning of the second list was somewhat easier than learning of the first when all syllables were different (A-B,C-D), much easier when the response syllables were the same (A-B, C-B), and a little harder when the stimulus syllables were the same (A-B, A-C) (see Table 2). The relative difficulties are compared using the A-B,C-D group as the norm.

Nonsense-syllable lists of low Glaze value and low intralist similarity were used when the experiment was replicated with EPAM. The normalized results are shown in the second line of Table 2. The effects in the simulated experiment were qualitatively the same as in the actual data. If we compare the conditions A-B,A-C and A-B,C-B with A-B,C-D, we find that identity of stimulus syllables impeded learning less, and identity of response syllables facilitated learning to the same degree in the simulation as for the human *S*s. The ratio of difficulty for the A-B,A-C compared with the A-B,C-B conditions, where total number of

TABLE 2   Comparison of EPAM with Bruce's Data
on Interlist Similarity[a]

	Condition of stimulus and response similarity		
	A-B, A-C	*A-B, C-B*	*A-B, C-D*
A-B, C-D condition = 100	130	75	100
EPAM-III A-B, C-D condition = 100	112	75	100

[a] Relative number of trials to criterion.

different syllables discriminated and learned was the same, was 1.73 for the human *S*s and 1.49 for EPAM.

From our analysis of the data of the Underwood and Bruce experiments, we conclude that EPAM provides satisfactory explanations for the main observed effects of intralist and interlist stimulus and response similarity upon the learning of paired-associate nonsense syllables. The effects predicted by EPAM-III are in the right direction and are of the right order of magnitude, although there is room for improvement in the quantitative agreement.

FAMILIARITY AND MEANINGFULNESS

Among the other independent variables that have been shown to have major significance for the ease or difficulty of learning nonsense syllables are familiarity and meaningfulness. A thorough discussion of the definition of these two variables can be found in Underwood and Schulz (1960).

The degree of familiarity of a syllable is usually not measured directly; instead, it is measured by the amount of *Familiarization training* to which *S* has been exposed with that syllable. In the following discussion, they are not synonymous. "Familiarization" will be used when reference is made to specific experimental conditions and operations. "Familiarity," on the other hand, will refer to a condition internal to an *S*: the state of information about a syllable in the memory of an *S* who has gone through some kind of familiarization training. Thus, familiarity is an intervening variable not directly observable. The use of an intervening variable hardly needs to be defended, since it is the rule rather than the exception in theory-building in the natural sciences as well as the behavioral sciences.

Familiarization training is accomplished by causing *S* to attend to the syllable in question in the context of some task other than the paired-associate learning task to be given him subsequently. It should be noted that there is no way of discovering, with this definition, how familiar a syllable may be for an *S* due to his experience prior to coming into the laboratory. Although the syllables are not meaningful words, the consonant-vowel combinations contained in them occur with varying frequency in English. The meaningfulness of a syllable, on the other hand, is generally determined by measuring the number of associations that *S*s make to it in a specified period of time. Nonsense syllables for learning experiments are generally selected from available lists that have been graded in this way for meaningfulness.

Since high familiarity and high meaningfulness both facilitate nonsense-syllable learning, there has been much speculation that the two phenomena might be the "same thing." This, in fact, is the central hypothesis examined in the Underwood and Schulz monograph. In one sense, meaningfulness and familiarity are demonstrably not the same, for a substantial amount of familiarization training can be given with low-meaningful syllables without significantly increasing their meaningfulness. However, Underwood and Schulz (1960) adduce a large body of evidence to show that there is a strong relation running the other way, i.e., that meaningfulness of words is correlated with their frequency of occurrence in English, and that ease of learning nonsense syllables is correlated with the frequency, in English, of the letters that compose them (for syllables of low pronounciability), or with their pronounciability.

The data are of course greatly complicated by the fact that *S*s may handle the material in either the visual or the aural mode, and that most *S*s probably encode into the latter, at least part of the time. Hence, for relatively easily pronounceable syllables, frequency of phoneme pairs in the aural encoding is a more relevant measure of frequency than frequency of the printed bigrams or trigrams. Thus, the finding by Underwood and Schulz that pronounciability is a better predictor than trigram frequency of ease of learning does not damage the hypothesis that familiarity of the component units is the critical variable, and that familiarity, in turn, is a function of previous exposure.

We conclude that high meaningfulness implies high familiarity, although not the converse. If this is so, then the correlation of meaningfulness with ease of learning may be spurious. Familiarity may be the variable that determines ease of learning, and meaningful syllables may be easy to learn only because they are highly familiar.

The idea that familiarity is the critical variable in learning rests on the idea, certainly not original with EPAM model, that there are two stages in paired-associate learning: (1) integration of responses, and (2) association of responses with stimuli. Underwood and Schulz (1960) have used this idea, and it plays an important role in their analysis. It also plays an important role in the structure of EPAM (Feigenbaum, 1959, 1960). From our earlier description it can be seen that these two stages of learning are also present in EPAM-III, but that both stages make use of the same pair of learning processes: image building and discrimination learning.

If response integration is the mechanism accounting for the relation between meaningfulness and familiarity, on the one hand, and ease of learning, on the other, then there should be a point of saturation beyond which additional familiarization will not further facilitate learning, i.e., the point at which the syllables are so familiar that they are completely integrated. In the EPAM-III mechanism this would be the point where the syllable images were complete and where the tests in the net were fully adequate to discriminate among them.

There is strong empirical support for the hypothesis of saturation. At the end of the meaningfulness scale, further increases in meaningfulness of syllables have relatively little effect on ease of learning, but the effects are large over the lower range of the scale. In fact, and this is the most striking evidence relevant to the issue, the experiments on meaningfulness in the literature reveal a remarkably consistent upper bound on the effect of that variable. Underwood and Schulz (1960) survey a large number of the experiments reported in the literature, of both paired-associate and serial learning of CVC syllables, and find rather consistently that the ratio of trials to criterion for the least and most meaningful conditions, respectively, lies in the neighborhood of 2.5. That is to say, syllables of very low meaningfulness take about two and one-half times as long to learn as syllables of very high meaningfulness (and about two and one-half times as many errors are made during learning).

Before the significance of this 2.5/1 ratio is considered further, it is necessary to discuss one difficulty with the hypothesis that familiarization and meaningfulness (via familiarity) facilitate learning primarily by virtue of responses being integrated prior to the associational trials. The effects reported in the literature with meaningfulness as the independent variable are generally much larger than the effects reported for familiarization. No one has produced anything like a 2.5/1 gain in learning speed by familiarization training.

There is now some evidence, primarily in a doctoral dissertation by Chenzoff (1962), that the main reason for this discrepancy is that the familiarization training in experiments has been too weak, has stopped too soon. It appears that no one has carried out familiarization training with his *S*s to the point where the syllable integration achieved is comparable to the integration of syllables of high meaningfulness.

Chenzoff's experiment can be summed up as follows. First, in his experiment he manipulated both meaningfulness and familiarization of both stimuli and responses. Thus, he had 16 conditions: all possible combinations of H-H, L-H, H-L, and L-L for stimulus and response meaningfulness[4] with F-F, U-F, F-U, and U-U for familiarity. Second, he employed a more thorough familiarization training technique for the F condition than had any previous investigator. The syllables were presented one at a time to *S* at about a 2.5-sec rate. The *S* was required to pronounce each syllable. After five trials, *S* was asked to recall the syllables in any order. If an incorrect syllable was given, *S* was told it was not a member of the list. If, within 30 sec, *S* could not perform completely correctly, five more familiarization trials were administered. This continued until *S* learned the list. The range of number of trials for the various *S*s was 10-30; the median and mode were 15 trials.

With this training, the effects of familiarization were qualitatively similar

---

[4]The two levels of the meaningfulness variable were constructed as follows (using CVC's):

H: 53-100 Glaze, 85-100 Krueger, 67-97 Archer, 2.89-3.66 Noble (m), 3.08-3.87 Noble (a).

L: 0-53 Glaze, 39-72 Krueger, 9-48 Archer, 0-92-1.83 Noble (m), 1.38-94 Noble (a).

to, and more than half as large in magnitude as, the effects of meaningfulness. Specifically:

(1) For the H-H (high meaningfulness) conditions, amount of familiarization of stimuli, responses, or both had no effect on ease of learning; the saturation was complete [Table 3, Column (1)].

(2) For the L-L (low meaningfulness) conditions, unfamiliarized syllables (U-U) took 1.8 times as long to learn[5] as familiarized syllables (F-F). Response familiarization (U-F) had a greater effect than stimulus familiarization (F-U); the ratios were 1.2 and 1.6, respectively [Table 3, Column (3)].

(3) When familiarization training was provided, the effects of meaningfulness upon ease of learning were reduced by about two thirds. In the F-F conditions, the L-L pairs took only 1.2 times as long to learn as the H-H pairs, the L-H and H-L pairs falling between the two extremes. Saturation was not quite complete but was clearly visible [Table 3, Column (2)].

(4) In the absence of familiarization training, the usual large effects of meaningfulness were visible. In the U-U conditions, the L-L pairs took 2.2 times as long to learn as the H-H pairs [Table 3, Column (4)].

Thus, except for the quantitative deficiency in the effect of familiarization, Chenzoff shows meaningfulness and familiarity to be equivalent. But they are not additive because of the saturation effect.

Further and very strong evidence for the syllable-integration hypothesis is obtainable from the predictions of **EPAM-III**. By presenting syllables with appropriate instructions, EPAM can attain familiarity with stimulus syllables, response syllables, or both. Amount of familiarity can be manipulated by varying the number of familiarization trials. In particular, familiarity can be carried to saturation—to the point where complete syllable images are stored in the discrimination net. The maximum effects predicted by EPAM-III for familiarization are of the same magnitude as the maximum effects of meaningfulness observed in the empirical studies. Table 3 shows, for the four conditions, and taking the L-L conditions as the norm, the relative rates of learning as predicted by EPAM-III [Column (5)], and as reported by Chenzoff's Ss [Column (4)]. Except for the rather high value for the H-L condition for Chenzoff's Ss, which is in disagreement with the other experiments in the literature on this point, the quantitative agreement with the EPAM-III predictions is remarkably close. In particular, EPAM predicts the 2.5 maximum ratio that has been so often observed. Since syllable integration is the mechanism that EPAM employs to achieve this result, this implication of the theory provides support for the hypothesis that syllable integration is the mediating mechanism in producing the effect of meaningfulness (and familiarization) upon ease of learning.

---

[5]Because of the form in which Chenzoff presented his data, the actual measure of speed of learning used here is the reciprocal of the number of correct responses between particular (fixed) trial boundaries, relative to the (H-H, F-F) condition taken as a norm of 1.0 (see Table 3).

TABLE 3   Effects of Familiarization and Meaningfulness

Meaningfulness (or familiarity)	Chenzoff's (1962) data[a]			EPAM-III[b]		
	(1) High meaning- fulness	(2) High familiar- ization	(3) Low meaning- fulness	(4) No familiar- ization	(5) No previous familiar- ization	(6) Previous familiar- ization
H-H or F-F	1.0	1.0	1.0	1.0	1.0	1.0
L-H or U-F	1.0	1.1	1.2	1.2	1.3	1.0
H-L or F-U	1.0	1.2	1.6	1.2	1.8	1.5
L-L or U-U	1.0	1.2	1.8	2.2	2.5	1.7

[a]Reciprocal of number of correct responses; H-H or F-F = 1.0.
[b]Relative number of errors to criterion; H-H or F-F = 1.0.

DEGREE OF FAMILIARIZATION

If the present interpretation of the mechanism of familiarity is correct, then the effects of a given amount of familiarization training will depend, in a sensitive way, upon how familiar the syllables were at the beginning of training. There is no way of knowing this exactly, although it is reasonable to assume that nonsense syllables of low association value are close to the zero level of familiarity. (See, however, the findings of Underwood and Schulz on differential pronounciability of such syllables.)

To examine the effects of varying amounts of familiarization training upon the ease or difficulty of paired-associate learning, EPAM was tested with various combinations of zero to five trials of stimulus- and response-syllable familiarization. The results are shown in Table 4 in terms of number of errors to criterion.

TABLE 4   Effects of Stimulus and Response Familiarization[a]

Response familiar- ization (trials)	Stimulus familiarization (trials)			
	0	1	2	3
0	52	44	38	38
1	48	35	32	32
2	39	24	24	24
3	27	21	21	21

[a]Number of errors to criterion in paired-associate learning.

Under the conditions employed in these experiments, the maximum possible effects of familiarization were obtained with a combination of three trials of response familiarization and one trial of stimulus familiarization; additional familiarization did not facilitate learning. The asymptote, 21 errors, for this amount of familiarization was not attainable with any amount of response familiarization in the absence of stimulus familiarization, or with any amount of stimulus familiarization without response familiarization. The asymptotes in the latter two cases were 27 errors and 38 errors, respectively, and were reached with three trials and two trials, respectively, of familiarization.

The detail of Table 4 shows some exceedingly complicated relations. For example, if syllables have received no prior familiarization, one trial of stimulus familiarization reduces errors more than one trial of response familiarization (reductions of eight and four errors, respectively) from 52 in the no-familiarization case. On the other hand, for syllables that had already received one trial of stimulus and response familiarization, an additional trial of stimulus familiarization reduced errors only by three, while an additional trial of response familiarization reduced errors by 11, from a level of 35. Other similar results may be read from Table 4. Many of the numerous small anomalies in the literature on familiarization training may be attributable to the lack of control over the amount of prior familiarity that Ss had with the syllables used in the experiments.

In Table 3, Column (6), we show the predicted effect, estimated from the EPAM data of Table 4, of familiarization training with syllables that were already somewhat familiar before the experiment began (i.e., that had previously received one simulated trial each of stimulus and response familiarization). Under the F-F condition, we would have 21 errors to criterion; under the U-F condition (one stimulus familiarization trial, three response trials), 21; under the F-U condition (three and one stimulus and response familiarization trials, respectively), 32; and under the U-U condition (one S and one R familiarization trial), 35. The resulting indexes of relative difficulty for the four conditions are 1.0, 1.0, 1.5 and 1.7, respectively, as shown in Column (6). These may be compared with the values 1.0, 1.2, 1.6, 1.8, for the actual data in Column (3). In other words, the fact that the effects shown in Column (3), and even in Column (4), are somewhat smaller than the predictions in Column (5) may be due simply to the fact that the syllables were already slightly familiar to the Ss at the beginning of the experiment.

CONCLUSION

In this paper we have compared the predictions of EPAM-III, a theory of human verbal learning, with data from the experiments of Bruce, Chenzoff, Underwood, and others, on the effects of intralist and interlist similarity, of familiarization, and of meaningfulness upon difficulty of learning. We find that there is good quantitative, as well as qualitative, agreement between the published data and the predictions of the theory. Finally, we have used our findings to discuss the relation between familiarity and meaningfulness, and have shown that most of the known facts can be explained by supposing that a symbolic structure necessarily becomes familiar in the process of becoming meaningful, but that the converse is not necessarily the case.

SUMMARY

Results obtained by simulating various verbal learning experiments with the Elementary Perceiving and Memorizing Program (EPAM), an information-processing theory of verbal learning, are presented and discussed. Predictions were generated for experiments that manipulated intralist similarity (Underwood, 1953); interlist similarity (Bruce, 1933); and familiarity and meaningfulness. The stimulus materials were nonsense syllables learned as paired-associates.

A description of the EPAM-III model is given.

The predictions made by the model are generally in good agreement with the experimental data. It is shown that the quantitative fit to the Underwood data can be improved considerably by hypothesizing a process of "aural recoding."

The fit of the EPAM predictions to data of Chenzoff (1962) lends support to the hypothesis that the mechanism by means of which a high degree of meaningfulness of items facilitates learning is the high familiarity of these items.

The effects of varying degrees of stimulus and response familiarization on ease of learning were studied, and are shown to be surprisingly complex.

REFERENCES

Bruce, R. W. Conditions of transfer of training. *J. Exp. Psychol.*, 1933, 16, 343-361.

Chenzoff, A. P. The interaction of meaningfulness with S and R familiarization in paired-associate learning. Unpublished doctoral dissertation. Carnegie Institute of Technology, 1962.

Feigenbaum, E. A. *An information processing theory of verbal learning.* Report P-1817. Santa Monica, Calif.: The RAND Corporation, 1959.

Feigenbaum, E. A. The simulation of verbal learning behavior. *Proceedings of the 1961 Western Joint Computer Conference,* 1961, 19, 121-129.

Feigenbaum, E. A., and Simon, H. A. Forgetting in an association memory. *Reprints of the 1961 National Conference of the Association for Computing Machinery,* 1961, 16, 2C2-2C5.

Feigenbaum, E. A., and Simon, H. A. A theory of the serial position effect. *Brit. J. Psychol.,* 1962, 53, 307-320. (a).

Feigenbaum, E. A., and Simon, H. A. Generalization of an elementary perceiving and memorizing machine. *Information processing 1962, Proceedings of IFIP Congress 62.* Amsterdam: North-Holland Publishing Co., 1962, 401-406. (b).

Newell, A., Tonge, F. M., Feigenbaum, E. A., Green, B. F., Jr., and Mealy, G. H. *Information processing language V manual.* (2nd ed.) Englewood Cliffs, N.J.: Prentice-Hall, 1964.

Underwood, B. J. Studies of distributed practice: VIII. Learning and retention of paired nonsense syllables as a function of intralist similarity. *J. exp. Psychol.,* 1953, 45, 133-142.

Underwood, B. J., and Schulz, R. W. *Meaningfulness and verbal learning.* Philadelphia: Lippincott, 1960.

3

Individual Behavior Simulations

– Personality

JOHN C. LOEHLIN

INTRODUCTION

Thirty-odd years ago, the late Gordon Allport listed at least fifty different senses in which the word *personality* had been used (Allport, 1937); the total will hardly have shrunk since then. This is not the place to try to disentangle all the conceptual complexities involved, but it may help to put our view of computer simulation of personality into perspective if we glance at three of the principal polarizations.

1. *Surface versus depth.* Does personality refer primarily to certain features of the behavior a person emits, or does it refer to some inferred underlying machery that generates or colors the behavior? Much popular usage inclines toward the former ("she has a pleasing personality"), as does some technical usage (e.g., when a personality trait is identified with the score on a personality test). But when Freud (for example) describes id, ego, and superego in the anatomy of the personality (Freud, 1933), he clearly has in mind something much more toward the depth end of the dimension.

2. *Uniqueness versus commonality.* Murray and Kluckhohn put it succinctly: "Every man is in some respects (a) like all other men, (b) like some other men, and (c) like no other man" (1948, p. 35). In some discussions of personality, the emphasis is on the first of these, the general common properties of that class of entities we call "persons"; in other discussions, it is on making discriminations within the class. (These discriminations may be in terms of subclasses of persons, "personality types," or dimensions of variation, "personality traits," or both.)

3. *Inclusion versus exclusion.* For some writers, personality is what you have after you put perception, motivation, emotion, learning, thinking, social behavior, and so forth, together in an individual—while for other writers, personality is what you have left over after you have taken all these other things out. From the first point of view, personality comprises the distinctive total behavior of the individual, whereas from the second point of view, personality is the residual variation left unaccounted for by simpler processes.

In the present chapter, we will be taking the following positions with respect to these three polarities. On the surface-depth dimension, we will clearly favor depth. The whole point of personality simulation is to construct models of the structures and processes though to underlie distinctive individual behaviors. On the uniqueness-commonality dimension, we will play it both ways— we will hold the computer models ought to be able to deal both with the distinctive and the shared aspects of personality, and we will be interested in seeing how existing simulations accomplish this. Finally, on the inclusion-exclusion dimension, we will incline toward inclusion. However, since simulations of cognitive processes are discussed by Krulee elsewhere in this book, we will largely ignore perception and thought: We simply note here that individual distinctiveness of processing style in such programs is relevant to our present topic. In addition, while the way in which the individual person relates to his social context is of great interest from the standpoint of personality, the simulation of this social context will not be discussed in the present chapter, since it is treated extensively throughout the remainder of the present volume. On the other hand, the modeling of motives, needs, emotions, attitudes, values, and conflicts of the individual will be held to be very much a part of "personality simulation."

The Nature of Personality Theory

Quite apart from the matter of defining the domain of phenomena comprising "personality," there have been various views on what a proper personality theory should look like. Traditional "personality theories" have been predominantly verbal and metaphorical in character. Some of the metaphors have been from the physical world—"charges," "pressures," "tension," "equilibrium," "mechanisms"; others have been from the social world—"conflict," "defense," "censorship," "regnancy," "cues." The great metaphorical resources of natural languages have made such theories rich and suggestive but at the same time have made it almost impossible to decide just what implications do or do not flow from a given set of statements in the theory.

At the opposite extreme, some theorists have sought to express the structures of personality mathematically. The most popular of such schemes, factor analysis, uses an extremely simple model of the person—the vector or profile: An individual is described by a set of scores on a series of abstract traits, or "personality factors." (Cattell, 1946; Eysenck, 1947; Guilford, 1959.) Some additional complexities in the way of interactions among traits are occasionally introduced, particularly by Cattell (see, for example, 1963), but so far there has not been much exploitation of these elaborations of the basic model.

Is there some convenient middle ground between representing the person by a complex and ambiguous verbal-metaphorical construction, and representing the person by a short linear string of numbers? One such possible middle ground is the computer model, which permits a considerable degree of dynamic and structural complexity while retaining the cardinal virtue of the mathematical model: the explicitness that permits one to generate a variety of specific implications from it.

Strategies in Personality Research

Whatever the formal nature of one's theory, it remains a mere speculation until given empirical support. The efforts of students of personality to give empirical substance to their theories fall fairly conveniently into three categories: naturalistic observation, psychometrics, and experiment.

In many ways, the most appealing of the three strategies is naturalistic observation. One simply observes man in his natural settings and sees whether his behavior is indeed as theory predicts. In fact, since we each carry around a fairly large supply of miscellaneous behavior records in our memories, we may often be able to reach conclusions about an inference from a theory without even having to make any new observations.

But naturalistic observation has its pitfalls. For one thing, it can be highly distorted by subjective biases. While the capacity of man to deceive himself may not be infinite, it is certainly impressive. Another problem is presented by the great complexity that characterizes natural situations. Exact behavior predictions can rarely be made because of the vast array of factors present that may influence the behavior but either are not specified in the theory or are difficult or impossible to assess in the situation. Even worse, the variables that one *can* assess are likely to be correlated with one another in such a way as to make inference difficult.

The psychometric tradition in personality research arose out of the first of the difficulties of naturalistic observation noted above. A personality test consists of a systematic sampling of behaviors or attitudes of the individual; in the construction of the better of such tests, considerable effort and ingenuity have gone into avoiding or compensating for various sources of potential bias.

Experimental approaches to the study of personality have arisen out of the second difficulty of naturalistic observation—the lack of control of variables. If persons can randomly be assigned to groups receiving different experimental treatments, many difficulties of interpretation disappear. Of course, many experiments on personality variables are impossible for ethical or practical reasons, but in some areas of personality study experimental methods can be applied, and indeed have been applied, with considerable success.

Any of the above research strategies may in principle be used to test the adequacy of a computer model. We will have some comments in a later section of this chapter concerning the ways in which such models have in fact been tested.

Modeling Persons in Computers

The idea of a mechanical simulation of a man has a long history, and the notion of such a contrivance developing motives, temperament, and emotions of its own, and turning against its creator, has been a familiar literary theme. The immediate stimulus for the development of digital computer models of personality processes seems, however, to have been the work on the computer simulation of cognitive processes that took place in the 1950s, particularly the efforts of Simon and Newell and their associates at Carnegie Tech and the RAND Corporation. This work is reviewed in the chapter by Krulee in the present volume. Also influential, both directly and by way of the cognitive area, were the cybernetic ideas of Norbert Wiener and his associates at M.I.T. (Rosenblueth, Wiener, and Bigelow, 1943; Wiener, 1948). An early expression of this influence in psychology came in E. G. Boring's presidential address to the Eastern Psychological Association under the title "Mind and Mechanism" (Boring, 1946). Boring discusses the various functions that would have to be incorporated in the design of a bright and attractive dinner companion or a perfect professor of psychology. He acknowledges a conversation with Wiener as having set him off on this theme. Boring concludes (p. 192):

> I believe that robotic thinking helps precision of psychological thought, and will continue to help it until psychophysiology is so far advanced that an image is nothing other than a neural event, and object constancy is obviously just something that happens in the brain. That time is still a long way off, and in the interval I choose to sit cozily with my robot, squeezing his hand and feeling a thrill—a scientist's thrill—when he squeezes mine back again.

In Britain, a somewhat similar analysis was undertaken by Donald MacKay, who in 1951 outlined a hybrid digital-analogue device and concluded: "In such an artefact analogues of concepts such as emotion, judgment, original-ity, consciousness, and self-consciousness appear" (MacKay, 1963, p. 240).

By 1960, summaries and reviews of the work on the computer simulation of cognitive processes were beginning to appear in the psychological literature, as well as speculative extension of these ideas to other aspects of psychology. Hovland (1960) reviewed the computer simulation of thinking. Uhr (1960) discussed perception by man and machine. Taylor (1960) proposed a treatment of motivation in these terms. And Miller, Galanter, and Pribram (1960) in a fascinating little book entitled *Plans and the Structure of Behavior* extended information-processing ideas to almost every aspect of psychology, including motivation, skills, personality, language, thinking, and memory. Shortly thereafter, Saunders (1961) treated the problem of individual differences in personality in his account of different variants of an imaginary computer named Zodiac.

By this time, several actual computer simulations were underway in the personality area. A conference held in Princeton, New Jersey, in June 1962 may perhaps be taken as the official launching of the field of computer personality modeling. Three simulations (Abelson's, Colby's, and the present writer's) were described, along with developments in related areas. There were critiques and discussions by well-known psychologists in the field of personality. The proceedings of the conference were published under the title *Computer Simulation of Personality* (Tomkins and Messick, 1963).

A short review of developments in the area occurs in a chapter in the 1968 *Annual Review of Psychology* (Hunt, 1968). The present writer has contributed a book and a briefer account of computer models of personality (Loehlin, 1968a, b). Some recent textbooks on personality theory (Mehrabian, 1968) and personality measurement (Kleinmuntz, 1967) include discussions of computer simulation. A highly pertinent analysis of motivation and emotion in information-processing terms has also appeared (Simon, 1967), and much relevant material is included in a chapter on simulation in social psychology by Abelson (1968a).

EXISTING COMPUTER MODELS
OF PERSONALITY PROCESSES

The major existing computer models of personality take a general form which can be characterized as follows:

The central component is an *attitude structure* that represents the individual's more-or-less-enduring dispositions to respond in characteristic ways to his environment and to his own actions. This attitude structure is built up of *concepts* (possibly linked together into more complex *beliefs)* which contain emotional or evaluative as well as cognitive components. In computer terms, this attitude structure is represented by a complex *data structure,* in array or list form.

Operating in conjunction with this attitude structure is a more-or-less-coherent set of *personality processes,* in the form of computer program *subroutines.* These processes generate appropriate behavioral or verbal outputs in response to inputs, or "spontaneously." They also add to or alter the attitude structure itself, in complex ways which may involve various biases, distortions, selective perceptions, and the like.

The operating subroutines contain certain adjustable constants or *parameters,* which govern their style of processing and can be considered to represent *personality traits* of the model. Personality differences thus may arise from differences in the form and content of attitude structures or from differences in the parameters of the personality processes that use or change these structures.

Abelson's Model
of an Individual Attitude Structure

Most of the features of the generalized description above are realized in the simulation of an individual attitude structure constructed by Robert P. Abelson (Abelson, 1963, 1968a, b; Abelson and Carroll, 1965; Abelson and Reich, 1969) using the programming language IPL-V and, more recently, SNOBOL.

Abelson's model is designed to input statements that it interprets in the light of its existing attitude structure. Statements consonant with its present beliefs and attitudes are added to the structure.   Dissonant statements either are rejected outright as incredible or are subjected to distortion by defense mechanisms until they can be definitely accepted or rejected. Suitable output sentences are generated to express agreement, disagreement, or reinterpretation of the input statement. Parameters in the various subroutines govern such properties of the system as its tolerance for dissonant inputs or its preferences among the processing strategies in its repertoire.

Let us consider a (fictional) example. Suppose the input sentence is "Fidel Castro likes TV Westerns." Let us assume the system already contains some relevant information: the concepts "Fidel Castro" and "TV Westerns"; the verb "likes"; and various combinations of these and other concepts forming the predicates "likes TV Westerns," "likes TV soap operas," the sentence "Fidel Castro likes TV soap operas," and so forth. Each of these elements is represented by an IPL list structure which identifies its subelements and the compounds into which it enters, as well as class-subclass relationships ("Fidel Castro" may be a member of the class "Communist leaders," and "TV soap operas" may include "As the World Turns"). In addition, each item has an evaluative component—the owner of this particular attitude structure may detest Fidel Castro and soap operas and may dote on TV Westerns.

When the sentence "Fidel Castro likes TV Westerns" is input to the system, the credibility of this assertion is first investigated. If the attitude structure already contains this sentence or its negation, an immediate judgment of credibility or incredibility is made. If not, a further exploration of related beliefs is undertaken. During this exploration, the discovery of connections indicating that "TV Westerns" and "TV soap operas" are both "escapist entertainment" would tend to count as favorable evidence for credibility (since we already know that Fidel likes soap operas); whereas the information to the effect that "Mao Tse-tung hates TV Westerns" would tend to count in the opposite direction.

If the exploration of the data structure indicates that the input statement is at least marginally credible, its evaluative consistency is examined. In the present instance a dilemma would be posed at this point: A negatively valued subject "Fidel Castro" is associated with a positively valued predicate "likes TV Westerns." When an imbalance of this nature is discovered, two mechanisms are called into play, one an evaluative transfer between the sentence elements (the person tends to think a little better of Fidel Castro and a little worse of TV Westerns), and the other a distortion of the cognitive content of the sentences by such mechanisms as rationalization or denial. (For example, the sentence "TV Westerns contain much bloodshed," if present in the system, might provide the basis for a successful rationalization attempt.)

Thus by the input of a sequence of belief statements from a social environment the model builds up a characteristic attitude structure, one that tends to resist information radically inconsistent with its present contents but is nonetheless subject to gradual alteration over time. Modification also occurs in the manner of processing information. The program contains built-in learning mechanisms which tend to develop preferences for using processes that work successfully. Thus on the basis of its experience the model may come to favor a characteristic method of rationalization or to emphasize inductive rather than deductive approaches to establishing credibility. By an appropriate sequence of inputs one should presumably be able to produce naive credulity, tough-minded rationality, or paranoia; however, only limited explorations along these lines have been undertaken so far with the model.

Abelson and his students have carried out several experiments with human subjects to test the plausibility of different inference-making mechanisms in the program (Gilson and Abelson, 1965; Abelson and Kanouse, 1966; Kanouse and Abelson, 1967). In general, these experiments have provided support for the model but have suggested a need for incorporating some additional variables into it.

In addition, Abelson has attempted to set up an attitude structure in the model corresponding to a real-life set of attitudes—the conservative foreign policy views of a well-known U.S. political figure. The concepts, beliefs, and evaluations expressed by this individual in his published writings and speeches were hand coded into the elements of the model's initial attitude structure. The responses of the model to suitable input sentences could then be simulated. The outputs of the model have been described by informal observers as often strikingly characteristic of the utterances of the individual in question.

Colby's Model of a Neurotic Personality

Perhaps the most elaborate computer model of personality is the one that has been developed by the psychoanalyst Kenneth Mark Colby and his associates (Colby, 1963, 1964, 1965; Colby and Gilbert, 1964). Colby's model was originally programmed in IPL; it is now expressed in a version of ALGOL. Since the model is described in some detail in the case-example reprinted following this chapter, only brief note of some major features will be taken here.

As in Abelson's model, the central data structure is an attitude structure. Concepts, and the attitudes formed using them, are listed in several tables, or *matrices.* These matrices contain both cognitive information (e.g., class-subclass relationships among the concepts) and evaluative information (e.g., the *charge,* or emotional value, that a given belief has for the person being simulated).

Again, Colby's model contains a number of subroutines, or processes, which operate on the data structures in the matrices. The model may be set up either to respond to inputs, as in Abelson's model, or simply to try to express its conflictual beliefs, like a patient free associating on a psychoanalyst's couch. Prominent in the operations of the model are *transforms* and *monitors.* The former are a set of defense mechanisms which distort conflictual beliefs so they can be expressed. The latter continuously record levels of conflict, anxiety, and so forth, within the system and redirect processing to less troublesome areas of the attitude structure if things get too unpleasant.

As described in the case-example, Colby has set up his model's matrices with contents corresponding to the beliefs and attitudes of a particular woman patient of his; the resulting output is "something like" the behavior of a real patient.

Colby's more recent work in this area has mainly concentrated on two problems. One is an attempt to build up a representation of a person's attitude structure in the computer by way of a teletype dialogue between the person and the machine (Colby, Watt, and Gilbert, 1966; Colby and Enea, 1967, 1968b). The other is the elaboration of the processing of beliefs to give the model more effective inférence-making abilities and hence to allow it to make occasional rational readjustments in its attitude structure which may decrease its long-run potential for conflict (Colby, 1967, 1968; Colby and Enas, 1968a; Tesler, Enea, and Colby, 1968). The work along both these lines looks promising, but the final outcome is not yet clear.

Aldous

The present writer has constructed a personality model called Aldous (Loehlin, 1962, 1963, 1968a), which is much simpler and more schematic than the preceding two models but resembles them in being based on a central data structure and a set of program subroutines that interact with it. The original version of Aldous was programmed in machine language for a small four-thousand-word computer; it has since been translated into FORTRAN and run on larger machines.

The central attitude structure of the model is built around *concepts,* which are locations in the computer memory indexed by combinations of defining properties. At each such location is stored an indication of the familiarity of the object or class of objects referred to by the concept, as well as the strengths of the model's three emotions—fear, anger, and attraction—toward the object or category represented. A concept with its associated emotional predispositions may be considered as constituting an *attitude*. Concepts are organized hierarchically: If Aldous has an unpleasant experience with an object in the class, say, of young red-haired females, his attitudes toward such categories as females, red-haired females, young people, and so forth, will also be affected in greater or lesser degrees.

Aldous reacts to objects that are input to him as coded descriptions of their properties. Three internal subroutines carry out this reaction: *recognition,* in which the input guides retrieval of relevant portions of the attitude structure from the model's permanent memory; *emotional reaction,* in which an emotional response to the object is generated from the retrieved attitudes, modified by the model's current mood; and *action selection,* in which a generalized reaction (approach, withdrawal, attack, indifference, conflict) appropriate to the model's emotional state is chosen according to simple fixed rules.

Once a response is emitted, an external subroutine—*consequences*—provides suitable feedback to the model, based on the likely outcome of the model's response, given the realistic properties of the object; if this feedback results in significant arousal of the model's emotions, a *learning* subroutine makes appropriate modifications in Aldous's attitudes. Various parameters (proportion of errors in recognition, threshold for action, relative weight given to recent and distant past, etc.) govern the style of operation of the various subroutines in the model and are conceptualized as representing *personality traits.*

Thus based on the interaction over time of its characteristic response style and its input history, each model acquires a distinctive and unique "personality."

Informal explorations with the model have included setting it up in a computer with another of its kind and permitting the two to interact—that is, the output of the one becomes the input to the other (Loehlin, 1965). In this case, attitudes are formed toward actions or sequences of actions of the two models. The flavor of such interactions may perhaps be conveyed by a "psychotherapy" example.

It had previously been observed that when one of a pair of models was set up with generally favorable attitudes toward its world, and another with generally unfavorable attitudes, and the two were allowed to interact, the relationship tended ultimately to degenerate into mutual hostility. A version of Aldous was modified to carry out the role of a (rather directive) psychotherapist. The therapist model has a fixed repertoire of responses—it attempts to counteract mild levels of fear or hostility on the part of its patient, withdraws in the face of stronger levels, and responds encouragingly to all neutral or positive behaviors. A model of Aldous with initially negative attitudes was permitted to interact with the "therapist" over a period of trials. These encounters were marked mostly by withdrawal at first, but strongly positive behavior toward the "therapist" eventually developed. The patient was then allowed to interact with a "normal" model. Things went along fairly promisingly for a while, but then hostility began emerging and the attitudes of the pair started to show mutual deterioration. The patient was returned for another period of interaction with the therapist. Considerable hostility was exhibited by the patient at the start of these sessions, but this in due course lessened, and his behavior again became strongly positive. Finally, the patient was started out with a new normal partner, and this time a stable positive relationship was eventually developed and maintained.

Other Models

While the three models that have been described constitute the most fully specified attempts so far to represent personality structures and processes in digital computer programs, a number of other modeling efforts involve the representation of personality to a greater or lesser degree. For example, simulations of social or interpersonal processes typically imply representations of the participants—representations that may range from the extremely simple to the quite elaborate. An example at the elaborate extreme is provided by the Gullahorns' HOMUNCULUS, which could quite well stand as a personality model in its own right. Other less-detailed modelings would include the interacting individuals in Coe's (1964) simulation of aggressive interaction, or in Pool and Kessler's (1965) simulation of elite decision-makers, or in many other examples cited in the chapters on social simulations. And, as noted earlier, any individual behavior simulation that incorporates distinctive individual parameters may be regarded as in some sense a personality model as well.

DISCUSSION

Now, how does the simulation of personality by computer models such as these fit into the broader context of theory and research on personality as we have outlined it earlier?

At a minimum, we can take such models as demonstrating the descriptive resources of the computer medium. Clearly one can represent structures of considerable complexity, and interacting processes of considerable subtlety, in a computer model—and yet have them explicit: The programs do run.

But the modelers have made somewhat stronger claims than this, namely, that computer models of personality—maybe not these particular computer programs, but programs more or less like them—will one day serve as touchstones for evaluating personality theories, will make useful practical predictions about individuals, will act as durable practice cases for novice psychotherapists, and so on. In short, the claim has been made that such descriptions have some degree of validity. How can this claim be evaluated?

In principle, the answer is simple. Set up the models in question to correspond to actual individuals; let the models behave, let the persons behave, and then examine the correspondence between the behaviors. In more formal terms, derive predictions from the theory by operating the model, then test these predictions by empirical observation.

I do not propose even to enumerate, much less explore, the complex questions that are raised by this apparently simple statement of the matter. Many general issues of evaluating computer simulations are discussed in some detail elsewhere in this volume, particularly in the concluding chapters. Instead, I will briefly comment on three points: first, some special difficulties relating to computer simulation of personality, as opposed to computer simulation in other areas of the social sciences; second, the ways in which computer models of personality have been evaluated; and third, some possible techniques for matching models to persons.

Special problems of personality simulation.   Perhaps the main problem inherent in computer simulation of personality is the difficulty of observing the events to be simulated. If one is dealing with complex structures and processes lying deep within the organism, in many cases not even available to the individual's introspections, major difficulties are implied both for representing these structures and processes in the first place and later for evaluating and improving the goodness of fit of the model to reality. The simulator of an economic system, say, is much more fortunate in this respect—the fine structure of his subject matter is open to direct inspection. If the economist needs to, he can always ascertain whether Joe Smith did in fact buy a twenty-one-thousand-dollar house in Des Moines in the spring of 1968. The personality psychologist, for many important events, is in the position of an economist forever limited to inferring Joe Smith's purchase from its effects on the national economy.

It should be made clear that this is not a criticism of the use of computer simulation in the personality area—the problem is basic to personality theory and exists whatever the form in which the theory is expressed. Putting the theory on a computer thus in no sense creates the difficulty, although it may make it harder to gloss over.

Tests of existing computer models.   Earlier in the present chapter three general strategies of personality research were described: naturalistic observation, psychometrics, and experiment. Overwhelmingly, tests of existing computer models have relied on the first of these, in the form of crude comparison of the model's output to casually observed human behavior. Abelson is a partial exception, since he conducted some formal experiments to test certain aspects of his model. The present writer has *simulated* a psychometric instrument, a personality questionnaire, to assess Aldous' attitudes but has not attempted to match scores on this questionnaire with any real-world questionnaire data.

Does this lack of formal verification of these models invalidate them as scientific or practical instruments? The writer would argue that at this stage in their development informal observation would be more than sufficient to suggest inadequacies in the models and that an elaborate apparatus of psychometrics and experiment would constitute theoretical overkill. But clearly, at *some* stage, if these models cannot interact profitably with the systematic gathering of empirical data, their scientific value must seriously come into question.

Matching models to persons.   It will be recalled that both Abelson and Colby have attempted to supply their models with attitude structures corresponding to those of actual individuals: Colby to a patient of his, Abelson to a political figure. This is not a trivial task. Most individuals may be presumed to have a vast array of interlocking beliefs, attitudes, and knowledges in any area of central concern to them. This cognitive-affective structure must be inferred from a large volume of assertions and actions of the individual. If there were a one-to-one relation between assertions and beliefs, this would not be too bad, but of course the output stream is both highly redundant and highly elliptical. Redundant, in the sense that there may be thousands of particular outputs from an individual that boil down to the one assertion, "I hate women." Elliptical, in that many assertions depend for their intelligibility on a large implicit store of general information. To interpret "the Soviet Union punts on second down" requires (a) a grasp of the system of ordinal numbers; (b) some knowledge of football rules and past and present fashions in football strategy; (c) the information that the Soviet Union is a country and that countries do not play football, hence that the statement is to be taken as metaphorical; (d) the knowledge that the football reference fairly strongly implies an American speaker and hearer, and that in the United States the Soviet Union is predominantly viewed in a context of geopolitical competition; and so on. The computer model will not know these things unless they are explicitly or implicitly provided for it.

How is one to extract and organize information from the output stream of a particular human in order to construct an accurate model of him? Perhaps one can enlist the data-processing power of the computer itself. One possibility is automated content analysis of the human's output by some such program as the General Inquirer (Stone *et al.,* 1966). Another involves extracting information from real-time man-computer dialogue. Efforts by Colby along this line have already been mentioned, and related work has been done by Weizenbaum (1966). Programs such as the above attempt to deal with the fairly free output of the person. An alternative strategy is to restrict the output by obtaining responses to specified inputs. The outputs may be as constrained as yes-no-? responses to computer-selected MMPI items (Kleinmuntz and McLean, 1968) or at the intermediate level of complexity of sentence completions (Goldberg, 1966; Veldman, 1967; Veldman, Menaker, and Peck, 1969). More-or-less-structured computer interviews are also under investigation (Starkweather, 1967; Bellman, Friend, and Kurland, 1966). While it is still too early to tell which line of approach will prove most profitable in the long run, it is clear that considerable relevant activity is in progress.

Review and Prospect

Computer models of personality processes exist; there are problems in constructing them; there are problems in validating them. These assertions can confidently be made today. Confident assertions about tomorrow are less easy.

Still, if one grants the complexity that many have seen in human personality, it is difficult to think of any *other* way in which it may be formally objectified than as a computer model. But exactly what such models will look like ten years from now—or twenty—only time will tell.

REFERENCES

Abelson, R. P., "Computer Simulation of 'Hot' Cognition," in S. S. Tomkins and S. Messick (eds.), *Computer Simulation of Personality*. New York: John Wiley & Sons, Inc., 1963.

———, "Psychological Implication," in R. P. Abelson, E. Aronson, W. J. McGuire, T. M. Newcomb, M. J. Rosenberg, and P. H. Tannenbaum (eds.), *Theories of Cognitive Consistency*. Chicago: Rand McNally & Company, 1968b.

———, "Simulation of Social Behavior," in G. Lindzey and E. Aronson (eds.), *The Handbook of Social Psychology* (2nd ed.), Vol. II. Reading, Mass.: Addison-Wesley Publishers, 1968a.

Abelson, R. P., and J. D. Carroll, "Computer Simulation of Individual Belief Systems," *American Behavioral Scientist*, 8:9 (May 1965), 24-30.

Abelson, R. P., and D. E. Kanouse, "Subjective Acceptance of Verbal Generalizations," in S. Feldman (ed.), *Cognitive Consistency*. New York: Academic Press Inc., 1966.

Abelson, R. P., and C. M. Reich, "Implicational Molecules: A Method for Extracting Meaning from Input Sentences," a paper presented at the International Joint Conference on Artificial Intelligence, Washington, D.C., May 7-9, 1969.

Allport, G. W., *Personality: Psychological Interpretation*. New York: Henry Holt & Co., 1937, 588 pp.

Bellman, R., M. B. Friend, and L. Kurland, "Simulation of the Initial Psychiatric Interview," *Behavioral Science*, 11 (1966), 389-99.

Boring, E. G., "Mind and Mechanism," *American Journal of Psychology*, 59 (1946), 173-92.

Cattell, R. B., *Description and Measurement of Personality*. Yonkers, N.Y.: World Book, 1946, 602 pp.

———, "Personality, Role, Mood, and Situation-Perception: A Unifying Theory of Modulators," *Psychological Review*, 70 (1963), 1-18.

Coe, R. M., "Conflict, Interference, and Aggression: Computer Simulation of a Social Process," *Behavioral Science*, 9 (1964), 186-97.

Colby, K. M., "A Programmable Theory of Cognition and Affect in Individual Personal Belief Systems," in R. P. Abelson, E. Aronson, W. J. McGuire, T. M. Newcomb, M. J. Rosenberg, and P. H. Tannenbaum (eds.), *Theories of Cognitive Consistency*. Chicago: Rand McNally & Company, 1968.

———, "Computer Simulation of a Neurotic Process," in S. S. Tomkins and S. Messick (eds.), *Computer Simulation of Personality*. New York: John Wiley & Sons, Inc., 1963.

———, "Computer Simulation of Change in Personal Belief Systems," *Behavioral Science*, 12 (1967), 248-53.

———, "Computer Simulation of Neurotic Processes," in R. W. Stacy and B. D. Waxman (eds.), *Computers in Biomedical Research,* Vol. I. New York: Academic Press Inc., 1965.

———, "Experimental Treatment of Neurotic Computer Programs," *Archives of General Psychiatry,* 10 (1964), 220-227.

Colby, K. M., and H. Enea, "Heuristic Methods for Computer Understanding of Natural Language in Context-Restricted On-Line Dialogues," *Mathematical Biosciences,* 1 (1967), 1-25.

———, "Machine Utilization of the Natural Language Word 'Good,'" *Mathematical Biosciences,* 2 (1968d), 159-63.

———, "Inductive Inference by Intelligent Machines," *Scientia,* 103 (1968b), 1-10.

Colby, K. M., and J. P. Gilbert, "Programming a Computer Model of Neurosis," *Journal of Mathematical Psychology,* 1 (1964), 405-17.

Colby, K. M., J. B. Watt, and J. P. Gilbert, "A Computer Method of Psychotherapy," *Journal of Nervous and Mental Disease,* 142 (1966), 148-52.

Eysenck, H. J., *Dimensions of Personality.* London: Kegan Paul, 1947, 308 pp.

Freud, S., *New Introductory Lectures on Psycho-Analysis.* New York: W. W. Norton & Company, Inc., 1933, 257 pp.

Gilson, C., and R. P. Abelson, "The Subjective Use of Inductive Evidence," *Journal of Personality and Social Psychology,* 2 (1965), 301-10.

Goldberg, J. B., "Computer Analysis of Sentence Completions," *Journal of Projective Techniques,* 30 (1966), 37-45.

Guilford, J. P., *Personality.* New York: McGraw-Hill Book Company, 1959, 562 pp.

Hovland, C. I., "Computer Simulation of Thinking," *American Psychologist,* 15 (1960), 687-93.

Hunt, E. B., "Computer Simulation: Artificial Intelligence Studies and Their Relevance to Psychology," *Annual Review of Psychology,* 19 (1968), 135-68.

Kanouse, D. E., and R. P. Abelson, "Language Variables Affecting the Persuasiveness of Simple Communications," *Journal of Personality and Social Psychology,* 7 (1967), 158-63.

Kleinmuntz, B., *Personality Measurement.* Homewood, Ill.: Dorsey Press, 1967. 463 pp.

Kleinmuntz, B., and R. S. McLean, "Diagnostic Interviewing by Digital Computer," *Behavioral Science,* 13 (1968), 75-80.

Loehlin, J. C., "A Computer Program That Simulates Personality," in S. S. Tomkins and S. Messick (eds.), *Computer Simulation of Personality.* New York: John Wiley & Sons, Inc., 1963.

———, *Computer Models of Personality.* New York: Random House, Inc., 1968a, 177 pp.

———, "'Interpersonal' Experiments with a Computer Model of Personality," *Journal of Personality and Social Psychology,* 2 (1965), 580-84.

———, "Machines with Personality," *Science Journal,* 4:10 (October 1968b), 97-101.

———, "The Personality of Aldous," *Discovery,* 23:7 (1962), 23-26.

MacKay, D., "Mindlike Behavior in Artefacts," reprinted in K. M. Sayre and F. J. Crosson (eds.), *The Modeling of Mind.* Notre Dame, Ind.: University of Notre Dame Press, 1963.

Mehrabian, A., *An Analysis of Personality Theories.* Englewood Cliffs, N.J.: Prentice-Hall, Inc., 1968, 217 pp.

Miller, G. A., E. Galanter, and K. H. Pribram, *Plans and the Structure of Behavior.* New York: Henry Holt, Inc., 1960, 226 pp.

Murray, H. A., and Kluckhohn, C., "Outline of a Conception of Personality," in C. Kluckhohn and H. A. Murray (eds.), *Personality in Nature, Society, and Culture.* New York: Alfred A. Knopf, Inc., 1948, 572 pp.

Pool, I. de S., and A. Kessler, "The Kaiser, the Tsar, and the Computer: Information Processing in a Crisis," *American Behavioral Scientist,* 8:9 (May 1965), 31-38.

Rosenblueth, A., N. Wiener, and J. Bigelow, "Behavior, Purpose and Teleology," *Philosophy of Science,* 10 (1943), 18-24.

Saunders, D. R., "How to Tell Computers from People," *Educational and Psychological Measurement,* 21 (1961), 159-83.

Simon, H. A., "Motivational and Emotional Controls of Cognition," *Psychological Review,* 74 (1967), 29-39.

Starkweather, J. A., "Computer Methods for the Study of Psychiatric Interviews," *Comprehensive Psychiatry,* 8 (1967), 509-20.

Stone, P. J., D. C. Dunphy, M. S. Smith, and D. M. Ogilvie, *The General Inquirer: A Computer Approach to Content Analysis.* Cambridge, Mass.: M.I.T. Press, 1966, 651 pp.

Taylor, D. W., "Toward an Information Processing Theory of Motivation," in M. R. Jones (ed.), *Nebraska Symposium on Motivation: 1960.* Lincoln: University of Nebraska Press, 1960.

Tesler, L., H. Enea, and K. M. Colby, "A Directed Graph Representation for Computer Simulation of Belief Systems," *Mathematical Biosciences,* 2 (1968), 19-40.

Tomkins, S. S., and S. Messick (eds.), *Computer Simulation of Personality.* New York: John Wiley & Sons, Inc., 1963, 325 pp.

Uhr, L., "Intelligence in Computers: the Psychology of Perception in People and in Machines," *Behavioral Science,* 5 (1960), 177-82.

Veldman, D. J., "Computer-Based Sentence-Completion Interviews," *Journal of Counseling Psychology,* 14 (1967), 153-57.

Veldman, D. J., S. L. Menaker, and R. F. Peck, "Computer Scoring of Sentence Completion Data," *Behavioral Science,* 14 (1969), 501-7.

Weizenbaum, J., "ELIZA—A Computer Program for the Study of Natural Language Communication between Man and Machine," *Communications of the Association for Computing Machinery,* 9 (1966), 36-45.

Wiener, N., *Cybernetics.* New York: John Wiley & Sons, Inc., 1948, 194 pp.

COMPUTER SIMULATION
OF NEUROTIC PROCESSES

KENNETH MARK COLBY

COMPUTER SIMULATION

In psychiatry there is a class of unsolved problems which now appear amenable to an information-processing approach using computer simulation. When a psychiatrist and his patient engage in therapeutic dialogue, they exchange a large number of messages containing semantic information. The purpose of this exchange is to achieve an enduring beneficial modification of the patient's mental suffering. Events of this clinical dyad are too complex to handle with a purely numerical approach. In view of its success in other complex simulations, computer simulation seems worth a try.

To simulate a domain of interest means to represent certain of its aspects in the form of symbols which can be manipulated much more easily and quickly than the real-life referents they represent. Simulation itself is an old analytic device. When the number of different symbols needed for satisfactory representation and their rules of manipulation becomes large, a computer is necessary in order to do a large amount of symbol-manipulating work in a short time. In principle, one could turn the desired work over to a person, but it would take months or years to do by hand what a computer can do in minutes, the hand results would be full of hidden errors, and one could never be sure the person had rigorously followed the rules of manipulation given him.

Computer simulation as a technique should be distinguished from other aspects of artificial intelligence using programs. Some programs imitate human performance but their mechanisms are not intended to represent a theory of what is going on in the mind-brain of persons performing the same tasks. Other programs attempt to surpass human performance, but again they do not claim to work the way persons work. In simulation programs, however, not only does the output match human performance but the program's processes are intended to represent a theory, at the information-processing level, of what is occurring in the mind-brain of persons whose output is similar to the program output. Such programs are quasi-explanatory and become justified by serving as auxiliary aids in understanding and coping with certain problems. Since they are combinations

Reprinted with permission of Academic Press, Inc., from *Computers in Biomedical Research*, Vol. I, by Ralph W. Stacy and Bruce D. Waxman (eds.). Copyright © 1965.

of hypotheses and heuristic fictions, testing them for their match with facts of observation constitutes a difficult problem.

When a psychiatrist treats a patient by means of therapeutic dialogue, he utilizes a theory of mental pathology (from the Greek word *pathos* = suffering) and a theory of remedial techniques to guide his efforts. These theories attempt in an explanatory way to account for mental suffering and its relief in psychological or, more precisely, in information-processing terms. This does not deny the role of physiological or biochemical explanations in the domains of psychopathology. But at the moment psychological theories have the greatest adequacy, scope, and utility. A practitioner is a practical man and he requires theories closely connectible to the operational terms he uses in thinking about and applying his skills. Progress in the field of psychotherapy will involve improvements in methods and techniques but most of all it will require better explanatory theories of what is going on in clinical dialogue.

PSYCHOPATHOLOGY AND PSYCHOTHERAPY

There is little difficulty obtaining consensus about signs and symptoms commonly found in persons who apply for psychotherapeutic help. Clinicians of various orientations agree at this level of description and definition. But when an attempt is made to classify these observations into "disease" categories, interjudge reliability disappears. Except for the distinction between organic brain disease and nonorganic brain disease disorders, our systems of diagnostic classification remain of low reliability (Zigler and Phillips, 1961). Although patients are dis-eased, i.e., discomforted, distressed, and suffering, there is considerable doubt whether these conditions should be viewed as diseases at all in the traditional medical sense. There may even be advantages to the view of asserting there is only one mental illness, it is called "mental illness," and it has a frequency in the population of one out of one (Menninger, 1963).

In the light of the currently unsatisfactory diagnostic situation, it would not be possible to simulate a specific disease entity since there would be no agreement as to what to include and what to exclude. A more feasible goal would be the simulation of characteristic phenomena observable in self-descriptions of patients and in actual clinical contexts. While these observation reports of patients and therapists are rough and qualitative, they capture cardinal facts of experience and they form the basis of reliable regularities. They include descriptions of chronically or intermittently high levels of negative affects such as anxiety, depression, anger, together with thoughts recognized by patient, therapist, or both as distorted. Such episodes occur repeatedly over time and are experienced by patients as painful, leading them to seek help and relief. In addition to providing immediate and repeated observations of such phenomena as they are communicated verbally and nonverbally, the clinical situation offers an opportunity to study attempts to modify them by therapeutic dialogue. A psychotherapist tries to relieve mental suffering through inputs of semantic information. All will agree his effectiveness is limited and that in order to increase this effectiveness we need a greater understanding of the processes involved in person-on-person influence.

It is the semantic and pragmatic aspects of information theory which are involved in therapeutic dialogue (Colby, 1961). Besides the familiar property of complexity we must struggle with the property of elusiveness which characterizes so many of the events of clinical discourse. Consider the following tape-recorded exchange between a therapist and a patient. The patient is a young woman unable to give up an attachment to an older man and fearful of love relationships with men her own age.

PT.—Over the weekend, when I'd feel lonely, I'd sit down and start figuring money. It interests me. My bills, mother's bills, budgets, income for next year. It's a big waste of time, but it gives me some sort of security. Whenever anything goes wrong I start counting my money.

TH.—Money is something you can count *on*. It doesn't go away like your father and mother did.

PT.—I guess so. It has to do with security. When I get low I go back to it. An enormous waste of time. It keeps me occupied. . . . On Friday I met a man. He's really not so dreadful (laughs nervously). It just went through my mind maybe you knew him.

TH.—Why would *I* know him?

PT.—I don't really think you would. But I seem to be defending myself on all sides today. I suppose he's not a complete idiot. I'm trying to think why he seems so ghastly and why you might know him.

TH.—Maybe the point is you would retract your criticism of him *now,* in case I knew him, because *now* when you're alone, you want to count *on* me—like counting on your money. You don't want to irritate me.

PT.—That's right. Ah (pause) I suppose that's very right. Well anyway, so I went shopping in the afternoon and . . . (drops the topic introduced by the therapist and avoids further descriptions of the man).

Clinicians can readily recognize in the above factors of transference and resistance and the patient's avoidant response to the therapist's interpretation. Such events compromise much of the interaction observable in clinical situations. To create experimental analogs of them is extremely difficult and the yield uncertain because of lack of control over crucial variables involved (Colby, 1964). If we could simulate this real-life situation, we could then conduct simulation experiments with perfect control and perfect observation. The hope would be to use a computer as an experimental tool, a type of laboratory instrument heretofore unavailable for research in this area.

A particular example will now be described in brief.

Example

This example of a computer simulation of a neurotic process will illustrate a version of a program running as of May, 1964. During the past 2 years we have developed several versions in an attempt at greater comprehensiveness, consistency, and fidelity.[1] The continuing improvements have been in the direction

[1] I am indebted to my collaborators, John Gilbert of the Center for Advanced Study in the Behavioral Sciences, Stanford, California, and Janet Kreuter of the Department of Psychology, Stanford University.

of enriching the data structures and providing for greater variety of relations in processing.

Aware of the many conflicting uses of the term "model" in mathematics and the sciences, I shall use it to mean a partial representation of a theory expressed in **natural** language sentences. The term "program" will be used to mean the realization or exemplification of these sentences in the form of a computer language.

The model assumes a mental system which, in the course of processing its knowledge content or informational input, runs into severe conflict. Conflict consists of a command to carry out an operation and a countercommand which prohibits it. Presence of conflict is signaled by a rise in anxiety. Attempts are made to reduce the conflict state by creating commands. The result is a new command which is a disguised and distorted substitute for the original. Through this means conflict is temporarily avoided and anxiety reduced but the solution is unsatisfactory in the long run since the original members of the conflict remain in the system unaltered and will clash again in the future. Repeated cycles of conflict, of episodes of anxiety, and of thought distortions characterize the process here termed as "neurotic."

The program which realizes this theoretical model is coded in SUBALGOL (Stanford University's version of BALGOL) and runs on an IBM 7090. Typical output is shown in Fig. 1. On the left are simplified English sentences representing verbal output and on the right are numbers representing concomitant nonverbal output. Each line is to be viewed as a thematic event whose patient output counterparts, because of the great redundancies of natural language, might be of sentence, paragraph, page, or even chapter length. The program output is not intended to simulate the syntax and grammar of patient output but rather to represent the semantic and emotional content of typical thematic concerns.

VERBAL OUTPUT	NON-VERBAL OUTPUT				
			MONITORS		
	DANGER	EXCITATION	PLEASURE	ESTEEM	WELL-BEING
1. SELF MUST NOT MARRY POOR MAN	0.10	6979	0.69	0.03	0.36
2. SELF MUST HAVE PRESTIGE	0.11	7958	0.68	0.04	0.36
3. SELF MUST BE GLAMOROUS	0.11	6916	0.69	0.04	0.37
4. MOTHER IDOLIZES FATHER	0.14	3667	0.68	0.05	0.37
5. AL OUGHT ADMIRE SELF	0.16	4833	0.67	0.05	0.37

FIGURE 1   Sample output (new beliefs underlined).

A few details of the program will be described to illustrate its workings. A complete account can be found in Colby and Gilbert, 1964.

The program is divided into two portions, DATA and PROGRAM (words referring to specific aspects of the program will be capitalized). DATA consists of four data structures, DICTIONARY, SUBSTITUTE MATRIX, BELIEF MATRIX, and REASON MATRIX. The DICTIONARY contains 275 English words along with their property lists. The SUBSTITUTE MATRIX contains instances, synonyms, and antonyms of DICTIONARY words. These words are used in constructing the semantic portion of a BELIEF. A BELIEF is a molecular unit for processing consisting of words and numerical weightings. A BELIEF is defined as a proposition accepted and held as true by the program. The BELIEF MATRIX contains a list of 114 starting BELIEFS while the REASON MATRIX lists those BELIEFS which justify other BELIEFS, i.e., serve as their reasons for being accepted. The data structures are made up of information collected from a specific patient mentioned briefly in Section II. We assume the DATA to be unique to a given patient while the processes of the PROGRAM are assumed to be general as dispositional properties.

The program can be self-activated or input-activated. Self-activated processing begins with the gathering together of a cluster of BELIEFS termed the POOL (see Fig. 2). The POOL is formed according to relevance criteria around a nucleus termed a CORE BELIEF which is a BELIEF of major concern and imperativeness at the moment. A REGNANT BELIEF is selected randomly (actually pseudorandomly) and matched against each member in the POOL in a research for severe conflict as defined by multiple command-countercommand criteria. If conflict above a certain degree is not found, the semantic content of the REGNANT is expressed as an output sentence. If a high degree of conflict is found the REGNANT must be transformed to yield a new BELIEF which is nonconflictual in the POOL and hence can give rise to an output sentence.

A TRANSFORM is a procedure which carries out the transformation of a BELIEF. TRANSFORMS are divided into two groups (see Fig. 3). The first group consists of three procedures which operate when the REGNANT is a BELIEF involving ESTEEM, namely a proposition which obeys or disobeys what the SELF "ought." (Although we use a number of anthropomorphic terms as labels, they refer to definite computations in the program.) When ESTEEM is not involved a second group of TRANSFORMS are called into operation. They are simple or compound in type, the latter representing combinations of the simple TRANSFORMS.

The bulk of the work done in transformation operations consists of computing analogs and finding synonyms or antonyms. Verbs in the DICTIONARY have stronger and weaker forms (unless they are at the top or bottom of lists) and synonyms or antonyms at the appropriate level are found by searching the SUBSTITUTE MATRIX. How analogs are computed is illustrated in the flow diagram of Fig. 4. These operations are made possible by the way in which the DICTIONARY is hierarchically organized. Each DICTIONARY noun and adjective constitutes a class having instances and is itself an instance of a class. If the word is a primitive, a class having only itself as a member (e.g., FATHER), its associated property list gives the superordinate classes it is an instance of. Subordinate classes are those instances of the class named by the DICTIONARY word. In this manner the common logical relations of set inclusion and set membership are represented.

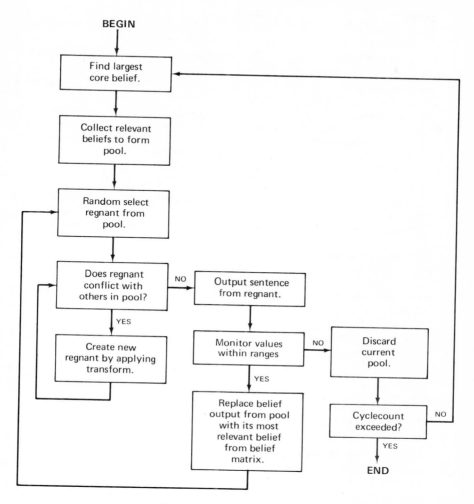

FIGURE 2　**Flow** diagram of self-activated processing.

New nonconflictual beliefs, created by the transforming operations and which can be expressed, are added to the BELIEF MATRIX thus giving them an opportunity to be selected in the formation of future POOLS.

During all this processing a system of MONITORS surveys what is going on to see if operations are being carried out within optimal ranges of certain numerical parameters. MONITORS affect the direction in which processing is to continue or whether it is to be discontinued. For example, the value of a DANGER MONITOR, which is a function of the amount of conflict in turn determined by numerical weight of the REGNANT, determines which TRANS-FORM is to be chosen among the Group I and Group II TRANSFORMS. When signals from MONITORS indicate that ranges are being exceeded, processing of the current POOL is switched off and a new POOL is formed. This process simulates the frequently observed decision of patients not to think about or

REGNANT='SELF ADMIRES FATHER'

	TRANSFORM	OPERATION	NEW REGNANT
GROUP I	ASCRIPTION	FIND ANALOG TO SUBJECT	'MOTHER ADMIRES FATHER'
	EXCHANGE	EXCHANGE SUBJECT AND OBJECT	'FATHER ADMIRES SELF'
	REVERSE EXCHANGE	EXCHANGE SUBJECT AND OBJECT AND FIND AN ANTONYM TO VERB	'FATHER REJECTS SELF'
GROUP II	ATTENUATION	FIND WEAKER SYNONYM FOR VERB	'SELF APPROVES FATHER'
	DEFLECTION	FIND ANALOG TO OBJECT	'SELF ADMIRES AL'
	REPLACEMENT	FIND WEAKER VERB SYNONYM AND ANALOG TO OBJECT	'SELF APPROVES AL'
	REVERSAL	FIND ANTONYM TO VERB	'SELF REJECTS FATHER'
	COMPOUNDS	FIND ANTONYM TO VERB AND THEN COMBINE WITH ATTENUATION, DEFLECTION AND REPLACEMENT TRANSFORMS	e.g. 'SELF REJECTS AL'

FIGURE 3   Transforms.

explore further an area of concern. A new topic of concern is selected but, as Leibniz recognized ("Nature does not make leaps"), continuities will be found between the old and the new topics.

When processing is input-activated the therapist's statement is represented in the program as format of a BELIEF which is a potential candidate for acceptance as true. A search is made through the BELIEF MATRIX for those BELIEFS in an identity, negation, similarity, and contrast relation to the input. Depending on MONITOR conditions, one of these is selected to serve as the NUCLEUS for a POOL. A response of affirmation or denial is made, giving the appropriate relation as output, and the POOL is then processed in the usual way. Here the program attempts to exemplify the theoretical notion that the patient does not respond directly to therapeutic input but to what he is able to find and to express from his own structures in connection with that input.

At this time the program has two limitations which we are attempting to remedy. First, the structure of the DICTIONARY is not thick enough to generate great variety. Besides properties, synonyms, and antonyms it needs predicates which procedures can operate on to find reasonable implications. Hence we are now including a type of higher-order predicate calculus in the data structures.

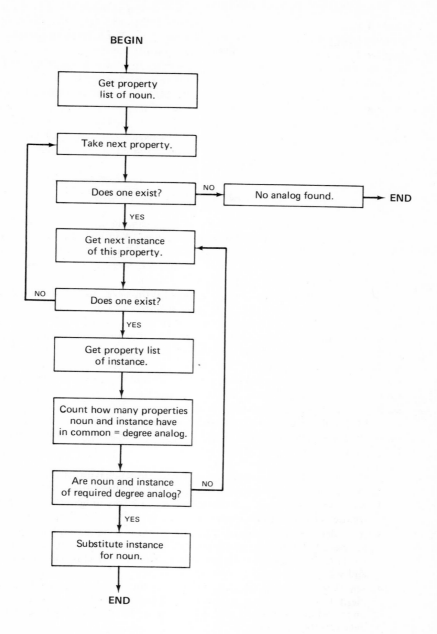

BEGIN

Get property
list of noun.

Take next property.

Does one exist? — NO → No analog found. → **END**

YES

Get next instance
of this property.

Does one exist?

NO

YES

Get property list
of instance.

Count how many properties
noun and instance have
in common = degree analog.

Are noun and instance
of required degree analog?

NO

YES

Substitute instance
for noun.

END

FIGURE 4   Flow diagram for computing analogs.

The second limitation of the program is that it has no way of correcting itself. Before the program can be expected to "benefit" from therapeutic input, it must have procedures for partially resolving conflict and undoing distortions. In theory, a therapist does not directly control therapeutic change. He offers opportunities for the patient's self-restorative capacities to apply themselves in key areas. However stunted, these capacities must already be present as potentialities. If such theoretical notions are plausible, then a program receiving therapeutic inputs must have some way of utilizing them in addition to simple responses of affirmation or denial. From among the many mechanisms which are probably at work in the beneficial change of psychopathology, we are currently trying to write procedures for a few to be represented in the program.

METHODOLOGICAL PROBLEMS

From the previous discussion and the example given, the general idea of this type of computer simulation should be clear. A model of theory attempting to explain cardinal phenomena of a field is first constructed. A program realizing the theory is written and tested on a computer. Since the term "test" is often used loosely, it deserves more exact specification.

To "test" a program can mean to run it on a computer (*a*) to make sure it is free of program errors, (*b*) to discover whether or not it is self-consistent, (*c*) to find out if the postulates are truly independent. This work differs from a fidelity test in which the output behavior of the program is compared to the output behavior of persons in an attempt to estimate how well the two streams of data fit. If there is reasonable agreement between the outputs we begin to think that the program offers an approximate and perhaps useful set of explanations for the observable field phenomena. But the difficulty lies in deciding what shall consist of "reasonable agreement."

In the absence of good quantitative measures at the moment, we must rely on qualitative judgments of experts. A series of increasingly difficult tests would be as follows:

1. Do experienced clinicians agree that output behavior of the program is roughly like that of patients in psychotherapeutic situations?
2. When shown two sets of outputs, can clinicians judge at above chance frequency which came from a program and which from a person?
3. A clinician might request—"say this to your program and tell me what happened. I know in general what would happen if you said this to patients like this." Is there agreement between the clinicians' prediction and what happened to the program?
4. Simulate a specific case known to the clinician up to a point in treatment time. The clinician asserts: "I now said this to the patient and I know what happened. Try it on the program," Is there agreement between the outcomes?
5. Simulate a specific case up to the current point in treatment. During a period when the patient and therapist are not in contact (say over a weekend) try various inputs on the program. Select those having greatest impact and suggest to the clinician he now try them on the patient. Is there agreement between the effects and outcomes?

These empirical tests are admittedly crude, qualitative, and full of pitfalls. But one must do the best one can with what is available. As we learn more about computer simulation of thought processes, better techniques for empirical matching will be developed. In the meantime to obtain qualitative agreements at the levels outlined above would be a worthwhile achievement. Failure to obtain them leaves it an open question whether the program or the reliability of expert judgments must be improved. In the case of the example desolved, a few clinicians have agreed the output is already "something like" the behavior of patients. Obviously a great deal more work must be done in developing both model and program before planning extended empirical tests.

Another issue concerns whether such programs should be considered theories, in the sense of testable hypotheses, heuristic fictions in the mathematical sense, or some combination of the two. A theory offers an explanation of a universal or statistical generalization. If the theory is adequate it predicts outcomes when the generalization is combined with initial individual conditions. A theory as a conjunction of hypotheses is expected to be confirmable in the external world with some degree of probability. A heuristic fiction, in contrast, is not expected to be confirmable. It may contain elements known to be false and contradictory to reality (of course it cannot be self-contradictory). It represents a thought invention rather than a discovery and it is justified or vindicated (not verified) by its usefulness as an auxiliary aid in understanding or solving some problem in the external world. A program such as the example cited contains both testable hypotheses and invented fictions serving as simplifying assumptions and boundary conditions. Hence perhaps it should be termed quasi-explanatory rather than explanatory (Beckner, 1959).

In time we should be able to tell whether programs simulating neurotic processes and their modification are useful instruments for dealing with subject matter problems of a psychotherapeutic situation. At the moment their main contribution has been to clarify theory by forcing the theorist to be explicit about his system, its entities, and their relations. A new desideratum for a scientific theory may be that it be programmable. If it cannot be programmed, then it will be a weak theory lacking in explicitness.

Work in this area has implications for the study of cognitive processes in general and in particular for the basic behavioral science of social psychology. We would like to know **more about** how people change their minds, how beliefs and belief structures change from thinking and talking both to oneself and in response to the talking of other persons which indicates (we think) what they are thinking.

APPLICATIONS

Besides aiding the practitioner by giving him better theories of pathology and therapy, there are a number of other applications of computer simulation techniques in psychiatry. A program can serve as a training device. For example, a simulation of the first psychiatric interview, which consists mainly of questions and replies, is being constructed (Bellman *et al.,* 1964). Psychiatric residents can practice on a computer and estimates made whether their interviewing technique is convergent or divergent to the task at hand. More difficult will be the problem of using a simulation of therapy for training. Here a time-sharing system will be required so that the therapist can be on-line with his computer "patient." Thus the continuous dialogue of a clinical discourse can be reproduced in real time. Life is too short to write a program which will cover all the contingencies of therapeutic interactions. A therapist must decide what to do next depending on what he knows of the patient and what has just happened. To simulate this type of "adaptive decision process" (Bellman *et al.,* 1964) will require close and rapid man-machine intercommunication in order to be feasible.

In the example program cited epigenetic or ontogenetic factors as to how the patient became that way are omitted. However, if one had the possibility of being in on-line communication with a large computer, one could approach the problem of raising a child in different environments to see what sorts of similarities and differences might be produced. Some work along these lines has already been done (Loehlin, 1963). One could test out various child-rearing hypotheses and try out alternatives in a much shorter time. This advantage of raising a "child" in the compressed time of a few minutes and having full control over crucial variables would have great significance for prevention of mental disorders. Already we are witnessing qualitative changes in the nature of disorders due to different child-rearing practices in our culture, e.g., the disappearance of catatonic stupor.

Developments of these applications will depend on collaboration between computer specialists, psychiatrists, and psychologists. Psychiatric research will turn increasingly towards the computer both for simulation and for data-analysis purposes. The current lag in this respect is not due to the unavailability of computers but to the inertia against learning a new technique. Psychiatrists and many behavioral scientists appear reluctant to learn programming. There may be two reasons for this: (1) writing programs represents too great a discontinuity from familiar and conventional methods; (2) programming is taught mainly for those interested in numbers using numerical examples and problems which behavioral scientists neither understand nor sympathize with. There is no solution to the first except to wait for the professional arrival of those undergraduates and graduate students to whom programming is becoming a familiar technique. The second problem can be solved by programming courses designed for behavioral scientists in which the approach is entirely nonnumerical. In the author's experience students can be writing small programs and running them on the machine after only 2 hours of instruction. This can be accomplished by ignoring hardware details, limiting the programming language to the simplest expressions which will get the job done, and operating only with alphanumeric symbols arranged in the form of lists and list structures.

In the author's view the psychiatrist or behavioral scientist involved in computer simulation must write the program himself or be so conversant with it that he can understand what is going on by reading the program. He cannot simply turn over the model to a programmer for coding. The way in which the model is realized in the program is a crucial step in the process. Writing and running a program not only teaches a theorist a great deal about what he is really saying in his model but also gives him an opportunity to learn how to improve it.

REFERENCES

Beckner, M., (1959). "The Biological Way of Thought." Columbia Univ. Press, New York.

Bellman, R., M. B. Friend, and L. Kurland. (1964). On the construction of a simulation of the initial psychiatric interview. Paper presented at a meeting of the American Psychiatric Association.

Colby, K. M. (1961). *Am. Scientist* 49, 358-369.

Colby, K. M. (1964). *Arch. Gen. Psychiat.* 10, 220-227.

Colby, K. M., and J. Gilbert (1964). *J. Math. Psychol.* 1 (2), 405-417.

Loehlin, J. (1963). In "Computer Simulation: Frontier of Personality Theory," S. S. Thomkins and S. Messick (eds.), pp. 189-211. Wiley, New York.

Menninger, K. (1963). "The Vital Balance," Viking Press, New York.

Zigler, E., and L. Phillips. (1961). *F. Abnorm. Soc. Psychol.* 63, 607-618.

4

Social and Cultural System

Simulations

JOHN T. GULLAHORN

JEANNE E. GULLAHORN

Theorizing about sociocultural systems probably dates back to the time when members of primitive hordes began speculating about the nature and consequences of their interdependence. By the seventeenth century, advances in science and mathematics had influenced the concepts and assumptions applied in interpreting man and society, leading to a "Social Physics" (cf. Sorokin, 1928, chap. 1) that Pareto later developed into a "rational mechanics" (cf. Henderson, 1935), which ultimately became represented in modern theories concerning equilibrium maintenance in a social system composed of mutually interrelated elements (cf. Parsons, 1951).

Along with mechanistic conceptions, organismic analogies have long characterized social thought, finding scientific expression in Spencer's Social Darwinism (1897) and more recently in functionalist theories concerning the "prerequisites" of social systems in maintaining their institutionalized structures by means of automatic, "homeostatic" mechanisms (cf. Radcliffe-Brown, 1952; Parsons, 1964). By focusing on the more stable, normative characteristics of social systems in a search for the social equivalent of an organism's relatively fixed structure, functionalists have tended to overemphasize structural representation at the expense of dynamic analysis of interpersonal and intergroup processes (cf. Buckley, 1967). As we shall note later, the recently developed social exchange theory represents a major overhaul in social system theorizing,

© John T. and Jeanne E. Gullahorn, 1971. Preparation of this chapter was supported as part of a larger project by PHS Research Grants MH14582 and MH16935 from the National Institute of Mental Health. This chapter may be reproduced in full or in part for purposes of the United States Government.

"bringing men back in" (Homans, 1964) by incorporating psychological factors in social explanation.

Except for recent reformulations, however, much contemporary social system theory is taxonomic and descriptive, possessing, as Homans incisively notes, "every virtue except that of explaining anything" (1961, p. 10). Perhaps as a consequence of the traditional emphasis on social structure—culminating in a welter of terminology and static categorization—most computer applications concerning sociocultural systems involve clique or network detection (cf. Hubbell, 1965; Gardin, 1965) or the systematic classification of linguistic, anthropometric, or cultural traits (cf. Hymes, 1965; Clarke, 1968). Such developments, of course, augment descriptive precision and help eliminate some of the redundance and ambiguity inherent in verbal categorization.

Elaborate verbal conglomerates of conceptual pigeonholes and x-fold tables, however, do not lend themselves to simulation, which requires well-defined statements of relationships among variables, with implications of *ceteris paribus* disclaimers clearly specified in order to actualize change processes (cf. Gullahorn and Gullahorn, 1965a). Thus, lacking an overall empirically linked explanatory theory, social system simulators have not developed macro-level simulations of total societies; instead they have simulated single universal processes affecting a social system or have modeled dynamic microbehavioral processes within sociocultural contexts.

Plan of the Chapter

This chapter will review a number of diverse simulations of sociocultural processes. Table 4-1 lists the models that are discussed in some detail; additional related simulations are mentioned in the chapter and cited in the References.

Since this chapter focuses on theory construction and research implementations of simulations, we have omitted such teaching applications as Coleman's recent gaming endeavors (1966, 1969) and Gamson's SIMSOC classroom simulation of a society (1969). Similarly, since the internation simulations are discussed in depth in the chapter by Smoker in this volume, they are not reviewed here. Obviously, other models of economic, political, business, and marketing processes have social system import, and the reader is referred to other chapters in this volume for further information.

In endeavoring to organize the varied types of simulations that were retrieved in our bibliographic search, we have distinguished between models that emphasize universal processes affecting a social system and those that focus on microbehavioral processes within sociocultural contexts. The distinction is by no means absolute. However, while not all the models simulating total system processes exhibit a "black box" approach to individual decision-making, they nevertheless tend to trivialize autonomous information processing. The simulations we have termed microbehavioral, on the other hand, incorporate social psychological considerations and tend to deemphasize total system processes. In

TABLE 4-1   Selected Sociocultural System Simulations [a]

Simulations of Total System Processes

1.	Social Organization	Kunstadter *et al.* (1963), Gilbert and Hammel (1965), Kasdan and Smith (1968)
2.	Population Dynamics	Lazer (1969), Beshers (1965)
3.	Policy Processes	Michelena (1967)
4.	Information Dissemination	Popkin (1965), Kramer (1969), Hagerstrand (1953), Tiedmänn and Van Doren (1964), Wolpert (1967), Deutschmann (1962), Hanneman (1969), Carroll (1969)

Simulations of Microbehavioral Processes

1.	Socialization Processes	Hanson *et al.* (1967), McPhee (1961)
2.	Other Reference Group Processes	Coleman (1962), Breton (1965), Smith (1969)
3.	Individual Processing of Social Communications	Pool and Kessler (1965), Boguslaw, Davis and Glick (1966), Abelson and Bernstein (1963), Bråten (1968)
4.	Social Exchange Considerations in Decision-Making	Gullahorn and Gullahorn (1963), Coleman (1962)

---

[a]   Dated by time of first publication, not necessarily time of completion of model.

some instances in the latter approach the lack of explicit programming of total system processes is deliberate, since the simulation is intended to explore whether system-level consequences can be generated by interaction of the individual units.

SIMULATIONS OF TOTAL SYSTEM PROCESSES

Change is one of the "constants" in social systems. Even relatively static classificatory schemes embody some implications of social change processes, particularly in assumptions regarding equilibrium maintenance. One of the attractive potentialities of simulation derives from the capability of programming hypothesized change processes within a represented social system so that computer-generated consequences can be compared with real-life data (cf. Gullahorn and Gullahorn, 1965b). In this section we shall describe simulations of changes within total social systems emanating from rules governing social relationships, from processes affecting population dynamics, from social conditions eliciting government response, and from dissemination of information via mass media or interpersonal interaction. Since these simulations focus on universal processes affecting the represented social systems on the whole, they tend to trivialize autonomous individual information processing.

Simulations of Social Organization

In interpreting ethnographic data concerning unilateral cross-cousin marriage systems, Levi-Strauss (1949) argues that a societal preference for (matrilateral) marriage with mother's brother's daughter occurs more frequently than does the alternative (patrilateral) marriage choice of father's sister's daughter because the matrilateral rule has inherent functional value, producing greater solidarity in social organization. The circularity of Levi-Strauss's functionalist position and the lack of causal explanatory propositions preclude any attempt at simulation. Homan's psychologically based explanation of marital choice behavior in societies practicing unilateral cross-cousin marriage (Homans and Schneider, 1955), on the other hand, could be modeled by means of routines specifying linearity, the locus of jural authority, and the pattern of interpersonal relations among kinsmen.

Ethnological simulators, however, have not undertaken this type of endeavor; instead they have concentrated on a more limited controversy regarding cousin marriages. Some social anthropologists contend that the incidence of cross-cousin marriages results from an explicit prescriptive rule or from a preferential bias regarding marital selections. Harrison White has developed mathematical models assuming the operation of such rules (1963). Other ethnologists, however, postulate that cousin marriages are a consequence of certain territorial and residential regulations and can thus be accounted for without invoking a kinship-phrased rule. Since these alternative formulations have been represented in separate computer simulations of marriage behavior, systematic comparison is not possible. Ideally, a modularized general computer model could be developed to simulate marital choices. In one simulation run the processes implied by prescriptive or by preferential kinship-choice rules would be operationalized; in another run these routines would be replaced by others implementing the alternative hypothesis concerning territorial residential preferences. The output of each run could then be compared with ethnographic data regarding rates of cross-cousin marriage.

While no computer model has yet been designed to compare the consequences of the alternative formulations regarding cross-cousin marriage, one simulation has investigated the rate of such marriages in computer runs incorporating a specific kinship-choice rule as contrasted with null-condition runs in which this rule was omitted (Kunstadter *et al.*, 1963; Kunstadter, 1965). The computer model developed by Kunstadter and associates determines the effects of variations in demographic variables on the operation of an ideal pattern of marriage preference. Programmed birth and death rates as well as age-specific marriage rates are derived from population statistics from India, which was chosen to be representative of world areas prior to the introduction of modern medicine. Monte Carlo routines apply these rates to individuals in the simulated population.

The model incorporating a specific matrilateral cross-cousin marriage preference rule includes the following assumptions:

1. All women marry according to age-specific probabilities.
2. Women can marry starting at age five, but they do not begin reproducing until age eleven. Marital fertility probabilities change with age, starting at 0.05 at age eleven and declining to 0 at age forty-five.
3. Only married females reproduce.
4. Marriages are monogamous; however, widowed individuals can remarry.
5. The ideal marriage partner for a man is his mother's brother's daughter; for a woman, her father's sister's son.
6. The search for an ideal mate begins early in childhood (age two for girls, five for boys). If no matrilateral cross-cousin is available at the time of marriage, a mate is randomly chosen from unmarried individuals with whom marriage is not restricted either by incest taboos eliminating primary relatives or by exogamy proscriptions eliminating patrilineal relatives up to three generations removed from the concerned party.

In the simulation, the initial population is input, with each individual identified by age, sex, marital status, and presumably kinship network. At the beginning of the simulated yearly cycle, appropriate matrilateral cross-cousin marriage partners are sought. For each unmarried unengaged woman of engageable age (two to fifty), a search is made for all her father's sisters' sons who are alive, unmarried, unengaged, and of marriageable age (five to fifty). If any are found, the girl is "engaged" to the eldest. After the search for ideal mates has been completed, the remaining eligible women are engaged to randomly selected males who are not primary relatives or patrilineal relatives up to three generations removed from the fiancée.

Following the engagement cycle, the program searches the population of engaged women, applies age-specific probabilities of marriage to them, and uses a Monte Carlo procedure to determine whether marriage occurs. If a couple thus is married, this fact is recorded. If they are not married, and if the engagement is of the ideal cross-cousin type, the engagement is continued. Otherwise, the engagement is broken and the partners return to the pools of prospective mates for the following year.

A similar Monte Carlo application of an age-specific birth rate to married women results in the birth of children, whose sexes are randomly assigned on the basis of sex ratio for live births. Population mortality results from processing age- and sex-specific probabilities of dying for each simulated individual. If the deceased were engaged or married, their partners reenter the marriage pools in the next yearly cycle.

At the end of an annual cycle, the population is updated and results of the year's operation are output, including the number of persons alive and dead, the number of males and females, the number and sex of newborn children, the number of ideal engagements and ideal cross-cousin marriages, and the proportion of ideal extant marriages.

Operating under the described maximal conditions for the ideal-type liaisons, the model generates an average rate of 27 percent for matrilateral cross-cousin marriages. This incidence is affected by variations in demographic parameters: The proportion of ideal marriages varies directly with population growth and inversely with age at marriage, and variability in the proportion of ideal marriages varies inversely with population size.

In the null-condition runs omitting the preferential cross-cousin rule, marriages occur at random within the population, except for the restrictions on age at marriage and the proscriptions concerning incest and exogamy. Under these conditions, the computer-generated incidence of fortuitous cross-cousin liaisons reaches only 2 percent; thus, given the specifications of the computer model, only a trivial proportion of cross-cousin marriages can be expected merely by chance.

Presumably, according to the experimental condition simulation runs, a population could achieve a 25-30 percent rate of cross-cousin marriages by consistently following the preferential rules incorporated in the model. Kunstadter and colleagues thus conclude that the simulation raises the question of why observed frequencies of cross-cousin marriages are so low in real societies whose members profess such marital preference. Additional field data and hypothesis formulation appear necessary to provide parameters and structure for further computer simulations.

Following a different theoretical approach in simulating marriage behavior over several generations, Gilbert and Hammel hypothesize that cousin marriage rates can be accounted for without specifying a preferential marriage rule phrased in kinship idiom (1965, 1966). Unlike the previously-described simulations, the Gilbert and Hammel model involves patrilateral parallel cousin liaisons in an Arab society, manipulating residence and territorial rules in an effort to demonstrate that the incidence of such cousin marriages constitutes an "epiphenomenon" of the interrelationships among the input variables.

Beginning with a population of specified size and age distribution, subdivided into a given number of local village groups, the Gilbert and Hammel computer model implements the following rules:

1. The life of simulated individuals is so divided that from birth to age fifteen they are ineligible for marriage; from fifteen to thirty they are eligible; from marriage until the wife's age is forty they reproduce; and after age forty they neither marry nor reproduce.
2. Marriages are monogamous.
3. Incest taboos prohibit marriage with male Ego's mother, daughter, sister, aunt, or niece.
4. The probability of making a group-endogamous marriage is specified (varying between 0 and 1).
5. Age-matching procedures in the selection of wives specify that spouses should be relatively close in age.
6. Residence is patrivirilocal; therefore, sons live in their father's village and women join their husbands at marriage.
7. On the average, three children are born to each married couple;

however, the number of children born to a particular woman varies inversely with her age at marriage.

8. Approximately one-third of the children "born" do not reproduce.
9. There is no simulated death; nonreproducing individuals over forty are retained on the population list for purposes of genealogical reckoning.

In the program, simulated marriages occur according to the following procedures: A village from which a husband will be chosen is randomly selected; then a village from which a wife will be drawn is selected according to the probability of the husband making an endogamous marriage. Thus, when this probability is zero, the husband's village is excluded from the selection set.

For each village, a list of eligible males and a list of eligible females are generated from the general population list. These lists of eligibles are age graded, comprising all unmarried individuals over the fifteen-year puberty point and below the forty-year senility limit. A potential husband is randomly chosen from the list of eligible males in the selected village. His age is retrieved, and then from the list of eligible women in the selected village for the wife, a potential bride is chosen from a set of five females—three younger and two older than the husband (thus creating a bias toward younger wives). The genealogical trees of the proposed spouses are compared, and at a predetermined level of genealogical depth, all possible relationships between them are computed. If any of the computed relationships violates the programmed incest taboos, the spouse pair is rejected, and the program cycles to a new random selection of potential partners.

If the marriage is permissible according to the implemented territorial and incest rules, any genealogical relationship between the spouses is recorded, and the two individuals are removed from the list of eligibles and placed onto a married couple list, ordered in terms of the woman's age. They remain on the couple list until the wife reaches age forty; thus their tenure on the list and consequent chance of reproducing are determined by the woman's age at marriage.

Randomly selected couples from the couple list are assigned children according to a Poisson distribution with a mean of 3. The sex of each child is assigned from a binomial distribution with a mean of 0.5, and each child is assigned to his father's village. Children are then listed at the bottom of the age-graded general population list. This simulated time passage changes the puberty and senility points on the population list, the eligibles lists, and the couples list.

The marriage process is repeated until one thousand marriages have been contracted, at which point the tabulated results are output. The program is flexible, allowing compilation of a variety of data concerning age and sex distributions, number of children per couple, number of cousins per person, and numbers and types of cousin marriages. These demographic and genealogical data can be compared for different probabilities of endogamous marriage and for different population sizes. For example, Gilbert and Hammel report that in both large and small populations, first-cousin marriages increase by a factor of about 1.2 over the previous level for each increase of 0.1 in the probability of endogamy, although the absolute level is higher in small populations. Thus the simulation indicates that parallel cousin marriage varies inversely with population size and directly with endogamic preference.

Operating with territorial preference parameters estimated from Arab ethnographic data, the model generates an average of only 4-5 percent of father's-brother's-daughter marriages, as contrasted with the reported Near Eastern rate of 9-12 percent. Given the estimated inputs, therefore, only about half of the frequency of patrilateral parallel cousin marriages can be accounted for on the grounds of territorial preference. In discussing possible reasons for the discrepancy between their model's output and the real-life data, Gilbert and Hammel suggest that finer territorial divisions may produce a higher incidence of cousin marriage. On the basis of their modeling efforts, they further note the necessity of additional numerical information, seldom available in published ethnographic reports.

The simulations just reviewed indicate the potential contribution from operationalizing in a computer program certain rules of social organization and then observing their generated effects on kinship and other social system relationships. As suggested previously, this modeling could be extended to provide more systematic testing of alternative theoretical formulations. At present it is impossible to compare the endeavors of Kunstadter and associates with those of Gilbert and Hammel, since separate programs were implemented involving different social systems and different types of cousin marriages.

Another possible extension involves combining such simulation approaches with other anthropological computer applications focusing on classification and grouping. For example, the Coult and Randolph computer representation of kinship networks produces empirical analyses of genealogical data that would otherwise be impossible, such as the sorting of complicated systems of relationships in the Bedouin Arab marriage system (Coult and Randolph, 1965; Randolph and Coult, 1968). Findler and McKinzie (1969) have also developed a program to generate and analyze kinship structures. Incorporating such data-processing classification procedures in a simulation of kinship behavior might permit more extensive comparisons between computer-generated and real-life findings.

Integrating simulations of change processes with classification applications might also further elucidate prehistoric sociocultural systems (cf. Clarke, 1968, chap. 13). For example, by simulating ecological changes and postulated cultural adaptive responses in a program incorporating archaeological classifications, one might test inferences such as those proposed by Hill in his investigation of Pueblo social organization (1966). Beginning with random samples of artifacts from a large Arizona Pueblo settlement, Hill applied factor analysis programs to identify clusters of artifact and pottery types and motifs according to their distribution in rooms and over time. This computer analysis differentiated the Pueblo rooms according to their functions (food storage and preparation areas versus living quarters) and also isolated five localized unit clusters (matrilineages) within two broader overall clusters (clans). Relating these findings to regional data for the period A.D. 1000-1300, Hill postulates that environmental changes during that time threatened subsistence for scattered lineage communities, leading to nucleation of the population into more highly organized settlement aggregates such as the one investigated.

To a varying extent, the marital choice simulations previously described incorporate structural elements and decision processes representative of actual social systems. Such a strategy facilitates validating a model through comparisons with real-life findings. Another possible application of computer simulation for developing and actualizing theoretical formulations, however, involves constructing a hypothetical social system and varying certain processes to observe their effects. Such is the approach followed by Kasdan and Smith in their ethnological simulation of a nonexistent society (1968).

To investigate the impact of primogeniture on land fractionalization, mobility, and population growth among different classes, Kasdan and Smith simulate a thousand-year span in their hypothetical society, under conditions where social organization and existing resources remain constant. The output indicates that among the intermediate classes, primogeniture facilitates upward mobility, and its absence results in downward mobility and heavy land fractionalization. For the upper classes, an elite class practicing primogeniture becomes numerically smaller and possesses less total land of the highest quality (and, by inference, wields less power) than a corresponding elite class not practicing primogeniture. From these results Kasdan and Smith conclude that the absence of primogeniture tends to polarize the class structure of a society into property owners and nonowners, thus encouraging centralization of power and providing opportunity for despotism. Where primogeniture exists, however, polarization does not tend to occur, and middle classes emerge. Under such circumstances power is hierarchically distributed, making despotism less likely. Such conditions, furthermore, appear conducive to industrialization. Further explorations with this type of model should yield interesting insights regarding societal development.

Simulations of Population Dynamics

Some consequences for a population as a whole were observed in the previously described simulations of regulations affecting social relationships in terms of marital choice, residence, and inheritance. Population dynamics are also modeled in the microanalysis of the American economy by Orcutt and associates (1961). This Monte Carlo simulation incorporates baseline census data in simulating stochastic changes in family life cycles, trends in the changing composition of the labor force, and assumptions about household economic behavior. Harper's simulation of a sociological survey (1968, 1969) can also incorporate birth and death processes, marriage, social mobility, and geographic mobility.

Still other computer analyses and syntheses involving census data are reported in a volume edited by Beshers (1965a).

Though many computer models include demographic variables in their inputs and outputs, actual simulations of the vital experiences of a population have not been reported. Perhaps this lacuna reflects the current state of demography, which lacks coherent theory to guide research and explain observed relationships. Thompson's empirical generalizations concerning Western European population growth (1929) have contributed to the development of a conceptual scheme concerning demographic transition (cf. Coale and Hoover, 1958), but explanatory propositions have not been well specified. As a step toward developing an explanatory theory of population growth, S. Charles Lazer has advanced a set of propositions concerning demographic transition and has specified processes implied by these propositions in his design for a computer simulation model (1969).

Among the propositions incorporated in Lazer's model are the following:

1. A primitive state of no significant mortality control will serve to maintain a high mortality rate.
2. Increased political stability will contribute to a declining mortality rate.
3. Continued declining mortality will eventually yield (when the limit of mortality control is achieved) a low stable mortality rate.
4. A primitive state of nature with no significant mortality control will favor a traditional value system predisposed to high stable fertility.
5. Urbanization will contribute to the establishment of a value system that is predisposed to a low stable fertility rate.
6. A value system predisposed to low stable fertility will contribute to a declining fertility rate.

The overall simulation includes two related submodels concerning mortality decline (flow charted in figure 4-1) and fertility decline. Since the improved transportation and communication networks (TRANCOM) generated in the mortality decline simulation lead to general modernization, including urbanization and industrialization which are assumed to affect fertility, fertility decline lags behind mortality decline in the model as it does in actual population change.

In the flow chart for processing mortality decline (figure 4-1), ORDER refers to political stability, and any process or event contributing to such stability can be programmed as a member of the class ORDER. Probabilities for the events included in this set can be derived from specific historical data when the model is used for prognostication purposes; or such probabilities can be assigned on the basis of assumptions about antecedent conditions peculiar to a population, such as cultural norms and technology.

Associated with each ORDER event is a numerical value representing the hypothesized impact of such an event on the mortality rate. For example, establishment of a strong central government where none existed before would

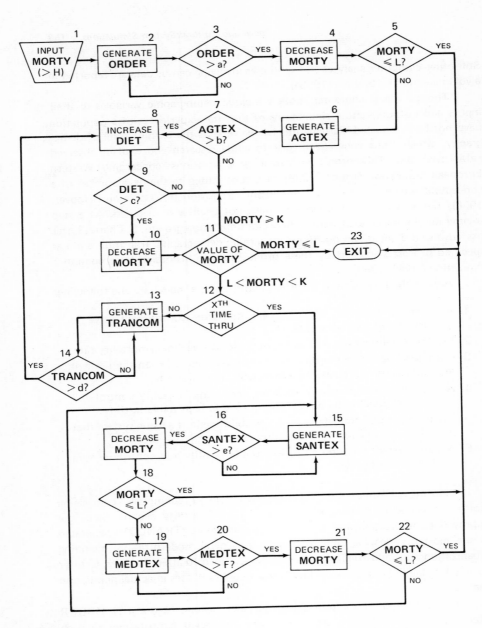

FIGURE 4-1   Mortality decline.

presumably reduce mortality more than would the establishment of a different feudal system. When the sum of the values affecting mortality exceed a predetermined threshold value (*a* in box 3, figure 4-1), the death rate is decreased.

Three mortality parameters, in conjunction with the processes specified in the theory, determine the operation of the mortality simulation. These are *H* (the level of the death rate above which mortality can still be assumed to be uncontrolled), K (the death rate, lower than H, above which the population is affected only by the earliest and simplest variables contributing to mortality decline), and L (the level at which the population has achieved a low stable death rate). The simulated population undergoes its most precipitous decline in the range between K and L.

The mortality parameters can be set from available data from Western European countries or they can randomly be assigned within an observed or theoretically interesting range to explore their consequences. Similarly, official indexes or theoretical assumptions can provide the threshold values of other variables including political stability (ORDER), improved resource extraction (AGTEX), improved food supply and diet (DIET), improved transportation and communication networks (TRANCOM), improved public sanitation (SANTEX), and increased medical technology and improved medical care (MEDTEX).

Lazer plans to tune his model for a best-fit simulation of Western European population dynamics, using procedures comparable to those followed by Pool and his colleagues in a campaign simulation that also depended on a plethora of survey data (cf. Pool *et al.,* 1965, chap. 3). Once the processes and parameters of the model are so adjusted that the output approximates the vital experiences of Sweden, this model will be used to simulate population change in England. Lazer faces a number of difficult problems in completing a running simulation and then endeavoring to validate it, but his work should contribute to the needed advancement of demographic theory.

Potential explorations with such a model have important policy implications in addition to their theoretical application in simulating population growth under varying conditions. For example, the model can generate the consequences of changing the sequence of modification of variables by introducing modern medical practice before political order and agricultural improvement. Data having policy relevance for underdeveloped countries can also be produced regarding the impact of compressing time in realizing a stage of the modernization process and the effects of introducing a fully developed improvement rather than cycling through successive stages of its development.

Besher's nonstationary stochastic model for birth projections also generates the consequences of alternative social policies (1965b, 1967a). Current experimentation with the model incorporates data from India to simulate the impact of large-scale introduction of birth control in underprivileged nations. Beshers is also developing a computer representation of migration in metropolitan areas (1967a, 1967b), which may later be used to simulate federal policy on jobs and housing.

The theoretical significance and policy ramifications of Lazer's and Besher's models hold great promise. But since these simulations are not yet fully operational, the extent to which they fulfill expectations remains to be seen:

Simulations of Polity Processes

Michelena's VENUTOPIA is yet another computer model intended to predict social system effects of different policy proposals (1967). Using data inputs from an extensive Venezuelan survey, VENUTOPIA assesses conflicts and antagonisms among various occupational groups, processes polarizing coalitions among groups, and elicits government responses to the state of the social system. Each simulated group is described in terms of background characteristics (age, occupational prestige, education, income, socioeconomic status, status congruence, mass media exposure, level of information, etc.); normative orientations (leftism or rightism and attitudinal modernity); psychological states (optimism or pessimism and heterogeneity—i.e., antagonism between members with differing levels of attitudinal modernity); and political capacity (participation and intergroup contact).

Programmed multiple regression analyses identify weightings of significant predictors of each group's evaluations of seven subsystems (family, community, polity, economy, education, church, and foreign influences) as well as of thirteen roles (including businessmen, labor leaders, government officials, police, military, mass media communicators, student leaders, and schoolteachers). These evaluations in conjunction with a set of programmed propositions regarding group relationships determine when conflicts and antagonisms polarize groups. In part, the political effect of polarization depends on the degree of congruence between the polarized groups' and the government's evaluations of subsystems and of roles. In responding to system states, the government can bargain, initiate reforms, or repress a group by reducing its income, its level of political activity, or its word-of-mouth communication.

With the vast number of variables involved in the multiple regression equations, assumptions of linearity and of no interaction appear implausible. Nevertheless, the model incorporates a fascinating array of data and some interesting propositions; therefore, further developments should prove noteworthy.

Other simulations of political processes and of international relations have obvious social system import. These endeavors are discussed in chapters 6 and 7 of this volume.

Simulations of Information Dissemination

That personal communication is an important determinant of opinion formation and ultimately of sociocultural change is a truism of social system theory whose ramifications have been incorporated in verbal theories, mathematical models, and more recently in computer simulations. With their long-standing interest in the process of cultural diffusion, anthropologists are among the early contributors to verbal theories concerning the diffusion process (cf. Linton, 1936). Similarly, a long tradition of research in rural sociology has focused on how the individual farmer obtains information about agricultural innovations through his network of friends and neighbors, as well as from the extension agent and the agricultural media. While viewing the potential adopter as enmeshed in a network of social relationships, rural sociologists generally have not systematically explored this network—an innovation in diffusion research introduced in the Coleman, Katz, and Menzel drug study (1957, 1966). Indeed, with the findings from the investigation on the adoption of new prescription drugs by physicians, Coleman developed accurate deterministic mathematical models to represent the aggregate behavior of the diffusion system for both the sociometrically integrated and the isolated doctors (1964). To account for additional complexities in actual diffusion processes, however, mathematical models become unwieldy; and as Coleman concludes, "to mirror the full complexity of such structures would require detailed measurement followed by a simulation with the structure mapped onto the memory of a computer" (1964, p. 495).

Another tradition of research on mass communication has also investigated the dissemination of new ideas, but from a different perspective, viewing the urbanized individual as directly subject, in almost hypodermic needle fashion, to the contagion of mass media influences. As Elihu Katz notes (1959), rural sociology and mass communications in a sense discovered each other in the late 1950s, and the expectation is that each may modify the other's research design, assisting rural sociologists in generalizing from studies of neighborhoods and communities and enabling mass communications researchers to take account of interpersonal contacts. Computer simulation is one methodological tool that can incorporate insights and findings from different social science traditions to converge on the general process of the diffusion of ideas over time within a sociocultural system.

Thus far, mass communication simulations have been less integrative than have recent computer models of diffusion processes. For example, in simulating information flow, Hartman and Walsh "inoculate" a demographically heterogeneous set of individuals only via exposure to newspapers, and not through interpersonal interaction (1969).

Media exposure is central also in mass communications simulations conducted at the Center for International Studies at M.I.T. The COMCOM (*Com*munist *com*munication) simulation models the flow of messages through the mass media to the populations of the Soviet Union and China (Popkin, 1965; Pool, 1967). Each simulated individual in the samples from each country is described by certain demographic, social, and attitudinal characteristics, such as age, sex, location of residence, occupation, party membership, attitude toward the regime, and trustfulness or distrustfulness regarding messages. Each person also has a probability of being exposed to each of the different media or message types. In China, where indexes comparable to the Nielsen ratings for electronic media audiences are unavailable, audiences for different media are estimated by combining various ancillary data. Each message type is described in terms of its source, certain credibility characteristics, volume (e.g., space in the newspaper), and time and place of appearance. Monte Carlo procedures determine the basic exposure of each individual to each message. Thus, incorporating assumptions about such processes as the frequency of exposure and duplication (the incidence of shared audience among different media), the simulation generates data regarding who is reached and how often by the news flowing through the media in a total society.

In a similar simulation of audience exposure in a mass media system, Kramer (1969) models the 1947 Cincinnati United Nations Information Campaign that had previously been investigated by a NORC panel survey. Computer inputs concerning the population and its media habits are derived from census statistics and from newspaper and radio audience studies. Simulated message themes relevant to the attitude measures in the NORC survey come from a content analysis of press coverage. For each of the twelve themes so identified, the simulation generates data concerning the most important media types and vehicles in producing exposure, the duplication of exposure across themes, and the growth and distribution of exposure in different population subgroups (described by sex, education, and socioeconomic status). Kramer reports that the themes producing the highest average frequency of exposure for the simulated population were the same as those that appeared most closely related to large changes in the NORC panel.

Klein and associates follow a different approach in modeling communicative behavior, incorporating conversational interaction within a speech community in order to simulate language change in Tikopia and Maori (1969). Interpersonal contacts, of course, are the principal fodder for models of innovation dissemination. Actually, the impetus for simulating diffusion processes has not come from sociologists, despite their discontent with the inherent limitations of cross-sectional studies, but from the pioneering efforts of Hägerstrand, a Swedish geographer who initiated Monte Carlo diffusion simulations in the early 1950s (1953, 1965a, 1965b). Since then Hägerstrand's approach has been extended in new applications by a coterie of quantitative geographers (cf. references cited in Hanneman *et al.*, 1969; Chapter 8 this volume).

In investigating the spatial-temporal distribution of an innovation, Hägerstrand notes that whereas acceptance "outposts" develop on the geographical map of the locale under investigation, most adoptions are concentrated around the initial possessors of information. Hägerstrand terms this observed spatial proximity in adoption the "neighborhood effect," which he quantifies in his simulations in terms of a "Mean Information Field."

Essentially, the neighborhood effect postulates an inverse relationship between distance and probability of adoption, based on the observation that contact among individuals decreases as the distance between them increases. Decreasing communication contacts lessen the occasions for learning of an innovation or for being persuaded to adopt it. It is interesting to note the similarity between Hägerstrand's assumption that geographical distance between adopting units is a major determinant of spatial diffusion and Linton's generalization from field studies of early anthropologists: "Other things being equal, elements of culture will be taken up first by societies which are close to their points of origin and later by societies which are more remote or which have less direct contacts" (1936, p. 328).

Quantitative geographers compute neighborhood effect probabilities from such spatial measurements as the distance of migration, the distance between people talking over the telephone, and the distance between the residences of bride and groom prior to their marriage (cf. Marble and Nystuen, 1963). To simulate diffusion processes, Hägerstrand and his followers input these contact probabilities in a composite matrix, called the Mean Information Field, in which the central cell is the orientation point for the origin of a communication act. Thus the Mean Information Field expresses the probability of communication between a source individual (in the central cell) and any other person in the geographical region (other matrix cells) as a function of the distance between them.

In Hägerstrand's early models, an innovation was adopted at once when heard of in a meeting between an informed and an uninformed individual. While acknowledging the invalidity of such an oversimplification, Hägerstrand notes that the diffusion of information about government subsidies for farmers provokes such low resistance that it almost approximates immediate adoption. To take account of psychological resistance to adopting an innovation, Hägerstrand later modified his diffusion rules so that adoption would occur only after a specified number of tellings (the threshold) that would differ among individuals. In applying this rule, Hägerstrand assumed that resistances varying from one to five tellings were normally distributed in the population; therefore, resistance scores were generated for simulated individuals so as to reproduce a normal distribution. Preliminary findings from this model suggest that the resistance factor leads to slower development over time and to further spatial concentration, since the more dispersed, erratic tellings have little chance to cause adoptions (Hägerstrand, 1965b). Pitts has programmed extensions of Hägerstrand's models and varied decision rules regarding the number of tellings per resistance category (1963).

In focusing on spatial variables, most published diffusion simulations virtually ignore other social structural and social psychological variables identified in past research as important determinants of diffusion processes. For example, though one of Hägerstrand's models includes communication barriers that block or diminish communication frequencies along certain cell boundaries, these barriers are designed to imitate the effect of lakes and roadless forests in the simulated region and thus do not reflect possible social constraints. Of course, the inclusion of resistance considerations contributes a cognitive element to what has heretofore essentially been a "black box" approach to decision-making in diffusion simulations. The implementation, however, involves random assignment of resistance scores and fails to relate resistance to demographic, attitudinal, or economic characteristics of the potential adopter.

One extension of a Hägerstrand-type model to incorporate some sociological diffusion research findings is the Tiedemann and Van Doren simulation of the diffusion of hybrid seed corn in an Iowa county (1964). A previous investigation indicated that farmers usually first heard of the innovation from commercial seed sellers; however, they were most likely persuaded to adopt the innovation through interactions with neighbors. In their model, therefore, Tiedemann and Van Doren construct two Mean Information Fields, one representing the influence of local seed stores and the other accounting for farmer-to-farmer contact. To simulate a two-step diffusion process, the innovation is first communicated from seed stores over a relatively large area and then by farmer-to-farmer diffusion over a relatively smaller Mean Information Field.

Wolpert's simulation of the diffusion of farm machinery in Sweden also extends Hägerstrand's approach to encompass some behavioral parameters of information flow (1967). Like Tiedemann and Van Doren, Wolpert incorporates a two-stage communication process involving an external flow of information from experts in a source region to farmers in the sample units and an internal diffusion among farmers within the areas. Though acknowledging the importance of Karlsson's proposals concerning the impact of source-receiver similarity on effective communication (1958), Wolpert operationalizes his similarity variable in terms of a spatial indicant (farm size) instead of a social-psychological measurement. Designating five size categories of farms, Wolpert regulates the internal flow of information by the following rules: There is a higher probability that farmers in one size category will direct information to others in the same category; there is a greater likelihood that information will be believed when received from a farmer in the same category; and large farmers need less contact with relevant new information before accepting it than do small farmers.

Thus, instead of following Hägerstrand's procedure of generating a psychological resistance score from a normal distribution for each contacted person, Wolpert assigns resistance probabilities to adopting units on the basis of farm size. Furthermore, Wolpert's model apparently does not make provision for "gestators" (partially persuaded persons requiring additional tellings before adopting an innovation), since he implements his resistance rule in terms of the probability of acceptance of information on the *first* telling for each farm size category. It would be interesting to have comparative data concerning the consequences of Hägerstrand's and Wolpert's operationalizations of resistance in what is otherwise the same diffusion model. Thus far, regrettably, diffusion simulators have not exploited this potentiality of computer modeling and have not programmed exchangeable modules to represent different hypothesized processes so that overall system effects of these modules can be compared and evaluated (cf. Gullahorn and Gullahorn, 1965b).

Recently Wolpert and Zillman have restructured the Hägerstrand-type approach, considering information diffusion a specific application of a more general and flexible model of the spatial context of decision-making (1969). In representing variables they note that whereas geographers may literally interpret a spatial concept like proximity, other social scientists may consider it an analogue to a cognitive process. Wolpert and Zillman's programmed decision processes incorporate behavioral findings, and the model's components include the following: a population subject to a spreading criterion (threat or appeal) that may alter the informational bases for decisions, barriers or channels to this spread, interdependence between population members who act cooperatively or competitively with respect to a limited set of response alternatives, poles of approach or avoidance, and nonadaptive as well as adaptive search procedures.

Preliminary applications of this model involve simulations of panic behavior in a theatre fire context and, on a more aggregate scale, in a city invasion crisis.

Diffusion theorists in the sociological tradition as well as quantitative geographers like Wolpert have been influenced by Karlsson's syntheses (1958) of Hägerstrand's spatial considerations with generalizations from interpersonal communication research. Taking account of the impact of social as well as geographical distance on communication probabilities, Karlsson's models deal with the consequences of selective perception and exposure to information, as well as the influence of reference groups, message characteristics, and other source credibility determinants. Some of Karlsson's suggestions were incorporated in Deutschmann's proposals to simulate information diffusion and attitude change in small social systems (1962). Assuming the geographical area containing the population to be relatively small, Deutschmann did not include spatial distance between individuals as a variable in his models but instead divided his population into socially homogeneous cliques, with message transmission more probable within cliques than between them. While some of the previously mentioned diffusion models incorporated two-stage processes in information flow, Deutschmann's proposals are unique in explicating the processing of the two-step flow of communication hypothesis (cf. Katz and Lazarsfeld, 1955), which asserts that "ideas often flow *from* radio and print *to* the opinion leaders and *from* them to the less active sections of the population" (Lazarsfeld, Berelson, and Gaudet, 1948, p. 151). In Deutschmann's models, therefore, messages from the mass media incite communications between individuals; however, within the simulated community there is a small group of opinion leaders with a high probability of transmitting information to others after they have received it, and all others have a low probability of passing information. This implementation markedly differs from the communication exchanges in other diffusion models which assume that all individuals who become knowers have equal chances of contacting nonknowers on a random, with-replacement basis.

Deutschmann's untimely death precluded his actualizing his models on a computer; however, Stanfield, Clark, Lin, and Rogers developed a computer model based on Deutschmann's proposals. This model was designed to simulate the diffusion of awareness about an innovation in a small, relatively isolated Latin American peasant community. The initial version of this model is sketchily described by Rogers (1969), who reports poor correspondence between the output and reality data from the village. Hanneman modified this computer model in a version called SINDI 1 (*Si*mulation of *IN*novation *DI*ffusion), a stochastic model programmed in FORTRAN (Hanneman, 1969; Hanneman *et al.*, 1969), using input parameters and probabilities derived from Rogers' study of a Colombian village (1969). In SINDI 1, an innovation enters the social system through external communication channels (the extension service agent or the schoolteacher). Mass media channels are not important in this model because the simulated villagers are illiterate and are not exposed to agricultural radio programs. Within the community, individuals are divided into cliques of highly interacting members, with information transmission flowing more frequently within cliques than between them. A small group of "tellers" (opinion leaders) in the community have a high probability of passing information they have received about innovations; all others have a low probability of transmission.

The input to SINDI 1 includes parameters designating the number of cliques, the number of members in each clique, and the number of potential tellers in each clique. The four interacting cliques and one group of isolates were identified by analysis of sociometric data collected in the village; and the nine potential tellers were those individuals who had received three or more sociometric choices from their peers as sources of agricultural information. Other parameters include the number of contacts allowed each external channel source per time period; the number of contacts allowed a teller once he becomes a knower; the probability of a nonknower becoming a knower through any external channel source; and the probability of a member of a clique becoming a knower through a contact with a teller from any clique.

Technically, SINDI 1 consists of a main program that monitors the simulation and five subroutines, including a random-number generator. The first subroutine inputs parameters and initializes arrays for the beginning of a run and a time period; the next routine processes external messages. Each external channel randomly contacts a specified number of individuals. Associated with each person is an information transfer probability based on his channel orientation (either to an external source or to local, interpersonal contacts). This probability is compared with a randomly generated number between zero and one: If the random number is equal to or less than the assigned probability, the person becomes a knower; otherwise, he remains a nonknower. The teller contact subroutine functions like the external message section, except that the information transfer probability depends on the person's clique membership and the clique affiliation of the contacting teller. The output subroutine prints out a summary of the information transfer events for the simulation.

While SINDI 1 represents an improvement over the earlier model reported by Rogers (1969), the simulation output only roughly approximates the real-world diffusion curve based on respondents' recall of when they first heard of a weed-spray innovation. Some shortcomings in the formulation of SINDI 1 may account for this lack of correspondence; unfortunately, there are no reported explorations with alternative subroutines to generate comparative outputs.

The modeled assumption of random contacts probably misrepresents actual communication behavior; in fact, empirical data from the peasant village indicate that some individuals have considerable contact with external sources while others have none. In addition, though the model simulates only the diffusion of information rather than adoption (and thus does not consider psychological resistance), it assumes that the tellers begin disseminating as soon as they become knowers. Such sequencing might initially accelerate the simulated diffusion by failing to account for the likelihood that tellers become active disseminators only after actually adopting the innovation and becoming convinced of its value.

Perhaps the crucial contributor to SINDI 1's failure to simulate the cumulative diffusion of information, however, is the extremely restrictive operationalization of the two-step flow of communication hypothesis. In the implementation, the tellers have a probability of 1.0 for disseminating information received; whereas nontellers have a probability of 0 for communicating to nonknowers. This exclusive dichotomization of opinion leadership (which actually might better be represented as a continuous variable) produces a large number of nonknowers, since the isolates as well as those individuals in cliques without tellers cannot learn of the innovation. Inadvertently, therefore, SINDI 1 appears to simulate diffusion in a social system comprising a significant proportion of isolates. Indeed, the model's diffusion curves resemble those produced by the Constant Source Model that Coleman fitted to the drug adoption data for physicians who were relatively isolated in the sociometric networks (Coleman, 1964). It is regrettable that the teller and nonteller probabilities in SINDI 1 were not modified to test whether such changes would better simulate the logistic type S-curve of diffusion expected for an integrated interacting community.

Another obstacle encountered in the SINDI 1 programming reflects a chronic impediment to would-be simulators: The original data were not collected for the purpose of providing inputs to a simulation model; thus some parameters and probabilities had to be crudely estimated from available data. In what appears to be the most sophisticated diffusion simulation model to date, Carroll avoided this pitfall in parameter estimation by conducting field research specifically designed to gather data inputs for his program (1969).

Carroll's model of the diffusion of dairy innovations in a Brazilian community (SINDI 2) simulates the communication process by which individuals are influenced to adopt an innovation through exposure to various types of persuasive messages. In SINDI 2, information and influence flow to potential adopters via the print media (newspapers and magazines), the electronic media (radio and television programs), and interpersonal communication with adopters encountered individually and at meetings of community organizations. The model incorporates the following assumptions:

1. In reacting to environmental influence, each simulated individual manifests a resistance to adopting an innovation. This resistance is a function of the individual's demographic and attitudinal characteristics and of the economic characteristics of his enterprise as well as the characteristics of the innovation. Thus, instead of arbitrarily assigning resistance scores according to an untested assumption that resistance approximates the normal distribution in the aggregate, Carroll utilizes a multiple-partial correlation to predict each individual's resistance to adopting a particular innovation.

2. Each potential adopter is influenced proportionally to his exposure to innovation information from the mass media and to his frequency of association with adopters in the community.

3. An individual becomes an adopter when the magnitude of the cumulative influence from various sources exceeds the magnitude of his resistance to adoption.

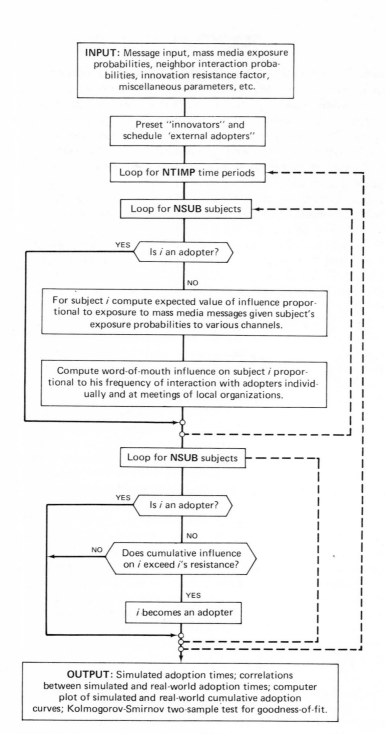

INPUT: Message input, mass media exposure probabilities, neighbor interaction probabilities, innovation resistance factor, miscellaneous parameters, etc.

Preset "innovators" and schedule 'external adopters'

Loop for **NTIMP** time periods

Loop for **NSUB** subjects

Is *i* an adopter? — YES

NO

For subject *i* compute expected value of influence proportional to exposure to mass media messages given subject's exposure probabilities to various channels.

Compute word-of-mouth influence on subject *i* proportional to his frequency of interaction with adopters individually and at meetings of local organizations.

Loop for **NSUB** subjects

Is *i* an adopter? — YES

NO

Does cumulative influence on *i* exceed *i*'s resistance? — NO

YES

i becomes an adopter

OUTPUT: Simulated adoption times; correlations between simulated and real-world adoption times; computer plot of simulated and real-world cumulative adoption curves; Kolmogorov-Smirnov two-sample test for goodness-of-fit.

FIGURE 4-2   Flow Chart for SINDI 2.

Figure 4-2 flow charts the processing in SINDI 2 (cf. Carroll, 1969, p. 40). Once the input data and required parameters are read in, the model initializes innovators (those who adopted the innovation prior to the simulation starting time) and schedules external adopters (those who adopted the innovation outside the community and subsequently migrated into the community after the simulation starting time). Innovators, therefore, are influential from the onset of the simulation; external adopters become information sources only after entering the social system.

Although some attributes are expressed as probabilities of exposure or interpersonal association, SINDI 2 is processed as a continuous, expected-value model rather than as a stochastic model that depends on random number generators to determine the course of events. The program cycles through a preset number of time periods within which there are two loops that cycle through the simulated subjects. The first of these loops computes the expected value of influence on each person who has not yet adopted the innovation. This expected value is proportional to each subject's fixed probability of exposure to innovation messages from the various print and electronic media and to the subject's probability of encountering adopters individually and at local organization meetings. Contacts with nonadopters in an individual's circle of discussion partners reduces the amount of influence on him resulting from association with adopters. SINDI 2 processes communication linkages between individuals in terms of sociometric dyads that represent potential one-way flows of information. Attributes associated with each dyad include the probability of daily contact between the two individuals and the credibility rating of the sender, as evaluated by the receiver.

The second subject loop at the end of each time period compares the cumulative influence on a subject with his resistance to adoption. When the total influence from the mass media and from contacted adopters exceeds an individual's resistance to adoption, he becomes an adopter and can begin influencing others in subsequent time periods. Carroll's implementation thus avoids the unrealistic dichotomization of knowers into tellers and nontellers incorporated in the SINDI 1 model.

After cycling through the designated time periods, SINDI 2 outputs a computer plot of the simulated and empirical cumulative adoption curves and performs the Kolmogorov-Smirnov two-sample test. This explicit programmed application of a goodness-of-fit test of the simulated output against empirical data is a marked improvement over earlier diffusion models in which comparisons were based on relatively nonrigorous inspections. Of course, good fit at the aggregate level does not necessarily validate a model. Carroll systematically reviews various validation considerations before concluding that SINDI 2 represents a reasonably valid model of the diffusion system to which it was applied (cf. Carroll, 1969, chap. 5).

While SINDI 2 avoids the "black box" approach to decision-making noted in previous diffusion models, it is not programmed as an active information-processing simulation. The probabilities of contact between the members and the source of credibility evaluations do not change on the basis of simulation events. In SINDI 2, a simulated person simply passively accumulates persuasive messages until they exceed his resistance threshold, whereupon he automatically adopts the innovation and begins recommending it to others. As Carroll acknowledges, there is no provision for selective retention of past communications or for purposeful delay in actual adoption because of other considerations. Like the early cognitive problem-solving computer models, diffusion simulations tend to represent "single-minded" behavior—an apparently necessary simplification at this stage (cf. Neisser, 1963).

SIMULATIONS OF MICROBEHAVIORAL PROCESSES

Individual information-processing mechanisms creating changes in simulated persons—and in the social systems in which they are embedded—are represented with varying degrees of complexity in another set of computer models we have termed simulations of microbehavioral processes. In this section we shall review some simulations of social influence processes, including adult socialization to an urban environment and political socialization; other simulations incorporating individual processing of social communications; and simulations involving social exchange considerations in decision-making.

Simulations of Socialization Processes

In simulating aspects of adult socialization to an urban setting, Hanson and associates (1967, 1968) share the communication and diffusion modelers' interest in the basic process of the dissemination of information. Rather than represent simulated individuals as relatively passive targets of the information that happens to hit them, however, the urbanization model simulates the active search behavior of rural migrants as they explore communication networks for available opportunities in the unfamiliar social structures of the city and as they modify their searching as a consequence of their experiences. Both impersonal and personal communication sources are included in the model, whose basic assumptions involve the following:

1. Men search their environment for opportunities to achieve goals arising from their current living situations and their recent past experiences (e.g., individuals may seek better-paying jobs, larger homes, new cars, etc.).
2. The "opportunity structure" of a city involves four major market sectors (housing, employment, consumer goods, and leisure activity).
3. Opportunities are advertised through three types of communication structures (the impersonal bureaucratic, natural proximity, and social network structures), and men search these structures for desired opportunities.

4. As a consequence of the interaction of advertised opportunities and searching men, a man may accept an opportunity that changes his current living situation as well as his goal orientation.

5. The changed living situation and goal orientation of a person alter his future search for opportunities.

Some of the model's major variable indexes are derived from factor analyses of extensive empirical data collected from rural migrants. These indexes include "on arrival constants and variables" such as age, family size, socio-economic status, IQ, self-esteem, and prudence; "current situation variables" such as degree of employment, degree of settled housing, total property value, earned income, current payments, organizational participaton, and current satisfaction variables; and "recent past behavior variables" such as propensity to get drunk and "rootless situation" (comprising recent unemployment, change of residence, and turnover in friendships). These background variables are input in multiple regression equations to predict the persistence of search, the range of search, and the attributes of each simulated individual. Parameters for these three dependent variables determine the character of each man's search for opportunities in the communication structures of the city's market sectors.

In the simulation, available opportunities are advertised in the impersonal, proximity, and social net structures in the four market sectors. Monte Carlo processing of programmed propositions determines the information flow of these opportunities. Origins of men in the communication structures depend on the location of opportunities they presently hold or on the locus of their search during the previous simulated time cycle. After the simulated individuals search the networks and match their attributes with the criteria associated with located opportunities, current data are compiled on changes resulting from accepted opportunities; and current situation attributes are input in new prediction equations to alter individual search parameters in the next time cycle. From the history tape that accumulates individual change records, summary statistics and time trend plots are generated for the migrants as a whole or for various subgroups within the simulated population. Though the model is still in the tuning stage, it promises valuable insights into the complicated socialization process involved in urbanization.

Although simulated migrants utilize their social contact networks in searching for opportunities, mutual social influence processes are not represented in the urbanization model. McPhee's individual voter simulations, however, do incorporate interpersonal influence considerations along with personal predispositions and external persuasion (1961, 1963; McPhee and Smith, 1962). Figure 4-3 reproduces Rosenthal's (1968) reconstructed flow chart of McPhee's model.

In McPhee's model, a voting population is input along with initial partisanship parameter values. A stimulation process then exposes simulated voters to political stimuli for the current election. The stronger a voter's predisposition toward a party, the weaker the campaign stimuli required to elicit a choice of that party. That is, if the election stimuli exceed the minimum threshold for a voter's party predisposition value (e.g., is $S_{dj} > 1 - P_d$ in figure 4-3), and if

Definitions:

P_d = Democratic predisposition $\qquad 0 \leqslant P_d \leqslant 1$
P_r = Republican predisposition $\qquad 0 \leqslant P_r \leqslant 1$
S_{dj}, S_{rj} = stimuli drawn at random from $\qquad 0 \leqslant S_{dj} \leqslant 1$
probability distribution for voter type $j \qquad 0 \leqslant S_{rj} \leqslant 1$
IP = initial preference
INT = interest in the campaign
C = final voting choice
DEM = Democratic preference or choice
REP = Republican preference or choice
NOV = No vote preference or choice
DD = Democratic aggregate learning weight
RR = Republican aggregate learning weight
NN = Nonvoting aggregate learning weight

Initialization: For each voter, values of P_4, P_r, DD, RR, and NN must be supplied. His contact must be indicated. For each voter type, the distribution of S_{dj}, S_{rj} must be supplied.

Note: For simplification, the limits on S_{dj}, S_{rj} have been altered. Also, in the actual simulation, P_d and P_r were constrained to the interval 0.1, 0.9.

1. Simulation — This occurs once for each voter and a second time for those voters reentered from the discussion process

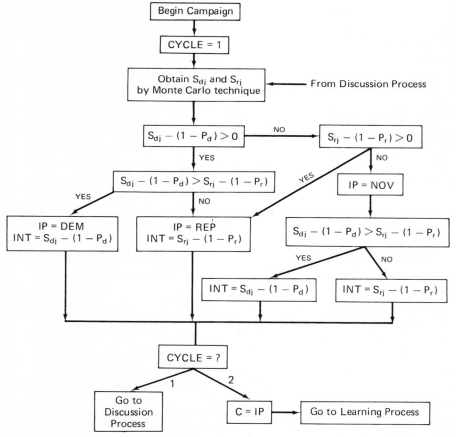

Note: McPhee not specific on action when equality obtains.

FIGURE 4-3 Detailed flow charts for the McPhee simulation.

2. Discussion — This operation is performed once for each voter

Definition: IP_r = initial preference of contact

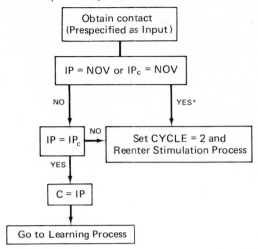

*Note: According to McPhee, "discussion takes place if the <u>average</u> of the two interests is above the indifference point or minimum to vote" (McPhee, 1963, p. 146). This condition is always fulfilled if both the voter and his contact have IP ≠ NOV (McPhee, 1963, pp. 137–138). It is never fulfilled if both have IP = NOV. On the other hand, if one member of the pair has a partisan initial preference while the other has a no-vote initial preference, they will automatically disagree and restimulation will occur (McPhee, 1963, pp. 146–147). Hence the simplicity of the above flow chart, which seems to counter McPhee's intention (McPhee, 1963, p. 146) that some "persuasion" should occur when a voter and nonvoter meet.

3. Learning — This operation is performed once for each voter

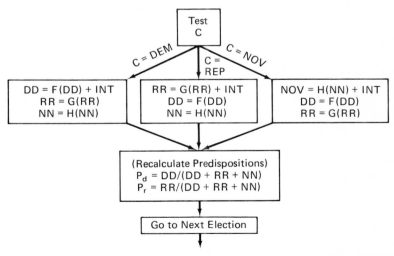

*Note: F(DD), G(RR), and H(NN) denote functional relationships used to weight past choices less heavily than current choices. In the model actually used,

$$F(DD) = 0.85\,DD$$
$$G(RR) = 0.85\,RR$$
$$H(NN) = 0.85\,NN$$

FIGURE 4-3 (Continued)

this excess is greater than any excess for the other party, then a voter develops a party preference for this election, and his interest in the campaign is defined by that excess. A voter's interest not only affects his subsequent discussion behavior but also influences his long-term partisanship.

After all voters have thus developed initial party preferences, a discussion process ensues. If the discussion partners' preference levels exceed the minimum indifference threshold for voting, they compare their choices. In the typical case, if a voter's initial preference agrees with that of his discussion partner, this preference becomes his final choice. In the event of disagreement, however, the voter and the discussion partner repeat the stimulation process; therefore, either or both might change party preference. Generally, the preference the voter develops during the second stimulation cycle becomes his final choice. Later in the program, nonetheless, another member of the sample might choose the original voter as discussion partner; and should they disagree, the voter is again exposed to the stimulation cycle and hence to further possible change. Thus there are complications not depicted even in Rosenthal's detailed flow diagrams.

A voter's final decision determines his vote on election day and also serves as input to the model's learning process, thus influencing his future partisanship behavior. In the learning processing, a voter's current interest in the campaign is combined with a time-discounted measure of his past choices, and his new predilection for each party is generated from the ratio of his selections of that party to the total choices. The new predispositions are then input in another time cycle simulating the next election.

As Rosenthal notes in his critical review (1968), the learning process of McPhee's model has several valuable features. Inclusion of the interest weighting, for example makes provision for simulating "critical" elections. Furthermore, the averaging method programmed in the learning processing provides a mechanism for simulating how old people become more "set in their ways" and thus more immune to counterappeals.

While McPhee's model has specifically been adapted to the study of single campaigns and has actually rather closely matched observed election results, it was developed for more theoretically interesting purposes than simple vote forecasting (in contrast to the Pool, Abelson, Popkin simulmatics project, 1964). McPhee's model simulates political socialization, producing a "politicized" body of voters who exhibit varying degrees of allegiance to one party. A report on this application appears as a case-example following this chapter.

Aspects of political socialization have also engaged the interests of other simulators. Levin (1962) reports preliminary work on a model of reference group influences on adolescents' party affiliation, and Browning (1968) has developed a partially data-based simulation of political recruitment.

Simulations
of Other Reference Group Processes

A different facet of adolescent socialization is modeled in Coleman's simulation of reference group behavior among high school students (1962,

1965). A computer analysis involving the sociometric network identified in a survey of adolescents revealed that while similarity or difference in smoking habits among friends influenced the incidence of smoking over the period of a school year, such behavioral concordance did not affect the maintenance of friendships. Estimating parameters of change in smoking habits from the survey data and inputing each simulated student's initial state, Coleman then simulated the course of the adolescents' smoking behavior as affected by the smoking practices of their friends. Comparing the simulation projections with the expected number of adoptions of the habit with age, Coleman concludes that friendship networks noticeably augment the overall number of regular smokers.

In another investigation of the impact of social influence on an individual's overt behavior, Breton simulates productivity in an industrial work group (1965). Instead of following the functionalist tradition of postulating the existence of a consciously endorsed output norm that helps maintain the social organization of the group, Breton programs his model so that any observed uniformity in productivity results from the structure of the work situation and the interaction among the workers. Thus, in Breton's simulation a worker's productive behavior depends on his rate of pay, his level of motivation for monetary rewards, and the constraints applied by co-workers for whom he initially has positive sentiments.

According to the model's output, pressures from fellow workers compress the variation over time among weakly motivated workers, making them more homogeneous in output—as if they were conforming to an informal norm. While constraints from co-workers also reduce the overall output of strongly motivated workers, among this group the sanctions have the opposite effect of increasing variability in productivity as some workers develop negative sentiments toward those attempting to restrict them. Inadvertently, therefore, Breton appears to have simulated the boomerang phenomenon observed in negative reference group processes. As he notes, the mechanisms involved in the simulation suggest that variables should be investigated in further empirical research.

While social comparisons alter smoking behavior in Coleman's model and produce sanctioning and sentiment change in Breton's model, neither simulation actually processes relative deprivation effects. Such considerations are incorporated in Smith's proposed simulation (1969) of Rossi's accounting scheme for family migration (1955). In representing a family's decision to move, Smith utilizes a dynamic stimulus-organism-response mechanism.

In Smith's model, a simulated resident's complaints about the neighborhood (the stimulus) result from an invidious comparison of his social acceptability with that of randomly selected neighbors. Social acceptability is a combination of ascribed rank (race and ethnicity) and achieved rank (socioeconomic status). If a resident's social acceptability exceeds his neighbor's, this relative deprivation in status associates elicits complaints about the neighborhood; otherwise, relative status gratification reduces his complaints. By combining a simulated resident's total complaints with his prejudice score (attitudinal predisposition), a value for anxiety about moving (reaction potential) is generated. If a person's anxiety value exceeds a threshold parameter, he moves; otherwise he remains in the neighborhood. Summing individual "decisions" produces an aggregate rate of moving behavior. Once this simulation is operational, Smith plans to investigate the differential racial segregation effects of restrictive covenants and open housing.

Smith has incorporated the same stimulus-organism-response mechanism in another proposed simulation of the effects of leadership (1969). This model, based on Selvin's analysis of army survey data (1960), follows the stimulation-discussion-learning format implemented in the McPhee voting simulations discussed previously. During the stimulation process, pressures toward disobedience (the stimulus) emanating from the leadership climate in an army company are combined with a recruit's attitude about the illegitimacy of army authority (the predisposition) to produce a score for complaints about army life (the reaction potential). Complaint values exceeding a preset threshold are coordinated with an intention to disobey; lower values are associated with obedient tendencies.

After all the simulated recruits thus make tentative decisions about disobedience, they are exposed to an influence process during which each discusses his intentions with a comrade if the average of their complaints exceeds an indifference threshold for disobedience. If the discussion partners agree in their plans to disobey, they reinforce each other's decision; if they disagree, both are restimulated by the pressures of their leadership climates and make new decisions about conforming behavior. Once the recruits have made final decisions as a result of the discussion or restimulation processes, they enter the learning process during which new predispositions toward disobedience are calculated on the basis of their past and current behavior. These new predispositions are then input in another time cycle of the simulation.

Smith's simulation scheme explicates the model of the individual recruit left implicit in the original survey research and also synthesizes the elements in the accounting scheme for behavior in different leadership climates. While it thus has great promise, it also shares some shortcomings associated with the original McPhee format on which it is based. That is, the discussion process formulated by McPhee and Smith lacks theoretical coherence and empirical verisimilitude because (1) a voter or a recruit is usually exposed to only one discussion partner, (2) agreement confirms an initial decision even though this intention may barely exceed the indifference threshold, and (3) disagreement results only in one reexposure to randomly generated (rather than selectively chosen) environmental stimuli and does not provoke attempts to gain social support from other discussion partners in order to reduce dissonance. But it is impossible to represent all facets of social psychological processes at once, and perhaps future developments with the McPhee and Smith models will remedy these apparent inadequacies.

Simulations of Individual Processing of Social Communications

Although the social influence models considered thus far involve some degree of individual information processing, as just noted they do not incorporate social psychological mechanisms regarding selective tendencies in perceiving and interpreting social communications. Such considerations are included, however, in several models of social behavior in situations of emergent controversy.

Using a scenario of messages representing communications during the historic week preceding the outbreak of World War 1, Pool and Kessler (1965) incorporate selective attention tendencies in simulating leaders' reactions to a crisis (Crisiscom). In processing incoming messages, the simulated Kaiser and Tsar pay more attention to news dealing with them and to news from trusted, liked sources; each attends more to facts he must act upon and to facts bearing on actions he already is involved in and less to facts that contradict his previous views. As a consequence of such information filtering, each simulated leader modifies his world image and originates new messages, which then serve as input for another program cycle and thus influence the state of the simulated world. In evaluating the configuration of affects, perceptions, and message trans- missions, Pool and Kessler deem the simulation "intuitively very satisfactory."

Similar assumptions about human distortion of information are in- corporated in the psychological submodel of the Technological, Economic, Military and Political Evaluation Routine (TEMPER), an ambitious four thousand variable simulation of cold war conflict (cf. Abt and Gorden, 1966; Clemens, 1968). Psychological variables and inertial memory also distort information as it is channeled between candidates and voters in an election simulation model by Coombs and associates (1968). While selective mechanisms may be operative in the man-machine interactions in the Romes' Leviathan model of hierarchical organization (1966), it is noteworthy that in one simula- tion of an intelligence communication control center (1964), performance improved during crisis conditions, ostensibly because increased commitment to common goals offset the effects of heavier information loads.

Still another investigation of human information processing is represented by the laboratory and computer simulations of national policy formation reported by Boguslaw and colleagues, (Davis, 1963; Boguslaw *et al.,* 1966; Boguslaw and Davis, 1969). In the original simulation, called PLANS, six human participants enact roles representing significant interest groups (business, mili- tary, labor, civil rights, internationalist, and nationalist). As background each is informed of his interest group, as well as the state of the cold war, comparative military strengths, unemployment rate, and trends of other economic and social indicators. Each participant can negotiate with the others to gain support for his policies, and at the end of each six-month cycle in the simulated five-year span, he allocates influence units to the policies he supports. Once policy proposals concerning disarmament, welfare programs, and so forth, are adopted, the socioeconomic environment is modified to reflect the consequences of such changes.

In one run of the simulation using professionals and administrators, the participants focused on a limited number of problems and adopted five policies during a three-year simulation period. When graduate students enacted the roles, they distributed their resources widely over a larger number of policies, actually adopting none during a three-year simulation span. On the basis of these preliminary findings, the investigators suggest that issue crystallization appears necessary for policy formulation and adoption.

To further explore the social processes of policy formation, Boguslaw and Davis (1969) have developed a computer model of their manned simulation, using separate programs to represent alternative hypotheses concerning the human participants' decision-making strategies. Three of the decision modes involving complete randomness, limited consistency, or absolute consistency generate decisions without reference to the interest groups' goals and thus appear to be straw men in the simulation, achieving no match with the human data. Most theory assumes that behavior is goal oriented, and some confirmatory results are generated by two simulated decision modes—undifferentiated goal behavior, approximating the diffuse behavior of the graduate students, and crystallized goal behavior, representing the professionals' decision-making. When cooperation is combined with crystallized goal behavior in the programmed implementations, more policies are adopted than in any other decision-making mode; furthermore, with this strategy the unemployment rate at the end of the five-year period is considerably lower than that in other computer runs and even lower than the rate observed in the human simulations.

Unfortunately, the report presents insufficient detail concerning the program and does not include data regarding the extent of fulfillment of interest group goals in the computer simulations; therefore, comparisons with the human data are limited. The interest group goals are central to another limitation in empirical validation: The computer programs automatically accept the input goals in processing decisions, whereas real individuals tend to espouse and promote additional values not specified by the formally prescribed interest group goals. While the manned and computer simulations generate important insights regarding social planning and action, their broader significance for social system theory remains unclear. Further explication of the research should be encouraged.

Perhaps the greatest tour de force in modeling social psychological influence processes is represented by the Abelson-Bernstein simulation of the diffusion of competing assertions in a community referendum controversy (1963; Alker, 1968). A general overview of the processing during a time cycle of the referendum campaign is presented in figure 4-4, a flow chart based on Abelson and Bernstein's verbal description. (A more detailed flow chart and review from a political science perspective appears in Francis's chapter in this volume.)

General mass communication media (radio, TV, newspapers, etc.) and particularized channels (town meetings, propaganda circulars, organized telephone campaigns, etc.) are represented in the model. At the beginning of a time cycle, channel assertions are updated on the basis of content analyses of editorials, cartoons, phone messages, and so forth. Each simulated citizen is then exposed to certain sources in certain channels, the exposure probabilities being

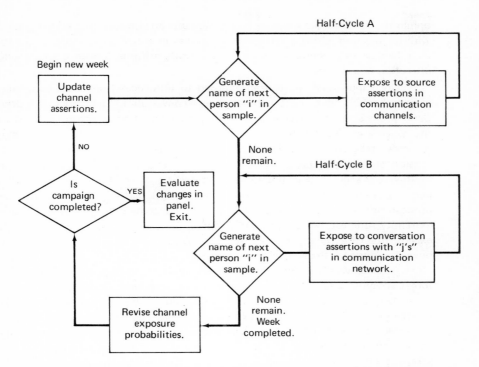

FIGURE 4-4   Weekly cycle in Abelson and Bernstein: community referendum
simulation.

determined by the individual's preset attraction toward a channel. If a citizen is
exposed to a particular channel, he is confronted in turn with each source using
that channel, and a weighted random decision routine determines whether he is
exposed to all the assertions made by a particular source. Whether the citizen
attends to any given source depends primarily on his interest in the issue. Thus,
while cartoons will be noticed by most regular newspaper readers, editorials will
be read only by those with relatively high levels of interest.

Other conditions affecting a person's receptivity to a source include the
extremity of his position, the attention value of the source's assertions, and the
"assertion match" (congruence) between the individual and the source. In
addition, an individual will be more likely to accept an assertion if he likes the
source, if he has not previously disagreed with the assertion, if he has already
accepted similar assertions, and if the assertion is consonant with his predispositions and with his present position on the issue. A negative reference source
has a boomerang effect on the receiver: When a person's attitude toward a
source is negative and his receptivity is very low, the source's assertions that the
person has not previously encountered and that are inconsistent with his position have the effect of promoting the person's acceptance of converse assertions.
In general, a person's interest increases when a source's assertions agree with his

opinions and decreases when he himself accepts an assertion that is inconsistent with his position. Other programmed propositions specify conditions for change in individual satisfaction, position on the issue, attitude toward a source, and attraction to a channel.

Once individuals have been exposed to influences from mass communication channels, they next contact their "conversation" networks. A Monte Carlo processing of conversational probabilities based on a person's interest in the issue determines whether a "conversation" will occur. Actually, the informaton flow programmed in this phase of the model is simply a one-way transmission of messages from one individual to another and thus fails to simulate the mutual exchanges and comparisons characteristic of real-life conversations. A person's reactions to the assertions of his contact partner are determined by propositions similar to those previously listed concerning the impact of attitude extremity, interest, receptivity, and so forth, on responses to a channel source, except that the parameters are set to make face-to-face encounter a more powerful influence than channel exposure.

After a week's exposure to the referendum campaign, a person tends to forget accepted assertions that are inconsistent with his position, but he always remembers those that are consistent with his predispositions. His level of interest is affected by the number of exposures he has encountered; if he has experienced no exposure, his interest decreases. The number of channel and conversational exposures, along with the number of counterassertions he has accepted and retained, determine whether the person changes his attitude position. If he does change, the probability increases that in a subsequent cycle he will attend to sources biased in favor of his new position and ignore those favoring the opposition.

Preliminary runs of the Abelson-Bernstein model with hypothetical data have produced plausible outputs. As the campaign progresses, attitudes tend to polarize (a different outcome from the consensus in the direction of stimulus bias noted in McPhee's model); interest in the issue slowly increases; and greater consistency tends to develop between the attitudes of a simulated citizen and those of his friends.

As is obvious from the general description of the model, the processing is exceedingly complex, incorporating some fifty interrelated propositions concerning social influence processes. This flagrant lack of parsimony has evoked criticism from Boudon (1965) and Rosenthal (1968), and Abelson and Bernstein themselves acknowledge that in simulation "perhaps Occam's Razor should be replaced by Occam's Lawnmower" (1963, p. 114). Nevertheless, the model possesses attractive features with its conceptual richness and its verisimilitude in programming a number of forces that complexly interact in producing a variety of outcomes.

Of course, validation of the model is fraught with difficulties, in part because of the interdependencies. For example, exposures to channels are processed as a direct function of interest, and interest varies directly with the number of exposures. Other problems might be anticipated in securing empirical data to set parameters and to validate portions of the model. That is, the very attempt to conduct a community survey during a school desegregation controversy or a fluoridation referendum might itself influence the course of the campaign, the attitudes and interest of the interviewed voters, and possibly even the outcome of the referendum. Abelson, however, indicates that validation investigations are underway (1968); further evaluation must await his report.

In much the same spirit, Bråten is developing a simulation of communication model (SIMCOM) to study the impact of mass communication and personal influence on the choice behavior of a target population (1968). One of the competing organizations represented in the program uses two mass media channels in its campaign; the other employs two agents who make personal contacts with potential clients. A complex system of social psychological mechanisms similar to those incorporated by Abelson and Bernstein determines the model's microlevel operating characteristics. For example, an individual's exposure probabilities (to a channel, agent, or another member of the population) depend on his evaluation state and his orientation to the source, and his attention to a message is a function of his evaluation and motivation states as well as the congruence between the message and his own preference. Once the parameter tuning phase is completed, Bråten plans to undertake such applications as the simulation of marketing communication, voting behavior, and diffusion of agricultural or health care information.

Simulations of Social Exchange Considerations in Decision-Making

Gullahorn and Gullahorn (1963, 1964, 1965a, 1970a, forthcoming) have developed a computer model of social man, HOMUNCULUS, based on the social exchange theory of George Homans (1961). The theory envisages humans as exchanging behavior as well as commodities. In face-to-face interaction, each person's rewards and costs tend to come from the behavior of others; hence, an individual's responses depend on the quantity and quality of the reward and punishment his actions elicit. To explain such elementary social behavior, Homans advances five propositions derived from experimental psychology and classical economics: consideration of recency and frequency of reinforcement, stimulus generalization and response generalization, value of reward, deprivation and satiation, and distributive justice (the norms requiring that a person's rewards should be proportional to his costs and his profits proportional to his investments).

Since Homans assumes that his explanatory propositions apply to all people, regardless of their cultural backgrounds or role relationships, the routines operationalizing the propositions are programmed to be constant for all simulated persons. Flexibility and individuality in the model stem from the unique list structures describing each simulated person (cf. figure 4-5).

Essentially, the HOMUNCULUS program simulates the decision-making behavior of each represented individual as he seeks to derive psychic profits from

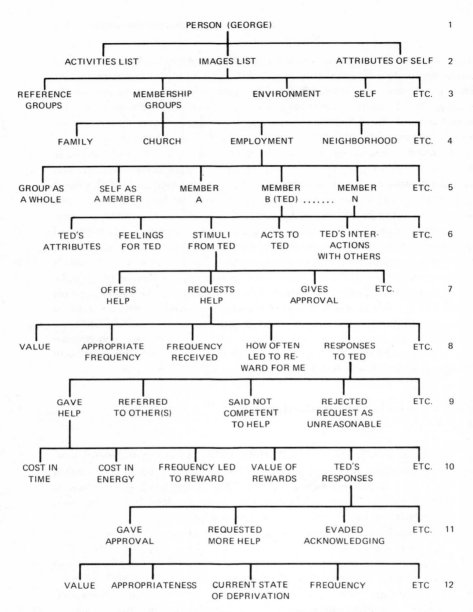

		PERSON (GEORGE)				1
ACTIVITIES LIST		IMAGES LIST		ATTRIBUTES OF SELF		2
REFERENCE GROUPS	MEMBERSHIP GROUPS		ENVIRONMENT	SELF	ETC.	3
FAMILY	CHURCH	EMPLOYMENT	NEIGHBORHOOD	ETC.		4
GROUP AS A WHOLE	SELF AS A MEMBER	MEMBER A	MEMBER B (TED)	MEMBER N	ETC.	5
TED'S ATTRIBUTES	FEELINGS FOR TED	STIMULI FROM TED	ACTS TO TED	TED'S INTER-ACTIONS WITH OTHERS	ETC.	6
	OFFERS HELP	REQUESTS HELP	GIVES APPROVAL	ETC.		7
VALUE	APPROPRIATE FREQUENCY	FREQUENCY RECEIVED	HOW OFTEN LED TO RE-WARD FOR ME	RESPONSES TO TED	ETC.	8
GAVE HELP	REFERRED TO OTHER(S)	SAID NOT COMPETENT TO HELP	REJECTED REQUEST AS UNREASONABLE	ETC.		9
COST IN TIME	COST IN ENERGY	FREQUENCY LED TO REWARD	VALUE OF REWARDS	TED'S RESPONSES	ETC.	10
	GAVE APPROVAL	REQUESTED MORE HELP	EVADED ACKNOWLEDGING	ETC.		11
VALUE	APPROPRIATENESS	CURRENT STATE OF DEPRIVATION	FREQUENCY	ETC		12

FIGURE 4-5   Memory structure of a person in HOMUNCULUS.

social interaction. Reference to figure 4-5 will help illustrate the programmed information retrieval and evaluation. The first versions of HOMUNCULUS involve a civil service office setting where a skilled worker, George, is initially asked for help by a less-competent colleague, Ted. In reacting to this input,

George first determines whether the stimulus is appropriate in the circumstances—that is, whether it violates social norms concerning the reciprocity of
rewards and costs. Checking his image of Ted (level 6 in figure 4-5), George may
find that Ted is less skilled than he and has generally not behaved inappro-
priately in the past. Since the request for assistance thus appears legitimate,
George next checks his activities list (level 2) to determine whether his responses
to such requests usually have been reinforced, and if so, whether they have been
rewarded by Ted (level 8). If such is the case, George reviews his own previous
responses to Ted's solicitations (level 9) and selects three that have been
rewarded (level 10). He then predicts Ted's expected response to each of the acts
under consideration (level 11), and for each of Ted's likely responses he determines (level 12) how frequently it has occurred, how valuable it has been as a
reward, and his current state of deprivation or satiation with respect to it.
Finally, George computes the expected psychic profit from each contemplated
response by subtracting costs (time and energy required, rewards foregone, etc.)
from the predicted reward value, and he rationally selects the behavior expected
to yield the greatest profit. George emits this optimal activity, which becomes
the stimulus for Ted to process, and the interaction cycles.

Additional processing and updating of list structures occur during the
decision-making sequence. Each stimulus and response modifies memory structures of the participants and affects later assessments of appropriateness.
Furthermore, if expectations regarding distributive justice are violated, certain
routines generate and process guilt or anger responses. Still other routines determine whether the respondent has stored an activity that he wishes to emit to the
stimulus agent or whether he has a response of free-floating aggression that is
awaiting release, and if so, whether it appears safe to emit it now.

Even with the limited social scenario represented in the early HOMUN-
CULUS runs, proliferating behavioral repertoires and memory trees of past
interactions threatened to overflow computer memory space. For this reason,
and to remove the situational dependency of the initial program, the activity
repertoire in HOMUNCULUS was modified to incorporate Bale's twelve-
category system (1950) describing interactions observed in small group research
(e.g., giving or *requesting* orientation, agreeing or disagreeing, expressing solidarity or antagonism).

With the modified version of HOMUNCULUS, simulation runs explored
the conditions under which triads remained a system of three members, devolved
into a dyad and an isolate, or segmented into three isolates who could not
interact in a mutually rewarding manner. In general, the computer-generated
outcomes depend on whether previous acquaintanceship provided individual
memory structures with evidence of rewarding contacts.

In the simulation, triads of previously acquainted, friendly members tend
to remain systems of three interacting positively. Though only two individuals
initiate conversation, the third receives a positive response when he interrupts,
and all tend to participate with some regularity. With unacquainted triads, on
the other hand, initial friendly exchanges between two participants lead to their

developing memory structures regarding the rewarding nature of their encounters, and each begins storing activities previously rewarded by the other. If
the third person does not interrupt early in the interaction, pathways of rewarding interaction with him do not develop, and the cumulative effect of the
profitable dyadic exchanges causes the three to devolve into an interacting pair
and an isolate. These simulation findings are compatible with Simmel's early
analyses (1950) and with small group research observation. The formalization of
Homans' reinforcement theory in HOMUNCULUS thus appears sufficient to
produce and explain coalition formation and other structural developments in a
triad.

Coleman (1962, 1964, 1965) reports similar simulation studies of triadic
interaction based on an even simpler formulation: (1) if two persons find interaction with each other rewarding they will tend to interact more frequently and
(2) time available for interaction is limited. His simulations typically produce a
dyad and an isolate, but another asymmetrical relationship also occasionally
develops in which A forms a strong bond with both B and C, but B and C are
weakly connected—suggesting the emergence of leadership.

While the early HOMUNCULUS simulations generated plausible output,
the total model proved intractable for more rigorous assessment; therefore,
empirical validation studies were undertaken with limited portions of the model
concerning the effects of value, cost, and profit on social behavior (Gullahorn
and Gullahorn, 1965b, 1970b). The abridged model simulates the decision-
making strategies of labor union members completing a questionnaire on role
conflict resolution. The computer program is modularized so that routines
representing each of three different hypothesized decision processes can be
incorporated in separate computer runs. Such modularization enables the *same*
programmed respondents (with identical list structures including personal
preference values and sensitivity values for pressuring reference groups) to enact
each decision process in resolving the role dilemmas presented in the programmed questionnaire. Since each run's output is thus exclusively generated by
the routines operationalizing one hypothesis, the relative explanatory power of
each strategy can be assessed by comparing the answers of simulated respondents
with those of the labor union members originally surveyed. In this situation, a
programmed decision strategy incorporating subjective estimates of the reward
values of consonance and integrity in optimizing social profit produced data that
did not significantly differ from the item responses of the original union
members.

While the role conflict simulations with the abridged model demonstrate
the utility of operationalizing different hypotheses so that their generated
output can be evaluated against human data, these efforts neglect the computer
model's capacity to simulate learning by modifying list structures on the basis of
experience. Therefore, further explorations with HOMUNCULUS (Gullahorn
and Gullahorn, forthcoming) again involve dynamic interaction among triads in
simulating Gerard's study of opinion anchorage in small groups (1954). In
Gerard's investigation, subjects first read a case about a union-management

dispute and predicted the outcome on a seven-point scale, also rating their degree of confidence in their predictions. Gerard then assigned them to three-man discussion groups according to their relative agreement, and he also manipulated cohesiveness through instructions to the subjects.

Though Gerard neglected considerations regarding subjects' confidence ratings in assigning them to experimental groups, social exchange theory suggests that a subject's confidence should vary directly with the extremity of his stated opinion and that resistance to opinion change should vary directly with confidence. Because Gerard's original data had been lost, this prediction could not be checked; however, since other small group studies supported the social exchange prediction, a series of sensitivity runs was undertaken with the simulation model, varying only the pattern of initial opinion distribution. These simulation runs reproduced Gerard's findings concerning opinion change toward agreement with another group member; hence the computer-generated data enhanced the credibility of an alternative hypothesis explaining Gerard's findings.

This application demonstrates a frequently observed sequence in simulation research. Designing the model required more detailed analysis of the original study and suggested alternative hypotheses concerning the findings. While simulation runs supported one rival hypothesis, sufficient data for further testing were not available from the original study (the simulator's chronic lament); new laboratory experiments were suggested to furnish needed data. Working back and forth between computer modeling and laboratory or field research thus appears to be a necessary strategy for adequately developing and testing sociocultural system simulations.

According to the theory represented in HOMUNCULUS, individuals are motivated to gain psychic profits through social behavior; therefore, they selectively reinforce each other according to the profits or losses resulting from social encounters. Through repeated interactions, simulated persons learn to predict the outcome of "contemplated" (tentatively selected) acts, and eventually those with similar values and need levels begin to respond relatively consistently in similar situations. When previously reinforced responses do not culminate in the expected rewards, these individuals "feel punished" (i.e., evaluate this contingency as inappropriate) and hence withhold reinforcement in turn. Even in the oversimplified interaction sequences developed in the early HOMUNCULUS simulations, something very much like norms seems to develop for the synthetic groups.

This finding is similar to Breton's report concerning the apparent output norm that emerged in his previously cited simulations of productivity in industrial groups. Taken together, these simulation developments help support Homans' contention (1961; 1964) that social system behavior can be explained in terms of psychological processes and that no principles beyond those necessary to explain individual behavior are required. The outputs of these programs were generated only by the decision-making routines implemented, and they involved no functionalist assertions regarding what norms would produce "organic solidarity."

Lest we appear too sanguine regarding these demonstrations, let us note several shortcomings. In the HOMUNCULUS studies, no groups larger than three have been simulated; thus extrapolation to more complex systems is unrealistic. In addition, the time periods represented in the simulation runs have been insufficient for complex role definitions to emerge and to be subjected to conflicting pressures over time. For both of these reasons, definitions of positions within a social system have not been abstracted independently from the expectations associated with specific role incumbents. Extension of HOMUNCULUS would thus increase its value in exploring social system theory.

Models like HOMUNCULUS provide a great deal of information about a theory, particularly about its logical structure, and they generate output that mirrors the hypothetical social system postulated by the theory. But because of their complexity and because generally accepted conventions and methods for their validation have not yet been developed, the truth value of such models is almost impossible to assess satisfactorily. (See discussions of Carroll and of Abelson and Bernstein in this chapter and other treatments of validation problems in Feldman, 1962; Reitman, 1965; Naylor *et al.*, 1966; Abelson, 1968; and Gullahorn and Gullahorn, forthcoming. An insightful comparison of the relative importance of a model's truth value versus its information value appears in Hanna, 1970.)

CONCLUSION

From the variety of topics, strategies, and levels of complexity incorporated in the simulations reviewed in this chapter, it appears that computer modeling—like verbal theorizing about sociocultural systems—is very much a do-your-own-thing type of enterprise. Innovations in verbal theories, however, offer exciting potentialities for simulation models.

As noted in the introduction, the taxonomic bias in much social and cultural system theory has led to a cataloging of analogical representations of social structure. Social psychology has also tended toward static classifications (e.g., of authoritarian personalities), possibly because in its early stages the field consisted mainly of reasonably sophisticated methods in search of a theory. Now, however, significant developments in both disciplines are effecting a dynamic convergence. The introduction of social exchange theory has reoriented sociology to real analysis of social processes by returning to social psychological basics. Fortunately, the evolution of distinctive middle-range social-psychological theories provides complementary propositions concerning social comparison processes as well as cognitive and behavioral consistency. By denoting the individual as the elemental acting unit in any social system and by specifying sequences of independent and dependent variables, the emerging social psychological exchange theory is particularly amenable to computer simulation in which dynamic changes in the interacting units can generate observable effects in the social systems in which they are embedded.

The potential contribution of further computer exploration of social exchange propositions is suggested by Coleman's reflections:

> The failure of social theories in the past arises from . . . mismatching between the human mind's abilities and the needs of the theory. The mind is well suited to development of a model or theory of purposive action at the level of the individual actor. But to construct from this theories of social organization or a social system is not an activity to which the human mind is well-suited. As a consequence, such action-theories have consisted in the past of an heroic leap from the individual level to the level of the system, with a vast empty gap in between, because the complexity outruns our mental abilities (1965, p. 105).

Whether simulation can really fulfill this challenge in illuminating how micro-behavioral processes effect social changes remains to be seen.

Perhaps some final reflections and caveats are in order regarding the state of the art (or science?) of social system simulation. Some of the models reviewed in this chapter are just moving from the flow chart design stage to programmed implementations. Although others are actually running, their parameters are still being tuned and their routines debugged for unanticipated branch processing. In some instances lack of ready access to computer facilities has deterred progress. Computer time limitations, for example, have precluded completion of the numerous runs necessary to provide data for estimating the central tendency and variance of the adopter distributions generated in early diffusion simulations. Aside from hardware considerations, many modelers have learned by bitter experience that programming theoretical processes should be a do-it-yourself operation, lest a zealous mathematical programmer implement elegant functions that inadvertently violate one's basic assumptions.

Problems involving machine availability and programming can thus prolong what is already a very slow process of translating a verbal model's qualitative description of entities, variables, and relationships into a format amenable to computer representation and operation. Since very few models have yet been sufficiently developed and tested to make the recommended experimentation with modules representing alternative theoretical hypotheses feasible, rigorous assessment has not been undertaken in this chapter; rather we conclude that evaluation of computer simulations of social system processes must "wait and see." What seems apparent from experience with such models thus far is that though the creative challenge of simulation is a powerful incentive, tremendous concentration and effort are necessary, and the expected rewards require considerable tolerance for delayed reinforcement. The endeavor, therefore, might prove very costly for an academician whose anxious dean wishes to count publications the way an Indian scalper once counted *coups.*

As noted previously, computer models involve difficult validation problems, many of which have yet to be encountered by the sociocultural system simulations reviewed here. Furthermore, since social system theory can be so complex, simulators tend to be fair game to critics who eagerly catalog the omissions and oversimplifications in the models and at the same time bemoan the lack of parsimony.

With reference to the parsimony issue, perhaps it is best to follow a strategy based on George Homans' advice: "Cut down as far as you dare the number of things you are talking about. 'As few as you may; as many as you must,' is the rule governing the number of classes of fact you take into account" (1950, pp. 16-17). On the other hand, computer simulation offers a tempting opportunity for the social cosmologist to represent and experiment with the complexities of his theoretical scheme in a way not possible in field or laboratory research. Indeed, the present authors plead guilty to succumbing to their latent Rube Goldberg tendencies in their modeling efforts. Thus, depending on the predispositions and discipline of future simulators, we might expect a continuum of complexities in emergent sociocultural system simulations.

REFERENCES

Abelson, R. P., "Simulation of Social Behavior," in G. Lindzey and E. Aronson (eds.), *The Handbook of Social Psychology*, II, pp. 274-356. Reading, Mass.: Addison-Wesley Publishers, 1968.

Abelson, R. P., and A. Bernstein, "A Computer Simulation Model of Community Referendum Controversies," *Public Opinion Quarterly*, 27 (Spring 1963), 93-122.

Abt, C., and M. Gordon, "Report on Project TEMPER." Paper presented at the Institute on Computers and the Policy Making Community, University of California, Lawrence Radiation Laboratory, April 1966.

Alker, H. R., Jr., "Computer Simulations: Bad Mathematics but Good Social Science?" Paper presented at the Australian UNESCO Seminar on Mathematics in the Social Sciences, May 1968.

Alker, H. R., Jr., and R. D. Brunner, "Simulating International Conflict: A Comparison of Three Approaches," *International Studies Quarterly*, 13 (Spring 1969), 70-110.

Bales, R. F., *Interaction Process Analysis.* Cambridge, Mass.: Addison-Wesley Press, 1950.

Beshers, J. M., "Birth Projections with Cohort Models," *Demography*, 2 (1965b), 593-99.

———, "Computer Models of Social Processes: The Case of Migration." Paper presented at the meetings of the Population Association of America, April 1967a.

———, "The Social Theorist: On Line Theory Construction and Concept Validation." Paper presented at the meetings of the American Sociological Association, August 1967b.

———, *Computer Methods in the Analysis of Large-Scale Social Systems.* Cambridge, Mass.: Joint Center for Urban Studies, 1965a.

Boguslaw, R., and R. H. Davis, "Social Process Modeling: A Comparison of a Live and Computerized Simulation," *Behavioral Science*, 14 (May 1969), 197-203.

Boguslaw, R., R. H. Davis, and E. B. Glick, "A Simulation Vehicle for Studying National Policy Formation in a Less Armed World." *Behavioral Science,* 11 (January 1966), 43-61.

Boudon, R., "Réflexions sur la Logique des Modèles Simulés," *European Journal of Sociology,* 6 (May 1965), 3-20.

Bråten, S., *Progress Report on the SIMCON Model.* Solna, Sweden: Institute of Market and Societal Communication, September 1968.

Breton, R., "Output Standards and Productive Behavior in Industrial Work Groups." Paper presented at the annual meeting of the American Sociological Association, Chicago, 1965, mimeographed.

Brightman, H. J., E. E. Kaczka, and B. H. Shane, "A Simulation Model of Individual Behavior in a Work Group," University of Massachusetts, Center for Business and Economic Research, Working Paper No. 69-821-1, 1969, mimeographed.

Browning, R. P., "Hypotheses about Political Recruitment: A Partially Data-Based Computer Simulation," in W. Coplin (ed.), *Simulation in the Study of Politics,* pp. 303-25. Chicago: Markham Publishing Company, 1968.

———, "Quality of Collective Decisions: Some Theory and Computer Simulations," Michigan State University, Department of Political Science, June 1969, mimeographed.

Buckley, W., *Sociology and Modern Systems Theory.* Englewood Cliffs, N.J.: Prentice-Hall, Inc., 1967.

Carroll, T. W., "Computer Simulation of the Diffusion of Dairy Innovations in a Rural Community of Brazil." Ph.D. dissertation, Massachusetts Institute of Technology, Department of Political Science, June 1969.

Clarke, D. L., *Analytical Archeology.* London: Methuen & Co., Ltd., 1968.

Clemens, W. C., Jr., "A Propositional Analysis of the International Relations Theory in TEMPER—A Computer Simulation of Cold War Conflict," in W. Coplin (ed.), *Simulation in the Study of Politics,* pp. 59-101. Chicago: Markham Publishing Company, 1968.

Coale, A. J., and E. M. Hoover, *Population Growth and Economic Development in Low-Income Countries.* Princeton, N.J.: Princeton University Press, 1958.

Coleman, J., "Analysis of Social Structures and Simulation of Social Processes with Electronic Computers," in H. Guetzkow (ed.), *Simulation in Social Science: Readings,* pp. 61-69. Englewood Cliffs, N.J.: Prentice-Hall, Inc., 1962.

———, "In Defense of Games," *American Behavioral Scientist,* 10 (October 1966), 3-4.

———, *Introduction to Mathematical Sociology.* New York: The Free Press, 1964.

———, "Simulation and Games," Johns Hopkins University, Department of Social Relations, 1969, xeroxed.

———, "The Use of Electronic Computers in the Study of Social Organization," *European Journal of Sociology,* 6 (May 1965), 89-107.

Coleman, J., E. Katz, and H. Menzel, *Medical Innovation.* Indianapolis: The Bobbs-Merrill Co., Inc., 1966.

———, "The Diffusion of an Innovation," *Sociometry,* 20 (1957), 253-70.

Coombs, S., M. Fried, and S. Robinovitz, "An Approach to Election Simulation through Modular Systems," in W. Coplin (ed.), *Simulation in the Study of Politics,* pp. 286-99. Chicago: Markham Publishing Company, 1968.

Coult, A. D., and R. R. Randolph, "Computer Methods for Analyzing Genealogical Space," *American Anthropologist,* 67 (1965), 21-29.

Davis, R. H., "Arms Control Simulation: The Search for an Acceptable Method," *The Journal of Conflict Resolution,* 7 (September 1968), 590-602; and *The Journal of Arms Control,* 1 (October 1963), 684-96.

Deutschmann, P. J., I: "A Machine Simulation of Attitude Change in a Polarized Community," II: "A Machine Simulation of Information Diffusion in a Small Community," III: "A Model for Machine Simulation of Information and Attitude Flow." San Jose, Costa Rica: Programa Interamericano de Informacion Popular, 1962. (Three mimeographed papers.)

Feldman, J., "Computer Simulation of Cognitive Processes," in H. Borko (ed.), *Computer Applications in the Behavioral Sciences,* pp. 336-59. Englewood Cliffs, N.J.: Prentice-Hall, Inc., 1962.

Findler, N. V., and W. R. McKinzie, "On a Computer Program That Generates and Queries Kinship Structures," *Behavioral Science,* 14 (July 1969), 334-40.

Gamson, W. A., *SIMSOC: Simulated Society.* New York: The Free Press, 1969.

Gardin, J. C., "The Reconstruction of an Economic Network of the Second Millenium," in D. Hymes (ed.), *The Use of Computers in Anthropology,* pp. 378-91. The Hague, Netherlands: Mouton, 1965.

Gerard, H., "The Anchorage of Opinions in Face-to-Face Groups," *Human Relations,* 7 (1954),313-25.

Gilbert, J. P., and E. A. Hammel, "Computer Simulation and Analysis of Problems in Kinship and Social Structure," *American Anthropologist,* 68 (February 1966), 71-93.

———, "Computer Simulation of Problems in Kinship and Social Structure," in D. Hymes (ed.), *The Use of Computers in Anthropology,* pp. 513-14. The Hague, Netherlands: Mouton, 1965.

Gullahorn, J. T., and J. E. Gullahorn, "A Computer Model of Elementary Social Behavior," in E. Feigenbaum and J. Feldman (eds.), *Computers and Thought,* pp. 375-86. New York: McGraw-Hill Book Company, 1963. Also printed in *Behavioral Science,* 8 (1963), 354-62.

———, "A Non-Random Walk in the Odyssey of a Computer Model," in M. Inbar and C. S. Stoll, *Social Science Simulations.* New York: The Free Press, 1970a.

———, " Computer Simulation of Human Interaction in Small Groups," American Federation of Information Processing Societies Conference Proceedings, 1964, Spring Joint Computer Conference. Baltimore, Md.: Spartan Books, 1964, pp. 103-13. Also reprinted in *Simulation,* 4 (1965), 50-61.

———, "Computer Simulation of Role Conflict Resolutions," in W. A. Starbuck and J. M. Dutton (eds.), *Computer Simulation in Human Behavior.* New York: John Wiley Sons, Inc., 1970b.

———, *Explorations in Computer Simulation.* New York: The Free Press, forthcoming.

———, "Some Computer Applications in Social Science," *American Sociological Review,* 30 (1965b), 353-65.

———, "The Computer as a Tool for Theory Development," in D. Hymes (ed.), *The Use of Computers in Anthropology,* 427-48. The Hague, Netherlands: Mouton, 1965a.

Hägerstrand, T., "A Monte Carlo Approach to Diffusion," *European Journal of Sociology,* 6 (May 1965b), 43-67.

———, *Innovationsforloppet ur Koroloqisk Synpunkt.* Lund, Sweden: Gleerup, 1953. Trans. by Alan Pred as *Innovation Diffusion as a Spatial Process.* Chicago: University of Chicago Press, 1968.

———, "Quantitative Techniques for Analysis of the Spread of Information and Technology," in C. A. Anderson and M. J. Bowman (eds.), *Education and Development.* Chicago: Aldine, 1965a.

Hanna, J., "Information-Theoretic Techniques for Evaluating Simulation Models," in W. A. Starbuck and J. M. Dutton (eds.), *Computer Simulation in Human Behavior.* New York: John Wiley & Sons, Inc., 1970.

Hanneman, G. J., "A Computer Simulation of Information Diffusion in a Peasant Community." Master's thesis, Michigan State University, Department of Communication, 1969.

Hanneman, G. J., T. Carroll, E. Rogers, J. D. Stanfield, and N. Lin, "Computer Simulation of Innovation Diffusion in a Peasant Village," *American Behavioral Scientist,* 12 (July-August, 1969), 36-45.

Hanson, R. C., W. N. McPhee, R. Potter, O. Simmons, and J. Wanderer, "A Simulation Model of Urbanization Processes." Paper presented at the American Sociological Association meeting, San Francisco, August 1967.

Hanson, R. C., and O. Simmons, "The Role Path: A Concept and Procedure for Studying Migration to Urban Communities," *Human Organization,* 27 (Summer 1968), 152-58.

Harper, D. H., "Simulating a Survey." University of Rochester, Department of Sociology, 1969, mimeographed.

———, "The Computer Simulation of a Sociological Survey," in *Calcul et Formulisation dans les Sciences de l'Homme.* Paris, France: Editions du Centre National de la Recherche Scientifique, 1968.

Hartman, J. J., and J. Walsh, "Simulation of Newspaper Readership: An Exploration in Computer Analysis of Social Data," *Social Science Quarterly,* 49 (March 1969), 840-52.

Henderson, L. J., *Pareto's General Sociology.* Cambridge, Mass.: Harvard University Press, 1935.

Hill, J. N., "A Prehistoric Community in Eastern Arizona," *Southwestern Journal of Anthropology,* 22 (Winter 1966), 9-30.

Homans, G. C., "Bringing Men Back In," *American Sociological Review,* 29 (December 1964), 809-18.

———, *Social Behavior: Its Elementary Forms.* New York: Harcourt, Brace & World, Inc., 1961.

―――, *The Human Group.* New York: Harcourt & Brace, 1950.

Homans, G. C., and D. M. Schneider, *Marriage, Authority, and Final Causes.* New York: The Free Press, 1955.

Hubbell, C., "An Input-Output Approach to Clique Identification," *Sociometry,* 28 (December 1965), 377-99.

Hymes, D. (ed.), *The Use of Computers in Anthropology.* The Hague, Netherlands: Mouton, 1965.

Karlsson, G., *Social Mechanisms: Studies in Sociological Theory.* Uppsala, Sweden: Almquist and Wilksells; and New York: The Free Press, 1958.

Kasdan, L., and C. W. Smith, "The Ethnology of a Non-Existent Society." Paper presented at the Ohio Valley Sociological Society meetings, Spring 1968.

Katz, E., "Communication Research and the Image of Society: Convergence of Two Traditions," *American Journal of Sociology,* 65 (March 1959), 435-40.

Katz, E., and P. F. Lazarsfeld, *Personal Influence.* Glencoe, Ill.: The Free Press, 1955.

Klein, S., M. Kuppin, and K. Meives, "Monte Carlo Simulation of Language Change in Tikopia and Maori." University of Wisconsin, Computer Sciences Department Technical Report, No. 62, June 1969.

Kramer, J. F., "A Computer Simulation of Audience Exposure in a Mass Media System: The United Nations Information Campaign in Cincinnati, 1947-1948." Ph.D. dissertation, Massachusetts Institute of Technology, Department of Political Science, 1969.

Kunstadter, P., "Computer Simulation of Preferential Marriage Systems," in D. Hymes (ed.), *The Use of Computers in Anthropology,* pp. 520-21. The Hague, Netherlands: Mouton, 1965.

Kunstadter, P., R. Buhler, F. Stephan, and C. Westoff, "Demographic Variability and Preferential Marriage Patterns," *American Journal of Physical Anthropology,* 21 (December 1963), 511-19.

Lazarsfeld, P. F., B. Berelson, and H. Gaudet, *The People's Choice* (2nd ed.). New York: Columbia University Press, 1948.

Lazer, S. C., *Toward a Formalization of Demographic Transition Theory.* Master's thesis, Michigan State University, Department of Sociology, 1969.

Levin, M. L., "Simulation of Social Processes," *Public Opinion Quarterly,* 26 (Fall 1962), 483-84.

Levi-Strauss, C., *Les Structures Élémentaires de la Parenté.* Paris: Presses Universitaires de France, 1949.

Linton, R., *The Study of Man.* New York: Appleton-Century-Crofts, 1936.

Marble, D. F., and J. D. Nystuen, "An Approach to the Direct Measurement of Community Mean Information Fields," *Papers, Regional Science Association,* 11 (1963), 99-109.

McPhee, W. N., *Formal Theories of Mass Behavior.* New York: The Free Press, 1963.

―――, "Note on a Campaign Simulator," *Public Opinion Quarterly,* 25 (July 1961), 184-93.

McPhee, W. N., and R. B. Smith, "A Model for Analyzing Voting Systems," in

W. N. McPhee and W. A. Glaser (eds.), *Public Opinion and Congressional Elections*. New York: The Free Press, 1962.

Michelena, J. A. S., "VENUTOPIA 1: An Experimental Model of a National Polity," in F. Bonilla and J. A. S. Michelena (eds.), *A Strategy for Research on Social Policy*, I, *The Politics of Change in Venezuela*, pp. 333-67. Cambridge, Mass.: M.I.T. Press, 1967.

Naylor, T., J. Balintfy , D. Burdick, and K. Chu, *Computer Simulation Techniques*. New York: John Wiley & Sons, Inc., 1966.

Neisser, U., "The Imitation of Man by Machine," *Science*, 139 (1963), 193.

Orcutt, G. H., M. Greenberger, J. Korbel, and A. Rivlin, *Microanalysis of Socioeconomic Systems: A Simulation Study*. New York: Harper & Row, Publishers, 1961.

Parsons, T., "Evolutionary Universals in Society," *American Sociological Review*, 29 (June 1964), 339-57.

———, *The Social System*. Glencoe, Ill.: The Free Press, 1951.

Pitts, F. R., "Problems in Computer Simulation of Diffusion," *Papers, Regional Science Association*, 11 (1963), 111-19.

Pool, I. de S., "Computer Simulations of Total Societies," in S. Klausner (ed.), *The Study of Total Societies*, pp. 45-65. New York: Doubleday Anchor Books, 1967.

Pool, I. de S., and A. Kessler, "The Kaiser, The Tsar, and the Computer: Information Processing in a Crisis," *American Behavioral Scientist*, 8 (May 1965), 31-38.

Pool, I. de S., R. Abelson, and S. Popkin, *Candidates, Issues, and Strategies: A Computer Simulation of the 1960 and 1964 Elections*. Cambridge, Mass.: M.I.T. Press, 1965.

Popkin, S. L., "A Model of a Communication System," *American Behavioral Scientist*, 8 (May 1965), 8-11.

Radcliffe-Brown, A. R., *Structure and Function in Primitive Society*. New York: The Free Press, 1952.

Rainio, K., *A Stochastic Model of Social Interaction*. Copenhagen, Denmark: Munksgaard, 1961.

———, "Social Interaction as a Stochastic Learning Process," *European Journal of Sociology*, 6 (May 1965), 68-88.

Randolph, R., and A. Coult, "A Computer Analysis of Bedouin Marriage," *Southwestern Journal of Anthropology*, 24 (Spring 1968), 83-99.

Reitman, W., *Cognition and Thought*. New York: John Wiley & Sons, Inc., 1965.

Ridley, J. C., and M. C. Sheps, "An Analytic Simulation Model of Human Reproduction with Demographic and Biological Components," *Population Studies*, 19 (March 1966), 297-310.

Rogers, E. M. (with L. Svenning), *Modernization among Peasants*. New York: Holt, Rinehart & Winston, Inc., 1969.

Rome, B., and S. Rome, "Communication and Large Organization." Santa Monica, Calif.: System Development Corporation, SP 1690/000/000, September 1964.

———, "Leviathan: An Experimental Study of Large Organizations with the Aid of Computers," in R. Bowers (ed.), *Studies on Behavior in Organizations,* pp. 257-311. Athens, Ga.: University of Georgia Press, 1966.

Rosenthal, H., "Voting and Coalition Models in Election Simulations," in W. Coplin (ed.), *Simulation in the Study of Politics,* pp. 237-85. Chicago: Markham Publishing Company, 1968.

Rossi, P. H., *Why Families Move.* Glencoe, Ill.: The Free Press, 1955.

Selvin, H. C., *The Effects of Leadership.* Glencoe, Ill.: The Free Press, 1960.

Simmel, G., *The Sociology of Georg Simmel.* K. H. Wolff (trans., ed.). Glencoe, Ill.: The Free Press of Glencoe, 1950.

Smith, R. B., "Simulation Models for Accounting Schemes," *American Behavioral Scientist,* 12 (July-August, 1969), 21-30.

Sorokin, P., *Contemporary Sociological Theories.* New York: Harper & Row, Publishers, 1928.

Spencer, H., *Principles of Sociology* (3rd ed.). New York: Appleton-Century-Crofts, 1897.

Thompson, W. S., "Population," *American Journal of Sociology,* 34 (1929), 959-75.

Tiedemann, C. E., and C. S. Van Doren, "The Diffusion of Hybrid Seed Corn in Iowa: A Spatial Simulation Model." Michigan State University, Institute for Community Development and Services, Technical Bulletin B-44, 1964.

White, H. C., *An Anatomy of Kinship.* Englewood Cliffs, N.J.: Prentice-Hall, Inc., 1963.

Wolpert, J., "A Rational Simulation Model of Information Diffusion," *Public Opinion Quarterly,* 30 (Winter 1966-67), 597-608.

Wolpert, J., and D. Zillman, "The Sequential Expansion of a Decision Model in Spatial Context." University of Pennsylvania, Department of Geography, 1969, mimeographed.

A THEORY OF
INFORMAL SOCIAL INFLUENCE

WILLIAM N. MCPHEE
JACK FERGUSON
ROBERT B. SMITH

INTRODUCTION

The purpose of this paper is to describe a social influence process that is not vulnerable to the objections raised by psychologists and political scientists to what they call excessive "sociologizing" in the explanation of choices.[1]

The choices here happen to be votes—although the process to be described is not limited to this application. What is meant by "social influence" is the observation that people in close contact for long periods, such as husbands and wives or parents and children, are found to have remarkably nonindependent preferences. At any given time, their choices are highly correlated. What is meant by objectionable "sociologizing" of this phenomenon, however, are interpretations of how these correlations come about over time. It is too easy to make the choice *itself* a function of deliberate social manipulation by others or conscious adaptation to others.

No one really advocates this kind of "social determinism," however. Nearly everyone admits the necessity, and most of us the primacy, of two other determinants of the immediate choice: namely, (1) external political stimuli such as current events and (2) internal psychological dispositions such as political party loyalties learned in the past. But if we admit the primacy of these brute facts, then how can responses to them nevertheless be so correlated with the subtle cues from others in the contiguous social environment?

This paper describes a social influence process that has the desired

Reprinted with permission of The Macmillan Company from *Formal Theories of Mass Behavior,* by William N. McPhee. Copyright © 1963 by The Free Press of Glencoe, a division of The Macmillan Company. The initial work on this paper drew on funds from a grant to the first author by the Behavioral Sciences Division of the Ford Foundation. It was revised and all runs replicated with funds from a grant by the Social Science Division of the National Science Foundation, NSG G No. 13045. Machine time on an IBM 650 was freely supplied by Watson Scientific Computing Laboratory at Columbia University.

[1]For example, V. O. Key, Jr., and Frank Munger, "Social Determinism and Electoral Decision: The Case of Indiana," in *American Voting Behavior,* Eugene Burdick and Arthur J. Brodbeck, eds. (New York: The Free Press of Glencoe, 1959), pp. 281-307.

subtlety. It leaves the determination of the choice at all times to external (that is, political) stimuli and individual (that is, psychological) dispositions. Working through these, however, such a process is capable of generating any of the social correlations found in choice data.

A THREE-PROCESS MODEL OF VOTING

The influence process to be described is one of three processes that make up a model for the study of voting behavior. It is no accident that the other two processes concern the primary determinants of voting referred to above: the response to external stimuli and the learning of internal dispositions. The complete model has been described in detail elsewhere.[2] It was designed and is being used for large problems of national scope and historical time spans.[3] Therefore, the processes by which a single individual forms a single vote in one election are necessarily too simple to be interesting as psychological theories. Here they will be briefly explained—and with a certain amount of literary license—by the flow diagram of Figure 1.

[2]William N. McPhee and Robert B. Smith, "A Model for Analyzing Voting Systems," in *Public Opinion and Congressional Elections,* William N. McPhee and William A. Glaser, eds. (New York: The Free Press of Glencoe, 1962).

[3]For example, William N. McPhee and Jack Ferguson, "Political Immunization," in McPhee and Glaser, *op. cit.*

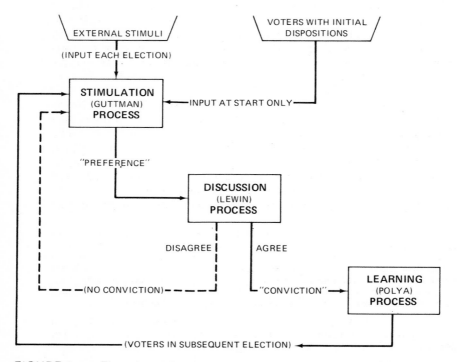

FIGURE 1   Flow chart of the model.

This model is operated in a computer. We shall assume that a large sample of "voters," representing a community in some political period, has already been put into the machine storage or memory of the computer. Each voter is represented by a limited series of characteristics, many of which will be modified by events during each election period. Let us assume that we are about to start an "election," as one of a sequence of them.

The model first takes from the input hopper an injection of "political stimuli" for the period since the last election. These are distributions of stimuli graded as to their attractiveness or cogency from strong to weak. The distributions differ for different groups in the sample electorate, determined by needs of the problem not relevant here.

The *stimulation* process is shown in the upper left-hand block of Figure 1. It assigns a particular stimulus to each individual voter by a Monte Carlo method. This method preserves the distributions of strong and weak stimuli from the party to the voter *group* affected. But within that restriction, for any individual voter the assignment is at random. This has an important consequence for present purposes. If for any reason this voter repeats the stimulation process, he may *not* get the same impression (level of stimulus). This difference is analogous to a voter's getting a different impression of a party on reconsideration in October from his first casual impression in May. With the exception of this stochastic flavor—the justification for which can be found in the work of Wiggins among others[4]—the stimulation process follows a rule most of us take for granted as a truism.

A. The stronger the individual's internal disposition towards a party, the *weaker* the external stimulus required to elicit a "yes" choice for that party.
B. Conversely, the weaker the internal disposition of an individual towards a party, the *stronger* the external stimulus required to elicit a "yes" choice for that party.[5]

Recognition that this is how people respond to questionnaire stimuli, if those are all on the same topic and differ only in cogency or attractiveness, has been made explicit by, among others, Louis Guttman's scaling procedures.[6] Therefore, for a label, call the stimulation process a "Guttman process." Its output is an initial preference for one of the two choices (parties), although under certain circumstances voters develop no preference for either party and become incipient nonvoters. This initial preference may be likened to a first impression of a situation that may be changed by later reflection.

[4] Lee M. Wiggins, "Mathematical Models for the Interpretation of Attitude and Behavior Change: The Analysis of Multi-Wave Panels," 1955, unpublished Ph.D. dissertation, Columbia University.

[5] See "A Campaign Simulator," *Formal Theories of Mass Behavior*, and Part III of "Attitude Consistency," in McPhee and Glaser, *op. cit.*, for a more formal treatment of the stimulation process.

[6] Louis Guttman, "The Basis for Scalogram Analysis," in *Measurement and Prediction*, Vol. IV of the series "Studies in Social Psychology in World War II," Samuel A. Stouffer, ed., Vol. IV (Princeton, N.J.: Princeton University Press, 1950). In psychology, like ideas date back to Thurstone.

Next, note the *discussion* process shown at the center of Figure 1. The initial preferences above are its input, and this process produces as its output surviving preferences, that is to say, preferences that survive discussion and the reflection it may prompt. A surviving preference can be interpreted as the final "conviction" about how to vote at election time; but here we will be less concerned with the vote than with how anything that helped reach that conviction also affects *learning* (modification of dispositions) and thus the individual's future behavior. Hence, returning later to discuss social influence—our main topic—let us first see how the learning process operates.

As suggested by its location at the lower right of Figure 1, the learning process takes as input the convictions surviving the discussion process and its output is some modification of the disposition that will affect choices in the subsequent elections. To give the name "learning" to this is perhaps a euphemism, although the mathematical models of learning in psychology are equally simple.[7] In its most elementary form, the gist of the present idea is identical to certain "urn models" worked out in mathematics to fit the growth of disease and similar self-developing or progressively intensifying phenomena. The rule is, in oversimplification adequate for present purposes, that

A. The more often one chooses a party now, the more likely he is to prefer the *same* party in the future.
B. Which is implemented by making the probability of preferring a party the same as the *frequency* of (weighted) preferences for it, each new one thus adding to the probability.

While this learning process was designed without knowledge of the above urn models of progressive phenomena, one by G. Polya most resembles the present case, which we therefore call a "Polya process."

The disposition that is modified is a general probability of choosing each of the alternatives. A high probability of choosing one political party *and* a low one of choosing the other, for example, would correspond to what we call "party loyalty." In any event, these modified dispositions then form the input, with new stimuli, for the stimulation process in the next cycle, for example, the next election.

THE INFLUENCE PROCESS

Let us now retrace steps to cover in more detail the process in the middle of Figure 1, whose inputs, to repeat, are initial impressions gained from the stimulation process and whose outputs are surviving convictions affecting learning.

The latter convictions may be reached by either of two routes. One we shall call "social-reality testing." This idea is associated with the name Kurt Lewin, and so we call (this aspect of) the discussion routine a "Lewin process."

---

[7]The present learning process has a resemblance to the models of William K. Estes, that is, the cumulative learning of sampled elements from a "stimulus" that is actually a distribution of such elements. Estes models can be shown, in turn, to be formally equivalent to that of Bush and Mosteller.

Lewin's idea was simple. Suppose that we want to test the truth of a belief. One can often do this by objective means. If he wants to find out whether glass will break if hit with sufficient force by an object, then he can actually try to break some. This will be a test against the criterion of "objective reality." But, suppose one gets the idea that, say, Governor Rockefeller would make a good President. There is no easy way to determine the truth of this proposition; and few of us, as voters, have the means to test it.

Consciously or not, many of us adopt an alternative kind of criterion: we *ask* somebody, preferably somebody from New York. What does he think about Governor Rockefeller? Or we notice how others react to the Presidential candidates and who people around us think is best. If others agree with our own impression, we retain it. And so this initial impression will tend to survive—our own judgment confirmed by others—and grow into a conviction. Lewin called this kind of cross-checking with others the testing of a belief against the criterion of "social reality." In ambiguous situations, where the cost of objective testing is prohibitive, everyone would agree that "social reality" is the only test criterion.

This cross-checking one's own impressions against the social reality of others' impressions is accomplished in the present model as follows. People in the sample (computer) population are tied together in sociometric nets representing intimate contact, for example, family members and close friends. If two such people have sufficient interest in politics (their joint interest adds to more than an indifference point), they "discuss" the topic, and their initial preferences now become visible to one another. Then the following happens:

If a voter's own initial impression *agrees* with that of his intimate friend or parent during the discussion process, then doubts are set at rest. The initial reaction is "confirmed" as an acceptable choice. The significance is that this will probably be the choice learned, that is, the one that modifies future dispositions towards a party. And this is as it should be in any learning process. For, from the standpoint of the person, this confirmed choice was the product of (1) his own reaction to objective stimuli and (2) the one socially rewarded as well.

The case of *disagreement* is handled in a way that provides a complement to Lewin's notion. One must always keep in mind that political stimuli are never wholly ambiguous. They have a rational reality that is often compelling despite contrary social advice (which we know is often wrong). When other persons disagree with our impressions of reality, we are just as likely to suspect that the others' opinion is wrong as we are to suspect that we ourselves interpreted reality incorrectly. If both are suspect then, it means that social reality has become ambiguous. In that event, who or what is the arbitrator?

Obviously, it is objective reality again. Just as social reality is the arbitrator when objective reality is ambiguous, so objective reality is the arbitrator when social reality is ambiguous.

In the model, this last effect is accomplished by renewed exposure (after such social disagreement) to the external stimuli again. In the model's version now running, both disagreers are simply sent back through the stimulation process again. Each voter again forms impressions by sampling the distributions of external stimuli, as these interact with his internal political dispositions exactly as before. A new preference comes out of this second exposure. The rule for deciding between the former and the new preference is:

A. If the new preference is the same as his former one, then the voter has "confirmed himself" and will retain his *old* preference despite contrary advice.

B. If his preference is different and now accords with the impression of his social intimate, he will adopt his new preference (because his *own* recheck against reality confirms the advice he received).

It happens that the preference retained is logically identical, in either event, to the one resulting from the voter's own impression on the second trip through the stimulation process.

Complications arise in how to handle repetitive occurrences of this process in the same election and other problems. One can therefore expect that many variants of the process will emerge in future simulation programs, especially in application to different fields, of which we discuss one in a Technical Note. But in any such variant there is one central idea—that the voter makes up his *own* mind and social influences merely assist him to determine which of his samplings or impressions of reality is stable in the face of cross checks.

SOCIALIZATION EFFECT

Let us turn now to some results illustrating how this influence process behaves. Because the model is a long-term one, designed especially to deal with political generations, a primary function we demand of any social influence process is to reproduce the facts of political *socialization.* That is, it should be able to make loyal Democrats and partisan Republicans out of youths who at first have unpredictable tendencies. Figure 2 gives an illustration of how the combination of discussion and learning processes work together to accomplish this.

The data of Figure 2 were generated with the following experimental (hypothetical) input:

1. Half of the machine population represented very young people, for example, ten to fifteen years old. "Very young" has a technical meaning here equivalent to little past experience in politics (which is therefore easily modified by new experience).[8]

2. This starting experience was such that these youths would, if left to themselves, have about *equal* probabilities of voting for both parties, as well as some probabilities of not voting.[9]

[8]The learning process above and as described in detail in McPhee and Smith, *op. cit.,* computes the probability that serves as disposition to vote for a party as a frequency roughly like, omitting weights, this:

$$\text{Pr} \left\{ \text{Dem.} \right\} = \frac{\text{No. of Democratic choices}}{\text{No. of all choice opportunities}}$$

The absolute magnitudes involved, that is, of the denominator, make the process unstable if they are small, stable if they are large. In the particular experiment, whereas the denominator was 28 units for adults (the "parents" of the text), it was 10 for youths.

[9]The youths were given, in footnote 8, four units of Democratic choice, four Republicans, and two neither party.

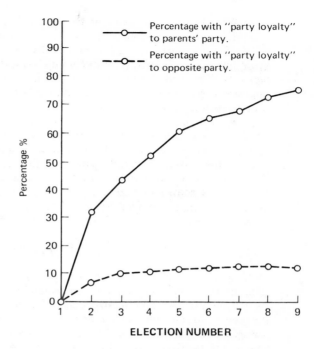

FIGURE 2   The growth of "party loyalty." (Party loyalty = high probability of voting for one party *and* low probability of voting for the other.)

3. Each such young voter was then tied, in a friendship or sociometric network, to the equivalent of a *parent* with much political experience. That parental experience was biased almost wholly to favor one party.[10]

This experimental setup means that when the youth develops initial preferences in ensuing elections during his teens and early twenties, he is discussing his choice with a strongly partisan parent who is, most of the time, unwavering in his convictions. With this approximation of reality in, say, Iowa Republican families and New York City Democratic families, what we want to know is whether the same discussion process by which adults influence each other will socialize children as well. This was determined by holding a sequence of elections corresponding to the decade or two during which youths are first taking some interest in politics, but are still not independent of their elders' coaching or example. Nine elections happened to be run, and in all elections here the distributions of stimuli for each party were equal and normal.[11]

[10]The parents were 24 units for one party, one for the other, and three neither.

[11]"Normal" refers to a rectangular distribution where all strengths of stimuli are equally likely, in McPhee and Smith, *op. cit.* If the individual were alone, it would simply translate party dispositions (probabilities) into manifest preferences (corresponding frequencies). For example, of those with 0.7 probability of voting for a party, 70 per cent of the cases would actually do so.

What Figure 2 shows is the percentage of youths who develop the kind of one-sided dispositions called "party loyalty." This means a high probability of voting for one party and a low probability of voting for the other. The upper trend line shows the percentage of youths who developed party loyalty for the *parents'* party and the lower trend line those who developed loyalty for the opposite party, all having started with equal tendencies in both directions and thus no such loyalty.[12]

If each election is considered to have taken place at the end of a two-year period, then it can be seen that the direction of the two trend-lines was well established after four or five elections, or a decade of political activity. If we had released the youths from their parental influence at the end of the first decade, they would have *continued* to vote for the same party during the second decade. For what is shown in Figure 2 is a measure of the growth of internalized dispositions. In this model, as in real life, after a time external guidance is no longer required, for its effect is now perpetuated by inner conviction.

ADULT DISPOSITIONS

So much for socialization. The power of the influence process seems so great in that function that one immediately fears it would be "too strong" for purposes of the less effective social influence among adults. This is not so. In the above socialization experiment we gave each youth equal dispositions towards both parties at the beginning of the period. And the distributions of stimuli for each party for all persons were also equal in all periods in order to show the effect of the discussion process when all other things were equal—all other things *except* the sustained influence exerted by partisan parents. But in adult political life all other things are not equal.

A main reason has just been shown. Precisely the effect of social influences themselves, via socialization, is to make dispositions toward the two parties unequal. Thus, we later find in an adult population relatively few people who are still equally "influenceable" in any direction. This finding is illustrated by Table 1.

Table 1 was prepared by classifying every adult respondent in a Roper survey in 1956. The classification was by an estimated probability of the person voting Democrat (top of table) and his estimated probability of voting Republican (left-hand side of table). This classification was done by relatively crude means, to serve as input to the model;[13] but with these survey respondents as its

[12]Population size for computer simulation shown in Figure 2—the number of *youths* involved—is approximately 400 for each election in the sequence of nine, in four independent replications of 100 each. Of course, a larger population would be necessary for precise quantitative study of such curves, not the purpose here.

[13]Thanks are due the Roper Opinion Research Center, Williams College, for this information, and for its processing to a research group at IBM, Inc., headed by Eugene Lindstrom. The index was constructed by first assigning to each person a "turnout" probability, or probability to vote, which was based on the relative *frequencies* with which he expressed interest, had previously voted and expressed intention to vote. Each person was then assigned a "directional" probability based upon party identification, previous vote for Senator or Governor, and previous Presidential vote. What is shown as the table entry here is the joint probability, multiplying probabilities of voting turnout and of directional or party choice. Equal weights for all items in the index is assumed here, but current work with latent-structure scaling that does not require that assumption does not seem to lead to significantly different results.

TABLE 1 Distribution of Cases in a National Sample by Estimated Probabilities of Voting for Each Party

		Probability of Voting Democrat[a]									
		.0	.1	.2	.3	.4	.5	.6	.7	.8	.9
Probability of Voting Republican	.0	522	142	158	58	62	84	71	70	61	296
	.1	65	75	21	21	5	1	—	—	—	—
	.2	100	6	67	6	25	51	24	57	12	—
	.3	54	13	3	17	—	29	—	17	—	—
	.4	50	—	6	—	3	1	8	—	—	—
	.5	95	—	25	12	—	33	—	—	—	—
	.6	64	—	5	—	—	—	—	—	—	—
	.7	75	—	33	16	—	—	—	—	—	—
	.8	73	—	14	—	—	—	—	—	—	—
	.9	229	—	—	—	—	—	—	—	—	—

[a] The table entry is the number of Roper interviews out of a sample of 2,936 cases, September, 1956.

"voters," the model behaves accurately for the 1956 election period.

The table shows the (raw) number of survey respondents for every combination. The main result for present purposes is that the bulk of the cases fall along the top and the left-hand margins. This means that most of the people have some probability of voting for one party but very little for the other party. Only those few individuals in the center cells around the main diagonal have approximately equal probabilities of voting for either party. We can see by the small numbers of the latter that in the adult world other things are *seldom* equal. Let us turn, therefore, to the effects of the discussion process where other things are unequal—the real-life circumstance.

INFLUENCE AND DISPOSITION
AT CROSS PURPOSES

A new experiment was designed using youths and adults of the same characteristics as before. The difference now is that the little experience in politics given the youths is biased. And it is biased towards the party *opposite* that of the parents. This experiment could be considered a simulation of the New Deal period, where, even before voting age, many youths were learning to like Roosevelt's party, despite Republican parents. Specifically, a probability of 0.2 was assigned for voting for the parents' party and 0.6 of voting for the opposite party. This design ensured that there would be disagreement between youths and parents ranging between 70 and 90 per cent in all variations of the experiment.

The variations mentioned were introduced to represent different political climates (stimulus situations). These aided or hindered the youths' tendencies to go against their parents' traditional party. Specifically:

A. One is a "control" variant. Here stimuli for the youths' party—party their dispositions favored (0.6 probability)—were *equal* to stimuli for the parents' party (for which youths had only 0.2 probability). This is identical, except for the youths' biased disposition, to the above experiment.

B. Another is a "strong" variant. Here the youths' own party—the party their dispositions favored—offered *stronger* stimuli throughout the period than did the opposite party, that is, the party of the parents, since parents and youths disagreed.[14]

C. Finally, in a "weak" variant, the youths' own party offered *weaker* stimuli throughout the period than did the opposite party for which they had weaker dispositions—again the party of the parents.

These three variations of the experiment were run for the same number of elections as before. Results are shown in tables and charts in much of the remainder of this paper.

A key result, in Table 2, is a measure of the effectiveness (when there is disagreement) of parents in *changing* the youths' preferences. Recall what happens after disagreement occurs in the voting model. The youth goes back through the stimulation process again. This is interpreted as reconsidering reality in the face of a social situation that is not supporting the original choice. The youth gets a new preference from repeating the stimulation process. It may be quite different from the first impression because the process is in part governed by chance events (sampling of grades of stimulus strengths). A count is kept by the computer of these changes. The direction of change after the disagreement is what is shown as the table entry of Table 2. For simplicity figures shown are averages for all elections in several independent replications on this problem.[15]

TABLE 2 The Outcome of Youths' Second Stimulation after Initial Disagreement with Parents

| | | *AFTER DISAGREEMENT WITH PARENTS YOUTHS' DIRECTION OF CHANGE WAS:* | | | |
		Towards Parents	*No Change*	*Farther Away from Parents*	
Youths' *own* party stimuli	Strong	13%	78%	9%	100%
	Normal	25%	60%	15%	100%
	Weak	40%	43%	17%	100%

For example, when the youths' own party was given strong stimulation, only 13 per cent of the youths changed to the parents' party. Actually, 100 per cent could have changed, since the table considers only those who initially

[14]"Strong" refers to a *distribution* of stimuli. If, for example, there were five grades of stimuli from the strongest to the weakest, then the probability of each occurring is, respectively, .30, .25, .20, .15, .10. (See McPhee and Smith, *op. cit.*) The "weak" distribution used is .10, .15, .20, .25, .30, as the frequencies of occurrence of strongest to weakest stimuli, respectively.

[15]N for each row of Table 2 is about 200, in four independent replications, for any one election, and we average nine elections.

disagreed with parents. A full 78 per cent of the youths kept their original choice. The remaining 9 per cent are those who were originally not intending to vote, but who now decided to vote for the party opposite that of the parents. So, almost nine out of ten youths *resisted* the parents' advice when political stimuli favored the party towards which their own internal dispositions also leaned.

The center row of Table 2 shows the normal or control case and the bottom row the situation where the stimuli of the youths' own party were weak. The latter illustrates a condition for effectiveness of social influence in causing changes against the direction of the youths' internal disposition. The condition for effective influence is when *objective* reality (the stimulus distribution) is contrary to internal disposition. Otherwise, other than when the youths' own party stimuli were weak, the parents' counsel was not effective against youths' dispositions that were now in the opposite direction.

Table 2 shows only about half of the potential influence, however. It does not show what happens when the youth happens to *agree* initially with the parent's party choice. Then, of course, the youth is encouraged to keep that preference. This positive encouragement in the case of agreement is discussed later as a "reinforcement" phenomenon.

The sizable joint effect of both of these types of influence is summarized in Table 3. The sample base is now all the youths (not only those who disagreed initially as in Table 2). The matter of concern is now the final choice regardless of how it was reached, that is, the vote.[16] It is averaged for the first decade in the upper table and for the second decade in the lower.

A set of additional runs is added, however, with which to compare the cases we have been considering. The latter, whose tendency was contrary to the parents, are the cases shown in the *right*-hand column. On the left are the new cases. They are identical in all respects except that they tend to *agree* with their parents. (Youth's probability of voting for parent's party is here 0.6, for the opposite party 0.2, the mirror opposite of the above.)

TABLE 3   Per Cent of Youths Voting for Their *Own* Party: The Party of Their Initial Disposition

| | | *FIRST DECADE* | |
		When Parents' Party Is the Same *as Youths' Party*	*When Parents' Party Is* Opposite *Youths' Party*
Youths' *own* party stimuli	Strong	97%	80%
	Normal	88%	59%
	Weak	71%	31%
		SECOND DECADE	
Youths' *own* party stimuli	Strong	98%	79%
	Normal	93%	57%
	Weak	75%	24%

[16]N is about 200 for each *cell* of Table 3, in independent replications as above.

The table entry in all cases is the rate of voting for the youth's "own party," the one his internal dispositions favor, 0.6 to 0.2. On the left side, where parents agree, this is also the parent's party, and influence is in the same direction as the disposition. On the right side, the vote for the party favored by the disposition is being cast *despite* contrary influence. Comparing the rows, then, gives the difference due to influence compatible with dispositions versus influence against dispositions, under different stimulus conditions down the columns.

To summarize the results: Social influence here makes only a minor difference in the extent to which youth votes for the party of its own inclinations, when that party is offering strong stimuli (top row).

When the stimuli are not so strong, however, the story is different. The middle row lists the situation where the stimuli for the youths' own party are normal and equal to the other party. This is the ambiguous case of no reliable guidance either way from external reality. The difference that an encouraging versus a discouraging parent makes, here, is quite sizable and increases still more in the second decade. In the third row, where the youths' party is offering weak stimuli in this period, a big vote for that party never materializes when the parents are in opposition (right side). The revolt, so to speak, "aborts" when objective or external reality fails to support it.

Here is something curious, however. When parents *do* support this objectively weak position, agreeing with the youth's own disposition (left side), the vote for it remains sizable despite the weak stimuli from external reality. It suggests the real-life counterpart when social influence "insulates" a compatible disposition from contrary reality, as, for example, a dense distribution of Republicans in Vermont tended to insulate youth there from political trends in the 1930s contrary to their dispositions.

CHANGE IN DISPOSITION VERSUS VOTE

Yet, in the actual world, does discussion alone "really" change people? We are all familiar with compliant behavior without change of feeling. To examine this, note that we have to this point been discussing single preferences, or discrete votes in one election. "Really" to change a person, however, might mean to change his basic dispositions toward the two parties. The latter are sluggish in this model. Many a borderline or chance vote can be changed, but to alter party dispositions requires (1) a sustained influence that is (2) working under ideal conditions. Figure 3 illustrates this feature of the model's behavior.

We again consider only the original sample, where all youths start with a disposition *contrary* to their parents' party. We also retain the three experimental variants, with the rebellious youth's party strong, normal, and weak in its stimulus offerings (for example, appeal of candidates). The table entry in Figure 3 is now different, however. It is not votes but the "party loyalty" concept. Specifically, it is the percentage of the youths with high probability of voting for "their own" party, the one we started them inclining towards, combined with a low probability of voting for their opposing parents' party. In effect, the table entry is the percentage of the incipient rebellions that are becoming *permanently* successful by a given time, the higher the proportion in Figure 3 the more

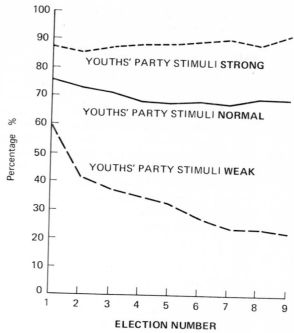

YOUTHS' PARTY STIMULI **STRONG**

YOUTHS' PARTY STIMULI **NORMAL**

YOUTHS' PARTY STIMULI **WEAK**

Percentage %

ELECTION NUMBER

FIGURE 3   Per cent of youths with "party loyalty" to that party favored by their initial dispositions. (Party loyalty=high probability of voting for one party *and* low probability of voting for the other.) Parents' party is opposite in all examples.

youths have become (almost irreversibly) committed to the party they inclined toward but their parents did not.[17]

In the model, at least, the outcome is clearcut: When the party of the youths' own disposition offers him strong stimuli, his loyalty to it is untouched by contrary social influence. In fact it grows. When the youths' own party's stimuli are normal—equal to those of the other party—the result is also a resistance to social influence (after an initial minor effect) on what "really" matters, the disposition. Earlier, we saw that substantial differences in *roles* in any given election were attributable to contrary social influence in this situation (see Table 3). But here the deeper party *dispositions,* once given a direction early in the socialization process, are not really reversed by the impressions and advice that, it is true, affect discrete votes. All this simulates a kind of "accommodation": overt behavior that conforms to the environment is more common than underlying disposition change.

The latter kind of fundamental reorientation, a reversal of a previous disposition, is rare in real life for the same reason as in the model. It occurs only in the extreme example, where not only a sustained social influence is con-

[17]*N* is 200 for each point (election) on each time path, each independently replicated in smaller samples.

sistently contrary over a substantial period, but the influence is also always supported by the objective reality of external events (lower time path in Figure 3). That is, it requires that, when the parent tells a youth that his rebellious party's offerings are actually weak, when the youth reconsiders these offerings, he finds that indeed they are weak. Thus, social direction of basic dispositions is only heeded, in this model, where it *should* be.

Such a situation so ideally favorable to the influence over many elections seldom happens, however, so the gist of the results as we approach adulthood, *after* socialization or other causes have started dispositions in one direction, is that thereafter social influence alone cannot reverse that direction. This realistic example completes the illustration that a process strong enough to produce the striking correlations in socialization data (for example, 70 per cent or more people voting for the same party as their parents did) is at the same time a *weak* enough process to accord with common-sense observations of adults. Namely, for all that they do and rationally should consult one another on specific choices, they resist one another's efforts to change deeper internal allegiances.

AN ILLUSTRATIVE IMPLICATION

The examples of this paper do not differ from those usually introduced in support of a verbal theory: they are already known from everyday experience or past research. That the model can reproduce them says something for their explanation and for its validity. The main hope for these methods, of *working* models and therefore working theory, is, however, that they will generate *new* implications not previously apparent from the static verbal statement of the same theory. At this early stage of experience little has been delivered on that promise, but we conclude with a small example:

The process above was designed with "reconsideration" as the central idea; then its similarity to Lewin's "social reality" was noted. But now note that the same formal logic can also be interpreted as "reinforcement theory," as in experiments on animal learning. There the environment (played by the experimenter) *selects* out of large number of randomlike behaviors of the animal that which is to be rewarded. The stimulation process in the present model generates the same kind of partially random behavior of subjects (youths' preferences). Then the influence process plays a role like that of the experimenter, selecting *which* of these partly random preferences it will choose to reinforce (pass on to the learning process).

So, here Lewinian and reinforcement theories turn out to be representable by the same logic (computer program). The difference is simply heuristic, namely, where one locates himself as observer. Lewin takes the introspectionist point of view of the subjects (animals, voters) who are discovering which behavioral choice is the valid one, that is, the best way to vote or get food. The behaviorist takes the point of view of the experimenter (environment, parent) waiting for the subjects to propose the choice he will select as the valid one that gets the reward.

The latter view of the model has an interesting implication in politics. An example is shown in Figure 4. The points along the time paths in Figure 4 plot the vote of youths for the party towards which originally they inclined in initial

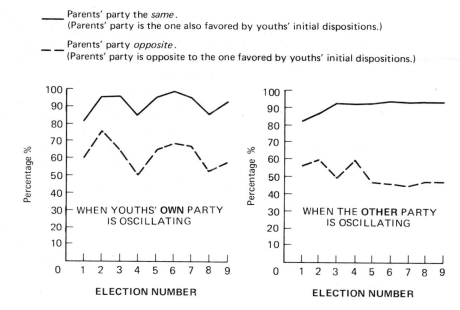

_____ Parents' party the *same*.
(Parents' party is the one also favored by youths' initial dispositions.)

_ _ _ Parents' party *opposite*.
(Parents' party is opposite to the one favored by youths' initial dispositions.)

FIGURE 4   Youths' vote for their *own* party, that is, the party favored by their initial disposition.

disposition, "their own party" as before (their probability of choosing it 0.6 versus 0.2 for the other). The top line on each side shows the vote in cases where the parents' party is the *same* as the youths'. The bottom line on each side plots the vote of youths for their party when the parents' party is opposite. Thus, the "gap" between the two lines is a measure of the difference that a consistently favorable versus a consistently unfavorable social environment makes in voting for the party of one's own initial inclinations.

The stimulus situation for the two parties now differs from previous experiments. One party is offering *constant* stimuli of the normal or control type of distributing, as before. But the other party is offering *oscillating* stimuli, first being favored and then later disfavored by "great historical events." For example, the second and sixth elections are "booms," like 1936 for the Democrats. The fourth and eighth elections are "disasters" of equal and opposite magnitude against the same party.[18]

The difference between the two diagrams of Figure 4 is that in the leftmost one it is the youths' own party that offers an opposite party that is oscillating. If we think of resistance to influence as depending on the "support" one gets from his *own* party's offerings, these are what vary in the left chart. On the other hand, if we think of the effectiveness of influence as depending on the "temptations" offered by the influencer's side, the other side, then the attractiveness of these temptations is what is variable in the right chart.

[18]Booms are created by introducing "strong" stimulus distributions as in footnote 14, disasters by "weak" stimulus distributions as in footnote 14. *N* is 200 for each point on each time path, as usual in several replications (and here illustrating the conclusions of other samples and experiments as well).

Magnitudes and timing are otherwise identical. So, what Figure 4 suggests is that, while response changes more in this case where "support" for one's own cause is varying, the effect of the *social* influence is greater when "temptations" toward the other cause are varying. This is seen by the large gap at the end of the experiments in the right-hand chart. Why is that?

First, consider the case when parents or other influences want the person to vote contrary to his disposition, and for the party he normally would have low probability of choosing. In that event, the influencer seldom sees an *instance* of the choice he wants to reward and reinforce. It is like trying to reinforce an animal in behavior for what he does not do in the first place (enough to "catch" and reward). Here, however, when the appeal of the influencer's cause is made extraordinarily strong from time to time, as when a Roosevelt runs, then a large number of unusual votes for that party arise, in this model either as the youth's own first impression or in his rechecks with reality during disagreement. The influencer thus has *instances to reward*—something to work with and encourage.

Next, consider the opposite case, when parents or other influencers agree with the youth's own party inclinations. In that event parents play a subsidiary role, because he is a "good boy" and votes their way most of the time by inner disposition. But suppose we introduce great variability in the strength of "temptations" offered the youth by the other party. Then the parent or influencer has more deviant *instances to correct*—which he can in this case because dispositions are in his favor. Again he has something to work with.

As animal experimenters and trainers of animal acts know, it is nice to have a good "random animal" who will more often display the unusual behavior one wishes to reinforce. And as every householder knows, one sometimes wishes a pet would display some unusual but obnoxious behavior more often, for example crossing a dangerous street, so that corrections could be administered and learning influenced. In politics, the same kind of opportunities may be produced by wide variations in the strength of external stimuli for the party opposite that which dispositions favor. For that will produce a richer variety of deviant behavior to correct or reinforce *selectively*.

The New Deal's rise and fall, for example, was the time to influence Republican youth and Eisenhower's rise and fall the time to influence Democratic youth—both for *and* against the cause in question—if this hypothesis applies in the real world. And if so, an unobvious implication follows. It is that compelling variations in the strength of objective stimuli from external sources can lead not to a less but a *more* decisive influence of the immediate social environment. For, a counterpart of animal variability that gives an experimenter rich opportunities for selection reinforcement is unusual variability in the deviant "temptations" that give the social environment rich opportunities to encourage or discourage selectively.

CONCLUSION

Serious analysis of the implications remains to be done, however, and purpose here has only been, to repeat, to illustrate a process that can reproduce the kind of "sociological" correlations found in voting and choice data, but not at the expense of the primary determinants of the choice: external stimulus and internal disposition, The portion of the model that accomplishes this can be interpreted as "selective reinforcement," but we have chosen to emphasize a less behavioristic interpretation which instead takes the point of view of the influencee. He is simply *cross*-checking impressions obtained from objective and "social reality" with one another, as all of us have to resort to doing in ambiguous situations.

The consequences at the aggregate level of social significance are generally not repugnant. Influence, and thereby tradition, counts when one has nothing else to go by—as in socialization prior to experience. Otherwise, influence is effective in this model only when it is facilitating something else, inner convictions to resist or compelling stimuli to change. If the real world is like this, it makes both sociological *and* rational sense.

TECHNICAL NOTE

As the introduction said, this is an example of a process that is simple enough—alone—to be tractable in formal analysis; but it illustrates the value of doing the (easy) computer simulation first, namely, as working theory to discover what is *worth* the (difficult) formal proof. This note is to sketch how one could attack the latter, in the hope it will get the serious work this paper suggests it warrants.

Let p be the probability of voting for the cause with which the other person *disagrees* (the other's choice for the moment being fixed). Then $p + q = 1$, where q is the probability of voting for the cause with which the other would agree. Now, if the latter (type q) choice is realized, there is no disagreement in the first place, no reconsideration, and thus no chance to persist in the dis-agreeable choice; but if the former (p) choices comes up, there is. So, if we let $P(p;k)$ be the probability of *persisting* in the choice disagreeable to the other even after k reconsiderations,[19] it is

$$P(p;0) = p$$
$$P(p;1) = p(p) = p^2$$
$$P(p;2) = p(p^2) = p^3$$

$$. \tag{1}$$

$$P(p;k) = p^{k+1}$$

A few examples give the flavor of how sensitively all this depends on P, that is on the *strength* of the inner disposition whose manifestation is now disagreeable to others. For example, with one reconsideration:

If $p = .20$, then $P(p;1) = (.20)^2 = .04$, or 1/5 of p alone
If $p = .50$, then $P(p;1) = (.50)^2 = .25$, or 1/2 of p alone
If $p = .80$, then $P(p;1) = (.80)^2 = .64$, or 4/5 of p alone

[19] A "reconsideration" is a new exposure to stimulation after social disagreement with the outcome of the previous stimulation. Earlier, we permitted only one $k = 1$, for each pair of persons (but a person could be involved in many different pairs interacting in other problems).

Now consider why the influence also depends sensitively on the strength of stimuli. Without going into details of the stimulation process that is discussed in "Note on a Campaign Simulator" (McPhee, 1963), the gist of the latter is that strong stimuli make for a *manifestation* of p with conditional probability p' that is substantially larger than p where p is moderate or small, for example, at .50 and .20 in the above example. So, if $p' > p$, then $(p')^2 \gg p^2$. Then people persist much better in their choices disagreeable to others. By the same stimulation process, when weak stimuli are present, the manifestation of p is a conditional, probably $p' < p$. Especially, p' is smaller where p is moderate or small. Then, since $p' < p$, it will be $(p')^2 \ll p^2$, and people will not persist well in the disagreeable choice.

Next note how the number of considerations matters. Suppose people were equally distributed over all p values. Then the *total* number of contrary choices when there are $1, 2, \ldots, k$ reconsiderations permitted would be, in contrast to no reconsiderations,

No reconsideration: $\quad \displaystyle\int_0^1 p \, dp = \tfrac{1}{2}$

One reconsideration: $\quad \displaystyle\int_0^1 p^2 dp = \tfrac{1}{3}$

$\hspace{9.5cm}$ (2)

Two reconsiderations: $\quad \displaystyle\int_0^1 p^3 \, dp = \tfrac{1}{4}$

$k - 1$ reconsiderations: $\quad \displaystyle\int_0^1 p^k dp = \dfrac{1}{k+1}$

Now, consider that people are not equally distributed over all p. Then the bimodal populations found in politics would make people *resistant* to influence, while the normally distributed populations found for other attitudes would make people *vulnerable*. The reason is that the number of choices changed is the initial proportion of disagreement minus the probability of persisting in that disagreement unchanged. With one reconsideration, it is

$$\text{Number of changes} = p - p^2 \qquad (3)$$

which reaches its maximum (derivative zero) at $p = \frac{1}{2}$, or where normal distribution would be dense.

Next, let the other person's choice no longer be fixed. If we let p_i and p_j be their respective probabilities of choosing one alternative, then q_i and q_j are the respective probabilities of choosing the other. Now let A $(p;k)$ be the probability they will *agree* on the p-type choice when k reconsiderations are permitted. It is for $k = 1$, one reconsideration as in the voting model:

$$\begin{aligned} \text{A } (p;1) &= p_i p_j + p_i q_j p_i p_j + q_i p_j p_i p_j \\ &= p_i p_j (1 + p_i q_j + q_i p_j) \end{aligned} \qquad (4a)$$

Since the expression in parenthesis would usually be much greater than 1.0, there is a good deal more *consensus* "in a world of reconsideration" than without it (where it would be $p_i p_j$ alone). And yet, choice being made by individuals in the light of their own dispositions (p), it is not an intolerable consensus. Indeed, since the chance of agreeing on the q-type choice is symmetrical,

$$A(q;1) = q_i q_j (1 + p_i q_j + q_i p_j) \qquad (4b)$$

with the expression in parenthesis the same as above, it depends only on $p_i p_j$ versus $q_i q_j$ which alternative is finally agreed upon. Thus, the decision satisfies the higher average disposition; for example, if one "really" wants his choice, the other gives in. But is also has the desirable property of a "built-in veto," for, no matter how high p_i is, if p_j is near zero, the joint probability is effectively the latter.

If we let any number, k, of reconsiderations take place until agreement, it is then a model of the problem when people *have* to agree finally on a common choice, for example, a television program they will all watch or an automobile purchased for common use. The algebra then becomes formidable and has only been investigated far enough to suggest it will fall into a sum of geometric series. But Ithiel Pool and Howard Raifa suggest that we already know the sums involved, on the following grounds. Suppose that agreement finally comes at the kth trial, where k is as large as we please. In theory, at least, the formal process has no "memory," no more than dice do. Therefore, the probability that the p type will win at time k, given that someone does win, is

$$A(p;k) = \frac{p_i p_j}{p_i p_j + q_i q_j}$$

the same as the conditional probability that the p type will win at *any* time, if one indeed does win.

5

Economic System Simulations

IRMA ADELMAN

Simulation is at once one of the most powerful and one of the most misapplied tools of modern economic analysis. "Simulation" of an economic system means the performance of experiments upon an analogue of the economic system and the drawing of inferences concerning the properties of the economic system from the behavior of its analogue. The analogue is an idealization of a generally more complex real system, the essential properties of which are retained in the analogue.

While this definition may be consistent with the denotation of the word (see the discussion of simulation's meanings in Chapter 1), the connotation of simulation among economists active in the field today is much more restricted. The term "simulation" has been generally reserved for processes using a physical or mathematical analogue and requiring a modern high-speed digital or analogue computer for the execution of the experiments. In addition, most economic simulations have involved some stochastic elements, either as an intrinsic feature of the model or as part of the simulation experiment field. Thus, while a pencil-and-paper calculation on a two-person Walrasian economy would, according to the more general definition, constitute a simulation experiment, common usage of the term would require that, to qualify as a simulation, the size of the model must be large enough, or the relationships complex enough, to necessitate the use of a modern computing machine. Even the inclusion of probabilistic elements in the pencil-and-paper game would not suffice, in and of itself, to transform the calculations into a bona fide common-usage simulation.

Reprinted in modified form with permission of the publisher from the INTERNATIONAL ENCYCLOPEDIA OF THE SOCIAL SCIENCES, David L. Sills, editor. Volume XIV, pp. 268-273. Copyright © 1968 by Crowell Collier and Macmillan, Inc.

In order to avoid confusion, the term "simulation" will be used hereafter in its more restrictive sense.

To clarify the concept of simulation, it might be useful to enumerate the steps involved in performing a simulation, and to indicate how the simulation process is related, on the one hand, to the analytic solution of mathematical models of an economy, and on the other, to the construction of mathematical and econometric models of economic systems.

The first step in performing a simulation is to construct a mathematical or physical model of the system. The model must specify the relationships between the relevant economic variables as well as the specific dynamics of their inter-action. The equations of the model or the rules of operation of the physical analogue must constitute an accurate representation of the behavioral processes (operating characteristics) of the part of the economy being modeled. In addition, they must also express the identities among the variables of the system. Some of the variables involved are *exogenous,* in the sense that their values are fixed outside the model; the *initial conditions* of the system are included here. The other variables are *endogenous,* their values being specified by the rela-tionships that govern the interactions within the model. Of the endogenous variables, some are *predetermined* by the solution of the model at a point in time prior to the one for which the calculations are being carried out; the rest are contemporary.

Second, specific numerical values must be assigned to the parameters of the model and to its exogenous variables. These numerical values are generally based on statistical estimates derived from samples of time-series or cross-section data describing the relevant features of the system being modeled. Some parameters are selected on a priori grounds and reflect theoretical or empirical judgments based on experience with similar systems.

Third, assumptions about the probability distribution of the stochastic structure of the model must be made. These assumptions relate both to the disturbance terms and to the parameters of the behavioral equations of the model.

Fourth, the equations must be put into computer soluble form. Some of the equations may be combined by simultaneous solution to make the system simpler. Some techniques to solve the nonlinear equations of the model (if any) must be introduced. A flow chart for the solution of the system, which specifies the order of the steps to be followed in the machine solution of the system, must also be devised.

Finally, the solution values of the endogenous variables are derived by combining, in the manner specified by the model equations and in the sequence previously devised, the assumed values of the exogenous variables with the stochastic disturbance terms.

It should be stressed that any simulation can only be as good as its inputs. In particular, one cannot emphasize enough the importance of formulating a realistic model of the system being simulated and of deriving good estimates of the parameters of the model. Otherwise, the simulation will merely offer a numerical solution to a system of equations with no empirical relevance. In fact, unless these criteria are met, the procedure followed will not constitute a simulation at all in the sense of our definition, since the correspondence between the system being simulated and its physical or mathematical analogue will be quite poor.

The initial phase of a simulation experiment therefore usually consists of a lengthy iterative process in which computer solutions of the model equations are examined and judged to be unrealistic in some respect. Either the model equations or particular parameter estimates, initial conditions, or values of exogenous variables are then modified in the hope of obtaining more "sensible" results for the endogenous variables and the systems solved once more. The previous sequence is then repeated until satisfactory convergence obtains between the properties of the simulation model and those of the real economy as the investigator believes it to be.

The relationship between simulation and econometric and mathematical formulations of economic theories is quite intimate. As indicated above, it is these descriptions of economic processes that constitute the basic inputs into a simulation. After a mathematical or econometric model is translated into language a computing machine can understand, the behavior of the model, as described by the machine output, represents the behavior of the economic system being simulated. The solutions obtained by means of simulation techniques are quite specific. Given a particular set of initial conditions, a particular set of parameters, and the time period over which the model is to be simulated, a single simulation experiment yields a particular numerically specified set of time paths for the endogenous variables (the variables whose values are determined or explained by the model). A variation in one or more of the initial conditions or parameters requires a separate simulation experiment which provides a different set of time paths. Comparisons between the original solution and other solutions obtained under specific variations in assumptions can then be used to infer some of the properties of the relationships between input and output quantities in the system under investigation. In general, only very partial inferences concerning these relationships can be drawn by means of simulation experiments. In addition, the results obtained can be assumed to be valid only for values of the parameters and initial conditions close to those used in the simulation experiments. By contrast, traditional mathematical approaches for studying the implications of an economic model produce general solutions by deductive methods. These general solutions describe, in functional form, the manner in which the model relates the endogenous variables to any set of initial conditions and parameters and to time.

The models formulated for a simulation experiment must, of course, represent a compromise between "realism" and tractability. Since modern computers enable very large numbers of computations to be performed rapidly, they permit the step-by-step solution of systems that are several orders of magnitude larger and more complicated than those that can be handled by the more conventional techniques. The representation of economic systems to be investigated with the aid of simulation techniques can therefore be much more complex; there are considerably fewer restrictions on the number of equations that can be utilized. Simulation, therefore, permits greater realism in the extent and nature of the feed-back mechanisms in the stylized representation of the economy contained in the econometric or mathematical model.

The impetus for the use of simulation techniques in economics arises from three major sources. First, both theory and casual observation suggest that an adequate description of the dynamic behavior of an economy must involve complex patterns of time dependencies, nonlinearities, and intricate interrelationships among the many variables governing the evolution of economic activity through time. In addition, a realistic economic model will almost certainly require a high degree of disaggregation. Since analytic solutions can be obtained for only very special types of appreciably simplified economic models, simulation techniques, with their vastly greater capacity for complexity, permit the use of more realistic analogues to describe real economic systems.

The second driving force behind the use of simulation, one that is not unrelated to the first, arises from the need of social scientists in general and of economists in particular to find morally acceptable and scientifically adequate substitutes for the physical scientist's controlled experiments. To the extent that the analogue used in the simulation represents the relevant properties of the economic system under study, experimentation with the analogue can be used to infer the results of analogous experiments with the real economy. The effects in the model of specific changes in the values of particular policy instruments (e.g., taxes, interest rates, price level) can be used to draw at least qualitatively valid inferences concerning the probable effects of analogous changes in the real economic system. Much theoretical analysis in economics is aimed at the study of the probable reactions of an economy to specified exogenous changes, but any economic model that can be studied by analytic techniques must of necessity omit so many obviously relevant considerations that little confidence can be placed in the practical value of the results. Since a simulation study can approximate the economy's behavior and structure considerably more closely, simulation experiments can, at least in principle, lead to conditional predictions of much greater operational significance.

Finally, the mathematical flexibility of simulation permits the use of this tool to gain insights into many phenomena whose intrinsic nature is not at all obvious. It is often possible, for example, to formulate a very detailed quantitative description of a particular process before its essential nature is sufficiently well understood to permit the degree of stylization required for a useful theoretical analysis. Studies of the sensitivity of the results to various changes in assumptions can then be used to disentangle the important from the unimportant features of the problem.

Past uses of simulations. The earliest applications of simulation approaches to economics employed physical analogues of a hydraulic or electrical variety. Analogue computers permit the solution of more or less complex linear or nonlinear dynamic systems in which time is treated as a continuous variable. They also enable a visual picture to be gained of adjustment processes. On the other hand, they are much slower than digital computers and can introduce distortions into the results because of physical effects, such as "noise" and friction, which have no conscious economic analogue.

Subsequent economic simulations have tended to rely primarily on digital computers. As the speed and memory capacity of the computers have improved, the economic systems simulated have become increasingly more complex and more elaborate, and the emphasis in simulation has shifted from use as a mathematical tool for solving systems of equations and understanding economic models to use as a device for forecasting and controlling real economies. In addition, these improvements in computer design have permitted the construction of microanalytic models in which aggregate relationships are built up from specifications concerning the behavioral patterns of a large sample of microeconomic units.

As early as 1892, Irving Fisher recommended the use of hydraulic analogies "not only to obtain a clear and analytical *picture* of the interdependence of the many elements in the causation of prices, but also to employ the mechanism as an instrument of investigation and by it, study some complicated variations which could scarcely be successfully followed without its aid" (1892, p. 44). It was not until 1950, however, that the first hydraulic analogues of economic systems were constructed. Phillips (1950) used machines made of transparent plastic tanks and tubes through which colored water was pumped to depict the Keynesian mechanism of income determination and to indicate how the production, consumption, stocks, and prices of a commodity interact. In the diagram which represents the Philips simulator (Figure 1) the production, stock and consumption flows of a commodity are represented by the flow of water into, out of, and away from the tank labeled stocks. Price is assumed to be determined at any instant by the quantity of stock and the demand curve for stocks, thus price is inversely related to the height of the water in the tank. Equilibrium conditions are determined by the regulation of the valve controlling the production and consumption flows, given a specific demand for stocks. The values equate the positioning of the production and consumption curves.

Electrical analogues were used (Strotz *et al.,* 1957) to study models of inventory determination and national income oscillation, both with and without stochastic shocks. In the inventory determination model, consumer behavior is described by a linear demand function. The industry is assumed to regulate its production in accordance with a linear output function, and to follow a long-run policy of keeping inventories at a constant level, although short-run deviations in inventories result from market disturbances. The flow pattern of this model is pictured in Figure 2. Market inertia reflects the difficulties of altering the rate at which goods are made available at retail outlets, and production inertia reflects the difficulty of promptly changing the rate at which goods are produced in the factories. The model is represented by the equations:

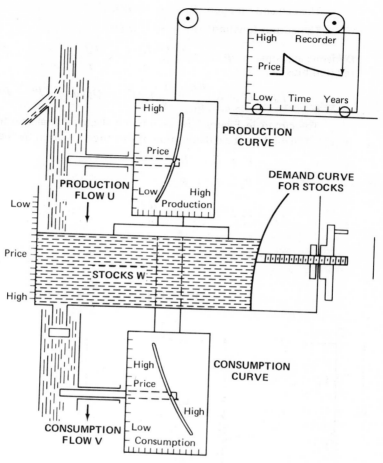

FIGURE 1 Diagram of the Phillips Model. Source: Phillips (1950), p. 285.

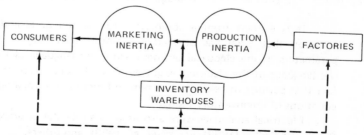

FIGURE 2 Economic flow pattern. Source: Strotz *et al.* (1957), p. 559.

The demand function $P_d = \alpha_1 - B_1 Q_e$

The output function $P_s = \alpha_2 + B_2 Q_p$

Dynamic adjustment relationships
$$P_d - P_s = \lambda_1 \frac{dQ_e}{dt} + \lambda_2 \frac{dQ_p}{dt}$$

$$P_d - P^o = \lambda_1 \frac{dQ_e}{dt} + \frac{1}{\gamma} \int_{t^o}^{t} (Q_e - Q_p)\, dt + (P^1 - P^o)[1 - \epsilon^{-\mu(t-t^o)}]$$

The circuit in Figure 3 represents the electrical analogue to the economic model. The mathematical description of this circuit is given by the equations:

$$V_1 = E_1 - R_1 i_1$$

$$V_2 = E_2 + R_2 i_2$$

$$V_1 - V_2 = L_1 \frac{di_1}{dt} + L_2 \frac{di_2}{dt}$$

$$V_1 - V^o = L_1 \frac{di_1}{dt} + \frac{1}{c} \int_{t^o}^{t} (i_1 - i_2)\, dt$$

$$+ (V^1 - V^o)[1 - \epsilon - \frac{R_3}{L_3}(t - t^o)]$$

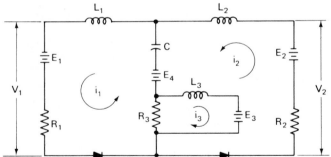

FIGURE 3   Electrical circuit analogue. Source: Strotz *et al.* (1957), p. 559.

It is evident from a comparison of the economic and electrical circuit equation that what has occured is a simple relabeling of the variables and parameters in the electrical analogue from the language of the economist into the language of the electrical engineer. The relationships remain intact. It is this fact that permits the electrical system to function as an analogue for solving the equations of the inventory model.

Electrical analogues were also used to study the business cycle models of Smith and Erdley (1952); Strotz, *et al.*, 1953); and others.

The shift to the use of digital computers began in the late 1950s. In a simulation study on an IBM 650, Adelman and Adelman (1959) investigated the dynamic properties of the Klein-Goldberger model of the U.S. economy by extrapolating the exogenous variables and solving the system of 25 nonlinear difference equations for a period of one hundred years. In this process no indications were found of oscillatory behavior in the model. The introduction of random disturbances of a reasonable order of magnitude, however, generated cycles that were remarkably similar to those characterizing the U.S. economy. On the basis of their study, they concluded that (1) the Klein-Goldberger equations may represent good approximations to the behavioral relationships in the real economy; and (2) their results are consistent with the hypothesis that the fluctuations experienced in modern highly developed societies are due to random perturbations.

Duesenberry, Eckstein, and Fromm (1960) constructed a 14-equation quarterly aggregative econometric model of the U.S. economy. Simulation experiments with this model were used to test the vulnerability of the U.S. economy to depressions and to assess the effectiveness of various automatic stabilizers, such as tax declines, increases in transfer payments, and changes in business savings.

A far more detailed and, indeed, the most ambitious macroeconometric simulation effort to date, at least on this side of the iron curtain, is being carried out at the Brookings Institution in Washington, D.C. The simulation is based on a 400-equation quarterly econometric model of the U.S. economy constructed by various experts under the auspices of the Social Science Research Council (Duesenberry et al., 1965). The Brookings model has eight production sectors. For each of the nongovernment sectors there are equations describing the determinants of fixed investment intentions, fixed investment realizations, new orders, unfilled orders, inventory investment, production worker employment, nonproduction worker employment, production worker average weekly hours, labor compensation rates per hour, real depreciation, capital consumption allowances, indirect business taxes, rental income, interest income, prices, corporate profits, entrepreneurial income, dividends, retained earnings, and inventory valuation adjustment. The remaining expenditure components of the national product are estimated in 5 consumption equations, 11 equations for nonbusiness construction, and several import and export equations. For government, certain nondefense expenditures, especially at the state and local levels, are treated as endogenous variables, while the rest are taken as exogenous. On the income side of the accounts there is a vast array of additional equations for transfer payments, personal income taxes, and other minor items. The model also includes a demographic sector containing labor force, marriage rate, and household formation variables and a financial sector that analyzes the demand for money, the interest rate term structure, and other monetary variables. Finally, aside from a battery of identities, the model also incorporates two matrices: an input-output matrix that translates constant dollar GNP (gross national product) component demands into constant dollar gross product originating by industry, and a matrix that translates industry prices into implicit deflators for GNP components. (See Figure 4.)

The solution to the system is achieved by making use of the model's block recursive structure. First, a block of variables wholly dependent on lagged endogenous variables is solved. Second, an initial simultaneous equation solution

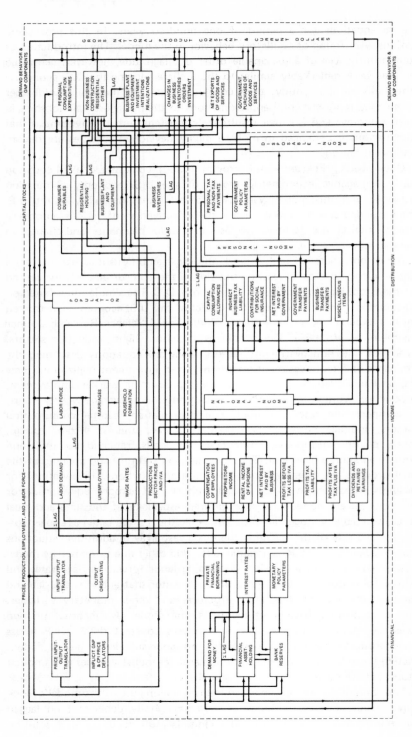

FIGURE 4   Condensed flow diagram: Brookings econometric model.
Source: Duesenberry *et al.* (1965), pp. 24, 25.

is obtained for variables in the "quantity" block (consumption, imports, industry outputs, etc.) using predetermined variables and using prices from the previous period in place of current prices. Third, a simultaneous equation solution is obtained for variables in the "price" block using the predetermined variables and the initial solution from the quantity block. Fourth, the price block solution is used as a new input to the quantity block, and the second and third steps are iterated until each variable changes (from iteration to iteration) by no more than 0.1 percent. Fifth, the three blocks are solved recursively.

The work on the model was initiated in 1961, drawing upon about 30 experts to formulate the structural equations in each behavioral subblock of the model. The initial version of the model became available in 1965, and efforts since then have been devoted to gaining information about the system by solving it simultaneously and subjecting it to policy and stochastic shocks. As expected, the first solutions indicated that the model was deficient in many respects. A substantial reformulation of various subsectors (inventories, consumer durables, labor demand, nonwage income components, tax receipts, state and local government expenditures, various other market variables) and of the process of converting GNP final bill of goods items and GNP prices into industry value added and industry prices was finished by 1968 to yield Brookings II. Further substantial efforts to improve the model have been undertaken since then. Thus, even though a great deal of effort and expertise have gone into the formulation of the model, as of 1969 the model is still in the iterative verification-reformulation stage.

Simulation experiments show that Brookings II yields predictions both within and beyond the sample period that lie well within the range of accuracy of other models. The average correlation between predicted and actual values for 14 variables and 8 quarters is 0.88. Simulation studies have also been undertaken to determine the potential impact of personal income and excise tax reductions, of government expenditure increases, and of monetary tightness and ease. The results of the simulations indicate that tax multipliers are less than the government expenditure multipliers. In addition the use of various welfare functions to compare the outcomes of the above alternative policy instruments suggests that the utility ranking of pure monetary policies is superior to that of any other combination of policy instruments tested. (The later conclusion must be qualified, however, by the particularly weak nature of the specification of the monetary and real sector interactions in the Brookings model.)

Other simulation experiments carried out with Brookings III relate to testing the properties of the model at past cyclical turning points for the U.S. economy. These tests indicate that the model solutions have lower amplitude than the economy. The model reproduced (qualitatively) the 1954 trough and the 1957-58 recession, but failed to reproduce the 1960-61 recession. Sample period simulations for 36 quarters were carried out with the model. These indicate that Brookings III is an improvement over Brookings I, in that real GNP and the GNP deflator both have the appropriate trend growth and some of the appropriate cyclical content. Twenty-five-year simulations beyond the sample period indicate that the dynamic properties of the Brookings model are similar to those of the Klein-Goldberger model: smooth time paths in the absence of exogenous shocks, and cyclical fluctuations in the stochastic, shocked solutions.

A rather different approach to the estimation of behavioral relationships for the entire socioeconomic system has been developed by Orcutt and his associates (Orcutt *et al.,* 1961). Starting with the observed behavior of micro-economic decision units rather than macroeconomic aggregates, they estimate behavioral relationships for classes of decision units and use these to predict the behavior of the entire system. For this purpose, functions known as operating characteristics are estimated from sample surveys and census data. These functions specify the probabilities of various outcomes, given the values of a set of status variables which indicate the relevant characteristics of the micro-economic units at the start of a period.

Their initial simulation experiments were aimed at forecasting demo-graphic behavior. For example, to estimate the number of deaths occurring during a given month, the following procedure was applied: an operating characteristic,

$$\log P_i(m) = F_1[A_i(m - \tfrac{1}{2}), R_i, S_i] + (m - m_0)F_2[A_i(m - \tfrac{1}{2}), R_i, S_i] + F_3(m),$$

was specified. Here, p_i (m) indicates the probability of the death of individual i in month m, A_i (m−½) is the age of individual i at the start of month m, R_i denotes the race of individual i, and S_i denotes the sex of individual i. The function F_1 is a set of four age-specific mortality tables, one for each race and sex combination, which describe mortality conditions prevailing during the base month m_0. The function F_2 is used to update each of these mortality tables to month m. The function F_3 is a cyclic function which accounts for seasonal variations in death rates.

This and other operating characteristics were used in a simulation of the evolution of the U.S. population month by month between 1950 and 1960. A representative sample of the U.S. population consisting of approximately 10,000 individuals was constructed. In each month of the calculation a random drawing determined, in accordance with the probabilities specified by the relevant operating characteristics, whether each individual in the sample would die, marry, get divorced, or give birth. A regression analysis was then used to compare the actual and predicted demographic changes during the sample period. Close agreement was obtained. This approach is currently being extended to permit analysis of the consumption, saving, and borrowing patterns of U.S. families, the behavior of firms, banks, insurance companies, and labor unions, etc. When all portions of the microanalytic model have been fitted together, simulation runs will be made to explore the consequences of various changes in monetary and fiscal policies.

So far, we have discussed simulation experiments for prediction purposes only. However, the potential of simulation as an aid to policy formulation has not been overlooked by development planners in either free-enterprise economies or socialist countries. A microanalytic simulation model has been formulated at the U.S.S.R. Academy of Sciences in order to guide the management of the production system of the Soviet economy (Cherniak, 1963). The model is quite detailed and elaborate. It starts with individual plants at a specific location and ends up with interregional and interindustry tables for the Soviet Union as a whole. In addition, the U.S.S.R. is contemplating the introduction of a 10-step joint man-computer program. The program is to be cyclic, with odd steps being man-operated and even steps being computer-operated. The functions fulfilled by the man-operated steps include the elaboration of the initial basis of the plan, the establishment of the criteria and constraints of the plan, and the evaluation of results. The computer, on the other hand, will perform such tasks as summarizing and balancing the plan, determining optimal solutions to individual models, and simulating the results of planning decisions. This procedure will be tested at a level of aggregation corresponding to an industrial sector of an economic council and will ultimately be extended to the economy as a whole. I have been unable to find any information on the progress of this work more recent than that in Cherniak (1963).

To take an example from the nonsocialist countries, a large-scale interdisciplinary simulation effort was undertaken at Harvard University, at the request of President John F. Kennedy, in order to study the engineering and economic development of a river basin in Pakistan. The major recommendations of the report (White House, 1964) are currently being implemented by the Pakistani government, with funds supplied by the U.S. Agency for International Development.

In Venezuela a 250-equation dynamic simulation model has been constructed by Holland, in conjunction with the Venezuelan Planning Agency and the University of Venezuela. This model, which is based on experience gained in an earlier simulation of a stylized underdeveloped economy (Holland & Gillespie, 1963), is being used in Venezuela to compare, by means of sensitivity studies, the repercussions of alternative government expenditure programs on government revenues and on induced imports, as well as to check the consistency of the projected rate of growth with the rate of investment and the rate of savings. An interesting macroeconomic simulation of the Ecuadorian economy, in which the society has been disaggregated into four social classes, is currently being carried out by Shubik (1966).

Monte Carlo studies. The small sample properties of certain statistical estimators cannot be determined using currently available mathematical analysis alone. In such cases simulation methods are extremely useful. By simulating an economic structure (including stochastic elements) whose parameters are known, one generates samples of "observations" of a given sample size. Each sample is then used to estimate the parameters by several estimation methods. For each method, the distribution of the estimates is compared with the true values of the parameters to determine the properties of the estimator for the given sample size. This approach, known as the Monte Carlo sampling method, has frequently been applied by econometricians in studies of the small sample properties of alternative estimators of simultaneous equation models.

Limitations and Potential. In assessing the present state of simulation of economic systems, one should distinguish among at least three classes of potential uses: forecasting, policy making, and serving as a research device for improving our understanding of economic structure. As of now, the a priori beliefs of structural estimation embodied in simulation models usually entail larger error variances than a nonstructural forecasting approach (Christ, 1951; Cooper, 1969). Simulation models formulated to date have therefore led to little improvement in forecasting ability. With respect to policy, most comprehensive econometric models are too coarse to indicate the effects of the highly specific detailed changes with which economic legislation is usually concerned. As pointed out by Kuh (1965, p. 365):

> The day is still far off, as it should be, when the President of the United States will request economic guidance from a computer without human mediation, although there have been times when better advice might have been elicited from our transistorized robots. At the same time, one might expect that governmental economic expertise will rely more on econometric capability for analyzing alternative policy implications instead of current primary dependence on the expert back-of-the-envelope calculation.

At present, the major use of simulation models is to educate the intuition of economists. The main benefits of simulation lie in exploring the properties of systems for which analytic solutions are difficult or impossible and in revealing incongruities in the behavior of a simultaneous system when individual equations are plausible. There is much to be learned from tracing through the reasons for and remedying these incongruities.

Why have results to date with simulation models been of so limited a value? To a large extent, the very strength of simulation is also its major weakness. As pointed out earlier, any simulation experiment produces no more than a specific numerical case history of the system whose properties are to be investigated. To understand the manner of operation of the system and to disentangle the important effects from the unimportant ones, simulation with different initial conditions and parameters must be undertaken. The results of these sensitivity studies must then be analyzed by the investigator and generalized appropriately. However, if the system is very complex, these tasks may be very difficult indeed. To enable interpretation of results, it is crucial to keep the structure of the simulation model simple and to recognize that, as pointed out in the definition, simulation by no means implies a blind imitation of the actual situation. The simulation model should express only the logic of the simulated system together with the elements needed for a fruitful synthesis.

Another major problem in the use of simulation is the interaction between theory construction and simulation experiments. In many cases simulation has been used as an alternative to analysis, rather than as a supplementary tool for enriching the realm of what can be investigated by other, more conventional techniques. The inclination to compute rather than think tends to permeate a large number of simulation experiments, in which the investigators tend to be drowned in a mass of output data whose general implications they are unable to analyze. There are, of course, notable exceptions to this phenomenon. In the water-resource project (Dorfman, 1964), for example, crude analytic solutions to simplified formulations of a given problem were used to pinpoint the neighborhoods of the solution space in which sensitivity studies to variations in initial conditions and parameters should be undertaken in the more complex over-all system. In another instance, insights gained from a set of simulation experiments were used to formulate theorems which, once the investigator's intuition had been educated by means of the simulation, were proved analytically. Examples of such constructive uses of simulation are unfortunately all too few.

Finally, in many practical applications of simulation to policy and prediction problems, insufficient attention is paid to the correspondence between the system simulated and its analogue. As long as the description incorporated in the simulation model appears to be realistic, the equivalence between the real system and its analogue is often taken for granted, and inferences are drawn from the simulation that supposedly apply to the real economy. Clearly, however, the quality of the input data, the correspondence between the behavior of the outputs of the simulation and the behavior of the analogous variables in the real system, and the sensitivity of results to various features of the stylization should all be investigated before inferences concerning the real world are drawn from simulation experiments.

In summary, simulation techniques have a tremendous potential for both theoretical analysis and policy-oriented investigations. If a model is chosen that constitutes a reasonable representation of the economic interactions of the real world and that is sufficiently simple in its structure to permit intelligent interpretation of the results at the present state of the art, the simulations can be used to acquire a basic understanding of, and a qualitative feeling for, the reactions of a real economy to various types of stimuli. The usefulness of the technique will depend crucially, however, upon the validity of the representation of the system to be simulated and upon the quality of the compromise between realism and tractability. Presumably, as the capabilities of both high-speed computers and economists improve, the limitations of simulation will be decreased, and its usefulness for more and more complex problems will be increased.

REFERENCES

General discussions of the simulation of economic processes can be found in Conference on Computer Simulation 1963; Holland & Gillespie 1963; Orcutt *et al.* 1960; 1961. A comprehensive general bibliography is Shubik 1960. On the application of analogue computers to problems in economic dynamics, see Phillips 1950; 1957; Smith & Erdley 1952; Strotz *et al.* 1953; 1957. Examples of studies using digital computers are Adelman & Adelman 1959; Cherniak

1963; Dorfman 1964; Duesenberry *et al.* 1960; 1965; Fromm & Taubman 1968; White House 1964.

Adelman, Irma; and Adelman, Frank L. (1959) 1965. The Dynamic Properties of the Klein-Goldberger Model. Pages 278-306 in American Economic Association, *Readings in Business Cycles.* Homewood, Ill.: First published in Volume 27 of *Econometrica.*

Cherniak, Iuri I. 1963 The Electronic Simulation of Information Systems for Central Planning. *Economics of Planning* 3:23-40.

Christ, Carl F. 1951 A Test of an Econometric Model of the United States 1921-1947 in *Conference on Business Cycles.* New York: National Bureau of Economic Research.

Conference on Computer Simulation, University of California, Los Angeles, 1961, 1963, *Symposium on Simulation Models: Methodology and Applications to the Behavioral Sciences.* Edited by Austin C. Hoggatt and Frederick E. Balderston. Cincinnati, Ohio: South-Western Publishing.

Cooper, Ronald L. 1969 The Predictive Performance of Quarterly Econometric Models of the United States. Paper presented at National Bureau of Economic Research Conference on Income and Wealth Nov. 14-15.

Dorfman, Robert 1964 Formal Models in the Design of Water Resource Systems. Unpublished manuscript.

Duesenberry, James S.; Eckstein, Otto; and Fromm, Gary 1960 A Simulation of the United States Economy in Recession. *Econometrica* 28:749-809.

Duesenberry, James S. *et al.* (editors) 1965 *The Brookings Quarterly Econometric Model of the United States.* Chicago: Rand McNally.

Duesenberry, James S. *et al.* (editors) 1969 *The Brookings Model: Some Further Results.* Chicago: Rand McNally.

Fisher, Irving (1892) 1961 *Mathematical Investigations in the Theory of Value and Prices.* New Haven: Yale University Press.

Fromm, Gary; Klein, L. R.; and Schinck, G. R. 1969 Short and Long Term Simulations with the Brookings Model. Paper prepared for National Bureau of Economic Research Conference on Income and Wealth Nov. 14-15.

Fromm, Gary; and Taubman, Paul 1968 *Policy Simulations with an Econometric Model.* Washington, D.C.: Brookings Institution.

Holland, Edward P.; and Gillespie, Robert W. 1963 *Experiments on a Simulated Under-developed Economy: Development Plans and Balance-of-Payments Policies.* Cambridge, Mass.: M.I.T. Press.

Kuh, Edwin 1965 Econometric Models: Is a New Age Dawning? *American Economic Review* 55:362-369.

Orcutt, Guy H. *et al.* 1960 Simulation: A Symposium. *American Economic Review* 50:893-932.

Orcutt, Guy H. *et al.* 1961 *Microanalysis of Socioeconomic Systems: A Simulation Study.* New York: Harper.

Phillips, A. W. 1950 Mechanical Models in Economic Dynamics. *Economica* New Series 17:283-305.

Phillips, A. W. 1957 Stabilisation Policy and the Time-forms of Lagged Responses. *Economic Journal* 67: 265-277.

Shubik, Martin 1960 Bibliography on Simulation, Gaming, Artificial Intelligence and Allied Topics. *Journal of the American Statistical Association* 55:736-751.

Shubik, Martin 1966 Simulation of Socio-economic Systems. Part 2: An Aggregative Socio-economic Simulation of a Latin American Country. Cowles Foundation for Research in Economics, *Discussion Paper* No. 203.

Smith, O.J.M.; and Erdley, H.F. 1952 An Electronic Analogue for an Economic System. *Electrical Engineering 71:362-366.*

Strotz, R. H.; Calvert, J. F.; and Morehouse, N. F. 1957 Analogue Computing Techniques Applied to Economics. American Institute of Electrical Engineers, *Transactions* 70, part 1:557-563.

Strotz, R. H.; McAnulty, J. C.; and Naines, J. B., Jr. 1953 Goodwin's Nonlinear Theory of the Business Cycle: An Electro-analog Solution. *Econometrica* 21:390-411.

White House—[U.S.] Department of Interior, Panel on Waterlogging and Salinity in West Pakistan 1964 *Report on Land and Water Development in the Indus Plain.* Washington: Government Printing Office.

SIMULATION OF AN ECONOMY WITH
DEVELOPMENT AND TRADE PROBLEMS

EDWARD P. HOLLAND

Economists cannot make laboratory experiments on national economic systems. If, however, we could devise and build a miniature national economy, complete in every detail, which behaved exactly like a real one but on a much-speeded-up time scale, and if we could adjust its parameters and then manipulate policies and exogenous events while recording time histories of its economic variables—then we could learn much by experimenting with it. By means of simulation with a computing machine, we can do something of that kind. Although we are forced by practical considerations to simplify the representation of the economy, using aggregate variables and omitting many details, we can, nevertheless, simulate and explore an economic-system model which behaves much more realistically than any model that can be dealt with by conventional analytic techniques.[1]

The purpose of this article is to illustrate how the technique of simulation can be used to study problems of economic development and foreign trade policy for an underdeveloped country. Graphical and numerical results from a few computer runs are presented and compared to show how simulation might actually be used as an aid in formulating a development investment program.

These runs are a small sample selected from about 200 made during experiments on the model to be described. The parameters and initial values of variables in the model were based largely on the Indian economy as of 1951. The development programs and policies which were tried out on the model, however,

[1] An over-all picture of the state of the art of simulation, with some of its history, rationale, methods, and typical applications, is given in a symposium of articles in the December 1960, *American Economic Review* by Orcutt (1960), Shubik (1960), and Clarkson and Simon (1960). Bibliographies with these articles cover a wide variety of simulations and background material, unfortunately omitting, however, the interesting analog work of A. W. Phillips (1950; 1957).

Reprinted with permission from *The American Economic Review,* 52 (June 1962), 408-30. Copyright © 1962 by the American Economic Association. This work was financed jointly by the Center for International Studies and two other M.I.T. units: the Research Laboratory of Electronics and the Sloan Research Fund of the School of Industrial Management. An earlier analog simulation project, out of which this project developed, was financed by a grant from the Rockefeller Foundation. Assistance in making the transition to the digital computer, using the DYNAMO program, was rendered by the Industrial Dynamics Group of the School of Industrial Management. Services of an IBM 704 computer were provided by the M.I.T. Computation Center.

were purely hypothetical and do not have any relation to actual experience or plans of the Indian government. No policy recommendations for India should be inferred from the results presented, because, for simplification, several considerations have been omitted which might be crucial for India. The work done was partly a process of developing a technique and partly a study of some hypothetical situations. Use of this method for policy guidance in real situations is a phase that is yet to come.

The objectives of a simulation study may include establishing a consistent and technically feasible development plan, exploring the effects of changes and alternatives in such a plan, and discovering what variables provide the best signals to tell how well the plan is succeeding or to warn of impending trouble. Other objectives include designing policy measures for handling problems that may originate either inside or outside the system, discovering what side effects are likely to be induced by any given policy, and finding out how to cope with such effects if they appear. For these purposes, simulation offers several advantages. It makes it possible to observe the dynamic interaction of aspects of the economy which are usually analyzed in isolation from each other, such as foreign trade consumers' behavior, and investment decisions. It also permits comparatively realistic representation of all sorts of nonlinearities, discontinuities, time delays, and irreversibilities. Inclusion of such characteristics at the appropriate points is usually crucial for reproducing the modes of dynamic behavior that can occur in the real system.

THE ECONOMIC SYSTEM MODEL

In the description that follows the aim is to give a picture of the main features of the system, including fundamental assumptions about what elements of the structure are constant, which variables are exogenous and which are endogenous, and what sort of relationships control each of the main variables, with due attention to nonlinearities and other special characteristics. It would not be practical, in a short article, to describe such a complex model, or even a part of it, in equations, because it would require so many of them, with so much explanation, to be meaningful.[2]

An actual developing economy would be observed and analyzed in terms of the time histories of such variables as gross national product at current prices, real gross national product, consumers' price index, price indexes for various categories of goods, investment in various sectors, total investment, production from various sectors, imports of various types, exports, and the balance of payments.

The model described here was formulated with the aim of making the counterparts of these variables behave realistically under dynamic conditions. This meant representing underlying processes of adjustment pertinent to nonequilibrium situations rather than merely relationships which must exist between variables when they are in equilibrium.

---

[2] A complete specification, for those who want it, is included in Holland, Tencer, and Gillespie (1960). Not counting about two and one-half chapters of introduction, it occupies 167 pages and takes more than 250 equations.

Thus, for example, long-run supply curves were not used, because the relation between price and supply quantity at any future time will be affected by the particular paths along which some of the other variables move during the process of change and because the occurrence of a long-run type equilibrium is extremely improbable under the conditions for which the model was designed to be used.

Instead, very-short-run supply functions for the various sectors of the economy were formulated, with parameters subject to continuous change, reflecting changes in available fixed capital, prices of intermediate goods, wage rates, and labor productivity. A short-run supply curve for the product of one of the industrial sectors would look at a particular time like curve *a* of Figure 1.

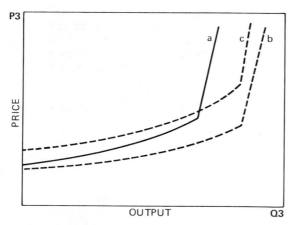

FIGURE 1   Short-run supply curves.

The part of the curve to the left of the kink is based on variable costs, with the assumption that only enough plants operate to produce the total output efficiently. To the right of the kink all plants are assumed to be operating at or above their normal output, and the price of their produce is affected by scarcity. If net capital formation were carried out with no change in wage rates or intermediate-goods prices, the supply curve at some later time would have shifted horizontally to a position like *b*. (The lowering of the intercept on the price axis reflects increased productivity in new, technically improved plants.) If the same capital formation had been accompanied by a 50 per cent increase in wages and other prices, the short-run supply curve would have shifted to *c* instead of *b*.

The model includes six domestic production sectors, of which four have short-run supply curves similar to that described above. The four are power-using consumer goods manufacturing (Sector 1), capital and intermediate goods production (Sector 3), nonpower-using consumer goods manufacturing (Sector 4), and transport, communication, housing, etc. (Sector 5). In addition, the supply of output from agriculture (Sector 2) is assumed to be completely inelastic in the short run, and the supply of personal services (Sector P) is assumed to be perfectly elastic on any time scale. The model also has three sectors concerned with imports and exports (Sectors 6, 7, and 8).

Associated with each of the first production sectors named above is a capacity-creating activity which can maintain or expand the sector's capacity or allow it to decay. At any given time decisions are being made about undertaking capital formation projects in each sector, both for replacement and for expansion. (The determination of these decisions is explained in Section II.) The units of productive capacity started during a particular time increment (the increments used were .05 year long), continue under construction throughout a gestation period and then become available to use for production. Factor payments and capital goods purchases, constituting investment expenditures, are determined at any time by the number of capacity units which are in the gestation process. The units of capacity completed and ready for production in a given time increment remain part of the sector's capacity throughout statistically distributed life spans. Thus, at any time, the rate at which old capacity is decaying, the rate at which new capacity is coming into being, the existing capacity, and the level of investment expenditure all depend on the past history of the rate of starting capacity-creating projects, and whenever this rate of starts is altered, the effects on investment and especially on capacity are delayed and smoothed out.

Demand for the product of any sector may be a combination of consumers' demand, demand derived from the current production of other sectors, and demand derived from capacity-creating projects. Not all of these are relevant to all sectors, however. The significant demand relationships are indicated diagrammatically in Figure 2.

Each block in Figure 2 represents a combination of several relations, and each arrow represents an important signal from one group of relations to another (i.e., a dependent variable from one group entering another group as an independent variable). The block labeled "power-using consumer goods manufacturing," for example, stands for a price-determining function like the short-run supply curve of Figure 1, a function for evaluating the potential return on investment on the basis of current conditions, a wage-negotiation function, which responds to profits and in turn affects costs and the supply price, and a set of input-output coefficients (fixed in real terms) determining the demand by this sector for intermediate goods from other sectors. All of the relations implied in this block are related to current production in the particular sector (Sector 1).

Touching the block described above to indicate a high degree of interdependence is another block representing the capacity-creating activity and the capital life cycle for this sector. Arrows from the current-production block (Sector 1) to Sectors 2 and 5 indicate the demand for products of those two sectors induced by production in Sector 1. Another arrow, going to Sector 3, originates in both the production and capacity-creation blocks of Sector 1, indicating that products from Sector 3 are demanded in proportion to the level of capital-formation activity as well as in a different proportion to current output. For each demand arrow shown it would be appropriate to have a price arrow going the other way, since the prices of intermediate goods used in any sector affect that sector's production cost and hence its supply price schedule. Price arrows, however, have been omitted for clarity.

The export sector, Sector 8, has no production or capital formation and is

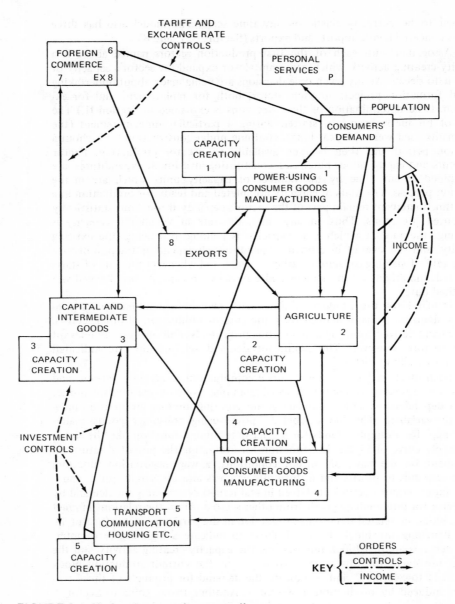

FIGURE 2   National economic system diagram.

merely an accounting relationship, combining products of Sectors 1 and 2 into a product-mix for export. The foreign commerce sector includes a fixed supply price (in foreign currency) for imported capital and intermediate goods, another price for consumers' imports, and a demand equation for exports. The price elasticity of export demand may be specified as desired for each run. Exchange rates and tariffs may be held fixed, made flexible, or manipulated in various ways.

The aggregate national income, earned by factors in production and capital-formation activities, is adjusted for taxes and business savings to determine disposable income, which, with a lag, becomes a principal independent variable of the consumers' demand function. The other independent variables are the prices of various goods, determined from the supply functions, and the population. Population grows exponentially at whatever rate is programed, independently of other variables.

The consumers' demand function has characteristics of a classic multi-dimensional indifference map, determining the allocation of expenditures in response to income and prices. The per capita demand for food increases less than proportionately as per capita real income rises, in keeping with Engel's Law, while savings and the demands for other goods and for services increase more than proportionately. Various goods (including imported ones) are substitutable for each other at diminishing marginal rates. This demand function and the supply functions for the various consumer goods interact continuously to adjust prices as conditions change.

SPECIFICATIONS FOR A RUN

Before going on to show how the economic-system model performed in a particular run, it is necessary to indicate what functions and constraints were included as aspects of policy, what variables were assigned exogenous values or time-paths, and what degrees of freedom were kept open.

A development plan was specified in terms of two sets of time-schedules. For Sectors 2, 4, and 5—agriculture, nonpowered consumer-goods production, and public overhead services (transport, communication, housing, etc.)—specific time-programs for starting capacity-creation projects were set up and rigidly followed. This does not imply that the enterprises in these sectors are necessarily government-operated, although some may be. The assumption is that it is government policy that determines the level of capital formation, but the actual investment may be carried out by private entrepreneurs. It has been observed in India, for example, that very little investment takes place in agriculture in regions where no government activity is carried out. Community development programs are an effort by the government to help and encourage villagers to undertake improvements in their facilities and in their methods. Irrigation dams and canals built by the government must, to be useful, be supplemented by local ditches and banks built by individuals or local groups. Thus the privately implemented activity tends to be determined by that of the government.

Development of the small-scale and cottage industries lumped together in Sector 4 is, in India, also induced largely by government incentive, even though the actual firms are privately owned. In Sector 5 there is a combination of private enterprises encouraged and regulated by government—such as part of housing—with purely government-owned utilities like power and railroads. For the purposes of this study we are not concerned with ownership, nor with the question of how control is effected, but only with the fact that in these three sectors it is reasonable to assume that government policies control the rate of capital formation.

For Sectors 1 and 3—the modern industrial sectors producing respectively consumers' goods and intermediate and capital goods—time-programs were specified for the *minimum* rates of starting capacity-creation projects. These sectors have private investment decision functions based on profits, and it is expected that these endogenous mechanisms will often sustain higher rates of project starts than the minimums. However, in the event that starts in either of these sectors should fall below the levels specified, it is assumed that some action will be taken to stimulate them—perhaps direct government investment, perhaps financial incentives to private investment; again we do not identify the means of carrying out the policy but assume that some action can be devised that will prevent capital formation in each of these sectors from falling below the specified minimum at any time.

The private investment decision function in each of these two sectors is primarily a response to the profit rate, determined from the current price and production costs in a new plant. This profit margin is reduced to a rate of return on the basis of the cost of building new capital facilities. Adjustments are made for overhead and financing costs and in some cases for optimism or pessimism of entrepreneurs. Near-future replacement requirements are taken into account, as well as the amount of new capacity already under construction but not finished. The outcome of all these considerations at any time is the rate at which private entrepreneurs want to start new projects in each sector. This will be the actual rate if it is above the scheduled minimum and below any ceiling that may be imposed.

Ceilings on the rate of starting capital projects were called for, under certain conditions in some runs, as measures to combat inflation or to reduce imports of capital goods. The means of effecting this control are not included in the simulation. They might be restriction of import licenses, direct limitation of construction, or indirect limitation through a tight money policy.

Besides the capital-formation profiles—definite ones for Sectors 2, 4, and 5, and minimum levels for Sectors 1 and 3—specifications of foreign-trade policy were included. The exchange rate was fixed, with no provision for changing it in this run. A tariff mechanism was set up to protect the capital-and-intermediate goods sector by raising the tariff on equivalent imported goods if any domestic capacity became idle. A tariff on consumers' imports was specified to be imposed in case of a crisis in the balance of payments (defined, for this run, as a current-account deficit greater than $500 million per year). The tariffs were increased periodically if the situation for which they were invoked persisted. A definite schedule was assumed for tapering off food imports during the first ten years as home production expanded.

Other assumptions were made, not as policies, but to tie down exogenous variables. The prices (in foreign currency) of imported goods of each type were set at fixed values, independent of quantity and without time variations. The demand for export goods was also devoid of time-trend or cycles but was made sensitive to price, with a value of -1.3 for elasticity. Population was assumed to grow exponentially with a growth coefficient of .02 per year.

No limit was set either on credit financing of investment (government and private) or on the deficit on current account in the balance of foreign payments. This does not imply that projections of these deficits would not be major considerations in planning or that their actual magnitudes would be ignored in practice. The deficits are determined, however, by the operation of mechanisms which have already been fully specified. As a technique of investigation, they are left open-ended; the resulting inflation and the foreign-exchange deficit are among the criteria that will be used later in this article for comparing different runs.

A SAMPLE HISTORY[3]

An economic history of the hypothetical country for a 20-year period is shown in Figures 3 and 4. This is part of the output generated in one computer run, referred to here as *Run A*. In addition to plotting the points through which these graphs were drawn, the computer tabulated values of about 80 variables by half-year intervals. (The computations were made at intervals of 1/20th of a year, approximating a continuous process, and involved many additional variables, but since it would not be worth while to record every step of the computations, selected values were actually tabulated only by half-year intervals.)

For this particular run a relatively intense program of investment in public overhead (Sector 5) was carried out, with considerable emphasis also on the capital and intermediate goods industries (Sector 3). Agriculture (Sector 2) was expanded enough to replace food imports and increase the food supply about 4 per cent per year, twice as fast as the population was growing. Relatively little emphasis was given to consumer goods manufacturing (Sectors 1 and 4). This program is of a type often advocated for starting a development process; similar programs are being followed in a number of developing countries. It is based on the theory that the low levels of public overhead capital and of capital-goods capacity are bottlenecks and that investment in those sectors will induce expansion of the consumer goods manufacturing both by opening the bottlenecks and generating demand.

The real investment carried out in each sector is shown in the bottom part of Figure 3. In Sectors 2, 4, and 5, as explained above, these real investments were rigidly programed. In Sector 3, the profit-motivated level of investment rarely came up to the programed minimum; thus the latter almost always became the actual value. In Sector 1, on the other hand, the private investment induced by profit expectations was almost always above the low minimum set by the program.

With this investment program as prime mover, and with unlimited credit and foreign exchange, the gross national product in current prices increased dramatically—at an average rate of 7.5 per cent per year (see the middle part of Figure 3). A large part of this increase, however, was not growth of real output but only a rising price level. Even so, real gross national product, as shown, became more than 2½ times as great in 20 years, an average growth rate of 4.8 per cent per year. Growth of real disposable income per capita averaged 2.8 per cent per year.

Gross investment is shown as a percentage of gross national product at the top of Figure 3. This percentage was a dependent variable, inasmuch as GNP was

---

[3]Most of the time-history graphs that are represented in this and the next section are from Tencer (in preparation).

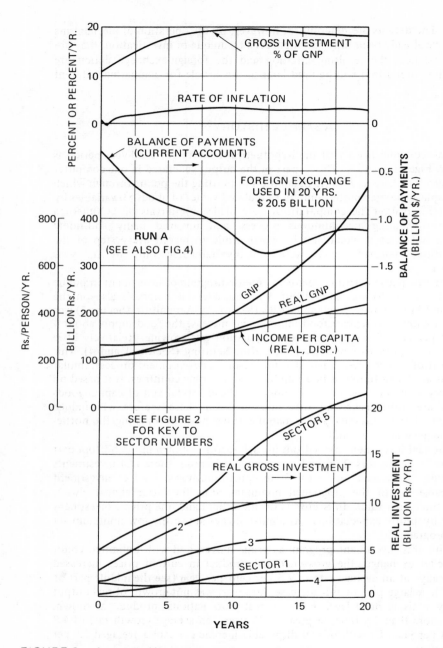

FIGURE 3   A sample History: part of results from Run A.

generated within the system while real investment was independently programed. Starting from about 10 per cent of national product at the beginning of
the run, gross investment increased to 20 per cent in the middle of the run and

then declined very slightly. Figure 3 also shows the annual rate of increase of the consumers' price index, which was in the neighborhood of 3 per cent per year throughout most of the run, and the balance-of-payments deficit on current account.

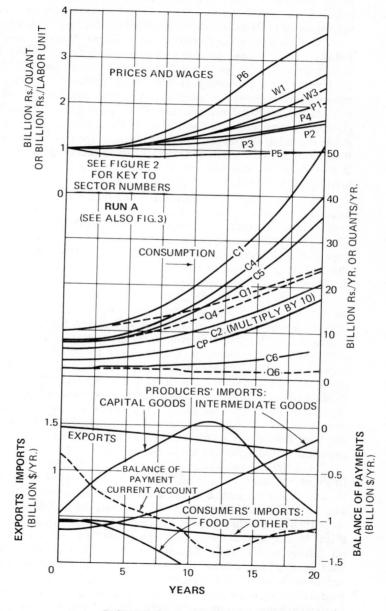

FIGURE 4   Additional results from Run A.

Figure 4 shows the behavior of prices and wages, the changing pattern of consumption, and, at the bottom, the various components of foreign trade and the resulting balance on current account. Except in Sector 5, where capacity was very rapidly expanding, prices rose generally. Industrial wages rose even faster, partly because of increasing labor productivity in the industrial sectors. The price of consumers' imports, P6, was rapidly elevated by an almost continuous succession of tariff increases, intended to help improve the balance of payments.

The growth of consumption of different kinds of goods and services followed Engel's Law, in that expenditures on food, initially high, grew less rapidly than other expenditures as per capita income grew. (This refers to basic foodstuffs; extra processing of food is in a different sector.) In addition to consumption measured in current values, the graph includes physical quantity indexes for industrial manufactures (Q1), small-scale and handicraft products (Q4), and consumers' nonfood imports (Q6). The choice among these three goods was relatively sensitive to price changes. Thus, as P1 climbed more steeply than P4, the consumption of Q4 caught up to that of Q1 (in terms of quantity). It is evident also that the ever-increasing tariff included in P6 was effective in holding down the quantity of Q6 consumed.

Although consumers' expenditures on nonfood imports, C6, doubled in 15 years and reached 3½ times the initial value in 20 years, the increase went into tariff revenue and not into payments abroad. In the bottom panel of Figure 4, where the balance-of-payments components are shown, it can be seen that the expenditure abroad for consumers' nonfood imports was actually reduced during the first 15 years, and in spite of some increase thereafter was still below the initial value by the twentieth year.

Imports of food were gradually tapered off during the first 9½ years as part of the program—not in response to market forces. Imports of capital goods rose rapidly during the first 12 years as required for the high rate of capital formation but declined thereafter as the domestic capital goods sector reached a sufficiently rapid rate of growth to begin taking over the supply. Imported intermediate goods, on the other hand, were assumed to be of types for which domestic substitutes were not developed within the 20-year period. Thus expansion of the economy induced ever-increasing imports in this category. With exports declining because of the rising price of exportable goods, and with imports of intermediate goods steadily rising, the import-replacement process in the food and capital goods sectors was not sufficient to bring the balance-of-payments deficit below the billion dollar a year level.

The initial assumption that this balance-of-payments deficit could be financed by long-term capital inflows is clearly open to question, not only because of the magnitude of the accumulated deficit over the 20-year span but also because even at the end of the 20 years the deficit continues at a rate of more than $1 billion per year. The rate of inflation, too, while not catastrophic, might be considered undesirably high in itself (in addition to its adverse effect on foreign trade). Thus, although the investment program and other policies used in Run A produced a significant rate of economic growth, the results cannot be accepted as satisfactory. Let us then see whether the program can be modified to keep the price level from rising so fast and to reduce the amount of foreign exchange used up by the deficit in the balance of payments.

EFFECTS OF SOME CHANGES IN PLANS AND POLICIES

The simulation was repeated a number of times with a series of cut-and-try changes to help in understanding the dynamics of the system and to seek an improved plan for the allocation and phasing of investment and improved policies for control of foreign trade. The last of the further runs described here is by no means an optimal solution, nor is it in fact an adequate solution for the long run, inasmuch as another crisis is imminent at the end of the period considered. Nevertheless, these runs show some improvement and illustrate the process of investigation. What is lacking is simply further application of the same approach to arrive at a still more satisfactory policy combination.

Run B: Considering that the price inflation and heavy imports of capital goods in Run A are both associated with a high level of investment, it is appropriate to try cutting back the investment program to see what improvement might be made in price stability and the balance of payments and at what cost in growth of real output. Accordingly, Run B was made with about 20 per cent less investment in Sector 2 (agriculture), Sector 3 (capital and intermediate goods), and Sector 5 (public overhead). The main results from this trial are shown in Figure 5, together with the corresponding results from Run A and from other runs yet to be discussed. The less intense program, as expected, reduced the balance-of-payments deficit significantly by reducing capital goods imports. It is noteworthy, however, that the reduction in investment did not reduce the rate of inflation at all. In fact, the final value of the consumers' price index was actually a shade higher for Run B than for Run A (1.65 against 1.62). Evidently the slower growth of consumers' demand was offset by the reduced rate of expansion of supplies.

The reduction in investment, while ineffective in abating inflation, caused a significant loss in growth of output. Starting from 106 billion rupees per year initially, real gross national product grew in 20 years to 237 billion rupees per year in Run B, as compared with 270 billion rupees in Run A. This is an average annual growth rate of 4.11 compared with 4.78 per cent per year.

These comparisons, and others to come, are summarized in the table at the bottom of Figure 5.

Run C: An attempt was made to alleviate inflation by applying the following policy, leaving everything else as it was in Run B: Whenever the rate of change of the consumers' price index reached 3 per cent per year, the rate of starting new capital projects in the industrial sectors (1, 3, and 5) was curtailed until such time as inflation of the price index fell below 3 per cent per year, whereupon the program of Run B was resumed.

This policy was first invoked in Run C about 8½ years after the start, when the inflation rate first reached 3 per cent per year. The resulting drop in investment is noticeable in the graph at the top of Figure 5, while in the one just below, it can be seen that the policy did indeed have the effect of reducing the rate of inflation to less than the chosen limit. This was not a major reduction, however, since it had not been very much above 3 per cent for much of the time in Run B. A slight drop in currently priced national product is noticeable, but real product was almost identical in the two runs. The most notable effect of

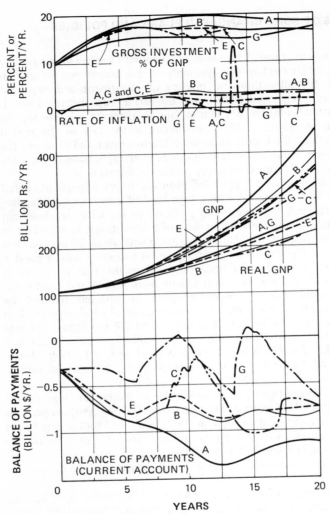

FIGURE 5   Effects of investment pattern and policy changes.

RUN	SYMBOL	GROWTH OF REAL GNP % PER YR.	PRICE INDEX AT 20 YRS.	FOREIGN EXCHANGE USED IN 20 YRS. $ BILLION
A	————	4.78	1.62	20.5
B	————	4.11	1.65	15.8
C	—·——	4.06	1.59	15.0
E	– – – –	4.56	1.36	14.2
G	—·—·—	4.86	1.18	5·4

this "anti-inflation" policy proved to be a reduction in the balance-of-payments deficit, through reduced imports of capital goods.

Run E:[4] The fact that reducing industrial investment during part of Run C saved foreign exchange and slightly reduced inflation without significantly hurting real growth suggested that the investment pattern was too heavily biased toward the industrial sectors and that some gain might be effected by shifting emphasis away from some of them. Therefore Run E was programed with 10 per cent less investment in public overhead capital than Run C but with a 25 per cent increase in agricultural investment. The graphs and table tell the story: significantly less inflation, higher real output, and another $800 million saving in foreign currency.

Run G: Between Run E and Run G a good many exploratory runs were actually made, and the program was changed in several respects. The investment pattern was changed further in the direction of less public overhead (Sector 5) and more agricultural expansion (Sector 2). The capital and intermediate goods program (Sector 3) was smoothed out a bit and built up a little earlier, while investment in nonfood consumer goods manufactures (Sector 1) was accelerated after the tenth year. Only in nonpowered consumer goods production was the program left unchanged. The new investment profiles (as they actually occurred, including some spontaneous investment) are compared with those of Runs A and E in Figure 6. The changes in the Sector 3 program, which were intended to smooth out the actual investment profile in that sector, did not have that effect, but instead simply changed the timing of the fluctuations in spontaneous investment.

As a measure to inhibit inflation, especially when a sudden policy change might set up a sharp transient stimulus (e.g., in case the currency should be devalued), the time lag in the wage increase function was doubled, implying some sort of wage control policy. This change does not alter the equilibrium or target level toward which wages move under any given set of conditions but slows down their adjustment toward that target.

For the first 10 years the program gave a much better result than the previous ones (see Figure 5). Real gross national product grew faster, with less inflation, and with a diminishing balance-of-payments deficit. With the main emphasis on sectors requiring less capital goods, capital goods imports were eliminated after 9½ years, at the same time that food imports were phased out and just after the balance-of-payments deficit had been brought to zero.

At that point the character of the process changed abruptly. The abruptness resulted from the crudeness of approximations used in the model formulation, but the nature of the change is indicative of a problem that is actually of concern in some developing countries. That problem is the difficulty of indefinitely continuing the import substitution process. As more and more imports are replaced, the remaining opportunities become less and less attractive or feasible. In this model, the effect of this decreasing flexibility was roughly approximated in the nonfood consumer goods category by the diminishing marginal rate of substitution in the demand function. For either food or capital goods, however, import substitution remained perfectly feasible until all such imports, and therewith, of course, the substitution possibilities were suddenly eliminated. The resulting shock was enhanced in this run by the accidental simultaneity of this event in the case of both food and capital goods imports.

[4]Runs D and F have been omitted from the sequence.

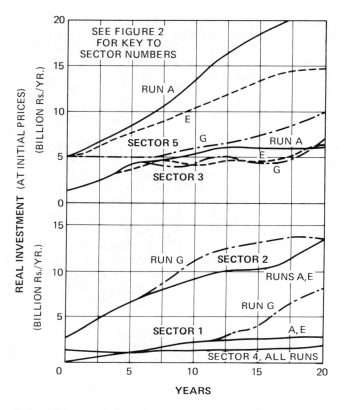

FIGURE 6   Time profiles of real investment, Runs A, E, and G.

Before the elimination of these imports 9½ years from the start of the run, a favorable trend in the balance of payments had been established by the reduction of these imports faster than imports of consumer goods and inter-mediate goods were increasing. (Exports were hardly changing.) Afterward, the continuing increase in consumers' and intermediate imports rapidly built up a new balance-of-payments deficit.[5] The ending of the import-substitution process affected consumers' prices favorably, largely because the rate of expansion of agriculture was maintained, with all of the increase being added to the supply instead of some being offset by reduced imports. Not only was inflation checked but from the twelfth year on the consumers' price index even declined.

By the time the current-account balance-of-payments deficit had grown from zero to $600 million per year in four years, and was still increasing at about the same rate, it seemed time for some drastic countermeasures. Several experiments were tried, only one of which is reported here. When the deficit

[5]Of course, it is just as unrealistic to assume no substitution possibilities for intermediate goods as to assume the possibilities are perfect for capital goods. Undoubtedly these features can be improved in a future model. However, except for the sharpness of the discontinuity, these approximations compensate each other to some degree. It is not the object of this article to draw general conclusions but to illustrate a technique of investigation.

reached $600 million per year (13¾ years from the start), the currency was devalued so that 50 per cent more rupees were required in exchange for a dollar, and at the same time a 20 per cent tariff was imposed on consumers' imports.

The short-run impact was an improvement in the current account of $750 million per year, as seen in Figure 5. Of the total, $600 million was from reduced consumer-goods imports and $150 million from increased exports. This improvement may be regarded as permanent, relative to the deficit that would otherwise have developed if it could have been financed and allowed to grow indefinitely. A more realistic interpretation, however, is that a crisis was impending and was deferred for a few years by the devaluation. Although the balance of trade deficit was greatly reduced, its rate of growth was not affected, as can be seen by comparing the slopes of the balance-of-payments graph of Figure 5 for the periods before and after the devaluation (before year 13½ and after year 14½). By the twentieth year the deficit had reached $700 million per year and showed no sign of leveling off.

This illustrates the need for some action outside the range of the assumptions used thus far in the investigation. With the fixed export demand scheduled assumed in these runs, and the requirement of intermediate-goods imports in proportion to the growing output of a domestic industrial sector, there is really no way to stabilize the trade balance, even if all consumer and capital goods imports are cut off. Although the limitations assumed are reasonably realistic for some countries at some stage in their development, they cannot be tolerated in the long run and must somehow be overcome.

With the assumed limitations, as long as growth is successfully maintained, measures acting through price effects (i.e., tariffs and devaluation) give only temporary relief. To effect a more permanent improvement it would be necessary either to develop domestic substitutes for all imports and work toward autarky or to expand export demand through successful promotional campaigns abroad and the development of production of new export goods. Attempts to solve the balance-of-payments problem without such measures may be observed in some real countries; perhaps this simulation helps to show why they seldom succeed.

Although it ended with a problem still to be solved, Run G was on several counts an improvement over the previous ones. The growth of output was a little faster than in Run A, and significantly faster than in the intervening runs, while the over-all rise in the price index was the least, and the total drain on foreign exchange reserves was only a fraction of that in the other runs. (See the table in Figure 5.) On the other hand, the slower accumulation of industrial productive facilities left the economy at the end of the run less well equipped for future expansion.

The great improvement made by reducing investment in public overhead capital could mean either of two things: It could mean that the original program was actually too heavily slanted toward public overhead capital. More plausible, however, is the possibility that the assumptions about requirements for overhead services were not realistic. A large part of the output of this sector, according to our statistical sources, was used by consumers. The assumption that their demand for these services (including housing) was neutrally price-elastic should surely be questioned. Also, it might be desirable to subdivide the sector into services primarily for industries and those primarily for consumers.

CONCLUDING OBSERVATIONS

From the series of examples presented here it is apparent that an evaluation of policy alternatives for an actual situation could well involve scores or even hundreds of runs, depending on the problem addressed, the freedom available for variations in different dimensions of policy, and the degree of certainty with which the economic system structure is known. Obviously no substantive conclusions could be drawn from these few runs. In fact, critical consideration of the results led to questions about the validity of assumptions underlying the balance-of-payments problem and the requirements for public overhead services, and to suggestions for trying to improve the formulation.

There is, of course, nothing unique about simulation that makes its results more dependent on assumptions than any other method. The assumptions in such a model as this are so explicit and so numerous as to emphasize the dependence. It is less noticeable in more simplified analyses where it is implicitly assumed that all sorts of things can simply be omitted from the model, but it is no less limiting. A simulation study easily can, and always should, include tests of the effects of changing questionable assumptions or parameter values, at the same time developing ways to make the success of policies relatively insensitive to such changes.

It is not to be expected that simulation, using an approximate model whose parameters are not precisely known, will produce an accurate forecast for a 10-, 20-, or 30-year period. Apart from shortcomings of the model, events exogenous to the real system may easily start cumulative processes leading to major deviations from what was otherwise the most probable outcome, so that no technique can yield reliable predictions. The purposes of a simulation study are to learn about possible problems and how to cope with them, and to make long-range projections for a variety of contingencies, in order to provide perspective for working out shorter range policies. Like an itinerary for a long trip, a long-run plan should be subject to revision whenever new knowledge or new opportunities warrant, or whenever unexpected obstacles arise. At any time it is desirable to have some vision of long-run goals, as well as more detailed targets for the shorter run together with plans for attaining them. Thus a 20-year projection provides a frame of reference within which to work out a 5-year plan, so that the 5-year planners will not, for example, underrate worth-while projects that take 6 years to become productive.

Either for working out development plans or for devising measures to deal with particular dynamic problems, simulation offers some powerful advantages over other techniques. One is the capability of handling a relatively complete system—one which provides for dynamic interactions between aspects of the economy that are usually analyzed in isolation from each other. The value of this is illustrated by some not entirely obvious conclusions from a study of exchange-rate devaluation which was made by Robert W. Gillespie (1961) using this same model. He found that for this model and the assumed conditions the most important determinants of the effectiveness of devaluation were not the frequently analyzed elasticities of demand but the induced effects on the pattern of investment allocation and the resistance of the system to inflation of prices. The resistance to inflation, in turn, proved upon further testing to be very sensitive to the speed with which consumers adjust their expenditures to changes of income.

These results cannot, of course, be taken as general, being based on a partial survey of a particular model. However, they do make it clear that some of the most innocent-appearing *ceteris paribus* assumptions can be treacherous, and that exchange-rate policy had better not be decided purely on the basis of comparative statics. Farsighted theoreticians and policy-makers have already described the processes involved here and how they pertain to the effectiveness of a devaluation (Alexander, 1959; Gutt, 1951). But until the development of simulation, the means of exploring the problem fully have been lacking, and little has been done about it on the theoretical level beyond discussion.

As compared with maximization techniques, simulation permits a clearer separation of the technical investigation from the defining of policy objectives. No formal welfare function or over-all performance index is needed in the analysis itself. Thus it is not necessary to formulate value judgments in a vacuum before observing the effects of alternative policies. Since the effects show up as changes in such incommensurable criteria as the time profiles of income, employment, and prices and the stability of the foreign exchange rate, any such formulation is extremely dubious. A better approach is to choose among alternative outcomes in terms of all of their dimensions. With simulation this can be the final step and can be done, if desired, by decision-makers other than the experimenter.

The formulation of the model for a simulation study is crucial and requires extensive research and good judgment. The techniques of computer programing have developed to such a point that almost anyone can set up some kind of a model and—after some initial difficulties—get it to work. The results, however, may be thoroughly misleading if the model does not correspond suitably to what it is supposed to represent. The importance of careful formulation of the model applies both to its qualitative structure and to the numerical magnitudes involved. Actually a good deal of insight can be gained—as has often been done with engineering systems—by experimenting with a model with the right qualitative features, such as nonlinearities and time lags, even without much knowledge of the numerical values of the parameters. At least, from such a hypothetical study one can learn what modes of dynamic behavior are possible and which parameters are critical in determining which mode will prevail.

In problems of dynamics it is probably more important to have these qualitative features represented in the right places than it is to have accurate numerical estimates of the parameters. It is not even clear what is meant by the idea of highly accurate estimates of the parameters of a model which is qualitatively wrong. On the other hand, many issues hinge on the quantitative values of parameters. Without some reasonably reliable statistical estimates we are likely to find that certain policy choices simply cannot be made one way or the other. Without some knowledge of the orders of magnitude of the quantities we are dealing with, it is not even possible to formulate a good model qualitatively, for any model must be a simplification, omitting many aspects of reality, and we need some basis for judging what can safely be omitted. It is not great precision but reliable approximations that are needed.

Simulation thus is no substitute for empirical research. In fact it sharpens the need both for statistical information and for accurate descriptions of relationships and dynamic processes. With models carefully formulated from such foundations, the possibilities for increasing our understanding of economic dynamics in general and for developing workable policies for particular situations are great.

REFERENCES

S. S. Alexander, "Effects of a Devaluation," *Am. Econ. Rev.,* March 1959, *49,* 22-42.

G. P. E. Clarkson and H. A. Simon, "Simulation of Individual and Group Behavior," *Am. Econ. Rev.,* Dec. 1960, *50,* 920-32.

R. W. Gillespie, *Simulation of Economic Growth with Alternative Balance of Payments Policies,* Ph.D. thesis, Massachusetts Institute of Technology, Cambridge 1961.

Camille Gutt, "Policies to Make Devaluation Effective," in *Money, Trade, and Economic Growth,* New York 1951.

E. P. Holland, Benjamin Tencer, and R. W. Gillespie, *A Model for Simulating Dynamic Problems of Economic Development,* Center for International Studies report C/60-10, Massachusetts Institute of Technology, Cambridge, 1960.

G. H. Orcutt, "Simulation of Economic Systems," *Am. Econ. Rev.,* Dec. 1960. *50,* 893-907.

A. W. Phillips, "Mechanical Models in Economic Dynamics," *Economica,* Aug. 1950, *17,* 283-305.

———, "Stabilization Policy and the Time-Forms of Lagged Responses," *Econ. Jour.,* June 1957, *67,* 265-77.

Martin Shubik, "Simulation of The Industry and The Firm," *Am. Econ. Rev.,* Dec. 1960, *50,* 908-19.

Benjamin Tencer, Ph.D. thesis, Massachusetts Institute of Technology, Cambridge (in preparation).

6

Political Process Simulations

WAYNE L. FRANCIS

The scope of this chapter is a bit more limited than its title. Political process simulations in the field of international relations are evaluated in the chapter by Paul Smoker. Bureaucratic processes are also reviewed elsewhere (e.g., chapters 4, 9, and 11). Thus it is the remaining types of political processes—electoral, legislative, and judicial at the national, state, and local levels—that are of concern here. Within these generous confines, the task of surveying the simulation literature is both manageable and frustrating. Reported simulations are few in number, and completed simulations are even more rare.

BACKGROUND

Ongoing political process simulation research no doubt has diverse origins. The enterprise benefits from twenty-years of intensive information gathering, storage, and statistical analysis; widespread exposure to attitudinal research; growing awareness of the need to represent cognition systematically; and increasing attention to the time function and formal models of political change. In some areas, repeated use of standard correlational analysis may be yielding a diminishing marginal utility, causing researchers to look for new ways of managing data. Also, the development of game or coalition theory seems to be reaching a point where more particularistic analogues are necessary. These conditions, trends, and influences appear to be setting the stage for an increasing amount of simulation work. Although the works described here focus upon legislative participants and voters, it would seem only a matter of time before simulations with different focuses, such as intercommunity decision-making or judicial behavior, were reported (see Goodman, 1967).

From a theoretical and methodological perspective, it is helpful to think of political process simulation as having two predecessors, game theory and multivariate statistics. While it is evident that game theorists have usually limited themselves to highly constrained small group experiments (Riker, 1967; Klahr, 1966; Browning, 1969), the thrust of their work does seem to call for greater explication of coalition processes. The Abelson-Bernstein interaction process (1963) and the coalition formation process have a note of striking similarity. So even though it is difficult to think of the repetitive and short-lived nature of many gaming experiments as process simulation, the elaboration of these experiments and theories into games of intermediate and more subtle payoffs with a large number of players will require heuristics similar to those developed in process simulation.

Voting decisions and legislative decisions have been subjected to a wide variety of statistical analyses for postdicting or explaining decision outcomes. Until very recently almost all multivariate models could be described by a single equation that involved no more than two points in time—time t, the predicted value for the dependent variable, and time $t-1$, the net weights of the independent variables. Relief from this valuable but wooden approach has come in some measure from applications of the Simon-Blalock technique; however, many data banks cannot approximate the demanding assumptions of the technique. The need for a more flexible approach to meet the data at hand has led some political scientists to political process simulation.

At this time, *participant-type* simulations are extremely difficult to evaluate, not because of their elusive qualities but because either they have been designed strictly for classroom learning or they are still in a highly exploratory stage. These simulations include Coleman's *Game of Democracy,* Gamson's *Simsoc: Simulated Society,* Lewis's community decision-making version of the "Game of Influence," and Goodman's "Intercommunity Simulation." All of these simulations emphasize role playing and competition among the participants, and the outcomes are heavily dependent upon the strategies and bargaining skills of the actors. These simulations probably constitute only a small proportion of such experiments, but it seems certain that most participant simulations will have the same general characteristics. Essentially, there must be some payoff, either the winning of office in elections or conventions, or the winning of preferential policy. It is precisely the significance of strategy in determining outcomes that makes these simulations interesting. To what extent do strategies vary from experiment to experiment? What are the winning strategies? What differences occur when the data inputs are altered—when the participants are given different information or when restraints upon behavior are introduced?

Because of the significance of strategy, participant simulations of political processes appear to be a closer relative of game theory than of multivariate statistics. It would be difficult to draw a fine line between game theory experiments and simulation experiments in this context, but in a cryptic vein, it might be argued that simulation experiments were close enough to reality to defy mathematical solutions. Mathematical neatness, most easily achieved with a small number of parameters, is sacrificed in favor of *perceived* "real-world" complexity, most easily achieved with a large number of parameters.

Computer simulations of political processes are at present more accessible for scholarly evaluation. They, too, are characterized by a large number of parameters, but as distinguished from the participant simulations, all the parameters are known. Unknown personality skills and motivations cannot affect the outcomes. In the simulations to be covered, strategy is more a by-product than it is the focus of the process. These simulations are in various stages of development and may be divided into two substantive areas, electoral behavior and legislative behavior, as follows:

1. Electoral simulations
    a. Pool, Abelson, and Popkin (1965)
    b. McPhee (1963) /
    c. Abelson and Bernstein (1963) /
    d. Browning (1968) /
2. Legislative simulations
    a. Cherryholmes and Shapiro (1968)
    b. Matthews and Stimson (1968)
    c. Francis (1969)

The symbol / indicates that the researchers, to this writer's knowledge, have not made available an empirical assessment of their simulation ideas except in a very preliminary way.

ELECTORAL SIMULATIONS

In 1963 Robert Abelson and A. Bernstein distinguished between two types of simulation, *process* simulation, which they applied to their model of community referendum controversies (Abelson and Bernstein, 1963, p. 121), and *prognostic* simulation, which they applied to the Simulmatics Project (Pool, Abelson, and Popkin, 1965). The Simulmatics Project was developed prior to 1960 to predict state-by-state voting for the two major presidential candidates, Kennedy and Nixon, and the basic procedures were applied to the 1964 Johnson-Goldwater election.

The model was constructed by accumulating the results of several national voter surveys taken long before the campaigns began. The socioeconomic characteristics of respondents from each state were assumed to constitute the actual distribution of characteristics for the entire state, and a large number of issue clusters were developed from the survey results. Respondents with the same socioeconomic characteristics were called a voter type, which was described by a distribution of opinion on various political issues—pro, anti, or undecided. It was then necessary to decide to what issues the presidential candidates would appeal. By analyzing the voter types in relation to candidate appeals, the percentage of vote for each candidate in each state could be estimated.

The 1960 and 1964 election models contained very few dynamics and could be described as three-stage simulations. In the first stage are the synthetic universes for the American states; in the second stage, the candidate appeals; and in the third stage, the voter outcomes. The equations they employ to describe the models, however, are only two-stage equations; that is, there is a direct leap from selected population characteristics to voter outcomes. Candidate appeals are defined by the variables in the equations, and no attention is given to possible variance in how different parts of the country may perceive campaign and media messages. Were New Hampshire citizens in 1964, for example, *more* aware of Goldwater's statements about social welfare than Tennessee citizens, regardless of voter type? Yet in 1964, by employing three campaign issues (civil rights, social welfare legislation, and nuclear responsibility), party identification, voter turnout, and two geographic considerations, they produced a ranking of states that correlated 0.90 with the actual outcome.

Two other features of the 1964 election model deserve mention. First, the seven-factor model was a preanalysis and not a postanalysis, meaning that the 0.90 correlation described a prediction and not a postdiction. As will be seen, the remaining simulation exercises described in this chapter have not been subjected to such close scrutiny. Second, the three basic equations, which included all but the two geographic adjustments, confirmed previous knowledge about Democratic, Republican, and Independent voter types. They are a confirmation in the sense that the long-standing distinction formed the crux of the model, one equation for Democratic, one for Republican, and one for Independent voter types. The equation for Independent voter types:

$$V_j = P_{36} \frac{P_1 + (1 - P_1 - Q_1)P_3P_4}{P_1 + (1 - P_1 - Q_1)P_3P_4 + (1 - P_1 - Q_1)(1 - P_3 - Q_3)P_4}$$

$$\frac{+ (1 - P_1 - Q_1)(1 - P_3 - Q_3)P_4}{+ (1 - P_1 - Q_1)P_3(1 - P_4 - Q_4) + Q_1 + (1 - P_1 - Q_1)Q_3Q_4}$$

$$\frac{+ (1 - P_1 - Q_1)P_3(1 - P_4 - Q_4)}{+ (1 - P_1 - Q_1)(1 - P_3 - Q_3)Q_4 + (1 - P_1 - Q_1)Q_3(1 - P_4 - Q_4)}$$

is much more complex than the equations for Democratic and Republican voter types:

$$V_j = P_{36}(1.0 - Q_2P_2) \quad \text{and}$$
$$V_j = P_{36}(P_1P_3 + P_1P_4(1 - P_3) + P_3P_4(1 - P_1)).$$

In contrast, the authors of *The American Voter* (Campbell *et al.,* 1960, p. 143) conclude that

... Independents tend as a group to be somewhat less involved in politics. They have somewhat poorer knowledge of the issues, their image of the candidates is fainter, their interest in the campaign is less, their concern over the outcome is relatively slight, and their choice between competing candidates, although it is indeed made later in the campaign, seems much less to spring from the discoverable evaluations of the elements of national politics.

The results of the two studies are not necessarily inconsistent, but they do point up the need for a greater elaboration of the communication processes that evolve in political campaigns.

A further evaluation of the meaning of the equations reveals that in each case voter turnout (P_{36}) modifies the extent to which the other variables will lead to democratic voting (V_j). Excluding the turnout effect (expressed in a percentage), we see that Democrats vote democratic unless:

$(1.0 - Q_1 P_2)$ they are against civil rights (Q_1) and feel strongly about the matter (P_2).

Republicans vote democratic when:

$(P_1 P_3)$ they are for both civil rights (P_1) and nuclear restraint (P_3).

or (+)

$(P_1 P_4 (1 - P_3))$ they are for both civil rights and social welfare (P_4), controlling for those who met the first requirement $(1-P_3)$,

or (+)

$(P_3 P_4 (1 - P_1))$ they are for both nuclear restraint and social welfare, controlling for those who met either of the first two requirements $(1-P_1)$.

The examples above indicate how the equations apply, remembering that all values are expressed in percentages and that the additivity (+) is accomplished by controlling for previously established conditions.

The equation for Independents is a bit more complex. The numerator is calculated by adding to those who favor civil rights, those neutral on civil rights but favoring nuclear restraint and social welfare, those neutral on civil rights and nuclear restraint but favoring social welfare, and those neutral on civil rights and social welfare but favoring nuclear restraint. The denominator duplicates the above sequence but then adds those against civil rights, those neutral on civil rights but against both nuclear restraint and social welfare, those neutral on civil rights and nuclear restraint but against social welfare, and those neutral on civil rights and social welfare but against nuclear restraint. Thus the equation is a symmetrical proportion wherein the numerator is modified by the percentage of voters having precisely the opposite characteristics, entered, of course, as part of the denominator. It is important to note that for any issue Q_j is not the complement of P_j. Rather, $1 - P_j = Q_j + (1 - P_j - Q_j)$, the last term identifying those people who are neutral on the issue.

In general, what do the equations convey? They tell us that for all *voters* in 1964, Independents were at least as sensitive to the issues as were Democrats and Republicans. The turnout rate (P_{36}), of course, will be lower for the Independents, but it is essentially the turnout that needs to be explained. Then the results of *The American Voter* research might seem more consistent. The following two studies (McPhee and Abelson-Bernstein) provide a framework for moving in this direction.

William McPhee reported a campaign simulator (McPhee, 1961, 1963) that incorporated the notion of an electorate making repeated responses to campaign stimuli. The voting population is represented through a sample survey, and the voters are then classified into a fairly large number of groups. Each group is demographically homogeneous and has a prior distribution of favorableness toward each candidate. The simulation is set in motion when a candidate makes an appeal, or an argument to attract votes. The researcher might then determine the strength of the appeal within each subgroup by a follow-up sample, asking each subject to indicate how convincing the argument was on behalf of the candidate. The appeal scale serves to modify the voting dispositions of the electorate. Many of those who were undecided might not favor the candidate, but it would take a very high appeal score to change the voting position of someone who was originally strongly opposed to the candidate. The simulation continues when a new argument is made by one of the candidates, resulting in another distribution of appeal and consequent modification of voting positions. The process continues until election day.

The McPhee campaign simulator, reported in 1963, must be considered a description of simulation ideas and not a fully implementable computer program. Some of the weaknesses in the simulation concepts probably resulted from reports that it was an unfinished experiment. The example of the 1964 Wisconsin primary, utilized in the McPhee study, was simply a cue-giving illustration, lacking most of the dynamics implied by the overall description. For instance, the study gives the reader no rule for determining voter turnout. It appears as if the author is dealing with only those who will vote, perhaps predetermined in the proper proportions for all subgroups. Does this then mean that the position-building process central to the simulation is independent of turnout? Would not farmers turn out in greater strength if intensive appeals to their economic interests were made?

The McPhee simulator is apt to be difficult for the reader to interpret because its description does not clarify the elaborate empirical base necessary for its implementation. Under ideal information conditions the researcher must have not only an original survey and classification of voters but successive opinion samples and a way of determining what appeals are communicated by the candidates. In each opinion sample respondents would convey their interpretation of an appeal made by a candidate, but it is quite possible that many of the respondents are totally unaware of the appeal. As McPhee suggests, the appeal can be visualized as a set of substimuli, perhaps possessing great diversity in the news media, and essentially voters sample from these substimuli. The model must then be tested by determining whether voting districts containing varying proportions of demographic voter types behave in a way that is consistent with the simulated results for voter types (called homogeneous demographic groups by McPhee). One empirical question is left open: How are the appeals identified? Does the researcher make educated guesses, using his own perceptions as the base? Does he simply ask the survey respondents to identify the appeals they perceive? Or does he make a systematic analysis of the news media?

In the absence of an elaborate empirical base, the implied parameters in the McPhee simulation design remain greatly underspecified. He does, however, assert a monotonic relationship between prior dispositions of voters toward candidates and the likelihood of accepting an appeal. If a voter strongly favors a candidate, he is much more likely to accept a weak appeal. This relationship can be expressed in Guttman fashion for discrete variables, instead of for continuous variables, as suggested by McPhee. People who strongly favor the candidate will accept even the weakest appeals. Those who favor the candidate, but not quite as strongly, do not accept the weakest appeals but accept any appeal of greater strength, and so on, until identified are those in the last group, who accept no appeal whatsoever.

Except for the above working assumption, the McPhee design is completely open. What is the relationship between sampling of stimuli by potential voters and objective measures of media transmission? To what extent do appeals, when accepted, modify dispositions toward candidates? Do voters reach a satisficing or saturation point after which they ignore political communication? How do voters influence each other? These are questions that need to be answered through precise simulation procedures. Since the original writing of this piece, Howard Rosenthal (1968) has provided an elaborate and imaginative interpretation of the McPhee simulator, giving answers to some of the above questions. Rosenthal, who also surveyed voting simulations, chose to operationalize the McPhee design but apparently felt that the Abelson-Bernstein model did not deserve the same attention. This writer's reaction was the converse.

Many of the questions left open by McPhee seem to have been more tightly managed in the parallel simulation of community referendum controversies by Abelson and Bernstein (1963), who developed a propositional system that borrows heavily from the literature of social psychology. In particular they dealt with a community fluoridation issue, but more important, their procedures and assumptions have wide social science applicability. The propositional system has the additional virtue of being sufficiently specific and interrelated to sustain simulative activity for an indefinite period of time.

The Abelson-Bernstein simulation is constructed as a cyclical process. Each cycle can represent a week of real time, and each cycle is a reasonable facsimile of every other cycle. Information input for the first cycle is derived from a sample survey of potential voters and an analysis of the news media. Questions in the sample survey elicit attitudinal positions, interest levels, socioeconomic characteristics of respondents and conversational friends of respondents, attraction to various communication channels (television, radio, telephone, newspapers, magazines, circulars, billboards, town meetings, etc.), and a variety of assertions relating to the topic under study. The many communication channels are simply analyzed for their assertion *content, attention value,* and *source* Ideally the communication channels would be examined each week, but the simulation would still operate if such information were gathered only at the beginning of the experiment.

Each cycle is governed by three sets of rules: (1) the effect of channel communications upon potential voters, (2) the effect of communications among potential voters through what the authors call the "method of pseudopartners," and (3) the nature of feedback, preparing each new cycle with modified information, making it theoretically unnecessary to take successive community surveys.

The propositional system central to the simulation is fairly difficult to interpret because the authors have not distinguished between those propositions that are limiting conditions and those that are bivariate monotonic relationships, nor have they explicitly highlighted those propositions that have regenerative feedback qualities. In some cases it is also difficult to know the form of the input data and whether certain variables have unique indexes or whether they are derived in a conceptual way from indexes given other variable names. An interpretation of their scheme is portrayed in figures 6-1 and 6-2.[1] The broken arrows represent feedback flow. Variables enclosed by rectangular boxes are based on operational definitions external to the flow of the simulation. Variables enclosed by circles are derived from other variables within the simulation flow. Limiting conditions are stated below each figure. The symbol i refers to the potential voter; c, the communication channel; s, the source of communication; and j, a conversation partner.

In figure 6-1 the potential voter is examined to determine whether he is exposed to any element of the set called communication channels. This may be accomplished by comparing a random number between 0 and 1.0 with a probability, x_3, derived from the empirical values of x_1 and x_2 (i's attraction to c and i's interest in the issue). Alternatively, the values of x_1 and x_2 can be summed, giving x_3, which is then compared to the random number y_1 which must fall within the range of possible values of x_3. If the potential voter is not exposed to a single communication channel he is then confronted by possible conversational partners in the manner described by figure 6-2. If he is exposed, however, the process described in figure 6-1 continues. From each exposure an assertion analysis is compared with the attitudes and predispositions of the potential voter to determine the extent to which the potential voter has altered the extremity of his attitude position on the issue and his level of interest in the issue (x_6 and x_2). The extremity of his attitude position is affected by his satisfaction with the source of communication (x_{12}), which is in turn determined by three other variables, x_4, x_8, and x_9 (also see conditions *C, D,* and *E*).

---

[1] This operationalization of the Abelson-Bernstein propositional system cannot be assumed to have a one-to-one correspondence to their computerized version, but it is hoped that the differences are minor and detract very little from the strength of their total concept.

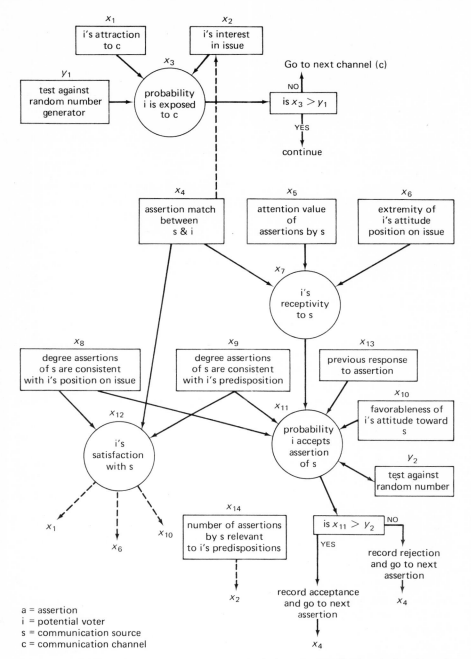

FIGURE 6-1   Interpretation of Abelson-Bernstein simulation, Part 1.

1. If $x10<0$, and if $x7$ is very low, and if $x13$ is empty, and if $x9<0$, calculate probability of $\sim a$ and test against random number.
2. If $x8<0$ and i accepts a, $x2$ decreases.
3. $x6$ changes only if s's attitude position, z, is more extreme than $x6$.
4. If condition 3 holds, $x6$ approaches z if $x10>0$, and moves away from z if $x10<0$.
5. $\Delta x_6 = z - x_6.$

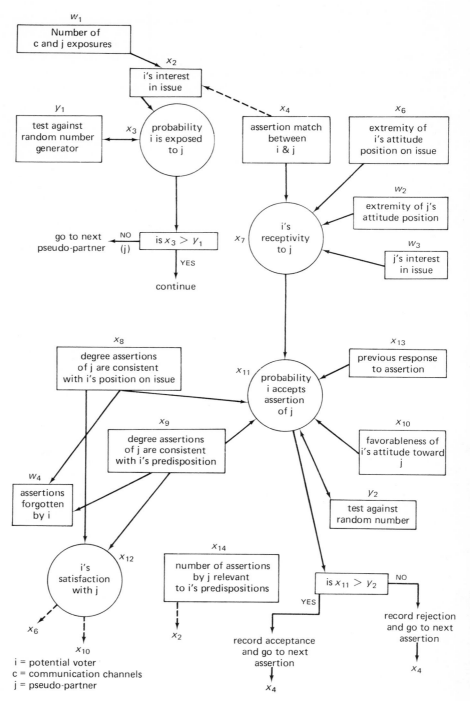

FIGURE 6-2   Interpretation of Abelson-Bernstein simulation, Part II.
1. Same as 1-4 in Figure 6-1.
2. J will not forget his assertion if accepted by i.

The potential voter's interest in the issue is altered in the first instance by the assertion match between s and i, which is in turn determined by the outcomes of prior exposures. The assertion pool of i is enlarged when the values of x_8, x_9, x_7, x_{10}, x_{13}, and x_{14} (see conditions A and B) yield a probability of acceptance (x_{11}) that exceeds a random number, Y_2.

After the possible channel exposures have been exhausted, the extremity of i's attitude position on the issue and his level of interest may further be modified by his conversations with other members of the public, described in figure 6-1. The process is quite similar to that developed for media transmission, and it should suffice to point out that there are only five important differences: (1) the level of interest of i is affected by the absolute number of exposures to elements of c and j, (2) j's interest in the issue and extremity of attitude position are taken into account, (3) assertions can be forgotten by i, (4) i may gain an attraction to the bias of an element of c, and (5) the pseudopartner, j will not forget an assertion if it is accepted by i.

Once every possible pseudopartner has been considered, the cognitive values representing i are stabilized, and a second member of the population is selected and simulated in the same way. After every member has been simulated, the cycle is complete, but only to be modified by subsequent cycles representing later periods of time. The general organization of the simulation may vary. It may occur as described above, but alternatively all members of the sample may first be considered in relation to media messages, treating conversational patterns as a distinct second stage of the cycle. Or possibly for each randomly selected member of the sample it may randomly be determined whether or not he is first tested by the media process or the conversational process, and for each randomly selected conversation partner it may randomly be determined whether he will also first be subjected to the media process.

When the last cycle has been completed, the simulation is tested by making two predictions. Voter turnout is predicted for any individual from his interest in the issue and the extremity of his attitude toward the issue. If he is predicted to show up at the polls, he is predicted to vote according to the direction of his attitude on the issue. The authors suggest that the simulation is designed to predict only general levels of turnout and preference and that there are too many unknowns to expect a high level of accuracy for individual members of the sample. Herein lies perhaps the most severe problem for political process simulators. What are the probabilistic guidelines for accepting or rejecting any given inference structure and any given inference within it? A rich data bank will always provide a large number of solutions for a group level prediction such as voter turnout.

In summary, the Abelson-Bernstein simulation may be said to have a number of strengths and weaknesses. On the positive side, it is a fully developed formal model, both operational and highly dynamic, drawing together strands of knowledge from social psychology, learning theory, and voting behavior. Sophisticated formal models, however, without extensive empirical testing, are highly tentative, and it would seem that some systematic means of altering the model would be necessary. The essential weakness of the Abelson-Bernstein simulation is not so much that when reported an empirical test had not been made or that some of the propositions may be contrary to fact as that it lacks a dynamic for altering its propositional content.

The simulations presented in the foregoing paragraphs focus upon processes that have an immediate effect upon voter choice. The simulations depend upon estimations of the interplay between candidate appeals and mass consumption during the campaign period. They do not describe an earlier step in the electoral process wherein people make choices to seek public office. In a recent study (Browning, 1968), a preliminary simulation model was developed for this purpose. The model suggests that certain political system characteristics (residency, age requirements, etc.) will limit the pool of possible office seekers by affecting both the recruiting practices of leaders and the perceptions of potential candidates. Within these limits, however, personality motivations and previous political socialization will have a strong impact upon the choice to seek office. Those who seek office on their own initiative are likely to have motivations different from those who are recruited by other leaders. Motivations will also influence the kind of office sought and the behavior of the individual once he is in office.

Except to the extent that Browning indicates how data values are assigned and altered for many variables, the exact dynamics of the computer program are not made clear. The simulation assumes that the researcher has a data bank on possible political aspirants, including social background information and personality motivation scores relating to "power," "achievement," and "affiliation," as developed in psychology by McClelland and others (see Browning, 1968, for citations). A second data bank examines the available political positions and the system characteristics that circumscribe each position. It is the interrelationship of these two data banks that yields predictions of who will seek what office. The simulation then adopts procedures similar to those already covered to predict winners and losers. Once the winners are determined, their motivation scores may be utilized to predict their style of conduct while in office—"passive," "status-oriented," "organization-oriented," "policy-concerned," or "policy-influencing," as specified by Browning (1968 , p. 103).

In essence, the Browning model sketches an evolution of political job filling, reaching from the recruitment stage to the "behavior in office" stage, and presumes that the entire process may be recycled for each election period. Verification of the utility of the model would require extensive interview documentation, as well as other types of record keeping. In a practical sense, the prominence of personality characteristics of elites injects a very high cost into the necessary research, even assuming the elites have been identified. It would be helpful to know whether the distribution of motivation scores was normally apportioned in selected populations (depending upon the level of office) and whether the scores on each of the three dimensions exhibited independence. Precise knowledge of distribution, and degree of independence, would for general purposes allow the simulation to operate in a probabilistic manner, perhaps giving adequate predictions over a large number of cases when controlling for political system characteristics.

LEGISLATIVE SIMULATIONS

The Abelson-Bernstein simulation has its counterpart in the study of legislative behavior, namely, in the work of Cherryholmes and Shapiro (1968). Their simulation is similar to the Abelson-Bernstein simulation in several respects: (1) the simulation is characterized by an elaborate communication-interaction inference structure, (2) this inference structure only modifies previously ascertained attitudes or dispositions, (3) they use a random number generator inherent in the method of pseudopartners, (4) they attempt to predict voting outcomes, and (5) many of the variables and inferences are similar to those of the previous study.

The Cherryholmes-Shapiro simulation is designed to account for the roll call voting behavior of members of the United States House of Representatives on bills that represent a *relatively* homogeneous subset of all bills that come out of committee for a roll call vote. They employ two highly related but separate simulations, one on "expansion of federal role" votes and one on "foreign policy" votes. At its inception, the simulation calculates the extremity of each member's prior position on a given bill by reference to his previous voting behavior on similar bills and by reference to several variables pertaining to the bill itself. A member's prior position is more favorable, for example, if he tends to vote for federal role bills (called memory), if the bill is sponsored by his party or state delegation, if he is a member of the originating committee or one of the bill's sponsors, and if the bill favors his region or constituency. His prior position is less favorable if he signs a committee minority report and for each of the above conditions that does not apply. In this phase of the simulation the effects of the above conditions are treated additively. Each condition yields a score of +1 or -1, except for the previous voting behavior dimension, which has a range of +2 to -2. The scores for each member are summed. At this point the final position of many members can be recorded. If the total score of any member does not fall within the range of +2 to -2, it is assumed that he has an extreme position and cannot be influenced. All members whose scores fall within the range can then be influenced by the communication process.

The communication process, too elaborate to describe fully here, is a variation on the method of pseudopartners, taking into account the structure of congressional business. The authors have acquired, from a broad search of legislative literature, estimates of (1) the number of interactions legislative participants are likely to have with their colleagues and (2) the types of legislators who are likely to interact (e.g., a Democrat is more likely to interact with a Democrat than with a Republican, or, as they estimate, four times more likely). These estimates make it possible to apportion probabilities among all possible interactions.[2] The simulation is organized so that there is a probability apportionment for president-representative communication, House leadership interaction, House rank-and-file interaction, interaction with House seniority leaders, and House leadership interaction with rank-and-file members.

When a representative enters the communication process he is confronted by the president and each of his 434 colleagues. As each pseudopartner is confronted, a probability is calculated in an additive way, determining one by one whether a series of conditions exist and adding the corresponding probabilities if appropriate until, finally, the conditions tests have been exhausted. The total probability is matched against a random probability to decide whether the interaction occurs. If the interaction occurs, the pseudo-partner's predisposition score, described above, is added to an influence total, which is the sum of the predisposition scores for all those with whom the representative has interacted. The final influence total is divided by the number of interactions (not confrontations), producing a mean score. The representative is then given a new position on the bill which falls halfway between his original position and the mean score. If this midpoint is negative he is predicted to vote against the bill, and, if positive, in favor of the bill.

The Cherryholmes-Shapiro simulation contains fewer dynamics than the Abelson-Bernstein simulation. A cyclical feedback process is not employed. For each bill the representative is confronted by each of his colleagues only once. Even though he changes his position on that bill, whatever he learns is forgotten before the next bill is considered. Perhaps this assumption of "forgetting" should explicitly be stated in the simulation. Although in the Abelson-Bernstein simulation several cycles are anticipated before a vote takes place, in the legislative situation each cycle can have its own vote; however, it would probably be necessary to chart exactly when the votes took place and how long they had been considered since being reported out by committee. Then a judgment could be made as to whether total forgetting was appropriate in each case and whether further inferences were needed.

Whatever could be done to the Cherryholmes-Shapiro simulation, it would still be difficult to improve upon its postdictive quality. The simulated total votes correlate at least 0.91 with the actual total votes, regardless of how the variables are defined. Furthermore, over the forty-eight bills, an average of 84 percent of the legislators vote consistently with their simulated vote. The simulation is also examined by systematically removing parts of its input. Most segments of the model, when removed, substantially reduced the performance of the simulation. The only omission that had little effect was that of "previous voting habits," called memory. In a sense this last finding simply strengthens the remainder of the simulation. The simulation can be run without recourse to earlier events produced by the same causes. One might guess that the remainder

[2] If, for example, there were one hundred members in a legislative chamber and a given legislator was said to interact with about twenty-five members on an important bill, the likelihood that he would interact with a member selected at random would be about 0.25. If he interacted with Democrats four times as often as with Republicans and the chamber had fifty members from each party, then the probability of interacting with a Democrat would be 0.4, and with a Republican, 0.1.

of the simulation would also postdict the earlier events with considerable accuracy.

In the ongoing study of other researchers (Matthews and Stimson, unpublished manuscript), this problem is being approached in a different way. Previous roll calls are characterized by the kinds of cues that were given (state delegation, party leadership, committee chairman, party majority, Democratic Study Group, etc.). Each member is given an average score for each cue which is based on the degree to which he votes with the cue. Each percentage is then subtracted from a chance criterion of 0.5 to produce a propensity index whose values range from –0.5 to +0.5. The index values constitute the legislators' predisposed responsiveness toward the cue givers. Subsequent votes are simulated by observing whether the cues are present, giving them a value (c) of +1, –1, or 0, depending upon whether they support a yes vote, no vote, or are not available. The legislator will respond to the cues in a precise order, responding first to the cue that for him has the highest negative or positive number on the propensity index, and last to the cue whose propensity value (p) most closely approximates 0. His vote is determined by whether Σ (pc) is positive or negative, with one limiting property. If at any point the sum is greater than 0.5 or less than –0.5, the legislator has acquired sufficient information to arrive at a position and the summation ceases, explaining, of course, why cues are confronted according to the absolute value of p.

The Matthews-Stimson technique has been applied to all roll call votes and has not been restricted to particular kinds of issues. Overall it has predicted *individual* voting with an accuracy of 85 to 90 percent, clearly a high degree of success, considering both the scope of their application and the relatively few assumptions they make. Their evidence seems to suggest that previous voting habits are highly significant for building cognitive decision maps regardless of issue content. Ideological inferences drawn from Guttman scaling procedures may be unnecessary.

There is a possible redundancy in the Matthews-Stimson simulation, one that can be empirically resolved but does not appear in their early report. In establishing the presence of cues in an upcoming vote, how is it possible to know whether certain cues are present on so many votes unless they are revealed by the vote itself? Such cues as "party majority" and "state delegation" appear to be taken from the actual vote, meaning that an individual's vote is partly predicted from his contributions to group voting patterns. The extent to which this redundancy will affect the prediction level may not be great, but it detracts from the validity of the assumption that legislators are actually observing cues. The degree to which their prediction level is biased can be tested by developing a set of random votes, each roll call corresponding in the total number of yea and nay responses to an actual vote, but any given legislator's vote is randomly determined. How well, given the same propensity scores, does the simulation predict?

Even though the simulations described above incorporate many of the same variables and yield predictions within the same range of accuracy, they do have radically different inference structures. The Cherryholmes-Shapiro simulation includes an elaborate interaction analogue, whereas the Matthews-Stimson simulation rests upon an ordered cognitive map. One weakness common to both techniques, and perhaps characteristic of most social science simulations, is found in their adherence to one-way causation. Reciprocal effects are not treated by direct analogy. In the Shapiro simulation a pseudopartner does not alter his position. In the second simulation a cue giver does not respond to a cue taker within the same policy area.[3]

Another similar ongoing study has focused upon postdicting whether or not the bills of a state legislature will be reported out of committee (Francis, 1970). It would seem appropriate here to describe only the simulation technique rather than the content. Essentially, the committee success of each bill is predicted from codified information about the characteristics of the bill and relevant legislative participants, including the governor and a few interest groups. All condition testing in the simulation is reduced to a binary path net structure whose termination points reflect whether a committee member opposes or supports a bill. The same variable may appear more than once in the net structure; however, different test values may be employed for continuous variables. In such cases, test values are usually either the mean or some multiple of the standard deviation. The basic statistical problem is to find the net structure that postdicts with the greatest accuracy. The solution requires a multicycle scanning process which proceeds within certain probability limits.

Since all condition tests result in a + or – value (e.g., Democrat or Republican, does the governor support the bill, is a sponsor on the committee, etc.) and the final result is either + or– (reported out of committee or withheld), the computer may be instructed to search for the variable containing a value that produces the fewest errors. If the governor supports 18 bills (+) and all 18 survive committee, then this criterion provides the first node and termination point in the net structure, providing, of course, that no other variable provides equal accuracy for a greater number of bills. The 18 bills are then removed from the simulation, and the process is repeated. Before accepting *any* inference, however, two requirements must be met: (1) the postdictions must be above a particular level of accuracy and (2) a sufficient number of bills must be accounted for by the inference. Shortly it will be shown how these two requirements are blended into one probabilistic criterion. For the moment let us say that only three variables are found wherein the two requirements are met— gubernatorial support, State Teacher's Association support, and committee chairmen sponsorship—and that 160 out of the original 460 bills are now "explained" by the three tests. The computer program then initiates another cycle, taking two variables at a time for the remaining 300 bills, searching for one of the following value arrays,

---

[3]This is *not* to say that the authors ignore reciprocity. Principal actors and pseudopartners do reverse roles in the Cherryholmes-Shapiro simulation; whereas in the Matthews-Stimson simulation, cue givers and cue takers reverse roles when the issues change (e.g., a member is the relevant chairman on one issue and is not on the relevant committee on a second issue).

```
+  +
+  —
—  +
—  —
```

that will postdict plus or minus bill outcomes with sufficient accuracy. When the utility of this search is exhausted, the variables can be taken three at a time, and so on, until the number of bills accounted for by any array are too few to meet the requirement. Actually, whenever a set of bills is removed by a successful test the simulation starts over, giving preference to a simpler explanation of the remaining bills.

The advantage of this procedure is that it allows for a probabilistic control for every logical inference. The control is implemented by determining what proportion of the bills in the *unexplained* pool actually pass or fail in committee and by subsequently calculating the likelihood that a postdiction will be achieved by chance. For example, if two-thirds of 460 bills pass committee and the gubernatorial test correctly predicts that 18 bills will pass, the chance calculation for each success is $(2/3)^{18}$. For more complex situations, either combinatorial probabilities or z transformations on the normal curve can be used; however, it is more complete and accurate to use the former by way of the following common equation:

$$_nC_rP^rQ^{n-r} + \frac{n!}{(n-r)!(r)!} P^rQ^{n-5}$$

The probabilities resulting from application of this formula are compared in the simulation with some desired departure from chance, say 0.05, to be considered for acceptance. The variable test giving the lowest probability is accepted, and the simulation is recycled for the remaining bills. As now constructed, the simulation has a definite bias toward parsimony. A single variable test yielding a sufficiently low probability will be accepted before multivariate tests are implemented. The degree of parsimony bias, however, is determined by the desired acceptance level. One feature of the procedure is that it allows a comparison of net structures for different acceptance levels. Net structures may also be compared for different time periods in the legislative session by simply entering time as a partitioned variable. One other feature of the program is that it will find postdictive relationships that are contrary to the researcher's preconceived notions.

SUMMARY

The simulation studies covered in this chapter may roughly be classified as best-fit and experimental heuristics. The Abelson-Bernstein, Browning, Cherryholmes-Shapiro, and McPhee simulations can be called *experimental,* primarily because the rules for changing the simulation content and flow are left open. To a somewhat greater degree, the Pool-Abelson-Popkin, Matthews-Stimson, and

Francis simulations push experimentation back to the derivation and selection of data. All the simulations make some assumptions about cognition; however, the Abelson-Bernstein and the Shapiro simulations make a definite analytic and empirical distinction between cognition and interaction. The Matthews-Stimson and Francis simulations treat interaction abstractly by the use of ordered cognitive net structures. The main advantage of the latter technique is that it absorbs the gradation between intense conversational interaction and low-cost observation of cues. The converse disadvantage is obvious, but if intense interaction is highly related to remote cue taking (i.e., the pseudopartner and cue giver tend to be one and the same), it will be difficult to establish which analogue is most appropriate. Further research may show that the interaction-cognition distinction provides greater explanatory power only on certain types of legislation, perhaps when the legislation is so significant that it requires considerable negotiation, well beyond what is normally required. To some extent this speculation is supported by the Cherryholmes-Shapiro simulation which found that in the "foreign policy" area, the "interaction" phase made a difference only on roll call votes over the final passage of bills.

Rather than dwell on the infancy of political process simulation, it is perhaps more relevant to point out briefly what will be necessary in the future for work of this type to gain acceptance in the scientific community. For the individual researcher it is extremely helpful to remain free of the inertia created by established modes of data analysis and verification. Exciting analogues of voting, legislative, and other forms of political behavior are probably more likely to flourish in the spirit of unconstrained inference. In the end, however, imaginative simulation procedures must be assessed not only by their ability to predict selected decision outcomes but by more complete inductive statistical procedures. The sensitivity testing implemented by Shapiro is a step in the right direction, and if carried out for every inference rather than segment of the model, will constitute a convincing form of confirmation. The Francis procedure, based upon inductive probabilities, is another alternative when the researcher has assigned dichotomous attributes to each condition test. These procedures, however, are clearly inadequate to account for the total problem of eliminating alternative models and specific inferences. Failure to manage this problem, it seems, will inordinately delay the success of the simulation enterprise.

REFERENCES

Abelson, Robert, and A. Bernstein, "A Computer Simulation Model of Community Referendum Controversies," *Public Opinion Quarterly*, 27 (1963), 93-122.

Browning, Rufus P., "Quality of Collective Decisions: Some Theory and Computer Simulations." Unpublished manuscript, 1969.

———, "The Interaction of Personality and Political System in Decisions to Run for Office: Some Data and a Simulation Technique," *Journal of Social Issues*, Vol. XXIV, No. 3 (1968).

Campbell, Angus, Philip E. Converse, Warren E. Miller, and Donald E. Stokes, *The American Voter* (Survey Research Center, University of Michigan). New York: John Wiley & Sons, Inc., 1960.

Cherryholmes, Cleo H., and Michael J. Shapiro, "Representatives and Roll-Calls: A Computer Simulation of Voting in the Eighty-Eighth Congress." Indianapolis: The Bobbs-Merrill Co., Inc., 1968.

Coleman, James S., *Game of Democracy*. Washington, D.C., National 4–H Club Foundation.

Francis, Wayne, L., "Simulation of Committee Decision-Making in a State Legislative Body," *Simulation and Games*, September 1970. Originally delivered at the annual meeting of the American Political Science Association, 1969.

Gamson, William A., *Simsoc: Simulated Society* (Instructor's and Participant's Manuals). New York: The Free Press, 1969.

Goodman, Robert F., "The Inter-Community Simulation: Participants' Manual," UCLA, Department of Political Science, 1967.

Klahr, David, "A Computer Simulation of the Paradox of Voting," Carnegie Institute of Technology, *The American Political Science Review*, Vol. LX (June 1966).

Lewis, Eugene, "Influence: A Simulation of Community Decisionmaking Processes," Unpublished manuscript based upon "The Game of Influence" developed by the High School Curriculum Center in Government, Indiana University.

McPhee, William N., "Note on a Campaign Simulator," in *Formal Theories of Mass Behavior*, Chap. 4, 169-83. London: Collier-Macmillan, Ltd.; New York: The Free Press of Glencoe, 1963.

Matthews, Donald R., and James A. Stimson, "Decision-Making by U.S. Representatives: A Preliminary Model." Unpublished manuscript, University of North Carolina at Chapel Hill, 1968.

Pool, Ithiel De Sola, Robert P. Abelson, and Samuel L. Popkin, *Candidates, Issues, and Strategies*. Cambridge, Mass.: The M.I.T. Press, 1965.

Riker, William A., "Bargaining in a Three Person Game," *The American Political Science Review*, Vol. LXI, No. 3 (September 1967).

Shapiro, Michael J., "The House and the Federal Role: A Computer Simulation of Roll Call Voting." Ph.D. dissertation. Northwestern University, Department of Political Science, 1966.

———, "The House and the Federal Role: A Computer Simulation of Roll-Call Voting," *The American Political Science Review*, No. 2 (June 1968), 495-517.

THE HOUSE AND THE FEDERAL ROLE:
A COMPUTER SIMULATION
OF ROLL-CALL VOTING

MICHAEL J. SHAPIRO

Much of the business of the U.S. Congress in the post war period has involved issues concerning the size and scope of activities of the federal government. The legislation in this area can be traced, for the most part, to measures which originated during the period of the New Deal in response to the Great Depression and to measures enacted during World War II to meet the short-run exigencies attendant to rapid economic and social mobilization. From the point of view of the expansion of the federal role, the Eisenhower years are of some moment. While they marked a lull in the expansionist trend witnessed under the Democratic presidencies of Roosevelt and Truman, their significance lies in the fact that despite the change in administrations, there was no reversal of the policies begun during the Roosevelt years. While most of the Republican legislators were on record in opposition to the expansion of the federal role, the failure of the Republican Party to introduce and enact legislation to reverse the trend of federal expansion resulted in a new plateau of federal activity from which the congressional dialogue was to proceed during the Kennedy and Johnson Administrations.

While the 87th Congress, meeting during Kennedy's first two years in the White House, did not enact the quantity of legislation expanding the federal role that Kennedy had called for in his inaugural, in the 88th Congress both parties supported a larger federal role to a greater extent than they had previously. In fact the first sessions of the 88th Congress as it bears on the federal role has been summed up as follows: "At no time did the majority of both parties reject a larger federal role." (*Congressional Quarterly Almanac,* 1963, p. 724). With two exceptions, the statement holds true for the second session in 1964.

Reprinted with permission from the *American Political Science Review,* 62 (June 1968), 495-517. Copyright ©1968 by The American Political Science Association. The author is indebted to Dr. Cleo Cherryholmes of the Department of Political Science at Michigan State University, who shared in the design of the simulation model and rendered valuable advice and assistance in its application reported here. The author wishes, in addition, to thank Dr. Harold Guetzkow and Dr. Kenneth Janda of the Department of Political Science at Northwestern University, who aided immeasurably in   conception and execution of the research upon which this article is based.

The research reported in this article is based upon a computer simulation study of roll-call voting on twenty-one bills in the 88th session of the U.S. House of Representatives dealing with the expansion of the Federal role. Its purpose is twofold: first, to elucidate and explicate the determinants of the success of legislation resulting in the expansion of the federal role; and second, to develop and test a model that represents a theoretical overview of the roll-call voting process in the House.

THE MODEL

This simulation model of roll-call voting in the U.S. House of Representatives focuses on two basic processes. The first is a cognitive process involving the predisposition toward voting on a bill that a representative develops when he assesses his past voting behavior on that type of bill, the individuals and groups in the House with positive and negative positions on the bill, and the substance of the bill in terms of its benefits and disadvantages for his constituency and region of the country. The second is a communication process in which representatives who do not develop strong predispositions on a bill in the first phase of the model confront the President and their colleagues and receive influence.

The Predisposition Phase

The predisposition phase of the model is depicted in the flow chart of Figure 1. In this phase a predisposition score is calculated for each representative, one at a time, as the attributes of the bill in question are compared with those of the representatives.[1]

The first determination made when a bill is entered into the model and confronts the first representative is whether or not the representative's past voting on that type of bill suggests a bias or predisposition for or against the bill. This aspect of the model is based upon the notion that representatives have ideological postures that provide cues for their roll-call voting behavior.[2] The use of past votes on federal role bills to determine a representative's ideological

---

[1] The attributes of the bills were determined from their descriptions in the Congressional Quarterly Almanac for 1963 and 1964. The bills chosen for the simulation study are essentially the same as those identified by the Congressional Quarterly Service as pertaining to the federal-role issue area. A few recommital motions on these bills have been added to provide a larger number of bills for purposes of analysis in general and to allow, in particular, for comparisons between bills introduced by the Democratic Administration and those introduced by Republicans (the latter having been reflected, for the most part, in recommital motions).

[2] For a number of different perspectives on the effects of personal ideology on legislative roll-call voting, see D. R. Brimhall and A. S. Otis, "Consistency of Voting in Our Congresses," *Journal of Applied Psychology*, 32 (1948), 1-14; Charles Farris, "A Method for Determining Ideological Groupings in the Congress," *Journal of Politics*, 20 (1958), 308-338; Duncan MacRae, *Dimensions of Congressional Voting: A Statistical Study of the House of Representatives in the Eighty-First Congress* (Berkeley: University of California Press, 1958); and Lewis A. Froman, Jr., *Congressmen and Their Constituencies* (Chicago: Rand McNally, 1963).

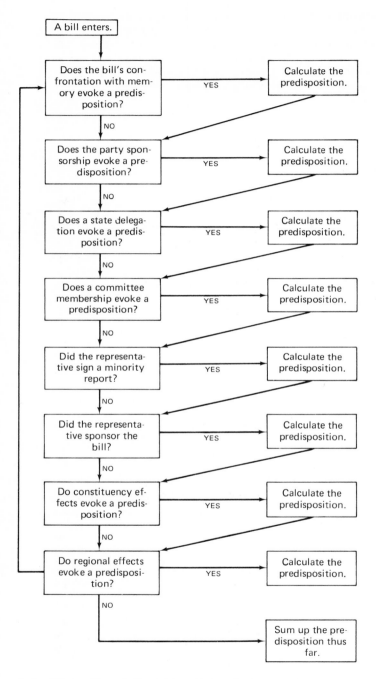

FIGURE 1   The predisposition phase of the model.

position with respect to present federal role bills is based upon the discovery of what constitutes, in effect, a federal role dimension in a number of roll-call analyses of a legislative voting.[3] Thus weights of two or one are added to or subtracted from the predisposition scores of each representative if he has voted consistently for or against bills dealing with the expansion of the federal role.[4]

The remaining variables in the predisposition phase of the model relate, first of all, to the individuals and groups that support or oppose the bill in question, the parties, state delegations, individual sponsors, committees that

[3]From the standpoint of face validity, the case for the unidimensionality of congressional attitudes on the federal role is a good one. In recent years it has been one of the few issue areas that has consistently distinguished the platforms of the two parties. This consistency has been reflected by the increasing tendency for both casual observers and systematic investigators of politics to employ this issue dimension as a liberal-conservative index. In one case, the use of representatives' past votes on the federal role as a liberal-conservative index has provided an instance of empirical validation of the issue area's unidimensionality (Froman, *op cit.*). It was found that congressmen from constituencies associated with conservative voting are low on the index (at the conservative end), while those from constituencies associated with liberal voting are high on the index (at the liberal end). This type of validation is not completely satisfying in as much as Froman's criteria for liberal and conservative voting with which the constituencies were purported to be associated are not entirely independent of the factors or bills that make up the liberal-conservative index employed. A better case for the unidimensionality of representatives' attitudes (inferred from voting behavior) towards the federal role is based upon inferences from the results of Guttman scaling techniques applied to congressional roll-call votes. This scaling technique, based upon rank-ordering procedures, was designed to tap the unidimensionality of attitude clusters. See Louis Guttman "The Basis for Scalogram Analysis," in S. A. Stouffer (ed.), *Measurement and Prediction* (Princeton, N. J.: Princeton University Press, 1950). MacRae, in his Guttman scale analysis of roll-call votes in the 81st Congress, found what he termed "Fair Deal" and "Welfare State" scales for Democrats and Republicans respectively (*op. cit.*). Within both of these scales were most of the roll-call votes concerned with the federal role. Anderson obtained similar scales in his analyses of four succeeding congresses. See Lee Anderson "Variability in the Unidimensionality of Legislative Voting," *Journal of Politics,* 26 (1964), 568-585. Finally, a Guttman scaling analysis was conducted on the 21 bills included in my simulation study. When two bills were eliminated from this analysis, a reproducibility coefficient of .90 was obtained. The two bills that did not scale were not coded for a memory effect in the simulation.

[4]After some testing of the model, predisposition weights of plus or minus two were assigned to representatives who had voted for the expansion of the federal role either eighty-five percent of the time of more or fifteen percent of the time or less. Predisposition scores of plus or minus one were assigned for percentages between sixty-five and eighty-five percent and between fifteen and thirty-five percent. Predisposition scores based upon past voting were calculated only for those representatives who had voted on at least ten bills in this issue area. In general our measure of past voting behavior on the federal role issue dimension was based on the previous two House sessions. All other weights for variables in the predisposition phase of the model are plus or minus one. It can be observed, then, that the predisposition phase of the model provides a theoretical framework for combining propositions about legislative voting behavior that have emerged from more inductive legislative studies. This theoretical framework is based upon the assumptions that predispositions are additive and that all the predisposing variables have approximately equal weight. It is important to note that this last assumption relates to initial predispositions and not to voting behavior. As will be noted below, these variables appear in the communication phase of the model where their effects are not equal.

have reported the bill out of the elected leadership,[5] and secondly to the substance of the bill. Thus each bill is coded in terms of the region or regions of the country and types of constituencies it might aid or hinder.[6]

The representative in this conceptualization is thought of as developing a predisposition on the basis of his ideological posture, and the audiences he may be expected to perceive as relevant to a bill. This predisposition phase, however, represents only part of this theoretical conception of the roll-call voting process in the House. After predispositions have been calculated for all the representatives for a given bill, the communication phase of the model is activated.

The Predisposition-Communication Linkage

Not all the representatives are influenced in the communication phase of the model on each bill, but all representatives are potential influences in the communication process. Each representative's potential influence over his colleagues on a given bill is a function of the predisposition that he develops during his confrontation with it in the predisposition phase of the model. Only those representatives whose predispositions fall within the range between plus two and minus two enter the communication process to be influenced. The final votes of all others are determined in the predisposition phase of the model.

The decisions on who enters the communication phase of the model are based upon propositions from both social-psychological findings of influence and conformity in groups in general and findings related specifically to the influence on representatives' voting behavior. From the former we get the proposition that individuals with more extreme attitude positions are less susceptible to influence than individuals with less extreme attitude positions, and from the latter the proposition that a legislator with two or more pressures

[5]Two of these types of predisposition calculations require some clarification. On the basis of our knowledge of the pervasiveness of party conflict in legislatures I have constructed the model such that a coding of a bill for party sponsorship which predisposes all the members on one party to vote for the bill (adding one to all their predisposition scores) has the effect of predisposing all the members of the other party in the opposite direction (subtracting one from all their predisposition scores). In the case of a bill coded from committees, all members of a committee through which the bill has passed are predisposed for the bill except those representatives who signed the minority report. These are predisposed against the bill.

[6]It should be noted that when bills are coded such that some representatives acquire a predisposition as a result of an attribute of their constituency, it is then determined whether or not their constituency is competitive. If the representative comes from a competitive constituency, he acquires an additional unit of predisposition in the same direction. Competitive constituencies were designated as those that had not returned a representative from the same party in the last four consecutive congressional elections. The constituency attributes included in this study were percent urban, percent rural farm, percent rural non-farm, percent negro, percent foreign stock, percent owner occupied dwellings, percent white collar, and degree of competitiveness. With the exception of the competitiveness variable, the raw data on these variables was obtained from the *Congressional District Data Book, Districts of the 88th Congress,* 1963.

such as party and constituency influencing his vote in the same direction is unlikely to be susceptible to further influence concerning his vote.[7]

The decisions involved when a representative approaches the communication phase of the model are depicted in the flow chart in Figure 2. It can be noted that we have employed the simplifying assumption that the communication process for all legislators is based upon the predispositions derived before the process is begun. No cycling of the process is carried out to take account of the possibility of a diffusion of influence in which predispositions generated as a result of communications among representatives become the basis for further influence in later communicative interactions.

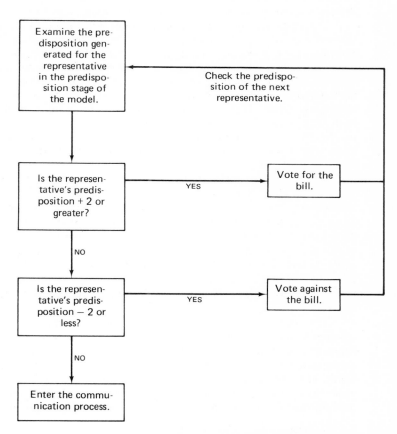

FIGURE 2   The communication entry criteria.

[7]For evidence with respect to the attitude extremity proposition see Carl Hovland, O. J. Harvey, and Muzafer Sherif, "Assimilative and Contrast Effects in Reactions to Communication and Attitude Change," *Journal of Abnormal and Social Psychology,* 55 (1957), 244-252. The proposition on immunity to influence of representatives with two influences predisposing them in the same direction is suggested by Julius Turner, *Party and Constituency: Pressures on Congress* (Baltimore: The Johns Hopkins Press, 1951), p. 125.

The Communication Phase

Bauer, Dexter, and Pool have asserted that "the U.S. House of Representatives is above all a communication node which serves to unify a large and heterogeneous land by bringing into confrontation a group of men representing its various parts."[8] It has been observed, on the other hand, that the patterns of interaction in legislative bodies tend to reinforce rather than moderate basic cleavages.[9] In constructing this communication process I have attempted, based upon investigations of interactions in legislative bodies, to simulate the types of confrontations that a legislator is likely to engage in when, because of either a lack of significant cues or a conflict among them, he has not evinced a strong predisposition to vote for or against a given bill.

The method employed to simulate these confrontations is an adaptation of the method of "pseudo partners" used by Abelson and Bernstein in their flouridation referendum simulation.[10] With this technique they matched conversation partners on the basis of potentiality for conversation, which was calculated as a function of social network information, a meshing of demographic characteristics and shared habitual loci of discussions about community issues. They based their conversation rates upon the levels of interest in the issue under discussion. I have adapted this procedure to the social network in the House by basing my conversation partner matching decisions on knowledge of the structure and processes of legislative systems in general and the House in particular.

The variables chosen to represent the structure of the communication network in the House are based upon observations of the frequency of interactions between representatives with attributes of their conversation partners and the general observations that representatives seek information from certain types of colleagues and that some representatives, particularly those in positions of leadership, seek to give information or exert influence on certain types of bills. For the purposes of communication, then, each representative is described in terms of his party, state, region, constituency characteristics, leadership positions, committee membership, rank on each committee, and seniority.

The first determination made when a representative enters the communication process is whether or not he has a conversation with the President. Such a conversation symbolizes, for the sake of simplicity, a conversation with either the President or one of his legislative liaisons. The inclusion of the President in the communication process is predicated on the observation that presidential-legislative relations of this type have been institutionalized in the post war period.[11] As is indicated in the flow chart of

[8]Raymond Bauer, Lewis Dexter, and Ithiel de Sola Pool, *American Business and Public Policy* (New York: Atherton Press, 1964).

[9]John Wahlke, Heinz Eulau, William Buchanan and Leroy Ferguson, *The Legislative System: Explorations in Legislative Behavior* (New York: Wiley, 1962).

[10]Robert Abelson and A. Bernstein, "A Computer Simulation of Community Referendum Controversies," *Public Opinion Quarterly*, 27 (1963), 93-122.

[11]See for example Richard Neustadt, "Presidency and Legislation," *American Political Science Review*, 49 (1955), 980-1021.

Figure 3, I have confined the President's communications with representatives to cases in which legislation that is a part of his program is concerned.

The probabilities of conversation generated when the President confronts a representative are depicted in Figure 3. To arrive at these probabilities we begin

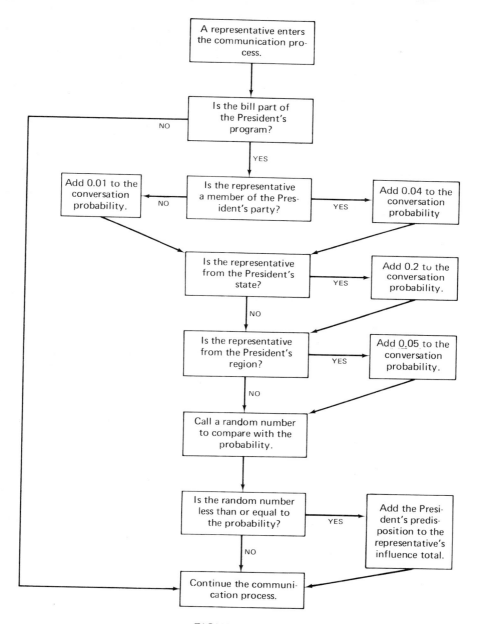

FIGURE 3 Communication with the President.

by setting a limit on the average number of representatives that the President might be expected to confront on bills that are a part of his program. Here we assume that his number of contacts would approximate those of the formal leaders of the majority party in the House (the Speaker, Majority Leader, and Whip). After setting such a limit (approximately 45 contacts) we can then apportion these contacts among the types of legislators he confronts in such a manner that his total number of contacts on a given bill will approximate our limit. For these approximations we assume that his frequencies of interaction with the various types of representatives are similar to those of a representative. He is thus given more contacts with the members of his party, representatives from his home state, and representatives from his region of the country.

When the probability of conversation between the representative and the President is calculated, a random number is called; if the random number is less than or equal to that probability, it is assumed that a conversation took place and the President's predisposition on the bill (plus two if he favors the bill and minus two if he opposes it) is added to the influence total for that representative.

The representative then enters that part of the communication process within which a determination is made as to the members of the House with whom he will converse. In this part of the communication process he confronts all 434 of his colleagues. For each confrontation a probability of conversation is calculated, based on a matching of attributes of the representative and his potential influencers. As in the case of confrontations with the President, a random number is called after each of these confrontations, compared with the probability of conversation generated, and, when it is less than or equal to that probability, results in the addition of the predisposition that the conversation partner developed in the predisposition phase of the model to the representative's influence total.

The formal leadership of the House is prominent in the communication process because, as Truman has observed, "The House has four and one half times as many members as the Senate. From this schoolboy fact comes the tendency for the formal leadership in the House to correspond closely to the actual. . . ."[12] Thus in the first part of the matching process (shown in the flow chart in Figure 4) it is determined whether the representative entering the communication process is an elected leader. If he is, it is then determined whether or not his potential conversation partner is an elected leader. The flow chart in Figure 4 indicates the probabilities of conversation generated when the various types of leadership confrontations take place. These probabilities are derived from the general observation that the leadership in the House interacts frequently, and from the findings of investigations of state legislative bodies which indicate that the presiding officer (speaker) and majority leader have a close relationship and that the interactions between these and the minority leaders are somewhat less frequent.[13] They are based, in addition, on the finding that frequent interactions occur between the whips and whip organizations, and between these and the other elected leaders.[14]

[12] David Truman, *The Congressional Party* (New York: Wiley, 1959).

[13] Wahlke *et al., op. cit.,* p. 225.

[14] Randall Ripley, "The Party Whip Organizations in the United States House of Representatives," *American Political Science Review,* 58 (1964), 561-576.

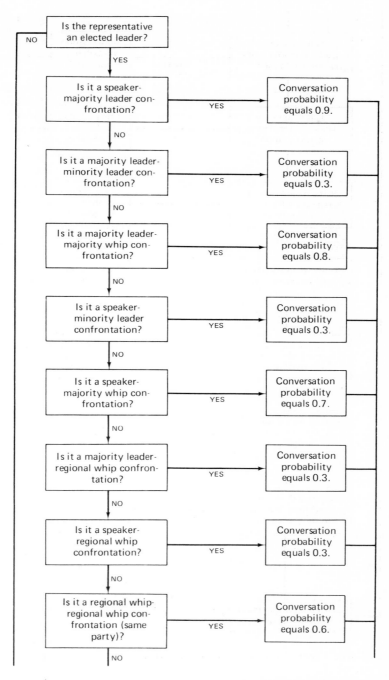

FIGURE 4   Leadership interactions.

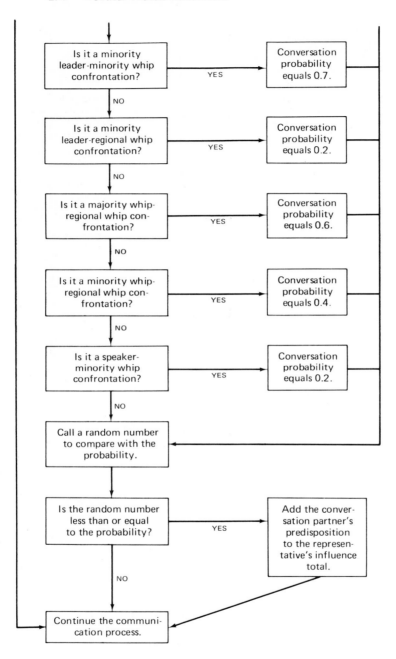

FIGURE 4 (cont.)   Leadership interactions.

The probabilities of conversation among the legislative leaders are based, in part, on the average total number of communications in which a non-leader becomes involved in legislative bodies. To distinguish formal elected leaders, I made use of the finding that the leaders become involved in twice as many communicative interactions on the whole as do the rank and file members of legislative bodies.[15] The probabilities appearing in Figure 4 are thus based upon an extrapolation from the total number of communications in which leaders become involved. The relative apportioning of probabilities for inter-party as opposed to intra-party confrontations is explicated in our discussion of non-leadership interactions below.

The flow chart in Figure 5 depicts the beginning of that part of the communication phase of the model in which a representative who holds no positions of elected leadership becomes involved after having confronted the President. The first determination made in this section is whether the potential conversation partner is a regional whip assigned to the region of the representative. The high probability of conversation generated when the representative confronts his regional whip is based upon the finding that such a conversation actually does take place on most bills.[16]

The next set of determinations depicted in Figure 5 constitutes the matching of attributes between the representative and his potential conversation partner. As in the case of the other types of confrontations discussed thus far, the probabilities of conversation generated are based on the limit set for the average total number of communications in which a representative tends to become involved. That average falls between twenty-five and thirty and is derived from investigations of communicative interactions in legislative bodies. It has been found that in general legislators in state legislative bodies exert personal influence over approximately a dozen colleagues. We have added another half dozen to take account of the larger size of the House which increases the possibility of a wider acquaintanceship, and the remainder has been added to represent the communications that a representative has with both elected and seniority leaders in the process of carrying out the task of legislating.[17]

The apportioning of the probabilities for confrontations between representatives involves a matching procedure. This procedure begins with a determination of whether the representative and his potential conversation partner are members of the same party. The relative size of the probabilities for inter- and intra-party interactions is based on the finding that a representative tends to have four times as many communicative interactions with members of his own party than with members of the opposition.[18]

The matching procedure then continues with a determination of whether the representative and his potential conversation partner are from the same state delegation. The additional probability of conversation added for intra-state

[15]Wahlke *et al., op. cit.,* p. 226.

[16]Ripley, *op. cit.*

[17]See Wahlke *et al., op. cit.,* p. 222 for data on the dozen basic interactions. The additional interactions are based upon the findings of Truman, *op. cit.,* pp. 145-275.

[18]See Garland Routt, "Interpersonal Relationships and the Legislative Process," *Annals of the American Academy of Political and Social Science,* 19 (1938), 129-136, and Wahlke *et. al., op. cit.,* p. 224.

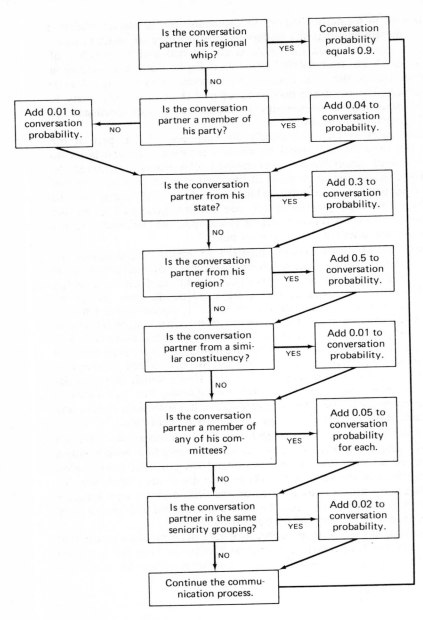

FIGURE 5   Rank and file interactions.

confrontations is based upon the observation that state delegations in Congress are somewhat cohesive in their voting behavior and upon congressmen's reports that their state delegations are important sources of their communications about legislation.[19] The size of the probability relates, again, to the limit set for the total number of communications a legislator can be expected to have and is weighted on the basis of inferences drawn from the above sources as to the importance of the state delegation as opposed to other factors in the communication process.

The next determination in the matching procedure is whether the representative and his potential conversation partner come from the same region of the country. The probability of conversation added for intra-regional confrontations is based upon both observations of interactions in legislative bodies and reports of conversation frequencies by members of the house.[20] The probability is based, in addition, on the limit set for total communications as in the case of the other probability increments.

The next matching determination depicted in the flow chart of Figure 5 is whether the representative and his potential conversation partner represent similar constituencies in terms of their urban-rural attributes. The probability of conversation added for a confrontation between representatives with like constituencies is based upon the finding that those with similar constituencies engage in a greater-than-average number of communications in a state legislature, and upon the finding that there is a slight tendency for those with similar constituencies to demonstrate cohesion on roll-call votes in the House.[21]

The next matching determination for confrontations among representatives relates to committee membership. The probability of conversation added when the confrontation is between representatives who share a committee assignment is based, first of all, on the common sense supposition that those who share committee assignments will have some communications in the course of carrying out those assignments, and secondly on the finding that those sharing committee assignments in the House demonstrate a slightly greater than average cohesiveness on roll-call votes.[22]

The last matching determination in this phase of the communication process relates to the effect of shared seniority status on communicative interactions of representatives. The probability of conversation added for shared seniority status is based upon the finding that interaction in legislatures occurs disproportionately within seniority groupings.[23]

---

[19]See Truman, *op. cit.*, p. 253; Donald Matthews *U. S. Senators and Their World* (Chapel Hill: University of North Carolina Press, 1960); John Kessel, "The Washington Congressional Delegation," *Mid-West Journal of Political Science*, 8 (1964), 1-21; Lewis Froman, Jr. and Randall Ripley, "Conditions for Party Leadership: The Case of the House Democrats," *American Political Science Review*, 59 (1965), 52-63; and Marc Ross, "Some Correlates of Voting Support for the Leadership in the House of Representatives" (Mimeo, Department of Political Science, Northwestern University, 1965).

[20]Routt, *op. cit.*, p.135; Samuel Patterson, "Patterns of Interpersonal Relations in a State Legislative Group: The Wisconsin Assembly," *Public Opinion Quarterly*, 23 (1959), 101-109; Ross, *op. cit.*

[21]Allan Fiellin, "The Functions of Informal Groups: A State Delegation," in R.L. Peabody and N. W. Polsby (eds.), *New Perspectives on the House of Representatives* (Chicago: Rand McNally, 1963), p. 66; and Truman, *op. cit.*, p. 210.

[22]Truman, *op. cit.*, p. 277.

[23]See Wahlke *et al.*, *op. cit.*, p. 224; and Routt, *op. cit.*, p. 135.

In addition to the criteria for the matching decisions enunciated above, the probabilities of conversation for these and other types of confrontations are based upon the finding that representatives in the majority party are more cohesive on roll-call votes than those in the minority party and on the finding that more interactions take place between members of the majority party.[24] The use of this criterion can be noted in the apportioning of probabilities of conversation for interactions between leaders and rank and file representatives to be described below.

The flow chart in Figure 6 depicts the determinants of communication that are included to account for interactions between rank-and-file representatives and the seniority leaders and technical experts in the House, the chairmen and members of the committees through which the bill that is the subject of communications passes. The probabilities of conversation added for these confrontations are apportioned on the basis of a matching of party between the representative and his potential conversation partner, the assumption being that the representative seeking technical advice on bills will display biases similar to those displayed in his choices of other conversation partners. In addition, in keeping with the finding that the seniority leaders (committee chairmen) are especially cohesive with their party on bills that have gone through their committees, the chairmen are weighted more heavily in the communication process.[25] The higher ranking members of the committees relevant to the bill in

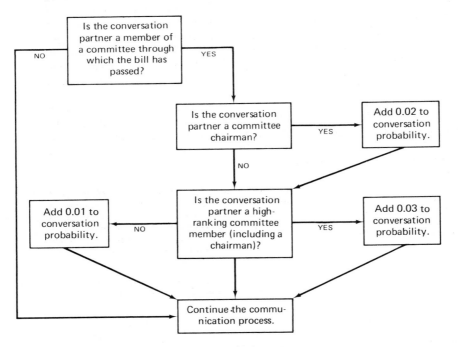

FIGURE 6   Interactions with the seniority leaders.

[24]Truman, *op. cit.*, p. 286; Routt, *op. cit.*, p. 135.

[25]*Ibid.*, p. 240.

question are given more weight in the communication process also, based upon the finding that technical experts receive more communications than rank and file legislators.[26]

The final determinations in the communication phase of the model relate to the probabilities of conversation added when a representative confronts an elected leader. The flow chart in Figure 7 depicts the matching procedure involved. The probabilities of conversation added are based, in part, on the finding discussed above that leaders are involved in twice as many communications as non-leaders and, again, on the limit set for total communications for each representative. The apportioning of probabilities for the various types of leaders is based on the reports of communicative interactions by congressional representatives.[27] As in the case of the seniority leaders, representative-leadership interactions are incremented only for intra-party confrontations.

An overview of the communication phase of the model is depicted in the flow chart of Figure 8. It can be noted that when the random number falls within the probability of conversation calculated for a given confrontation, the predisposition of the conversation partner is added to the influenced representative's influence total. When a representative has confronted all 434 of his colleagues, his new predisposition is calculated on the basis of the influence he has received in the communication in which he has become involved. This calculation is carried out by dividing the sum of the predispositions with which the representative is confronted by the number of conversations in which he has been involved to obtain the average predisposition or attitude toward the bill with which he has been confronted. The representative's new predisposition then becomes one that is halfway between his old one and the average one that he has confronted. This influence calculation is based upon research findings of attitude change and conformity in situations of communicated influence in small groups.[28]

The outputs of the communication process are thus the new predispositions of those representatives who receive influence in that phase of the model. It is these new predispositions that determine their votes on the bill in question.

THE PERFORMANCE OF THE MODEL

Because this simulation model aggregates the voting choices of each individual representative to obtain roll-call results, the basic evaluation of its

[26]Matthews, *op. cit.*, p. 252.

[27]Ross, *op. cit.*

[28]See S. C. Goldberg, "Three Situational Determinants of Conformity to Social Norms," *Journal of Abnormal and Social Psychology*, 50 (1954), 325-329; C. I. Hovland and H. A. Pritzker, "Extent of Opinion Change as a Function of Amount of Change Advocated," *Journal of Abnormal and Social Psychology*, 54 (1957), 257-261; C.I. Hovland, O. J. Harvey, and M. Sherif, *op. cit.*; S. Fisher and A. Lubin, "Distance as a Determinant of Influence in a Two-Person Serial Interaction Situation," *Journal of Abnormal and Social Psychology*, 56 (1958), 230-238; P. G. Zimbardo, "Involvement, Communication Discrepancy, and Conformity," *Journal of Abnormal and Social Psychology*, 60 (1960), 86-94.

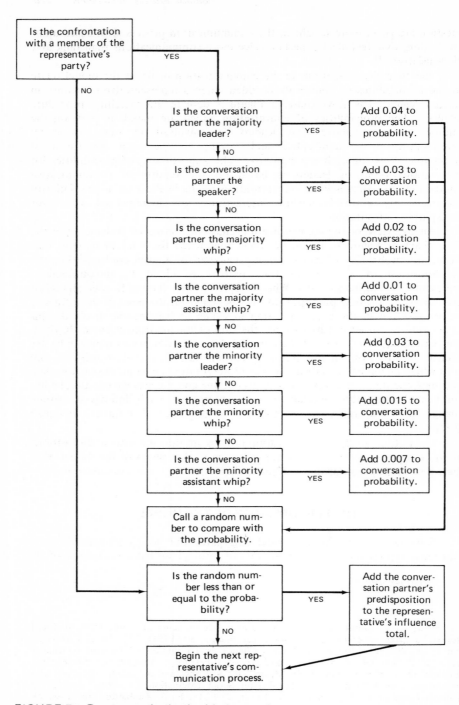

FIGURE 7   Representative-leadership interactions.

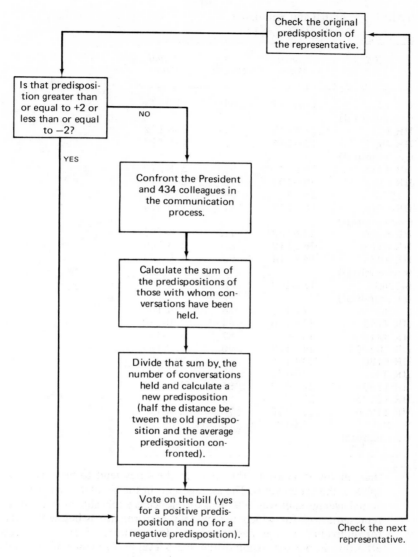

Check the original predisposition of the representative.

Is that predisposition greater than or equal to +2 or less than or equal to −2?

NO

YES

Confront the President and 434 colleagues in the communication process.

Calculate the sum of the predispositions of those with whom conversations have been held.

Divide that sum by the number of conversations held and calculate a new predisposition (half the distance between the old predisposition and the average predisposition confronted).

Vote on the bill (yes for a positive predisposition and no for a negative predisposition).

Check the next representative.

FIGURE 8   An overview of the communication phase.

performance has been carried out at both the macro and micro levels. The former involves the extent to which the voting outcomes in the simulation approximate the actual voting outcomes in the Eighty-Eighth Congress, and the latter considers the extent to which each individual representative's voting behavior in the Congress is approximated by his voting behavior in the simulation.

Considering, first of all, the macro level or voting outcomes as a whole we represent the simulated and actual voting outcomes in Table 1. It can be noted

TABLE 1   Actual and Simulated
Voting Outcomes

Bill	Simulated Voting	% Correct	Actual Voting
HR 12 (recommittal)	171-239	84	171-239
HR 12	280-128	77	288-122
HR 4955 (recommittal)	153-244	92	181-217
HR 4955	382-13	92	378-27
HR 6143	294-104	70	287-113
S 1576	353-3	94	335-18
HR 6196 (recommittal)	167-237	81	179-224
HR 6196	232-168	78	216-182
HR 6518	262-119	82	273-109
HR 4955 (recommittal)	159-214	90	180-193
S 2265 (recommittal)	153-209	86	174-188
S 2265	241-121	76	254-107
HR 7152	323-100	84	290-130
HR 8316	362-0	87	317-43
HR 10222	245-173	90	229-189
HR 6196	238-175	89	211-203
HR 3881	230-172	82	212-189
HR 11377	247-172	85	226-185
HR 12175	371-3	82	308-68
HR 11926	221-170	78	218-175
S 3060 (recommittal)	148-197	76	107-237

that, in the simulation, the splits in the votes tend to be larger than were the splits in the actual votes. The average split in the simulated votes is 141 while the actual average split was 109. This difference is explicable in terms of the types of representatives for which the model's performance is below its overall average for individual voting performances. The sources of error will receive more attention in our discussion of the micro level performance of the model. Despite this tendency for the simulation to over-emphasize the splits in the votes, it can be noted that the correct outcomes are obtained by the simulation for all the bills. Using the splits in the simulated and actual votes as raw scores, moreover, we obtain a product-moment coefficient of correlation between the simulated and actual voting outcomes of .97.[29] While this suggests a very positive

[29]If the sums of the yes and no votes are used as raw scores instead of the splits between the yes's and no's on each vote, correlation coefficients become .95 and .96 respectively. It should be noted, however, that because the product-moment correlation coefficient adjusts for the situation in which the range of the values for one variable is higher than that for the other, the coefficient we have obtained slightly overstates the degree of fit between the simulated and actual roll-call voting outcomes.

evaluation of the simulation model, the assessment of its performance must be made in the context of a consideration of what kinds of performances are obtained with variations or alternative formulations of the model.

Before turning to this type of assessment, however, the micro level performance of the original model remains to be considered. The percentages of individuals voting the same way in the simulation as they actually voted are presented for each bill in Table 1. The average percentage of individuals correctly simulated across the twenty-one bills is 84. Breaking that figure down for the three different categories of party sponsorship we find that for the four bipartisan bills the average percentage of individual representatives correctly simulated is 89. The percentages obtained for the eleven Democratic Administration bills and the six Republican sponsored recommital motions are 82 and 85 respectively. The assessment of these differences and all others reported must be made in the context of the amount of variation in results that can be obtained because of the stochastic variation of the communication phase of the model.[30]

These percentages obtained for the three party-sponsorship categories are explicable in terms of the types of individuals for which the simulation performs best. The most significant difference is between Democrats and Republicans. Averaging the percentages across all the bills yields scores of 86 and 79 correctly simulated Democratic and Republican representatives respectively. All of this difference is accounted for by the Democratic Administration-sponsored bills, for representatives in both parties voted correctly in the simulation: an average of 83 percent on the Republican-sponsored recommittal motions and an average 85 percent on the four bills with bipartisan support. The simulation thus tends to over-emphasize the cohesiveness of the Republicans on Democratic-sponsored bills despite the attempt to build greater cohesiveness for Democrats than Republicans into the communication phase of the simulation model described above.[31] It may be that it is this poorer micro level performance of the simulation for Republican representatives that accounts for part of the lower percentage of correctly simulated representatives on Democratic-sponsored bills.

Turning to the regional breakdown of the extent to which individual representatives are correctly simulated, we find that the poorest performance for regions obtained is for the southern representatives, with an average percentage correct across the twenty-one bills of 78. The next lowest percentage correct is for the eastern representatives at 83. The midwestern, mountain and far western representatives are simulated correctly with percentages of 88, 86, and 86 respectively.

It would appear that the simulation model confronts the same difficulty as the Democratic Whip organization, predicting the voting behavior of the Southern Delegation.[32] Because of the difficulty of discerning those bills on

---

[30]To determine this variation two bills were run ten times each, one being a bill with a large number of representatives entering the communication phase and the other with relatively few. In each case, the amount of variation in the percentage of representatives correctly simulated did not exceed one percent. Thus any difference in percentages that is two percent or greater can be treated as an actual difference.

[31]In keeping with previous findings cited in a number of the works noted above, the structure of the communication process provides for more communicative interactions within the majority party.

[32]See Froman and Ripley, *op. cit.*

which the conservative coalition of Republicans and southern Democrats takes shape, we must code the southerners as regular Democrats and thus predispose them in the same manner as the other Democratic representatives, with respect to party affiliation. By so doing, we miss a sizable percentage of them on occasions when they side with Republicans against "liberal legislation" such as bills proposing the expansion of the federal role. It should be noted in addition that it is undoubtedly the incorrect simulating of the voting of some southern representatives that accounts, in part, for the greater average split in the simulated votes as opposed to the actual votes, for the switching over of southerners has the effect of moderating the heavily party-oriented cleavages that account for the larger splits in the simulation.

The slightly lower average percentage correctly obtained for eastern representatives is probably explicable in terms of the traditional liberalism of eastern representatives. Several eastern Republican representatives usually side with Democrats on liberal legislation, and in general, eastern representatives from both parties tend to have more liberal voting records than their colleagues from other parts of the country. Our simulated voting results on the communications regulation bill, HR 8316, illustrate this phenomenon quite clearly. Coding the bill as a bipartisan measure resulted in an average percentage of representatives correctly simulated of 87. Because this bill results in the diminuition of the federal role, however, a number of liberal easterners who had voted against the bill ended up for it in the simulation as a result of the bipartisan effect. Thus a below-average percentage of eastern representatives of 84 was obtained. Again, this slightly lower performance for eastern representatives would have the effect of over-emphasizing the voting splits as is the case with the performance on southern representatives.

Looking at the performance of the simulation model for the different representatives in terms of the types of constituencies they represent reveals a range of correctly simulated individual representatives from 86 per cent for representatives with predominantly urban constituencies to 81 per cent for representatives with predominantly rural non-farm constituencies. The representatives with predominantly rural farm constituencies were correctly simulated on an average of 82. These differences are probably a result of the distribution of Republicans and southern Democrats in terms of constituency types. A disproportionate share of non-southern Democrats represent constituencies with a high proportion of urban residence while Republicans and southern Democrats disproportionately represent rural farm and non-farm constituencies.

One interesting constituency-type variation in per cent correctly simulated is for competitive versus non-competitive districts. Representatives from the former were simulated correctly on an average of 83 per cent.[33] While this can be interpreted as evidence of the relatively stronger hold that the parties (the strongest determinant in our model) have over representatives who are less firmly established, the fact that competitive districts are not equally allocated on a regional or party basis confounds the clear interpretation of this result.

Moving to the consideration of a representative categorization that is closely related to the competitiveness of constituencies we can look at the

---

[33]These percentages are computed for only those twelve bills on which constituency effects are coded.

extent to which the voting behavior of representatives with different seniority positions is correctly simulated. The findings from legislative research suggest that representatives with high seniority rankings are less likely to be loyal to their party and are thus far less predictable in terms of their roll-call voting.[34] The results of the simulation tend to support this proposition: it yields an average per cent correct figure of 87 for the low-seniority members of the House while the average per cent correct figure obtained for representatives with high seniority standing is 82. Again, this finding must be assessed in the context of possible confounding factors, not the least of which is the fact that southern representatives possess a disproportionate share of the high-seniority positions in the House.[35]

We have seen thus far that the model as originally constituted simulates the voting on federal role issues in the 88th Congress with results that match the outcomes of the actual voting quite closely at both the macro and micro levels. We have, in addition, gained some insights into the extent to which the voting behaviors of different types of representatives are explicable in terms of our theoretical formulation of the voting process. In the final section I turn to a consideration of the validity of my theoretical formulation by examining the extent to which alternative formulations of the simulation model can replicate the results obtained with the original formulation.

THE VALIDITY OF THE SIMULATION MODEL: ALTERNATIVE FORMULATIONS

Pool, Abelson, and Popkin have addressed themselves to the problems of evaluating a complex simulation model:

> A complex model can predict real-world outcomes and yet be wrong in many details. It may predict accurately because the main effects are correctly represented and yet the model may contain many irrelevances. So one must always question the details of a complex model, even if it passes the test of good prediction.[36]

Simulation is a modeling strategy that is perhaps ideally suited for the questioning of the details of a model. Assumptions can be changed, and in a matter of minutes the data can be processed on the reformulated model. In this section we turn to a consideration of the details of our simulation model of roll-call voting in the U. S. House of Representatives in order both to examine the validity of the model as it was originally formulated and to test the

[34] See Oliver Garceau and Corrine Silverman, "A Pressure Group and the Pressured: A Case Report," *American Political Science Review,* 48 (1954), 672-691; Truman, *op. cit.;* Froman, *op. cit.*

[35] If this inquiry were directed toward a rigorous examination of such effects as seniority, we could control for confounding effects and operate the model with these controls to obtain these more specific insights into the effects of individual mechanisms.

[36] Ithiel de Sola Pool, Robert Abelson, and Samuel Popkin, *Candidates, Strategies, and Issues: A Computer Simulation of the 1960 and 1964 Presidential Elections* (Cambridge: MIT Press, 1964), p. 64.

sensitivity of the determinants or effects programmed into the model to represent those processes and propositions that we had deemed to be significant factors in the roll-call behavior of the members of the House.

The Party Effect

There is little doubt that on the American legislative scene, party is the most significant factor in roll-call voting cleavages. Recent quantitative analyses of roll-call voting in the Congress have explored this phenomenon systematically.[37] Because party is a prominent determinant in our simulation model, appearing as both a predisposing factor and a communications parameter, a systematic assessment of the performance of the model requires the consideration of the extent to which the party effect contributes to that performance.

We thus examine, first of all, the performance of a simulation model in which party is the only effect. Maintaining the same coding rules for party sponsorship and omitting all other determinants we obtain the voting results shown in Table 2. The actual results and the results obtained with the original model are also shown for purposes of comparison. Again using the voting splits as raw scores, we obtain a product-moment coefficient of correlation between

TABLE 2   The Results of the "Just Party" Model

Bill	Original Model	% Correct	Party Model	% Correct	Actual Voting
HR 12 (recommittal)	171-339	84	171-239	84	171-239
HR 2	280-128	77	240-170	77	288-122
HR 4955 (recommittal)	153-244	92	163-235	93	181-217
HR 4955	382-13	92	399-0	95	378-27
HR 6143	294-104	70	237-163	60	287-113
S 1576	353-3	94	357-0	95	335-18
HR 6196 (recommittal)	167-237	81	169-235	81	179-224
HR 6196	232-168	78	232-168	78	216-182
HR 6518	262-119	82	220-163	78	273-109
HR 4955 (recommittal)	159-214	90	160-213	90	180-193
S 2265 (recommittal)	153-209	86	153-209	86	174-188
S 2265	241-121	76	210-152	74	254-107
HR 7152	323-100	84	423-0	69	290-130
HR 8316	362-0	87	362-0	87	317-43
HR 10222	245-173	90	242-176	90	229-189
HR 6196	238-175	89	237-177	89	211-203
HR 3881	230-172	82	235-167	75	212-189
HR 11377	247-172	85	245-167	84,	226-185
HR 12175	371-3	82	374-0	81	308-68
HR 11926	221-170	78	0-0	00	218-175
S 3060 (recommittal)	148-197	76	148-197	76	107-237

[37]See Turner, *op. cit.;* McRae, *op. cit.;* Truman, *op. cit.;* and Froman, *op. cit.*

the actual voting splits and the "just party" simulated voting splits of .80 as compared with the .97 coefficient obtained with the original model. The micro level performance or average percentage of representatives correctly simulated by the "just party" model is 78 as compared with the 84 per cent obtained with the original model. It is thus clear, on the one hand, that party plays an important role in the performance of the simulation model, but on the other, it is also clear that the remaining effects built into our model are not irrelevant, for the results obtained with only the party effect are not as good as those obtained with the original model at either level.

Another way to assess both the model as a whole and the effect of the party variable is to run the model in its original form without including the party effect. When this is done the performance criteria are markedly similar to those obtained when just party is used to simulate the voting. The voting results obtained with this variation of the model are presented in Table 3 along with the results obtained with the "just party" model and the actual votes for purposes of comparison. The product-moment coefficient of correlation between the actual voting splits and those obtained with party removed from the model is .68, somewhat lower than that obtained with just party as the model. But the average percentage of representatives correctly simulated is 78, the same figure obtained for the "just party" simulation.

Two types of interpretations suggest themselves from the sensitivity testing of the model with the party effect the subject of our manipulations. The

TABLE 3   The Results of the "Just Party" and "Without Party" Models

Bill	"Just Party" Model	% Correct	Actual Voting	"Without Party" Model	% Correct
HR 12 (recommittal)	171-239	84	171-239	158-250	84
HR 12	240-170	77	288-122	353-55	78
HR 4955 (recommittal)	163-235	93	181-217	120-175	83
HR 4955	399-0	95	378-27	285-113	70
HR 6143	237-163	60	287-113	301-96	82
S 1576	357-0	95	335-18	272-18	75
HR 6196 (recommittal)	169-235	81	179-224	127-272	75
HR 6196	232-168	78	216-182	257-137	76
HR 6518	220-163	78	273-109	315-66	80
HR 4955 (recommittal)	160-213	90	180-193	121-243	84
S 2265 (recommittal)	153-209	86	174-188	171-188	88
S 2265	210-152	74	254-107	288-72	75
HR 7152	423-0	69	290-130	165-243	64
HR 8316	362-0	87	317-43	265-97	75
HR 10222	242-176	90	229-189	265-149	87
HR 6196	237-177	89	211-203	247-164	86
HR 3881	235-167	75	212-189	209-190	85
HR 11377	245-167	84	226-185	255-156	86
HR 12175	374-0	81	308-68	297-75	79
HR 11926	0-0	00	218-175	224-170	79
S 3060 (recommittal)	148-197	76	107-237	130-213	79

relatively similar performance of the model with just party and without party suggests that the original model is somewhat overdetermined in terms of party. The absence of party in the second reformulation of the model is partly compensated because of the extent to which the communications process is structured by a party effect, but the communications is also structured by the preceding predisposition phase of the model. This suggests that there may be a party surrogate in this prior phase that accounts for the similar performances. We examine this possibility below.

The other type of interpretation afforded by manipulation of the party effect relates to the types of issues processed in the model. Included in Table 3 along with the juxtaposed voting results of the "just party" and "without party" simulations are the percentages of representatives correctly simulated on each bill. If we examine the three different categories of party sponsorship, bipartisan, Democratic, and Republican, separately, we can determine the extent to which the party effect is significant in each. These are broken down into the party sponsorship categories in Table 4.

TABLE 4   Comparison of the Original "Just
Party," and "Without Party" Models
on Three Party Sponsorship Categories

	Original Model % Correct	"Just Party" Model % Correct	"Without Party" Model % Correct
Bipartisan Bills	89	90	75
Republican Sponsorship	85	85	71
Democratic Sponsorship	82	70	79

Considering, first of all, the bills coded as bipartisan we obtain an average percentage of correctly simulated representatives of 90 with the "just party" model, slightly better than the 89 per cent obtained over the same bills with the original model. The percentage obtained with the "without party" model is only 75, considerably lower. Similarly, we obtain an average percentage of correctly simulated representatives on the Republican-sponsored recommittal motions of 85 with the "just party" model, the same figure obtained with the original model. The percentage obtained with the "without party" model is 71, again considerably lower.

When we examine the eleven remaining Democratic Administration bills, however, we find that the percentage of correctly simulated representatives is only 70 for the "just party" model as compared with the 82 per cent figure obtained with the original model. The "without party" model, on the other hand, correctly simulates an average of 79 per cent of the representatives on these bills. This reversal in performances between the "party only" and "without party" models suggests some validity for our presupposition about treating recommittal motions as incidents of party conflict rather than mediating the party effect with substantive coding for region and constituency variables as in the case of the Democratic Administration bills. At a general level, this result tends to validate the inclusion of constituency and regional effects in the model, for these enhance its performance. To assess the inclusions of regional and constituency effects more critically, however, we must directly manipulate them with further sensitivity testing of the model.

The Constituency Effect

Investigations of the relationship between the demographic and political characteristics of constituencies and the roll-call behavior of their representatives have been extensive. The findings from these studies all point in the direction of such a relationship.[38] While these investigations have not systematically explored the constituency effect from issue to issue in terms of the extent to which different types of issues result in different degrees of constituency loyalty on the part of representatives, Miller and Stokes' recent investigation explored the effect in this manner.[39] They found, among other things, that representatives tend to heed the desires they perceive their constituents to have as well as the ones they actually do have, particularly on domestic welfare legislation. The close similarity between the welfare legislation included in the Stokes and Miller study and our bills dealing with the federal role is the basis upon which the decision to code constituency effects was made.

In order to examine the validity of constituency coding in general and the specific coding decisions made in particular, I have run the model on the twenty-one bills without the constituency coding that had been included in the original run. The voting results obtained with this alternative formulation of the model are presented in Table 5 along with the actual voting results and the voting results obtained with the original model for purposes of comparison. The coefficient of correlation between the actual splits and those resulting from the "without constituency" simulation is .80 as compared with the .97 coefficient obtained between the original model and the actual votes.[40] It would appear that the constituency effect is not an irrelevant part of the simulation model.

To examine the extent to which the specific constituency coding decisions were well advised, we consider the micro level performance of the model without the constituency coding. The average percentage of representatives

---

[38]Turner, *op. cit.;* Duncan MacRae, "The Relation Between Roll-Call Votes and Constituencies in the Massachusetts House of Representatives," *American Political Science Review,* 46 (1952), 1046-1055; MacRae, *op. cit.,* 1958; J. Roland Pennock "Party and Constituency in Postwar Agricultural Price Support Legislation," *Journal of Politics, 18* (1956), 167-210; Froman, *op. cit.*

[39]Warren Miller and Donald Stokes, "Constituency Influence in Congress," *American Political Science Review,* 57 (1963), 45-56.

[40]When we compute the correlation between the actual splits and the "without constituency" splits across only the twelve bills on which constituency was originally coded we obtain a coefficient of .78 as compared with a .95 obtained with the original model for the same bills.

TABLE 5   The Results of "Without Constituency" Model

Bill	Original Model	% Correct	Without Constituency	% Correct	Actual Voting
HR 12 (recommittal)	171-239	84	170-239	84	171-239
HR 12	280-128	77	242-168	77	288-122
HR 4955 (recommittal)	153-244	92	149-249	91	181-217
HR 4955	382-13	92	380-17	91	378-27
HR 6143	294-104	70	169-230	66	287-113
S 1576	353-3	94	352-4	94	335-18
HR 6196 (recommittal)	167-237	81	168-236	81	179-224
HR 6196	232-168	78	233-166	78	216-182
HR 6518	262-119	82	224-159	79	273-109
HR 4955 (recommittal)	159-214	90	158-215	90	180-193
S 2265 (recommittal)	153-209	86	150-212	86	174-188
S 2265	241-121	76	213-149	74	254-107
HR 7152	323-100	84	297-126	88	290-130
HR 8316	362-0	87	362-0	87	317-43
HR 10222	245-173	90	243-175	90	229-189
HR 6196	238-175	89	238-176	88	211-203
HR 11377	230-172	82	236-166	76	212-189
HR 12175	247-172	85	245-167	84	226-185
HR 12175	371-3	82	351-20	83	308-68
HR 11926	221-170	78	317-0	42	218-175
S 3060 (recommittal)	148-197	76	147-198	76	107-237

correctly simulated on the twelve bills for which constituency was coded is 78 for the run with the "without constituency" model. This compares with an average of 85 per cent obtained for those twelve bills with the original model. While this is not conclusive evidence of the validity of the specific coding decisions made for the constituency effects, it argues strongly for that validity. It tends to validate the model as a whole and to support previous findings on the effect of constituency factors on legislative roll-call voting behavior.

The Region Effect

Sectionalism or regional cleavages in congressional voting behavior have been found in many investigations.[41] In our simulation of bills dealing with the federal role, only four were coded as having regional effects. All of these were coded to suggest a negative predisposition on the part of southern representatives.

To test the extent to which my regional coding aided in the performance of the simulation model we processed the twenty-one bills without the regional coding. The voting results obtained with this "without region" model are

[41]See for example H. G. Roach, "Sectionalism in Congress (1870-1890)," *American Political Science Review*, 19 (1925), 500-526; G. Grassmuck, *Sectional Biases in Congress on Foreign Policy* (Baltimore: John Hopkins Press, 1951); V. O. Key, *Southern Politics* (New York: Alfred A. Knopf, 1949).

presented in Table 6 along with the actual results and those obtained with the original model for purposes of comparison. The product-moment coefficient of correlation between the splits on the actual bill and those resulted from the "without region" simulation is .89 as compared with the .97 coefficient obtained between the actual voting splits and those resulting from the original simulation model.[42] The regional coding thus substantially improves the performance of the simulation model even though only four bills are coded as subject to a regional effect.

TABLE 6   The Results of the "Without Region" Model

Bill	Original Model	% Correct	Without Region	% Correct	Actual Voting
HR 12 (recommittal)	171–239	84	171–139	84	171–239
HR 12	280–128	77	287–123	75	288–122
HR 4955 (recommittal)	153–244	92	165–233	92	181–217
HR 4955	382–13	92	385–13	92	378–27
HR 6143	294–104	80	338–61	70	287–113
S 1576	353–3	94	354–3	94	335–18
HR 6196 (recommittal)	167–237	81	168–236	81	179–224
HR 6196	232–168	78	232–167	78	216–182
HR 6518	262–119	82	252–129	81	273–109
HR 4955 (recommittal)	159–214	90	159–213	90	180–193
S 2265 (recommittal)	153–209	86	154–208	86	174–188
S 2265	241–121	76	239–121	77	254–107
HR 7152	323–100	84	423–0	70	290–130
HR 8316	362–0	87	265–97	75	317–43
HR 10222	245–173	90	244–174	90	229–189
HR 6196	238–175	89	238–176	89	211–203
HR 3881	230–172	82	244–158	78	212–189
HR 11377	247–172	85	246–166	85	226–185
HR 12175	371–3	82	371–3	82	308–68
HR 11926	221–170	78	225–169	78	218–175
S 3060 (recommittal)	148–197	76	147–198	76	107–237

To test the extent to which my specific regional coding was well informed, we examine the micro level performance of the "without region" model. The average percentage of representatives correctly simulated with this variation of our original model is 77 as compared with an average of 83 correctly simulated with the original model for the four bills on which region effects were coded. We thus have partial validation of the regional coding decisions. It would appear that the substantive provisions of the simulation model (including both constituency and regional effects) is an important part of the legislative process involved in roll-call voting. The inclusion of these effects has aided the overall performance of the model at the macro level and provided better individual performance of the model on the bills for which coding makes the substantive provisions relevant.

---

[42]When we compute the correlation between the actual splits and the "without region" splits for only the four bills on which region was originally coded we obtain a coefficient of .62 as compared with a .99 obtained with the original model for the same bills.

The Memory Effect

On all but two of the bills, memory is coded in terms of whether the bill proposes an expansion or diminution of the federal role. Each representative thus has his predisposition affected if his past votes on the federal role issue dimension suggest a positive or negative attitude toward the expansion of the federal role. To examine the extent to which memory is an important effect in our model I have run the bills through the model without including the memory effect. The results of the "without memory" simulation are presented in Table 7 along with the actual and original simulation results for purposes of comparison.

Looking at the macro level performance of the "without memory" model, we obtain a product-moment coefficient of correlation of .97, the same coefficient obtained with the original model. The lack of contribution of memory can be further highlighted by viewing the results obtained when just memory is used to simulate the votes. The product-moment coefficient of correlation obtained between the actual voting splits and those resulting from the "just memory" simulation is only .21.

These findings suggest that the memory effect does not aid in the macro level performance of the simulation model. The presence of the memory effect may, however, account for the findings obtained when the party effect was manipulated. When we correlate the splits resulting from the "just party" simulation with those resulting from the "just memory" simulation we obtain a coefficient of .62. This suggests that memory tends to coincide to some extent with party membership and is thus somewhat of a party surrogate which may explain part of the similarity of the results obtained with the "just party" and "without party" simulations.

TABLE 7   The Results of the "Without Memory" Model

Bill	Original Model	% Correct	Without Memory	% Correct	Actual Voting
HR 12 (recommittal)	171−239	84	171−239	84	171−239
HR 12	280−128	77	336−72	71	288−122
HR 4955 (recommittal)	153−244	92	153−249	93	181−217
HR 4955	382−13	92	399−0	95	378−27
HR 6143	294−104	70	314−80	73	287−113
S 1576	353−3	94	357−0	95	335−18
HR 6196 (recommittal)	167−237	81	169−235	81	179−224
HR 6196	232−168	78	232−168	78	216−182
HR 6518	262−119	82	280−102	81	273−109
HR 4955 (recommittal)	159−214	90	160−213	90	180−193
S 2265 (recommittal)	153−209	86	166−196	83	174−188
S 2265	241−121	76	271−89	71	254−107
HR 7152	323−100	84	322−100	84	290−130
HR 8316	362−0	87	362−0	87	317−43
HR 10222	245−173	90	246−171	90	229−189
HR 6196	238−175	89	237−177	89	211−203
HR 3881	230−172	82	172−229	78	212−189
HR 11377	247−172	85	245−167	84	226−185
HR 12175	371−3	82	374−0	81	308−68
HR 11926	221−170	78	226−168	78	218−175
S 3060 (recommittal)	148−197	76	148−197	76	107−237

Despite the small role that memory has been shown to have in the macro level performance of the simulation model, the micro level performance of the "without memory" simulation model indicates that the memory effect is not entirely superfluous. The average percentage of representatives correctly simulated by the "without memory" model is 82 as compared with the 84 per cent figure obtained when memory is included. The memory effect is thus a slight aid in the micro level performance of the model.

Of some interest is the performance of the memory effect on the different types of bills. Comparing the average percentage of correctly simulated representatives obtained with the "without memory" model and the original model for the four bipartisan bills yields the same figure of 89. Similarly, the "without memory" model results in the same average percentage of representatives correctly simulated across the six Republican sponsored recommittal motions as in the original model, 85. It is on the Democratic Administration bills that memory appears to aid the micro level performance of the model.

While the original model correctly simulated the voting behavior of 82 per cent of the representatives on the Democratic Administration bills, the "without memory" model correctly simulated only 79 per cent of the representatives on these bills. Because it is these bills that deal with the substantive issues involved in legislation, the memory effect probably conveys the impact of ideological postures that have been found to have some effect on legislative roll-call behavior.[43] In the case of recommittal motions and bipartisan bills, the content of the bill that tends to evoke ideological effects is obfuscated by procedural concerns and the bipartisan effect respectively.

The Communication Effect

We turn finally to a consideration of the contribution made by the communications phase of the model. Again this assessment is made by testing the sensitivity of the effect by removing it. The voting results obtained when the model is run without the communications process are presented in Table 8 along with the actual and original simulation model voting results for purposes of comparison. Examining, first of all, the macro level performance of this alternative formulation of the model we obtain a product-moment coefficient of correlation between the actual voting splits and those resulting from the "without communications" simulations of .88 as compared with the .97 coefficient obtained between the actual and original model voting splits.

The micro level performance of the model is also enhanced by the communication process. The average percentage of representatives correctly simulated by the "without communications" simulation model is 78 as compared with the 84 per cent correctly simulated by the original model. We thus have a basis for believing that the communications process is not an irrelevant part of our overall model.

Several important questions can be raised with respect to internal aspects of the communications phase of the model. In constructing the process I found

[43]See footnote 2, page 265.

TABLE 8.   The Results of the "Without Communications" Model

Bill	Original Model	% Correct	Without Communication	% Correct	Actual Voting
HR 12(recommittal)	171−239	84	173−222	84	171−239
HR 12	280−128	77	254−119	72	288−122
HR 4995 (recommittal)	153−244	92	155−233	90	181−217
HR 4955	382−13	92	303−34	74	378−27
HR 6143	294−104	70	188−133	62	287−113
S 1576	353−3	94	278−13	77	335−18
HR 6196 (recommittal)	167−237	81	170−221	79	179−224
HR 6196	232−168	78	220−168	78	216−182
HR 6518	262−119	82	219−120	76	273−109
HR 4955 (recommittal)	159−214	90	158−200	88	180−193
S 2265 (recommittal)	153−209	86	164−169	84	174−188
S 2265	241−121	76	207−121	72	254−107
HR 7152	323−100	84	298−100	83	290−130
HR 8316	362−0	87	265−96	68	317−43
HR 10222	245−173	90	237−170	89	229−189
HR 6196	238−175	89	226−176	87	211−203
HR 3881	230−172	82	169−177	77	212−189
HR 11377	247−172	85	228−168	85	226−185
HR 12175	371−3	82	285−25	74	308−68
HR 11926	221−170	78	214−167	77	218−175
S 3060 (recommittal)	148−197	76	147−183	76	107−237

that changes in key mechanisms such as the salience of the party effect resulted in a dimunition of the performance of the model as a whole. Changes in the average number of communications of representatives had similar effects on the performance of the model as a whole. A thorough examination of the sensitivity of all the important effects in the communication process has not been conducted as yet, but the indications derived from the examinations thus far suggest that the communications phase of this model is quite sensitive to changes in parameters and mechanisms.

CONCLUSION

The fact that every alternative formulation of the model undertaken thus far has decreased some aspect of the model's performance lends evidence in support of the contention that the relatively good match obtained between the actual roll-call voting in the House on federal role issues and in our simulation is not a result of a number of errors or ill-advised hypotheses that cancel each other out to make the model right for the wrong reasons. What is especially significant is the fact that the theoretical model based upon propositions from legislative and face-to-face group research can be used to simulate the roll-call voting on an issue area in two sessions of the House with good performances at the macro and micro levels.

The results of the research just reported can be assessed from various perspectives. From the point of view of legislative behavior research, the performance of the model lends support to the propositions from legislative research upon which it is based. In addition, the model represents, in itself, a theoretical formalization of legislative voting behavior. To the extent that it was validated with the sensitivity testing reported in this chapter, that formalization has been shown to perform in a manner that suggests acceptance of the conceptualizations which comprise it.

In terms of the manner in which theories can be evaluated, this theoretical model has performed well on the basis of a prediction criterion. In addition, the ability of the model to simulate legislative voting effectively with few mechanisms grouped into its predisposition and communication phases suggests a positive evaluation in terms of a theoretical organizing criterion discussed in a recent article.[44] To the extent to which the model performs similarly on other types of issues and for other sessions of the House, its organizing power will be further enhanced.[45] The manipulability of the model has been demonstrated, in part, by the sensitivity testing reported in this paper. While it cannot be increased without sacrificing organizing power, its potential has not been exhausted here.

The performance of the model can also be assessed in terms of its successful simulation of a large decision-making body. It can be noted that the evaluation of the performance of processes in large groups depends upon the ability to discern outputs from the group processes in order to compare the functional equivalence of outputs in the simulation and thereby obtain evaluation criteria. The public character of roll-call voting provides an easily obtained quantifiable output and thus renders legislatures susceptible to the simulation mode of investigation. The success of this simulation is encouraging from the point of view of its implications for the application of the simulation technique to the study of social and political processes in large groups as well as its illumination of the determinants of the legislative process involved in the roll-call voting decisions of representatives.

Finally, a number of directions in which investigation of the model might proceed suggest themselves. In terms of the predisposition phase of the model, we could investigate the extent to which the inclusion of personal characteristics of representatives as predisposing effects would affect the simulation of roll-call votes. The communications phase of the model could be reformulated in two ways. First, a number of parameters could be added to affect conversation probabilities, and second, the communication process could be cycled to allow for a diffusion of influence whereby those whose predispositions have been changed as a result of interactions with colleagues could interact with other colleagues as influencers on the basis of their newly acquired predispositions.

What is perhaps most significant from the point of view of future research possibilities with the model is the relative similarity of state legislatures to the Congress. In order to develop our model to the point where it simulates legislative voting in a broader sense, we hope to employ it in simulations of roll-call voting in state legislatures as well as in several other sessions of the U.S. House of Representatives. In this way we hope to enhance the organizing power of a theoretical formulation, the potential of which we have begun to explore in the research reported here.

---

[44]K. W. Deutsch, J. D. Singer, and K. Smith, "The Organizing Efficiency of Theories."

[45]For a more extensive treatment of the model which includes an assessment of its performance on foreign affairs legislation, see Cleo H. Cherryholmes and Michael J. Shapiro, *Representatives and Roll-Calls: A Computer Simulation of Voting in the Eighty-Eighth Congress* (Indianapolis, Indiana: Bobbs-Merrill, 1968).

7

International Relations Simulations

PAUL SMOKER

INTRODUCTION

The study of politics and affairs of state dates back many centuries—the use of simulation in international relations dates back to just over a decade. Although it is possible to trace the evolution of international relations simulations from primitive war games, the starting point for modern simulations is marked by constructions such as Oliver Benson's Simple Diplomatic Game (Benson, 1961, 1963), Harold Guetzkow's Inter-Nation Simulation (INS) (Guetzkow, 1959, 1965; Guetzkow et al., 1963), and Lincoln Bloomfield's Political Military Exercise (PME) (Bloomfield and Padelford, 1959; Bloomfield and Whaley, 1965b). These pathbreaking constructions inspired further developments in the field, including the all-computer Technological, Military, Political Evaluation Routine (TEMPER) (Raytheon, 1965; Gorden, 1967, 1968), and two second-generation constructions: the World Politics Simulation (WPS) and the International Processes Simulation (IPS). William Coplin's WPS (Coplin, 1969) is a man-computer construction that uses on-line computing facilities and is currently in its third format (ICAF, 1969a, 1969e). Paul Smoker's IPS (Smoker, 1968) also uses interactive computing (1970c) and is in its second format (1970b). This review will discuss each of these simulations and its place in the field and will also refer to modifications of these basic models and to other lesser-known simulations of international relations.

Copyright 1970 by Paul Smoker. The author wishes to thank the Northwestern University Research Coordination Committee for partial support in writing this chapter.

DEFINITION

As Schultz and Sullivan note in the opening chapter of this book, each discipline defines simulation according to its own purposes; and even within disciplines, opinions differ. Guetzkow, one of the pioneers of international relations simulation, has progressively defined simulation as

> . . . an operating representation of central features of reality . . . (Guetzkow, 1959, p. 183)

and

> . . . a theoretical construction consisting not only of words, not only of words and mathematical symbols, but of words, mathematical symbols *and* surrogate or replicate components all set in operation over time to represent the phenomena being studied. (Guetzkow, 1968, p. 203.)

While some social science simulators have distinguished *games* from *simulations* using the criteria that simulations usually become games when human decision-makers are involved (Ackoff, Gupta, and Minas, 1962; Shubik, 1960), Guetzkow does not make this distinction. A "surrogate or replicate component" in Guetzkow's framework could be a human being or a subroutine in a computer program.

For some purposes, it might be considered desirable to replace all human decision-makers by computer routines so that many options could be considered relatively quickly. From this perspective, the use of human beings as surrogates in simulations such as INS, WPS, and IPS might be considered an admission of failure with regard to theories of the situation. Thus faced with a lack of adequate explicit theory concerning the processes involved, human beings can be used as "black box" components on the assumption that their behavior in the simulated environment will at least, in part, correspond to "the reality."

For other purposes, such as training foreign service officers or Peace Corps personnel, human decision-makers use the simulation environment—an environment that includes other human decision-makers—as an experience that may at some later date be relevant to a practical problem. Such a procedure enables the trainee to accumulate a "personal experience bank" in a relatively short time. Some simulators of international relations have seen simulation as a model, whereas others stress experimentation on a model. Brody (1963) defines a model as a collection of assertions about some reality—past, present, or predicted—and classifies models into four distinct types: pictorial, verbal, mathematical, and simulational. MacRae and Smoker (1967), on the other hand, adopt Ackoff's view that models are representations of systems, objects, and events, are idealized because they are simplifications that consider only relevant properties of reality, and hopefully demonstrate the nature of a reality. They argue that simulations may be considered as experimentation on models and that it is useful to distinguish between a model and an experimentation on a model.

While it may be possible to achieve consensus in defining simulations of international relations, it is questionable whether such consensus is desirable at this time. Simulation is still creatively evolving as a technique, and any premature closure upon a single formulation in the simulation of international processes could be disfunctional (Guetzkow, 1968, p. 257).

SOME USES OF
SIMULATION IN INTERNATIONAL RELATIONS

Because simulation can involve relatively complex models with a number of different "component parts" operating at many levels, it has been used in a wide variety of ways in the study of international relations. Some researchers have used simulation as an environment in which to study the behavior of individuals, a classic concern being the relationship between various personality characteristics and decision-making in crisis situations. Two good reviews of simulation crisis research have been published (Hermann, 1969; Robinson, Hermann, and Hermann, 1969). Charles F. Hermann (1965, 1967b) has considered crisis decision-making, threat, time, and surprise in an Inter-Nation Simulation; Margaret G. Hermann (1965, 1966) has used INS to study stress, self-esteem, and defensiveness; David Schwartz has analyzed threat, hostility, and behavior preferences in crisis decision-making, again using INS (Schwartz, n.d.). Brody, Benham, and Milstein have analyzed hostile international communication, arms production, and perception of threat (1966); Shapiro has considered cognitive rigidity and moral judgments (1966); Druckman has studied ethnocentrism and intergroup relations in an Inter-Nation Simulation (1968); whereas Driver and his associates have developed from structural studies of aggression, stress, and personality in an Inter-Nation Simulation (1962, 1965) to a long and detailed series of studies using the Tactical and Negotiations Game (TNG) (Streufert et al., 1965; Streufert, Castore, and Kilger, 1967; Streufert, 1968a, b). The Streufert-Driver TNG experiments use a relatively simple all-man format where the use of nation teams is a device for studying complex decision-making in individuals and groups (Streufert, Suedfeld, and Driver, 1965; Streufert and Streufert, 1969; Streufert, Streufert, and Castore, 1969; Streufert, 1969a). While the TNG has been used to compare perceptual differences between Chinese and American nationals (Streufert and Sandler, 1969), the central focus of this in-depth series of games has been complex decision-making (Streufert, Streufert, and Castore, 1969) and testing complexity theory with regard to individual personality (Streufert and Castore, 1968) rather than international relations theory. Of course, personal characteristics of real-world decision-makers appear to be important (Raser, 1965), and certainly there is ample evidence to suggest important differences of behavior in international relations simulations that are associated with individual characteristics (Crow and Noel, 1965; Smoker, 1968a, 1970a; Coplin, 1969). However, for the purposes of this review, the TNG is regarded as a small group simulation or experiment rather than as an international relations simulation. It would require a separate review to include all the relevant work in social psychological and psychological experimentation. Moreover, although many of these studies cover relevant topics such as pacifist strategies (Shure and Meeker, 1965), mediation and arbitration (Pruitt and Johnson, 1969; Johnson and Pruitt, 1969), negotiation (Mushakoji, 1968, 1971; Pruitt, 1968) boredom (Pilisuk *et al.*, 1967), and cooperation (Pruitt, 1966, 1969) none of these simulations focuses directly on international relations and international relations theory. As Streufert recognizes (1968b, p. 17), the results from these experiments should be interpreted cautiously even if face validity (Hermann, 1967a) is present. This is not to suggest that such studies are less important than simulations of international relations, as defined here, but to limit the scope of this article to those constructions that attempt to construct model world systems or to relate directly psychological characteristics to international relations in such systems.

For example, Charles and Margaret Hermann, in their attempt to simulate the outbreak of World War I by using a modified Inter-Nation Simulation, selected participants on the basis of the California Psychological Inventory in such a way that personality characteristics matched those of the real-world decision-makers at the time of the crisis (Hermann and Hermann, 1963, 1967). Real-world personality characteristics were inferred by using a complex content analysis procedure. Dina Zinnes's comparative study of hostile behavior of decision-makers in simulate and historical data, again the World War I crisis, provides a second example of international relations simulation study that includes personality variables as central to its design (Zinnes, 1966).

The studies of Cappello (in press), Crow and Raser (1964), and Ruge (1969) represent attempts to use the Inter-Nation Simulation as a cross-cultural research device and to relate cultural and personality factors to aspects of international relations. This use of simulation may, as Ruge points out (1969, p. 21), reveal relevant factors for inclusion in future theory concerning international relations.

A second use of simulation in international relations has been to study particular situations such as the outbreak of World War I (Hermann and Hermann, 1967; MacRae, 1967), nuclear proliferation in the 1960s (Brody, 1963), inspection for weapons production (Singer and Hinomoto, 1965), the Vietnam War (MacRae and Smoker, 1967; Milstein and Mitchell, 1968), the Taiwan Straits crisis (Pelowski, in press), the East-West disarmament negotiations of the mid-1950s (Bonham, 1967), the Middle East situation (DeWeerd, 1968), international events of the early 1960s (Meier and Stickgold, 1965), and the patterns of cohesion of NATO (Forcese, 1968). All-computer simulations such as Crisiscom (Pool and Kessler, 1965) have been of particular value in this use of simulation.

A third use of simulation has been as an educational device (Guetzkow, 1964a). Many scholars have used simulation in this manner. On occasion, systematic research has been undertaken to check the validity of educational simulations, be they in elementary schools (Targ, 1967, 1968a, 1968b), high schools (Cherryholmes, 1965), or colleges and universities (Robinson *et al.*,

1966; Cohen, 1962). More often than not, the educational value is hypothesized on the grounds of participant involvement and face validity (Judge, n.d.; Burton, 1966; Smoker and Martin, 1968). Simulations such as the PSW simulation developed by John Parker, Clifford Smith, and Marshall Whithed (Parker and Whithed, 1969; Smith and Whithed, 1968b) are justified by the designers because of the practical experience gained by students of international relations who participate. Again, if the focus is broadened from international relations simulations to educational simulations in general, a massive literature is available for instructional uses of simulation. This area is beyond the scope of this review, but an interested reader may consult some standard works (Twelker and Wallen, 1967; Boocock and Schild, 1968; Tansey, 1970).

A fourth use of simulation in international relations is the attempt to consider a particular theoretical or functional aspect of the international system. For example, collective decision-making and coalition behavior have been important topics in the study of international relations, and a number of simulation endeavors have attempted to investigate these phenomena (Alker, 1969; Browning, 1969). Another central focus of much international relations theory has been the concepts of balance of power and deterrence. Deterrence has been studied by Raser and Crow (1969), and balance of power is being investigated by Dina Zinnes and her associates (Zinnes, Van Howeling, and Van Atta, 1969) using a Simscript format. Other nonnational aspects of international relations, such as multinational corporations and international organizations, have been receiving increasing attention in the theoretical literature, and this focus has been reflected in the work of simulation researchers. Smith and Whithed have tried to study international business decision-making in a political setting by using a number of game-type simulations (Smith and Whithed, 1968a, 1969); Frank Hoole has simulated budgetary decision-making in an international organization by using a computer model (Hoole, 1970); and Pelowski has considered the problems of an all-computer simulation module of multinational corporate futures (Pelowski, 1969). The Pelowski approach is typical of the growing commitments to modular simulations in international relations research. Scholars with this perspective argue that full-blown simulations of international relations can be developed by progressively linking an increasing number of modular simulations as they develop. Each modular is a representation of some component part of international relations theory and can be inserted into the nest of modulars or can be replaced by an improved modular. In this way, fine-grain computer simulations of localized revolutionary processes, such as the Cornblit all-computer model of short-run political change (Cornblit, in press), might in principle be related to coarse-grain modules of international processes, such as Oliver Benson's Simple Diplomatic Game in its new modular format (Krend, in press), through a common data base.

A central use of simulation is described by Harold Guetzkow in "Simulations in the Consolidation and Utilization of Knowledge about International Relations" (Guetzkow, 1969). This perspective views simulation as a format for theory and distinguishes at least three thrusts for the simulation enterprise. To begin with, simulations are seen as vehicles for differentiating and amalgamating theories of international relations. Islands of theory, for example, might progressively be linked through a dynamic modular approach. A second aspect stresses simulations as vehicles in the validation of theories of international relations. Here empirical findings are organized so that the validity of their theoretical contents may be assessed. Finally, the role of simulations in the utilization of knowledge for policy-making in international affairs can be considered, including the complex questions associated with exploring alternative policy futures. The policy aspect of simulation has attracted a number of scholars (Ausland and Richardson, 1966; Bailey, 1967; Bobrow and Schwartz, 1968; Bobrow, 1969), and this trend is likely to continue.

Other reviews of simulation in international relations adopt alternative typologies for considering various uses of simulation (Hermann, 1968; Krend, 1968; Crow, 1966; Abt Associates, 1965; Guetzkow and Jensen, 1966). The multiple uses of simulation in international relations (SIP, 1966a, 1966b, 1968) and the wide variety of theoretical perspectives (Guetzkow, 1964b; Coplin, 1966; Chadwick, 1969) have led to considerable conceptual diversification among simulators of international relations. This diversification is also apparent in such documents as the Gomer master list of simulation variables, which itemizes hundreds of simulation variables currently in use (Gomer, 1968). Many of these variables from different simulation experiments have been recorded on IBM cards and microfilm, and a considerable volume of textual and numerical data is stored in a simulation data bank (Busse, 1969). In addition, a number of bibliographic and abstracted bibliographic sources include material dealing with simulation in international relations (Myers, 1968), thus complementing existing information sources in international relations (Park, 1968).

The variety of uses of simulation is, to some extent, related to the theoretical diversification of the users of simulation. In fact, the theoretical underpinnings of simulations in international relations can be used to differentiate important aspects of the major simulation models. Given a concern with military and power factors, it is not surprising to find that the ensuing simulation, as in TEMPER, has relatively fine-grain representation of military characteristics and relatively coarse-grain political characteristics. Similarly, given a concern with the influence of domestic political questions on foreign policy decisions, the World Politics Simulation is relatively fine grain for domestic political structures of nations and relatively coarse grain for the international system (ICAF, 1969d). The International Processes Simulation, because of a concern with the international system and its consequences on national, subnational, and transnational behavior, has relatively fine-grain modeling at the international system level (Pfaltzgraff, 1969).

At least two procedures may in the future be used to overcome this problem of partial modeling. The most obvious, but perhaps least practical, is to somehow construct one simulation that is homogeneous in terms of multilevel theory. This is to argue for multilevel theory that is well- (or poorly) developed for every relevant component part and relationship. At the present time, this appears to be a rather difficult proposition because theory is still relatively fragmented and insular and there are considerable—and legitimate—differences of opinion concerning basic concepts and frameworks.

An alternative approach would nest simulations in a modular fashion in such a way that only "undistorted" output from one modular would be used as input to the total network. Given the partial and lumpy nature of theory in international relations, this procedure is appealing. As theory develops in a particular section, modulars can be replaced. As new theory evolves in previously unmapped areas, new modulars can be added to the nest. As metatheory changes, theory concerning the relationships between modulars, the executive procedures in the computer programs, can be adapted.

The problem of metatheory is, of course, very important. Such "theory" might be developed by using Monte Carlo approaches to "fit" output from the nest of modulars to some data set alleged to be representative of "reality." But implicit in this approach are the assumptions that social reality is constant in terms of laws of human behavior, that the data set is somehow representative of "reality," and that correspondence of output is a necessary condition for correspondence of structure and process. The first and last of these assumptions require a discussion of validity; the second requires a consideration of relevance.

ON VALIDITY OF
SIMULATIONS OF INTERNATIONAL RELATIONS

Hermann, in his now classic discussion of validation problems in simulations of international relations (Hermann, 1967a), discusses the relationship between the purpose of a simulation and the question of validity. While other theorists had considered the problems in some detail (Campbell and Fiske, 1959; Kress, 1966; Crow and Noel, 1965), the validity criteria developed by Hermann have become important referents in considering performance characteristics of simulations of international relations. The five criteria suggested by Hermann use different aspects of simulation performance to evaluate its validity:

1. Internal validity, after Campbell (1957), demands replication of performance characteristics over repeated runs.
2. Face validity results from surface realism from the point of view of experimenter or participants.
3. Variable parameter validity involves comparisons of the simulation variables and parameters with their assumed counterparts in the referent system.
4. Event validity requires prediction of discrete events using simulation.
5. Hypothesis validity compares the performance characteristics of simulations to theoretical or hypothesized relationships.

Internal validity has had limited use (Elder and Pendley, 1966a) and some criticism (MacRae and Smoker, 1967, p. 5), but the other formulations have been used more often. Recently other scholars have developed alternative perspectives (Evans, Wallace, and Sutherland, 1967; Martin, 1968; Grant, 1970), some of them specifically related to simulation of international relations (Guetzkow, 1968; Smoker, 1969) and some to problems of organizational research and design (Sullivan, 1969; Vertinsky, in press).

Guetzkow sees validity as homomorphism of theory and empirical analyses (Guetzkow, 1968, p. 206) and suggests

> ... by using some systematic rigor in making comparisons between simulations and 'realities,' by taking reference data largely from extant international systems rather than laboratory and field research about non-international phenomena, and by finding in simulations internal processes and outputs which correspond to reference processes as well as reference outcomes, a convergence of evidence is gained which increases the credibility of the theoretical constructions of simulations. (Guetzkow, 1968, p. 208.)

Smoker adds an additional dimension when he argues:

> All of the studies so far surveyed define simulation validity as a function of the predicted and actual correspondences between simulation and 'reality.' The 'real world' is regarded as given, and the 'model world' is regarded as an attempt to demonstrate or show or reveal aspects of reality. If there is a lack of validity, the model world is altered in an attempt to increase the validity of the simulation.

> It is possible to take the complementary position and to evaluate the 'real world' relative to the 'model world' incorporated in a simulation. With this perspective the model world becomes an attempt to demonstrate or show or reveal the way parts of reality could or should be, and differences between the two worlds are rectified by changing aspects of reality through social and political action. (Smoker, 1969, p. 11.)

This definition of complementary validity raises more questions than it answers, but it may be of value as simulations of international relations are increasingly being used for practical military and political ends.

In the following sections of this paper where a number of international relations simulations are evaluated, problems of validity constantly arise. The last section considers problems of relevance and possible future developments.

EVALUATIONS OF
INTERNATIONAL RELATIONS SIMULATIONS

The Technological, Economic, Military, Political Evaluation Routine

Relatively little is published in book or journal form describing the TEMPER simulation, considering the voluminous mimeographed material available; but the best single article in generally available format is Abt and Gorden's report on project TEMPER (1969). This article describes the structure of TEMPER in terms of the actors, the nation groups, and the way they act. Of particular interest is the section dealing with the intellectual basis for TEMPER and the resulting set of theoretical assumptions. A central, and subsequently critical, theoretical assumption is that

> Nation actors are treated as highly aggregated decision making units that attempt to minimize the discrepancy between their ideal state and reality by allocating political, military, and economic resources within acceptable costs. The two super powers and their allies, the United States and its allies and the USSR and its allies, compete for the political loyalty, economic cooperation and military alliance of the neutral nations. (Abt and Gorden, 1969, p. 248.)

Elsewhere (Raytheon, 1965) considerable detail is given regarding the initial TEMPER model. According to this documentary source, TEMPER attempts "to describe global cold war conflict and simulates it with a digital computer program." The TEMPER simulation program is written in a subset of Fortran common to Fortran IV, CDC Fortran 62, and Fortran 63. To run the simulation, the minimum computer system must have a thirty-two-thousand-word memory, a card reader, three magnetic tape drives, an on-line printer, and, of course, a Fortran compiler. A more efficient system could use an off-line printer and about six magnetic-tape drives, but the use of all features of the simulation would require eleven magentic-tape drives.

The overall flow diagram for the simulation is shown in figure 7-1. This diagram is extracted from an analysis of TEMPER (MARK II) undertaken at the Industrial College of the Armed Forces in Washington, D. C. (De La Mater *et al.,* 1966). Four sets of components make up the computer program: the psychological submodel, four subroutines; the economic submodel, six subroutines; the war submodel, seven subroutines; and the decision-making submodel, twelve subroutines. Each submodel is time dependent, some subroutines being weekly, some quarterly, some semiannual, and some annual.

The psychological submodel consists of a threat subroutine (THREAT), a perception subroutine (PERCEP), an experience subroutine (XPERCE), and a cultural subroutine (KULTUR). PERCEP and XPERCE simulate "the distortion that would occur in the receipt of tactical, strategic and budget information from one country by another through diplomatic, intelligence and communication channels." (De La Mater *et al.,* 1966, p. 3-1.) Similarly, it is argued that THREAT computes the various elements of military and political threat that tend to modify normal cultural motivations, whereas KULTUR introduces and updates the cultural motivations that are considered most significant in influencing a nation in its relations with others.

As an example of TEMPER logic, it is of interest to consider KULTUR and the variables involved in defining these "most significant" cultural motivations. Six alleged motivations are included in the model: internal initiative,

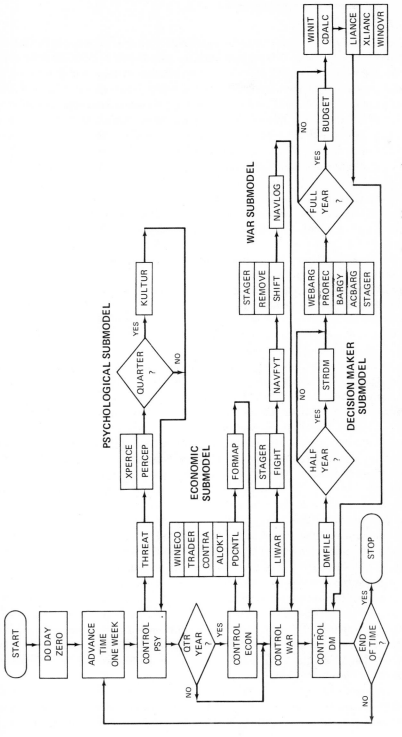

FIGURE 7-1   Sequencing of TEMPER subroutines. Source: De La Mater *et al.*, (1966).

305

military coercion, military initiative, propensity to tax, external dynamism, and military power ratio. These motivations are affected only by the various components of THREAT. THREAT itself has two major components: military threat and political threat—military threat being subdivided into tactical threat and strategic threat; and political threat being subdivided into global threat, threat from allies, and threat from neutrals. At the lowest level of aggregation, military threat uses concepts of counterforce units, nuclear credibility, level of military operations, population loss, and maximum military operations; whereas political threat boils down to alliances (East, West, or Neutral), alignment, land desire, and external dynamism.

To start the simulation at "Day Zero," the cultural motivation values are set by using the data base values modified slightly in four of the six cases by tactical threat. After "Day Zero," KULTUR is not reentered until the end of a quarter year (simulation time). Then, as initially, Military Power Ratio is set between its no-threat and its maximum-threat limit, depending on the level of threat (which, in turn, depends upon the relative military power of East and West). Similarly, internal initiative, military coercion, military initiative, and propensity to tax are updated by a function where threat drives the motivation either up or down toward its maximum threat limit. External dynamism is computed by using a formula that lowers it if confronted by a military threat and raises it for a political threat.

A casual consideration of "the cultural motivations which are considered most significant in influencing a nation in its relations with others" suggests that the theory underlying TEMPER assumes great importance to military and threat factors in international relations and international relations theory. This observation is supported when the nature of the variables used in TEMPER is considered. Approximately 160 variables are listed as having unique representation in the Fortran program as a whole. Of these variables, 90 are directly concerned with military factors; 30 with economic-resources-population factors; 13 with geographic factors; 12 with escalation, threat, and bargaining factors; 9 with political factors; and the rest with such things as recording simulation time. When it is realized that nearly all the geographic factors are military in the sense that they are variables (such as fraction of land held), this means that approximately 66 percent of the variables in TEMPER are military variables and 6 percent are political, most of the remainder being economic.

Since the simulation is alleged to be more than a war game and since "the fundamental theoretical assumption about world conflict on which the TEMPER Model is based is that the overall nature of the current world conflict and the national strategies designed for dealing with it are best defined in the comprehensive terms of global military, political and economic interactions" (Abt and Gorden, 1969, p. 251), a more detailed consideration of the international relations theory in the TEMPER simulation follows.

A scouring of the voluminous mimeographed literature concerning TEMPER reveals a number of international relations assumptions that, even if they were the only assumptions in the simulation, place grave doubts on the hypothesis validity of the model as a vehicle to study international relations. As a vehicle for war gaming, it may be extremely valid; but this review is not concerned with war games. The assumption is that TEMPER's reputation as a simulation of international relations, as propounded by Abt and Gorden—two of the major architects of TEMPER—must be weighted by various validity standards for international relations simulations.

CONFLICT REGION	WEST	EAST	NEUTRAL
1	S. VIET	N. VIET	LAOS
2	THAILAND	CAMBODIA	BURMA/CEYLON
3		RED CHINA	
4	PAKISTAN	MONGOLIA	INDIA
5	S. KOREA	N. KOREA	
6	JAPAN		
7		USSR	
8	USA/CANADA		
9	AUSTRALIA		AFRICA NATIONS
10	TURKEY/IRAN		MIDDLE EAST
11	NATO EUROPE	RED EUROPE	NEUTRAL EUROPE
12	BRAZIL/VENEZUELA	CUBA	NEUTRAL AMERICA
13	MALAYSIA	INDONESIA	

FIGURE 7-2 Key to TEMPER map. Source: De La Mater *et al.* (1966).

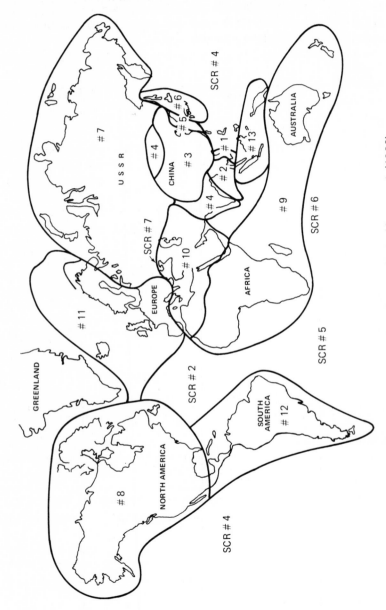

FIGURE 7-3 TEMPER map. Source: De La Mater *et al.* (1966).

The TEMPER data base includes data for 117 nations, and these data are recorded and aggregated by three TEMPER routines (TEMPO, TEMPET, and T-PUNCH), so that there are 39 nation groups and 20 conflict regions (13 land and 7 sea) on the TEMPER map. The 39 nation groups are further subdivided into an East Block and a West Block. The leftover nation groups, such as Israel and Egypt, are placed in a neutral block whose power ratio is zero, since power ratio is calculated by dividing East by West and vice versa. Thus TEMPER assumes there is one Western block, one Eastern block, and a neutral block. An actual TEMPER map of the world as used in the 1966 evaluation of TEMPER is shown in figures 7-2 and 7-3. An almost identical map has been used in subsequent years in the man-computer format discussed below (ICAF, 1966a,b, 1967, 1968b, 1968c). In addition, TEMPER includes just nations in its model and assumes that each nation group acts with a common purpose because no internal dissention or subversion is allowed in the model and no political or other interest groups are present. Similarly, no international organizations, multinational businesses, or other components of the international system are included.

This set of assumptions runs counter to many international relations theorists. It is difficult to grant much hypothesis validity to these assumptions, given the impressive array of theorists who do not share the view of TEMPER (McClelland, 1966; Burton, 1967; Waltz, 1967; Deutsch, 1968; Galtung, 1968; Angell, 1969; Alger, 1970). Of course TEMPER, in its newer man-computer format, is mainly an educational device for members of a United States military establishment, and it may be argued that the question of hypothesis validity is of less value than the educational usage. An alternative perspective will argue that it is not very desirable for many hundreds of military officers to be exposed to an image of international relations that is theoretically not very credible. If international relations were viewed as a war game, TEMPER might be seen as a means toward that goal, as individuals exposed to it might tend to act in accordance with the socializing experience involved. This is to take the notion of complementary validity seriously and to try to create a reality in the image of a model.

Other reviews of TEMPER have concluded that:

Indirect benefits notwithstanding, in our opinion TEMPER is a failure. TEMPER could not succeed at the present state of knowledge; it cannot be repaired. Our recommendation, therefore, is to abandon TEMPER. This should be done quickly since any further time or money spent on TEMPER is wasted.

The reasons why we have arrived at this conclusion are spelled out in detail in subsequent sections. Briefly, they are:

1. In comparison with the real world of the cold war, the TEMPER world is fatally distorted.

2. The relation between TEMPER's inputs, design features and outputs renders it invalid as a simulation.

3.   The design of TEMPER makes it practically useless as a subsidiary tool for the conduct of political gaming.

4.   The distortions in the TEMPER world render it unsuitable for any attempt to construct a new simulation or gaming model. (Balinski *et al.,* 1966, p. 2.)

And elsewhere:

The TEMPER map, around which the theory is structured, is too restrictive for any significant improvement to the present model. Further development will require an approach that permits more realistic and flexible international relations simulation.

The data base is inadequately described. No standard of measurement is provided for the dynamic simulation variables. There exists no theoretical justification for the methods of aggregation. Much of the 'theory' of the model is embodied in the values assigned to the data base variables, but this 'theory' has not been discussed in any of the documentation.

It is concluded that the data base must be adequately described to the player and researchers before the simulation outputs will gain any respect.

Research into TEMPER must begin with a study that results in acceptance or rejection of the basic assumptions. Subsequent studies must determine if the basic assumptions have been, or in fact, can be programmed. Any research into the micro aspects of the simulation, however interesting, will be of little value unless it can be correlated to the basic assumptions of the model, and viewed in the light of a specific use; eminence of the researchers notwithstanding. (Draper, 1966, p. 39.)

Draper's comments concerning the basic assumptions in the TEMPER model support the earlier discussion of inadequate international relations theory. Unfortunately, it is not at this time possible to give any objective evaluation of TEMPER in terms of empirical comparisons of real-world performance to TEMPER. No empirical validity studies have been undertaken on the model and, objectively speaking, its performance characteristics are unknown. Some output from TEMPER has been published (ICAF, 1968a) giving print-out from runs of the simulation conducted during 1968. Many of the variables published are strictly military; but some, such as trade, provide an opportunity for at least an "eyeball" validity study of this variable. The first thing that becomes apparent when inspecting the trade tables is that trade is defined in an unorthodox fashion. Two types of trade exist: political trade (or exports), which is used in TEMPER to regain declining friendship, and residual trade (or imports), which is defined as trade to meet unsatisfied demand. When the trade tables are inspected for these total categories, it becomes apparent that most of the trade links between nations in TEMPER are zero. The greatest amount of trade is manifested by the USACAN group in TEMPER, where trade links exist with about 12 percent of the other nations. The average for USACAN, NATOEU, and JAPAN is nearer 10 percent. In the real world for these nations with the others in the sample, the average is over 90 percent. Something is wrong, it would seem, with the performance characteristics of TEMPER with regard to trade; and this is, in part, acknowledged in the previously mentioned ICAF report (De La Mater *et al.*, 1966). On pages 4-14 of this report, the authors point out that there can be no trade between blocs in the TEMPER model because West and East Blocs do not award Ally Value (the degree to which a nation group values its alliance with another nation group) to each other. The authors then point out that because, in the real world, such trade does exist, "provision probably should be made to accommodate this real world situation" through either a modification of two of the subroutines or a major modification of TEMPER. It seems reasonable to suppose that this particular flaw in performance is associated with the archaic conceptual framework incorporated in the model. It does not seem unreasonable to suppose that many other performance characteristics of the model are likely to be similarly distorted. Before leaving the trade discrepancy, it should be pointed out that not all the discrepancy occurs because of the peculiar assumptions associated with the bipolar structure of the simulation. An inspection of the print-out reveals that, for example, in more than half the cases, the USACAN nation group has no trade at all with South Vietnam and has either no imports or no exports with the European NATO countries. Even if TEMPER is just a war game, this is a little hard to take.

The behavior of TEMPER has been compared with that of INS and of DETEX 11, a variant of PME (Alker and Brunner, 1969), and some criticism is levied at the debatable economic relationships and the aggregation procedures in TEMPER's bipolar world:

> Even if its implications need to be explored by many more operating runs the TEMPER model is especially suggestive of debatable relationships, partly because of its explicit nature. Thus we have already noted the considerable overlap of DETEX and TEMPER problem components, but the rather implausible relationships governing the aggregation in TEMPER. The projection of cost effectiveness decision making rules onto the Soviet Union, East and West Germany is also controversial, especially when various kinds of deterrent capabilities are considered. A lot of psychological variables, such as misperception tendencies, are included in TEMPER, but their impact on problem definitions and magnitudes in TEMPER does not correspond to their impact in DETEX 11 or INS. (Alker and Brunner, 1969, p. 109.)

The explicit nature of the TEMPER equations undoubtedly renders the simulation vulnerable to any criticism where less formal all-man or man-computer representations may avoid embarrassing questions. However, the basic criteria of adequate international relations theory apply to all representations, and it is on this point—not on the particular details of particular equations—that the present reviewer considers the TEMPER simulation not only inadequate but dangerous as a teaching device for military officers. In the absence of any validity studies of real-world performance characteristics, the only remaining criteria applicable to evaluating TEMPER are the criteria of hypothesis validity, and therefore the fundamental assumptions about internationals incorporated in the model are sufficiently in error to throw the validity of the whole simulation into serious question. It does not seem unreasonable to demand that those who use a simulation of international relations over a number of years as part of an educational experience for some hundreds of military officers present evidence as to the validity of the model and the resulting image of international relations it incorporates. Even if the simulation is used, from the parochial perspective of "the management of our (USA) national resources and the interrelation of economic, military, and political factors on an international level, from our (USA) national perspective" (Westfall and Draper, 1966, p. 1), some of the conclusions of an earlier review still appear to hold, notably:

> TEMPER has design deficiencies which make it unsuited for other than experimental use without major modification . . .

and

> The interplay of many variables used to describe a country and its relations with other countries has not been tuned to the point where results are valid and convincing. This tuning, to be effective, will require reforming of many of the equations, restructuring of the relationships between subroutines and extensive testing of results of computer runs. (ICAF, 1966a, p. 11.)

In summary, TEMPER may represent a brave first attempt at all-computer simulation of international relations when considered from the technical point of view; but from the theoretical and substantive point of view, it is seriously invalid unless it is being used deliberately, or otherwise, as a socializing device to demonstrate the way things can be or will be. Its validity might then be measured relative to actions of the United States military in the real world. This is a complex and controversial question, best left to the final discussion of relevance.

The RAND/M.I.T.
Political Military Exercise (PME)

The PME has its roots in the work of Herbert Goldhamer at the RAND Corporation (Goldhamer and Speier, 1961, p. 499). Goldhamer's games incorporated political decision-making and policy formulation into limited war games and were played by the RAND Corporation from 1955 onward (Averch and Lavin, 1964; Giffin, 1965, pp. 65-68). Using this experience, Lincoln Bloomfield and his associates developed the Political Military Exercise in its various forms (Bloomfield and Padelford, 1959; Bloomfield, 1960). The POLEX

1 exercise was run in 1958 using academic and United States government experts as players; subsequently, other series were run by Bloomfield, including the POLEX-DAIS and DETEX series (Bloomfield and Whaley, 1965a, 1965b). The EXDET series of games used only student decision-makers (Whaley and Seidman, 1964; Whaley, Ordeshook, and Scott, 1964), and several United States government agencies have used variants of the PME for training, policy, and research purposes (Barringer and Whaley, 1965, p. 456). More recently, the CONEX-EXCON series of games has been run, using a modified form of PME (Bloomfield, 1968) known as GENEX. Similar experiments have been run by Michael Banks, A. J. Groom, and A. N. Oppenheim in Europe (1968).

The PME is an all-man exercise in which there are a number of nation teams and a control team. The nation teams represent particular political entities within a nation (such as the National Security Council for the United States team and the Politburo for the Soviet Union team) and make moves simulating actions of the team they represent. Initially, all interteam communications are written and pass through the control team, the only exceptions being face-to-face conferences between particular decision-makers or groups of decision-makers which are held subject to the credibility evaluations of the control team. Teams can only take action during the well-defined move periods. The control team has several roles. It checks for credibility and only allows moves that it believes to be possible; it acts as "nature" in deciding physical outcomes of particular actions of particular teams; it simulates all important or relevant groups not represented by teams in the simulation, such as particular international organizations or domestic pressure groups. Thus, the teams represent particular decision-making groups in the situation, and control tries to represent all other relevant factors. A control team might, for example, introduce an accidental nuclear explosion or an earthquake or an assassination. Certainly the control team is able to manipulate the direction of the game to some extent and can specifically be selected for the particular goals of the research experiment. A PME crisis simulation of the Middle East situation might, for example, include regional experts on Middle Eastern affairs, military experts on conventional and guerrilla strategies, economic experts on important industries such as oil, and experts on crisis behavior. Some aspects of the control team's work can be preprogrammed, such as all or nearly all the essential messages sent from control in the early stages.

Nation teams are sometimes small—two or three people in particular roles, and sometimes large—up to about eight people. For some situations (such as simulating a Middle Eastern crisis), domestic pressure groups (such as Palestinian guerrillas) can be represented separately by a team. Decision-making positions within the team can be assigned by control prior to the game or can be self-selected by individuals in the teams. Move periods usually last about one to one and one-half hours, although much depends on the research interest of the exercise. Similarly, the relationship of game time to real-world time can be varied, as can the duration of the game. Features such as intelligence facilities and leaks are often introduced in the game, and "hot line" facilities are sometimes provided, again depending on the topic being simulated. The scenario, or world history, is particularly important in these games when some possible future is being simulated. Each participant is given some time to study the scenario before the game starts.

Because of the wide variety of formats in which PME has been used and because of the policy-oriented perspective of many of its users, relatively little concern has been given to questions of validity. Since PME is, to a large extent, seen by its users as a crisis game where the focus is on "crudely simulating the decision making process at the top governmental level" (Bloomfield and Whaley, 1965b, p. 857), evidence of validity is focused on actors and the decision-making process rather than on longer-term aspects of the international system. Lucas Fischer has considered the hypothesis validity of PME in various aspects of the decision-making process (Fischer, 1968). He argues, for instance, that the question of rationality in decision-making is dealt with more adequately in the PME than it has been in many verbal models since—although the game rules encourage rationality, great scope is given for irrational behavior in crisis situations, especially in the form of selective perception of issues, information, and alternatives. Moreover, certain structural aspects of the game, such as time pressure, incomplete and ambiguous information, and high threat to values, tend to impede rationality in an analogous way to real-world decisions. Fischer also concludes that although time is extremely important in both PME and reality in crisis decision-making, the sharp contrast between domestic and external politics differs from some theorists who see mutual penetration of domestic and foreign policies (Rosenau, 1966; Scott, 1967).

Fischer also argues that a great variety of organizational structures can be adopted within each team, and this makes possible many different forms of decision-making. On the other hand, organizational norms, sanctions, and rewards are less well represented, although at crisis times, such incentives may be less important. Fischer also suggests that factors related to time and uncertainty can produce tension in players, which is related to that hypothesized for real-world situations.

Fischer discusses instruments of national policy in PME, including alliance patterns and the use of violence, and patterns of international action, including power and international organization. Unfortunately, all the comparisons are conjectural, and no supporting data from the simulation is presented in any systematic manner. Nevertheless, Fischer argues that the PME is a valuable heuristic, and many important theoretical assumptions have been built into the exercise.

Alker and Brunner, in their comparison of TEMPER, PME, and INS, point out that of these three simulations, PME is the least likely to undergo major long-term changes in the international system and, as suggested by Fischer, has some built-in tendencies operating against going to war (Alker and Brunner, 1969, p. 104).

Any evaluation of performance characteristics of PME at this time is purely subjective, since no rigorous empirical attempt has been made to compare its behavior with real-world behavior. Many of the arguments presented by Fischer are quite appealing but are not necessarily valid. For example, he argues that PME players are a relatively uniform group and that this can be interpreted either as a statement of the game model to the effect that cultural and other sociological differences are not important in determining individual and group behavior or as a statement that foreign policy decision-making elites, during crisis, tend to behave similarly in similar circumstances, regardless of their personal or national background (Fischer, 1968, p. 22). There are a number of empirical studies of real-world behavior (McClelland *et al.,* 1965, 1968; Corson, 1970) and simulation (Smoker, 1968a, 1969; Coplin, 1969) that suggest that cultural, national, and personal differences are important in conflict behavior in general and crisis behavior in particular, in both simulations and realities. It is to Fischer's credit that he acknowledges that alternative theories exist and also that the Stanford crisis studies (Holsti, North, and Brody, 1968), for example, have assumptions opposite to that of the PME in this regard (Fischer, 1968, p. 24). But this does not resolve the problem that most, if not all, the empirical studies go against the particular set of theories Fischer chose for validation purposes.

On another point, the control team is central to the processes in the PME, since the explicit and implicit assumptions they hold about reality are critical in the subsequent course of events. Some research might be undertaken with PME using different types of control teams for the same scenario. It does not seem an unreasonable hypothesis to suggest that the various outcomes from such a series of runs may strongly be related to the particular theoretical positions of those in control.

Until empirical analyses of performance are undertaken with output from the PME, it is difficult, if not impossible, to evaluate its validity. It can be that the nature of the control team and the nature of the crisis problems tackled are such that PME acts as a reinforcer of basically incorrect perceptions of reality and incorrect theory in much the way that TEMPER does. Certainly there is no compelling empirical evidence to suggest that the results from PME be taken seriously, and there is some apprehension that the exercise has been used by United States government agencies with, apparently, no particular concern as to its real world validity as a policy device other than subjective untestable propositions about output from the simulations.

To some extent, this lack of concern with evaluating reality performance is understandable, given the foreign policy focus of the simulation and the relative difficulty of undertaking rigorous empirical analyses in such an area (Jensen, 1966a, 1966b). However, methodological developments in the quantification of events data have been made (Azar *et al.,* 1970) so that definition and measurement of a foreign policy event are possible in some detail (Hermann, 1970; McGowan, 1970) and with some measure of reliability (Sigler, 1970; Leng and Singer, 1970). Given these developments, relatively "hard" analyses might now be made in the comparison of the output and real-world events. Such comparisons could assist considerably in providing an empirical base for evaluating reality performance.

The Inter-Nation Simulation (INS)

The INS is perhaps the best-known simulation of international relations. It was developed by Harold Guetzkow and his associates at Northwestern University and was first reported on in book form early in the decade (Guetzkow *et al.,* 1963). It has subsequently been reviewed from several perspectives (Verba, 1964; Raser, 1969) and has been subjected to a wide range of validity studies (Guetzkow, 1968).

INS is a man-computer model—the computations being sometimes conducted by hand during the pause between game periods, sometimes by computer (Pendley, 1966; Smoker, 1966). Figure 7-4 gives the decision and information cycle for a typical INS. For this diagram, it has been assumed that the calculations are performed by hand and not by a computer program. Each period lasts roughly seventy-five minutes, although, of course, great variation exists in different formats of INS. For example, the INSKIT for use in schools has one time schedule for a ninety-minute simulation period and another for a forty-five-minute period (Guetzkow and Cherryholmes, 1966a, p. 19). In this situation, the decision forms used by participants are designed in such a way that they are self-calculating.

Typically, the participants are divided into nation teams, each team comprising a number of decision-making positions. In the INSKIT, for example, each prototypic nation may have a Head of State, a Foreign Policy Adviser, an Official Domestic Adviser, a Foreign Affairs Diplomat, and a Domestic Opposition Leader. An International Organization is often included under the direction of a secretary-general (Guetzkow and Cherryholmes, 1966b).

In figure 7-4, the schedule calls for the main set of economic and political decisions to be made after half an hour in each period. These decisions are used in the programmed environment to calculate the state of each nation at the start of the next period. Of course, during the period, participants can communicate with each other by messages or in conferences and can undertake a number of political activities, such as forming alliances or signing cultural agreements. If an International Organization is present, then it has, as a rule, a regularly scheduled meeting at the end of each period.

The unprogrammed part of INS is very analogous to the PME. Great flexibility exists for adapting the model to suit particular research interests. For example, controlled intervention has been used on a number of occasions to test certain hypotheses, such as the spread of nuclear weapons and its effect on alliance structure (Brody, 1963) and the consequences of implementing Osgood's graduated independent tensions-reduction initatives (Crow, 1963). Unlike PME, INS has been used to consider both short-term crisis situations such as the outbreak of World War I (Hermann and Hermann, 1967) and long-term patterns in international relations (Nardin and Cutler, 1969; Chadwick, 1969).

The programmed part of INS is analogous to the control team in PME in certain aspects. The programmed part consists of a series of equations relating various political and economic factors (Guetzkow *et al.,* 1963, Guetzkow,

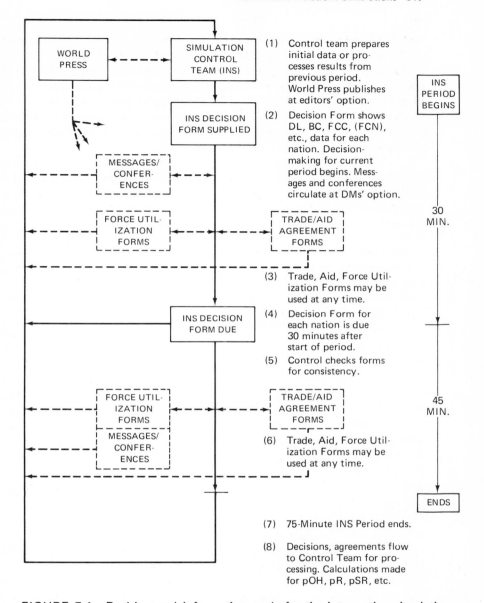

FIGURE 7-4 Decision and information cycle for the inter-nation simulation. Messages, conferences and other interactions occur throughout the cycle in an unprogrammed manner. Unrestricted messages flow through Simulation Control to recipient and to World Press. Restricted messages are leaked to press by Control by systematic sample. Unprogrammed elements are indicated by the broken line. Programmed decision-points and information flows are denoted by the solid line. Prepared by Jeffrey A. Krend, Simulated International Processes project, Northwestern University, October, 1969.

1965). For example, in the original INS model, the satisfaction of consumers with the standard of living depended upon the minimum level of consumer goods necessary and the amount of consumer goods in excess of this minimum figure in a diminishing-returns manner. Thus, successive increases in the quantity of consumer goods available result in decreasing amounts of extra consumer satisfaction. The equation for expressing this set of assumptions in INS is a quadratic equation (Guetzkow *et al.,* 1963, p. 125); however, it was subsequently found to be in error by Elder and Pendley (1966a).

In effect, the programmed assumptions in INS represent a simple computer simulation of aspects of international relations. These assumptions were originally made by expressing aspects of theory about international relations in terms of simple equations. Coplin has compared INS theory with theories of international relations and concludes that

> If there is one obvious conclusion from the preceding discussion, it is that the Inter-Nation Simulation is a theory building exercise conducted not by a number of unsophisticated psychologists who have run out of experiments, but rather by a group of scholars who are well acquainted with the verbal theories of international relations. Moreover the creators of the simulation have been eclectic in their approach. Their model has as much in common with "traditional" as with "nontraditional" theoretical positions. (Coplin, 1966, p. 577.)

INS differs from most other simulations of international relations in one extremely important manner. Apart from comparisons of the assumptions incorporated in the model with assumptions about international relations (Coplin, 1966; Gottheil, 1966), a considerable number of empirical studies have been undertaken in which reality and simulation are compared. Some of these studies consider one or a few variables (Caspary, 1965; Pendley and Elder, 1966), some focus on specific events (Targ and Nardin, 1965; Hermann and Hermann, 1967), and others consider the associations between a large number of variables (Chadwick, 1966, 1967).

The various studies of INS have independently been reviewed by Harold Guetzkow and George Modelski (Guetzkow, 1968; Modelski, 1969), and an evaluation of the validity of INS should be based on the critical work of these two scholars. Modelski and Guetzkow independently considered twenty-four specific studies using INS, the studies being written by twenty-nine investigators. From these twenty-four studies, Guetzkow selected fifty-five instances for which he made a judgment of correspondence between simulation and international realities. Guetzkow, on the basis of the empirical evidence, assigned a rating of *much, some, little, none,* or *incongruent* for the correspondence in each of the fifty-five cases. He discovered that thirty-eight out of fifty-five comparisons carried a rating of *much* or *some,* thirteen a rating of *little,* and four an assigned rating of *incongruent* (Guetzkow, 1968, p. 253).

Modelski independently concluded that thirty of the fifty-five cases merited *much* or *some,* and in only four cases did the Modelski scores differ by more than one degree. An independent study of the Modelski and Guetzkow comparisons (Krend, 1969) gave a correlation of 0.95 between the two sets of ratings, the agreement being highest for studies of the use of humans as surrogate decision-makers and lowest for studies of relations among nations.

In his critique of INS, Modelski argues that INS is essentially a nation state model of world politics and is, as such, one special case of world politics. He discusses four basic assumptions of INS in some detail, namely:

1. That nations are the basic unit of analysis
2. That nations are essentially self-sufficient
3. That validators of national decision-making are individuals and groups in the nation's political system
4. That central decision makers control the basic national capabilities and their allocation

Modelski's penetrating analysis concludes that a generalized simulation system of world politics should be able to cope with a universe of possible world politics that is multiorganizational, significantly interdependent, and equipped with global validation procedures and at the same time with politicians sensitive to various publics. All of his suggestions argue for greater complexity in terms of role positions and organizational structures. Modelski observes that the second generation IPS goes some way toward meeting his objections.

Although the reality performance of INS is unsatisfactory, the willingness of INS researchers to subject their results and models to rigorous empirical analyses has led to the creation of two second-generation man-computer simulations of international relations. Each of these simulations is based on INS but attempts to rectify errors and omissions in the former INS model. A review of these two models—the World Politics Simulation (WPS) and the International Processes Simulation (IPS)—follows.

The World Politics Simulation (WPS)

The World Politics Simulation, like INS, is a man-computer representation of aspects of international relations. The decision and information cycle for WPS is illustrated in figure 7-5. WPS is currently in its third format and is designed to create an operating environment somewhat similar to that in which national foreign policy makers act and interact (ICAF, 1969e, p. 1). Up to nine nations are included in WPS and each is represented through an economic submodel, a demographic submodel, and a political submodel. From the point of view of foreign policy, the political submodel is most innovative and contains the most interesting set of theoretical assumptions. The model assumes that a number of domestic pressure groups (known as policy influencers) exert pressure on decision-makers and that decision-makers are motivated by a desire to stay in

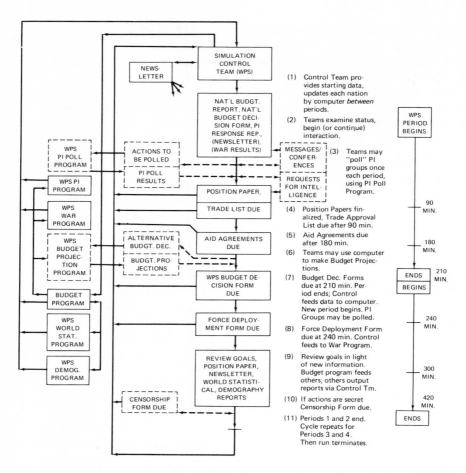

FIGURE 7-5 Decision and information cycle (part 1) for world politics simulation III/69. Prepared by Jeffrey A. Krend, Simulated International Processes **project**, Northwestern University, October 1969.

office. The participant's manual for WPS places considerable emphasis on the policy influencers (ICAF, 1969f), and the background materials for the simulation give examples of particular policy influencers for particular nations (ICAF, 1969c). In Israel, for example, the policy influencers are assumed to be the Center socialist parties, Center right parties, Center left parties, Foreign Ministry, Public Opinion, Organized Labor, Agrarian Federations, Industrial Managers, Military Moderates, and Military Radicals.

Figure 7-5 shows how WPS is organized in groups of two periods and illustrates some of the novel man-computer interactions introduced into the simulation. WPS uses an on-line terminal to speed up the man-machine interface and provides a limited look-ahead capability for trying out policies and gaining an estimate of the consequences on the policy influencers. Feedback to participants, as in INS, is between periods when consequences of human decisions on the simulated international environment may be calculated (ICAF, 1969b).

Like INS, the United Nations is often represented by participants; but like PME, other nations not represented by teams and the U.N. can be simulated by control if necessary to the purpose of the experiment. WPS in this regard is sometimes more vulnerable to the criticism of Modelski concerning other organizations in world politics, although its relatively fine-grain modeling of domestic processes is quite superior to the programmed domestic model in INS or IPS in terms of hypothesis validity.

As in PME and TEMPER, it is at this time difficult to evaluate the model on other than hypothesis validity grounds. No empirical studies of its performance characteristics relative to international relations have been undertaken, although a comparative study of diplomat and student behavior in WPS has been published (Coplin, 1969). This study reports on a series of four runs with WPS, two with United States State Department personnel and two with high school students. Strong differences were found for investment and performance in economic growth, foreign aid patterns, meeting frequency, message frequency, alliances, and activities involving the international organization. A mixture of differences and similarities were found for satisfaction of the policy influencers, uses of military force, foreign trade patterns, use of communications media, and patterns of hostility and friendship. Strong similarities of behavior were found for economic interest in trade and domestic versus foreign policy actions as sources of policy influencer satisfaction.

Coplin's analysis leads him to the conclusion that rather than view INS or WPS as a theory of political phenomena, it is more appropriate to say that each run represents a theory of political phenomena. To some extent, this disaggregated view of particular outcomes is consistent with the outlook of many foreign policy analysts and is complementary to the social scientific view of within- and between-run variance and its distribution. The various uses of simulation in the study of politics reflect these perspectives (Coplin, 1968, 1970).

In the absence of further evidence concerning the performance characteristics of WPS, it seems reasonable to suggest considerable caution in interpreting results from any of the runs with WPS, although hypothesis validity seems to suggest that the linkage between policy influencers and foreign policy is clearly superior to any other existing simulation of international relations. The international system component of the model is severely underdeveloped, and the focus on foreign policy may well have introduced distortions at other levels of analysis.

Certainly, it does not seem unreasonable to suggest that because of its continuous use as an educational device for officers in the armed forces of the United States of America, some effort should be made by those using it to present evidence as to its reality performance. While the theoretical assumptions incorporated in WPS seem reasonable—unlike those in TEMPER—the relatively strong emphasis of domestic influence on foreign policy and the relatively underdeveloped nature of the rest of the simulation theory suggest the possibility of lack of correspondence in, for example, the interaction between governments and the political aspects of the international system. Hopefully, those placing credence on WPS as an educational device for military officers will begin to require objective evidence concerning the performance characteristics of international relations simulations. In the absence of such evidence, a healthy skeptic might be forgiven for suspecting a manipulative and training goal rather than an educational motive in the use of TEMPER and WPS by the Industrial College of the Armed Forces in Washington. There is, after all, some evidence to support the assumption that a currently dominant military view equates international relations with a war game, tempered by the constraints of domestic policy influencers.

The International Processes Simulation (IPS)

The International Processes Simulation was developed by Paul Smoker at Northwestern and Lancaster universities. Like the WPS, it represents a second-generation man-computer simulation; but unlike WPS, it is particularly focused on the international system. The basic model is described elsewhere (Smoker, 1968b) and differs from all the other models so far in its overall image of international relations. After Galtung, Alger, and others, an attempt is made in IPS to introduce an international system containing International Governmental Organization, International Non-Governmental Organization, Multi-National Corporations, Nations, and National Corporations. The decision-making positions in the human part of the simulation reflect this structure as in addition to governmental decision-makers (such as Heads of State and Foreign Affairs Ministers), nongovernmental decision makers (such as citizens, managing directors of nation-based and multinational corporations) and officers and representatives of international organization are included. The actual structure of a particular simulation run can be changed by varying a nation-state component and an international component. Thus, using the same model, a variety of worlds can be generated changing from a nation-state type structure of the early 1900s to an international type structure of the 1970s and 1980s. The decision and information cycle for IPS is shown in figure 7-6.

It must be stressed that, although multinational business is included in IPS, IPS is not a business game after the style of the International Operations Simulation (Thorelli and Graves, 1964); similarly, the International Non-Governmental Organization component is present in IPS as a part of an international system rather than as a valid representation of such organization (Judge, 1967).

The computer component of IPS, like the human component, is developed from INS. There are, however, a number of differences which are fully explained elsewhere (Smoker, 1968b). The programs are modular and contain approximately 130 variables, including a number of additional political variables such as

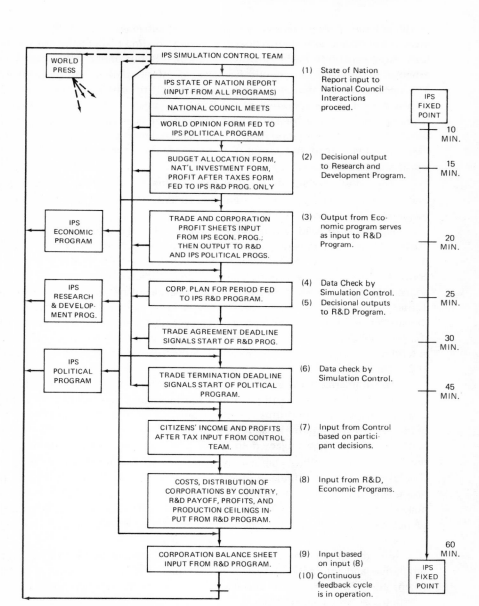

FIGURE 7-6 Decision and information cycle (part 2) for international processes simulation. As in other simulations in this series, communications and conferences are assumed to occur at the option of the decision-makers, hence may exist anywhere in the noncomputerized parts of the cycle. Prepared by Jeffrey A. Krend, Simulated International Processes project, Northwestern University, October 1969.

world opinion, public opinion, and political effectiveness. Like WPS, the program is interfaced with the participants through on-line equipment. There is no look-ahead capability in the present version of IPS; but unlike WPS, there is an attempt to simulate continuous feedback of information to the participants. This can be seen by the fairly regularly spaced left-to-right arrows in figure 7-6. This is critical to the operation of the simulation as an impression of continuity is created by fast and frequent feedback from the environment to the participants. As a result, there is no need to halt the simulation as in the case of INS or PME, and the end of the period is no more significant than the end of a real-world year.

The latest version of IPS uses two on-line terminals and two separate systems of interlocking programs (Smoker, 1968c). A choice of two data bases is included with the programs.

A number of validity studies have been conducted on IPS (Smoker, 1968a, 1969, 1970a), and comparative studies with INS have been undertaken whenever possible. The validity studies have been selected to test the model in terms of conflict behavior by replicating some of the studies of Rummel and Tanter concerning domestic and foreign conflict properties of nations (Rummel, 1963; Tanter, 1966); alliance and cooperative behavior of governments by replicating the studies of Singer and Small concerning over time alliance behavior of nations (Singer and Small, 1966); and general systemic properties by replicating and extending the Chadwick studies of INS and the international system (Chadwick, 1966, 1967).

To summarize these studies for thirty-three discrete types of domestic and foreign conflict behavior aggregated into fourteen categories using the Rummel coding scheme, the performance characteristics of IPS show strong correspondence to the international system of 1955-60 in terms of generating the same structures when subjected to factor analytical procedures. Additionally, the future world exhibits significantly different patterns of behavior to the past simulated world, and these patterns are consistent with the predictions of prominent peace theorists (Galtung, 1968). Similarly, professional decision-makers exhibit patterns of behavior significantly different from those of high school decision-makers. The high school and future worlds are also less conflictual.

For the Singer-Small alliance studies, when comparisons of 130 separate simulation and international system correlations of alliance patterns are undertaken, the IPS correlations correspond to approximately two-thirds of the international system correlations.

For the Chadwick replication of the intercorrelations between seventeen separate variables ranging from economic and trade variables to political, alliance, and conflict variables, roughly two-thirds of the 136 correlations were similar in the IPS and referent systems, whereas less than half were similar in the INS.

The evidence from these studies suggests that IPS demonstrates stronger correspondence to the referent international systems considered than does INS for these variables. However, care is required in the use of IPS as an educational technique, since on hypothesis validity grounds it seems reasonable to suggest that the focus of IPS is almost exclusively at the international system level. While domestic processes within nations are represented in the programmed and unprogrammed parts, relative to WPS, the model is poorly developed here. IPS has been used as an educational device (Biery, 1969), but until more evidence is accumulated (Busse, 1968; Soroos, 1969), care should be taken to explain its as yet limited confirmed validity.

On the other hand, relatively few theorists have models that transcend the "ping-pong-ball" view of planet Earth, and an exposure to IPS may be of value from this point of view. Relative to other simulations of international relations, the performance characteristics are at least tolerable, given the primitive state of the art; and in addition, rudimentary multilevel theory is incorporated in the model.

Within the context of future requirements, however, there are serious errors in structure and theory. No structural components in the simulation relate the ecological and technological global environment to economic, political, and social variables. No built-in procedure enables participants to change the rules of the environment under certain conditions. And an inadequate understanding of social time and over time processes makes it seem possible to have different parts of the simulation working on different time scales.

Each of these problems requires a considerable amount of work and some drastic overhauling of international relations theory. The next generation of man-computer simulations will, hopefully, make some progress in overcoming them.

SUMMARY AND PROSPECTS

Guetzkow has conceptualized the INS as a self-organizing system (Guetzkow, 1963), and to a large extent the community of international relations simulators can be seen in a similar way. Particular models, such as the INS, are taken and adapted in extremely innovative ways by different scholars in different parts of the world for different research purposes (Laulicht, 1967; Seki, 1966; Capello, 1971). As a vehicle for theory and development of theory in a cumulative fashion, the simulation endeavor, despite many problems (Schwartz, 1965), has been among the most rigorous and sustained during the last decade of international relations research (Guetzkow, 1970). Basic models such as INS have progressively been developed into constructions such as WPS (Coplin, 1969) and IPS (Smoker, 1968b). At the micro level, individual decision-makers in INS have been computerized (Bremer, n.d.), and detailed analyses of individual equations have led to revisions and improvements (Elder and Pendley, 1966a).

The problem of validity is still one of the most difficult questions facing all models in international relations, and simulators seem to be increasingly concerned with this question. It is no longer enough to have a model unless some sort of validation is considered a part of the continual process of model creation. Similarly, the question of relevance is likely to become increasingly important as different views on what is relevant confront one another. For example, both WPS with its focus on policy influencers and IPS with its concentration on the international system can be conceived as theoretical statements of relevance. An objective observer might conclude that both viewpoints should be incorporated in a next-generation simulation.

Of course, relevance considerations bring many important value questions into focus and, in this way, raise issues about the validity of aspects of reality. Those simulators concerned with practical policy may increasingly attempt to change reality as a result of some experience with a relevant model. A lively and hopefully productive debate between simulators and practitioners may be anticipated in the next decade of activity.

Early in the past decade, it seemed useful to differentiate between all-computer and man-computer simulations; but the advent of on-line equipment and almost immediate feedback has in some way blurred this previously clear distinction. An all-computer simulation has now become a situation with one experimenter interacting with one computerized environment, often modifying his experiments as a result of the response (or should we say "stimulus") from the computer. A man-computer simulation simply involves more than one experimenter, with the experimenters themselves interacting with each other. As we know from the beginning effect in simulations such as IPS, it takes a little while before the participants' behavior is conditioned by the computerized environment, and the same may be true for the all-computer experiment.

Whatever distinctions subsequently emerge in a typology of international relations simulations, increasing use will undoubtedly be made of computer approaches in the study of international relations, and a number of third-generation constructions are likely to emerge in the next decade. These constructions may incorporate whole new ranges of variables, such as those relevant to global ecosystems and survival, and may explore many more properties of world politics with a concern for validity and a sensitivity for relevance far in excess of all our present simulations.

REFERENCES

Abt Associates, Inc., *Survey of the State of the Art: Social, Political, and Economic Models and Simulations.* Cambridge, Mass.: Abt Associates, Inc., November 1965.

Abt, Clark C., and Morton Gorden, "Report on Project TEMPER," in Dean G.

Pruitt and Richard C. Snyder (eds.), *Theory and Research on the Causes of War,* pp. 245-62. Englewood Cliffs, N.J.: Prentice-Hall, Inc., 1969.

Ackoff, Russell, Shiv K. Gupta, and J. Sayer Minas, *Scientific Method: Optimizing Applied Research Decisions.* New York: John Wiley & Sons, Inc., 1962.

Alger, Chadwick F., "Trends in International Relations Research," in Norman D. Palmer (ed.), *A Design for International Relations Research: Scope, Theory, Methods, and Relevance.* Philadelphia: The American Academy of Political and Social Science, 1970, 7-28.

Alker, Hayward R., Jr., "Computer Simulations, Conceptual Frameworks and Coalition Behavior," in Goennings, Kelly, and Leiserson (eds.), *The Study of Coalition Behavior: Theoretical Perspectives and Cases from Four Continents.* New York: Holt, Rinehart, & Winston, Inc., 1969.

———, and Ronald D. Brunner, "Simulating International Conflict: A Comparison of Three Approaches," *International Studies Quarterly,* XIII, 1 (Spring 1969), 70-110.

*Angell, Robert C., *Peace on the March: Transnational Participation.* New York: Van Nostrand Reinhold Company, 1969.

Ausland, John, and Hugh Richardson, "Crisis Management: Berlin, Cyprus and Laos," *Foreign Affairs,* 44, 2 (1966), 291-303.

Averch, H., and M. M. Lavin, *Simulation of Decision Making in Crises: Three Manual Gaming Experiments,* RAND Report RM-4202-PR. Santa Monica: Calif.: RAND Corporation, August 1964.

*Azar, Edward, Stanley Cohen, Thomas Jukam, and James McCormick, "Methodological Developments in the Quantification of Events Data," East Lansing: Michigan State University, Cooperation/Conflict Research Group, April 1970.

Bailey, Gerald, *Utilizing Simulation of International Behavior in Political-Military Affairs: A Preliminary Analysis.* Arlington, Va.: Human Sciences Research, Inc., April 1967.

Balinski, Michel, Klauss Knorr, Oskar Morgenstern, Francis Sand, and Martin Shubik, *Review of TEMPER Model* (final report draft). Princeton, N. J.: Mathematica, September 1966.

Banks, Michael, A.J.R. Groom, and A. N. Oppenheim, "Gaming and Simulation in International Relations," *Political Studies,* XVI, 1 (February 1968), 1-17.

Barringer, Richard E., and Barton Whaley, "The MIT Political Military Gaming Experience," *Orbis, A Quarterly Journal of World Affairs,* IX, 2 (Summer 1965), 437-58.

Benson, Oliver, "A Simple Diplomatic Game," in J. N. Rosenan (ed.) *International Politics and Foreign Policy: A Reader in Research and Theory,* pp. 504-11. New York: The Free Press of Glencoe, Inc., 1961,

———, "Simulation of International Relations and Diplomacy," in Harold Borko

---

*These items are not directly relevant to simulation but are referred to in the text.

(ed.), *Computer Application in the Behavioral Sciences,* pp. 574-95. Englewood Cliffs, N. J.: Prentice-Hall, Inc., 1963.

Biery, James, "Playing Games with the Future," *Northwestern Review,* Evanston, Ill.: Northwestern University, Winter 1969, 18-23.

Bloomfield, Lincoln P., "Political Gaming," *United States Naval Institute Proceedings,* LXXXVI, 9 (September 1960), 57-64.

_____, "Report on CONEX 1," Cambridge Mass.: Massachusetts Institute of Technology, Center for International Studies, 1968.

_____, and N. J. Padelford, "Three Experiments in Political Gaming," *The American Political Science Review,* LIII, 4 (December 1959), 1105-15.

_____, and Barton Whaley, "POLEX: The Political Military Exercise," *The Military Review: Professional Journal of the U. S. Army,* Fort Leavenworth, Kansas, November 1965a, 65-71.

_____, "The Political Military Exercise: A Progress Report," *Orbis, A Quarterly Journal of World Affairs,* VIII, 4 (Winter 1965b), 854-70.

Bobrow, Davis B., "Computers and a Normative Model of the Policy Process." Paper presented at the 1969 Meetings of the American Association for the Advancement of Science, Boston, December 26-31, 1969.

_____, and Judah L. Schwartz, *Computers and the Policy-Making Community.* Englewood Cliffs, N. J.: Prentice-Hall, Inc., 1968.

Bonham, Gaylor, "Aspects of the Validity of Two Simulations of Phenomena in International Relations." Ph.D. dissertation, Massachusetts Institute of Technology, Department of Political Science, 1967.

Boocock, Sarane, and E. O. Schild, *Simulation Games in Learning.* Beverly Hills, Calif.: Sage Publications, 1968.

Bremer, Stuart A., "The Validation and Evaluation of the Siper Computer Simulation Model," in "National and International Systems: A Computer Simulation," Chap. 7. Draft, Ph.D. dissertation, Michigan State University, Department of Political Science, n.d.

Brody, Richard A., "Some Systemic Effects of the Spread of Nuclear Weapons Technology: A Study through Simulation of a Multi-Nuclear Future," *Journal of Conflict Resolution,* VII, 4 (December 1963), 663-753.

_____, Alexandra H. Benham, and Jeffrey S. Milstein, "Hostile International Communication, Arms Production, and Perception of Threat: A Simulation Study," Stanford, Calif.: Stanford University, Institute of Political Studies, July 1966.

Browning, Rufus P., "Quality of Collective Decisions, Some Theory and Computer Simulations," East Lansing: Michigan State University, June 1969.

*Burton, John W., *International Relations.* Cambridge: Cambridge University Press, 1967.

_____, "International Relations Simulations on the Cheap," Evanston, Ill.: Northwestern University, Simulated International Processes project, July 1966.

Busse, Walter R., "Negotiation in the International Processes Simulation." Dissertation Proposal, Northwestern University, November 1968.

———, "The Northwestern Simulation Archives: Man-Computer Models of International Relations," Evanston, Ill.: Northwestern University, Simulated International Processes project, 1969.

Campbell, Donald T., "Factors Relevant to the Validity of Experiments in Social Settings," *Psychological Bulletin,* 54 (1957), 297-312.

———, and D. W. Fiske, "Convergent and Discriminant Validation by the Multitrait-Multimethod Matrix," *Psychological Bulletin,* 56 (1959), 81-105.

Cappello, Hector, "International Tension As a Function of Reduced Communication (A Simulation Study)," in J. Laponce and P. Smoker (eds.), *The Vancouver Papers on Experimentation and Simulation,* in press.

Caspary, William, "The Causes of War in Inter-Nation Simulation." Evanston, Ill.: Northwestern University, 1965.

Chadwick, Richard W., "An Empirical Test of Five Assumptions in an Inter-Nation Simulation about National Political Systems," *General Systems,* XII (1967), 177-92.

———, "An Inductive, Empirical Analysis of Intra and International Behavior, Aimed at a Partial Extension of Inter-Nation Simulation Theory," *Journal of Peace Research,* VI, 3 (1969), 193-214.

———, "Developments in a Partial Theory of International Behavior: A Test and Extension of Inter-Nation Simulation Theory." Ph.D. dissertation, Northwestern University, Department of Political Science, 1966.

Cherryholmes, Cleo, "Developments in Simulation of International Relations in High School Teaching," *Phi Delta Kappa* (January 1965), 227-31.

Cohen, Bernard D., "Political Gaming in the Classroom," *Journal of Politics,* XXIV, 2 (May 1962), 367-80.

Coplin, William, "Approaches to the Social Sciences through Man-Computer Simulations," *Simulation and Games,* Vol. I. Beverly Hills, Calif.: Sage Publications, December 1970.

———, "Inter-Nation Simulation and Contemporary Theories of International Relations," *The American Political Science Review,* LX, 3 (September 1966) 562-78.

———, "Man-Computer Simulation As an Approach to the Study of Politics: Implications from a Comparison of State Department and High School Runs of the World Politics Simulation," *Proceedings, National Gaming Council, Eighth Symposium,* 1969.

——— (ed.), *Simulation in the Study of Politics.* Chicago: Markham Publishing Co., 1968.

Cornblit, Oscar, "A Model of Short Run Political Change," in J. Laponce and P. Smoker (eds.), *The Vancouver Papers on Experimentation and Simulation,* in press.

*Corson, Walter, "Conflict and Cooperation in East-West Crises: Measurement and Prediction." Ann Arbor: University of Michigan, Institute for Social Research, April 1970.

Crow, Wayman J., "A Study of Strategic Doctrines Using Inter-Nation Simulation," *Journal of Arms Control,* I, 4 (October 1963), 674-83.

———, "Simulation: The Construction and Use of Functioning Models in International Relations," in K. R. Hammond (ed.), *Egon Brunswiks Psychology,* pp. 340-48. New York: Holt, Rinehart & Winston, Inc., 1966.

———, and Robert C. Noel, *The Valid Use of Simulation Results.* La Jolla, Calif.: Western Behavioral Sciences Institute, June 1965.

———, and John Raser, "A Cross Cultural Simulation Study," La Jolla, Calif.: Western Behavioral Sciences Institute, November 1964, pp. 12-18.

De La Mater, Stephen T., Jack V. Dunham, Robert W. Granston, Herbert H. Schulke, Chester R. Smith, and Frederick R. Westfall, *ICAF Analysis of TEMPER MARK II.* Washington, D. C.: Industrial College for the Armed Forces, 1966.

*Deutsch, Karl W., *The Analysis of International Relations.* Englewood Cliffs, N. J.: Prentice-Hall, Inc., 1968.

DeWeerd, Harvey, "An Israeli Scenario for a Laboratory Simulation." Santa Monica, Calif.: System Development Corporation, April 1968.

Draper, George, "Technological, Economic, Military and Political Evaluation Routine (TEMPER)—An Evaluation." Washington, D. C.: National Military Command Systems Support Center, July 1966.

Driver, Michael J., "A Structural Analysis of Aggression, Stress, and Personality in an Inter-Nation Simulation." Lafayette, Ind.: Purdue University, January 1965.

———, "Conceptual Structure and Group Processes in an Inter-Nation Simulation. Part One: The Perception of Simulated Nations." Princeton, N. J.: Princeton University and Educational Testing Service, 1962.

Druckman, Daniel, "Ethnocentrism in the Inter-Nation Simulation," *Journal of Conflict Resolution,* XII, 1 (March 1968), 45-68.

Elder, Charles D., and Robert E. Pendley, "An Analysis of Consumption Standards and Validation Satisfactions in the Inter-Nation Simulation in Terms of Contemporary Economic Theory and Data." Evanston, Ill.: Northwestern University, November 1966a.

———, "Simulation As Theory Building in the Study of International Relations." Evanston, Ill.: Northwestern University, July 1966b.

Evans, George W., Graham F. Wallace, and Georgia L. Sutherland, *Simulation Using Digital Computers.* Englewood Cliffs, N. J.: Prentice-Hall, Inc., 1967.

Fischer, R. Lucas, "The Rand/MIT Political-Military Exercise and International Relations Theory." Evanston, Ill.: Northwestern University, 1968.

Forcese, Dennis, "Power and Military Alliance Cohesion: Thirteen Simulation Experiments." Ph.D. dissertation, Washington University, Department of Sociology, St. Louis, Mo., 1968.

*Galtung, Johan, "Entropy and the General Theory of Peace," *Proceedings,* Second International Peace Research Association General Conference, Vol. I. Assen: Van Gorcum & Comp. N.V., 1968.

Giffen, Sydney F., *The Crisis Game: Simulating International Conflict.* Garden City, N. Y.: Doubleday & Company, Inc., 1965.

Goldhamer, Herbert, and Hans Speier, "Some Observations on Political Gaming," in James N. Rosenau (ed.), *International Politics and Foreign Policy,* pp. 498-503. New York: The Free Press of Glencoe, Inc., 1961.

Gomer, Louise C., "Master List of Variables in the Simulation of International Processes." Evanston, Ill.: Northwestern University, Simulated International Processes project, March 1968.

Gorden, Morton, "Burdens for the Designer of a Computer Simulation of International Relations: The Case of TEMPER," in Davis B. Bobrow and Judah L. Schwartz (eds.), *Computers and the Policy-Making Community,* pp. 222-45. Englewood Cliffs, N. J.: Prentice-Hall, Inc., 1968.

———, "International Relations Theory in the TEMPER Simulation." Evanston, Ill.: Northwestern University, 1967.

Gottheil, Diane Levitt, "A Method for Comparing Verbal and Simulation Theory, with Its Application to an Essay on Alliances by Wolfers." Evanston, Ill.: Northwestern University, April 1966.

Grant, Lawrence, "The Problem of Validity for Computer Simulation." Ames: Iowa State University, January 1970.

Guetzkow, Harold, "A Decade of Life with the Inter-Nation Simulation," in Ralph G. Stogdill (ed.), *The Process of Model-Building in the Behavioral Sciences.* Columbus: Ohio State University Press, 1970.

———, "A Use of Simulation in the Study of International Relations," *Behavioral Science,* 4, 3 (July 1959), 183-91.

———, "Evaluation of the Inter-Nation Simulation As a Teaching Aid," *Proceedings of the American Society of International Law,* 58 (1964a), 78-79.

———, "Inter-Nation Simulation: An Example of a Self-Organizing System," in Marshall C. Yovits, George T. Jacobi, and Gordon D. Goldstein (eds.), *Self-Organizing Systems,* pp. 79-92. Washington, D. C.: Spartan Books, 1962.

———, "Simulation in International Relations," *Proceedings of the IBM Scientific Computing Symposium on Simulation Models and Gaming.* Yorktown Heights, N. Y.: Thomas J. Watson Research Center, December 7-9, 1964b, pp. 249-78.

———, "Simulations in the Consolidation and Utilization of Knowledge about International Relations," in Dean G. Pruitt and Richard C. Snyder (eds.), *Theory and Research on the Causes of War,* pp. 284-300. Englewood Cliffs, N. J.: Prentice-Hall, Inc., 1969.

———, "Some Correspondence between Simulations and 'Realities' in International Relations," in Morton A. Kaplan (ed.), *New Approaches to International Relations,* pp. 202-69. New York: St. Martin's Press, 1968.

———, "Some Uses of Mathematics in Simulation of International Relations," in John M. Claunch (ed.), *Mathematical Applications in Political Science,* pp. 21-40. Dallas, Tex.: The Arnold Foundation (Southern Methodist University), 1965.

———, Chadwick F. Alger, Richard A. Brody, Robert C. Noel, and Richard C.

Snyder, *Simulation in International Relations: Developments for Research and Teaching.* Englewood Cliffs, N. J.: Prentice-Hall, Inc., 1963.

———, and Cleo Cherryholmes, *Inter-Nation Simulation Kit: Instructor's Manual.* Chicago: Science Research Associates, Inc., 1966a.

———, *Inter-Nation Simulation Kit: Participant's Manual.* Chicago: Science Research Associates, Inc., 1966b.

———, and Lloyd Jensen, "Research Activities on Simulated International Processes," *Background: Journal of the International Studies Association,* IX, 4 (February 1966), 261-74.

Hermann, Charles F., *Crises in Foreign Policy: A Simulational Analysis.* Indianapolis: The Bobbs-Merrill Co., Inc., 1969.

———, "Crises in Foreign Policy Making: A Simulation of International Politics." Ph.D. dissertation, Northwestern University, Department of Political Science, 1965.

———, "Critique and Comment: Validation Problems in Games and Simulations with Special Reference to Models of International Politics," *Behavioral Science,* XII, 3 (May 1967a), 216-31.

———, "Games and Simulations of Political Processes," in *International Encyclopedia of the Social Sciences,* Vol. 14, pp. 247-81. New York: The Free Press, 1968.

———, "Threat, Time and Surprise: A Simulation of International Crisis." Princeton, N. J.: Princeton University, April, 1967b.

*———, "What Is a Foreign Policy Event?" A paper prepared for the Events Data Measurement Conference, Michigan State University, East Lansing, April 15-16, 1970.

———, and Margaret G. Hermann, "An Attempt to Simulate the Outbreak of World War I," *The American Political Science Review,* LXI, 2 (June 1967), 400-16.

———, "Validation Studies of the Inter-Nation Simulation." China Lake, Calif.: U. S. Naval Ordnance Test Station, December 1963.

Hermann, Margaret G., "Stress, Self-Esteem, and Defensiveness in an Inter-Nation Simulation." Ph.D. dissertation, Northwestern University, Department of Psychology, 1965.

———, "Testing a Model of Psychological Stress," *Journal of Personality*, 34, 3 (September 1966), 381-96.

*Holsti, Ole R., Robert C. North, and Richard A. Brody, "Perception and Action in the 1914 Crisis," in J. David Singer (ed.), *Quantitative International Politics: Insights and Evidence,* pp. 123-58. New York: The Free Press, 1968.

Hoole, Frank, "Decision Making in the World Health Organization: The Budgetary Process." Draft, Ph.D. dissertation, Northwestern University, Department of Political Science, April 1970.

Industrial College of the Armed Forces, *A Study and Evaluation of the Technological, Economic, Military and Political Evaluation Routine (TEMPER).* Washington, D. C.: Resident School 1965-66—TEMPER Committee Report, Industrial College of the Armed Forces, 1966a.

————, *Briefing Manual: World Politics Simulation III/ICAF.* Washington, D. C.: Industrial College of the Armed Forces, 1969a.

————, *TEMPER ICAF 1968 Reports,* Vols. 1 and 2. Washington, D. C.: Industrial College of the Armed Forces, 1968a.

————, *The Player Handbook: Addendum TEMPER–68.* Washington, D. C.: Industrial College of the Armed Forces, 1968b.

————, *The Player Handbook: TEMPER-66.* Washington, D. C.: Industrial College of the Armed Forces, 1966b.

————, *The Player Handbook: TEMPER-67.* Washington, D. C.: Industrial College of the Armed Forces, 1967.

————, *The Player Handbook: TEMPER-68.* Washington, D. C.: Industrial College of the Armed Forces, 1968c.

————, *World Politics Simulation: Administrative Manual.* Washington, D. C.: Industrial College of the Armed Forces, 1969b.

————, *World Politics Simulation: Background Materials.* Washington, D. C.: Industrial College of the Armed Forces, 1969c.

————, *World Politics Simulation: Computer Program Listings.* Washington, D. C.: Industrial College of the Armed Forces, February 1969d.

————, *World Politics Simulation: Description of Model.* Washington, D. C.: Industrial College of the Armed Forces, 1969e.

————, *World Politics Simulation: Participant's Manual.* Washington, D. C.: Industrial College of the Armed Forces, 1969f.

*Jensen, Lloyd, "Foreign Policy Elites and the Prediction of International Events," *Papers, Peace Research Society (International),* V (1966a).

*————, "United States Elites and Their Perceptions of the Determinants of Foreign Policy Behavior." Evanston, Ill.: Northwestern University, April 1966b.

Johnson, Douglas F., and Dean G. Pruitt, "Pre-Intervention Effects of Mediation versus Arbitration." Buffalo: State University of New York, Department of Psychology, August 1969.

Judge, A. J. N., "Management Game Techniques and International NGO's," *Associations Internationales,* 10 (1967), 659-65.

————, "Proposal for the Development of a 'World Game' As a Long-Term Education Technique." Brussels: Union of International Associations, n.d.

Krend, Jeffrey A., "A Comparison of Guetzkow-Modelski Correspondence Ratings." Evanston, Ill.: Northwestern University, Simulated International Processes project, 1969.

————, "A Reconstruction of Oliver Benson's 'Simple Diplomatic Game,'" in J. Laponce and P. Smoker (eds.), *The Vancouver Papers on Experimentation and Simulation,* in press.

————, "A Typology for Ordering the Simulation and Gaming Literature" (draft). Evanston, Ill.: Northwestern University, March 1968.

Kress, Paul, "On Validating Simulation: with Special Attention to Simulation of International Politics." Evanston, Ill.: Northwestern University, April 1966.

Laulicht, Jerome, "A Vietnam Peace Game: Computer-Assisted Simulation of Complex Relations in International Relations," *Computers and Automation,* XVI, 3 (March 1967), 14-18.

*Leng, Russell J., and J. David Singer, "Toward a Multi-Theoretical Typology of International Behavior." A paper prepared for the Events Data Conference, Michigan State University, East Lansing, April 1970.

*McClelland, Charles A., "The Access to Berlin: The Quantity and Variety of Events, 1948-63," in J. David Singer (ed.), *Quantitative International Politics: Insights and Evidence*, pp. 159-86. New York: The Free Press, 1968.

*———, *Theory and the International System.* New York: The Macmillan Company, 1966.

*———, *et al.*, "The Communist Chinese Performance in Crisis and Non-Crisis: Quantitative Studies of the Taiwan Straits Confrontation, 1950-64." China Lake, Calif.: Behavioral Sciences Group, Naval Ordnance Test Station, 1965.

*McGowan, Patrick, "The Unit of Analysis Problem in the Comparative Study of Foreign Policy." A paper prepared for the Events Data Measurement Conference, Michigan State University, East Lansing, April 1970.

MacRae, John, "Bosnian Crisis Simulation" (Participants' Manual and Scenario). Lancaster: University of Lancaster, 1967.

———, and Paul Smoker, "A Vietnam Simulation: A Report on the Canadian/ English Joint Project," *Journal of Peace Research,* IV, 1(1967), 1-25.

Martin, Francis F., *Computer Modeling and Simulation.* New York: John Wiley & Sons, Inc., 1968.

Meier, Dorothy, and Arthur Stickgold, "Progress Report: Event Simulation Project (INS-16)" (draft). St. Louis, Mo.: Washington University, 1965.

Milstein, Jeffrey S., and William Charles Mitchell, "Computer Simulation of International Processes: The Vietnam War and the Pre-World War I Naval Race," Sixth North American Peace Research Conference, November 1968.

Modelski, George, "Simulations, 'Realities' and International Relations Theory." Evanston, Ill.: Northwestern University, Simulated International Processes project, 1969.

Mushakoji, Kinhide, "Negotiations between the West and the Non-West— Cultural Problems in Conflict Resolution," *Proceedings,* Second International Peace Research Association General Conference, Vol. I, N. W., Assen: Van Gorcum & Comp. N.V., 1968.

———, "The Strategies of Negotiation," in J. Laponce and P. Smoker (eds.), *The Vancouver Papers on Experimentation and Simulation*, in press.

Myers, Mary Lynn, "Bibliographic and Abstracted Bibliographic Sources: An Annotated Listing of Sources Which Include Materials Dealing with Simulation in the Social Sciences, Especially Those Concerned with the Simulation of International Processes." Evanston, Ill.: Northwestern University, Simulated International Processes project, 1968.

Nardin, Terry, and Neal E. Cutler, "Reliability and Validity of Some Patterns of

International Interaction in an Inter-Nation Simulation," *Journal of Peace Research,* VI, 1 (1969), 1-12.

*Park, Tong-Whan, "A Guide to Data Sources in International Relations: Annotated Bibliography with Lists of Variables." Evanston, Ill.: Northwestern University, 1968.

Parker, John, and Marshall Whithed, "Developing an International Mutual Response System through Political Simulation Techniques." Troy, N. Y.: Rensselaer Polytechnic Institute, Department of Political Science, 1969.

Pelowski, Al, "Multi-National Corporate Futures: An Approach to Simulation Module Construction in International Relations." Evanston, Ill.: Northwestern University, Simulated International Processes project, 1969.

———, "Preliminary Simulation of the Taiwan Straits Crisis: An Event-Based Probability Model," in J. Laponce and P. Smoker (eds.), *The Vancouver Papers on Experimentation and Simulation,* in press.

Pendley, Robert E., "INSCAL: A Fortran Program for Performing the Calculations for the Inter-Nation Simulation." Evanston, Ill.: Northwestern University, 1966.

———, and Charles Elder, "An Analysis of Officeholding in the Inter-Nation Simulation in Terms of Contemporary Political Theory and Data on the Stability of Regimes and Governments." Evanston, Ill.: Northwestern University, 1966.

Pfaltzgraff, Robert L., "Simulation and International Relations Theory: A Comparison of Simulation Models and International Relations Literature." Philadelphia: University of Pennsylvania, 1969.

Pilisuk, Marc, Paul Skolnick, Kenneth Thomas, and Reuben Chapman, "Boredom vs. Cognitive Reappraisal in the Development of Cooperative Strategy," *Journal of Conflict Resolution,* XI, 1 (March 1967), 110-16.

Pool, Ithiel de Sola, and Allan Kessler, "The Kaiser, the Tsar and the Computer: Information Processing in a Crisis," *American Behavioral Scientist,* VIII, 9 (May 1965), 31-38.

Pruitt, Dean G., "Negotiation As a Form of Social Behavior." Buffalo: State University of New York, October 1968.

———, "Reward Structure and Cooperation, Part II: Motivational Processes in the Decomposed Prisoner's Dilemma Game." Buffalo: State University of New York, Department of Psychology, February 1969.

———, "Reward Structure and Its Effect on Cooperation," *Papers, Peace Research Society (International),* V (1966), 73-85.

———, and Douglas Johnson, "Meditation As an Aid to Face-Saving in Negotiation." Buffalo: State University of New York, Department of Psychology, January 1969.

*Raser, John R., "Personal Characteristics of Political Decision Makers: A Literature Review." La Jolla, Calif.: Western Behavioral Sciences Institute, 1965.

———, *Simulation and Society.* Boston: Allyn & Bacon, Inc., 1969.

———, and Wayman J. Crow, "A Simulation Study of Deterrence Theories," in Dean G. Pruitt and Richard C. Snyder (eds.), *Theory and Research on the*

Causes of War, pp. 136-49. Englewood Cliffs, N. J.: Prentice-Hall, Inc., 1969.

Raytheon Company, *Technological, Economic, Military and Political Evaluation* Routine (TEMPER), Vol. I *Orientation Manual,* July 1965; Vol. II *The Theory of the Model,* August 1965; Vol. III *Game Handbook,* July 1965; Vol. IV *Technical Manual,* September 1965; Vol. V *Operations Manual,* July 1965; Vol. VI *Reference Manual,* July 1965; Vol. VII *Data Collection Manual,* July 1965. Bedford, Mass.: Raytheon Company.

Robinson, James A., Lee F. Anderson, Margaret Hermann, and Richard C. Snyder, "Teaching with the Inter-Nation Simulation and Case Studies," *The American Political Science Review,* LX, 1 (March 1966), 53-65.

———, Charles F. Hermann, and Margaret Hermann, "Search Under Crisis in Political Gaming and Simulation," in Dean G. Pruitt and Richard C. Snyder, *Theory and Research on the Causes of War,* pp. 80-94. Englewood Cliffs, N. J.: Prentice-Hall, Inc., 1969.

*Rosenau, James N., "Pre Theories and Theories of Foreign Policy," in R. Barry Farrell (ed.), *Approaches to Comparative and International Politics,* pp. 27-92. Evanston, Ill.: Northwestern University Press, 1966.

Ruge, Mari Holmboe, "Decision Makers As Human Beings: An Analysis of Perception and Behavior in a Cross-Cultural Simulation Experiment." Oslo: University of Oslo, Institute of Political Science, and International Peace Research Institute, 1969.

*Rummel, Rudolph J., "Dimensions of Conflict: Behavior within and between Nations," *General Systems Year Book,* 8 (1963), 1-50.

Schwartz, David C., "Problems in Political Gaming," *Orbis: A Quarterly Journal of World Affairs,* IX, 3 (Fall 1965), 677-93.

———, "Threat, Hostility and Behavior Preferences in Crisis Decision Making: A Further Comparison of Historical and Simulation Data." Philadelphia: University of Pennsylvania, n.d.

*Scott, Andrew M., Jr., *The Functioning of the International Political System.* New York: The Macmillan Company, 1967.

Seki, Hiroharu, "Kokusai Taikei No Simulation—INS-J-4 No Modern Ni Tsuite" (The Simulation of the International System—on the Model of INS-J-4), *KodoKagaku Kenkyu (The Journal of Behavioral Science),* 1, 2 (1966), Tokai Daigaku Kiso Shakaikagaku Kenkyujo.

Shapiro, Michael, "Cognitive Rigidity and Moral Judgments in an Inter-Nation Simulation." Evanston, Ill.: Northwestern University, September 1966.

Shubik, Martin, "Bibliography on Simulation, Gaming, Artificial Intelligence and Allied Topics," *Journal of the American Statistical Association,* December 1960, pp. 736-51.

Shure, Gerald H., and Robert J. Meeker, "The Effectiveness of Pacifist Strategies in Bargaining Games." Santa Monica, California: System Development Corporation, September 16, 1964, Sp-1588/000/00, multilithed, 14 pp.

*Sigler, John, "Reliability Problems in the Measurement of International Events in the Elite Press." A paper prepared for the Events Data Measurement Conference, Michigan State University, East Lansing, April 1970.

Simulated International Processes project, "Another Partial Bibliography on Simulation in International Relations." Evanston, Ill.: Northwestern University, 1968.

Singer, J. David and Hirohide Hinomoto, "Inspecting for Weapons Production: A Modest Computer Simulation," *Journal of Peace Research*, 2, 1 (1965), 18-38.

*———, and Melvin Small, "Formal Alliances, 1815-1939," *Journal of Peace Research*, 3, 1 (1966), 1-32.

———, "Simulations and Gaming of International Military Political Behaviors: Survey of Activities." Evanston, Ill.: Northwestern University, 1966a.

———, "Studies Related to the Simulation of International Political and Military Processes: Collection of Abstracts." Evanston, Ill.: Northwestern University, 1966b.

Smith, Clifford Neal, and Marshall Hale Whithed, "International Business Decision-Making in a Political Setting: The Model SW-2 Simulation," *Proceedings, National Gaming Council, Eighth Symposium*, 1969.

———, "International Business Decision-Making in a Political Setting: The SW-1 Simulation." DeKalb, Ill.: Northern Illinois University, 1968a.

———, "PSW-1 Simulation." DeKalb, Ill.: Northern Illinois University, 1968b.

Smoker, Paul, "Analyses of Conflict Behaviours in an International Processes Simulation and an International System 1955-60." Evanston, Ill.: Northwestern University, 1968a.

———, "Fortran Program for Vietnam Simulation." Lancaster: Peace Research Center, 1966.

———, "International Processes Simulation: A Man-Computer Model." Evanston, Ill.: Northwestern University, 1968b.

———, "International Processes Simulation: An Evaluation," *Peace Research Reviews*, November 1970a.

———, "International Relations Simulations," in Pat Tansey (ed.), *Aspects of Simulations in Education*. London: McGraw-Hill, 1970b.

———, *IPS Program Pack*. Evanston, Ill.: Northwestern University, 1968c.

———, "Simulating the World," *Science Journal*, London, July 1970c.

———, "Social Research for Social Anticipation," *American Behavioral Scientist*, XII, 6 (July-August 1969), 7-13.

———, and John Martin, *Simulation and Games: An Overview*. Chicago: Modern Trends in Education Series, Science Research Associates, 1968.

Soroos, Marvin, "International Involvement and Foreign Conflict in the International Processes Simulation and the 'Real' World." Ph.D. dissertation proposal, Northwestern University, May 1969.

Stewart, Edward C., "The Simulation of Cultural Differences," *The Journal of Communication*, XVI, 4 (December 1966), 291-304.

Streufert, Siegfried, "Complexity and Complex Decision Making." Lafayette, Ind.: Purdue University, March 1969a.

———, "Increasing Failure and Response Rate in Complex Decision Making," *Journal of Experimental Social Psychology*, 5 (1969b), 310-23.

———, "The Components of a Simulation of Local Conflict: An Analysis of the Tactical and Negotiations Game." Evanston, Ill.: Northwestern University, 1968a.

———, "The Tactical and Negotiations Game: A Simulation of Local Conflict: An Analysis of Some Psychopolitical and Applied Implications of TNG Simulation Research." Lafayette, Ind.: Purdue University, January 1968b.

———, and Carl H. Castore, "Information Search and the Effects of Failure: A Test of Complexity Theory." Lafayette, Ind.: Purdue University, December 1968.

———, Carl Castore, and Susan Kilger, "A Tactical and Negotiations Game: Rationale, Method and Analysis." Lafayette, Ind.: Purdue University, June 1967.

———, M. Clardy, M. Driver, M. Karlins, M. Schroder, and P. Suedfeld, "A Tactical Game for the Analysis of Complex Decision Making in Individuals and Groups," *Psychological Reports,* 17 (1965), 723-29.

———, and Sandra Sandler, "Information Availability and Complex Perceptions of Americans and Chinese." Lafayette, Ind.: Purdue University, July 1969.

———, and Susan Streufert, "Effects of Conceptual Structure, Failure and Success on Attribution of Causality and Interpersonal Attitudes," *Journal of Personality and Social Psychology,* II, 2 (1969), 138-47.

———, Susan Streufert, and Carl Castore, "Complexity, Increasing Failure and Decision Making," *Journal of Experimental Research in Personality,* III, 4 (1969), 293-300.

———, Peter Suedfeld, and Michael Driver, "Conceptual Structure, Information Search and Information Utilization," *Journal of Personality and Social Psychology,* II, 5 (1965), 736-40.

Sullivan, Edward M., "Simulation As a Tool for Organizational Research and Design." Evanston, Ill.: Northwestern University, Department of Industrial Engineering and Management Sciences, January 1969.

Tansey, Pat (ed.), *Aspects of Simulation in Education.* London: McGraw-Hill, 1970.

*Tanter, Raymond, "Dimensions of Conflict Behavior within and between Nations, 1958-60," *Journal of Conflict Resolution,* X, 1 (1966), 41-64.

Targ, Harry, "Children's Developing Orientations to International Politics." Lafayette, Ind.: Purdue University, 1968a.

———, "Impacts of an Elementary School Inter-Nation Simulation on Developing Orientations to International Politics." Lafayette, Ind.: Purdue University, 1968b.

———, "The Inter-Nation Simulation: An Elementary School Exercise." Evanston, Ill.: Northwestern University, 1967.

———, and Terry Nardin, "The Inter-Nation Simulation As a Predictor of Contemporary Events." Evanston, Ill.: Northwestern University, Simulated International Processes project, August 1965.

Thorelli, Hans B., and Robert L. Graves, *International Operations Simulation, with Comments on Design and Use of Management Games.* New York: The Free Press (Collier-Macmillan), 1964.

Twelker, Paul A., and Carl J. Wallen, *Instructional Uses of Simulation; A Selected Bibliography*. Portland, Oregon: Teaching Research Division, Oregon State System of Higher Education in cooperation with Northwest Regional Educational Laboratory, 1967.

Verba, Sidney, "Simulation, Reality, and Theory in International Relations," *World Politics,* XVI, 3 (April 1964), 490-519.

Vertinsky, Ilan, "Methodology of Simulation and Experimentation for Social Planning," in J. Laponce and P. Smoker (eds.), *The Vancouver Papers on Experimentation and Simulation*, in press.

*Waltz, Kenneth N., *Foreign Policy and Democratic Politics*. Boston: Little, Brown and Company, 1968.

Westfall, F., and G. Draper, "The Initial Use of TEMPER." Washington, D. C.: Industrial College of the Armed Forces, April 1966.

Whaley, Barton, Peter C. Ordeshook, and Robert H. Scott, "EXDET III: A Student-Level Experimental Simulation on Problems of Deterrence." Cambridge: Massachusetts Institute of Technology, Center for International Studies, 1964.

———, and Aaron Seidman, "EXDET II." Cambridge: Massachusetts Institute of Technology, Center for International Studies, November 1964.

Zinnes, Dina A., "A Comparison of Hostile Behavior of Decision-Makers in Simulate and Historical Data," *World Politics,* XVIII, 3 (April 1966), 474-502.

———, Douglas Van Howeling, and Richard Van Atta, "A Test of Some Properties of the Balance of Power Theory in a Computer Simulation." Evanston, Ill.: Northwestern University, Simulated International Processes project, Summer 1969.

AN ATTEMPT TO SIMULATE
THE OUTBREAK OF WORLD WAR I

CHARLES F. HERMANN

MARGARET G. HERMANN

Political games and simulations are models or representations of particular political systems and their associated processes. They are techniques for reproducing in a simplified form selected aspects of one system, *A,* in some independent system, *A'.* Games and simulations have a dynamic quality produced by the complex interaction of properties in the model. This feature enables them to generate states of the system that differ radically from those present originally. The kinds of transformations that may occur between the initial and final states of a simulation or game are difficult to represent by other means, despite a diversity in modeling procedures ranging from verbal descriptions to differential equations. Because of their apparent applicability to many problems of politics, as well as their novelty, games and simulations have been developed in a variety of areas in political science.[1] They have been used in research,

Reprinted with permission from the *American Political Science Review,* 61 (June 1967), 400-16. Copyright © by The American Political Science Association. This research was conducted under Contract N123(60530)25875A from Project Michelson, U. S. Naval Ordnance Test Station, China Lake, California. An earlier report on this project was distributed by the contractor as *Studies in Deterence X: Validation Studies of the Inter-Nation Simulation,* NOTS Technical Paper 3351, December, 1963. The authors wish to acknowledge their indebtedness to Harold Guetzkow, principal investigator and mentor; Thomas W. Milburn, director of Project Michelson; and Robert C. North and colleagues at the Stanford Studies in Conflict and Integration who generously shared their document collection and data analysis on the outbreak of World War I. The Center of International Studies at Princeton University supported the first author during the preparation of the present article.

[1] In addition to the studies cited elsewhere in this paper, the variety is suggested by the following illustrations: Oliver Benson, "A Simple Diplomatic Game," in James A. Rosenau (ed.), *International Politics and Foreign Policy* (New York: Free Press, 1961), 504-511; William P. Davison, "A Public Opinion Game," *Public Opinion Quarterly,* 25 (1961), 210-220; Robert P. Abelson and Alex Bernstein, "A Computer Simulation Model of Community Referendum Controversies," *Public Opinion Quarterly,* 27 (1963), 93-122; Lincoln P. Bloomfield and Barton Whaley, "The Political-Military Exercise: A Progress Report," *Orbis, 8 (1965), 854-870;* Andrew M. Scott with William A. Lucas and Trudi M. Lucas, *Simulation and National Development* (New York: Wiley, 1966); J. David Singer and Hirohide Hinomoto, "Inspecting for Weapons Production: A Modest Computer Simulation," *Journal of Peace Research* (1965), 18-38; James A. Robinson, Lee F. Anderson, Margaret G. Hermann, and Richard C. Snyder, "Teaching with Inter-Nation Simulation and Case Studies," *American Political Science Review,* 60 (1966), 53-65.

instruction, and policy formation. Although the application of these techniques has been increasing, systematic evaluation of their performance is only now beginning. This essay reports one type of evaluation.

The researchers sought to structure a simulation of international politics so it would reproduce features of the political crisis that preceded the beginning of the First World War. Two separate trials or runs of the simulation were performed as a pilot project. With two runs, the data are sufficient only to illustrate what might be done in an expanded research program.[2]

The study was undertaken to investigate the use of a historical situation as a means of validating simulations. The problems of model validity are critical in determining the value of the simulation-gaming technique not only to political science, but to all the social sciences. In a fuller discussion of simulation validity elsewhere,[3] one of the authors has indicated that model validity is always a matter of degree and is affected by (1) the purpose for which the model is used, (2) whether or not human participants are involved, and (3) the types of criteria employed. The World War I simulation explores the third area—criteria for estimating validity. It focuses on possible standards or criteria for establishing the goodness of fit between the simulation and the system represented. To what extent do features of a political system or its processes correspond to their simplified representation in a model? One means of investigating this question is to ascertain if a simulation produces events similar to those reported in a historical situation. Another approach is to determine whether the simulation supports more general hypotheses about political phenomena which previously have been confirmed by independent methods. Both events and hypotheses are used as validity criteria in the simulation of the 1914 crisis.

Although the validity issue is the primary reason for conducting the simulation of World War I, several other purposes are served by the exercise. First, it provides a milieu in which to explore the relative effect on political actions of personality characteristics as compared to variables more frequently associated with political analysis. Second, the simulation of past events offers a possible device for teaching and studying history.

---

[2]The simulation runs were conducted in the summer of 1961 at Northwestern University. The exploratory nature of these runs led the authors to question whether the pilot study should be published. The supply of the original Navy report, however, is now exhausted. Moreover, no more complete set of historical runs has been conducted to date. Because a number of other published materials have discussed these pilot runs, it seems appropriate to make a fuller description of the World War I simulation more widely available. In doing so, the authors wish to caution that the work is primarily an examination of a means of evaulating simulations rather than a direct validation of the Inter-Nation Simulation. For examples of how this pilot project has been discussed elsewhere, see Arthur Herzog, *The War-Peace Establishment* (New York: Harper, 1963), esp. 183-184; Sidney Verba, "Simulation, Reality, and Theory in International Relations," *World Politics,* 16 (1964), esp. 507-515; James A. Robinson and Richard C. Snyder, "Decision-Making in International Politics," in Herbert C. Kelman (ed.), *International Behavior* (New York: Holt, Rinehart and Winston, 1965), 445, 512; and J. David Singer, "Data-Making in International Relations," *Behavioral Science,* 10 (1965), p. 77.

[3]Charles F. Hermann, "Critique and Comment: Validation Problems in Games and Simulations with Special Reference to Models of International Politics," *Behavioral Science,* 12 (May 1967), 216-31.

PROCEDURE

The Inter-Nation Simulation

Researchers differ as to the distinction between games and simulations. A number of experimenters, however, have associated "games" with operating models that involve human participants and "simulations" with models which do not.[4] Usually human participants are involved when the procedures or rules for designating the interplay of all components in the model have not been explicitly determined. When the model's relationships are incompletely programmed, human players and administrators are required to make judgments during the game. If the relationships are programmed, the need for human decision makers is reduced. In this essay an operating model will be defined as a simulation rather than a game, if a separate staff or computer is required to execute the programmed features. Thus, it is possible to have a simulation that is partially programmed and partially determined by human participants.

The model of international politics used in this study is such a hybrid. Developed by Harold Guetzkow and his associates,[5] the Inter-Nation Simulation incorporates both human participants and programmed calculations. In its usual format the Inter-Nation Simulation involves five or more nations. The government of each simulated nation is represented by human participants who assume one of several decision-making positions. During the 50-to-70 minute periods into which the simulation is divided, the decision makers allocate the military, consumer, and natural-industrial resources available to their nation. These various types of resources have different functions in domestic and international affairs. Using their resources the participants make decisions about internal matters such as economic growth, government stability, defense preparations, and research and development programs. At the international level nations may enter alliances, negotiate trades or aids, engage in various kinds of hostilities, and participate in international organizations.

Every period, which represents approximately one year of "real" time, the decision makers record their actions on a standardized decision form. Then, either a calculation staff or a computer applies the programmed rules to the decision form to determine the net gain or loss in the various types of resources. The structured part of the model also establishes whether the decision makers have maintained the support of the politically relevant sectors of the nation whose endorsement is required for them to remain in office. The calculated results are fed back to each nation, thus beginning a new period of interactions and decisions by the participants.

Adapting the Simulation

Five nations were represented in the simulation runs of the 1914 crisis. Each government was staffed by two decision makers. A third participant in

---

[4]For a discussion of the distinctions made between games and simulations, see Charles F. Hermann, "Games and Simulations of Political Processes," *International Encyclopedia of the Social Sciences.*

[5]Harold Guetzkow, Chadwick F. Alger, Richard A. Brody, Robert C. Noel, and Richard C. Snyder, *Simulation in International Relations* (Englewood Cliffs, N. J.: Prentice-Hall, 1963).

each nation acted as a messenger. The five simulated nations were intended to replicate features of Austria-Hungary, England, France, Germany, and Russia. Italy was excluded altogether and Serbia was represented symbolically by the researchers without participants. Several reasons can be offered for this treatment of Italy and Serbia. Reliable records of their diplomatic communications (a major input in the simulation of the other nations) were not available. Secondly, one can argue that although Serbia seemingly precipitated the immediate conflict and Italy was a member of the Triple Alliance, both nations were on the periphery in determining the question of world war when compared to the five other countries. Their exclusion, therefore, did not hinder the purposes of the exploratory runs. Although frequently other runs of the Inter-Nation Simulation have included more nations and more participants per nation, these changes are not as fundamental as several others.

Two major modifications were made in the basic simulation model. The first alteration established the initial conditions which the experimenters deemed important to characterize the international situation in the summer of 1914. Participants were introduced to some of the attributes of the historical setting by means of (1) a brief history of selected international affairs prior to the beginning of the crisis, (2) a statement of the current domestic and foreign policies of the participant's nation and the reasons they were being pursued, (3) a sketch of several personality traits of the historical policy maker whose role the participant occupied, and (4) a set of relevant historical diplomatic messages, conversations, and newspapers for the time between the assassination of the Austro-Hungarian Archduke on June 28 and the Serbian reply to the ultimatum on July 25, 1914. In addition, an effort was made to fit the programmed parameters and variables of the simulation to the national profiles of the countries involved.[6]

Several of these inputs require elaboration. Values for most of the components in the programmed part of the simulation were based on 1914 statistical indices (e.g., population, gross national product, size of armed forces) that approximated the meaning of the model's parameters. The 1914 indices were multiplied by an arbitrary constant to convert them to amounts convenient for use in the simulation. Individuals familiar with recent European history were asked to estimate decision latitude, the parameter that indicates how sensitive policy makers must be to the politically-relevant segments of their nation. These judges rated the decision latitude of each nation on a 10-point scale. Higher values were assigned to nations whose policy makers enjoyed greater freedom of action. Table 1 displays the values of each nation's basic parameters as they were reported to the simulation participants at the beginning of both runs. The historical diplomatic materials used in the simulation were compiled from hundreds of communications and documents which had been translated, edited, and verified by the Stanford Studies in Conflict and Integration under the

---

[6]The international history, the statements of domestic and foreign policy, and the personality sketches appear as appendices in *Studies in Deterrence X: Validation Studies of the Inter-Nation Simulation,* Technical Paper 3351, December, 1963. A complete set of the diplomatic messages in the form in which they were used in the simulation is on file with the contracting agency and with the International Relations Program, Northwestern University. For values assigned the basic parameters, see Table 1.

TABLE 1   Four Simulation Parameters
for 1914 National Profiles

Nation	Basic Capability Units[a]	Force Capability Units[b]	Validator Satisfaction[c]	Decision Latitude[d]
Austria-Hungary	45,540	14,560	4	7
England	86,940	25,000	6	3
France	62,100	20,800	4	3
Germany	120,000	24,500	7.5	8
Russia	78,660	23,000	4.5	9
Serbia	less than 4,140	less than 1,700	not given	not given

Note.—With the exception of decision latitude, the 1914 data and procedures for estimating these parameters were derived from James A. Winnefeld, "The Power Equation Europe, 1914," Stanford University (1960).

[a] Basic capability units represent the human, natural, and industrial resources available in a country. For the 1914 period the following indices were combined: steel production, national income, and total population weighted by the rate of male literacy.

[b] Force capability units are the military component in the simulation and were calculated by combining two indices, regular peacetime armies and capital ships.

[c] Validator satisfaction is the degree to which a decision-maker's policies are acceptable to those elite groups with power to authenticate his office holding. A "crisis coefficient" composed of the frequencies of certain types of events (e.g., civil disturbances and insurrections, assassinations) in a given country in the 50 years preceding 1914 was combined with an indicator of national security (relative military strength). The integrated estimates of satisfaction were placed along a 10-point scale. Higher values represent more satisfaction with the government.

[d] Decision latitude is an ideological element. It is defined as the degree to which probability of office holding is responsive to changes in validator satisfaction. To estimate decision latitude, judges rated the nations on a 10-point scale. Higher values represent greater latitude for the government (i.e., less sensitivity to the demands of validators).

editorship of Howard Koch.[7] The procedures for determining the personality characteristics are described in the section on participants and their historical counterparts.

All of the structured inputs were masked to avoid revealing to the participants that an actual historical situation was being modeled. Proper nouns, e.g., the names of individuals, countries, and alliances, were falsified. Misleading cues casting events in the future were introduced into the world history. In addition, after a pretest the assassination at Sarajevo was modified to avoid disclosing the identity of the historical setting. In the revised simulation account, several major Austro-Hungarian officials were killed by a strafing aircraft while they were on a reviewing stand in Serbia. These precautions were taken in order that the participants' knowledge of history would not bias their responses. The introduction

[7] Howard E. Koch is with the staff of the Stanford Studies in Conflict and Integration, *Documentary Chronology of Events Preceding the Outbreak of the First World War: 28 June-6 August, 1914* (Stanford University, mimeographed, 1959).

of an extensive amount of structured material in the initial phase of the operating model represented an important modification of the usual practice in the Inter-Nation Simulation.

The second major alteration in the Inter-Nation Simulation affected the time units represented in the model. Usually the simulation is divided into 50-to-70 minute periods which constitute the equivalent of a year in the "real" world. These intervals are associated with a year because of the programmed calculations. Every period the policy makers allocate their national resources. The consequences of their allocations are determined by the relationships between a number of variables in the programmed part of the simulation. Examples of these variables include the rate of depreciation in existing military equipment, the amount of lead time required for a new research program, and the extent of shift in popular support for the government. The equations used to calculate these and other variables are designed to reflect changes that might occur on roughly an annual basis. As a result, decisions taken by the participants and submitted for calculation normally represent the allocation of resources for approximately twelve months.

The present exercise, however, required the representation of not years, but the few critical days in late July, 1914. This reduction in time necessitated several seemingly contradictory changes. On the one hand, the existing programmed calculations had to be made relevant to the participants. If the basic model was to be maintained, persons involved in the exercise had to experience the constraints and demands imposed on their immediate behavior by the programmed features of the simulation as it is usually constituted. On the other hand, participants required a time framework that would allow them to deal with the kinds of decisions that policy makers might encounter on a daily rather than annual basis. In sum, the individuals in the simulation had to be able to make short-term decisions, while being aware of the long-term consequences as represented in the programmed calculations.

To meet these requirements only individuals who had previous experience with the Inter-Nation Simulation were invited to participate in the exercise. These experienced participants were told that the first few simulation periods would represent days. Moreover, there was the implication that these short time frames would be followed by periods which would be equivalent to years as in previous simulations. To further the impression that the initial periods were to be embedded in a series of longer time units, the participants received an annual decision form and were instructed to submit an updated version for calculations when the simulation periods began to represent years. These arrangements were made to encourage the participants to take account of the basic programmed variables and parameters in conducting their immediate interactions. A decision maker is more likely to act in a short-term situation so as not to damage such programmed variables as the probability of his continuing in office or the annual amount of consumer goods available to his nation, if he believes that the simulation is going to continue for "years." Notwithstanding the information given the participants, both simulation runs were terminated before the anticipated conversion to the longer time units had occurred. Hence no calculations were made in any of the programmed components of the model.

In addition to designating the periods as days, the shortened character of the time units was promoted by the diplomatic messages. These historical documents reflected events and decisions that were developing day by day, if not hour by hour. Finally, as an aid to short-term decision-making, a new form was introduced into the simulation. On the new instrument each nation's decision makers were able to indicate more immediate changes in the intensity of their action toward other nations. Participants were advised that daily variations in their nation's level of commitment would influence such annually computed variables as total available resources and the likelihood of office-holding. Thus, the intensity scale provided an explicit link with the long-term elements present in the programmed calculations of most Inter-Nation Simulation exercises.

Historical Figures
and Their Simulation Counterparts

An attempt was made to select participants with personality traits similar to some of those manifested by political leaders in the crisis of 1914. This task involved three subproblems. First, a restricted number of historical figures who were active in the crisis had to be selected. Second, a judgment had to be made regarding which personality characteristics of these men were salient in their political behavior. Finally, a method had to be devised for selecting simulation participants with similar personality profiles.

The resources available to the researchers necessitated that the total number of decision makers in each run be limited to 10—two participants for each nation. Consequently, we sought the two policy makers in each of the five European nations who were major contributors to the critical decisions during the crisis. More specifically, three criteria guided the selection: (1) which persons had a dominating influence on the foreign policy decisions of their nation at the time of the crisis? This criterion recognizes that the loci of decision making may not correspond with the "legitimate authority" to make such decisions. (2) Which persons received and dispatched (or at least read) diplomatic cables and related foreign policy documents? The historical figures whom the simulation participants represented should have occupied a reasonably central position in the diplomatic communication net because diplomatic messages acted as a major source of simulation inputs. (3) About which persons was autobiographical and/or biographical material available to help in the assessment of personality traits? Utilizing these criteria, the following historical personages were chosen for representation in the simulation:

Austria-Hungary
Berchtold (Minister for Foreign Affairs)
Conrad (Chief of General Staff)

England
Grey (Secretary of State for Foreign Affairs)
Nicolson (Permanent Under-Secretary of State for Foreign Affairs)

France
Poincare (President of the Republic)
Berthelot (Acting Political Director of Foreign Ministry)

Germany
    Wilhelm II (Kaiser)
    Bethmann-Hollweg (Chancellor)
Russia
    Nicholas II (Czar)
    Sazonov (Minister for Foreign Affairs)[8]

Once the political leaders judged to have assumed critical roles in the crisis were selected, it was necessary to establish which personality characteristics were salient to their political behavior. This .determination was made by a cursory content analysis of personal letters, autobiographies, and biographies of the chosen policy makers.[9] Each document was content analyzed for psychological characteristics or traits identified by one of several tests for measuring personality.[10] On the basis of the content analysis, dominance, self-acceptance, and self-control appeared to be characteristics which differentiated among all 10 of the political leaders. Furthermore, the personality of one or more of the selected individuals was strongly characterized by his attitude toward such concepts as fate, frankness, making decisions, his own country, peace, self-confidence, success, suspicion, and war. No claim is made that these traits provide a complete personality profile of the historical individuals under examination. Undoubtedly some important characteristics have been overlooked. This list of traits, however, yielded a distinctive profile for every leader which was consistent with features stressed in documents describing that individual.

    Two psychological tests were used to measure the personality charac-

---

[8]Some illustration of how the criteria were employed is appropriate. The selection of Berthelot provides a good example of the application of the first two criteria. Although Berthelot did not have the legitimate authority to make binding decisions for his government, he nevertheless extensively influenced French foreign policy during July of 1914. With the president, premier (who was also the foreign minister), and political director of the foreign ministry on a mission to Russia in July, Berthelot was placed in charge of the foreign ministry. He became the chief advisor during this time to Bienvenu-Martin, officially the Minister of Justice, who was a novice at foreign affairs, and relied heavily on Berthelot. For further evidence on the role of Berthelot, see Richard D. Challener, "The French Foreign Office: The Era of Philippe Berthelot," in Gordon A. Craig and Felix Gilbert (eds.), *The Diplomats: 1919-1939* (Princeton, N. J.: Princeton University Press, 1953), 49-85. The third criterion was important in the choice of Conrad. Far less autobiographical and biographical material is available on the Austro-Hungarian political leaders than on those from the other countries. Most information concerns Emperor Franz Joseph, a logical selection in addition to Berchtold for the simulation. However, because Franz Joseph was quite old and recovering from a serious illness, he was not as influential on the decisions as other officials. Among the key figures, Conrad had the most available material.

[9]No inter-coder reliability was performed in the content analysis of the personality traits. For this reason, as well as because of the very limited sample of materials that could be examined for each figure, the selected traits must be considered only as tentative approximations. A list of the sources used in the content analysis appears in Technical Paper 3351, *op. cit.*

[10]Sources for the personality categories were Raymond B. Cattell, *The Sixteen Personality Factor Questionnaire*, rev. ed. (Champaign, Ill.: IPAT, 1957); Allen L. Edwards, *Edwards Personal Preference Schedule* (New York: Psychological Corp., 1953); Harrison G. Gough, *California Psychological Inventory* (Palo Alto, Calif.: Consulting Psychologists Press, 1956).

teristics that the researchers associated with the selected historical figures. The California Psychological Inventory (CPI)[11] not only measured the three traits judged to be relevant to all the policy makers, but it also contained measures of some secondary characteristics identified in the content analysis. The second instrument was the semantic differential.[12] The nine attitudes were estimated with this testing device. Utilizing a suggestion made by Gough,[13] one of the researchers responded to both instruments as if she were the historical policy makers. The tests were completed after the biographical and autobiographical material on each of the chosen figures had been read. Test profiles for all 10 individuals were made in this way. These 10 profiles provided a standard against which to compare the responses of potential simulation participants on the same personality tests.

Some 101 high school students, who had participated in the Inter-Nation Simulation experiment in the summer of 1960,[14] were tested. As previously noted, persons already acquainted with the operation of the simulation were used in order to facilitate the shift in the time dimension. Furthermore, experienced participants freed the experimenters from training participants in basic simulation skills. The CPI profile of each prospective participant was compared with that of each historical figure. Particular attention was paid to profiles that matched exactly on dominance, self-acceptance, and self-control and were within one standard deviation of the other traits measured by the CPI. The semantic differential was utilized as a final selection step. In other words, those individuals chosen through the CPI matching were screened further by means of the semantic differential until three individuals per role were selected. Of the 10 individuals best matched to the historical figures, six were able to participate in the runs—five in the first run and one in the second. The balance of the participants in each run were second and third choices. Because of the interest in using individuals whose profiles most closely corresponded to those prepared for the actual leaders, the participants were not controlled on sex. Four women participated in the first run and three in the second.

Conducting the Revised Simulation

To provide an overview of the procedures used in the modified Inter-Nation Simulation, we will describe the operations in the two attempted replications of the pre-World War I crisis. Both simulation exercises were conducted in two days. On the first day, the participants in the runs assembled together for a general introduction and review of simulation procedures. After these activities, the remainder of the first day was used to introduce the structured input for each run.

[11] Harrison G. Gough, *ibid.*

[12] Charles E. Osgood, George J. Suci, and Percy H. Tannebaum, *The Measurement of Meaning* (Urbana, Ill.: University of Illinois Press, 1957).

[13] Personal correspondence from Harrison G. Gough, dated July 25, 1961.

[14] For a description of this earlier simulation research, see Richard A. Brody, "Some Systemic Effects of the Spread of Nuclear Weapons Technology: A Study through Simulation of a Multi-Nuclear Future," *Journal of Conflict Resolution,* 7 (1963), 663-753.

At the beginning of the input phase, each participant was assigned to a separate cubicle. There they started by reading a disguised statement of pre-1914 "world history" and a description of the individual whose role they would assume. Upon completing this material, every participant was given a set of masked diplomatic communiques, newspapers, and memoranda which might reasonably have come to the attention of his historical counterpart. To the extent that such information was available, messages were ordered in the sequence in which they were received by the historical figure. For example, although the simulation began with an incident representing the assassination of the Archduke on June 28 (known to the participants as Day 1), some participants learned about that event the same day; others did not. In general, a participant received an incoming message and then the recorded reaction, if any, which his counterpart had made to that message. If evidence indicated that the historical figure had been aware of responses made to a message by other members of his government, the simulation participant also received this information. Thus, the Kaiser's marginal notes on diplomatic communiques were seen by the individual assuming the role of the German Chancellor, Bethmann-Hollweg. The input phase terminated with the Austro-Hungarian ultimatum on July 25 (Simulated Day 28). As the historical events became more rapid and complex in the latter part of the input phase, the order in which participants received information took on considerable importance.

The participants were allowed to read the set of messages available to them at their own pace. They were encouraged to write their reactions to each communication in a space opposite the message. The researchers believed that the participants would become more involved in the situation if they recorded their thoughts about the messages as they read. Moreover, these reactions provided a source of data. For the same reasons, after Day 15 and again after Day 28, every individual was required to write a summary of the events that had occurred. On these two simulated days, the participants also were provided sheets of paper and asked to draw a map of the world as they conceived it at that time.[15] After completing the second history and map, the participants were dismissed until the following day. They were cautioned not to discuss the material they had read.

The first run was concluded on the morning of the second day (hereafter designated the M-run) and the second was finished that afternoon (the A-run). When the participants returned to their respective simulations, they were informed that the world would continue with the situation as it had evolved up to that point. They were told that there would be no more structured messages. Thereafter, the situation could be handled in whatever manner they chose. Each nation's decision makers were given an annual decision form and an intensity of action scale to indicate their country's resources and commitments at the beginning of the free activity phase of the simulation. They were instructed to designate each 50-minute period as one simulated day, but the number of such

[15] Geographical features are not incorporated in the Inter-Nation Simulation and no explicit geographical statements were included in the masked historical communiques. Therefore, the participants' maps provided data about the changing conception of relationships among nations. The distance between allied and hostile nations as well as the relative size of the nations represented in the maps were analyzed. Although the measuring device is worthy of further exploration, the results from these pilot runs proved to be quite ambiguous and are not reported here. They are included in Technical Paper 3351, *op. cit.*

periods that would occur remained unspecified. Following these initial instructions, the two participants in each nation were assigned to a room in which
they were separated by a partition. A messenger or courier sat at one end of the
partition to prevent unauthorized conversations, to relay messages, and to
operate a tape recorder during conferences between the decision makers. Written
messages and periodic conferences were used as means of communication both
within and between nations.

At the end of three periods (representing July 26-28, 1914), each participant was asked to draw another map and update his statement of world events.
Much to the surprise of the participants, upon completing this task they were
informed that the simulation was over. A post-mortem or debriefing session was
held with each simulation run. The participants first completed a questionnaire
and then described their plans and reactions to events in the simulation.

RESULTS

Two different standards or criteria were employed in estimating the
validity of actions taken by the decision makers in the free activity phase of the
simulation runs. First, both macro and micro level events in the 1914 crisis were
used as standards with which to compare incidents that occurred in the simulation. Second, two general hypotheses, previously tested with documents from
the outbreak of the First World War, were explored using simulation data.
Findings on the hypotheses from the two data sources were checked for
comparability.

Macro and Micro Events

In the present context, "macro event" refers to the occurrence of general
war. Did war break out in the simulation runs as it had in Europe in 1914? In
neither run did war—or more accurately, the representation of war—occur during
the last three periods. Historically, Austria-Hungary declared war on Serbia on
July 28th, the final day represented in the simulation runs. In 1914, the declarations of general war between the other European nations did not occur until
after the last time period portrayed in the simulation.[16] Although no war had
been declared when the simulations were terminated, hostilities were imminent
in the M-run.

In the researchers' opinion, if the M-run simulation had been continued for
another 50-to-100 minutes (one or two more simulated days), war would have
been declared along lines similar to the historical situation. This position is
confirmed by 10 of the 15 M-run participants and messengers in their debriefing

[16]The chronology of hostilities during this critical period in 1914 was as follows: July
28—Austria-Hungary declares war on Serbia; July 29—Russia orders and then cancels general
mobilization; July 30—Russia again orders general mobilization, France alerts troops along
German border; July 31—Austria-Hungary begins general mobilization, Germany issues
ultimatum demanding Russia stop mobilizing within 12 hours or Germany will mobilize;
August 1—France and Germany start general mobilization, Germany declares war on Russia.
After August 1, formal declarations of war follow in quick succession from the other major
European states.

questionnaires. Throughout the free activity phase of that run the two alliance structures held. Germany[17] was prepared to give secret aid to Austria-Hungary for an attack on Serbia while signing neutrality pacts with England and France to keep them out of the war. This effort was intended to localize the conflict and to assure victory for the Dual Monarchy. If Russia then considered assisting Serbia, she would lack the support of her allies—a fact which Germany and Austria-Hungary believed would deter such action. France and England, however, had plans to go to war if Austria-Hungary attacked Serbia. That attack by Austria-Hungary was being planned at the close of the simulation run. An internal conference between the German participants on the next to last simulated day illustrates the direction in which the M-run was moving.

> Central Decision Maker [Kaiser]: Is their war started yet? [Austria-Hungary's attack on Serbia]
> External Decision Maker [Bethmann-Hollweg]: I don't think so.
> CDM: Here's the problem—we can certainly give them [Austria-Hungary] the aid but the aid must be kept secret because if monetary aid isn't kept secret then the promises—written promises—which I am giving to Bega [England] and Colo [France] will mean absolutely nothing; total war is bound to break out.... If Enuk [Russia] enters it, we have to send secret aid; if Enuk [Russia] doesn't enter it, then he [Austria-Hungary] can defeat Gior [Serbia] by himself and that will be a thorn out of our side. We have already driven a wedge into the Tri-Agreement [Triple Entente].
> EDM: Another question: Suppose Colo [France] enters, then we go in?
> CDM: If Colo [France] enters the war we go in....

In the A-run the participant representing Lord Grey of England called for an international conference on the first simulated day of the free activity phase. Thereafter, most of the simulation time was spent in obtaining agreement from all nations to attend this meeting. At the conference Austria-Hungary was charged with making far too extreme demands against Serbia and was pressured into withdrawing them. This retraction resulted in substantial conflict among the decision makers in Austria-Hungary as well as a bitter dispute between Germany and the Dual Monarchy. Germany had pledged complete support to Austria-Hungary and was highly irritated at the failure of Austria-Hungary to consult privately before changing its policy.

Two observations should be made about these developments in the A-run. First, some historians have expressed the opinion that England should have taken a stronger position more quickly, forcing moderation upon the Triple Alliance.[18] Second, and perhaps of greater importance, if the input phase had been extended for one more historical day, the participants would have received England's diplomatic communiques proposing an international conference and the prompt rejection of this suggestion by several nations. Thus, an alternative

---

[17] It is to be understood that the references to nations here are to the student decision makers in the simulated nations.

[18] See, for example, Luigi Albertini, *The Origins of the War of 1914,* edited and translated by Isabella M. Massy (London: Oxford University Press, 1953), Vol. 2, p. 514; and Sidney B. Fay, *The Origins of the World War* (New York: Macmillan, 1930), Vol. 2, 556.

actually considered and subsequently excluded by the historical figures provided the avenue which the simulation participants followed for the resolution of the imposed situation.

Several major divergences from the outbreak of World War I developed in the A-run. Least parallel to historical events is the indication in some messages that England was considering the initiation of war on Germany while the advantage appeared on her side. This war was represented in communications as a defensive strategy resulting from the military and economic threat of Germany. A second significant variation was the agreement of Austria-Hungary to withdraw her ultimatum. The Austro-Hungarian decision maker, intended to represent Conrad, revealed pacifistic tendencies and readily accepted the objections to his nation's militaristic actions. This behavior suggests a need to match more closely the social-political attitudes of the historical figures and the simulation participants.

This account of the A-run's divergences fittingly introduces the analysis of smaller, more specific events that occurred in the simulation and in the last days of July, 1914. By comparing numerous micro occurrences, validity depends not on the correspondence between isolated events (war or no war) but on the overall goodness of fit between a distribution of events. In other words, we are interested in whether the overall pattern of simulation occurrences is more or less like the pattern of reported incidents prior to World War I.

To illustrate this approach a sample of micro events was drawn from the simulation (the M-run) that at the macro level displayed a higher degree of correspondence to 1914. In this analysis an action mentioned by the participants in their written communication constituted a micro event. Eighteen separate micro events were identified in the M-run messages during the first simulated day of the free activity phase. A somewhat longer series of events would probably have been discovered for that day if transcripts of conferences had been examined in addition to the written communications. No reason, however, has been found to suggest that the types of events produced in face-to-face inter-actions differ for validity purposes from those indicated in written messages.

After the simulation events had been abstracted, a major historical study of the beginnings of the First World War was examined for comparable events.[19] In Table 2 the simulation events are indicated in the left-hand column and the reported historical events are listed in the corresponding space in the right-hand column.

Several alternative means of comparing the events were identified. Comparable activity is one way to match events in both the pre-World War I crisis and the simulation. We describe this as occurrence similarity. A second basis of comparison involves intent or purpose. Men's purposes in initiating or manipulating events may be similar, regardless of differences in format or activity. For example, in Item 16 in Table 2 the simulation action is the reverse of that reported in the observable world of 1914, but the intentions of the decision makers in both actions were probably similar. Germany was concerned that any war between Austria-Hungary and Serbia be kept localized without the intervention of other countries. To achieve this objective the simulation decision

[19]Luigi Albertini, *ibid.*, Vol. 2.

TABLE 2   Comparison of Sample of Simulation Events
with Historical Events

Simulated July 26, M-run	Estimated Score[a]	Reported Historical Events[b]
1. Germany requests Rus- to demobilize.	T = 0 I = 1 O = 1 — 2	July 29: German note to Russia: " . . . further progress of Russian mobilization measures would compel us to mobilize and then European war would scarcely be prevented" (p. 491).
1. Russia notifies Germany that no demobilization will occur until safeguards are established for Serbia.	T = 11 I = 1 O = 0 — 2	July 23: Russian foreign minister informs German ambassador that Russia is ready to assist in "procuring legitimate satisfaction for Austria-Hungary without abandoning the standpoint, to which Russia must firmly adhere, that Serbia's sovereignty must not be infringed" (p. 404). There is no explicit mention of making demobilization conditional on such safeguards.
3. In a conference Russia asks England and France to mobilize; both refuse.	T = 0 I = 1 O = 0 — 1	July 24: Conversation between Russian foreign minister and ambassadors from France and England. France agrees to fulfill all obligations of her treaty. England more evasive but conciliatory. (Pp. 294-96.)
4. In a conference England and France inquire as to Russian military needs.	T = 0 I = 1 O = 1 — 2	
5. Germany notifies Russia that the latter's support of Serbia would be aggression.	T = 0 I = 1 O = 0 — 1	July 28: No evidence found of direct communication between Germany and Russia on this matter; however, Albertini concludes that this was a German objective. In support he cites several telegrams emanating from the German foreign office on July 28 including a note from the Kaiser: "It is an imperative necessity that the responsibility for a possible extension of the conflict to the Powers not immediately concerned should in all circumstances fall on Russia" (p. 474 and p. 476).
6. Austria-Hungary warns Russia not to mobilize further.	T = 0 I = 1 O = 1 — 2	July 29: There is some suggestion that the Austro-Hungarian ambassador warned the Russian foreign minister against Russian intervention on this date, although the conversation was interrupted by information that Austria-Hungary had bombed Belgrade. (Pp. 552-53.) On July 26 Austria-Hungary's alliance partner, Germany, issued warnings through her ambassador to the Russian foreign minister: "I made detailed and urgent representations to the Minister about how dangerous it seemed to me to attempt to strengthen diplomatic action by military pressure" (p. 481).
7. Discussion on need for attack on Serbia in Austro-Hungarian internal conference.	T = 0 I = 1 O = 1 — 2	July 7: Austro-Hungarian Council of Ministers for Joint Affairs: one of a number of internal conferences before delivery of the Serbian ultimatum in which plans for attack on Serbia and the expected Russian reaction were discussed. No explicit program to make Russia appear the aggressor is mentioned although this appears to have been part of the Triple Alliance strategy. (P. 166.)
8. Discussion in Austro-Hungarian internal conference on making Russia appear the aggressor in an expansion of hostilities.	T = 0 I = 1 O = 0 — 1	
9. Austro-Hungarian note to Russia charges Russia's action is aggressive.	T = 0 I = 1 O = 1 — 2	July 30: Austro-Hungarian foreign minister tells Russian ambassador: "Austria-Hungary had mobilized solely against Russia . . . By the fact that Russia is obviously mobilizing against us we should also have to extend our mobilization . . . " (p. 662).

[a] "T" stands for temporal equivalence, "I" for intention, and "O" for occurrence similarity. Each of the three categories was assigned a "1" if the simulation and 1914 event were judged as similar on that criterion; or a "0" if they were judged dissimilar. Simulation and historical events rated as similar on each of the criteria received an overall score of three.

[b] The source for the reported historical events is Luigi Albertini, *the Origins of the War of 1914*, edited and translated by Isabella M. Massy (London: Oxford University Press, 1953), Vol. 2. All page references are to this volume.

TABLE 2–(Continued)

Simulated July 26, M-run	Estimated Score[a]	Reported Historical Events[b]
10. Austro-Hungarian note to Russia states Austria-Hungary must hold Serbia accountable.	T = 0 I = 1 O = 1 —— 2	July 30: On the basis of above conversation with Berchtold, the Russian ambassador telegraphs his own foreign minister: "[Berchtold] is of the opinion that it is impossible for Austria to stop her operations against Serbia without having received full satisfaction and solid guarantees for the future" (pp. 662-63).
11. In note France assures England of friendship.	T = 0 I = 1 O = 0 —— 1	July 29: France appears to have repeatedly probed England on the issue of unified action and support. For example, the July 24 three-way conversation in St. Petersburg (pp. 294-96) or the inquiry to the British ambassador by his French colleague in Berlin, July 29 (p. 520n). A major discussion on this point took place between the British foreign minister and the French ambassador on July 29. In a follow-up conversation on July 31, the ambassador asked Grey if England would "help France if Germany made an attack on her." (P. 646.) However, the type of military aid was not specified.
12. France requests military aid in note to England.	T = 0 I = 1 O = 1 —— 2	
13. France urges a united front in a note to England.	T = 0 I = 1 O = 1 —— 2	
14. Germany makes internal decision to give secret aid to Austria-Hungary.	T = 0 I = 1 O = 1 —— 2	July 5–6: The Hoyos mission occurred, and Germany agreed to the so-called "blank check." Subsequent conversations between the Kaiser, his Chancellor, and military staff also reveal a discussion of military assistance to Austria-Hungary. (Pp. 133–59.)
15. Germany's internal decision to seek neutrality pact with France and England to keep them from aiding Russia should latter intervene in Austria-Hungary's war against Serbia.	T = 0 I = 1 O = 0 —— 1	While no direct documentation was found for verifying this internal decision, Albertini concludes: "The study of the German documents shows beyond all shadow of doubt (1) that in allowing Austria to attack Serbia, Germany started from the assumption that, if the attack developed into a European war, England would remain neutral; (2) that Grey's conduct until the afternoon of July 19 . . . strengthened the German leaders in this opinion. . . . ." (p. 514).
16. Germany urges Austria-Hungary not to attack until neutrality of France and England is assured.	T = 1 I = 1 O = 0 —— 2	July 26: Austria receives Germany's message urging military operations against Serbia as quickly as possible: "Any delay in coming military operations is regarded here as a great danger because of the interference of other Powers" (p. 453).
17. Decision in Russian internal conference not to reveal, at present, their plan to attack Austria-Hungary if she attacks Serbia.	T = 1 I = 1 O = 0 —— 2	
18. Decision in Russian internal conference to announce their intention to give Serbia full military aid.	T = 1 I = 1 O = 0 —— 2	July 25: Russian crown council decides on mobilization to be followed by war if Serbia is attacked. (P. 762.)

makers chose to elicit pledges of neutrality from two important nations. In actuality, the German decision makers decided that quick action, before commitments and reactions could be made, was the preferable policy. Another way of comparing events is through temporal equivalence. The timing of some simulation events closely matches that of similar events in reality; others deviate sharply in this temporal aspect.

One of the researchers rated the 18 simulation events on each of the three bases of comparison. For each simulation event, temporal equivalence was given a score of one (i.e., assumed similarity) if it occurred within one day on either side of the reported 1914 event to which it was compared. If the time disparity was greater than the arbitrary threshold, it was coded as nonsimilar and assigned a zero. The same zero or one scoring was used for comparing events on intention and on occurrence similarity. The estimated scores for each event are presented in the center column of Table 2.

If the scores for the three categories are combined for the 18 events, the highest possible goodness of fit score is 54. The overall score for the distribution of events in the first simulated day of the M-run is 31. More specifically, all 18 events were judged similar in intent to developments that occurred in the 1914 crisis. Weighed against the significance of this finding must be the recognition that it is difficult to ascertain the intent of simulation participants, to say nothing of historical policy makers. Rules for guiding such coding are exceedingly hard to devise. For half of the simulation events (9 of 18), the researchers were able to find historical actions which seemed to be equivalent in physical format. Four of the 18 simulation events took place at approximately the point in time that the simulation was intended to replicate. One possible explanation for the low correspondence on temporal equivalence involves the problem of equating intervals in the simulation to actual days. The simulation may not have been adequately structured to provide the participants with the impression that each 50 minute period represented 24 hours of "real" time.

Hypotheses

In one sense the comparison of simulation and historical events includes the investigation of a hypothesis. The hypothesis is that a macro event or a distribution of events in system A is comparable in a specified characteristic to a given event or distribution of events in system A'. Such a proposition, however, is oriented exclusively to the two systems involved—in the present case, the outbreak of World War I and the simulation of those events. In this section, attention is directed to more general hypotheses. These statements of relationship are intended to apply not only to the situation as it existed in Europe in the summer of 1914, but to other configurations of nations in international politics. Two hypotheses which have been tested with data from the 1914 crisis will be explored with the simulation data.

The three researchers who have studied the first hypothesis in the 1914 context have stated it as follows:

If a state's perception of injury (or frustration, dissatisfaction, hostility, or threat) to itself is "sufficiently" great, this perception will offset perceptions of insufficient capability, making the perception of capability much less important a factor in a decision to go to war.[20]

To explore their hypothesis, the members of the Stanford Studies in Conflict and Integration content analyzed 1,165 communications exchanged between the European states in the weeks prior to the beginning of the First World War. The purpose of this analysis was to determine the frequency of perceptions of capability and hostility. In the content analysis a perception of capability was defined as an assertion "concerning the power of another state or a coalition of states, or a statement with regard to the changing power of either a state or a coalition of states." A perception of hostility was defined as "a statement by one country about the hostility directed toward it by a second country, or about the hostility directed by a second country toward a third country.. .. Statements of hostility are defined to include statements of threat, fear and injury."[21]

The number of hostility and capability statements that the Stanford group found for the nations subsequently represented in the simulation are reported in the first columns of Table 3. Perceptions of hostility exceed those of capability for every nation. The overall difference is statistically significant ($p < .05$) by a sign test. Using the same definitions of hostility and capability, the messages written during the two simulation runs were content analyzed in a similar fashion.[22] As shown in the second and third sections of Table 3, the perceptions

TABLE 3   Number of Hostility and Capability Statements for Nations in 1914 and in Simulation

Nation	1914[a]		M-Run		A-Run	
	Hostility	Capability	Hostility	Capability	Hostility	Capability
Austria-Hungary	179	26	91	38	63	28
England	32	16	17	13	24	7
France	26	18	41	36	21	3
Germany	138	34	53	35	107	26
Russia	50	28	30	16	69	20

[a]The 1914 data are in Dina A. Zinnes, Robert C. North, and Howard C. Koch, "Capabilities, Threat and the Outbreak of War," in James A. Rosenau (ed.), International Politics and Foreign Policy (New York: Free Press, 1961), p. 476.

20Dina A. Zinnes, Robert C. North, and Howard E. Koch, "Capabilities, Threat and the Outbreak of War," in Rosenau, op. cit., p. 470.

21Ibid., p. 472.

22The inter-coder reliability for the simulation content analysis of hostility was .82 and for capability .83. For a description of the statistics used in this paper (the sign test, rank-order correlation, and Mann-Whitney U test) see Sidney Siegel, Nonparametric Statistics for the Behavioral Sciences (New York: McGraw-Hill, 1956).

of hostility were significantly greater ($p < .05$) than those of capability in both runs. Thus, the simulation runs and the 1914 data produce comparable results on this general hypothesis.

A further check was made on the goodness of fit between the simulation and the 1914 data. The difference between the number of hostility and capability statements was determined for each nation. Within each of the three sets of data these differences were placed in rank order, that is, with the nation having the largest difference being given the first rank, etc. Then the order of the ranks for each of the simulation runs was compared with the order produced by the 1914 data. The resulting rank-order correlation (.90) between the 1914 data and that of the M-run was statistically significant ($p = .05$). The correlation for the A-run was not significant, however.

The second general hypothesis asserts that when opposing alliances or

TABLE 4   Effects of Alliance Structure on Communication in 1914 and in Simulation

Source of Data	Number of Intra-Bloc Dyads (n_1)	Number of Inter-Bloc Dyads (n_2)	Mann-Whitney U^a	Significance
1914	8	12	15	*< .01*
M-Run	8	12	10.5	*< .01*
A-Run	8	12	43	—

[a]It should be noted that for a given n_1 and n_2, the smaller the U the greater its likelihood of being significant.

blocs emerge in international politics, the communication between blocs will be much less than that among alliance partners. Zinnes, who has tested this hypothesis with the 1914 data, states the relationship as follows: "Frequency of interaction within the bloc will be greater than the frequency of interaction between blocs."[23] To explore this hypothesis, a count was made of the number of communications that each nation dispatched to the other four nations primarily involved in the 1914 crisis. The number of messages exchanged between each possible combination of two nations was established. Rates of communication in dyads composed of nations in the same bloc were compared

[23]Dina A. Zinnes, "A Comparison of Hostile Behavior of Decision-Makers in Simulate and Historical Data," *World Politics*, 18 (1966), p. 477. This article is one of the few efforts to explore simulation validity by what we have described as the hypothesis approach. Statistical tests are conducted on hypotheses using data from World War I and from another series of Inter-Nation Simulation runs which made no attempt to replicate those historical events.

with those in dyads consisting of nations from opposing blocs. A Mann-Whitney *U* test applied to the rank-order positions of these dyads supports the hypothesis.[24] The results are shown in the first row of Table 4. Identical procedures were followed for the M- and A-runs using the messages sent by the decision makers to other nations. The results for the M-run, shown in the second row of Table 4, are statistically significant. By contrast, as indicated in the third row of Table 4, the findings for the A-run are not significant. Once again, the data from the M-run appear to be a better fit with that drawn from 1914 than do the data produced in the A-run.

EVALUATION

Replication of the 1914 Crisis

Can the Inter-Nation Simulation replicate occurrences in the observable universe such as the outbreak of World War I? Regardless of how intriguing this question is, for a number of reasons we cannot fully answer it with the present data. The exploratory nature of this initial research has been emphasized. A more complete study would require many more runs. It would be desirable to engage more mature participants whose previous experiences and backgrounds might lend themselves to the simulation problem. Moreover, the calculations used to set the beginning values of the parameters and the procedures used to code the simulation data need further refinement.

Even if one were tempted to make rough judgments on validity based exclusively on this pilot study, the divergence between the two runs imposes constraints. Using both events and general hypotheses to check the "fit" between 1914 and simulation data, the M-run approximates the political crisis prior to the First World War more closely than the A-run. At the macro event level, the M-run appeared on the verge of war with the same alliance commitments as observed in 1914. On the other hand, the A-run averted the immediate threat of war and involved several incidents in sharp contrast to those reported in the historical situation. In data from both runs, as in the material from 1914, perceptions of hostility were significantly more frequent than perceptions of capability. Only in the M-run, however, did the differences between these two kinds of perceptions correlate with similar data from the actual nations of 1914. Furthermore, the hypothesis that communication would be greater within alliances than between them was supported both by the 1914 data and the M-run data; but it was not confirmed in the A-run. These differences occurred despite a common introductory briefing, identical materials in the input phase, and the same set of initial values for the simulation parameters.

The two runs, however, did not share the same human participants. Although an effort was made to select participants in both runs with similar

---

[24]The 1914 data for this analysis were obtained from Zinnes, *ibid*. She performed a similar statistical test using Serbia as well as the five nations represented in the simulation. It is interesting that with the addition of Serbia, the result is not significant.

personality characteristics, it will be recalled that the matching was limited to a few traits. Moreover, on those characteristics for which an attempt at correspondence was made, the personality profiles of some individuals more closely matched their historical counterparts than others. As we noted previously, five of the six individuals whose profiles corresponded most completely with those designed to represent the historical figures participated in the M-run and only one in the A-run. In other words, half of the participants in the run that best approximated the 1914 crisis were first choices on the personality matching, whereas only one of the 10 decision makers in the more divergent run was a first choice. (In the A-run, six participants were second choices and three were third choices. Of the other five participants in the M-run, three were second choices and two were third choices.) Correspondence between a simulation and its reference system appears to have been facilitated by closer matching of the personalities of the participant and the historical figure.

Procedural Issues

These initial runs do not establish whether the Inter-Nation Simulation can replicate aspects of historical events, but they do uncover several procedural issues that a more complete study would confront. One problem is the participants' awareness of the historical events being simulated. We have described the efforts to mask the historical clues to avoid biasing participant responses. Despite these attempts, the disguise was not fully successful. On the debriefing questionnaire, the participants were asked to check from a list of historical situations those incidents that appeared "somewhat similar" and those that seemed "almost exactly like" their simulation run.[25] Table 5 indicates that five of the 10 participants in the M-run and two of the seven who completed the questionnaire in the A-run perceived some degree of similarity between their simulation and World War I. No other historical situation was identified by seven participants, but four others were each chosen by six individuals.[26]

Even though the results from this questionnaire item are not conclusive, they suggest the need for more vigilance in making past events unrecognizable. When, however, the content of historical situations is changed or misleading clues are added, there is the danger that essential attributes of the actual events will be so distorted they will not provide a validity check. This difficulty can be handled by selecting a less well-known situation—providing sufficient informa-

[25]In retrospect, this questionnaire item was not totally satisfactory. Estimating the difference between situations rated "exactly alike" and those rated "somewhat alike" is difficult. Because most participants checked several historical situations, it is not clear how much more the simulation resembled the 1914 crisis than, for example, Hitler's ultimatum to Poland. Furthermore, participants were not asked to state when they became aware of the apparent similarity between the simulation and a past event. Some evidence indicates that a messenger suggested the parallel to World War I to several individuals near the end of the M-run. An unstructured question, which provided no list of historical events, might have reduced the number of references to World War I. The alternative situations in the questionnaire item were selected from events mentioned by high school students who pretested the material for the input phase.

[26]The messengers in both runs also completed this questionnaire item. Six out of nine messengers reported a similarity to World War I. It may be that the overall view of events provided by this role increased their awareness of the similarities.

TABLE 5   Situations Viewed as Similar to the Simulation on the Debriefing
Questionnaires

Situation	M-Run Decision Makers (N = 10)		A-Run Decision Makers (N = 7a)		Totals (N = 17)		
	Similar	Almost Exact	Similar	Almost Exact	Similar	Almost Exact	Overall Total
World War I	3	2	1	1	4	3	7
Hitler's Ultimatum to Poland	1	2	2	1	3	3	6
Berlin Crisis	2	2	1	1	3	3	6
Hitler's Ultimatum to Sudetenland	2	1	2	1	4	2	6
Israeli-Arab Conflict	3	0	2	1	5	1	6
French-Algerian Conflict	1	1	2	1	3	2	5
Korean War	3	0	1	1	4	1	5
19th Century Colonialism	4	0	0	1	4	1	5
Spanish-American War	1	0	2	1	3	1	4
Attack on Pearl Harbor	1	0	0	0	1	0	1
World War II	1	0	0	0	1	0	1

a   Three of the A-Run participants were unable to attend the debriefing session.

tion exists on it to construct the necessary inputs. Alternatively, means of comparing the simulation and the actual event can be chosen which are unlikely to be affected if the participants identify the general situation. For example, in the present runs even those participants who saw a similarity to World War I were less likely to know the relationship between perceptions of hostility and capability, or to be aware of the micro events that occurred on July 26, 1914.

Another issue is the distinction between "self-structured activity" and "role playing." Should a simulation participant be required to play the role of a historical decision maker whose characteristic behaviors have been described to him or should the participant be free to structure his own activity? This problem is illustrated by an incident in one of the runs. A participant confronted a decision which he felt his assigned historical figure would answer one way, but he personally would answer another. Should he assume the role or play himself? Given difficulties in constructing profiles of historical figures and problems in matching participants on numerous characteristics, role playing may supplement incomplete personality correspondence. On the other hand, without detailed information one individual's interpretation of the probable behavior of another individual is likely to produce major distortions. Supplying such information would disclose the identity of the historical figure. Investigations probably should be limited to a few personality traits that are part of the natural disposition of the participant. In the event, however, that participants are asked to "play" a historical personality, selection procedures might include some indicators of the empathic qualities required for role playing or acting.

Some explicit reference should be made to the procedural problem of selecting the kind of historical situation that the simulation is intended to represent. All models—including games and simulations—are simplifications of the systems they are designed to replicate. This simplification is achieved in part by completely excluding certain properties of the referent system from the model or by combining numerous detailed elements in an aggregated form. Obviously, a simulation cannot be validated by comparing its output with historical phenomena the model was not designed to represent.

In several ways the effectiveness of the present research for validating the Inter-Nation Simulation is reduced because the 1914 crisis was selected as the historical situation to be replicated. The conversion of the basic time units in the simulation from years to days furnishes one example. With the time units scaled-down, the components in the programmed part of the model acted as constants, though many of them normally operate as variables. In these exploratory runs, then, the part of the simulated world contained in the structured programs affected the behavior of the participants only insofar as they recalled its impact on their previous simulation experience. The 1914 crisis did not provide a means of investigating the programmed relationships as they operate in the usual Inter-Nation Simulation. Although such short-term crises may be excellent for determining the validity of other simulation models, they seem somewhat less appropriate for the Inter-Nation Simulation.

The micro event analysis provides a second illustration of the problem. The variables in the programmed portion of the simulation are broad representations of properties within a nation (e.g., the sum of all human and natural resources, or overall military capability). The events in the micro analysis, however, were at the more specific level of a diplomatic conference or a decision to mobilize ground forces in a certain district. The Inter-Nation Simulation was able to produce events at this level of specificity, but it is not clear that they are produced by the aggregate variables that compose the model.[27]

HISTORICAL SITUATIONS AS AN
APPROACH TO VALIDITY

The procedural problems that emerged in the exploratory attempts to replicate the 1914 crisis are not insurmountable obstacles to the use of historical situations as validity tests for simulations. With careful attention to the selection of both participants and the past occurrences to be simulated, these difficulties can be minimized. Broader concerns about the use of historical data for verifying models, however, also must be considered. One challenge to this validity technique is raised by the developers of a simulated underdeveloped economy.

> When a simulation of a particular economy has been formulated, it should be subjected to shocks and exogenous trends like those that have impinged on the actual economy in the past so that its responses may be compared

[27]For further discussion of the level of specificity issue, see Harold Guetzkow, "Simulation in International Relations," in *Proceedings of the IBM Scientific Computing Symposium on Simulation Models and Gaming,* December, 1964, Thomas J. Watson Research Center, Yorktown Heights, New York, esp. pp. 264-267.

with historical records. It would seem that such tests would yield an independent verification of the model. Actually, in most cases it is unlikely that this ideal can be fully realized. . . . The difficulty, fundamentally, is that the information available in almost all countries is insufficient to establish a model without using all the relevant historical data in the formulation and in the adjustment of parameter values.[28]

The economic model mentioned above is a completely programmed computer simulation and, accordingly, required more data for establishing its initial values. Nevertheless, the problem is applicable to all historical validation attempts. Clearly, a simulation cannot be validated against the same historical material used to determine its parameters and beginning variable values. Historical replication always will be limited by the record of past events. Not only must there be a detailed account of decisions and actions, but the sequence of events must continue over a sufficient period of time without major changes which require resetting the model's parameters. For this reason historical verification may be more feasible for some types of simulations than others. The smaller the number of variables and parameters which require historical data, the more uncommitted historical material is available for establishing the goodness of fit. Moreover, if the content of the simulation deals with the kind of phenomena that recur frequently in the observable world, then historical validation is more applicable. Thus, a simulation of American judicial decision making is more readily verified with historical data than an operating model of disarmament processes.

Let us assume for the moment that sufficient historical data exist and that the content of a given simulation model permits the use of the validity approach described in this essay. Furthermore, let us assume that a comparison of events in simulation and in history reveals a high degree of correspondence. This correspondence does not demonstrate that the simulation correctly represents the structure and processes that were operative in the historical occurrence. We are speculating on the similarity between the historical and simulated inputs on the basis of the similarity of their outputs. Different relationships among various combinations of properties in the simulation conceivably could produce outcomes like those in the historical situation.

A simulation of the 1960 national Presidential election predicted the percentage of the vote for each candidate—the outcome—with considerable success. The designers of that simulation observe, however, that "it may legitimately be asked what in the equations accounted for this success, and whether there were parts of the equations that contributed nothing or even did harm."[29] Further analysis of the equations in the simulation revealed that the outcome was predicted despite the fact that at least one equation misrepresented aspects of voter turnout. Part of the structure of the simulation was incorrect, but the simulated result still matched the actual outcome. Despite this difficulty, our confidence that that simulation has captured some aspects of the voting process

---

[28]Edward P. Holland with Robert W. Gillespie, *Experiments on a Simulated Underdeveloped Economy* (Cambridge, Mass.: MIT Press, 1963), 207-208.

[29]Ithiel Pool, Robert P. Abelson, and Samuel L. Popkin, *Candidates, Issues, and Strategies* (Cambridge, Mass.: MIT Press, 1965), rev. ed., p. 64.

is greater than it would have been if the simulation had failed to replicate the campaign outcome. Confidence in the simulation would increase further as the operating model demonstrated ability to produce outcomes that corresponded with various elections. In sum, the similarity between simulation and historical events can provide at best only indirect and partial evidence for the correctness of the simulated structures and processes that produced the outcome.

Historical material can be used for validity purposes in other ways than by providing events for simulations to reproduce. In the exploratory runs, two validation techniques were tried—event comparisons and hypothesis testing. Although general hypotheses can include events as variables, they also can involve the processes by which events are produced. When events do become variables in a hypothesis, they tend to be more the micro events that occur with sufficient frequency to permit an adequate test. Hypothesis testing, therefore, is less susceptible to the criticisms of event comparisons made above. Verba has argued that difficulties in validly simulating a macro event (such as the outbreak of war) may exceed a model's potential contribution to theories of international politics. "Even if one could design a successful simulation in that respect, it might not be useful for future situations, which would not match the historical one in many important ways." Instead, he proposes that "if the situation can be decomposed into many subprocesses, such as communications flows, emotional states of decision-makers, and so forth . . . it may be possible to develop more widely applicable principles that can deal with many political situations."[30] Historical material may prove more useful for simulation validity explorations if it yields frequency distributions of events and processes that can be employed in hypothesis testing.

We have described some important procedural problems as well as two major limitations to using the replication of historical situations as a means of validating political simulations. In addition to event comparisons and hypothesis testing, alternative ways of verifying simulations are available. But they also have substantial liabilities.[31] In part because no one approach can fully establish the correspondence between a simulation and its intended referent system, simulation validity is always a matter of degree. Yet we cannot abandon the efforts to determine the goodness of fit between verifiable empirical observations and our conceptualizations—be they stated as verbal theories, mathematical equations, or simulation models. To improve our estimates of simulation validity, a strategy is required that includes multiple methods for discerning the degree of correspondence. In such a multimethod strategy one approach is historical replication. Until more validation exercises are conducted, it is premature to accept or reject simulation as an important new tool for studying political phenomena.

[30]Sidney Verba, *op. cit.,* 511, 513.

[31]For a survey of various validity approaches and a discussion of their assets and liabilities, see Charles F. Hermann, *Behavioral Science, op. cit.*

8

Human Geography Simulations

DUANE F. MARBLE

In the last fifteen years the discipline of geography has undergone a significant restructuring. The field has changed from one in which research activities were largely based upon description and intuitive analysis to one in which the development of theory and the use of modern analytical tools play central roles (Burton, 1963; Curry, 1967; Gould, 1969). Much of the present quantitatively oriented research in geography reflects the geographer's interest in spatial morphology and pattern. A great deal of the early quantitative work dealt with optimal patterns of activities or with statistical description of observed spatial patterns. Research on optimal spatial patterns has mainly been motivated by an interest in the efficient structuring of human activities in space or by attempts to use optimal solutions as a basis for estimating the efficiency of observed patterns of human spatial behavior. The analytical tools most commonly utilized have been mathematical programming and game theory, and more recent work has emphasized problems of optimal system design (e.g., transportation networks) that satisfy specified efficiency criteria.

Early research on spatial patterns made extensive use of tools and techniques developed primarily by statistical ecologists for the study of plant communities and was used to summarize the patterns of cultural and physical features (retail establishments, cities, river junctions) and to test if observed patterns corresponded to theoretical patterns. More recent work in this area has tended to emphasize the construction of stochastic point processes that operationalize various theories pertaining to the location of cultural features.

These two approaches have provided the basis for a wide variety of mathematical and statistical studies of spatial patterns, the forces that generate these patterns, and the consequences of these patterns for various types of human behavior. A great deal of this work has been statistical and major thrusts include the use of factor and components analysis for constructing regional partitions; contiguity studies for examining the presence of serial and auto-correlation in spatial data; use of trend surface, double Fourier, and other methods for describing surfaces; and analysis of networks and other linear patterns (Berry and Marble, 1968). Although some of this work has been applied directly to the abstract morphology of patterns, much of the present emphasis is on the interdependence between spatial patterns and human behavior (Cox and Golledge, 1969).

It is difficult to evaluate the significance of much of this research, since much of it has been done in relative isolation from the other social sciences; however, it is clear that there are no completely solved problems in geography. Many research thrusts have been frustrated because the analysis of spatial data presents problems that are quite different from the analysis of the data structures encountered in other social sciences. A good analogy is to physics, which structures problems at the one-, two-, and three-dimensional spatial levels; many physical problems that are readily analyzed in a one-dimensional space become exceedingly difficult when the jump is made to two dimensions. While geographers lack the mathematical sophistication of physicists, it is not clear that our two-dimensional problems are any simpler to resolve.

The limited use of simulation, both digital and analogue, in geographic research has been a response designed to bypass, in part, some of these complexities, which are multiplied when a researcher attempts to include considerations relating to temporal as well as spatial aspects of the problem. For many years simulation models, particularly digital simulations of the direct Monte Carlo type, have presented the only viable approach to modeling dynamic spatial systems. The researcher in this case attempts to derive an explanatory model not of a static spatial distribution but rather of the time path of development of a series of spatial distributions. Conceptually, the problem is a direct and simple one: Given a series of maps such as those shown in figures 8-1 and 8-2, describe the process that transforms the first spatial distribution into the second, the second into the third, the third into the fourth, and so forth.

THE DIFFUSION OF INNOVATIONS

The basic work in geographic simulation was undertaken by Professor Torsten Hägerstrand of the University of Lund in Sweden (Hägerstrand, 1953, 1967). Hägerstrand hypothesized that the nebulalike clusters seen in many spatial distributions (i.e., a concentrated core surrounded by a border zone of decreasing densities) were the result of a process related to the diffusion of the techniques and ideas that spread largely through the network of social contacts. He noted in several empirical studies that many innovations, such as those exemplified in figures 8-1 and 8-2, spread in generally similar fashions, that is, a start was made by a rather concentrated cluster of carriers which then expanded step by step in such a way that the probability of a conversion appeared to be

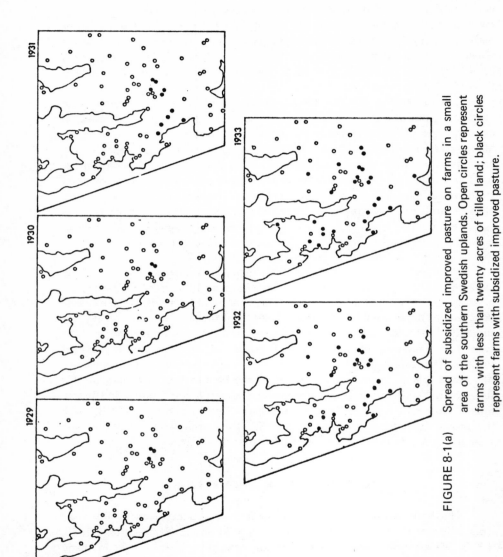

FIGURE 8-1(a)   Spread of subsidized improved pasture on farms in a small area of the southern Swedish uplands. Open circles represent farms with less than twenty acres of tilled land; black circles represent farms with subsidized improved pasture.

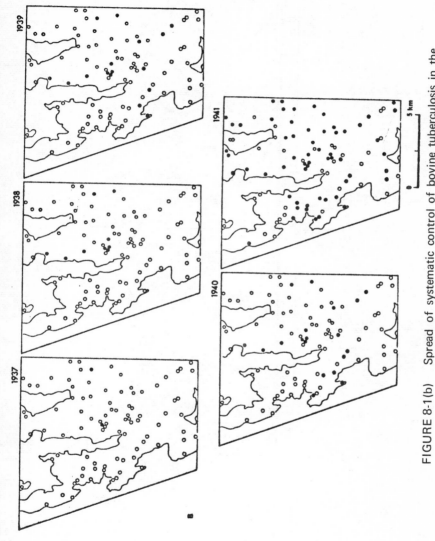

FIGURE 8-1(b)   Spread of systematic control of bovine tuberculosis in the same area as in Fig. 8-1(a). Open circles represent farms with cattle; black circles represent farms with organized control of bovine tuberculosis. Source: Hägerstrand, 1967.

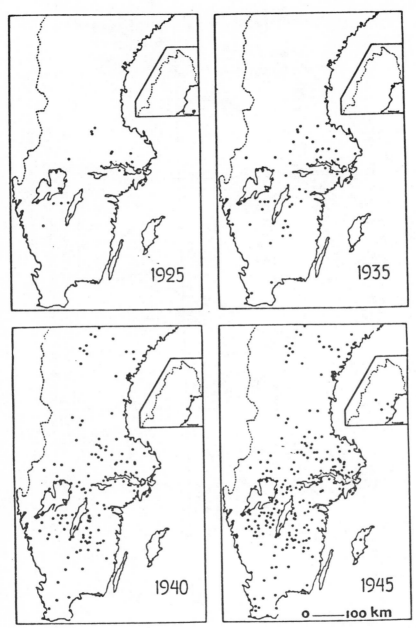

FIGURE 8-2   Spread in Sweden of a voluntary laymen's organization supporting the Swedish church. Each dot represents a primary group of the organization. Source: Hägerstrand, 1967.

higher among those who lived nearer the carriers than among those who lived farther away. He assumed that this was the result of spatial contiguity and borrowed a physical term, *neighborhood effect,* to describe it. Hägerstrand felt that the temporal changes in the observed spatial structures were directly related to the adoption of the innovation, and he reasoned that a central factor in the adoption process was that of information transfer. He then addressed himself to an examination of the geographic bias in social communications, and based upon an analysis of the empirical data from several areas in Sweden, concluded from the observed neighborhood effects that the importance of mass media was very small and could generally be ignored. This major simplification reduced the problem to one dealing only with the spatial aspects of interpersonal communications. He hypothesized that the spatial extent of private communications would significantly differ between social and economic groups and that certain groups would have communication structures that were spatially concentrated in the local range, whereas others would operate at regional, national, or even international scales. Hägerstrand's research was largely concentrated upon the diffusion of innovations at the local level and hence dealt only with the spatial pattern of interpersonal communications over rather short (10-15 km.) distances. In an attempt to determine the spatial bias in interpersonal communication at this scale he examined two surrogate patterns—local telephone traffic and local migration (see figure 8-3). The distribution of destination points about a common origin was studied, and an examination of the directionally smoothed spatial patterns of these phenomena (see table 8-1) indicated that the communications links of the average individual within the local area decreased very rapidly in number with increasing distance or, in the sample region, approximately with the square of the distance.

Having reinforced his notion that local communication patterns were severely restricted in space, Hägerstrand then proceeded to construct a simple Monte Carlo model of information transfer among individuals in the local area. This model may be equated with a diffusion model if certain restrictive assumptions pertaining to the nature of the adoption process are accepted.

Hägerstrand's spatial simulation proceeded according to a simple set of rules:

1. All action takes place upon an isotropic model plane which is divided into square cells of equal area which are inhabited by the same number of individuals. Each individual constitutes a potential adopter of the new item.
2. Only one person carries the item at the start.
3. The item is adopted as soon as information is passed.
4. Information is spread only by telling at pairwise meetings of individuals.
5. The tellings take place only at constant intervals (generation intervals) when every carrier tells exactly one other person, carrier or noncarrier.
6. The probability of being paired with a carrier depends upon the

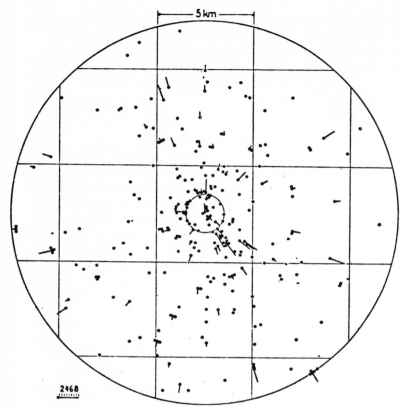

FIGURE 8-3   The Local Migration Field for the Asby, Sweden, Area,
1935-1939. Note: Dots represent one individual migrant and
dots with stems represent migratory groups with the length of
the stem proportional to the number of members in the group.
Source: Hägerstrand, 1953.

geographical distance between teller and receiver in a way determined
by empirical estimate.

These rules place the Hägerstrand model in the class of simple discrete
time, discrete space simulations. The simulation rules were severely limited as
were the number of areas covered by the simulation—modern computer technology not being available at the time the model was developed, all runs of the
simulation were carried out by hand. The severe restriction in the number of
rules for the operation of the simulation represented an attempt to examine
explicitly the role of spatial relationships in structuring the diffusion process.

A central feature of the model is the spatial autocorrelation function
which is implicit in Rule 6. Hägerstrand entered this in his model as a fixed size

but floating set of grid cells, each of which was the size of one of the cells in the study area. The carrier was assumed to be located in the central cell and the figures represented probabilities of contact with the cell in question. The figures reflect the strong distance decay functions derived from studies of empirical communication and movement patterns such as those discussed earlier. Information of the type presented in table 8-1 was reduced to a simple Pareto function of the form:

$$y = aD^b$$

where *y* equals the number of contacts/unit area and *D* is the distance between cell centroids. This function was used in turn to estimate total contacts and contact probabilities within the twenty-four cells immediately surrounding the

TABLE 8-1   Local Migration in the Asby Area

Distance (km)	Zonal Area (km^2)	Number of Migrants	Number of Migratory Units	Number of Migratory Units per km^2
0 - 0.5	0.79	13	9	11.39
0.5 - 1.5	6.28	69	45	7.17
1.5 - 2.5	12.57	72	45	3.58
2.5 - 3.5	18.85	44	26	1.38
3.5 - 4.5	25.14	37	28	1.11
4.5 - 5.5	31.42	34	25	0.80
5.5 - 6.5	37.70	33	20	0.53
6.5 - 7.5	43.99	45	23	0.52
7.5 - 8.5	50.27	36	18	0.36
8.5 - 9.5	56.56	18	10	0.18
9.5 -10.5	62.82	28	17	0.27
10.5 - 11.5	69.12	9	7	0.10
11.5 - 12.5	75.41	14	11	0.15
12.5 - 13.5	81.69	8	6	0.07
13.5 - 14.5	87.98	2	2	0.02
14.5 - 15.5	94.26	8	5	0.05

Source: Hägerstrand, 1953.

central cell (see figure 8-4). At each generation period the Mean Information Field was centered over each carrier, and a recipient was picked who then in turn became a carrier starting with the next generation. This procedure was repeated for every carrier location during a given generation. The result of these operations was a near geometric growth in the number of knowers in the system

0.0096	0.0140	0.0168	0.0140	0.0096
0.0140	0.0301	0.0547	0.0301	0.0140
0.0168	0.0547	0.4431	0.0547	0.0168
0.0140	0.0301	0.0547	0.0301	0.0140
0.0096	0.0140	0.0168	0.0140	0.0096

|←———————— 25 km ————————→|

FIGURE 8-4   Mean information field. (Carrier is always assumed to be located in center cell.) Asby data from Hägerstrand (1953).

and eventual saturation (see figure 8-5). Subsequent modifications of the basic model permitted extensions in which more than one individual carried the information or innovation at the start of the simulation and there were uneven population levels in the cells. In the latter instance the contact probabilities in the Mean Information Field were adjusted in accordance with the following formula:

$$Q_i = \frac{p_i \cdot N_i}{\sum\limits_{i=1}^{25} p_i \cdot N_i}$$

where Q_i represents the adjusted probability of a contact in cell i which has a population N_i and an original contact probability of p_i.

An interesting development in another of Hägerstrand's models was the introduction of barrier effects which represented localized discontinuities, and contact probabilities were assumed to be reduced as the result of intervening physical and cultural barriers to communication. Barriers in the model were confined to specific cell edges and were represented as a second probability process in the system. Each barrier has associated with it a probability that the

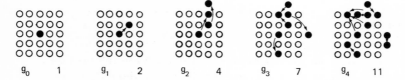

FIGURE 8-5   Spread of information by pairwise tellings in an even population with distance restraints on the probability of meetings. Source: Hägerstrand (1967).

communication will be permitted to pass. A 1.0 barrier represents completely open communication over the cell boundary in question, a 0.5 barrier would permit, on the average, one of every two communications to pass, whereas a 0.0 barrier would permit no communications to pass. When these barrier effects were included in the model, selection of a communication recipient involved two probability processes: (1) the selection of a cell for contact where the probabilities of contact were biased according to the Mean Information Field figures and (2) a determination if any barriers existed between the cell of origin and the cell of destination. If one or more barriers were found to intervene, a second random number draw was made to determine the impact upon the communication attempt. (Figure 8-6 shows observed versus simulated patterns.)

In his last model, Hägerstrand moved from the pure communication models (which could be held to be diffusion models only under the strict assumption of "instant" adoption) to an explicit consideration of resistance to adoption. He argued the existence of a "bandwagon" effect, as well as differential resistances to innovation based upon individual behavior. These were incorporated into the basic simulation model through an additional rule:

Resistance to the innovation decreases in direct proportion to the number of contacts with other individuals who have already adopted the innovation. The number of contacts necessary before adoption occurs is a function of the individual's characteristics.

This additional rule has frequently been misinterpreted because Hägerstrand, in the interests of computational simplicity, posited an identical distribution of resistances among the population within each cell. When no differences exist between areas in resistance to innovation, the net effect upon the output of the simulation is to introduce general temporal lags (e.g., 20 percent of saturation is reached in thirteen generations instead of seven, etc.). In actuality, different patterns of resistance may be assigned to specific areas representing, say, a cluster of traditionally oriented individuals who are highly resistant to new ideas. As Yuill (1964) has pointed out, the spatial impact of differential resistances to adoption is quite similar to that of barriers.

Several attempts have been made to extend and utilize the Hägerstrand simulation models, such as the two-level communications model proposed by Tiedemann and Van Doren (1964) to include explicitly the differential spatial role of change agents in the diffusion of hybrid seed corn in Iowa; the use of the models by Ramachandran (1967) to study the spread of new technology in portions of rural India; and the development of operational versions for modern, high-speed computers (Pitts, 1963 and 1965). Pitts has also developed an operational program for the generalization of the models to a more useful irregular lattice case. This more general program has replaced the original computational form. (Pitts, 1967; Marble and Bowlby, 1968a.)

The importance of the Hägerstrand models to modern geography should not be underestimated. Despite their simple conceptual structure, they isolated a number of critical problem areas (spatial dynamics, spatially biased communi-

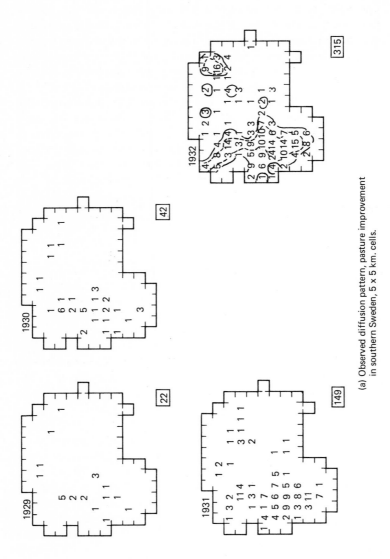

(a) Observed diffusion pattern, pasture improvement in southern Sweden, 5 x 5 km. cells.

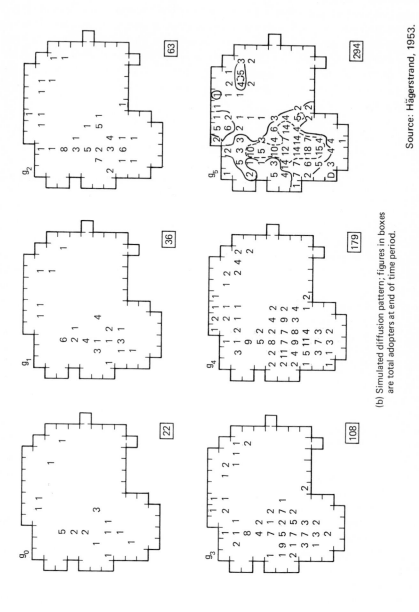

(b) Simulated diffusion pattern; figures in boxes are total adopters at end of time period.

Source: Hägerstrand, 1953.

FIGURE 8-6. Simulated versus actual diffusion of innovation.

375

cations patterns, behavioral factors in spatial analysis, etc.) at a time when the discipline was beginning a major redirection of its efforts, and they have indirectly exerted considerable influence upon current conceptual approaches in the field. Outside of the simulation approach they have encouraged a number of investigators to examine diffusion process within a spatial framework (Brown, 1968a; Beckmann, 1969; Day, 1970). As will be noted later, they also defined basic operational and verification problems that are now known to be common to all spatial simulations and, despite nearly two decades of work, remain largely unresolved today.

SPATIALLY BIASED COMMUNICATIONS SYSTEMS

Closely allied with the work on diffusions of innovations have been the studies of patterns of interpersonal communication in space. Most of these studies have been stimulated by Hägerstrand's Mean Information Field notion and have been empirical in nature (Marble and Nystuen, 1963; Morrill and Pitts, 1967; Marble and Bowlby, 1968b). Recently Marble *et al.* (1970) developed an experimental simulation model of an interpersonal communication system which eliminates many of the temporal reference problems encountered in Hägerstrand's original development and permits the introduction of social as well as geographical distance biases.

This model, known as SIMCOM, is based on the notion that the spatial diffusion of innovation may be viewed primarily as a result of interpersonal communication. Distance, both geographic and social, is recognized as an important factor influencing interpersonal communication. Geographic distance is incorporated in the model in the form of a distance decay function in which the probability of contact decreases as the distance between two individuals increases. However, unlike Hägerstrand's, this model also takes into account social distance by dividing the population into social groups or cliques and specifying the probabilities of communication within and between groups. The approach used is similar to the one proposed by the Swedish sociologist Karlsson (1958) in his "Model for Pure Interpersonal Communication." Other departures from Hägerstrand's work include (a) the treatment of time, (b) the inclusion of a simple death process, or what epidemiologists have termed *removal rate,* and (c) the application of more sophisticated computer technology. The consideration of time is different in that Hägerstrand's model tellings take place only at certain times with *constant* intervals when *every* adopter talks to one other person, whereas in the present model a clock mechanism is used to endogenously determine for each individual not only the time to the next communication but also the duration of each communication. A simple death process is incorporated into the model by allowing a portion of the adopters to become "passive" after a specified interval.

SIMCOM is operated on the following basis:

Rule 1. Communication takes place only during pairwise meetings of individuals. MAN1 is the person who attempts to initialize a communication, MAN2 is the person MAN1 contacts. If MAN1 is communicating with MAN2, no other person can communicate with MAN1 or MAN2 until the communication is finished.

Rule 2. If a person completes a communication at time t_0 the probability that he will attempt his next communication at t_1 is given by:

$$1/n \quad \text{if} \quad (t_o < t_1 \leq t_o + n)$$

$$0 \quad \text{if} \quad (t_1 > t_o + n)$$

The probability that contact takes place in the interval between t_0 and $t_0 + n$ is the same for each interval of time between t_0 and $t_0 + n$ where n is some prespecified time period.

Rule 3. Communication between MAN1 and MAN2 covers k time periods, where k is a function of both geographic and social distance.

Rule 4. The probability that MAN1 communicates with MAN2 at time t_0, given that MAN1 is scheduled to talk at t_0 and that MAN2 is not busy, is a function of both geographic distance and social distance. The effects of geographic and social distance are determined exogenously.

Rule 5. The number and location of each initial adopter is specified exogenously.

Rule 6. If MAN1, an active adopter, contacts MAN2, a nonadopter, MAN2 becomes an active adopter. An *active adopter* is a person who communicates the innovation in every two-way communication in which he participates. If MAN1, a passive adopter, contacts MAN2, a nonadopter, MAN2 remains a nonadopter. A *passive adopter* is a person who communicates the innovation only if contacted by another person.

The use of a clock and the corresponding finer division of the time dimension eliminates the problem of variable time scales encountered in the Hägerstrand models of the spatial diffusion process. This attempt to model the interpersonal communications portion of the diffusion process in some detail is still continuing, but it appears that it can be utilized as a basis for larger and more complex simulations of the spatial diffusion process. As all good simulations should, it also raises a number of questions which can only be answered through additional empirical research. One of the more interesting notions is incorporated in Rule 4 above; very little quantitative information appears to be available in this area.

PATTERNS OF URBAN GROWTH

Over a decade ago it was suggested that it might be possible to build computer simulation models of the growth and development of large-scale urban systems (Garrison, 1962). Garrison pointed out that there were two alternative approaches to the simulation depending upon the scale level that the investigator desired to examine: (1) the examination of the city as a functional unit within the overall system of cities and (2) the spatial arrangement of land uses and activities within an individual city. Models in both these areas could be exceedingly useful in theoretical investigations of urban relationships and also as practical planning tools. Planners could utilize these urban simulations to conduct sensitivity tests of alternate operating rules, such as variations in zoning regulations, or to examine the system-wide impact of major shocks, such as construction of a major urban freeway.

It was evident in 1960 that such comprehensive simulations could not be developed on the basis of current levels of knowledge relating to urban dynamics and within the severe restrictions on size and speed of digital computers. It was hoped that by 1970 such a simulation would become feasible. However, even with the development of significantly more advanced computer technology, we still find ourselves some distance away from the capability that was envisioned by Garrison in his original proposal. During the intervening period there have been numerous investigations pertaining to city size and structure and the way in which these change with time. The conceptual background for the construction of such a model is now on a firmer basis than in 1960, and several attempts have been made to construct partial simulations of restricted segments of the urban system. However, the full urban simulation envisioned by Garrison still seems to lie beyond our capabilities.

Systems of Cities

Geographers have been interested in the spatial pattern of cities for many years. A well-developed body of theory, central place theory, exists in this area, and it has produced a wide variety of theoretical and empirical investigations. However, central place theory, as currently structured, is basically static, and its equilibrium solutions provide little or no information on the growth of the system of cities over time. Some five years after Garrison made his original suggestion, Richard Morrill produced an ambitious monograph entitled "Migration and the Spread and Growth of Urban Settlement" which attempted to develop both simple and complex growth models for the spread of settlement in a developing area (Morrill, 1965a). The general model included three basic processes: (1) assignment of locations for transportation routes, for manufacturing activities, and for other noncentral place activities, (2) assignment of central place activities such as retailing, and (3) local migration. Some form of spatial control was exerted over all these processes with central place locations being subject to hinterland area requirements and transport route locations being dependent upon existing locations of cities and space-penetrating goals. Manufacturing was assumed to be dependent upon existing locations of resources, markets, and transportation, whereas migration was dependent upon the differential attractiveness of areas and was strongly affected by distance relationships.

Figure 8-7 illustrates an example of the output of the model. It begins with an infinite region divided into hexagonal subareas, a place of origin (*A*) through which all migration from outside the area is funneled, and a set of time periods, or generations. The state of technology, migration volumes, natural growth levels, and conditions for development at various levels of the central place hierarchy are specified exogenously. The model begins with one central area containing a specified number of persons, including a very small number in an agglomeration. During the first time period a number of migrants appear in *A*; some stay in the area while others migrate elsewhere. After assigning the migrants to specific locations, the model then checks to see whether sufficient population is present to justify the assignment of any central place activities. Population levels in this case (generation 1) are below the threshold level, so no activities are assigned. In the second generation further migration takes place through *A*, smaller agglomerations and transport routes linking them are developed, and the original point of settlement (*A*) has acquired enough residents to qualify as a low-level central place; a factor that also influences future patterns of development.

The example presented in figure 8-7 demonstrates the manner in which migration from a central threshold point leads to the gradual spread and filling up of a settlement, as well as the way in which stochastic elements in the system lead to the development of an irregular settlement frontier. The rise of the central place system in this example is marked by the domination of the city of origin, a general lag in the total number of centers developed, and a sectorial effect upon the hinterland demonstrating the impact of the direction of early migrants.

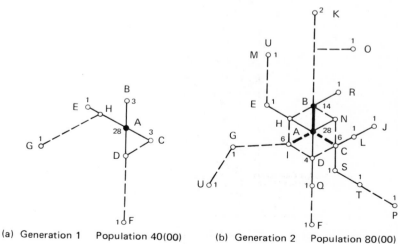

(a) Generation 1   Population 40(00)   (b) Generation 2   Population 80(00)

FIGURE 8-7.   Simulation of the growth of a system of central places and transportation routes. Source: Morrill (1965a).

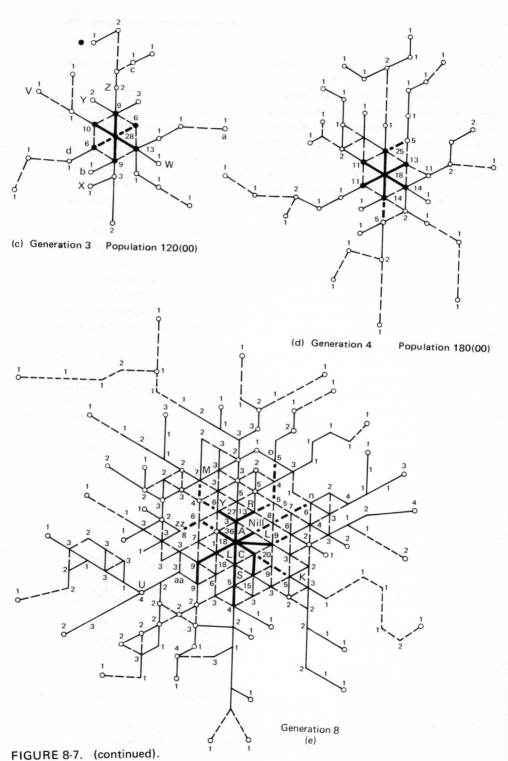

(c) Generation 3   Population 120(00)

(d) Generation 4   Population 180(00)

Generation 8
(e)

FIGURE 8-7. (continued).

Figure 8-8 represents the outcome of an application of a slightly modified form of this model to an area in southern Sweden during the period 1880 through 1960. The geometric restrictions of central place theory were relaxed in this instance, and the initial pattern of development of urban places and transport routes was taken as given. Morrill noted that his simulation did accomplish his general purpose of generating a pattern of urban centers comparable to the actual one, and through the migration mechanism built into the model, reproduced the radical redistribution of population that occurred in the actual case. The redistribution predicted by the simulation model differed in some locational details from the actual population development, but the general patterns were similar.

Morrill's attempt to simulate the pattern of development in an area represents perhaps the most ambitious geographic attempt at simulation of a large spatial system. Much of his work centered upon attempts to model the spatial pattern of local and interregional migration which, in turn, was stimulated by the early work of Hägerstrand in this area (Hägerstrand, 1957). Other work in the general area of simulation of settlement and urban structure has been undertaken by Byland (1960) and Malm, Olsson, and Warneryd (1966).

The Internal Spatial Structure of the City

Because of the intense interest arising from practical planning problems, geographers and others have given a great deal of attention to the development of models of the internal structure of the city. While the literature abounds with attempts to model the spatial structure of the city on both a static and a dynamic basis, relatively few of these models have utilized computer simulation techniques. Again, it has been Richard Morrill who has undertaken much of the geographic work in this area. In one article (Morrill, 1965), he presented a study that moved directly toward a simulation of the expansion of land uses on the urban fringe. The model considered the amount of available land, effects of accessibility to major developments, site quality variations, proximity to mixed uses or blighted areas, existence of neighborhood amenities, kind and size of residential developments, density of development, succession of land uses, and lag in the development of schools and major shopping centers. The model was applied to a study area north of Seattle, Washington, and new home and subdivision development was simulated for the period from 1957 to 1962. The result of the simulation was, of course, a map distribution of predicted development, and the summary in table 8-2 compared actual and simulated patterns of development. The correspondence seems quite high, and the model is being utilized to study expansion patterns in the same area over an extended period of time. The interaction between successive forecasting and observations of the developing spatial structure will hopefully lead to the development of a more accurate simulation model and also increase the level of understanding of the process of urban growth on the fringe. Morrill's work on the Negro ghetto (reprinted here as a case-example) is another example of this approach (Morrill, 1965a).

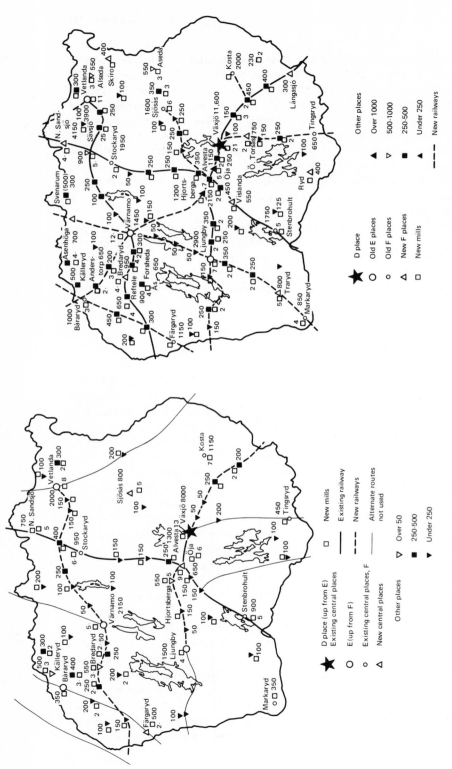

(b) SIMULATION ASSIGNMENT 1900-1920

(a) SIMULATION ASSIGNMENT 1880-1900

FIGURE 8-8. Simulation of the growth of the Smaland area in southern Sweden, 1880-1960. Source: Morrill (1965a).

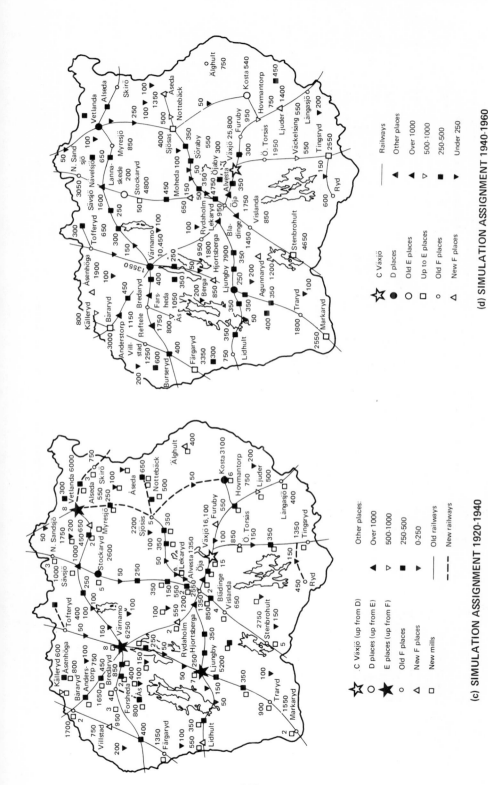

(c) SIMULATION ASSIGNMENT 1920-1940

(d) SIMULATION ASSIGNMENT 1940-1960

FIGURE 8-8. (continued).

383

TABLE 8-2 Comparison of Locational Characteristics, Actual and Simulation Development, Study Area Sohomish County, 1957-62

Size Distribution of Developments

	Actual	%	Simulated	%
1	300	.62*	229	.59*
2	55	.73	51	.72
3	22	.78	16	.76
4	21	.82	20	.815
5	12	.845	8	.84
6	7	.86	3	.845
7	7	.87	1	.846
8	7	.89	6	.86
9	4	.895	4	.87
10-15	17	.93	10	.90
16-20	5	.94	9	.92
21-25	9	.96	17	.96
26-30	3	.97	1	.965
31-40	5	.98	5	.98
41-50	1	.98	4	.99
51-60	5	.99	0	
61-70	3		1	
71-80	0		1	
80	2	1.00	2	1.00
	485*		338*	

Distribution of Distances of Larger New Developments to:

Major Existing Developments

	Actual	%	Simulated	%
Adjacent	24	.37	22	.33
1-2 blocks	18	.645	15	.55
3-4 blocks	9	.785	15	.78
5-6 blocks	7	.89	6	.87
7-8 blocks	3	.94	4	.925
9-10 blocks	2	.97	2	.96
11-12 blocks	1	.98	2	.99
13-14 blocks	2	1.00	1	1.00

no significant difference

Major Arterials

	Actual	%	Simulated	%
Adjacent	25	.38	21	.31
1 blocks	14	.59	15	.54
2 blocks	15	.82	17	.79
3 blocks	7	.92	10	.92
4 blocks	5	1.00	4	.98
5 blocks	0		1	1.00

no significant difference

Center of gravity

Actual: $\bar{x} = 2.35 \pm .069$, $\bar{y} = 4.71 \pm .216$

Simulated: $\bar{x} = 2.44 \pm 0.77$, $\bar{y} = 4.74 \pm .255$

no significant difference

Standard Distance

Actual: $3.17 \pm .228$ Simulated: $3.42 \pm .265$

no significant difference

Maximum difference in cumulative proportion: .033
not significant†

SOURCE: Morrill (1965).
*Number of developments differ because the actual array is for 1957-63 rather than for 1957-62.
†Kolmogorov-Smirnov tests are used to compare cumulative distributions, see Sidney Siegel, *Nonparametric Statistics for the Behavioral Sciences*, McGraw-Hill, 1956.

Closely related to geographic studies of migration have been studies of the pattern of nonmigratory movements of the population. These movements represent the recurrent contacts between the individual's residence and centers of employment, retail establishments, friends, and so forth. Two simulation models have been proposed covering subsets of these movement activities. Taafe, Garner, and Yeates (1963) examined a series of Monte Carlo models dealing with the journey to work to suburban locations in West Chicago. The simulations were used to isolate factors believed to underlie the observed distribution of trip origins. Nystuen (1967a) proposed a simulation model of customer movement to retail establishments. This was designed to incorporate explicitly the multiple-purpose shopping trip which, until that time, had largely been ignored in studies of the spatial structure of urban retailing. The model attempted to predict the total duration (measured in terms of the number of stops) of the trip as well as the mix of retail functions that the customer would encounter. The model structure was relatively simple, and its predictions were of disappointingly low accuracy. Nystuen's model has, however, stimulated a number of investigations into the structure of multipurpose trips and their relation to urban retail structure.

PROBLEMS OF GEOGRAPHIC SIMULATION

The simulation models discussed here have a number of common bonds. They all deal with a predefined set of geographic locations, much of their internal structure relates to various aspects of the spatial autocorrelation function, and their output is usually a map of the spatial distribution in question. Aside from Morrill's work, there appears to be a general lack of articulation with other simulation models. The models, once presented, are frequently not used again, and there is little effort to develop valid empirical tests with a view toward the construction of improved models. Underlying the geographic uses of simulation are a series of common problems, some operational and some methodological, which explain many of the facets of this pattern.

Conceptual Approach

Many geographic researchers are loath to utilize simulation approaches, since they are generally held to lack conceptual elegance. Researchers who do use simulation also feel that much of the value of a simulation model lies in the development of explicit statements of the internal relationships in the model rather than in its operation and manipulation. Many of the internal relationships turn out to be rather poorly known (see the comments on Rule 4 of SIMCOM), and efforts are then directly toward in-depth investigations of these areas rather than toward more explicit structuring and operation of the simulation.

Computation

Spatial simulations, even in their more simple forms, normally encounter massive problems of storage and computational speed brought on by the large number of persons and areas to be accounted for in each situation. Nearly all the models to date deal with interaction patterns on the individual level, and very large numbers of individuals and subareas must be incorporated into the model before "interesting" spatial results can be obtained. An example of this is a researcher attempting to utilize the Hägerstrand models to study the development of wave patterns in the diffusion of a new innovation. A fairly fine spatial grid is needed to define the wave fronts with any precision, and this may require breaking up the area studied into, say, two hundred cells on a side for a total of forty thousand cells each containing, say, fifty individuals. Problems of this size—together with associated graphic output—are not readily handled by existing computers. The cost factor is also a major one, since the simulation must be run many times to establish a reasonable variance structure. A simple spatial simulation may only use up, say, two minutes of central processor time, but the fifty to eighty passes needed to estimate effectively the parameters of the output distribution quickly raise the time needed for each problem from a few minutes to several hours. Existing simulation languages, for example, SIMSCRIPT, also do not readily lend themselves to the type of bookkeeping operations necessary in spatial models.

Verification

Perhaps the most difficult problem in spatial simulation lies in the area of verification of the model. Geographic simulations usually represent an attempt to define a process that will replicate some stated real-world distribution. By running the simulation model a number of times, the researcher may obtain what can be viewed as a series of sample observations drawn from some unknown frequency distribution. The sample observations may be used to estimate the parameters of the parent distribution and the question posed: "Does the observed real-world distribution represent a 'likely' member of this population?" This is a fairly common question in statistics, but it is considerably more difficult to answer when the frequency distributions are two dimensional rather than one dimensional.

Several early users of simulation in human geography either argued that only one pass through the simulation was necessary or utilized tests on the cells (such as x_2) that were aspatial in nature and, in the regular lattice cases, invariant under row and column operations on the cells. This latter attribute is unacceptable because the primary purpose of the spatial simulation is to generate a distribution with clearly defined locational properties. The results reported in Morrill's urban fringe study (table 8-2) represent a good example of this. Apparently only one pass was made with the model, and then generally aspatial tests were applied to the results. This approach was common in the early sixties but has now generally been discarded. The slow development of truly spatial tests combined with the large blocks of computer time needed for verification runs has generally acted to suppress geographic uses of simulation during the last few years.

Although some work has been undertaken on the verification problem (e.g., Anderson, 1965), for the most part it remains intractable and currently represents one of the major obstacles to further use of simulation approaches in geographic research. Research in this area has been undertaken based upon spatial filtering and the use of optical computers, but the results to date have not significantly reduced the magnitude of the verification problem.

CONCLUSIONS

Simulation modeling in human geography enjoyed a fair amount of popularity in the early sixties with the publication in English of some of Hägerstrand's early work. These early applications identified many critical research topics in an explicit fashion, but as a rule failed to produce results of the power and level of generality originally envisioned by their developers. Problems of size, computer capability, and severe difficulties in verifying the results of the modeling efforts have led during the last half of the decade to a sharp decline in the use of Monte Carlo simulations in geographic research. In the last several years, the only large geographic simulation to be constructed was the SIMCOM model outlined earlier, and Ramachandran's (1967) work represents the only large-scale empirical application of the Hägerstrand models.

Within the next three to five years, the problems posed by computer capability should generally be overcome, and several current theoretical studies seem to promise that satisfactory, if not elegant, solutions to the verification problem will be available at about the same time. The combination of these two factors should result in a renewed interest in simulation approaches in human geography.

REFERENCES

Anderson, David, *Three Computer Programs for Contiguity Measures,* Technical Report No. 5, Spatial Diffusion Study, Northwestern University, Department of Geography, 1965.

Bailey, T. J., *The Mathematical Theory of Epidemics.* New York: Hafner Publishing Co., Inc., 1957.

Beckmann, Martin J., *The Analysis of Spatial Diffusion Processes,* Research Paper No. 13. Brown University, Department of Economics, 1969.

Berry, Brian, J. L., and Duane F. Marble (eds.), *Spatial Analysis: A Reader in Statistical Geography.* Englewood Cliffs, N. J.: Prentice-Hall, Inc., 1968.

Bowden, L. W., *Diffusion of the Decision to Irrigate,* Research Paper No. 97. University of Chicago, Department of Geography, 1965.

Brown, Larry, *Diffusion Dynamics.* Lund, Sweden: C. W. K. Gleerup, 1968a.

———, *Diffusion Processes and Location: A Conceptual Framework and Bibliography.* Philadelphia: Regional Science Research Institute, 1968b.

Burton, Ian, "The Quantitative Revolution and Theoretical Geography," *The Canadian Geographer,* 7 (1963), 151-62.

Byland, Erik, "Theoretical Considerations regarding the Distribution of Settlement in Inner Northern Sweden," *Geografiska Annaler,* 42 (1960), 225-31.

Cox, Devin R., and Reginald G. Golledge (eds.), *Behavioral Problems in Geography: A Symposium.* Northwestern University Sutdies in Geography, No. 17, 1969.

Curry, Les, "Quantitative Geography," *The Canadian Geographer,* 9 (1967), 265-79.

Day, Richard H., "A Theoretical Note on the Spatial Diffusion of Something New," *Geographic Analysis,* 2 (1970), 68-75.

Garrison, William L., "Toward Simulation Models of Urban Growth and Development," *IGU Symposium on Urban Geography,* Lund Studies in Geography, Series B, No. 24. Lund, Sweden: C.W.K. Gleerup, 1962.

Gould, Peter, "Methodological Developments Since the Fifties," in C. Board, R. Chorley, P. Haggett, and D. Stoddart (eds.), *Progress in Geography,* Vol. 1. London: Edward Arnold, 1969.

———, *Spatial Diffusion.* Resource Paper No. 4, Commission on College Geography, Association of American Geographers, 1969.

Hägerstrand, Torsten, "Aspects of the Spatial Structure of Social Communication and the Diffusion of Information," *Papers,* Regional Science Association, 16 (1966), 27-42.

———, *Innovationsförloppet ur Korologisk Synpunkt,* 1953, trans. by Allan Pred as *Innovation Diffusion As a Spatial Process.* Chicago: University of Chicago Press, 1967.

———, "Migration and Area," in Hannerberg, Hägerstrand, and Odeving (eds.), *Migration in Sweden: A Symposium,* Lund Studies in Geography, Series B, No. 13. Lund, Sweden: C. W. K. Gleerup, 1957.

———, "On Monte Carlo Simulation of Diffusion," in W. L. Garrison and D. F. Marble (eds.), *Quantitative Geography,* Part I: Economic and Cultural Topics. Northwestern University Studies in Geography, No. 13, 1967.

———, "Quantitative Techniques for Analysis of the Spread of Information and Technology," in Anderson and Bowman (eds.), *Education and Economic Development.* Chicago: Aldine Publishing Co., 1965.

———, *The Propagation of Innovation Waves,* Lund Studies in Geography, Series B, No. 4, 1952.

Karlsson, Georg, *Social Mechanisms.* Stockholm: Almquist and Wiksell, 1958.

Malm, Roger, Gunnar Olsson, and Olaf Warneryd, "Approaches to Simulation of Urban Growth," *Geografiska Annaler,* Series B, 48 (1966), 9-22.

Marble, Duane F., and Sophia R. Bowlby, *Computer Programs for the Operational Analysis of Hägerstrand Type Spatial Diffusion Models,* Technical Report No. 9, Spatial Diffusion Study. Northwestern University, Department of Geography, 1968a.

———, and Sophia R. Bowlby, *Direct and Indirect Measurement of Urban Information Fields: Some Examples,* Technical Report No. 10, Spatial Diffusion Study. Northwestern University, Department of Geography, 1968b.

———, Perry O. Hanson, James O. Huff, Ashraf S. Manji, and Elias Pacheco, *A Monte Carlo Model for the Simulation of a Distance Biased Communications Net,* Technical Report No. 11, Spatial Diffusion Study. Northwestern University, Department of Geography, 1970.

———, and John D. Nystuen, "An Approach to the Direct Measurement of Community Mean Information Fields," *Papers,* Regional Science Association, 11 (1963), 99-109.

Morrill, Richard L., "Expansion of the Urban Fringe: A Simulation Experiment," *Papers,* Regional Science Association, 15 (1965), 185-202.

———, *Migration and the Spread and Growth of Urban Settlement,* Lund Studies in Geography, Series B, No. 26. Lund, Sweden: C. W. K. Gleerup, 1965a.

———, "Simulation of Central Places Over Time," in K. Norborg (ed.), *IGU Symposium on Urban Geography,* Lund Studies in Human Geography, Series B, No. 24, Lund, Sweden: C. W. K. Gleerup, 1962.

———, "The Negro Ghetto: Problems and Alternatives," *Geographical Review,* 55 (1965b), 339-62.

———, and Forrest R. Pitts, "Marriage, Migration and the Mean Information Field: A Study in Uniqueness and Generality," *Annals,* Association of American Geographers, 57 (1967), 401-22.

Nystuen, John D., "A Theory and Simulation of Urban Travel," in W. L. Garrison and D. F. Marble (eds.), *Quantitative Geography,* Part I: *Economic and Cultural Topics.* Northwestern University Studies in Geography, No. 13, 1967a.

———, "Boundary Shapes and Boundary Problems," *Papers,* Peace Research Society, 7 (1967b), 107-28.

Pitts, Forrest R., *HAGER III and HAGER IV: Two Monte Carlo Computer Programs for the Study of Spatial Diffusion Problems,* Technical Report No. 4, Spatial Diffusion Study. Northwestern University, Department of Geography, 1965.

———, *MIFCAL and NONCEL: Two Computer Programs for the Generalization of the Hägerstrand Models to an Irregular Lattice,* Technical Report No. 7, Spatial Diffusion Study. Northwestern University, Department of Geography, 1967.

———, "Problems in the Computer Simulation of Diffusion," *Papers,* Regional Science Association, 11 (1963), 111-22.

Ramachandran, R. "Technological Change and Spatial Diffusion of Innovations in Rural India." Ph.D. dissertation, Clark University, Department of Geography, 1967.

Taaffe, Edward J., Barry J. Garner, and Maurice H. Yeates, *The Peripheral Journey to Work: A Geographic Consideration.* Evanston: The Transportation Center at Northwestern University, 1963.

Tiedemann, C. E., and C. S. Van Doren, *The Diffusion of Hybrid Seed Corn in Iowa: A Spatial Simulation Model,* Technical Bulletin B-44. Michigan State University, Institute for Community Development, 1964.

Tornquist, G., *TV Agandets Utveckling I Sverige, 1956-65.* Stockholm: Almquist and Wiksell, 1967.

Yuill, Robert S., *A Simulation Study of Barrier Effects in Spatial Diffusion Problems,* Technical Report No. 1, Spatial Diffusion Study. Northwestern University, Department of Geography, 1964.

THE NEGRO GHETTO:
PROBLEMS AND ALTERNATIVES

RICHARD L. MORRILL

"Ghettos," as we must realistically term the segregated areas occupied by Negroes and other minority groups, are common features of American urban life. The vast majority of Negroes, Japanese, Puerto Ricans, and Mexican-Americans are forced by a variety of pressures to reside in restricted areas, in which they themselves are dominant. So general is this phenomenon that not one of the hundred largest urban areas can be said to be without ghettos.[1]

Inferiority in almost every conceivable material respect is the mark of the ghetto. But also, to the minority person, the ghetto implies a rejection, a stamp of inferiority, which stifles ambition and initiative. The very fact of residential segregation reinforces other forms of discrimination by preventing the normal contacts through which prejudice may be gradually overcome. Yet because the home and the neighborhood are so personal and intimate, housing will be the last and most difficult step in the struggle for equal rights.

The purpose here is to trace the origin of the ghetto and the forces that perpetuate it and to evaluate proposals for controlling it. The Negro community of Seattle, Washington, is used in illustration of a simple model of ghetto expansion as a diffusion process into the surrounding white area.

From the beginning of the nineteenth century the newest immigrants were accustomed to spend some time in slum ghettos of New York, Philadelphia, or Boston.[2] But as their incomes grew and their English improved they moved out into the American mainstream, making way for the next group. During the nineteenth century the American Negro population, in this country from the beginning but accustomed to servitude, remained predominantly southern and rural. Relatively few moved to the North, and those who did move lived in small clusters about the cities. The Negro ghetto did not exist.[3] Even in southern cities the Negroes, largely in the service of whites, lived side by side with the white majority. Rather suddenly, with the social upheaval and employment oppor-

Reprinted with permission from *The Geographical Review*, Vol. 55, (1965). Copyright ©1965 by the American Geographical Society of New York.

[1] Census Tract Reports, 1960, *Ser. PHC(1)*, selected cities. Subject Reports (Census of Population, 1960, Vol. 2), 1960, *Ser. PC(2):* Nonwhite Population by Race; State of Birth. (U. S. Bureau of the Census, various dates.)

[2] Oscar Handlin: The Newcomers (New York Metropolitan Region Study [Vol. 3]; Cambridge, Mass., 1959).

[3] Charles Abrams: Forbidden Neighbors (New York, 1955), p. 19.

tunities of World War I, Negro discontent grew, and large-scale migration began from the rural south to the urban north, to Philadelphia, New York, Chicago, and St. Louis, and beyond.

The influx was far larger than the cities could absorb without prejudice. The vision of a flood of Negroes, uneducated and unskilled, was frightening both to the whites and to the old-time Negro residents. As the poorest and newest migrants, the Negroes were forced to double up in the slums that had already been created on the periphery of business and industrial districts. The pattern has never been broken. Just as one group was becoming settled, another would follow, placing ever greater pressure on the limited area of settlement, and forcing expansion into neighboring areas, being emptied from fear of inundation. Only in a few cities, such as Minneapolis-St. Paul and Providence and other New England cities, has the migration been so small *and* so gradual that the Negro could be accepted into most sections as an individual.

America has experienced four gigantic streams of migration: the European immigration, which up to 1920 must have brought thirty million or more; the westward movement, in which from 1900 to the present close to ten million persons have participated; the movement from the farms to the cities, which since 1900 has attracted some thirty million; and the migration of Negroes to the North and West, which has amounted since World War I to about five million, including some three million between 1940 and 1960 (Table I). The pace has not abated. Contributing also to the ghetto population have been 900,000 Puerto Ricans, who came between 1940 and 1960, largely to New York City; about 1,500,000 Mexicans, descendants of migrants to the farms and cities of the Southwest; and smaller numbers of Chinese, Japanese, and others.[4] Economic opportunity has been the prime motivation for all these migrant groups, but for the Negro there was the additional hope of less discrimination.

TABLE I   Major Destinations of Net 3,000,000 Negroes
Moving North, 1940-1960[a]

New York	635,000	Washington, D. C.	201,000
Chicago	445,000	San Francisco	130,000
Los Angeles	260,000	Cleveland	120,000
Detroit	260,000	St. Louis	118,000
Philadelphia	255,000	Baltimore	115,000

[a]*Estimates only*

The rapidity and magnitude of the Negro stream not only have increased the intensity and size of ghettos in the North but no doubt have also accelerated the white "flight to the suburbs" and have strongly affected the economic, political, and social life of the central cities.[5] In the South, too, Negroes have participated in the new and rapid urbanization, which has been accompanied by increased ghettoization and more rigid segregation.

[4]*Ibid.*, pp. 29-43.
[5]Davis McEntire: Residence and Race: Final and Comprehensive Report to the Commission on Race and Housing (Berkeley, 1960), pp. 88-104.

TABLE II Minority Populations of Major
Urbanized Areas, United States, 1960

City	Minority Population	Total Population	Minority %
1. New York City	2,271,000	14,115,000	16
Negro	1,545,000		
Puerto Rican	671,000		
2. Los Angeles	1,233,000	6,489,000	19
Negro	465,000		
Mexican	629,000		
Asian	120,000		
3. Chicago	1,032,000	5,959,000	17
4. Philadelphia	655,000	3,635,000	18
5. Detroit	560,000	3,538,000	16
6. San Francisco	519,000	2,430,000	21
7. Washington, D. C.	468,000	1,808,000	26
8. Baltimore	346,000	1,419,000	24
9. Houston	314,000	1,140,000	28
10. San Antonio	303,000	642,000	47
11. St. Louis	287,000	1,668,000	17
12. Cleveland-Lorain	279,000	1,928,000	15
13. New Orleans	265,000	845,000	31
14. Dallas-Fort Worth	252,000	1,435,000	18
15. Atlanta	207,000	768,000	27
16. Birmingham	201,000	521,000	38
17. Memphis	200,000	545,000	37

Sources: Census of Population, 1960: Vol. 1, Chap. C, General Social and Economic Character-istics; Vol. 2, Subject Reports: Non-white Population by Race.

As a result of these migrations, the present urban minority population consists in the North and West, of 7.5 million Negroes and 4 million others, together 12.5 percent of the total regional urban population; in the South, of 6.5 million Negroes, 20 percent; in total, of 18 million, 14 percent.[6] The proportion is increasing in the North, decreasing in the South. Minority populations in large American cities are presented in Table II.

THE NATURE OF THE GHETTO

If we study the minority population in various cities, we can discern real differences in income, education, occupational structure, and quality of homes.[7] For example, median family income of Negroes ranges from $2600 in Jackson, Mississippi, to $5500 in Seattle; and as a proportion of median white family income, from 46 percent to 80 percent respectively. The United States median family income for Negroes in urban areas is only $3700, as compared with $6400 for whites, but it is more than double the figure for Negroes still living in rural areas, $1750. It is not hard, therefore, to understand the motivation for

[6]Nonwhite Population by Race [see footnote 1 above].
[7]Census Tract Reports [see footnote 1 above].

Negro migration to the northern cities, where striking progress has really been made.

But the stronger impression is of those general characteristics which are repeated over and over. The ghetto system is dual: not only are Negroes excluded from white areas, but whites are largely absent from Negro areas. Areas entirely or almost exclusively white or nonwhite are the rule, areas of mixture the exception. The ghettos, irrespective of regional differences, are always sharply inferior to white areas; home ownership is less and the houses are older, less valuable, more crowded, and more likely to be substandard.[8] More than 30 percent of Negro urban housing is dilapidated or without indoor plumbing, as compared with less than 15 percent for whites. The ghetto is almost always in a zone peripheral to the central business district, often containing formerly elegant houses intermingled with commercial and light industrial uses. As poor, unskilled labor, Negroes settled near the warehouses and the railroads, sometimes in shacktowns, and gradually took over the older central houses being abandoned by the most recently segregated groups—for example, the Italians and the Jews—as their rise in economic status enabled them to move farther out. More than one ghetto may appear on different sides of the business district, perhaps separated by ridges of wealthy, exclusive houses or apartments.

The Negro differs fundamentally from these earlier groups, and from the Mexicans and Puerto Ricans as well. As soon as economic and educational improvements permit, the lighter-skinned members of the other groups may excape the ghetto, but black skin constitutes a qualitative difference in the minds of whites, and even the wealthy Negro rarely finds it possible to leave the ghetto. Color takes precedence over the normal determinants of our associations.[9]

In the southern city Negroes have always constituted a large proportion of the population and have traditionally occupied sections or wedges, extending from the center of the city out into the open country. Indeed, around some cities, such as Charleston, South Carolina, the outer suburban zone is largely Negro. Figure 1 depicts the ghetto pattern for selected cities.

The impact of the ghetto on the life of its residents is partly well known, partly hidden. The white person driving through is struck by the poverty, the substandard housing, the mixture of uses, and the dirt; he is likely to feel that these conditions are due to the innate character of the Negro. The underlying fact is, of course, that Negroes on the average are much poorer, owing partly to far inferior educational opportunities in most areas, but more to systematic discrimination in employment, which is only now beginning to be broken. Besides pure poverty, pressure of the influx into most northern cities itself induces deterioration: formerly elegant houses, abandoned by whites, have had to be divided and redivided to accommodate the newcomers, maintenance is almost impossible, much ownership is by absentee whites. Public services, such as street maintenance and garbage collection, and amenities, such as parks and playgrounds, are often neglected. Residential segregation means de facto school segregation. Unemployment is high, at least double the white average, and delinquency and crime are the almost inevitable result. A feeling of inferiority

[8]McEntire, *op. cit.* [see footnote 5 above], pp. 148-156.
[9]Abrams, *op. cit.* [see footnote 3 above], p. 73.

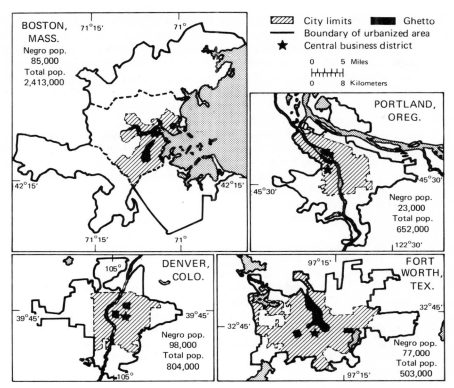

FIGURE 1   A group of representative ghettos. The dashed-line boundary on the Boston map indicates the inner urbanized area. Source: 1960 census data.

and hopelessness comes to pervade the ghetto. Most important is the enormous waste of human resources in the failure to utilize Negroes to reasonable capacity. The real cost of maintaining the ghetto system is fantastic. In direct costs the city spends much more in crime prevention, welfare payments, and so forth than it can collect.[10] The ghetto is the key to the Negro problem.

What are the forces that operate to maintain the ghetto system? Four kinds of barriers hinder change: prejudice of whites against Negroes; characteristics of the Negroes; discrimination by the real-estate industry and associated financial institutions; and legal and governmental barriers. Naked prejudice is disclaimed by a majority of Americans today. Today's prejudice is not an outright dislike; it is, rather, a subtle fear, consisting of many elements. The typical white American may now welcome the chance to meet a Negro, but he is afraid that if a Negro moves into his neighborhood it will break up and soon be all Negro. Of course, on a national average there are not as many Negroes as that—only one or two families to a block—but the fear exists because that is the

[10] John C. Alston: Cost of a Slum Area (Wilberforce State College, Wilberforce, Ohio, 1948).

way the ghetto has grown. A greater fear is of loss in social status if Negroes move in. This reflects the culture-bred notion that Negroes are inherently of lower standing. Some persons are terrified at the unlikely prospect of intermarriage. Finally, people are basically afraid of, or uncertain about, people who are different, especially in any obvious physical way. These fears combine into powerful controls to maintain segregation: refusal to sell to Negroes, so as not to offend the neighbors; and the tendency to move out as soon as a Negro enters, in order not to lose status by association.

The Negro himself contributes, however unwillingly, to ghettoization. It is difficult to be a minority as a group, but more difficult still to be a minority alone. Consequently the desire to escape the ghetto and move freely in the larger society is tempered by a realization of the problems in store for the "pioneer" and hesitancy to cut neighborhood ties with his own kind. Few people have such courage. In most cities, even if there were no housing discrimination, the ghetto would still persist, simply because a large proportion of Negroes could not afford, or would be afraid, to leave. Most Negroes achieve status and acceptance only within the Negro community. Usually Negroes who leave the ghetto prefer Negro neighbors; the risk is that this number, however small, is enough to initiate the conversion to full-scale ghetto.[11]

The Negro today suffers from his past. The lack of initiative and the family instability resulting from generations of enforced or inculcated subservience and denial of normal family formation are still present and are a barrier to white acceptance. The far lower levels of Negro income and education, no matter how much they are due to direct neglect and discrimination by the white majority, are nevertheless a strong force to maintain the ghetto. Studies show that whites will accept Negroes of equivalent income, education, and occupation.[12]

The strongest force, however, in maintaining the ghetto may well be real-estate institutions: the real-estate broker and sources of financing. It has always been, and continues to be, the clear-cut, official, and absolute policy of the associations of real-estate brokers that "a realtor should never be instrumental in introducing into a neighborhood a character of property or occupancy, members of any race or nationality, or any individuals whose presence will clearly be detrimental to property values in that neighborhood."[13] Many studies have attempted to resolve this problem. In the long run, property values and rents exhibit little if any change in the transition from white to Negro occupancy.[14] Sale prices may fall temporarily under panic selling, a phenomenon called the

---

[11]Chester Rapkin and William G. Grigsby: The Demand for Housing in Racially Mixed Areas . . . : Special Research Report to the Commission on Race and Housing . . . (Berkeley, 1960), pp. 27-30.

[12]Nathan Glazer and Davis McEntire, edits.: Studies in Housing & Minority Groups: Special Research Report to the Commission on Race and Housing (Berkeley, 1960), pp. 5-11.

[13]McEntire, *op. cit.* [see footnote 5 above], p. 245.

[14]Luigi Mario Laurenti: Property Values and Race: Studies in 7 Cities: Special Research Report to the Commission on Race and Housing (Berkeley, 1960); [Homer Hoyt:] The Structure and Growth of Residential Neighborhoods in American Cities (Federal Housing Administration, Washington, D. C., 1939); Lloyd Rodwin: The Theory of Residential Growth and Structure, *Appraisal Journ.*, Vol. 18, 1950, pp. 295-317.

"self-fulfilling prophecy"—believing that values will fall, the owner panics and sells, and thus depresses market values.[15]

The real-estate industry opposed with all its resources not only all laws but any device, such as cooperative apartments or open-occupancy advertising, to further integration. Real-estate and home-building industries base this policy on the desirability of neighborhood homogeneity and compatibility. Perhaps underlying the collective actions is the fear of the individual real-estate broker that if he introduces a Negro into a white area he will be penalized by withdrawal of business. There is, then, a real business risk to the individual broker in a policy of integration, if none to the industry as a whole. Segregation is maintained by refusal of real-estate brokers even to show, let alone sell, houses to Negroes in white areas. Countless devices are used: quoting excessive prices, saying the house is already sold, demanding unfair down payments, removing "For sale" signs, not keeping appointments, and so on. Even if the Negro finds someone willing to sell him a house in a white area, financing may remain a barrier. Although his income may be sufficient, the bank or savings institution often refuses to provide financing from a fear of Negro income instability, and of retaliatory withdrawal of deposits by whites. If financing is offered, the terms may be prohibitive. Similar circumstances may also result when a white attempts to buy a house—for *his* residence—in a heavily minority area.

Through the years many legal procedures have been used to maintain segregation. Early in the century races were zoned to certain areas, but these laws were abolished by the courts in 1917. The restrictive covenant, in which the transfer of property contained a promise not to sell to minorities, became the vehicle and stood as legal until 1948, since when more subtle and extralegal restrictions have been used.

Until 1949 the federal government was a strong supporter of residential segregation, since the Federal Housing Administration required racial homogeneity in housing it financed or insured. As late as 1963, when the President by Executive order forbade discrimination in FHA-financed housing, the old philosophy still prevailed in most areas. Finally, many states, and not just those in the South, still encourage separation. Even in the few states with laws against discrimination in housing, the combined forces for maintaining segregation have proved by far the stronger.

THE PROCESS OF GHETTO EXPANSION

The Negro community in the North has grown so rapidly in the last forty years, almost doubling in every decade, that even the subdivision of houses cannot accommodate the newcomers. How does the ghetto expand? Along its edge the white area is also fairly old and perhaps deteriorating. Many whites would be considering a move to the suburbs even if the ghetto were not there, and fears of deterioration of schools and services, and the feeling that all the other whites will move out, reinforce their inclination to move. Individual owners, especially in blocks adjoining the ghetto, may become anxious to sell.

[15]Eleanor P. Wolf: The Invasion-Succession Sequence as a Self-Fulfilling Prophecy, *Journ. of Social Issues,* Vol. 13, 1957, pp. 7-20.

Pressure of Negro buyers and fleeing white residents, who see the solid ghetto a block or two away, combine to scare off potential white purchasers; the owner's resistance gradually weakens; and the transfer is made.

The role of proximity is crucial. On adjacent blocks the only buyers will be Negroes, but five or six blocks away white buyers will still be the rule. In a typical ghetto fringe in Philadelphia the proportion of white buyers climbed from less than 4 percent adjacent to the ghetto itself to a 100 percent five to seven blocks away.[16] Figure 2 illustrates the great concentration of initial entry of new street fronts in a band of two or three blocks around a ghetto. The "break" zone contains 5 percent or fewer Negroes, but 60 percent of the

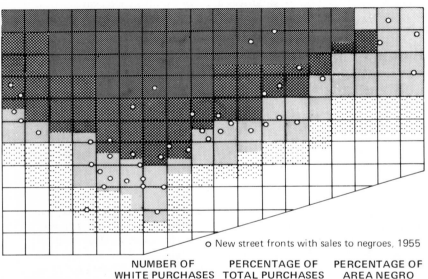

o New street fronts with sales to negroes, 1955

		NUMBER OF WHITE PURCHASES	PERCENTAGE OF TOTAL PURCHASES	PERCENTAGE OF AREA NEGRO
▨	Area 1	8	3.9	32
▨	Area 2	26	4.3	16
▨	Area 3	65	40.6	5
▨	Area 4	72	98.7	1
▨	Area 5	112	100.0	< 1

FIGURE 2   Distribution of Negro purchases on the edge of the ghetto, showing initial entry of street fronts, 1955. Adapted from diagram in Rapkin and Grigsby, The Demand for Housing in Racially Mixed Areas (see text footnote 11 for reference), p. 76.

purchases are by Negroes. Typically, a white on the edge does not mind one or two Negroes on the block or across the street, but if a Negro moves next door the white is likely to move out. He is replaced by a Negro, and the evacuation-replacement process continues until the block has been solidly transferred from

[16]Rapkin and Grigsby, op. cit. [see footnote 11 above], pp. 56-58.

white to Negro residence. Expansion of the ghetto is thus a block-by-block total transition.

In this process the real-estate agent is also operative. If the demand for Negro housing can be met in the area adjacent to the ghetto, pressure to move elsewhere in the city will diminish. The real-estate industry thus strongly supports the gradual transition along the periphery. After the initial break the real-estate broker encourages whites to sell. The transition is often orderly, but the unscrupulous dealer sometimes encourages panic selling at deflated prices, purchasing the properties himself and reselling them to Negroes for windfall profits. The probability of finding a white seller is high in the blocks adjacent to the ghetto but falls off rapidly at greater distances, as whites try to maintain familiar neighborhood patterns and conceive this to be possible if the Negro proportion can be kept small. The process of transition is destructive to both groups, separately and together. Whites are in a sense "forced" to sell, move, and see their neighborhoods disband, and Negroes are forced to remain isolated; and total transition reinforces prejudice and hinders healthy contact.

Spread of the Negro ghetto can be described as a *spatial diffusion* process, in which Negro migrants gradually penetrate the surrounding white area. From some origin, a block-by-block substitution or diffusion of a new condition—that is, Negro for white occupancy—takes place. The Negro is the active agent; he can move easily within the ghetto and can, though with difficulty, "pioneer" outside it. The white is passive, an agent of resistance or inertia. Resistance against escape of Negroes from the ghetto takes two forms: rebuff of attempts to buy, and diminishing willingness to sell with increasing distance from areas or blocks that already have Negroes. On the average the Negro will have to try more than once to consummate a sale, or, conversely, the owner will have to be approached by more than one buyer. Once the block is broken, however, resistance falls markedly, and transition begins. Although a complete model would take into account that a few whites continue to purchase in transition areas, the rate is insufficient, the net flow clear-cut, and the transition inevitable.

The proposed diffusion model is of the probabilistic simulation type.[17] It is probabilistic rather than deterministic for several reasons. We do not have sufficient definite information concerning the motivations for specific house-to-house moves of particular persons, but only general ideas concerning the likelihood of movement and how far. We are not dealing with a large aggregate of migrants, but with only a few individuals in a short period of time in a small area. If we had a thousand migrants, we could safely predict how many would move how far, but at the micro-level a probabilistic approach is required to evaluate individual decisions in the face of a complex of possible choices. Rather than determine that a specific migrant moves from one particular house to another, we find the probability of a typical migrant's move from a block to any and all other blocks, and we use random numbers to decide which destination, among the many possible, he chooses. We thus obtain a spatial pattern of moves, which spreads settlement into new blocks and intensifies it in old blocks.

[17]Herbert A. Meyer, edit.: Symposium on Monte Carlo Methods, Held at the University of Florida . . . . March 16-17, 1954 (New York and London, 1956); Everett M. Rogers: Diffusion of Innovations (New York, 1962); Warren C. Scoville: Minority Migrations and the Diffusion of Technology, *Journ. of Econ. History*, Vol. 11, 1951, pp. 347-360.

The model is simulated rather than "real" because it does not purport to predict individual behavior of actual people, but to simulate or pretend moves for typical households. Simulation is a valuable technique in science and technology, in which a model is constructed to depict artificially certain *major* features of some real process.

The simulation of diffusion model is important in biology, in rural and general sociology, and in communications, and has been used in geography.[18] It is an ideal vehicle for the characteristics of ghetto expansion—a process of growth in time, concerning behavior of small groups in small areas in small units of time, in which a powerful element of uncertainty remains, even though the general parameters of the model tend to channel the results. This randomness is evident in the real situation, since we observe that the ghetto, like a rumor or an innovation, does not progress evenly and smoothly in all directions but exhibits an uneven edge and moves at different rates in different directions, here advancing from block to block, there jumping over an obstacle.

We do not expect the simulated patterns to match precisely the actual patterns. We do want the model to generate a pattern of expansion that corresponds in its characteristics to the real pattern, and we can satisfy ourselves of the correspondence by visual and statistical tests. The purpose and hope are to discover and illustrate the nature of the ghetto expansion process, in full knowledge that the detail of the ultimate step is omitted—how the actual individual decides between his specific alternatives. The omission is justified, because we know that the combined effect of many individual decisions can often be described by a random process. The real test here is whether the spread, over a period of time, has the right extent, intensity, solidity or lack of it, and so on.

THE MODEL

A model of ghetto expansion must incorporate several elements: natural increase of the Negro population; Negro immigration into the ghetto; the nature of the resistance to Negro out-migration and its relation to distance; land values and housing characteristics; and the population size limits of destination blocks.

Beginning with the residential pattern at a particular time (in the Seattle example, 1940), migration and the spread of Negro settlement are simulated for ten two-year periods through 1960. The steps are as follows.

A. Taking into account natural increase for each period of the Negro population resident in the Seattle ghetto, at the observed rate of 5 percent every two years.

B. Assigning immigrants who enter the study area from outside at the observed mean rate of 10 percent every two years of the Negro population at the beginning of a period. These are assigned by random numbers, the probability that an area will be chosen being proportional to its present Negro population. Presumably, immigrants entering the

[18] Torsten Hägerstrand: On Monte Carlo Simulation of Diffusion, *in* Quantitative Geography (edited by William L. Garrison; in press); Forrest R. Pitts; Problems in Computer Simulation of Diffusion, *Papers and Proc. Regional Science Assn.,* Vol. 11, 1963, pp. 111-119.

area will find it easier to live, at least temporarily, and will find opportunities in houses or apartments or with friends, in approximate reflection of the number of Negro units available. After initial residence in the ghetto, the model allows these immigrants to participate in further migration.

C. Assigning internal migrants, at the rate of 20 percent of the Negro households (including natural increase and immigration) of each block every two years, in the following manner:

1. Each would-be migrant behaves according to a migration probability field (Fig. 3) superimposed over his block. This migration probability field can be shifted about so that each would-be migrant can in turn be regarded as located at the position indicated by X. The numbers in the blocks show where the migrant is to move, depending on which number is selected for him in the manner described below. Blocks adjoining position X have three numbers (for example, 48-50); more distant blocks have two numbers (for example, 54-55); and the most distant have one number (for example, 98). Since 100 numbers are used, the total number of these numbers used in any one block may be regarded as the probability, expressed as a percentage, that any one migrant will move there. Thus a movable probability field, or information field, such as this states the probabilities of a migrant for moving any distance in any direction from his original block. Probability fields are often derived, as this one was, from empirical observations of migration distances. That is, if we look at a large number of moves, their lengths follow a simple frequency distribution, in which the probability of moving declines as distance from the home block increases. Such probabilities reflect the obvious fact of decreasing likelihood of knowing about opportunities at greater and greater distances from home. Thus the probability is higher that a prospective migrant will move

1	2	3	4	5	6	7	8	9
10	11	12	13	14–15	16	17	18	19
20	21	22	23	24–25	26	27	28	29
30	31	32	33–34	35–37	38–39	40	41	42
43	44–45	46–47	48–50	X	51–53	54–55	56–57	58
59	60	61	62–63	64–66	67–68	69	70	71
72	73	74	75	76–77	78	79	80	81
82	83	84	85	86–87	88	89	90	91
92	93	94	95	96	97	98	99	00

FIGURE 3   The migration probability field.

to adjacent blocks than to more distant ones. The probability field provides a mechanism for incorporating this empirical knowledge in a model.

2. Randomly selected numbers, as many as there are migrants, are used to choose specific destinations, according to these probabilities, as will be illustrated below. The probability field as such makes it as likely for a Negro family to move into a white area as to move within the ghetto. A method is needed to take into account the differential resistance of Negro areas, and of different kinds or qualities of white areas, to Negro migration. Modification of the probability field is accomplished by the following procedures.

a) If a random number indicates a block that already contains Negroes, the move is made immediately (no resistance).

b) If a random number indicates a block with no Negroes, the fact of contact is registered, but no move is made.

c) If, however, additional numbers indicate the same block contacted in *b,* in the same or the next two-year period, and from whatever location, then the move is made. This provides a means for the gradual penetration of new areas, after some persistence by Negroes and resistance by whites. Under such a rule, the majority of Negro contacts into white areas will not be followed by other contacts soon enough, and no migration takes place. In the actual study area chosen, it was found that resistance to Negro entry was great to the west, requiring that a move be allowed there only after three contacts, if the simulated rate of expansion was to match the observed rate. This is an area of apartments and high-value houses. To the north and east, during this period, resistance varied. At times initial contacts ended in successful moves and transition was rapid; at other times a second contact was required. These facts were incorporated into the operation of this phase of the model.

D. There is a limit (based on zoning and lot size) to the number of families that may live on a block. Thus when the population, after natural increase and immigration, temporarily exceeds this limit, the surplus must be moved according to the procedures above. Obviously, in the internal-migration phase no moves are allowed to blocks that are already filled. The entire process is repeated for the next and subsequent time periods.

HYPOTHETICAL EXAMPLE OF THE MODEL

Immigration (A and B)

Let us assume at the start that the total Negro population—that is, the number of families—including natural increase is one hundred, distributed spatially as in Figure 4. Here the numbers indicate the number of families in each block. Ten immigrant families (10 percent) enter from outside. The probability of their moving to any of the blocks is proportional to the block's population and here, then, is the same in percentage as the population is in number. In order that we may use random numbers to obtain a location for each

		2	2		
1	5	10	10	5	1
5	10	15	15	15	4

FIGURE 4   Negro residents at start of period.

		1–2	3–4		
5	6–10	11–„20	21–30	31–35	36
37–41	42–„51	52–„66	67–,81	82–„96	97–00

FIGURE 5   Distribution of immigrants. Tally marks indicate entry into appropriate blocks.

immigrant family, the probabilities are first accumulated as whole integers, from 1 to 100, as illustrated in Figure 5. That is, each original family is assigned a number. Thus the third block from the left in the second row has two of the one hundred families, identified by the numbers 1 and 2, and therefore has a 2 percent chance of being chosen as a destination by an immigrant family. The range of integral numbers 1-2 corresponds to these chances. The bottom left-hand block has a 5 percent probability, as the five numbers 37-41 for the families now living there indicate. If, then, the random number 1 or 2, representing an immigrant family, comes up, that family will move to the third block in the second row. For the ten immigrant families we need ten random numbers. Assume that from a table of random numbers we obtain, for example, the numbers 91, 62, 17, 08, 82, 51, 47, 77, 11, and 56. The first number, 91, falls in the range of probabilities for the next to the last block in the bottom row. We place an immigrant family in that block. The second number, 62, places an immigrant family in the third block from the left in the bottom row. This process is continued until all ten random numbers are used. The final distribution of immigrant families is shown by the small tally marks in various blocks in Figure 5. The population of blocks after this immigration is shown in Figure 6. Here the large numerals indicate the number of families now in the blocks. It should be made clear that the migrants could not have been assigned exactly proportional to population, because there are not enough whole migrants to go around. The first two blocks, for example, would each have

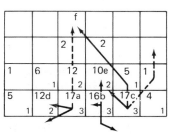

FIGURE 6   Movement of migrants from three sample blocks. Large figures, resident Negroes; italic figures, number of migrants; broken lines, contact only; solid lines, actual moves.

required two-tenths of a migrant. In the probabilistic model, however, this difficulty does not exist.

Local Migration (C)

Twenty percent of the Negro families of each block, rounded off to the nearer whole number, are now taken as potential migrants. The rounding off yields a total of nineteen families who will try to migrate from the blocks, as indicated by the italic numerals in Figure 6. To illustrate, let us consider migration from the three blocks identified by *a, b,* and *c* in the bottom row. Random numbers are now needed to match against the migration probability field, Figure 3. Let the random numbers now obtained from the table of random numbers be 49, 75, 14, 50, 36, 68, 26, 12, and 33. The first migrant from *a* is represented by the random number 49. This provides a location one block to the left of the migrant's origin, *X,* to *d.* The second migrant's random number, 75, provides a location two blocks down and one to the left, which is beyond the study area. We interpret this as moot, as though he were replaced by another migrant from outside the area. The third migrant's number, 14, provides a location three blocks up, location *f.* Since this block has no Negroes, this is only a contact, and no move is made at the time. This is indicated by a dashed line. Now let us proceed to migration from block *b.* The first migrant's number, 50, provides a location one block to the left, in block *a,* and the move is made. The second migrant's number, 36, provides a location one block up, in block *e,* and the move is made. The third migrant's number, 68, provides a location beyond the area. From block *c* the first migrant's number, 12, provides a location three blocks up and two to the left. This location coincides with the contact made earlier by the third migrant from block *a,* and the move is made. The third migrant's number, 33, provides a location one block up and one to the left, or block *e* again, and the move is made. The net result of all this migration is the opening of one new block to settlement, the reinforcement of three blocks, and two lost contacts.

NORTHWARD EXPANSION OF THE GHETTO IN SEATTLE

The ghetto in Seattle, with only 25,000 residents, is of course smaller than those in the large metropolises, and it may seem less of a threat to the surrounding area.[19] Nevertheless, the nature of expansion does not differ from one ghetto to another, though the size of the ghetto and the rate of expansion may vary.

The expansion of the Seattle ghetto is shown on Figure 7, on which the study area is indicated. From 1940 to 1960 the Negro population in the study

[19]Calvin F. Schmid and Wayne M. McVey, Jr.: Growth and Distribution of Minority Races in Seattle, Washington ([Seattle] 1965); Walter B. Watson and E. A. T. Barth: Summary of Recent Research Concerning Minority Housing in Seattle (Institute for Social Research, Department of Sociology, University of Washington, 1962); John C. Fei: Rent Differentiation Related to Segregated Housing Markets for Racial Groups with Special Reference to Seattle (unpublished Master's thesis, University of Washington, 1949).

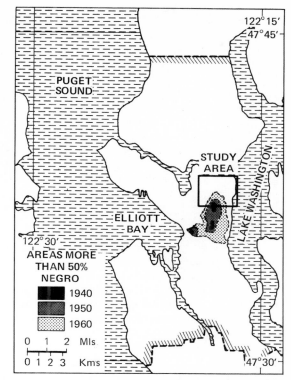

FIGURE 7   The ghetto area of Seattle. Source: Census data for the relevant
years.

area more than quadrupled, from 347 families to 1520. Except for a few blocks
just north and east of the 1940 Negro area, expansion was into middle-class
single-family houses. To the west, where expansion was least, apartments offer
increasing resistance, and to the northwest and along the lake to the east houses
reach rather expensive levels. Expansion was easiest along the major south-north
and southwest-northeast arterial streets, and northward along a topographic
trough where houses and land were the least valuable. The solidity of the ghetto
core, the relatively shallow zone of initial penetration, and the consequent
extension of the ghetto proper are shown on Figures 8 to 10. As the ghetto
became larger and thus more threatening, transition became more nearly solid.

The model was applied to the study area for ten two-year periods, beginning with the actual conditions of 1940 and simulating migration for twenty
years. For each two-year period the natural increase of the Negro population was
added to the resident population at the beginning of the period. Immigrants
were assigned as in the model. Migrants were assigned according to the
probability field (Fig. 3) and the rules of resistance. One example of the simulation of migration is shown on Figure 11, for 1948-1950. Typically, out of 147
potential migrants, 131 were successful and 16 made contacts, but only 8 of the

FIGURE 8 Blocks predominantly Negro in the northern part of Seattle's ghetto. Source: Census of Housing, 1960. (Block statistics for Seattle.)

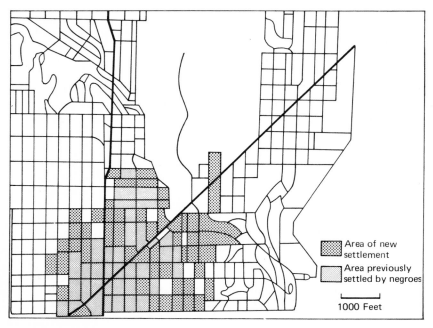

FIGURE 9 Expansion of Negro population in Seattle's ghetto, 1940-1950. Source: Census of Housing, 1950. (Block statistics for Seattle.)

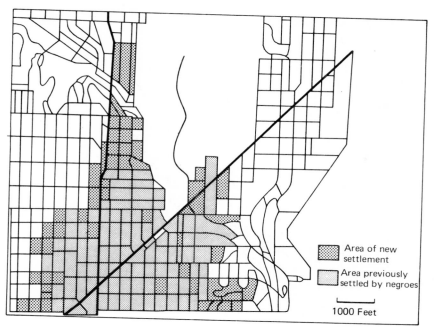

FIGURE 10   Expansion of Negro population in Seattle's ghetto, 1950-1960.
Source: Census of Housing, 1950. (Block statistics for Seattle.)

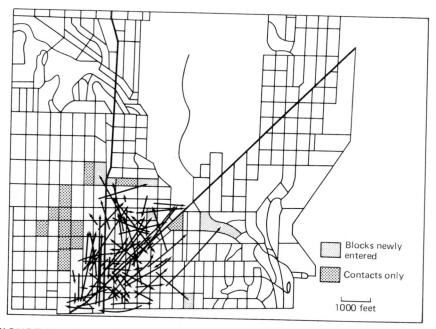

FIGURE 11   Simulation of migration in Seattle's Negro ghetto, 1948-1960.

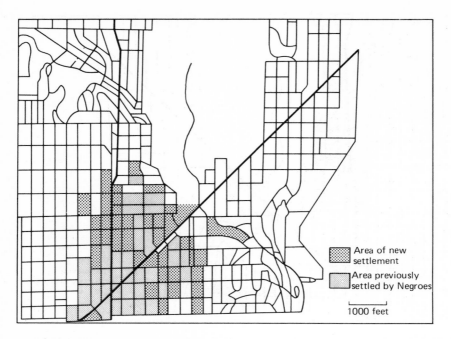

FIGURE 12   Simulated   expansion   of   Seattle's   Negro   ghetto,   1940-1950.

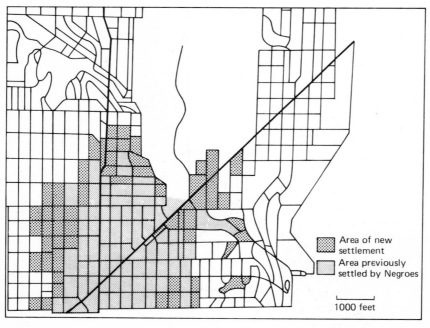

FIGURE 13   Simulated   expansion   of   Seattle's   Negro   ghetto,   1950-1960.

movers pioneered successfully into new blocks. The results of the simulation are illustrated by Figures 12 and 13, which summarize the changes within two larger periods, 1940-1950 and 1950-1960.

EVALUATION OF THE RESULTS

A comparison of Figures 9 and 12, and 10 and 13, showing actual and simulated expansion of the Seattle ghetto for 1940-1950 and 1950-1960 respectively, indicates a generally close correspondence in the patterns. The actual pattern extended more to the north and the simulated pattern more to the northwest. A field check revealed that neither the quality nor the value of homes was sufficiently taken into account in the model. Topography, too, was apparently crucial. By 1960 the Negroes were rapidly filling in the lower-lying, nonview land. The ridge and view properties remained more highly resistant. The model did not recognize the rapid movement northward along the topographic trough.

According to the most stringent test of absolute block-by-block conformity the model was not too successful. Less than two-thirds of the simulated new blocks coincided with actual new blocks. However, the model was not intended to account for the exact pattern. Sufficient information does not exist. The proper test was whether the simulated pattern of spread had the right extent (area), intensity (number of Negro families in blocks), and solidity (allowing for white and Negro enclaves), and in these respects the performance was better. The number of blocks entered was close, 140 for the simulation to 151 for the actual; the size distribution of Negro population was close; and similar numbers of whites remained within the ghetto (with the model tending toward too great exclusion of whites). This similarity, rather than conformance, indicated that both the actual and the simulated patterns *could have occurred* according to the operation of the model. This is the crucial test of theory.

A predictive simulation, as a pattern that could occur, using as the base the actual 1960 situation, was done for the periods 1960-1962 and 1962-1964 (Fig. 14). A limited field check showed that this pattern is approximately correct, except, again, with too much movement to the northwest and not enough to the north. No prediction from 1964 has been attempted, because of risk of misinterpretation by the residents of the area.

ALTERNATIVES TO THE GHETTO

The model attempted merely to identify the process of ghetto expansion and thus helps only indirectly in the evaluation of measures to control the ghetto. We know that such a diffusion process is common in nature—the growth from an origin or origins of something new or different within a parent body. Reduction of this phenomenon would seem to require a great weakening of the distinction between groups, here Negroes and whites, either naturally through new conceptions of each other or artificially by legal means.

In ghetto expansion the process is reduced to replacement of passive white "deserters" by active Negro migrants. Is there an alternative that would permit

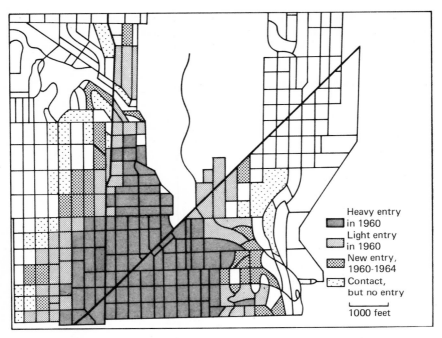

Heavy entry
in 1960

Light entry
in 1960

New entry,
1960-1964

Contact,
but no entry

1000 feet

FIGURE 14   Simulated   expansion   of   Seattle's   Negro   ghetto,   1960-1964.

the integration of minorities in the overall housing market and prevent the
further spread and consolidation of ghettos? Is it possible to achieve stable
interracial areas, in which white purchasers, even after Negro entry, are sufficiently numerous to maintain a balance acceptable to both? Three factors have
been found crucial: proximity to a ghetto; proportions of white and nonwhite;
and preparation of the neighborhood for acceptance of Negro entry.[20]
Proximity to a ghetto almost forbids a stable interracial situation. Fear of
inundation either panics or steels white residents. Only wealthy areas can
maintain any interracial character in such a location, since few, if any, Negroes
can afford to enter. Negroes entering areas remote from the ghetto are more
easily accepted (after initial difficulties), because the great body of Negroes does
not "threaten" neighborhood structures.

The proportion of Negroes in an area is critical for continued white
purchasing. Whites are willing to accept 5 percent to 25 percent (with a mean of
10 percent) Negro occupancy for a long time before beginning abandonment—
depending on such factors as the characteristics of the Negroes moving in, the
proximity of the ghetto, and the open-mindedness of the resident white population. On the other hand, although the Negro is accustomed to minority status,
he usually prefers a larger proportion of his own group nearby than the critical

[20]Eunice Grier and George Grier: Privately Developed Interracial Housing: An
Analysis of Experience: Special Research Report to the Commission on Race and Housing
(Berkeley, 1960), pp. 29-30.

10 percent. Thus a fundamental dilemma arises, and there are in fact few inter-racial neighborhoods. For cities with low Negro ratios, say less than 10 percent, the long-run possibilities are encouraging, especially with the rise of Negro education and income, increased enforcement of nondiscrimination laws, and the more liberal views of youth today. For urban areas with high Negro ratios, such as Philadelphia, with 20 percent (40 percent in the city proper), it is difficult to imagine an alternative to the ghetto. The same conclusion holds for southern cities. No spatial arrangement, given present levels of prejudice, will permit so large a proportion of Negroes to be spread throughout the city without serious white reaction.

Private interracial projects have begun integration and have been successful and stable, if few in number.[21] From these experiments it has been learned that white buyers in such developments are not usually liberal but are a normal cross section. Also, the spatial arrangement that permits the largest stable proportion of nonwhites has been found to be a cluster pattern—small, compact colonies of a few houses—rather than dispersed isolates.[22] This makes possible easy contact within the minority group, but also good opportunity for interaction with the white group, while minimizing the frequency of direct neighbors, which few whites are as yet able to accept.

Integrated residential living will become more acceptable as Negroes achieve equality in education and employment, but housing integration will probably lag years or decades behind. At most we may expect an arrest of the extension of existing ghettos, their internal upgrading, and prevention of new ones. Experience certainly indicates a long wait for goodwill to achieve even internal improvement; hence a real reduction in ghettoization implies a governmental, not a voluntary, regulation of the urban land and housing market—that is, enforced open-housing ordinances. Everything short of that has already been tried.

The suggested model of diffusion-expansion still describes the dominant ghettoization pattern. In the future we may be able to recognize an alternative "colonization" model, in which small clusters of Negroes or other minorities break out of the ghetto and spread throughout the urban area under the fostering or protection of government.

[21]*Ibid.*, p. 8.

[22]Reuel S. Amdur: An Exploratory Study of 19 Negro Families in the Seattle Area Who Were First Negro Residents in White Neighborhoods, Of Their White Neighbors and of the Integration Process, Together with a Proposed Program to Promote Integration in Seattle (unpublished Master's thesis in social work, University of Washington, 1962); Arnold M. Rose and others: Neighborhood Reactions to Isolated Negro Residents: An Alternative to Invasion and Succession, *Amer. Sociol. Rev.,* Vol. 18, 1953, pp. 497-507; L. K. North-wood and E. A. T. Barth: Neighborhoods in Transition: The New American Pioneers and Their Neighbors (University of Washington, School of Social Work, Seattle), pp. 27-28.

PART THREE

Simulation in Administrative Science

9

Business System Simulations

WILLIAM T. NEWELL
ROBERT C. MEIER

Simulation has evolved as one of the most interesting and potentially powerful tools available for analyzing administrative problems.[1] The structure of these administrative problems is similar in business and nonbusiness organizations. This chapter focuses on simulations in business systems; the following two chapters focus on the use of simulation in marketing and public systems. While much of the work in administrative science simulation may seem "applied" in comparison with the social science simulations discussed so far, the purpose of the model-building effort is generally the same—that is, to delineate and understand the behavior of systems, whether social or administrative.

DEVELOPMENT OF SIMULATION IN BUSINESS

Simulation emerged as a major tool of analysis in business as digital computers started to become widely available in the mid-1950s, and since that time there has been a rapid growth of interest in digital simulation methods. Prior to the introduction of the digital computer, digital simulation techniques were little used in business analyses. The techniques were not well known, and the volume of computations usually required by simulation models of practical interest made manual computations too cumbersome and expensive. Occasionally, analogue devices had been used in simulation studies, but the difficulty

[1]The material in this chapter has been adapted in part from Meier, Newell, and Pazer (1969).

of finding suitable analogues precluded any widespread use of analogue methods.[2]

Because of programming difficulties, slow speed, and limited memory capacity of the first digital computers, as well as lack of familiarity with the technique, early simulation studies generally involved uncomplicated phenomena such as simple inventory systems, simple waiting line problems, and small economic models. As digital computation techniques have matured with the development of computers with greatly increased capacity and sophisticated software packages, simulation models of more complex processes and larger systems have become practical. In addition, some programming problems have been alleviated by the development of computer programs and language designed particularly for digital simulation of the types of systems likely to be encountered in business and economic studies. Specialized languages and programs such as GPSS (General Purpose Systems Simulation), SIMSCRIPT, and DYNAMO often substantially reduce the time and cost of writing simulation programs. Improvements in both hardware and software in recent years now make it feasible to simulate systems of greater complexity than could have been dealt with in the past. Consequently, the range of problems for which digital simulation is a practical tool of analysis has been increased manyfold.

SIMULATIONS OF BUSINESS PROBLEMS

The many types of simulation models that have been developed may be classified by the type of problem, the scope of the system under study, or the methodology employed. Most of the models developed to date deal with specific business problem areas such as inventories, scheduling, and distribution systems. Other models include the scope (if not the detail) of an entire enterprise, some focus upon the interactions of feedback processes, and some simulate individual decision processes. Operational gaming models fall into a special category because they are primarily used for education and training rather than for experimentation and investigation.

Inventory Systems

Inventory control was one of the first business areas to be examined mathematically, and it was therefore one of the first areas in which simulation was utilized. Not only has the large body of theoretical work assisted in clearly conceptualizing inventory problems, but inventory problems inherently involve relatively well-defined flows over time which lend themselves to simulation. The

[2]Irving Fisher, for instance, used a hydraulic model to simulate economic phenomena in the late nineteenth century. Later, electrical analogues were suggested for use in economic analysis. See Morehouse, Strotz, and Horwitz (1950).

justification for using simulation is readily apparent upon examination of the theoretical literature, since analysis of one-item inventory problems can be a challenging mathematical task even when simplifying assumptions are made regarding the characteristics of customer demand, characteristics of lead time, and cost behavior.

An early instance of the use of simulation to analyze an inventory system was reported by Patrick J. Robinson of Imperial Oil Limited in 1957 (Report of the System Simulation Symposium, 1957). One application of simulation discussed by Robinson was to determine the feasibility of using central warehouses in conjunction with field warehouses. The problem was to investigate working characteristics of a proposed central warehouse to ensure that no bottlenecks would occur in its operation. A simulation was performed on a digital computer using a program written especially for the project. The program was capable of accepting data regarding initial inventory levels of items stocked at the central warehouse and simulating operation of the system when recapitulations of actual daily orders from field warehouses were given as input data. The computer printed out periodic reports of stock levels and also provided information on shortages and waiting lines for facilities.

Robinson reported the following conclusions from the simulation studies:

... It is clear that in many cases where complex relationships, both of predictable and random natures, occur, it is easier to set up and run through a simulated situation than it is to develop and use a mathematical model representing the entire process under study. In many cases an activity can be affected by numerous random influences. The probabilities involved for each type of influence can be separately examined. However, the calculation of the probability of the combined sequence of activities spilling over into each other and interacting, in what is sometimes referred to as a cascading effect, leads into very deep mathematical waters in the field of Stochastic Processes.

To avoid an impossible or unprofitable attempt to solve a complex operation using equations to seek so-called optimal answers, we turn to a simulation of a system such that we may repeatedly experiment and obtain statistically reliable empirical results. . . .

Where no analytic solution is available, the search for an ever-improving answer through the sequential solution of alternate trials, until finally running and rerunning additional cases doesn't produce any material improvement, brings us near to what we can, with confidence rely on as being something approximating an optimal solution. . . .

Summing up, you may agree that through simulating reality on a modest scale, we can do some research that might not be feasible otherwise; while at the same time we may provide management with a dramatic demonstration piece to help appraise situations and make profitable decisions (Report of the System Simulation Symposium, 1957).

To illustrate how an inventory problem common to logistics, distribution, production, and materials management systems can be simulated, let us look at some of the characteristics of an inventory simulation developed by the authors.[3] A single item of inventory is to be replenished under an order point-order quantity system.

We will assume that demand and lead time vary according to certain probability distributions and that only one demand for the item occurs each day, although the number of units demanded varies randomly according to the distribution in table 9-1. Lead time varies according to the distribution in table 9-2. We will also assume that there is a certain probability, as shown in table 9-3, of losing orders that are backordered due to a shortage of inventory. The table is interpreted to mean that there is a 30 percent probability of losing an order by cancellation on the day it is received if inventory is not available, a 40 percent chance of losing an order that is held over to the next day, and so on up to the

TABLE 9-1   Probability Distribution of Daily Demand

Number of Units Demanded	Probability of Occurrence
5	0.01
6	0.03
7	0.06
8	0.11
9	0.19
10	0.31
11	0.17
12	0.07
13	0.03
14	0.02
	1.00

Average demand = 9.72 units/day

TABLE 9-2   Lead Time Distribution

Lead Time	Probability of Occurrence
2	0.15
3	0.20
4	0.30
5	0.20
6	0.15
	1.00

Average lead time = 4.0 days

[3] A complete discussion of the computer program and its use may be found in Meier, Newell, and Pazer (1969).

TABLE 9-3   Probability of Losing Backorders

Number of Periods Backordered	Probability of Loss
0	0.30
1	0.40
2	0.55
3	0.75
4	1.00

fifth day when there is a 100 percent chance that an order not filled by that day will be lost.

Mathematical analysis of this complex inventory problem to determine the optimum order point and order quantity is not an easy task because of interaction between the demand, lead time, and backorder loss distribution.

Figure 9-1 shows output from a simulation run of the system for one hundred periods, or days. Data used in the run are shown in the first part of the printout. In the computer run the number of periods was set at one hundred, stock level was reviewed every period to see whether the order point was reached, price was set at five dollars per unit, inventory was given an initial value of 100 units, an order point of 40 units and an order quantity of 316 units were used, cost of placing and receiving a replenishment order was set at twenty dollars, and an interest rate or carrying cost of 0.20/250, or 0.0008 per time period, was used. Let us also assume that it costs ten dollars to enter a backorder when demand cannot be filled and that the cost of losing orders because of unavailability of stock is the five dollar-unit profit margin.

The transactions section of figure 9-1 traces the behavior of the system through one hundred periods and lists each transaction as it occurred. In the run summary section is shown the results of the simulation run. These results are specific for this particular simulation run because the cost of operating this probabilistic system depends upon the way in which the probabilistic elements happen to interact during the run. Consequently, it is possible to obtain different costs of operation on different runs of the model.

Simulation of the stochastic problem makes it possible to observe the interaction of three probabilistic elements in the system and to see effects of this interaction in terms of total cost, service level, and length of time to fill orders. Additional runs of longer duration would be required to establish confidence in the results of the simulation. In actual practice, such a simulation could be used to search for optimum values of the order point and order quantity by experimenting with pairs of values other than the one pair used in the demonstration run.

Note that the simulation model does not locate the optimum value of the order quantity and order point. Most simulation models have no capacity to optimize; they simply represent what will happen if a system is set up to operate in a certain way with whatever values are chosen for the decision variables. As a consequence, when optimum values are unknown and the objective of a simu-

INVENTORY SIMULATION
DEMONSTRATION

RUN INPUT DATA

```
X  =   107.00
Y  =    97.00
INITIAL RANDOM NUMBER =  0.338383
NUMBER OF TIME PERIODS = 100
INTERVAL BETWEEN REVIEWS =   1
PRICE OF ITEM =        5.00   PER UNIT
BEGINNING INVENTORY =     100   UNITS
ORDER POINT =      40   UNITS
ORDER QUANTITY =     316   UNITS
COST OF PLACING REPLENISHMENT ORDER =     10.00
COST OF RECEIVING REPLENISHMENT ORDER =     10.00
COST OF ENTERING BACKORDER =     10.00
COST OF LOST DEMAND =       5.00   PER UNIT
INTEREST COST PER TIME PERIOD = 0.000800
```

LEAD TIME	CUMULATIVE FREQUENCY
2	0.1500
3	0.3500
4	0.6500
5	0.8500
6	1.0000

DEMAND	CUMULATIVE FREQUENCY
5	0.0100
6	0.0400
7	0.1000
8	0.2100
9	0.4000
10	0.7100
11	0.8800
12	0.9500
13	0.9800
14	1.0000

PERIODS BACKORDERED	PROBABILITY OF LOSS
0	0.3000
1	0.4000
2	0.5500
3	0.7500
4	1.0000

TRANSACTIONS

PERIOD	ON HAND	ON ORDER	ORDERED	DUE IN	RECEIVED	DEMANDED	ON BACKORDER	BACKORDERED	LOST	COST
1	91	0				9	0			
2	82	0				9	0			
3	74	0				8	0			
4	64	0				10	0			
5	51	0				13	0			
6	41	0				10	0			
7	36	0				5	0			
7	36	316	316	9						10.00
8	26	316				10	0			
9	342	0			316					10.00
9	332	0				10	0			
10	319	0				13	0			
11	310	0				9	0			
12	301	0				9	0			
13	291	0				10	0			
14	282	0				9	0			
15	271	0				11	0			
16	261	0				10	0			
17	254	0				7	0			
18	246	0				8	0			
19	237	0				9	0			
20	224	0				13	0			
21	213	0				11	0			
22	199	0				14	0			
23	189	0				10	0			
24	182	0				7	0			
25	170	0				12	0			
26	160	0				10	0			
27	154	0				6	0			
28	144	0				10	0			
29	138	0				6	0			
30	131	0				7	0			
31	122	0				9	0			
32	112	0				10	0			
33	102	0				10	0			
34	90	0				12	0			
35	80	0				10	0			
36	70	0				10	0			
37	60	0				10	0			

FIGURE 9-1   Computer simulation of probabilistic inventory problem.

TRANSACTIONS

PERIOD	ON HAND	ON ORDER	ORDERED	DUE IN	RECEIVED	DEMANDED	ON BACKORDER	BACKORDERED	LOST	COST
38	49	0				11	0			
39	43	0				6	0			
40	32	0				11	0			
40	32	316	316	44						10.00
41	20	316				12	0			
42	9	316				11	0			
43	0	316				10	1			10.30
44	316	0			316		1	1		10.00
44	315	0					0			
44	306	0				9	0			
45	300	0				6	0			
46	288	0				12	0			
47	277	0				11	0			
48	270	0				7	0			
49	261	0				9	0			
50	254	0				7	0			
51	242	0				12	0			
52	235	0				7	0			
53	228	0				7	0			
54	215	0				13	0			
55	201	0				14	0			
56	192	0				9	0			
57	184	0				8	0			
58	173	0				11	0			
59	163	0				10	0			
60	153	0				10	0			
61	142	0				11	0			
62	134	0				8	0			
63	123	0				11	0			
64	113	0				10	0			
65	106	0				7	0			
66	96	0				10	0			
67	88	0				8	0			
68	78	0				10	0			
69	68	0				10	0			
70	58	0				10	0			
71	51	0				7	0			
72	43	0				8	0			
73	35	0				8	0			
73	35	316	316	79						10.00
74	26	316				9	0			
75	15	316				11	0			
76	4	316				11	0			
77	0	316				10	0		6	30.00
78	0	316				10	10	10		10.00
79	316	0			316		10			10.00
79	306	0					0			
79	295	0				11	0			
80	284	0				11	0			
81	273	0				11	0			
82	264	0				9	0			
83	253	0				11	0			
84	243	0				10	0			
85	234	0				9	0			
86	223	0				11	0			
87	213	0				10	0			
88	201	0				12	0			
89	190	0				11	0			
90	181	0				9	0			
91	174	0				7	0			
92	165	0				9	0			
93	156	0				9	0			
94	147	0				9	0			
95	137	0				10	0			
96	130	0				7	0			
97	118	0				12	0			
98	110	0				8	0			
99	99	0				11	0			
100	94	0				5	0			

RUN SUMMARY

```
TOTAL DEMAND =        960.
AVERAGE INVENTORY =   155.68         CARRYING COST    =     62.27
NUMBER OF ORDERS =    3               ORDERING COST    =     30.00
NUMBER OF RECEIPTS =  3               RECEIVING COST   =     30.00
DEMAND LOST =         6               LOST DEMAND COST =     30.00
SERVICE FACTOR = 0.99
NUMBER OF BACKORDERS =  2             BACKORDERING COST =    20.00
AVERAGE NUMBER OF PERIODS TO FILL BACKORDERS =   1.00
AVERAGE NUMBER OF PERIODS TO FILL ALL ORDERS =   0.01

                                      TOTAL COST       =    172.27
```

FIGURE 9-1   (Continued).

lation study is to locate optimum values of one or more decision variables in a system, a simulation model is essentially used as a vehicle for search. A simulation model does not in itself produce optimum values of decision variables in the sense that a linear programming model, for example, produces optimum values.

Job Shop Scheduling

A job shop consists of a collection of machines, some of them similar and some with different characteristics, each machine operating independently and routing jobs or units of work in sequence through a set of machines that must perform job operations.

In scheduling jobs through a shop, the objective is to ensure that operations are done in proper sequence and that jobs are scheduled on machines without conflict while attempting to meet criteria such as minimizing late deliveries, or minimizing in-process inventories, or maximizing utilization of equipment. Scheduling is complicated by different routing, sequencing, and time requirements for the jobs, by breakdowns and other delays, and by changes made in jobs after processing has begun. Because these characteristics are found in many situations other than machine shops, job shop scheduling has come to be considered a general type of scheduling problem.

Theoretical work on the question of optimal control of job shop operation has been of fairly recent origin, utilizing in many cases newly developed operations research techniques. As a consequence of inherent theoretical and computational difficulties, interest in the use of simulation as an alternative approach to job shop problems developed at an early date, and discussions of job shop simulation are among the earliest applications of simulation found in the literature (Rowe, 1957; Jackson, 1957).

Simulation approaches are oriented toward representing dynamic behavior of the shop as jobs pass through the shop in such a way that various operating policies can be tested under reasonably realistic conditions and without the gross simplifications that are usually necessary to make mathematical formulations tractable. The basic scheme in job shop simulation is to release jobs to the shop over time and follow them as they wait for machines to become available, be processed, and finally be completed by the shop. By gathering appropriate statistics about queue length, waiting times, equipment utilization, labor utilization, and so forth, behavior of the shop may be evaluated with regard to alternative operating policies. In addition, effects of different load conditions can be observed together with effects of machine breakdowns, probabilistic processing times, rush orders, and the like. Because of the complexity of job shop operations and interactions between the many parts of the system, analytical approaches have been successful in dealing only with portions of the total system, and then usually with very restrictive assumptions. In contrast, figure 9-2 shows the characteristics of a job shop simulator and the wealth of realistic detail that can be included.

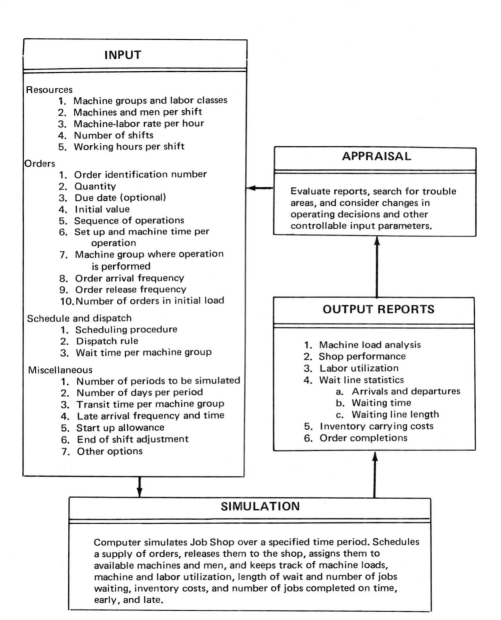

INPUT

Resources
1. Machine groups and labor classes
2. Machines and men per shift
3. Machine-labor rate per hour
4. Number of shifts
5. Working hours per shift

Orders
1. Order identification number
2. Quantity
3. Due date (optional)
4. Initial value
5. Sequence of operations
6. Set up and machine time per operation
7. Machine group where operation is performed
8. Order arrival frequency
9. Order release frequency
10. Number of orders in initial load

Schedule and dispatch
1. Scheduling procedure
2. Dispatch rule
3. Wait time per machine group

Miscellaneous
1. Number of periods to be simulated
2. Number of days per period
3. Transit time per machine group
4. Late arrival frequency and time
5. Start up allowance
6. End of shift adjustment
7. Other options

APPRAISAL

Evaluate reports, search for trouble areas, and consider changes in operating decisions and other controllable input parameters.

OUTPUT REPORTS

1. Machine load analysis
2. Shop performance
3. Labor utilization
4. Wait line statistics
    a. Arrivals and departures
    b. Waiting time
    c. Waiting line length
5. Inventory carrying costs
6. Order completions

SIMULATION

Computer simulates Job Shop over a specified time period. Schedules a supply of orders, releases them to the shop, assigns them to available machines and men, and keeps track of machine loads, machine and labor utilization, length of wait and number of jobs waiting, inventory costs, and number of jobs completed on time, early, and late.

FIGURE 9-2   Characteristics of job shop simulator (by permission from Earl LeGrande, "The Development of a Factory Simulation Using Actual Operating Data," *Management Technology*, Vol. 3, No. 1, May 1963).

Of the many aspects of job shop operations that could be investigated through simulation, the one that has been examined most extensively is the proper rule to use in assigning priorities to jobs as they arrive at machines for processing. These rules, called dispatch rules, determine which job will be placed on the machine when it becomes available. Some dispatch rules that have been suggested are:

1. Choose the job with shortest processing time on the machine.
2. Choose the job that has the smallest amount of slack time, i.e., scheduled completion date minus the sum of remaining processing times.
3. First come, first served.
4. Choose the job with earliest scheduled date for the current operation.
5. Choose the job with earliest scheduled completion date.
6. Choose a job at random from jobs waiting to be processed.

Evaluation of these and other rules, and combinations of the rules, has frequently been a primary objective of job shop simulation.

Results of one evaluation through simulation of the six rules mentioned above are summarized in table 9-4. They were obtained using data from the fabrication shop of El Segundo Division of Hughes Aircraft Company. As such, they are indicative of the relative merit of the dispatch rules under conditions in existence at this particular plant and not of the merit of the rules in general.[4] A direct outgrowth of job shop simulation in this case was development of a simulation scheduler which was incorporated into the actual control system of the fabrication facility.[5]

Table 9-4 shows the breadth of data that are obtainable by using simulation to approach the job shop problem. These results were obtained by varying only the dispatch rules from simulation run to simulation run and without changing any other factors or data. Similar data could be obtained to analyze, for instance, effects of varying the quantity of equipment in the shops while keeping all other factors unchanged, and countless experiments could be performed to explore effects of changing various factors or even combinations of factors. This aspect of job simulation points up what is frequently a significant difference between strictly theoretical and simulation approaches to problems. In theoretical approaches we often find that a portion of the problem is treated in the abstract with significant simplifying assumptions. The results, however, of such theoretical calculations are generally fairly compact and interpretable. In simulation many more factors and their dynamic interactions are usually included in the model and a variety of statistics regarding operation of the system may be obtained. However, the quantity of data, of which table 9-4 is an

---

[4] Some efforts have been made to evaluate dispatch rules and combinations of them independently of any particular job shop setting. For a description of this work, the reader is referred to Allen (1963), Fischer and Thompson (1963), and Heller (1960).

[5] A description of the control system is given in Steinhoff (1966) and Bulkin, Colley, and Steinhoff (1966).

TABLE 9-4 Summary of Results of Job Shop Simulation
Using Various Dispatch Rules[a]

Criteria	Dispatch rule[b]					
	1	2	3	4	5	6
Number or orders completed	1446	1044	1115	1078	1323	1030
Percentage of orders completed late	24.5	20.4	37.8	42.0	33.0	30.1
Mean of completion distribution[c]	−6.56	−4.16	−3.57	−3.04	−4.20	−5.20
Standard deviation of completion distribution	9.9	2.0	10.0	9.2	8.2	10.4
Average number of orders waiting in the shop	961.4	1313.4	1320.4	1416.9	1148.9	1432.1
Average wait time of orders	0.360	0.697	0.949	1.003	0.710	0.544
Yearly cost of carrying orders in queue	128,909	102,800	117,836	108,903	98,502	122,372
Ratio of inventory carrying cost while waiting to inventory cost while on machine	13.292	12.136	12.391	12.092	12.239	13.024
Percentage of labor utilized	0.632	0.580	0.587	0.574	0.548	0.579
Percentage of machine capacity utilized.	0.268	0.246	0.249	0.243	0.232	0.245
Total relative rank[d]	8.70	8.54	6.93	6.77	7.52	7.40

[a]Source: Earl LeGrande, "The Development of a Factory Simulation Using Actual Operating Data," Management Technology, Vol. 3, No. 1, May 1963.

[b]Numbers correspond to numbers of rules discussed in text.

[c]Negative values are days early.

[d]Based on the performance of each rule with respect to the best rule for each criteria and with equal weighting of the criteria. Highest total is the best rule.

example, places an additional burden of interpretation on the experimenter, since selection of appropriate criteria for evaluating operation of the system is often not much easier in a simulation than it is in the real world.

PERT Networks

Since their development in the late 1950s, critical path and PERT methods for control of projects consisting of many interrelated activities have achieved widespread publicity and acceptance. The basic concept in an analysis of the PERT or critical path type is to view a project as a series of activities that must be accomplished according to certain precedence relationships. These relationships may be represented by an arrow diagram, or network, as shown in figure 9-3, in which arrows represent activities that take time and circles represent events that are points in time. Precedence relationships in a network are indicated by establishing the convention that all activities leading into an event must be completed before activities leading out of an event can begin. Estimates are made of the time required for each activity, and these are the numbers shown next to the arrows in figure 9-3.

The time required and sequence of each activity are examined to locate one or more paths through the network that effectively determine the shortest time in which the project can be completed. This path(s) is known as the critical path(s) which in Figure 9-3 passes through events 1-2-3-4-5.

The PERT technique elaborates on this basic critical path procedure by requiring an activity in various lengths of time. To simplify making the probabilistic estimates and data processing requirements, PERT procedures require only an estimate of optimistic time, pessimistic time, and most likely time for each activity.

The PERT system then uses the mean activity times in computing the critical path. No use is made of the probabilistic element in the activity time estimates until after the critical path is found, and at that time variances of activities on the critical path only are used to compute the variance of the distribution of completion time for the end event.

It can be shown that PERT calculations consistently bias the mean of completion time for a network toward values that are too small. This occurs because PERT does not consider the time distributions of activities that are off the critical path in computing parameters of the project completion time distribution. As an example of how this occurs, all events on a critical path may be completed in their optimistic times, whereas events on a "noncritical path" may be in their pessimistic times. This situation could result in a noncritical path requiring longer to complete than the critical path.

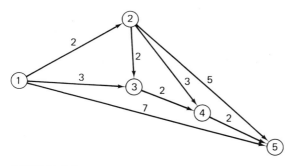

FIGURE 9-3   Project network (by permission from K.R. MacCrimmon and C.A. Ryavec, An Analytical Study of the PERT Assumptions, Memorandum RM-3408-PR, The RAND Corporation. Santa Monica, California, December 1962).

Obviously, interaction of the various activity time distributions both on and off the critical path can produce situations where noncritical paths can become critical and the critical path can become noncritical. As a result the PERT-calculated project completion time distribution is biased on the low side. The degree of bias is dependent upon whether there are any paths that are close to the PERT-calculated critical path in length and upon the variability of activity times both on and off the critical path.

Various methods for determining approximately the amount of bias and some methods for determining analytically the exact distribution have been suggested.[6] Simulation affords a straightforward method of determining, to any degree of accuracy required, the distribution of the project completion time. The simulation procedure consists of selection of one time at random from each of the activity time distributions in the network, and calculation of the critical path and project completion time based on this sample of activity times. The process is repeated over and over and results in the generation of the project completion time distribution. In addition, the frequency with which different activities appear on the critical path can be determined, if desired.

To demonstrate the use of simulation as outlined above, a computer program was written as a seminar project at the University of Washington.[7] The program was used to compare simulation results with analytical results obtained by K. R. MacCrimmon and C. A. Ryavec for the network in figure 9-3 (MacCrimmon and Ryavec, 1962). Times shown on the activities in table 9-5 are the means of the activity time distributions. The discrete activity time distributions used by MacCrimmon and Ryavec to obtain analytically the true project completion time distribution are shown in table 9-5. Results of the simulation are compared with the PERT-calculated mean and analytically calculated mean in table 9-6. After ninety samples the simulation results are very close to the actual values of μ and σ and are more accurate than the PERT-calculated results.

In the preceding discussion we have referred to the procedure as a simulation of the network. The Monte Carlo method, of course, is used to select times from the probability distributions of activity times, but the technique does not involve observing the operation of the network through time in quite the same sense that behavior of systems such as inventory systems are observed through time by simulation. Rather, times are selected by the Monte Carlo method for each of the activities, and calculation of the project completion time is then a straightforward arithmetic computation. This use of the Monte Carlo method to determine through multiple trials a probability distribution that would be difficult or impossible to get by analytical methods is more properly termed model sampling.

[6] An approximate method is given by Fulkerson (1962).

[7] Several different programs were developed at the University of Washington in the spring of 1963. The results discussed here were obtained by Messrs. Ahlers, Daniels, Plotke, and Tinius.

TABLE 9-5   Activity Time Distributions

Activity time distribution for activities 1, 2; 2, 3; 3, 4; 4, 5.

Activity time	1	2	3
Probability	$\frac{1}{5}$	$\frac{3}{5}$	$\frac{2}{5}$

Activity time distribution for activities 1, 3; 2, 4.

Activity time	1	3	5
Probability	$\frac{1}{5}$	$\frac{3}{5}$	$\frac{1}{5}$

Activity time distribution for activity 2, 5.

Activity time	2	5	8
Probability	$\frac{1}{5}$	$\frac{3}{5}$	$\frac{1}{5}$

Activity time distribution for activity 1, 5.

Activity time	3	7	11
Probability	$\frac{1}{5}$	$\frac{3}{5}$	$\frac{1}{5}$

Source: K.R. MacCrimmon and C. A. Ryavec, *An Analytical Study of the PERT Assumptions, Memorandum RM-340S-PR* (Santa Monica, Calif.: The RAND Corporation, December 1962).

TABLE 9-6   Comparison of Pert, Analytical, and Simulation Results

Method of computation	Project Completion Time	
	μ	σ
PERT	8.00	1.26
Analytically	9.23	1.39
Simulation		
40 samples	9.67	1.27
50 samples	9.12	1.55
90 samples	9.36	1.41

Risk Analysis of Capital Investments

Another example of an application of model-sampling techniques to analysis of a complex problem with probabilistic elements is in the risk analysis of capital investment projects. Such an approach as suggested by David B. Hertz would express the uncertainties surrounding forecasts of the important factors as probability distributions in a manner similar to that explained earlier in the inventory simulation example (Hertz, 1964). These factors include forecasts of the market, service life, and operating costs of the proposed investment.

The Monte Carlo technique is used to sample from these distributions to obtain rates of return or present values for various combinations of factors. Repetition of this sampling procedure generates a distribution of likelihood of returns resulting from a large number of combinations. Used in this fashion, model-sampling techniques generate an expected return and variability of expected returns for the investment based upon the interactions of the probability distributions of the various factors.

Waiting Lines

Waiting line, or queuing, problems arise in processes and systems in which customers or transactions arrive at a service facility where they must wait in line to be served. In these systems it is often of interest to determine the effects of such things as number of servers, service time distribution, arrival distribution, and queue discipline on the distribution of queue length, distribution of time in queue, total transit time through the facility, utilization of the facility, and so forth.[8] Queuing problems are of significant interest in the design of service facilities in the telephone industry, in the operation of toll-collecting and checkout facilities, in the design of data-processing facilities, and in the determination of repair facility requirement—to name a few major applications.

Waiting line problems have been investigated analytically since the early 1900s, and Erlang's *Theory of Probabilities and Telephone Conversations,* published in 1909, is the pioneering work in the field. In the 1920s T. C. Fry and E. C. Molina made significant contributions which have been followed, particularly since World War II, by many further theoretical advances. The general direction of this work has been toward extending theoretical results to a wider variety of systems and arrival and service characteristics. Although much analytical progress has been made, it is not difficult to propose system configurations, operating procedures, and input characteristics that are not susceptible to analysis even by the powerful methods that have been developed. Consequently, simulation has been found to be a very powerful and useful tool for analyzing systems in which queues are a significant feature to be investigated. Besides making tractable problems that could not otherwise be investigated, simulation permits easy observation of the dynamic behavior of the system under study, permits tracing of movement of individual customers or transactions through the system, and can include the use of complex logical decisions in directing and regulating the flow of traffic through the system. No similar flexibility of analysis is possible with a strictly theoretical approach, and this is a principal

---

[8] Queue discipline is the particular rule governing the operation of the queue. For example, queues may operate on a first-in, first-out basis or the customers may have various priorities that govern the order of service.

reason for the extensive application of simulation to such problems in recent years.

The general procedure in simulating these systems is to generate customer arrivals or transactions at intervals corresponding to the actual arrival distributions in the system under study. These transactions are then entered into the system, and statistics regarding their progress in time through the system are tabulated together with statistics regarding queue lengths and utilization of facilities within the system. Transactions, for instance, may be customers arriving at a service facility such as a checkout stand, or units of product moving through a manufacturing process, or customer orders flowing through an order-processing system.

It is apparent that much record keeping is involved in creating customer arrivals, moving them at the proper time and in the proper sequence through the system, and keeping necessary statistics. Since queuing formulas are available to determine statistics of interest in simple systems, they will probably be used in preference to simulation. With more complicated systems, queuing formulas may not be available, and simulation will then be the proper tool of analysis. Simulation of even a small system demands a substantial amount of computation and record keeping, and large systems would be impractical, if not impossible, to simulate without a digital computer.

A complex system that has been analyzed through simulation is an instrument calibration system in a large manufacturing organization. [9] Movement of instruments through the system, waiting lines and times that develop, and utilization of calibration technicians were of principal interest in the study. Because of the complexity of the system and desire to explore a variety of aspects of the system, simulation using GPSS was chosen as the tool for analysis.

There are four calibration laboratories in the company, each of which has several thousand pieces of general purpose test equipment assigned to it for periodic repair, calibration, and certification. When an instrument is sent to the calibration laboratory, it is logged in and cleaned by a receiving clerk and stored until a calibration technician who has the necessary equipment and skills is available to begin calibration on the particular instrument. If the calibration is not complete at the end of the shift, a technician on the next shift completes the work and sends the instrument to the receiving clerk.

The model has three basic parts: the instrument flow, a loading routine, and a shift clock. The instrument flow represents the calibration system itself and is a closed loop of all processes that instruments can go through. Transactions move through this loop, shown in figure 9-4, in simulated time, just as instruments move in the calibration system in real time. The loading routine is utilized at the beginning of a simulation run to distribute realistically throughout the system the total number of instruments to be simulated.

[9]This study was conducted by Charles E. Carpenter and Gary W. Dickson. The description of the simulation that follows is adapted from Charles E. Carpenter (1964). See also Stephen K. Didis and Charles E. Carpenter (1966).

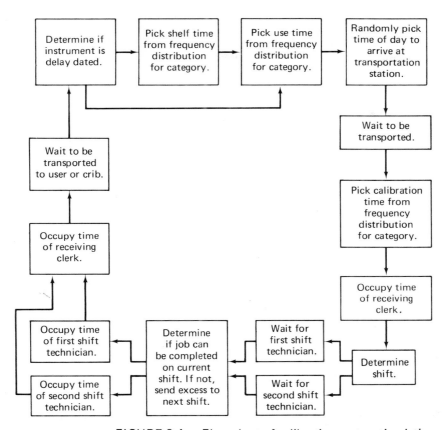

FIGURE 9-4 Flow chart of calibration system simulation.

Since the calibration system operates on two shifts with differing numbers of technicians on each shift, the shift clock is necessary to keep track of the shift and show when it is due to end. In addition, the shift clock is used to tabulate daily statistics about operation of the system.

The model can provide the following information on any run:

1. A history of each run giving the daily value of
   a. Quantity of instruments in the calibration lab at the end of each day
   b. Average flow time of calibrations completed during the day
   c. Average utilization of calibration technicians during the day
   d. Quantity of instruments arriving at the calibration lab
   e. Quantity of instruments leaving the calibration lab
2. A frequency distribution with mean and standard deviation of
   a. Quantity of instruments in the lab at the end of the day and at the end of the first shift

  b. Flow time of instruments through the lab
  c. Utilization of technicians each day
  d. Total quantity calibrated in each category and total work load in hours in each category
  e. Total time between calibrations for each instrument
  f. Time between removals of instruments from the shelf

One use of the model was to study the effects of a proposed combination of two laboratories, the larger of which calibrated about three times as many instruments as the other. The proposed combination entailed routing all the work through the large laboratory and setting up a transportation system to carry instruments to and from the small one. The procedure used was to simulate the operation of the two laboratories separately and then to simulate the combined operation with the same total number of technicians and with the same work load.

The simulation showed that in the combined laboratory average quantity of equipment in calibration or waiting for calibration would decrease from 154 to 121 instruments. This saving plus savings in administration cost, facilities, and better management control were felt to more than offset the one hour increase in average flow time, especially since users serviced by the small laboratory said that the increase in flow time would not seriously affect their operations. The combination of the two laboratories was then implemented.

Forecasting

A relatively simple direct application of simulation is in the evaluation of various statistical forecasting schemes. Such schemes are used to make forecasts, say of sales, for future time periods based on sales data from past time periods. Moving averages and exponential smoothing are two of the more widely used methods. One problem in determining whether a statistical forecasting method may be useful is the large number of alternatives. Even if other methods are ignored, moving averages and exponential smoothing alone offer many possibilities, since moving averages of different lengths may be tried and, in exponential smoothing, different values of the smoothing constant can be tested. In addition, the effects of including seasonal and cyclical factors may be investigated.

One of the principal ways of validating any statistical forecasting method is to try the method on past data and observe the amount of forecast error. In other words, a simulation is performed in which a forecast is made for each time period using only the past data that would have been available had the forecasting method actually been in use. Then, the simulated forecast is compared with actual data for the time period, and the procedure is repeated for succeeding time periods. By comparing distributions of forecast error for different forecasting methods, that is, various lengths of moving averages, different values of a etc., some evidence is obtained of expected performance of the methods. While simulated performance of a method is no guarantee of future performance, it is one of the best tests available.

If these analyses were frequently undertaken, the development of special computer programs would be appropriate. One such program reported in the literature has the capability of simulating the following forecasting schemes from Gross and Ray (1965):

1. Single moving average.
2. Single moving average with least squares trend.
3. Double moving average.
4. Single moving average with least squares trend and exponential seasonals.
5. Single exponential smoothing.
6. Single exponential smoothing with trend and seasonals.
7. Second order exponential smoothing.
8. Double exponential smoothing.
9. Triple exponential smoothing.

SIMULATIONS OF THE TOTAL FIRM

The preceding discussion of the use of simulation in analyzing various business problems naturally gives rise to the question of whether it is feasible to simulate operation of the total firm rather than just portions of it at a time. If such a total firm simulator could be constructed, observation and analysis of the interactions of the subsystems within a firm would be possible and problems of suboptimization, which inevitably arise when parts of the firm are individually examined, would be avoided.

The possibility of such a total firm simulator is dependent on the level of aggregation that the analyst is willing to accept. As shown by several of the preceding illustrations in this chapter, it is quite possible with existing techniques and data-processing equipment to simulate in fairly minute detail the events that occur in a limited area. However, it is not possible, even with the very powerful computing equipment that is becoming available, to simulate *all* the activities in detail in *all* the subsystems of a firm. Even if no technological constraints were present, the desirability of constructing an all-encompassing microscopic simulator of the total firm would be open to question, since the analyst would be faced with an impossible task of determining what experiments to perform, what data to collect, and what interpretation to make of the results. In other words, a model that is not a significant abstraction from the real world suffers from many of the same difficulties of interpretation and analysis that are found in the real world. As a consequence, any meaningful approach to total firm simulation involves a considerable amount of abstraction or aggregation, and just how this is done is a function of the point of view and objectives of the model builder.

Functional Model

The functional model is exemplified by the simulation of the hypothetical Task Manufacturing Corporation.[10] The philosophy underlying this simulator, designed by R. C. Sprowls and M. Asimow at UCLA, is to create models of the subsystems or functional areas that comprise the total firm and its environment. Figure 9-5 shows the various subsystems that comprise the total model. Models of each of the subsystems are designed to describe the behavior of individual

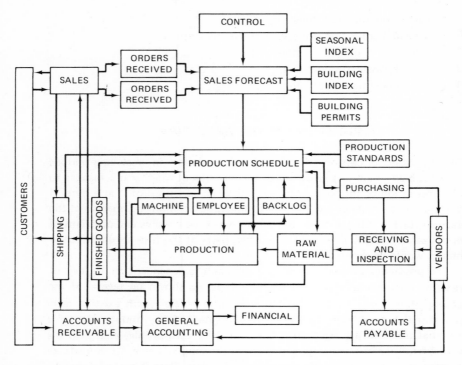

FIGURE 9-5   Subsystems of Task Manufacturing Company (by permission from R. Clay Sprowls and Morris Asimow, "A Computer Simulated Business Firm," in *Management Control Systems*, edited by Donald G. Malcolm and Alan J. Rowe, New York: John Wiley & Sons, Inc., 1960).

[10]This study was conducted by Charles E. Carpenter and Gary W. Dickson. The description of the simulation that follows is adapted from Charles E. Carpenter (1964). See also Stephen K. Didis and Charles E. Carpenter (1966).

units within that subsystem. In other words, the Task Manufacturing Corporation simulator is a collection of micromodels of the various functions of the business and its surroundings. To provide a specific basis for construction of the various models, it was assumed that Task was a small manufacturing firm employing four hundred persons and manufacturing five related products in the building hardware line. Sales are in the range of $3 million to $3.5 million annually. The manufacturing facility is made up of seven departments, and raw materials consisting of castings, steel strip, and steel bars are assumed to be available locally.

Sprowls and Asimow have described the simulator as follows:

> . . . . Each of the subsystems is sufficiently general, self-contained, and complete that it can be dealt with as an entirety. In a sense, each model of a subsystem is analogous to a "black box" and if certain inputs are specified, outputs will appear. Some of the outputs are uniquely determined and some are determined only in a stochastic sense.
>
> Just as a collection of subsystems does not comprise a business firm, neither does a collection of models comprise a representation of a business firm. The subsystems must be coupled together to permit inputs and outputs to come from and exit to both the external world and other subsystems. Formal policies, managerial decisions, and informal policies which have developed from customs and traditions determine the ways in which these couplings are allowed to occur. The set of human and material subsystems and the couplings conditioned by formal and informal policies comprise the business firm. Correspondingly, the set of separately programmable models of subsystems coupled by interconnecting programs comprises a representation of a business firm—a simulated firm which can be manipulated on a computer. Some of these interconnections—but by no means all of them—are shown in [fig. 9-5] (Sprowls and Asimow, 1960).

Information and Decision Model

Charles P. Bonini's model of the firm, although it deals with functional areas, as does the Task model, emphasizes the information and decision system within the firm (Bonini, 1963). Like the Task model, Bonini's model is designed to simulate a manufacturing firm. The firm produces four products, has five manufacturing departments, and sells through seven district sales offices with a total of forty salesmen.

One of the distinguishing features of Bonini's model is the inclusion in the model of behavioral factors, notably indexes of pressure on individuals within the organization which affect their performance and decisions within the model. For instance, the index of pressure on a salesman for a certain month is the average of five factors with weights as shown:

1. Index of pressure of his superior   (25%)
2. His quota relative to his sales in the past month   (40%)

3. Sales of the "average" salesman in his district relative to his sales   (10%)
4. Seventy-five one hundredths plus the fraction of his products less than 75 percent of quota   (10%)
5. His total quota for the past quarter relative to his total sales for the last quarter   (15%)

The index, in turn, determines whether any changes are made in the mean and standard deviation of the sales distribution from which the salesman's sales rate for the month is drawn at random. Bonini includes four types of salesmen in his model, and for each type there is a schedule showing the effects that various levels of the index of pressure have on the mean and standard deviation of the sales distribution which is used to determine the salesman's performance during the month. Similar procedures are used to affect the behavior or performance of the foremen, district sales managers, general sales manager, and vice-president of manufacturing.

The simulation model is organized around the three major activities of planning, control, and operations. Plans are made quarterly, or every three periods, since the month is the basic time increment in the model, and they involve the preparation of a sales forecast, sales administration budget, manufacturing administration budget, manufacturing cost estimate, and, finally, an overall company plan. The sales forecast, sales and manufacturing administrative budgets, and manufacturing cost estimates are functions of performance in previous periods. The overall company plan for the next quarter is determined by combining the cost and sales estimates with the product price estimates (assumed to be the prices from the previous period) to give an estimate of the expected profit. This profit is then compared with the profit goal (the average of profits from the preceding ten quarters), and if the goal is not attained by the plan an iterative procedure is followed to adjust the cost, sales, and price estimated until the profit goal is reached. This adjusted plan is then used to determine the target level of production operations for the next quarter.

Control in the firm is exerted through the establishment of sales quotas and manufacturing standard costs and through the use of the indexes of felt pressure. As mentioned previously, various indexes of felt pressure affect the simulated behavior and performance of the components of the firm. In each index of pressure a contagion factor reflects the pressure from the immediate level above in the organization and other factors which relate information about performance of an individual or in a specific area to the index of pressure. The various indexes of pressure then indirectly control the firm's operations by their inclusion in the computer programs that simulate the operations of the firm.

Actual operations of the firm are simulated on a period-by-period basis by computing for each period actual production, manufacturing costs, actual sales, and administrative expenses. In these calculations, the indexes of felt pressure have an effect on manufacturing costs, administrative expenses, and performance of the salesmen through alteration of the means and standard deviations of distributions from which the actual period values are chosen. The level of sales is also influenced by an assumed long-term upward trend and three-year cycle in the market for the firm's product and by certain assumed price elasticities for each of the products.

To provide the opportunity for experimentation, provision was made for altering certain portions of the model between simulation runs. The areas in which changes were made for purposes of experimentation with the model are:

1. Inventory valuation
2. Amount of contagious pressure
3. Sensitivity to pressure
4. Sales force knowledge of inventory
5. Variability of sales and cost distribution
6. Amount of market growth trend
7. Tightness of industrial engineering department
8. Utilization of past versus present information on control

To determine the effects of changes in these eight areas on the model and their interaction, Bonini used a fractional factorial experimental design involving sixty-four simulation runs, each of which was for 108 periods, or nine years. Of the many variables that could have been traced during the simulation runs, Bonini chose to record and analyze time series for six variables: price, cost, inventory, sales, profit, and index of pressure.

While the specific results of Bonini's experimentation are not of particular interest here, this simulation study illustrates quite well the amount of detail that must be included in any total firm simulation model. And it also illustrates how the interests of the analyst have much to do with the construction of the model. In Bonini's model, the treatment of the functional areas and the firm's information and decision system is greatly influenced by his emphasis on the behavioral aspects of decision-making.

Budget Model

Richard Mattessich has developed still another approach to total **firm** simulation using a budget model of the firm (Mattessich, 1964). This model uses the conventional accounting structure of a firm as its framework, and the output from the model is in the form of various period-by-period budgets and projected financial statements. Mattessich calls models such as the Task Manufacturing Company's and Bonini's model "control models," whereas he considers his own approach to be very close to conventional budgeting systems. The argument for this approach is that the control models that have been developed are too cumbersome and expensive for practical application in most firms, but the budget model is sufficiently similar to present accounting systems and practices to offer the promise of some practical application with relatively little reorientation of thinking within the firm.

Like the two models described previously, the budget model is designed to simulate a manufacturing firm. The model is so constructed that the dimensions, that is, number of products, number of raw materials, number of departments, and so forth, are variable and determined by the input data. Mattessich's model does not simulate the behavior or flow of individual entities or transactions within the firm. Instead, the behavior of the system is determined in an aggregate way through sets of coefficients, such as standard labor hours, operating expense rates, and overhead rates, which are inputs into the simulation program. Other data inputs determine the period-by-period levels of sales and production for the simulated firm. Outputs from the computer program are rather complete sets of period-by-period sales, production, material, labor, overhead, expenses, and cash budgets together with income statements and balance sheets. Since the model contains no stochastic elements, the outputs from the computer program are uniquely determined by whatever input data and coefficients are given to the program. A flow chart of the program is shown in figure 9-6.

INDUSTRIAL DYNAMICS

Industrial dynamics is one approach to analyzing the behavior of large-scale systems suggested by Jay W. Forrester of the Massachusetts Institute of Technology (Forrester, 1961). Analysis of the decision processes of the manager and his role as a decision-maker in an interacting environment under dynamic conditions is the central focus of industrial dynamics. A primary tool for analysis is system simulation in which events are considered in the aggregate. To facilitate construction of these types of models, the DYNAMO computer simulation language was developed.

The industrial dynamics philosophy proposes that effective management of an organization should recognize the character of the organization as a series of interconnected feedback networks with dynamic interactions among the components of the organization. As Forrester has defined it:

> Industrial dynamics is the study of the information-feedback characteristics of industrial activity to show how organizational structure, amplification (in policies), and time delays (in decisions and actions) interact to influence the success of the enterprise. It treats the interactions between the flows of information, money, orders, materials, personnel, and capital equipment in a company, and industry, or a national economy.

> Industrial dynamics provides a single framework for integrating the functional areas of management—marketing, production, accounting, research and development, and capital investment. It is a quantitative and experimental approach for relating organizational structure and corporate policy to industrial growth and stability (Forrester, 1961).

Forrester has not proposed a single model to represent industrial activity. Instead, he has proposed the development of particular models to represent

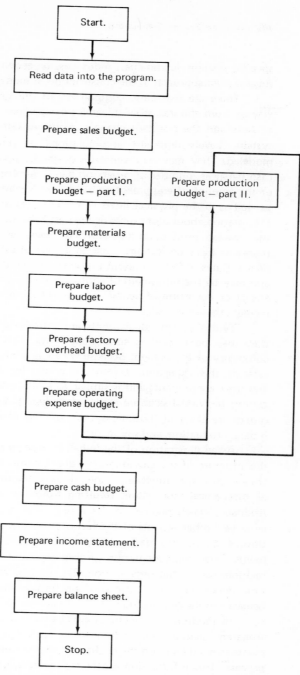

FIGURE 9-6   Budget simulation program flow chart (reprinted by permission from Richard Mattessich, *Simulation of the Firm Through a Budget Computer Program*, Homewood, Illinois: Richard D. Irwin, Inc., 1964).

specific systems (companies, industries, or economies). In other words, he has suggested an approach to the construction of models.

There are two basic components of the structure of a feedback system viewed from the industrial dynamics perspective: the varying activities or *rates of flow* and the resultant changes in system states or *levels of variables* in the system. Levels represent accumulations of resources in the system being modeled. They may be inventories of goods, amounts of information or ideas, cash balances, number of orders on hand, number of people in the work force, or the amount of equipment on hand. Levels may exist in the information flows as well as in the physical flows in the organization. They may take the form of information about past events or the past behavior of rates of flow—for example, the average production rate for last week or last month. Levels may also represent ideas or feelings, such as a level of satisfaction or a degree of confidence. Rates of flow represent the activities and decision functions in the system and may be the movement of goods, payments of money, expenditure of effort, arrival or departure of people, or acquisition of equipment. Decision functions regulate these flow rates.

Typically, individual events are considered in an aggregate sense so that they may be treated as continuous flows. This permits concentrating on the continuity of the system. Because we are dealing here with models of large-scale systems, this aggregation permits us to examine the overall behavior of systems and their components and is similar to a top-level manager's view of his organization. Individual sales are made, individual items are produced, and individual orders are shipped; but he is primarily concerned with the rates of flow of orders, production, and shipments.

As an illustration of the industrial dynamics point of view, let us consider the structure of the closed-loop feedback control system in figure 9-7. There are three primary functions: the decision activity, which may be human or mechanical; the action resulting from the decision; and the information feedback, which reports on the action. The process of control is continuous. However, other aspects of the feedback loop must be considered. Decisions go through an implementation process which transforms them into actual achievements. The implementation process is complicated by complex structural relationships in the organization. The process may further be complicated by time delays between a decision and its implementation and noise or random behavior in the organization.

In addition, a distinction exists between actual and apparent achievements. Apparent achievements will be more or less representative of actual achievements depending upon the quality of information received through the feedback process. This information is affected by delays, noise, and bias in the information channels. The decision process itself is characterized as goal-seeking behavior that responds to any difference between the organization's objectives and its apparent achievements. It is the industrial dynamics point of view that managers can operate most effectively if they recognize that this control system structure permeates each organization and its components.

The importance of the effect of time lags on the behavior of dynamic

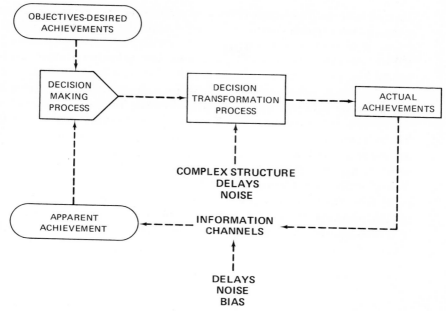

FIGURE 9-7   Control system structure (adapted from Edward S. Roberts, "Industrial Dynamics and Design of Management Control Systems," *Management Technology*, Vol. 2, No. 3, December 1963).

systems is stressed in industrial dynamics literature, because time lags determine many aspects of system behavior. Delays may be found in the flows of both physical quantities and information. In the former instances, lags would represent delays in movements of goods, people, money, and so forth. In the latter instance, they would represent such things as information transmission delays and smoothing of time series of information. Smoothing of information flows is closely related to the concept of delaying the flow of physical quantities in a system. Smoothing delays the impact of transient fluctuations on the system and filters out higher frequency fluctuations, thereby altering the sensitivity of the system to them. Thus, in information smoothing there is a tradeoff to be evaluated between reducing the impact of noise and other short-term fluctuations and reducing time lags in recognizing significant changes in the data being received.

Dynamics of Steady-State Systems

Two major classifications of much of the work done to date in industrial dynamics are (1) the dynamics of steady-state systems and (2) the dynamics of growth systems. The purpose of models of steady-state systems is generally to stabilize fluctuations in activities at steady-state levels.

One of the earliest steady-state system models developed was that of a production-distribution inventory system. A detailed discussion of such a system comprises a major part of Forrester's *Industrial Dynamics*. We will examine here a model developed by the authors of a simple production-distribution inventory system.[11]

The configuration of such a system would include the rates of orders, production, and shipments; levels of inventories; ordering decision; and delays in the flows of orders and shipments. The basic structure of the model is illustrated in figure 9-8. In this system the manufacturer's products are sold directly to dealers, and in turn to customers. The company maintains a factory warehouse which is adjacent to, but is managed more or less autonomously from, the factory production organization. Orders from dealers are from the factory warehouse. The warehouse places orders with the factory. The factory cannot respond instantly to changes in the rate of flow of incoming orders but is restricted by an adjustment delay representing the time required to adjust production rates.

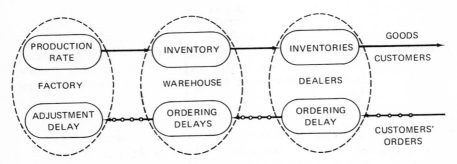

FIGURE 9-8   Basic structure of production-distribution system.

There are two major physical flows in this system—the flow of goods from the factory through the warehouse inventory through the dealers' inventories to the customers and the flow of orders from customers to dealers, from dealers to factory warehouse, and from warehouse to factory. Other physical flows may be included, such as money, personnel, and capital equipment, but will be omitted from this example. Interconnecting these flows is a complex network of information flows. Delays in the adjustment of factory production rates, delays in the response of the dealer and warehouse sector occasioned by their re-ordering decision rules, and delays in the transportation of goods from the factory warehouse to the dealers are included in the model.

Figure 9-9 depicts plotted output from a simulation of fifty-two weeks'

11The structure of the model is explained in detail in Meier, Newell, and Pazer (1969).

operation of this system. The model was started in an equilibrium state with the rates (sales, ordering, shipping, and production) at ten thousand units per week and inventories at forty thousand units. An increase of two thousand units per week in dealers' sales rate was injected at the eighth week to observe transient response of the system.

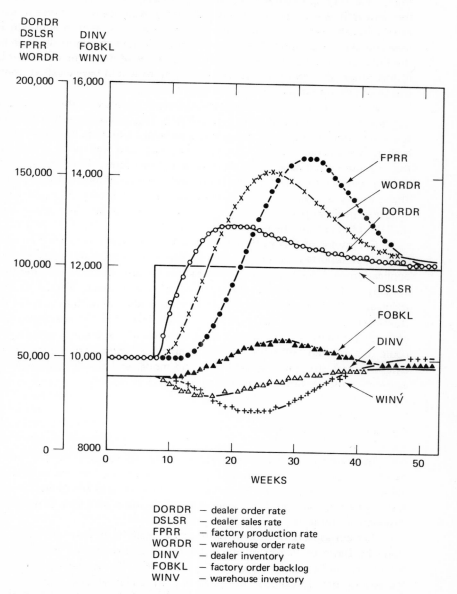

DORDR	— dealer order rate
DSLSR	— dealer sales rate
FPRR	— factory production rate
WORDR	— warehouse order rate
DINV	— dealer inventory
FOBKL	— factory order backlog
WINV	— warehouse inventory

FIGURE 9-9 Plotted DYNAMO output from production-distribution model.

The system responds with the type of amplification and time lags that we have come to associate with this type of system. Following a 20 percent increase in sales, dealers' order rate increased 29 percent by the twentieth week (a twelve-week lag), warehouse order rate rose 40 percent by the twenty-sixth week (an eighteen-week lag), and factory production rate was up 44 percent by the thirty-first week (a twenty-three-week lag). Dealer inventories fell 25 percent by the sixteenth week, and warehouse inventory was down 49 percent by the twenty-fourth week. It required nearly a year for the system to approximate a new equilibrium position. The degree of amplification shown in figure 9-9 and the time lags in the system are primarily functions of the particular delays used in this run of the model. Other behavior patterns would have resulted from the choice of different parameter values.

Another example of steady-state system analysis is the study of the Sprague Electric Company (Forrester, 1961). This project, the first to explore the application of industrial dynamics to an actual situation, was undertaken to evaluate the interaction of production, inventory, and employment policies for the purpose of reducing costly fluctuations. One product line was selected for detailed examination, and a simulation model was developed to test hypotheses about causes of aggregate system behavior patterns. Insights into the system's behavior, new priority rules for scheduling incoming orders, and new inventory policies resulted from the project.

Extensions of this work to include similar applications to other companies have been reported in the literature. Kenneth J. Schlager reported on the application to three companies in the Milwaukee area which were found to have systems and problems not unlike those of the Sprague Electric Company (Schlager, 1964). One company was a manufacturer of instruments for liquid flow measurement and was experiencing significant fluctuations in production and employment. It was found that some parts of the original Sprague model could be adapted, substantially reducing model construction time. A more extensive cost structure was added to the model and certain sectors were disaggregated, such as production-inventory control, personnel, and purchasing.

Dynamics of Growth Systems

In contrast to steady-state systems, growth systems exhibit long-term trends over time. Characteristics of this type system are the growth of product markets, enterprises, industries, urban communities, and economies, as well as systems dealing with projects having discernible life cycles, such as research and development projects.

Certain aspects of growth patterns of small firms were investigated in a study by David W. Packer in which the impact of two sets of basic policies on growth of the firm were examined (Packer, 1964). These policies governed the acquisition of professional manpower, both technical and managerial, and acquisition of productive capacity. An outline of the basic model is shown in figure 9-10.

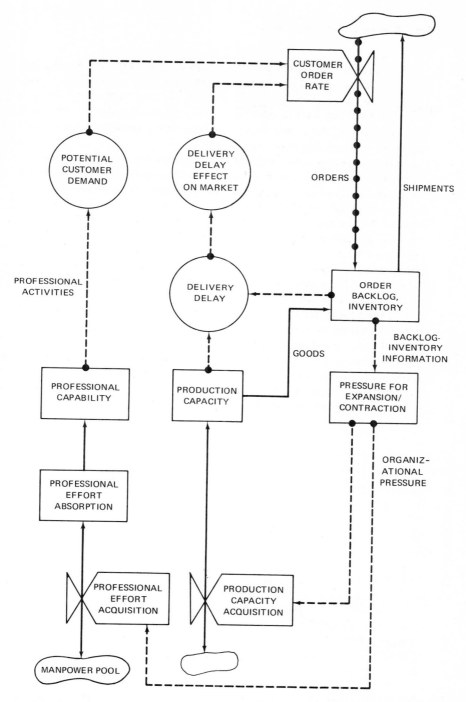

FIGURE 9-10   Model of corporate growth system (by permission from David
W. Packer, *Resource Acquisition in Corporate Growth*, Cambridge, Massachusetts: The Massachusetts Institute of Technology Press, 1964).

Growth of the system occurs through the action of two primary positive feedback loops. The first relates the positive effect of productive capacity expansion on increase in customer order rate. An increase in order rate raises the order backlog and lowers inventory. This creates pressure for expansion of capacity, which when realized reduces the firm's delivery delay—one factor working to increase the order rate.

The second positive feedback loop relates the effect of acquisition of professional manpower and the firm's ability to absorb additional manpower on the customer order rate. As in the first feedback loop, increases in customer order rate cause order backlog to rise and inventory to fall, thus creating pressure for expansion of the firm's professional effort. Acquisition of professional manpower results in an increase in the firm's professional capabilities, but this increase is restrained by its ability to absorb additional professional effort. The expanded professional capability expended on activities affecting the firm's market has a positive effect on potential demand—another factor working to increase the order rate.

Several other feedback loops interact with these to produce the fluctuating growth behavior indicated in figure 9-11. According to the study, this behavior resulted from the firm's inability to absorb professional employees into the organization rapidly. Attention devoted to training and fitting in newcomers detracted from activities necessary to create demand for the firm. As productive capacity grew faster than demand, cutbacks were necessary. These interactions resulted in cycles of expansion and contraction.

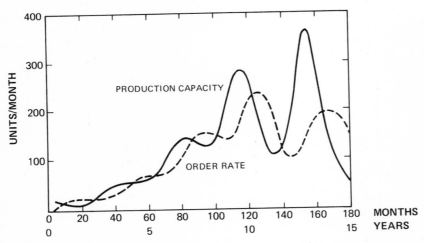

FIGURE 9-11   Corporate growth system behavior (by permission from David W. Packer, *Resource Acquisition in Corporate Growth*, Cambridge, Massachusetts: The Massachusetts Institute of Technology Press, 1964).

HEURISTIC METHODS

The preceding sections have dealt with simulation of business systems that are principally physical or economic, but in recent years increasing attention has been given to problem-solving techniques and procedures, usually computerized, which are similar to, or simulate, those that might be employed by intelligent problem solvers. The word *heuristic* has often been employed to describe methods of this type. Heuristic, however, is a broad term also used to describe search methods that would probably not be thought of as simulations of human behavior or thought processes. The social science research tradition on problem solving is reviewed by Gilbert K. Krulee in chapter 2.

Heuristic Problem Solving

Alfred A. Kuehn and Michael J. Hamburger have developed a heuristic approach to solving the problem of locating warehouses where it is desired to equate the "marginal cost of warehouse operation with the transportation cost savings and incremental profits resulting from more rapid delivery" (Kuehn and Hamburger, 1963). A flow diagram of the heuristic program used to solve the problem is shown in figure 9-12.

The program utilizes three problem-oriented heuristics.

1. Most geographical locations are not promising sites for a regional warehouse; locations with promise will be at or near concentration of demand (Kuehn and Hamburger, 1963).

The problem is reduced by this rule from one with nearly an infinite number of possibilities to a problem that is finite. In fact, in a trial problem it was found necessary to consider only twenty-four concentrations of demand as potential warehouse sites.

2. Near optimum warehousing systems can be developed by locating warehouses one at a time, adding at each stage of the analysis that warehouse which produces the greatest cost savings for the entire system. (Kuehn and Hamburger, 1963).

The heuristic above permits analysis of the problem as one of step-wide minimization. Resulting from this heuristic is a reduction of the number of required cost evaluations from 2^m to M^2 where M is the number of potential warehouse sites being considered.

3. Only a small subset of all possible warehouse locations need be evaluated in detail at each stage of the analysis to determine the next warehouse site to be added (Kuehn and Hamburger, 1963).

An operational statement of this heuristic is to evaluate the M potential

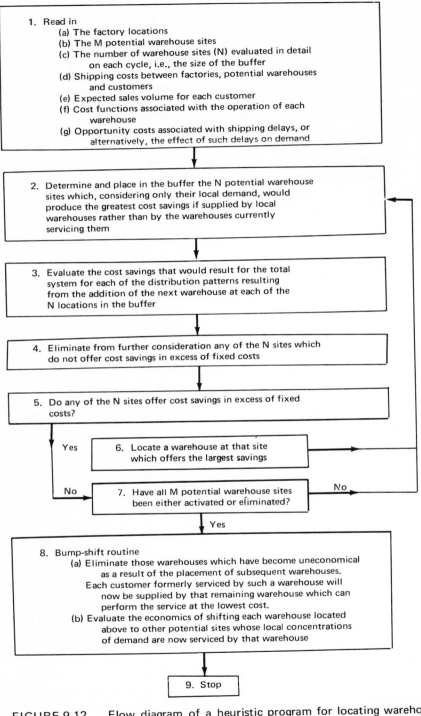

1. Read in
    (a) The factory locations
    (b) The M potential warehouse sites
    (c) The number of warehouse sites (N) evaluated in detail
         on each cycle, i.e., the size of the buffer
    (d) Shipping costs between factories, potential warehouses
         and customers
    (e) Expected sales volume for each customer
    (f) Cost functions associated with the operation of each
         warehouse
    (g) Opportunity costs associated with shipping delays, or
         alternatively, the effect of such delays on demand

2. Determine and place in the buffer the N potential warehouse
    sites which, considering only their local demand, would
    produce the greatest cost savings if supplied by local
    warehouses rather than by the warehouses currently
    servicing them

3. Evaluate the cost savings that would result for the total
    system for each of the distribution patterns resulting
    from the addition of the next warehouse at each of the
    N locations in the buffer

4. Eliminate from further consideration any of the N sites which
    do not offer cost savings in excess of fixed costs

5. Do any of the N sites offer cost savings in excess of fixed
    costs?

Yes

6. Locate a warehouse at that site
    which offers the largest savings

No

7. Have all M potential warehouse sites
    been either activated or eliminated?

No

Yes

8. Bump-shift routine
    (a) Eliminate those warehouses which have become uneconomical
         as a result of the placement of subsequent warehouses.
         Each customer formerly serviced by such a warehouse will
         now be supplied by that remaining warehouse which can
         perform the service at the lowest cost.
    (b) Evaluate the economics of shifting each warehouse located
         above to other potential sites whose local concentrations
         of demand are now serviced by that warehouse

9. Stop

FIGURE 9-12 Flow diagram of a heuristic program for locating warehouses
(by permission from Alfred A. Kuehn and Michael J. Hamburger, "A Heuristic Program for Locating Warehouses,"
Management Science, Vol. 9, No. 4, July 1963).

warehouse locations considering only the cost savings if *local* demand were supplied by that warehouse. Sites are ranked in order of potential local saving and *N* selected of the locations that rank the highest. The *N* locations are called the intermediate buffer and are analyzed in detail to evaluate the cost savings that would accrue to the total system. A reduction in search is accomplished from M^2 combinations to $N \times M$.

The second stage in the program is utilized after all *M* potential warehouse sites have either been assigned a warehouse or eliminated. This stage is called the Bump-Shift Routine. Warehouses placed early in the procedure may become uneconomical as the result of later site selection. These warehouses are eliminated from the solution. A second function accomplished by the Bump-Shift Routine is determining if a cost savings can be accomplished by moving a warehouse to another site within its territory (*territory* is defined as those warehouse sites that are served by the warehouse being evaluated). If savings can be accomplished, the move is made.

Kuehn and Hamburger tested their program on twelve sample problems involving location of warehouses in a distribution system with fifty markets, or concentrations of demand, and either one or two factories. In these test problems costs associated with the initial "no warehouse" configuration were lowered by adding or adding and deleting warehouses through application of the heuristics outlined above. Kuehn and Hamburger noted that in four of the test problems a minor adjustment in the final warehouse configuration would improve the solution slightly, but in the other eight problems no improvement on the solution could be found. Since computationally feasible optimizing techniques were not available, there was no way to determine whether better solutions were possible.

Artificial Intelligence

Some computer approaches to assembly line balancing may be considered types of artificial intelligence, particularly in pattern recognition and organization planning. The line-balancing problem typically occurs in mass production industries where work elements must be performed on each unit processed by the line within a given set of sequencing limitations. The problem is to combine elements into a series of work stations capable of meeting a desired rate of production. The objectives are to create stations that do not exceed the maximum amount of time that may be taken at each station, to equalize the amount of work assigned to each station, and to minimize the number of work stations on the line.

As yet no computationally feasible mathematical algorithms have been developed for line balancing, nor are exhaustive search procedures possible in large problems because usually too many possible combinations of elements would have to be considered.

Fred M. Tonge has developed a computer program for line balancing using heuristics which avoids the problem of searching all possible combinations of elements into work stations (Tonge, 1961). On the assumption that the sequencing and time limitations have been established, the program operates in three major phases.

In the first phase the program factors the heterogeneous environment of work elements into recognizable patterns. Output of the first phase serves as a starting point for activities of the second phase, which generates solution through modification of the first phase output. Components of the second phase include an allocation routine and regrouping routines. Allocation is accomplished through a recursive procedure for matching work stations and groups of tasks. When the assignment routine falls, regrouping procedures are used by the program.

Once a satisfactory solution is obtained in the second phase, an attempt is made in the third phase to improve this solution by equalizing the work load between stations. This is accomplished by using transfer and trading heuristics to reduce the time requirement of the largest work station. As reduction is accomplished, a new station will now become the largest, and the procedure is repeated.

Tonge solved three sample problems by the program—an eleven-element problem and a twenty-one-element problem previously considered as examples by other authors and a seventy-element problem representing actual appliance industry data. An industrial engineer's solution of the seventy-element problem required twenty-seven workmen, which exceeded by one the twenty-six determined by Tonge's program. Although learning was not included, Tonge commented that a learning routine could be incorporated to determine which heuristics tended to require less computing effort.[12]

Simulation of Human Thought

In simulation of human thought, an efficient solution to a problem is not necessarily the goal, but it is duplication of the thought process. Consequently, such heuristics are oriented around the human thought process rather than around the problem or computer. An excellent example of the simulation of human thought processes in business is the work done by Geoffrey P. E. Clarkson on simulation of the decision-making process of a trust investment officer (Clarkson, 1962). In his study, Clarkson noted in detail the processes through which the trust officer arrived at investment decisions. These processes, or protocols, were then translated into a computer program which simulated the trust officer's decision-making process. The model of the trust investment process is shown diagrammatically in figure 9-13. Three major subdivisions exist in the model.

[12]For a discussion by Tonge of subsequent work on the utilization of learning in the selection of heuristics, see Tonge (1965).

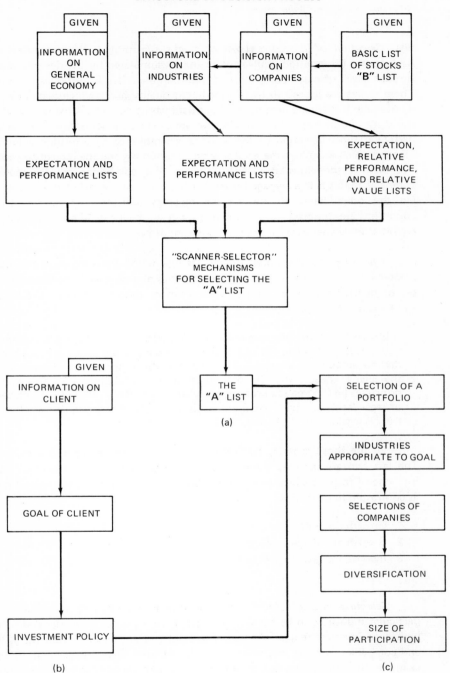

STRUCTURE OF DECISION PROCESS

FIGURE 9-13   Model of the trust investment process (by permission from Geoffrey P.E. Clarkson, *Portfolio Selection: A Simulation of Trust Investment*, Englewood Cliffs, New Jersey: Prentice-Hall, Inc., 1962).

Selection of the current list of stocks. To provide the computer with background information equivalent to that accumulated by the decision-maker through years of experience, a list of stocks, in the *B* list, is introduced as input. These stocks are judged to be of investment quality and are included in this list if, and only if, they are held by leading trust institutions. The *A* list is a subset of the *B* list and includes those stocks that appear to be a good investment at this particular time. Selection of the *A* list is accomplished by a routine called the scanner which examines the economy expectation list and the industry expectation list. If the expectation lists indicate that conditions are favorable, a company that is below average for its industry may still be included in the *A* list. On the other hand, if the expectation lists indicate that environmental conditions are doubtful, some companies will not be included even though they are rated somewhat above average for their industry.

Choosing the investment policy. It was found from the protocol that a complex of information regarding characteristics of each client was factored by a set of heuristics into four types of investment goals, each with a different emphasis on growth and income.

Information on the client is fed into the model in the form of a list which contains the following attributes: (i) the desired amount of growth, (ii) the desired amount of income, (iii) whether current income is sufficient for the client's needs, (iv) the desired amount of stability of income and principal, (v) income tax bracket, (vi) client's profession, (vii) client's place of legal residence, (viii) whether trust is revocable or not, and (ix) whether trust is legal or not (Clarkson, 1962).

Of the above criteria, the first seven are used in this phase of the program. The last two are used by the portfolio selection process. Each client is placed into one of four classifications on the basis of the relative desire for appreciation as contrasted to dividends:

1. Growth account
2. Growth and income account
3. Income and growth account
4. Income account

Selecting the portfolio. The investment policy will determine the particular industries to be considered, since there is a set of industries associated with each investment goal. Appropriate industries are selected from the *A* list and companies in other industries are excluded from further consideration. For example, the following group of industries is associated with an income-oriented account: auto and equipment, banks, container, food, insurance, metals (copper), oil (domestic), retail trade, railroad, and utilities.

After selecting the appropriate list of industries, companies in each of the pertinent industries are ranked on the basis of the dominant attribute of the investment policy. If a portfolio were being selected for an income account, companies within each of the above industries would be ranked on the basis of yield. The Selector then picks from an industry the company that is at the end of this list and applies a set of tests to determine if it is in line with the investment goal. Unless the value of some attribute is considerably out of line, the selection heuristic will pick that company and move to the next industry. If, however, that company is rejected, the next company on the list is subjected to the same tests and so on until a selection has been made or until all companies in the industry have been evaluated. The next industry is then considered. If, after processing all the industries, funds remain to be invested, the process is repeated with a relaxed criterion.

Table 9-7 provides a comparison of output of the simulation with actual decisions made by the trust investment officer. One difference exists between the two portfolios. General Public Utilities was considered by the model to be overpriced and was not included in list *A* and this had not been noticed by the trust investment officer. Clarkson remarks that the model was at fault for not allowing for possible time lags in assimilation of information. In general, however, the program did very well in modeling both the results and the observable thought process of the decision-maker.

GAMES AND GAMING

Although games are used both as educational devices and as tools for analysis and research, the greatest use of the gaming technique in business has been for education. Objectives of gaming for educational purposes vary from developing decision-making and organizational skills in general to imparting knowledge regarding specific techniques of analysis and optimization. Research uses of games have mainly been to investigate behavioral and organizational aspects of the play of the game, but games also have research uses in the development of dynamic theories of the firm and economic theories of oligopoly behavior. Programmed play game structures offer an approach to the investigation of effective long-range strategies and automated decision-making.

Business games may be used either to improve decision-making skills of the players or to improve performance of the system being simulated, or both. The term *operational gaming* is often used to further distinguish games used to study a process with the objective of finding and imparting optimal solutions (Thomas and Deemer, 1957).

It is generally agreed that the earliest business management game was the Top Management Decision Simulation, developed by the American Management Association in 1956 (Riccardi, 1957; Bellman et al., 1967). This game, the operation of which is typical of current models, is a computerized general management or total enterprise game, as opposed to a functional or industry game, since all major functions of the enterprise are represented and decisions are those made at top management levels. The game is designed for team play in

TABLE 9-7 Comparison of Simulated and Actual Investment Decision

1 Goal of account: high income and principal safety.
2 Funds available for investment and in common stocks: $28,000.

The TRUST INVESTMENT MODEL selected the following portfolio for Account 3

Number of Shares	Stock	Price	Total	Estimated Dividends
100	American Can Company	$38	$ 3,800	$ 200
100	Continental Insurance	53	5,300	200
100	Duquesne Light Company	24	2,400	116
100	Equitable Gas	37	3,700	185
100	Pennsylvania Power and Light	27	2,700	125
100	International Harvester	45	4,500	240
100	Libbey Owens Ford	50	5,000	250
100	Socony Mobil Oil Co.	38	33,800	200
			$31,200	$1,516

Estimated Yield 4.9 per cent

The TRUST INVESTMENT OFFICER selected the following portfolio for Account 3

Number of Shares	Stock	Price	Total	Estimated Dividends
100	American Can Company	$38	$ 3,800	$ 200
100	Continental Insurance	53	5,300	200
100	Duquesne Light Company	24	2,400	116
100	Equitable Gas	35	3,500	185
100	General Public Utilities	26	2,600	112
100	International Harvester	45	4,500	240
100	Libbey Owens Ford	50	5,000	250
100	Socony Mobil Oil Co.	37	3,700	200
			$30,800	$1,503

Estimated Yield 4.9 per cent

By permission from Geoffrey P. E. Clarkson, *Portfolio Selection: A Simulation of Trust Investment*, Englewood Cliffs, N. J.: Prentice-Hall, Inc., 1962.

an interactive, competitive environment which is characteristic of other general management games. Participants are assigned to a team comprising the management of a hypothetical company producing a single product and competing with the other teams in a single market.

At each stage of the game each firm possesses information about its own condition, and it has certain information about the industry and market in which it is selling. On the basis of this information players make decisions for the next time period. These are submitted to the computer, which has been programmed to calculate results of the decisions. Financial statements reflecting results of the period's operations are delivered to the players, who then proceed with the next quarter's decisions, and the cycle is repeated until the desired number of periods has been completed. The A.M.A. game and the manual game designed by G. R. Andlinger had a considerable influence on the development of business games (Andlinger, 1958).

One of the first uses of a management game on a university campus was the Top Management Decision Game developed by Albert N. Schrieber at the University of Washington. This game, patterned after the A.M.A. game, was first used in the summer of 1957. At about the same time management games were constructed by a number of universities and companies, including UCLA, Carnegie Tech, IBM, Remington Rand UNIVAC, General Electric, Westinghouse, and Pillsbury.[13]

Besides general management games, which initially received the most attention, a wide variety of games oriented to particular functional areas of business management have been developed. Marketing management, inventory control, scheduling, and finance are among the areas for which games are now available.

Following the introduction of business games, their number and use expanded so rapidly that by 1961 it was estimated that there were over one hundred games in existence and that over thirty thousand executives had played them. Paul S. Greenlaw, Lowell W. Herron, and Richard H. Rawdon catalogued eighty-nine games (Greenlaw, Herron, and Rawdon, 1962); Joel M. Kibbee, Clifford J. Craft, and Bert Nanus, eighty-five games (Kibbee, Crafts, and Nanus, 1961); and a study by Alfred G. Dale and Charles R. Klasson in 1962 disclosed that two-thirds of the major collegiate schools of business were using business games (Dale and Klasson, 1964).

One of the early, and still one of the most complex, games is the Carnegie Tech Management Game. It is modeled after a specific industry—packaged detergents—and provides for three companies marketing in four regions. The game's complexity is demonstrated by the facts that players are permitted to make over three hundred decisions each period and that one to two thousand items of information are provided. Decision areas include production schedules, work force, inventory planning, plant investment, purchase of seven raw

---

[13] A description of several of these games may be found in Kibbee, Craft, and Nanus (1961).

materials, pricing, product improvement and development, distributing and advertising each product in each region, market research, and a number of financial decisions. Its complexity is such that the teams of five to ten persons must organize themselves to assign functional responsibilities to each member.

Besides these university games, a number of companies have built and used general management games, often designed to simulate the firm's operations in its industry. Understandably, information on many of them is unavailable either because they have not been successful or because they are kept confidential if they have been successful.

CONCLUSION

Many examples of applications of simulation in business can be given in addition to those in this chapter. Most early applications have been related to production and inventory problems, two areas in which much theoretical work, mathematical and statistical, has previously been done—work that on the one hand has provided a clear understanding of the structure of the problems and on the other hand has revealed that certain processes inherently difficult to attack by mathematical analysis are amenable to simulation analysis. As simulation techniques become more widely known and as problems in other areas, such as long-range planning, marketing, and distribution, are formulated and understood more precisely, applications of simulation in those areas will expand substantially.

Few applications have been repeated frequently enough to be called standard applications. "One of a kind" simulations, which do not closely resemble any preceding work either in subject matter or in structure, are perhaps more the rule than the exception. Of the areas in which simulation has been used, the analysis of inventory problems, job shop scheduling problems, and various types of waiting line phenomena are those in which the greatest amount of experience has been gained. In most other areas the investigator who chooses to use simulation is likely to find few precedents for his work.

Taken on balance, simulation as a practical tool of analysis has developed enormously. With growing understanding of the basic structure of business and economic problems and continued expansion of computing capability, an increasing number of proposed procedures for dealing with business and economic problems will be tested out by simulation before they are applied to the real world. It is not entirely fanciful to envision the day when many managers of business and economic affairs will leave a significant portion of the day-to-day work of making routine decisions to computer programs while they occupy themselves with devising new strategies, systems, and decision rules for the computer. These new systems and decision rules will in turn be tested through simulation before they are integrated into operating procedures.

REFERENCES

Allen, Morton, "The Efficient Utilization of Labor Under Conditions of Fluctuating Demand," in John F. Muth and Gerald L. Thompson (eds.), *Industrial Scheduling.* Englewood Cliffs, N. J.: Prentice-Hall, Inc., 1963.

Andlinger, G. R., "Business Games—Play One!" *Harvard Business Review,* 36, No. 2 (March-April 1958).

Bellman, Richard, Charles E. Clark, Donald G. Malcolm, Clifford J. Craft, and Frank M. Riccardi, "On the Construction of a Multi-Stage, Multi-Person Business Game," *Operations Research,* 5, No. 7 (August 1967).

Bonini, Charles P., *Simulation of Information and Decision Systems in the Firm.* Englewood Cliffs, N. J.: Prentice-Hall, Inc., 1963.

Bulkin, Michael H., John L. Colley, and Harry W. Steinhoff, Jr., "Load Forecasting, Priority Sequencing, and Simulation in a Job Control System," *Management Science,* 13, No. 2 (October 1966).

Carpenter, Charles E., "Use of General Purpose System Simulator II in System Analysis." Unpublished MBA Research Report, University of Washington, Seattle, 1964.

Clarkson, Geoffrey, P. E., *Portfolio Selection: A Simulation of Trust Investment.* Englewood Cliffs, N. J.: Prentice-Hall, Inc., 1962.

Dale, Alfred G., and Charles R. Klasson, *Business Gaming: A Survey of American Collegiate Schools of Business.* Austin: The University of Texas, Bureau of Business Research, 1964.

Didis, Stephen K., and Charles E. Carpenter, "Simulation Study of a Test Equipment Calibration and Certification System," *The Journal of Industrial Engineering,* 27, No. 8 (August 1966).

Fischer, H. C., and Gerald L. Thompson, "Probablistic Learning Combinations of Local Job-Shop Scheduling Rules," in John F. Muth and Gerald L. Thompson (eds.), *Industrial Scheduling.* Englewood Cliffs, N. J.: Prentice-Hall, Inc., 1963.

Forrester, Jay W., *Industrial Dynamics.* Cambridge, Mass: The M.I.T. Press, 1961.

Fulkerson, D. R., "Expected Critical Path Lengths in PERT Networks," *Operations Research,* 10, No. 6 (1962).

Greenlaw, Paul S., Lowell W. Herron, and Richard H. Rawdon, *Business Simulation in Industrial and University Education.* Englewood Cliffs, N. J.: Prentice-Hall, Inc., 1962.

Gross, Donald, and Jack J. Ray, "A General Purpose Forecast Simulator," *Management Science,* 11, No. 6 (April 1965).

Heller, J., "Some Numerical Experiments for an M X J Flow Shop and Its Decision-Theoretical Aspects," *Operations Research,* 8, No. 2 (March-April 1960).

Hertz, David B., "Risk Analysis in Capital Investment," *Harvard Business Review* (January-February 1964), 95-106.

Jackson, James R., "Simulation Research on Job Shop Production," *Naval Research Logistics Quarterly,* 4, No. 4 (December 1957).

Kagdis, John, and Michael R. Lackner, "A Management Control Systems Simulation Model," *Management Technology,* 3, No. 2 (December 1963).

Kibbee, Joel M., Clifford J. Craft, and Bert Nanus, *Management Games.* New York: Reinhold Publishing Corp., 1961.

Kuehn, Alfred A., and Michael J. Hamburger, "A Heuristic Program for Locating Warehouses," *Management Science,* 9, No. 4 (July 1963).

MacCrimmon, K. R., and C. A. Ryavec, *An Analytical Study of the PERT Assumptions,* Memorandum RM-3408-PR. Santa Monica, Calif.: The RAND Corporation, December 1962.

Martin, J. J., " Distribution of Time through a Directed, Acyclic Network," *Operations Research,* 13, No. 1 (January-February 1965).

Mattessich, Richard, *Simulation of the Firm through a Budget Computer Program.* Homewood, Ill.: Richard D. Irwin, Inc., 1964.

Meier, Robert C., William T. Newell, and Harold L. Pazer, *Simulation in Business and Economics.* Englewood Cliffs, N. J.: Prentice-Hall, Inc., 1969.

Morehouse, N. F., R. H. Strotz, and S. J. Horwitz, "An Electro-Analog Method for Investigating Problems in Economic Dynamics: Inventory Oscillations," *Econometrica,* 18, No. 4 (October 1950).

Packer, David W., *Resource Acquisition in Corporate Growth.* Cambridge, Mass.: The M.I.T. Press, 1964.

Report of the System Simulation Symposium. Sponsored by A.I.I.E., T.I.M.S., and O.R.S.A., New York, May 1957.

Riccardi, Frank M., *et al., Top Management Simulation: The AMA Approach,* Elizabeth Marting (ed.). New York: American Management Association Incorporated, 1957.

Rowe, Alan J., "Computer Simulation Applied to Job Shop Scheduling," in *Report of the System Simulation Symposium.* Sponsored by A.I.I.E., T.I.M.S., and O.R.S.A., New York, May 1957.

Schlager, Kenneth J., "How Managers Use Industrial Dynamics," *Industrial Management Review,* 5, No. 2 (Fall 1964).

Steinhoff, Harry W., Jr., "Daily System for Sequencing Orders in a Large-Scale Job Shop," in Elwood S. Buffa (ed.), *Readings in Production and Operations Management.* New York: John Wiley & Sons, Inc., 1966.

Sprowls, R. C., "Business Simulation," in Harold Borko (ed.), *Computer Applications in the Social Sciences.* Englewood Cliffs, N. J.: Prentice-Hall, Inc., 1962.

———, and M. Asimow, "A Computer Simulated Business Firm," in Donald G. Malcolm and Alan J. Rowe (eds.), *Management Control Systems.* New York: John Wiley & Sons, Inc., 1960.

Thomas, C. J., and W. L. Deemer, Jr., "The Role of Operational Gaming in Operations Research," *Operations Research,* 5, No. 1 (February 1957), 6.

Tonge, Fred M., *A Heuristic Program for Assembly-Line Balancing.* Englewood Cliffs, N. J.: Prentice-Hall, Inc., 1961.

———, "Assembly Line Balancing Using Probabilistic Heuristics," *Management Science,* 11, No. 7 (May 1965).

A SYSTEMS STUDY OF POLICY FORMULATION IN A VERTICALLY-INTEGRATED FIRM

EDWARD B. ROBERTS

DAN I. ABRAMS

HENRY B. WEIL

Vertically-integrated companies often provide unusual opportunities for conflicts among organizational performance measures, especially as the vertical interdependency of product flow interacts with organizational separation of profit objectives. The resulting problems and their potential resolution by policy reformulation are discussed in this paper for the case of a large manufacturer-retailer of perishable food products.

The methods of Industrial Dynamics constitute the systems analysis approach employed in the described study. Emphasis is placed upon the closed-loop feedback systems that interrelate the company's divisions and that provide the conceptual basis for development of a company simulation model. Even without computer testing of the simulation model, analysis of the feedback systems structure can be used in this case to highlight the general conflict between the company's sales and profit objectives.

Extensive computer simulation of the company model is utilized to test a variety of areas of possible policy change, including among others transfer pricing, market stimulation, budgeting, and performance measurement. A consistent finding from the simulations was that several sources of intensified organizational pressure (including negotiated transfer prices and ambitious budgets) contributed importantly to company growth.

Although the model formulation and simulation results were accepted as valid by company managers, little implementation of the recommended policy changes has yet occurred. Instead the principal benefits in the short run are seen as largely educational in nature, with new managerial perspectives the key outcome. This result appears reasonable and possibly necessary if management is to be educated effectively for later acceptance of policy change recommendations based on sophisticated management science approaches.

Organizations formulate policies, make decisions, and undertake various activities as responses to gaps between their objectives and their measured performance. As a firm's objectives shift relatively slowly, changes in its per-

Reprinted with permission from *Management Science*, 14 (August 1968), B-674-B-694. Copyright © 1968 by The Institute of Management Sciences.

formance gaps, hence in its behavior, can be created simply by changes in the measures used to monitor the firm's performance. Thus one of top management's more important tasks is the selection of a set of performance measures that will lead to behavior consistent with achievement of company goals.

This is not an easy task. The typical business enterprise has multiple goals that are changing in relative importance over time. Organizational responses to different types of performance measurement and reward schemes are not immediately obvious, especially when multiple (and often conflicting) measures are applied simultaneously.

The opportunity for conflict among performance measures, and therefore the lack of clarity as to how the measures mold behavior, is probably greatest in vertically-integrated companies. Not only do these firms offer the usual possibility of individual measures conflicting to some extent (e.g., sales and profits), but also the opportunity exists for interdivisional conflict as a result of the measurement process (e.g., the frequent outcome of applying the profit center concept).

This paper describes the use of Industrial Dynamics for investigating, understanding, and experimenting with the process of goal achievement in a vertically-integrated firm. Industrial Dynamics is a feedback systems-oriented approach to dynamic socio-economic problems. Its techniques include feedback systems analysis and computer simulation of complex models.[1] Based on the study of a vertically-integrated manufacturer and retailer of food products, the paper will discuss the desirability of interdivisional competition, the effect of pressure on performance, and the relative merits of a number of alternative performance measures.

METHODOLOGY

Although other approaches might have been adopted, Industrial Dynamics was considered appropriate for this study for several reasons.

1. As shall be indicated, the interaction of different levels in the vertical organization forms a closed-loop feedback-type system of relationships.

2. With Industrial Dynamics techniques it is relatively easy to model qualitative behavioral relationships such as reactions to organizational pressures and incentives.

3. Empirical data were not available sufficiently to permit use of statistical techniques for deriving some critical market relationships (e.g., retail price elasticity, consumer response to advertising). The Industrial Dynamics methodology does not insist upon such data availability, although added confidence in the model formulation does result when derivation of relationships can be enhanced by statistical analysis methods. Given these data unknowns, however, Industrial Dynamics does offer the advantage of providing an easy vehicle for testing the sensitivity of company performance to deviations from the initial estimates made of the relationships. (Any other approach that utilizes efficient simulation techniques would also provide this benefit.)

---

[1] The basic background text in the field is Forrester (1961). An early general discussion of the principles of the approach is Roberts (1963).

4. The relatively low mathematical sophistication required by the Industrial Dynamics approach enhances (but does not assure) the chances of management understanding of the system analysis effort. In the area of policy formulation it is likely that such enhanced understanding will for long be the principal direct and immediate result of a system study. Much more time than is consumed by the study will usually be required before any change in managerial understanding is reflected clearly in company policy and performance.

THE COMPANY STUDIED

The company studied is a large diversified corporation with activities in the supermarket, department store, food manufacturing and restaurant fields. The findings presented resulted from a several months' study conducted within the vertically-integrated retail food and manufacturing divisions accounting for sales of several hundred million dollars.

Retail Food Division

The company owns and operates a large chain of supermarkets in North America, enjoying an excellent growth record. The number of outlets has been expanded ten-fold since 1945, while retail food sales have grown by a factor of about 25 in the same period.

This division is characterized by two rather dissimilar activities:

1. operation and control of the supermarkets, each of which is a profit center; and
2. performance of a centralized merchandising procurement, and distribution function.

Included in the latter activity is responsibility for promotion and retail pricing, as well as the planning of each week's program of featured special merchandise items.

For certain classes of products, corporate policy dictates that most or all of the goods sold by Retail be "bought" from the company's Manufacturing Division. Under current company practices this means that the negotiation of transfer prices for these items is also an important activity.

The Retail Division's performance is measured in terms of both sales volume and profitability. More specifically, performance is compared with a monthly dollar sales budget and a desired retail gross margin percentage. The Division's performance is monitored quite closely by the company president.

Manufacturing Division

Within the past five years, the company has become heavily engaged in the manufacture of private label products for sale primarily in its stores. Although the manufacturing activities were first operated by the Retail Division, they soon grew to such proportions that financial control considerations dictated the establishment of a separate division.

As the Retail Division is Manufacturing's sole customer for most of its output, and because Manufacturing is capital-intensive, performance of the Manufacturing Division is measured almost exclusively in terms of profitability. Specifically, the Division's income contribution (i.e., its accounted profits before allocation of corporate overhead charges) has been used as its principal performance measure.

The Product

A new bakery facility, requiring substantial capital investment, had recently been added to Manufacturing. In light of this it was decided to limit the scope of the study to the manufacture, distribution, and sale of one class of product—bakery goods. The major portion of the company's manufacturing investment is contained in this area.

Bakery products are characterized by high perishability (i.e., they must be sold within one to two days of production), impulse sales, and high substitutability within commodity groups (e.g., cakes). The latter characteristics of impulse sales and substitutions are prevalent in many consumer goods (e.g., detergents, apparel, gasoline). The perishable nature of these products precludes extensive use of inventories as buffers to uncouple either production from orders by the supermarkets (for the company) or demand from actual product availability (for consumers). Thus, management is faced with the classic "Christmas tree" or "newspaper" problem.

A PRODUCTION-SALES
MODEL OF AN INTEGRATED FIRM

The development of a model of a complex organization is usually a task of moderate-to-significant difficulty. In addition to overcoming the problems involved in finding a purposeful focal point for the organizational model, the model builder must effectively conceptualize the model's overall structure and carefully define the model's functional relationships. A variety of data gathering and analysis approaches are employed in these several stages of organization model development. But regardless of the care exercised by the model builder the resulting model can always be subjected to question on the grounds of model validity.

In developing the model of the production and sales activities of the company described in this paper, several different data sources were used. Most important among them were the personal observations of the several active participants in the model development, the plentiful information gained through extensive interviewing, and the results of analyses of a variety of company reports covering a several year period. Despite the concerns of the model builders, the resulting model is admittedly still a simplification of reality, but does represent the authors' best conceptualization of the major behavior-determining decisions and the ways in which they interact. When completed the model included equations for 110 variables written in the DYNAMO compiler language.

A model of this scope is far from "intuitively obvious," particularly to the manager not experienced in management science endeavors. Thus special attention was given throughout the model development to the gaining of managerial acceptance of the model. Although it was not possible to derive every equation rigorously from empirical evidences, the assumptions underlying each relationship were documented and discussed with participating managers. One satisfying measure of model validity is that the model was wholly accepted by company management as being sufficiently realistic for policy testing before actual computer simulations began.

Presentation of the model centers around a discussion of two related flow streams: the flow of orders and goods, and the flow of cash. Continual reference will be made to the accompanying diagrams of these flows.

The Flow of Goods and Orders

Figure 1 shows the flow of goods to the market place in response to orders generated at two points within the system. The discussion begins at the top of the figure and follows the flows downward.

One set of orders for bakery goods originates at the store level; these are orders for non-special items. "Non-specials" are items that during a particular week are not being featured at special low prices or with bonus stamp offerings. The specials are changed each week. In placing these orders, the person in charge of each store's bakery department relies quite heavily on historical data. This individual is called "the dairyman," as in addition to the bakery department he is responsible for the store's dairy and frozen foods sections.

Orders for items to be featured as weekly specials are generated by the Merchandising Group based on its decisions on how much to allocate to the stores. Historical data again form the basis for ordering, but Merchandising deviates from the amounts so indicated in response to pressures from higher up in the Retail Division. A substantial body of evidence supports the existence of these pressures, including monthly Merchandising Meeting minutes and statements by Retail Division managers. When under pressure to increase sales, Merchandising tends to order more than is indicated by past data. Based on interviews, it was conservatively estimated that increased allocations of up to 10% might occasionally occur. This behavior is motivated by a desire to make certain that there are sufficient goods in the stores to capitalize on the increased market potential created by sales-boosting efforts. It represents an attempt by Merchandising to increase sales by pushing goods into the stores. On the other hand, Merchandising is under pressure to be very conservative in its store allocations of special items whenever gross margin percentage is low. In such situations Retail's top management gets concerned about the problem of "reductions," i.e., those sales at 1/3 off that are necessitated when the perishable goods remain unsold at the expiration date marked on the merchandise.

Thus, two sets of orders flow into the Bakery manufacturing organization and serve as a basis for production planning. But this does not mean that the amount produced and shipped exactly equals the amount ordered. Rather,

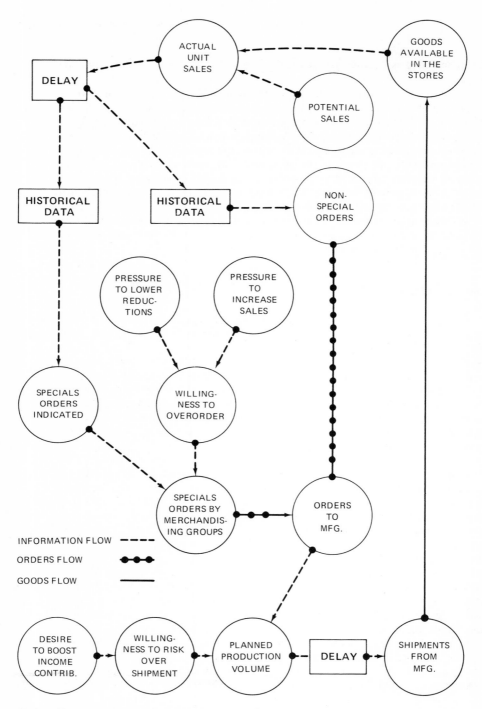

FIGURE 1 The flow of goods and orders.

Manufacturing has the option to modify orders, and both the magnitude and direction of these modifications are affected by the desire to increase Manufacturing's accounted income contribution. Specifically, Manufacturing has tended to overship in an effort to improve its income contribution through increased volume. Interviews disclosed that 10% overshipment occurred infrequently, but that 5% was not uncommon. After lead-time and production delays, Manufacturing ships its goods to the stores.

Referring to the upper left-hand corner of Figure 1, one sees that the flows of goods and orders really form a closed loop. Orders based on historical data determine production, which in turn limits sales, as Retail cannot sell more than Manufacturing has produced and shipped. Today's sales, of course, become part of tomorrow's historical data base.

Potential Sales

Figure 2 shows the six major determinants of potential retail unit sales:

(a) the amount of effort devoted by dairymen to bakery products;
(b) dollars spent on advertising and in-store promotion;
(c) retail prices;
(d) the number of bakery specials;
(e) the price elasticity of bakery items, especially of the specials;
(f) total market size.

It is hypothesized that a dairyman has a definite effect on sales through his ordering and by his handling of displays and in-store promotion. Interviews disclosed an opinion among Retail and store operations personnel that the difference between a bad dairyman and one doing an outstanding job can be as much as 25-30% in unit sales.

It is the opinion of several of the company's executives that bakery items are often bought on impulse, and as a result unit sales are very responsive to increased advertising expenditures. For example, market reactions to "blockbuster" (or key feature, hard-sell approach) advertising supports this. Given the fact that there is some (but unfortunately an unknown) degree of responsiveness to advertising expenditures, several alternative linear market response curves were used in model simulation experiments.

Experience with specials has provided ample evidence that unit sales rise when prices are cut. Again, the exact shape of the response curve was unknown, but our assumption (Alternative A in Figure 3) appears to be quite reasonable in light of Retail's experience with specials. As examples, one variety of bread at half price sells three times the normal amount; a popular pie at sixty percent of its regular price sells over five times the normal amount, but partially at the expense of other items.

Normally five bakery products are selected as specials each week, and past records indicate that these consistently contribute a substantial portion of total bakery goods sales. From the data available, it seems reasonable to assume that a sixth special might boost sales 20% if it were a normal one and 20% if Retail

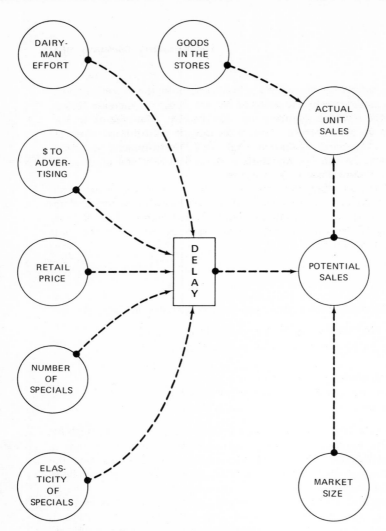

FIGURE 2   Factors affecting potential sales.

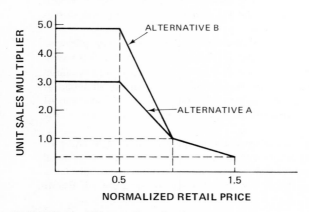

FIGURE 3   Effects of retail price.

offers a "super-special." In supporting this formulation Bakery Merchandisers gave several examples of how "super-specials," such as 29-cent apple pie (regularly 49 cents), have been effective in the past.

The last sales multiplier under management's direct control is the price elasticity of items featured as specials. The model includes an estimated 50% difference between maximum and minimum elasticities, which seems quite conservative in the context of the data.

Cash Flows

Starting near the upper-left hand corner of Figure 4, the diagram indicates that dollar sales are merely unit sales (see the flow of goods and orders, Figure 1) times average retail price. When dollar sales compare unfavorably with the budgeted level, the interviews and Merchandising Meeting minutes indicate that pressure built up within the Retail Division to take corrective action. It should be recognized that this desire to remove an unfavorable sales variance can be amplified by a general pressure for the whole company to boost sales.

Under pressure to increase sales, Retail Division's management responds through the five sales multipliers discussed in the preceding section. These responses, as evidenced in the company records, are:

(a) through personal contact and by its weekly newsletter, the Merchandising Group exerts pressure on the dairymen to stimulate a higher level of attention to bakery products;

(b) expenditures on advertising and in-store promotion are increased, and one or more "blockbuster" ads may be featured;

(c) the Merchandising Group selectively lowers retail prices, through lower pricing of new products or greater price cuts for specials;

(d) the number of weekly specials may temporarily rise to six;

(e) items with large price elasticities are chosen as specials.

After lead-time delays in achieving their effects, the impacts of these stimuli are felt in the market place. After reporting delays, an improvement in sales becomes apparent; the unfavorable variance is reduced. Thus, sales performance and the management decisions that control it form a definite closed loop.

Referring again to Figure 4, there is a second control loop involving Retail gross margin percentage (center of the diagram). Under pressure to improve an unsatisfactory gross margin, Retail is quite unwilling to lower prices. Prices are cut only when there exists simultaneously a significant desire to boost sales. On the other hand, the interviews showed that Retail rarely increased regular prices, even to improve gross margin. Further, Retail top management becomes quite concerned about the problem of reductions on "expired date" merchandise when gross margin is low. The Merchandising Meeting minutes clearly show that these concerns encourage the Merchandising Group to be conservative in its allocations to the stores.

It is apparent that management's efforts to boost sales conflict with its efforts to increase gross margin, especially in the areas of pricing and production policies. To increase sales, one lowers prices (among other things) to create

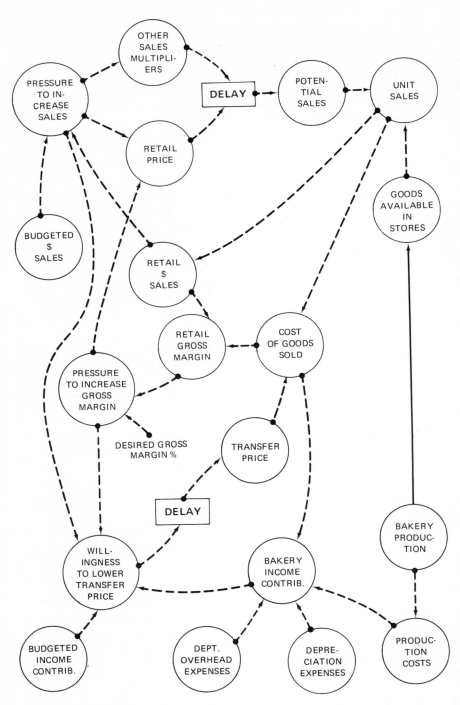

FIGURE 4   Cash flows.

expanded market potential and ships more to the stores to capitalize on the additional demand. To improve gross margin, however, management does the opposite. There are no price cuts and Merchandising is under pressure to be quite conservative in its allocations to the stores so as to avoid excessive reductions. In the section on the system's dynamic behavior, it will be shown that this conflict (and its resolution) is a critical determinant of how well this part of the company performs.

Turning now to the last of the three performance measures, Manufacturing income contribution equals dollar sales (at transfer prices) to Retail minus production costs, depreciation, and departmental overhead. When income contribution is low, Manufacturing takes the initiative by offering lower transfer prices to Retail in exchange for expected (but not guaranteed) larger volume. Further (as indicated earlier in this section), Manufacturing ships more than is ordered to increase its volume even more. So long as transfer prices are not reduced too much, the increased volume so generated improves Manufacturing's income contribution position.

THE SYSTEM'S DYNAMIC BEHAVIOR

The System's Control Loops

The preceding section describes a production and distribution system in terms of its structure and the rationale behind it. Now, let us turn to the ways in which Manufacturing, the centralized Retail activities and the stores interact through the system's three major control loops (Figure 5).

It is apparent from this diagram that the interactions among the three performance measures (i.e., manufacturing income contribution, retail dollar sales, and retail gross margin percentage) form a complex feedback system that determines company behavior.

The first control loop represents performance measurement of the Manufacturing Division. In response to pressure to correct an unfavorable income contribution variance, Manufacturing attempts corrective action. It ships more to the stores than was ordered and endeavors to stimulate more aggressive sales promotion at Retail by offering lower transfer prices. The longer lead time required for stimulating increased retail sales suggests that the effects of the overshipment are felt first. Overshipments result in increased "reductions" caused by quantities above historical levels being shipped to the stores without attendant sales-boosting efforts by the Retail Division. The immediate result of this, at worst, is a drop in the gross margin percentage that causes Retail to be more conservative in its allocations to the stores. This causes manufacturing income contribution to decline even further than before (because of the lower transfer prices). At best, Retail uses the lower transfer prices to maintain an acceptable gross margin percentage at the same sales level as before the concerted overshipping began, leaving Manufacturing with about the same income contribution variance as it faced prior to lowering transfer prices and overshipping.

Turning now to the second control loop, consider an instance in which

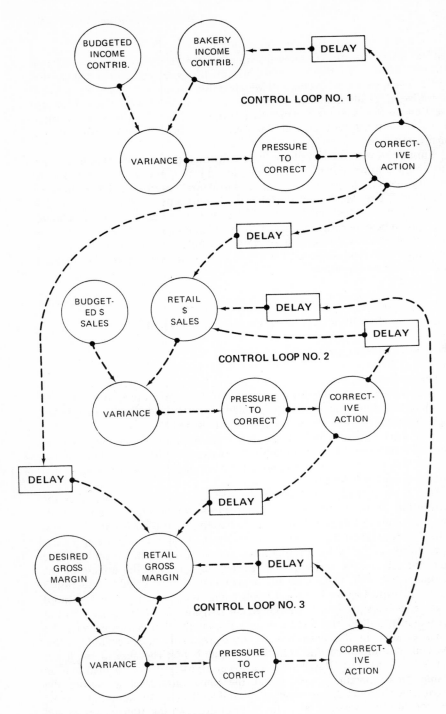

FIGURE 5   The system's control loops.

sales have fallen below their budgeted level. Retail's response is a relaxing of prices, an increase in advertising expenditures, and the possible offering of more specials. Pressure increases on the manufacturing group for more sales results in specials with larger price elasticities, allocation and ordering in excess of historical amounts, and pressure on the dairymen. A diagram of this process is shown in Figure 6.

However, none of these actions has an immediate impact on sales because all require many weeks of lead-time. In fact, the unfavorable variance becomes worse as built-in growth in the budget causes the budget targets to rise each week. Especially when under pressure from the President, Retail intensifies its corrective measures. This is to be expected, since for many weeks management does not have an opportunity to measure the success of its first attempts.

After some weeks the first effects of Retail's efforts are apparent as sales begin to recover. The market-place's full response follows shortly, and everyone seems much happier.

Sales are rising, but lower prices and over-ordering already have had a depressing effect on gross margin. Thus, with sales in relatively good shape, Retail's top management naturally turns its attention to this second area, improvement of profits, shown as Loop No. 3 in Figure 5.

Retail's natural reaction to a low gross margin is unwillingness to lower prices below their normal level. In addition, pressure is exerted on the Merchandising Group to boost gross margin and cut reductions. Thus, Merchandising responds by being quite conservative in its choice of specials and its allocations of merchandise to the stores. Figure 7 shows this in diagram form.

The tendency to continue in this pattern of behavior is intensified by Manufacturing income contribution's close relation to dollar sales, both of them being functions of unit sales. Manufacturing's usual corrective action in this area has a definite multiplier effect on Retail's efforts to raise sales. Manufacturing tends to over-ship, puts pressure on Retail for increased sales, and often lowers transfer prices as an incentive.

A Simulation Run of the Basic Model

In this section of the paper a simulation run is used to illustrate the results of the behavior described previously. For the purpose of the simulation, budgeted growth was set at 10% per year and annual growth of the potential market was set at 5%. In addition, there was no new investment over the 5 year simulated period. Initial conditions of the model were set to correspond with 1962 company data.

From examination of the graphical computer output in Figure 8, two general observations can be made. First, there are definite short-term fluctuations in dollar sales, retail gross margin dollars and manufacturing income contribution. Second, none of these performance measures attains its budgeted level, but all are able to grow at a rate in excess of the market's 5% growth. These five year results were compared with the actual company performance during the 1962-1966 period. Management's satisfaction with this comparative test provided further validation of the company model.

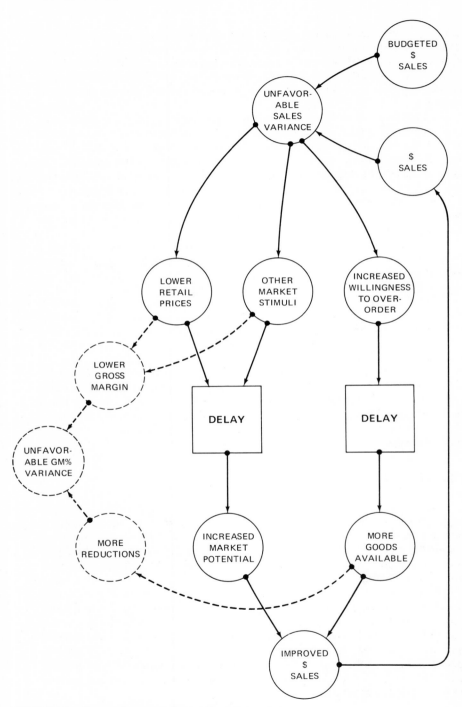

FIGURE 6   Correcting an unfavorable sales variance.

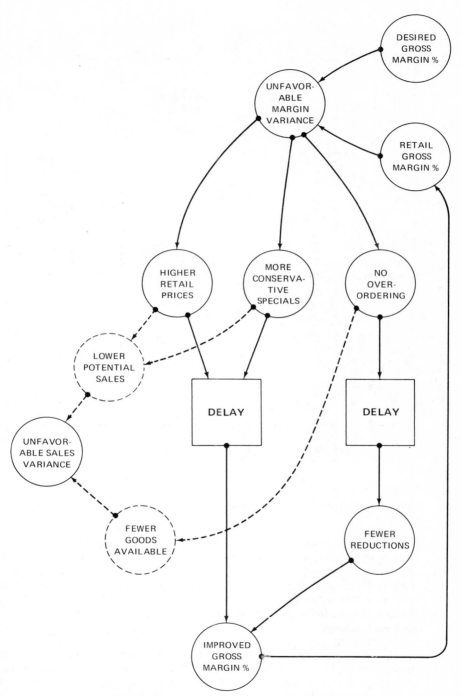

FIGURE 7   Correcting an unfavorable gross margin variance.

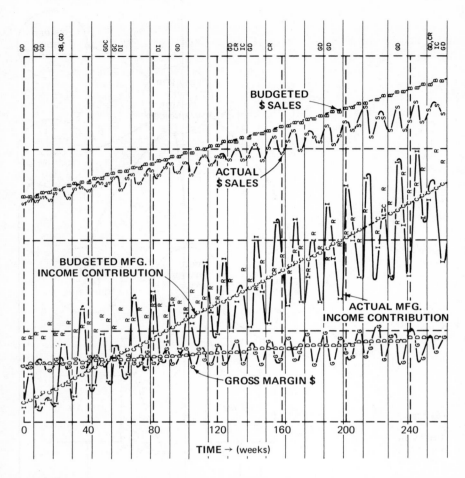

FIGURE 8   Basic simulation run, 1962-1966 (budgeted growth = 10% per year, market growth = 5% per year).

The short-term oscillations result from the conflict in behavior caused by employing dollar sales and gross margin percentage as performance measures as discussed in the previous section (see Figures 6 and 7). Therefore, let us now turn to the second phenomenon. Why is it impossible to grow at the budgeted rate when this rate is faster than the growth of the market? Again, the cause is the conflict between dollar sales and gross margin percentage as Retail performance measures.

Assuming that prices, advertising expenditures, and the other sales multipliers (Figure 2) stay at their normal levels, sales cannot grow faster than the market. In the model these "normal" levels are, by definition, those which are just sufficient to maintain constant market share. Thus, growth faster than the market can only be achieved with lower than normal prices, greater than normal advertising, or a combination of these and other stimuli.

In the simulations, the market is responding to price much more than it is responding to advertising. This is in line with the disclosure from the interviews that Retail seemed much more willing to use price as a stimulus than to increase advertising expenditures. The latter variable remains relatively fixed in current practice. Hence, if gross margin were ignored, Retail could progressively lower prices to keep up with its sales budget.

One cannot, however, ignore gross margin. After all, sales volume would grow amazingly if goods were given away. Thus, there must be—and is—a compromise: prices are cut to a point where pressure to raise them as a help to gross margin exactly equals pressure to lower them as a stimulus to sales. The result of such a compromise is growth faster than the market, but slower than the budgeted rate.

Summary

From the preceding discussion, one concludes that performance measures are important determinants of behavior within the system. If there is inherent conflict between performance measures, then organization behavior reflects that conflict. While the resulting overall performance may be "satisfactory," it falls short of the firm's potential.

Finally, it is important to review exactly how growth is achieved. Lowering prices, as well as using other stimuli (Figure 2), increases the potential market for bakery goods. However, no manufacturer can sell more than he produces. Thus, the key to growth is simultaneous market stimulation and increased output to capture the newly-created demand.

In the company studied, production is based on orders from Retail. These orders, however, seem often to be intuitive straight-line extrapolation of historical growth.

There must be more to it than that, however, or the company could never grow out of a stagnation period. The keys to accelerated growth in production are the pressures within the system that cause the Merchandising group to abandon historical guidelines and allocate more goods to the stores and the pressures that cause the Manufacturing Division to ship more than is ordered. Thus, perhaps unfortunately, only the violations of present formal company policy enable the company to break out of the trap of past performance and to produce enough to capitalize on a more rapidly growing future potential. It seems worth noting that similar breaking of corporate internal rules, such as by "bootlegging" research projects, have in other companies also provided the key to achieving growth objectives.

COMPARATIVE SIMULATION RESULTS

In addition to utilizing the model to understand and investigate system behavior, it was employed to investigate alternative policies in a number of areas. These policy areas include:

(a) transfer pricing;
(b) retail pricing;

(c) promotional expenditures;
(d) performance measuring;
(e) pressures within the organization;
(f) budgeting.

Rather than discussing each simulation run individually, this section summarizes the general points concerning policy formulation and goal achievement.

Interdivisional Competition

The principal way in which the Retail and Manufacturing Divisions compete is for shares of the total profit that results from making and selling an item. The split is, to a large extent, determined by transfer prices.

Three possible transfer pricing policies were investigated:

(a) negotiated transfer prices;
(b) costs + % of costs;
(c) costs + constant $ mark-up.

The percentage and constant mark-ups were chosen so that they agreed with the initial negotiated price, thereby making the simulations directly comparable.

Despite its aura of randomness and disconnection with costs, the policy of negotiated transfer prices proved to be superior in both 5 year cumulative $ sales and 5 year cumulative total profits (see Table 1). This is true even though Costs + Constant Mark-up (the next best policy) produced higher unit sales.

Both of the alternatives to negotiated transfer prices give Retail an increasingly larger share of total company profits as unit sales grow. The reason for this is that Bakery's mark-up is a constant while manufacturing costs per unit fall. Thus, transfer price drops as unit volume grows, causing Retail costs to decline substantially. This is not unlike the effects produced by a quantity discounting scheme. However, under these alternatives Retail receives an amazingly high gross margin, which (in both cases) it proceeds to "give away" by cutting prices.

In periods of stagnation, these three policies are virtually identical, but under conditions of rapid growth, negotiated transfer prices make more money for the company. With a fixed or nearly fixed mark-up, the Manufacturing Division has little trouble meeting its income contribution budget. Thus, it has

ABLE 1   Effects of Alternative Transfer Pricing Policies

	Normalized Cumulative 5 year SALES $	Normalized Cumulative 5 year PROFIT $*	Normalized % of 5th Year Profit to Retail
egotiated Transfer Prices	1.00	1.00	1.00
osts + Constant Mark-up	0.98	0.96	1.31
osts + % of Costs	0.97	0.93	1.39

*Profit = Retail Gross Margin $ + Bakery Income Contribution

no reason to over-produce, and production tends to lag the market potential created by Retail's low prices.

In this context, interdivisional competition is not the ogre conjured by those who believe that attempts to optimize on the Divisional level inevitably lead to sub-optimization of corporate achievement. On the contrary, in this type of vertically-integrated organization, interdivisional competition for profits stimulates volume-consciousness in Manufacturing and prevents excessive promotional price cutting by the Retail Division.

Performance Measures

The introduction to this paper contained the assertion that performance measures strongly influence behavior. The computer simulations provide excellent illustrations of this phenomenon.

Three basic performance measures are currently being used in the system:

(a) retail dollar sales;
(b) retail gross margin percentage;
(c) manufacturing income contribution.

Two of these, sales and income contribution, are accounting quantities that reflect the level of the company's activities. They are dynamic, growing as the company grows.

Gross margin percentage, on the other hand, gives only a very incomplete picture of Retail's profit performance. One can only spend gross margin dollars, not gross margin percentage. By concentrating on gross margin percentage, management does not take advantage of the possibility of reducing gross margin but realizing higher total profit through greatly increased sales.

In addition, Retail gross margin percentage has evolved into something more than just a performance measure. It is a tradition, an institution, not only in the company studied but throughout the retailing industry. While budgeted sales and income contribution are accepted as products of the accounting system, the expected gross margin percentage is different. It is a magic number (or sacred cow). As a result, it takes years to change people's idea of what a proper gross margin should be.

In the simulated replacement of negotiated transfer prices with costs plus a fixed percent mark-up, it was shown that at one point Retail realized a gross margin considerably in excess of its traditional performance. Yet, slowly, this increased profit was lost because Retail lowered prices. Gross margin in excess of the historic target was spent in attempts to boost sales.

Similarly, simulated price reductions of 10-20% created a vastly greater potential market for bakery goods. However, the potential went unrealized because Retail, disturbed by gross margins below traditional levels, was too conservative.

The simulation runs demonstrated that the use of a performance measure that was decoupled from sales volume had a definitely negative effect on both sales and profit growth. Thus, it was recommended to the company management that gross margin dollars be substituted as a measure of Retail profitability.

Effects of Pressures on Goal Achievement

Two types of pressures affect behavior within the bakery goods manufacture and sales system:

(a) internally generated pressures arising from management's desire to perform well in terms of sales, gross, margin, and income contribution;
(b) pressures generated by sources outside the bakery products system (the President, for example).

The objective of this portion of the paper is to show that these pressures are an important determinant of corporate growth.

The first series of simulations concentrated on the impact of pressures on the Merchandising Group. As shown in Table 2, the amount of pressure has a definite effect on both five-year sales and five-year profits. While the initial differences are not huge, they do grow, and the correlation is inescapable between pressure and profits.

These results are strong argument against any form of automated merchandising that eliminates the Merchandising Group's ability to feel and respond to pressures.

TABLE 2   Effects of Various Forms of Pressure

	Normalized Cumulative 5 Year SALES $	*Normalized Cumulative 5 Year PROFIT $*
No Pressure	0.98	0.98
Normal Level of Pressure	1.00	1.00
Sales Pressure Increased by 100%	1.05	1.08
Sales Pressure Increased by 300%	1.11	1.17
Profit Pressure Increased by 100%	1.02	1.07

NOTE: Negotiated transfer prices were used in all of the above simulations.

The Merchandising Group's willingness to allocate more to the stores and its willingness to choose high-elasticity specials are the most direct ways by which it contributes to accelerated growth in the simulation runs. While increasing pressures by 100% to 300% is quite drastic, additional simulations showed that the same improvements can be realized if incentives of some sort make the Merchandisers more responsive to current pressures.

A second set of simulations investigated the effects of Manufacturing overshipment. Despite wide-spread feelings in the Retail Division that Manufacturing should merely produce what is ordered, simulations show that overshipment is a significant growth stimulus. It permits the company to take advantage of the market potential created by advertising and low prices, and its elimination reduces five-year profits by about 10%.

An earlier part of this section discussed how negotiated transfer prices led to better performance than a fixed mark-up above costs. This, too, was traced to pressures. The fixed mark-up made it easy for Manufacturing to achieve its budgeted income contribution and, thereby, removed from Manufacturing its incentive to increase volume.

The simulation results in Table 3 highlight the effects of pressure as a stimulus to growth:

It can be seen that as each pressure is eliminated, five-year performance grows worse. Under conditions of constant mark-up, no Manufacturing over-shipment, and no pressure on the Merchandising Group, it becomes impossible for sales to grow faster than the market.

In the same way that removal of pressures has a detrimental effect on performance, the application of more pressure through higher budgeted rates of growth leads to significant improvements. With the market growing at 5% per

TABLE 3   Impacts of Pressures are Additive

	Normalized Cumulative 5 Year SALES $	Normalized Cumulative 5 Year PROFIT $
Negotiated Transfer Prices; Other Pressures Normal	1.00	1.00
Constant Mark-up; Other Pressures Normal	0.98	0.96
Constant Mark-up; No Manufacturing Over-shipment	0.93	0.86

annum, the budgeted growth rates in Table 4 were used.

In every case, higher budgeted growth leads to higher actual growth. Although these very high growth-rate budgets were never achieved, they stimulated growth several times faster than the market.

As a result of these findings, incentive schemes were recommended to make critical areas of the system more pressure-sensitive.

TABLE 4   Budgets as Stimuli to Performance

Budgeted Annual Growth Rate	Normalized Cumulative 5 Year SALES $	Normalized Cumulative 5 Year PROFIT $
15%	1.00	1.00
23%	1.08	1.09
30%	1.19	1.20

NOTE: Negotiated Transfer Prices; Gross Margin $ used.

RESULTS TO DATE

At various points during the study several meetings were conducted with key company executives. These sessions covered several topics: (1) the feedback systems conceptualization of the company, shown earlier as Figures 1 and 4; (2) the detailed assumptions, including parameter values, that were included in the model equations; (3) the principal simulation results, emphasizing those covered in Tables 1 through 4; and (4) the implications of the results with accompanying recommendations for policy. It is reasonable to point out that some aspects of the study effort, in particular some parts of the system conceptualization, were not accepted immediately by all the affected management personnel. However, an encouraging fact is that by the end of the study general agreement did exist (among the company president and the vice presidents of retailing, manufacturing, and finance) that the feedback concepts and the model did embody the essential characteristics of the organization.

The response to the study's completion (at least, temporary completion) has not been a rush to implement the specific recommendations derived from the computer results. This appears to be too much to hope for from an initial use in a company of management science tools aimed at policy formulation problems. Longer familiarity with the concepts, tools, and analysts is probably needed before specific policy changes will directly follow simulation outcomes.

However, the study has resulted in an ongoing series of meetings, both formal and informal, in which the company's basic operating practices and performance pressures have been reexamined. Furthermore, in addition to encouraging this new dialogue, the systems study described in this paper has supplied a new conceptual framework within which to conduct the examination. The study apparently has contributed significantly to a greater understanding of the complexities and interdependencies of the firm's activities in this vertically integrated sector. Finally the study has underlined the need for divisional executives to consider the company-wide effect of decisions previously regarded as of divisional importance only.

The experiences encountered in this study reemphasize the challenge to management scientists posed by the policy formulation (and other) problems of top management. They also point a way by which eventual success may be attained. If management science is to alter the way by which the corporation is directed, early efforts must be launched at issues critical to top management. But the objectives of these early studies in a firm need not be immediate policy change—such an objective is probably unrealistic in most organizations. Rather the effort should be aimed at top management education in the management science process and possibilities via exposure to studies of real company problem areas. Direct impact on managerial attitudes and understanding will pave the way to later policy change.

REFERENCES

Forrester, J. W., *Industrial Dynamics,* The M.I.T. Press, Cambridge, 1961.

Roberts, E. B., "Industrial Dynamics and the Design of Management Control Systems," *Management Technology,* Vol. 3, No. 2 (December 1963).

10

Marketing System Simulations

PHILIP KOTLER
RANDALL L. SCHULTZ

The American marketing system is a huge, complex network of marketing firms and facilities organized to carry out the task of distributing over one trillion dollars of goods and services each year. It is a system that accounts for 30 percent of the gainfully employed persons in the United States and costs the consumer over forty cents of each retail dollar (Cox, 1965). It operates within an economic system characterized by a high rate of product innovation, multiple instruments and channels of marketing communication and persuasion, and constantly evolving patterns of buyer wants, attitudes and behavior. The institutional and theoretical details of the American marketing system are being studied with increasing rigor by economists, behavioral scientists, and marketing scholars. The two questions most often asked are, How does the system work? and How can it be improved?

These questions are being researched in a variety of ways. Many of the early marketing studies were descriptive, taking the form of case studies, taxonomies, and statistical tabulations of marketing activity in the United States. The last decade, however, has witnessed an increasing amount of analytical and theoretically oriented research. The credit belongs largely to the rapid diffusion of *behavioral concepts* and *quantitative tools* in this and other business fields. Behavioral concepts that were developed in sociology, psychology, and cultural anthropology brought new life and understanding to the study of buyer behavior and the organizational behavior of marketing agents and institutions (Howard, 1965).

Quantitative tools came into marketing largely as a potential means for helping marketing executives make decisions in areas of high uncertainty, such as launching products, setting advertising budgets, and designing sales force strategies. Early model-building in marketing was characterized by an effort to impose standard operations research tools—calculus, linear programming, queuing theory, Markov processes—on highly simplified versions of marketing problems (Bass *et al.,* 1961). For example, consumer brand selection was modeled as a simple first-order Markov process even though this left out competitive marketing strategy, in-store purchase factors, mass communications, and word-of-mouth influence in the buying process (Maffei, 1960). The problem of optimal advertising media was modeled in straightforward linear programming terms even though this meant overlooking media duplication and replication of advertisements and the prevalence of nonlinear media discounts (Day, 1962). In time it was recognized that the simplified problems bore insufficient resemblance to the real problem structures facing marketing executives. Marketing problems are characterized by lagged effects, nonlinear relations, interacting variables, and stochastic processes that often call for original and complex models (Kotler, 1965b).

It is not surprising that a growing number of marketing scholars have turned to computer models and simulation to seek improved understanding and control of complex processes in the marketing system. Many marketing simulations have already appeared and more are under development. They all proceed from the conviction that too much is lost in macro models of marketing relationships. There is a strong desire to delineate marketing problems on a micro-process level and then seek their macro-model equivalents. The emergence of complex models of the marketing system has greatly been aided by the progress made in computer and simulation science.

Overview

This chapter will undertake to review and evaluate a selected sample of marketing simulations. The term *simulation* is broadly interpreted to reflect the variety of ways it is used by persons inside and outside of marketing to describe their work. At least four different usages of the term *simulation in marketing* can be distinguished.

1. *Simulation as behavioral modeling.* The most popular connotation of simulation in marketing is that it amounts to constructing a model that appears to imitate the real-time process of a system. Simulation and imitation are here synonymous. This meaning of simulation is exemplified in Amstutz's model of the shopping behavior of consumers, Alba's model of the adoption of touch-tone telephones, and Mayer's simulation of the field interviewing process. All these models attempt to imitate the essential real-time behavioral characteristics of a system.

2. *Simulation as a way of introducing and handling uncertainty.* Marketing behavior and response are characterized by uncertainty, and simulation can serve as a vehicle for expressing and studying this un-

certainty. This use of simulation is found in the Pessemier model, where the future history of new produce prices, sales, and profits is generated stochastically, and in the Simulmatics model, where households are exposed to media on a probabilistic basis. While stochastic modeling is not coextensive with simulation, many simulations include stochastic elements.

3. *Simulation as a computational technique for measuring parametric sensitivity.* A common problem in all models is to determine the effect of variations in input data and/or structural parameters on output results. One reason for doing this is to compensate for inaccurate data by seeing how the output will be affected by alternative estimates; another reason is that initial conditions in a simulation are often arbitrary and it is important to determine whether the output is significantly affected by different starting conditions. In simple analytic models, it is often possible to solve directly for the effect of altered parameters on output. In more complex models, it may be necessary to run the model for each possible systematic variation in the parameters. Tuason used simulation in this way to determine the performance of his strategy program in a variety of different environments, and Cloonan used it to vary the characteristics of territories to test the efficacy of different sales call decision rules. Many of the other models unfortunately did not get to the stage of examining how the results were affected by variations in input data or structural parameters.

4. *Simulation as a heuristic technique for finding an approximately optimal solution.* Simulation can be used as a systematic technique in searching for improved solutions to marketing decision problems. When a model is too complex to express or solve analytically, the space of possible solutions can be searched at discrete intervals to explore the profit response surface. This use of simulation was made by Urban in his new product decision model and consisted of investigating the payoff characteristics of a sample of different marketing programs and then examining the neighborhood of the best solution through a finer probing interval.

Given this broad interpretation of simulation, a choice had to be made of what to include in this review. The decision was to include (1) at least one example of simulation dealing with each major marketing subsystem and marketing function, (2) a variety of different modeling and experimentation techniques, and (3) the less-known over the better-known simulations. For example, while Forrester's (1959) model of advertising is well known, his general approach is discussed elsewhere (chapter 9). And several of our examples are largely unpublished, such as simulations developed by Alba, Tuason, Gensch, and others. The simulations ultimately selected for review are listed in table 10-1.

Marketing simulations can conveniently be divided into three classes: (1) computer models of the behavior of *marketing system components,* (2) computer models of the effect of different *marketing instruments* on demand, and (3) man-machine marketing simulations, that is, *marketing games.* A fourth

TABLE 10-1   Selected Marketing Simulations [a]

Marketing System Simulations	Examples
1. Distribution systems	Cohen (1960), Balderston and Hoggatt (1962), Shycon and Maffei (1960)
2. Customer systems	Amstuts (1969), Alba (1968)
3. Competitive systems	Kotler (1965), Tuason (1965)
Marketing Instrument Simulations	
1. Price decisions	Cyert and March (1962), Morgenroth (1962)
2. Advertising decisions	Simulmatics (1962), Gensch (1969)
3. Sales force decisions	Cloonan (1969)
4. New product decisions	Pessemier (1966), Urban (1968)
5. Marketing research decisions	Mayer (1964)
Marketing Games	Carnegie Tech MATE (1965), M.I.T. Marketing Game (1961

[a] Year refers to date of a convenient published version rather than date of completion of model or preprint publication.

category, all-man marketing simulations, has not been prominent in marketing.

Marketing system simulations attempt to reproduce the behavior of distributors, customers, and competitors. *Distribution simulations* seek to describe the behavior of institutions that affects the flow of product from producers to final buyers. *Customer simulations* seek to describe the formation of attitudes and buying behavior of actual and potential customers. *Competitive simulations* seek to describe the marketing actions and reactions of competitors in an on-going marketing process.

In contrast, marketing instrument simulations seek to model the effect on demand of marketing instruments such as *price, advertising, personal selling, new product development,* and *marketing research.* Included here are computer models that purport to describe how marketing executives actually make their marketing decisions as well as models that investigate improved solutions to marketing problems.

The man-machine simulations, or marketing games, do not presume to specify the actual decision behavior of marketing executives but instead seek to transform decisions of marketing "players" into specific market results. However, some of the market models have become extremely interesting in themselves, particularly those in which the game designer sought to develop authentic simulations of specific industries.

The major part of this chapter will review the simulations listed in table 10-1. The final section will evaluate the main areas of future opportunity for simulation as a methodology for the study of marketing phenomena.

DISTRIBUTION SIMULATIONS

The field of marketing primarily got its start in the early part of this century as a study of distribution channels. Academic economists tended to neglect the complex operations of middlemen and facilitating agencies in their formal models of price and output determination. This area attracted institutional economists, and their early studies of distribution were largely descriptive with some broad speculations about institutional evolution. More recently, marketing scholars have begun to study analytically such questions as the optimal number of middlemen, the stability of a distribution channel, the bargaining strategies of producers and middlemen, and so forth (Mallen, 1967). In addition to straightforward analytical studies, some interesting distribution simulations have recently appeared.

Three distribution simulations will be examined here, each dealing with a different problem. The first simulation was developed by Kalman Cohen to attempt to explain the behavior of various channel members in the vertical structure of the shoe leather industry. The second simulation was developed by Balderston and Hoggatt to explore the mechanisms of price and sales determination in the context of the West Coast lumber industry. The third simulation was developed by Shycon and Maffei to serve as a tool for estimating the cost of alternative physical distribution arrangements.

Cohen Simulation

The question considered by Cohen was, How well could simulation methodology represent and illuminate the complex behavior of various "typical" channel members in the shoe industry? (Cohen, 1960.) The vertical flow in the shoe industry starts with *hide dealers* (primarily meat producers). Their product goes to *tanners,* then to *shoe manufacturers,* then to *retailers,* and then to *final buyers.* (Wholesalers and jobbers also operate in this industry, but data were insufficient to permit modeling their behavior.) Each period every member of the distribution channel makes key decisions that affect the sales and prices of other channel members. The key decisions that Cohen chose to analyze are shown in figure 10-1—mainly price, order level, and production level decisions.

Using data previously collected for the industry (Mack, 1965), Cohen

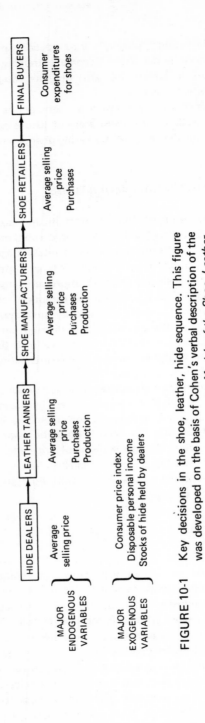

FIGURE 10-1   Key decisions in the shoe, leather, hide sequence. This figure was developed on the basis of Cohen's verbal description of the model. Source: Cohen, *Computer Models of the Shoe, Leather, Hide Sequence.*

proceeded to build a model which consisted of over sixty equations when it was completed. For example, one of the equations in the system attempted to account for monthly consumer expenditures on shoes from 1930 through 1940. Cohen tested several different independent variables and found that the following formulation worked best (Cohen, 1960):

$$S_t = -69,800,000 + .0147\ Y_t + 910,000\ I_t$$

where S_t = retailers' shoe sales in month t (in constant dollars)

Y_t = disposable personal income in month t (in constant dollars)

I_t = seasonal index in month t for retail dollar shoe sales from 1930 through 1940 (in percent)

The regression coefficients in this equation were all statistically significant at the 5 percent level and the equation's R^2 = 0.97. Variables such as deflated shoe prices and a time-trend factor were tried in this equation but were found statistically nonsignificant. Cohen proceeded to fit and test equation forms for other variables in the industry in the same manner.

The system of equations was too complex to solve directly for the time paths of various key variables in the system: There were too many interrelationships, nonlinearities, and lagged values. Cohen resorted to simulating the equation system over time. He ran two different types of simulation, which he called the *one-period* approach and the *process* approach, respectively. The one-period approach used the actual historical data of each preceding period to generate the simulated values for each following period. The process approach used the simulated results at the end of each period to generate the next period's simulated values. The latter approach risks compounding earlier errors and is not expected to give as good a fit to the historical data as the one-period approach.

Cohen's simulated results varied in their goodness of fit to the actual historical data. Figure 10-2 shows the actual and simulated time series of factory shoe prices. Although the fit appears good enough to capture the long price decline from 1930 to 1933 and the subsequent relative stability, it misses or severely misplaces most of the minor cyclical movements after 1932. Figure 10-3 shows the actual and simulated time series of manufacturers' shoe production and the fit appears to be very poor. The simulated series fluctuates too much and is not consistent in its timing relation to the actual series. Cohen's explanation of the poor fit is that aggregate sector variables do not satisfy the same functional forms as the individual firm's behavioral mechanisms. Important differences in the timing of retail orders and in the size distribution of inventories among individual firms are lost in models using the "typical firm" approach.

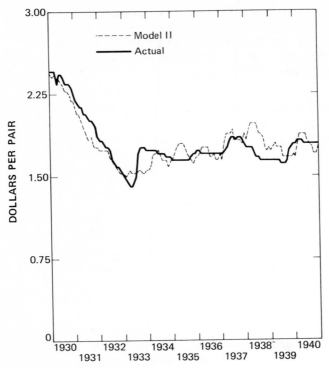

FIGURE 10-2   Factory shoe price. Source: Cohen, *Computer Models of the
Shoe, Leather, Hide Sequence,* p. 97.

Comment. Cohen's was a pioneering attempt to simulate the complex,
interrelated behavior of multilevel agents in a *real* vertical market. The researcher
who attempts to build a model of channel behavior on the basis of historical
data will face the following problems:

1. Choosing appropriate functional forms connecting the endogenous and
   exogenous variables. Economic theory supplies hypotheses about
   directions of relationships among economic variables, and the analyst
   must conceive and test alternative functional forms.
2. Achieving unbiased and consistent estimates of the coefficients of the
   equations despite the complexity of the equation system, the weaknesses of the historical time series data, and the rigid statistical assumptions of the underlying estimation procedure.
3. Validating the overall model when many time series are involved and
   individual goodnesses of fit vary widely.

From the point of view of marketing simulation, it would have been
desirable to see more marketing factors explicitly included, such as advertising

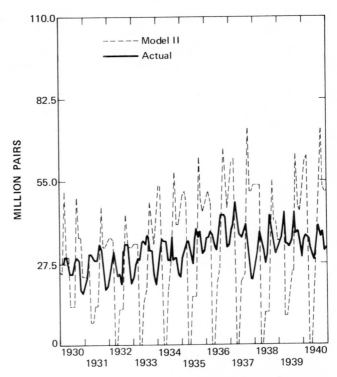

FIGURE 10-3   Manufacturer's shoe production. Source: Cohen, *Computer Models of the Shoe, Leather, Hide Sequence*, p. 107.

outlays by the shoe manufacturers, the roles and effect of wholesalers and jobbers in the system, and so on. If data permitted, differences in the decision behavior of different-size establishments could have been introduced to get around the "typical firm" assumption. Still, Cohen's model remains a prototype for econometric modeling in the area of vertical market structures. Despite its lack of strong results, it is disappointing that in the years since the publication of this model, no other effort of this kind has been reported. Rich data banks exist for various industries, and important contributions may be forthcoming from attempts to simulate the vertical marketing systems in other industries.

Balderston and Hoggatt Simulation

The Balderston and Hoggatt simulation is also oriented around an actual industry—the West Coast lumber industry (Balderston and Hoggatt, 1962). The authors here are less concerned with the whole vertical structure of the lumber industry than with the critical role played by the lumber brokers in searching and trying to match potential sellers (timber growers) and potential buyers (lumber retailers) for profitable transactions. The authors concentrated on the information and decision behavior of the lumber broker and did not attempt to use real data. They developed a variety of plausible decision and information rules to govern the participants' behavior and plausible initial levels of inventory, demand, and price to start the simulation. Their interest lay in exploring, through the medium of microbehavioral simulation, the role of particular factors in affecting market efficiency.

Balderston and Hoggatt created an initial market consisting of twenty growers, ten brokers, and forty retailers. At the beginning of a *market period,* each grower posts a bid price and bid quantity, and these prices are not negotiable. To begin the *first cycle* of a market period, each broker sends a "search message" to *preferred* buyers and sellers. The broker's preferences are determined either from previous experience with individual buyers and sellers or from random selection. (The difference made by these two methods is one of the questions studied in the simulation.) The broker can reach potential buyers and sellers by personal contact, telephone, or letter, each method involving a different cost per message. A "reply," containing a bid or offer price and quantity, is then received by the broker. All brokers have the opportunity to search for one buyer-seller pair during a cycle. If a potential transaction seems profitable to the broker (i.e., the total revenue from the transaction exceeds the cost of completing the transaction plus a profit margin), the broker sends an "order" message to the grower and the retailer involved. If not, the broker waits for another opportunity to search at the beginning of the next cycle. Since a particular buyer or seller can receive any number of orders (or none), preference ordering is also used to determine which they will accept. If an order is accepted, the broker sends a "confirm message" to the parties involved. After every broker has had the opportunity to complete or otherwise terminate a transaction, the first cycle is ended. A summary of the steps in the typical transaction is shown in figure 10-4.

A second cycle within the market period begins with brokers moving to their next preferred buyers and sellers and repeating the communication and transaction process described above. The last cycle occurs when there is no broker remaining who wishes to search for another possible transaction. At this point, payments are settled between the market participants and the retailers selling their goods in the final market. The working capital positions and profits are calculated for each firm, and firms with negative working capital exit from the market. This sets the stage for a new market period.

The chief experimental parameters in this simulation are *preference orderings* and *unit message costs.* Two alternate preference order rules are tested: (1) a preference order developed by each firm on the basis of its specific experience with other market participants and (2) a preference order generated randomly for each firm at the end of each market period. Unit message costs are set at one of four levels ($10, $12, $48, and $192) and remain the same throughout any given run. Each run is an observation of the market's time path

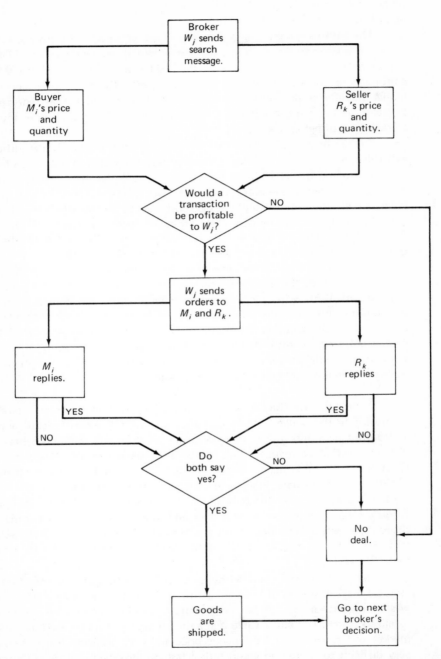

FIGURE 10-4   Steps in a Typical Transaction. Source: Balderston and Hoggatt, *Simulation of Market Processes.*

under specified parametric conditions and lasts sixty market periods, the equivalent of five years of market operation.

The authors sought to test the *viability* and *efficiency* of the market under different combinations of the two experimental parameters. The viability requirement was met as in each of the eight runs of the model the market continued to operate throughout the sixty periods. The authors recognize that model viability is no evidence of the validity of the model in representing behavior in the lumber industry. It merely indicates that they have found a workable set of behavioral rules.

The authors also tested the efficiency of the market under different experimental conditions. Market efficiency was computed as the ratio of total realized revenue for all firms divided by total potential revenue for all firms, where unity indicates the most efficient market. Market efficiency was found to fall very gradually until the unit message cost reached the extremely high level of $192. But even for high message costs, efficiency seemed to rise in later periods. Furthermore, market efficiency was not significantly affected by whether the brokers' preferences were determined by experience or at random.

Contrary to most traditional economic interpretations of the size distribution of firms, the simulation results indicated that market pressures and successes of firms with identical endownments could result in skewed size distributions. Furthermore, these distributions depended on preference rules and unit message cost. Another important finding concerns *market segments* defined here as groupings of firms that trade with each other but not with other firms. With greater trading loyalties and increased message costs, greater segmentation occurred. Other hypotheses were similarly tested.

Comment. The Balderston and Hoggatt simulation forms an interesting contrast with the Cohen simulation. Cohen started with a set of historical input and output data and sought to reconstruct the economic decision behavior of different market participants that led to the observed results. Balderston and Hoggatt, on the other hand, started with a set of postulated economic behavior of different market participants and used artificial data to determine how such a market worked under different conditions. Cohen's simulation exemplifies *analytic simulation,* centering as it does on trying to construct a model that will account for actual data. Balderston and Hoggatt's exemplifies *synthetic simulation,* centering as it does on trying to understand the properties of a hypothetical model.

The market system studied by Balderston and Hoggatt departed from the classical market models of pure competition and pure monopoly. Their market was characterized by limited information, varying information costs, and localized search. It is conceivable that a formal model could be developed of this market and analyzed for its dynamic price and sales behavior. But it would be very difficult to explore in an analytical fashion the effect of different information costs and preference rules on the ultimate number, size, and profitability of firms. Balderston and Hoggatt's choice was to create hypothetical buyers, sellers, and brokers and synthesize their behavior. The runs generated under different experimental conditions could be analyzed to learn about the effects of pure variations in marketing elements on market efficiency. These effects could be compared with those that would be predicted from classical theoretical propositions in economics. The methodology could also be used to predict the effects of complex changes in marketing channels on channel efficiency. Unfortunately, subsequent studies that apply simulation to examine the effect of different channel arrangements on the efficiency of a particular marketing channel have not appeared.

Shycon and Maffei Simulation

The Shycon and Maffei simulation seeks to provide a tool for estimating the costs and customer service times of alternative physical distribution arrangements serving a market. Companies have to decide how many plants and warehouses to operate and where to locate them in relation to the retail centers they serve. Each extra warehouse adds costs of various kinds and saves costs of other kinds. In the last decade, several analytical models have been proposed (Kuehn and Hamburger, 1963). The analytical models, however, require many simplifying assumptions that often make the results a weak approximation of reality. This led Shycon and Maffei to develop a simulation approach to the problem (Shycon and Maffei, 1960).

The Shycon and Maffei simulation grew out of attempts to improve physical distribution systems of the H. J. Heinz Company and the Nestlé Company. These companies seek to distribute their products at *minimum cost* through a system of geographically dispersed factories, warehouses, and customers. The large number of combinations of factories, warehouses, and customers, as well as a need for a management-oriented model with minimal abstraction from the real world, led the researchers to the technique of simulation. Simulation provides a means of dealing with a complex problem from a perspective that management can easily understand and use.

The approach requires management to specify a new or revised physical distribution system consisting of a proposed set of warehouses (their number, location, and sizes), inventory policies, and modes of transportation. The simulation model is used to estimate the total distribution costs under this system. The original computer program used information on the longitudes and latitudes of the factories, warehouses, and retail centers to calculate approximate shipping costs for the various potential product flows. The use of air miles to estimate shipping costs substantially reduces the computer storage requirements of the program. The computer then proceeds to identify the five warehouses nearest each customer. Customers are finally assigned to specific warehouses on the basis of a total cost analysis. Up to this point, capacity limitations of factories are ignored, but this is subsequently handled through a linear programming solution that assigns warehouses to factories.

Management now considers whether the estimated cost of the new physical distribution system makes it appear attractive enough to implement. This is likely to occur if it promises considerable cost savings (or better customer service) compared with the existing system. Otherwise, management might construct another physical distribution plan whose costs could readily be estimated through the computer program. Shycon and Maffei have included an optional routine that can generate new distribution networks by shifting existing warehouses in fixed or random directions and distances. However, this procedure does not possess any properties that would necessarily and efficiently lead to improved distribution networks. Simulation here is essentially a methodology for testing the relative efficiencies of given plans and not an algorithm for discovering a near-optimal plan.

Comment. The use of simulation to evaluate realistic representations of physical distribution systems allows management to develop and test complex plans for distribution systems in a laboratory setting before committing millions of dollars to them. Distribution systems are difficult to change once they are established, and this underscores the importance of finding the best techniques for their prior evaluation.

The Shycon and Maffei model did not fully exploit the potential of simulation as an alternative to purely mathematical approaches to distribution network evaluation. Its reliance on air distances for estimating transportation costs between two locations overlooked the gross transportation rate distortions that exist in practice because of different sizes of shipments, classes of goods, directions of travel, and so forth. The simulation does not incorporate a routine for reflecting seasonal cost variations, although this is important to many shippers. Shycon and Maffei developed one of the first physical distribution simulators, and now others can seek to extend its main advantage, that is, its capacity to allow the study of physical distribution systems in realistic detail.

CUSTOMER SIMULATIONS

Effective marketing planning requires well-constructed models for analyzing and forecasting market behavior. Traditionally, companies have relied on extremely simple models of market behavior. Customers' demand would be visualized as having a simple relationship to national income, population, product price, and possibly a few other variables. Types of customers would rarely be distinguished, and little effort was made to incorporate the large variety of marketing and psychological factors that affect buyer behavior.

More recently, several models have been constructed to describe brand choice and switching behavior over time in the category of frequently purchased consumer packaged goods (Frank, 1962; Kuehn, 1962; Howard, 1963; Massy, 1966; Morrison, 1966; Kotler, 1968c; Montgomery, 1969). These models take various positions on whether household brand-switching probabilities are influenced by zero, one, two, or more past brand choices of the household. Most of these models deal with total market or market segment brand-switching tendencies, and their properties as models are investigated through analytical methods. In contrast, two recent models of customer behavior follow the mode of microbehavioral simulation.[1] They revolve around an artificial population of consumers, generate behavior as a probabilistic function of each individual's characteristics and exposures, and cumulate the behavior into resulting market shares each period. This section will examine first Amstutz's model and then Alba's model.

Amstutz Simulation

Markets are made up of individual customers. This led Amstutz (1967) to examine the feasibility of using the computer to create a representative sample of hypothetical customers and to simulate their consumer action through time, as it is affected by a host of environmental, marketing, and psychological factors. He designed a computer program capable of performing four important tasks: (1) breeding a consumer population in any desired number, (2) developing and assigning product and brand awarenesses and attitudes to each individual consumer, (3) putting the consumer through a set of experiences, and (4) updating the consumer's attributes as a result of these experiences.

Figure 10-5 shows the characteristics and experience of one such consumer in the week starting February 19, 1962. According to line 2, this person lives in a suburb (SU) located in the Northeast (NE). He is between twenty-five and thirty-five, with an income of eight to ten thousand dollars, and college educated. These characteristics could describe an actual person recorded in a sample survey. Alternatively, this person could have been created artificially through a random number generator applied to basic census data. In the latter instance, the computer program randomly draws an *age bracket* from an age distribution table stored in the computer, where the chance of drawing a particular age bracket is proportional to its relative frequency in the population. Next, the computer program draws an *income* for the person from a frequency distribution of incomes within that age bracket. Finally, the computer program randomly draws an *educational level* from an education distribution. In this way, a string of characteristics is generated to create a plausible person. There is no limit to the size of the artificial population that can be bred; furthermore, the population that is bred can be expected to reflect the distribution of characteristics found in the real population.

Turning to line 3, the person owns brand 3, which is now six years old. The product is assumed to be a consumer durable such as an electric floor

---

[1]An interesting third model has recently appeared which describes customer shopping and brand choice behavior using a combination of analytical formulation and Monte Carlo procedures. See J. Herniter and V. Cook, "The NOMMAD Simulator: A Multidimensional Stochastic Model of Consumer Purchase Behavior," a paper presented at the 36th National Meeting, Operations Research Society of America, Miami Beach, Florida, November 10, 1969.

SIMULATION APP-03 TEST RUN APRIL 4, 1965   1400 HOURS

-- CONSUMER 0109 NOW BEGINNING WEEK 117 --   FEBRUARY 19, 1962

  - REPORT MONITOR SPECIFIED. TO CANCEL PUSH INTERRUPT.

  - CHARAC - REGION NE SU, AGE 25-35, INCOME 8-10K, EDUCATION COLLEG

  - BRANDS OWN 3, 6 YEARS OLD. RETAILER PREFERENCE 05, 11, 03

  - MEDIA AVAILABLE 1 0 0 1 0 0 0 0 1 1 1 1 0 0 0 0 0 0 0 0 0 0 0 0

  - ATTITUDES . 1   2   3   4   5   6   7   8   9   10   11   12

     ....

```
PROD CHAR .  0  +1  +1   0  -3  -1   0  +5   0  +3   0   0
APPEALS   . -3   0  +1  +5   0  -3  +3   0   0   0  +5   0
BRANDS    . +2  +1  +3  +2
RETAILERS . +1  -5  +3  +1  +5  -5  -5  +1  -1  -3  +5  +1
          . -3  +1  -1  +3  +1  +1
AWARENESS .  1   0   0   0
```

  - MEMORY DUMP FOLLOWS. BRANDS LISTED IN DESCENDING ORDER 1 TO 4

PRODUCT CHARACTERISTIC MEMORY   APPEALS MEMORY

```
:  :  :  :  :  :  :  :  :  :  :  :  :  :  :  :  :  :  :  :  :  :  :  :  :
1  2  3  4  5  6  7  8  9 10 11 12 . 1  2  3  4  5  6  7  8  9 10 11 12
.  .  .  .  .  .  .  .  .  .  .  .     .  .  .  .  .  .  .  .  .  .  .  .

2  3 15  0  5  5  4 14  8  7  1  3 . 8  9  7  3  1 11  7  4  4  3  9  3
8  0  6  4  9  5  4 13  0  3  6  7 . 6  8  0  7  0  9  2  4  3 10  3  1
0  6 15  7  0  3 11  3  5  2  5  7 . 0  4  8 10  9  2 14  3  9  7  9  5
7  9  3  7  3  2  7  2  6 12 14  2 . 0  9  7  8 13  9 11  6  0  2  5  9
```

  - MEDIA EXPOSURE INITIATED

      - MEDIUM 003 APPEARS IN WEEK 117 -- NO EXPOSURES
      - MEDIUM 004 APPEARS IN WEEK 117
        - EXPOSURE TO AD 013, BRAND 3 -- NO NOTING
        - EXPOSURE TO AD 019, BRAND 4
          - AD 109, BRAND 4 NOTED. CONTENT FOLLOWS
          - PROD. C 11 P = 4, 4 P = 2,
          - APPEALS 5 P = 2, 7 P = 2, 12 P = 2,
      - MEDIUM 007 APPEARS IN WEEK 117 -- NO EXPOSURES
      - MEDIUM 012 APPEARS IN WEEK 117
        - EXPOSURE TO AD 007, BRAND 2
          - AD 007, BRAND 2 NOTED. CONTENT FOLLOWS
          - PROD. C 8 P = 3, 12 P = 1,
          - APPEALS 2 P = 1, 4 P = 1, 6 P = 1, 10 P = 1,
        - EXPOSURE TO AD 013, BRAND 3 -- NO NOTING
        - EXPOSURE TO AD 004, BRAND 1 -- NO NOTING
      - MEDIUM 016 APPEARS IN WEEK 117 -- NO EXPOSURES
      - MEDIUM 023 APPEARS IN WEEK 117 -- NO EXPOSURES

FIGURE 10-5. One Simulated Consumer in Week 117. Source: Amstutz, *Computer Simulation of Competitive Market Response,* pp. 394-95.

```
- WORD OF MOUTH EXPOSURE INITIATED

        - EXPOSURE TO CONSUMER 0093 -- NO NOTING
        - EXPOSURE TO CONSUMER 0104 -- NO NOTING
        - EXPOSURE TO CONSUMER 0117 -- NO NOTING

- NO PRODUCT USE IN WEEK 117

- DECISION TO SHOP POSITIVE -- BRAND 3 HIGH PERCEIVED NEED
                             -- RETAILER 05 CHOSEN

- SHOPPING INITIATED

    - CONSUMER DECISION EXPLICIT FOR BRAND 3 -- NO SEARCH
    - PRODUCT EXPOSURE FOR BRAND 3
        - EXPOSURE TO POINT OF SALE 008 FOR BRAND 3
            - POS 008, BRAND 3 NOTED.  CONTENT FOLLOWS
        - PROD. C   3 P = 4,   6 P = 4,
        - APPEALS   5 P = 2,   7 P = 2,   10 P = 2,   11 P = 2,
    - NO SELLING EFFORT EXPOSURE IN RETAILER 05

- DECISION TO PURCHASE POSITIVE -- BRAND 3, $ 38.50

    - DELIVERY IMEDAT
    - OWNERSHIP = 3, AWARENESS WAS 2, NOW 3

- WORD OF MOUTH GENERATION INITIATED

    - CONTENT GENERATED, BRAND 3
        - PROD. C   3 P = +15,   8 P = +15,
        - APPEALS   4 P = +50,   11 P = +45

- FORGETTING INITIATED -- NO FORGETTING

-- CONSUMER 0109 NOW CONCLUDING WEEK 117   --   FEBRUARY 25, 1962

-- CONSUMER 0110 NOW BEGINNING WEEK 117    --   FEBRUARY 19, 1962

    QUIT,
    R 11.633+4.750
```

FIGURE 10-5 (continued)

polisher, and the person prefers to deal with retailers 5, 11, and 3, in that order. He has attitudes toward twelve different product characteristics and twelve different appeals, the attitudes ranging from a−5 (very negative) to a +5 (very positive). He also has attitudes toward the four brands and the eighteen retailers. Finally, he has been exposed in the past to a varying number of communications on different product characteristics and appeals for each of the four brands.

How was this information developed? Once again, the information could describe the attitudes, store preferences, brand owned, and communications recall of a real person who was interviewed in a survey. Or these characteristics could artificially have been generated from various frequency distributions compiled in earlier surveys.

Next, this consumer passes through a set of events and experiences resembling those that members of the real population may have passed through during the week beginning February 19, 1962. Consumer 109 is exposed to media and may or may not note advertisements pertaining to this product; he may be exposed to word-of-mouth influence, use the product during the week, make a decision to shop, deal with a salesman, and decide to purchase. Afterward he may generate favorable or unfavorable word of mouth to others. Some forgetting of his past exposures will also occur.

Each of these events in the consumer's week is regulated by mathematical functions and/or probability distributions. For example, the probability that the consumer shops in this week is related to his perceived need, which is an increasing function of his brand attitude score, opportunity for product use, and time since last purchase; it is also dependent on the level of his income. These factors are combined in a nonlinear mathematical expression.

By generating experiences for each member of the artificial population week after week during the period under study, Amstutz is able to aggregate and summarize their purchases in the form of a simulated time series of brand shares.

Amstutz specified a multilevel validation procedure for his model. *Function-level validation* involves two steps: first, a sensitivity analysis indicating the relative sensitivity of total system performance to various functions within the system structure, and second, a chi-square test of the null hypothesis that observed relationships are due to random variation. *Cell-level validation* establishes that the behavior of an artificial consumer within the simulated population cannot be differentiated by an expert from that of a similar member of the real-world population (Turing, 1950). *Population-level validation* is undertaken to insure that the behavior of members of the simulated population does not significantly differ from the behavior of corresponding members of the real-world population. Finally, *performance validation* is the ability of the model to duplicate historical real-world population behavior. This level of validation calls for initializing parameter settings in the model to duplicate those existing in the real-world system at a particular point in time and then comparing simulated and actual results.

In a reported application of this methodology to another product area, a pharmaceutical product, Amstutz and Claycamp indicate that a good fit to the data was obtained (Claycamp and Amstutz, 1968). They simulated the prescribing behavior of one hundred simulated doctors toward the ten brands in this product class over a year's time. There was some change in the market share ranks of the ten brands between the beginning and the end of the year. Their simulated results correctly predicted all the year-end ranks, as shown in table 10-2, which also shows that the same data according to market shares and their average error in predicting year-end values was 0.7 percent in market share.

A more detailed account of Amstutz's work appears in the case-example following this chapter.

Comment. The overriding question raised by Amstutz's micro-behavioral model is whether the extensive data gathering that this type of model requires produces enough in extra understanding or predictive ability to be

TABLE 10-2   Actual and Predicted Brand Shares

| Brand Identification | Rank Order Brand Share Comparisons | | | Absolute Brand Share Comparison | | | |
| | Rank as Initialized | Year End Rank | | Initialization Value | Year End Value | | Difference (Magnitude) |
		Simulated	Actual		Simulated	Actual	
1 — Y	4	2	2	13.7%	15.0%	16.1%	− 1.1%
1 — 0	5	6	6	9.7	9.1	8.7	+ .4
1 — X	6	5	5	7.3	9.3	9.0	+ .3
1 — +	8	8	8	5.0	3.2	2.8	+ .4
1 — □	10	10	10	0	0	0	—
2 — □	1	1	1	23.2	27.6	28.8	− 1.2
2 — 0	2	3	3	18.1	13.0	12.7	+ .3
2 — X	3	4	4	15.6	13.9	14.4	− .5
2 — +	7	7	7	6.2	5.9	5.5	+ .4
2 — Y	9	9	9	1.0	2.5	2.0	+ .5
				99.8%	99.5%	100.0%	.7%

Source: Amstutz, *Computer Simulation of Competitive Market Response*, pp. 409-11.

worth the cost. Amstutz's model undoubtedly allows considerable details of the consumer buying process to be brought in and interrelated, and this can be a real contribution to theory construction. The analyst is challenged to specify the relationships between demographic characteristics, media exposure probabilities, word-of-mouth exposure probabilities, responses to appeals and product characteristics of various kinds, and so forth. These are the kinds of variables that are worth thinking about both by the theorist searching for understanding and by the marketing practitioner searching for clues to promotional strategy. And they are worth thinking about in the operationally defined way required by simulation. But it is a different matter to justify this effort and cost as a route to developing useful forecasting and marketing planning models. In the first place, there is the question of whether the various frequency distribution data on consumer characteristics can be obtained and, if so, at what cost.[2] There is the question of how many consumers should be created for a representative sample population for trying out marketing strategies. Each consumer must separately be processed through many events and many weeks, and the computer cost of maintaining and processing all this detail is high. There is the question of the many types of errors that can creep into the data and the program and affect the results without easily being spotted because of the considerable model detail and complexity. There is the question of accurately measuring the marketing actions

[2] Amstutz does not describe how he develops or obtains the underlying conditional probability distributions for generating the demographic attributes. We can only guess whether he goes back to original census data (not census summaries), conducts original surveys, or uses the marginal distribution of traits (not the conditional distributions).

of competitors' weekly advertising media expenditures and in-store promotions. Competitive intelligence is extremely difficult and costly to collect, and this model demands more detail about competitive behavior than the typical aggregative model does. There is the question of validating the many functions used in the model. So many possible sensitivity tests can be performed on different functions in the system that the cost of sensitivity testing can rapidly get out of hand. There is the question of performance validity, and so far only one real application has been made. Here the performance validity appeared good, but more applications are needed before it can be determined whether this was a chance result or an intrinsic accomplishment from this level of model-building.

As for performance validity, could a simpler model have done as well? Amstutz did not test his elaborate model against possible simpler models—and such tests seem desirable. We are not doubting the value of microbehavioral simulation for stimulating theory construction and firing the imagination of practitioners but only raising the question of its promise in the area of practical forecasting and planning. Does Amstutz's microbehavioral model increase the degree of predictability over simpler models and, if so, to an extent justifying the substantially larger investment required? It is too early to answer this question.

Alba Simulation

Alba simulated consumer behavior toward a "new" product, the "touch-tone" telephone (Alba, 1968). He used data collected earlier from one hundred residents of Deerfield, Illinois, who had adopted this new phone (Robertson, 1967). The data described the traits of each adopter (seven traits were measured: interest polymorphism, venturesomeness, cosmopoliteness, social integration, social mobility, privilegedness, and status concern), their sources of information about the phone (personal influence and mass communication), and the date of its adoption. Alba's aim was to see whether he could simulate the adoption process in this community accurately enough to:

1. Reproduce the actual number of new adopters each period
2. Reproduce the order in which the one hundred households in the sample had adopted the touch-tone telephone

To do this, he developed the probabilities of each individual being exposed to touch-tone telephone advertising and word-of-mouth communications and, in turn, exposing others. The probabilities were derived from statistical regressions on the traits of individuals. For example, the probability that an individual would be exposed to a touch-tone phone advertisement was given by:

$$PE_i = .120 \, (VE_i/7 + .359 \, (IP_i/7) - .150 \, (CO_i/7) +$$

$$.200 \, (SI_i/7 + .202 \, (SM_i/7) + .184 \, (PR_i/7)$$

$$+ .085 \, (SC_i/7)$$

where:

PE_i = probability that individual i will be exposed to a touch-tone advertisement

VE_i = individual i's score on "venturesomeness" (on a 7-point scale from 1 to 7)

IP_i = individual i's score on "interest polymorphism" (on a 7-point scale from 1 to 7)

CO_i = individual i's score on "cosmopoliteness" (on a 7-point scale from 1 to 7)

SI_i = individual i's score on "social integration" (on a 7-point scale from 1 to 7)

SM_i = individual i's score on "social mobility" (on a 7-point scale from 1 to 7)

PR_i = individual i's score on "privilegedness" (on a 7-point scale from 1 to 7)

SC_i = individual i's score on "status concern" (on a 7-point scale from 1 to 7)

If an individual scored seven (the maximum score) on all the traits that correlated with the tendency to adopt a touch-tone telephone, his probability would be one. Otherwise PE_i would be between zero and one.

After the individual's PE is determined, a random number is drawn to determine whether he saw a persuasive communication that period. If he did, another random number is drawn to determine whether he underwent an attitude change. Alba defined an attitude change as tantamount to adopting the phone, although he notes that in most empirical studies (including his data) adoption refers to actual purchase. The probability of an individual undergoing an attitude change is affected by the level of the persuasiveness of the ad. The persuasiveness of the particular ad is rated on the basis of the message, source, transmitter, and content.

The individual who did not adopt the phone might still be exposed to a neighbor who talked about the phone. Alba formulated a model that generated a probability that any two residents of Deerfield would discuss the touch-tone phone. The probability varied directly with the degree to which the two residents shared the same traits.

Alba's steps in simulating the word-of-mouth process are shown in figure 10-6. Alba distinguishes between non-knowers, knower non-adopters, and knower-adopters, and he employs different probabilities of confrontation between any two categories of individuals. As confrontations take place each

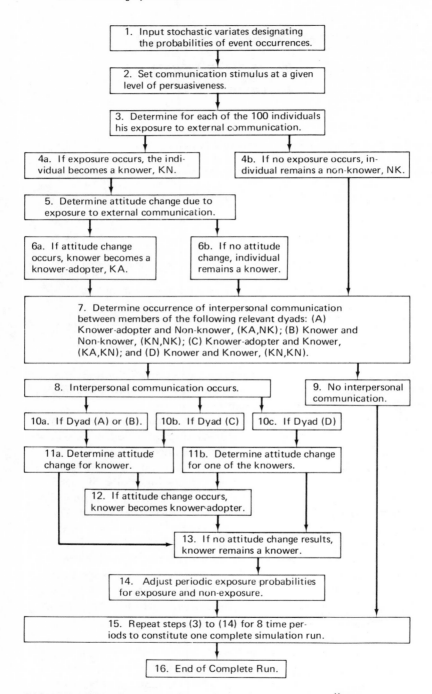

1. Input stochastic variates designating the probabilities of event occurrences.

2. Set communication stimulus at a given level of persuasiveness.

3. Determine for each of the 100 individuals his exposure to external communication.

4a. If exposure occurs, the individual becomes a knower, KN.

4b. If no exposure occurs, individual remains a non-knower, NK.

5. Determine attitude change due to exposure to external communication.

6a. If attitude change occurs, knower becomes a knower-adopter, KA.

6b. If no attitude change, individual remains a knower.

7. Determine occurrence of interpersonal communication between members of the following relevant dyads: (A) Knower-adopter and Non-knower, (KA,NK); (B) Knower and Non-knower, (KN,NK); (C) Knower-adopter and Knower, (KA,KN); and (D) Knower and Knower, (KN,KN).

8. Interpersonal communication occurs.

9. No interpersonal communication.

10a. If Dyad (A) or (B).

10b. If Dyad (C)

10c. If Dyad (D)

11a. Determine attitude change for knower.

11b. Determine attitude change for one of the knowers.

12. If attitude change occurs, knower becomes knower-adopter.

13. If no attitude change results, knower remains a knower.

14. Adjust periodic exposure probabilities for exposure and non-exposure.

15. Repeat steps (3) to (14) for 8 time periods to constitute one complete simulation run.

16. End of Complete Run.

FIGURE 10-6   Summary flow chart. Source: Alba, "Microanalysis of the Socio-Dynamics of Diffusion of Innovation," p. 111.

period, those who undergo an attitude change are considered adopters, and the computer program totals the number of adopters each period.

Alba defined a simulation run as covering eight periods, each period lasting approximately two months. He replicated the process fifty times and compared the mean result with the actual adoption curve. The actual and simulated curves are shown in Figure 10-7. The fit is not highly satisfactory: Alba's one hundred simulated residents of Deerfield started out more slowly in their adoption rate and later adopted more quickly than the actual residents. By period 5, almost all the simulated residents had adopted the phone, although only 60 percent of the real residents had in the sample. These differences were shown to be highly significant in both chi-square and Kolmogorov-Smirnov tests of goodness of fit. The fault may be that Alba did not vary the intensity of mass communications or personal influence on a seasonal basis, whereas in the real world both probably varied seasonally. If the telephone company did not advertise the new phone during periods 3 to 5, whereas Alba continued the advertising in the simulation, this would account for some of the discrepancy.

Alba fared better in his predictions concerning the earlier and the later adopters. The rank correlation coefficient (Spearman's *rho*) for the times of adoption of the real and simulated residents of Deerfield was high and statistically significant. His model was also able to discriminate accurately between residents who underwent attitude change because of mass communication influence and those who changed because of personal influence.

Alba also tested the influence of a higher versus a lower level of advertising in accelerating the adoption rate and found no significant difference. His higher level of advertising may not have been sufficiently high. Furthermore, as mentioned earlier, he did not vary the intensity of advertising seasonally.

Comment. Alba's simulation of the diffusion of an innovation was a

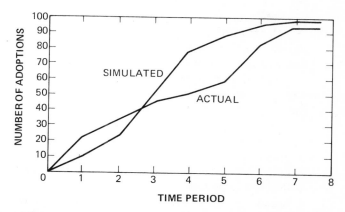

FIGURE 10-7   Alba's simulation results. Source: Alba, "Microanalysis of the Socio-Dynamics of Diffusion of Innovation," p. 209.

pioneering attempt to determine the relative influence of *social factors* on the one hand and *chance* on the other in explaining the rate at which a sample population learned about and decided to adopt a new product. He worked against the limitations of an existing data bank that was not originally designed for his needs, and yet he managed to develop a complex set of events and generate a plausible set of outputs that matched some of the actual occurrences fairly well. If the data had originally been collected for this project, more would have been gathered on the actual timing of advertisements, the media habits of the sample of residents, and the actual friendship or acquaintance patterns in the area. These data would have permitted the community to be modeled more concretely, and the diffusion process, still a highly stochastic one, could have been modeled more faithfully. The whole area of diffusion research begs for more knowledge on the relationship between product interest levels, consumer traits and neighbor interaction patterns, and conversations about products. The main contribution of Alba's work is that it shows that the complex processes of individual and community adoption of new products can be studied insightfully through the medium of microbehavioral simulation.

COMPETITIVE SIMULATIONS

A major aim of economic theory has been to evolve an explanation of the nature and consequences of different forms of business competition. Economists have developed an elaborate classification of market structures revolving around differences in the number of competitors and the degree of product homogeneity (monopoly, duopoly, oligopoly, monopolistic competition, and pure competition). Economists have primarily been interested in explaining *market prices and output,* and their analytical success has been highest for the two polar states of monopoly and pure competition, which, unfortunately, are the least descriptive of actual market structures in the United States.

Most business competition takes place in markets composed of few or several competitors offering differentiated products or services and using a variety of competitive instruments in addition to price. It was not until 1933 that Edward Chamberlin formally introduced into microeconomic theory such competitive instruments as product quality and selling effort (Chamberlin, 1933). These instruments still receive little formal attention from economists compared with the time spent in refining the pure theory of the price mechanism.

Simple analytical models of competition, such as the Cournot model, game theory, and so forth, do provide insights into competitive behavior, but their empirical relevance still has to be demonstrated. On the other hand, simulation models of competitive behavior often start with observed data and *then* proceed to specify the firm and market interactions without severe abstraction from actual decision-making processes. The model builder is more free to incorporate the various criteria and competitive tools used by firms without subjecting them to undue simplifications such as complete information or profit maximization. Unlike the simple analytical models, however, the more complex (and presumably more realistic) simulation models cannot be "solved" for optimality, but through systematic experimentation on the simulated system the nature of the real system can often more fully be revealed. While the results may not lead to a general theory of competition, the simulations can produce empirically testable hypotheses to which the rules of validity can then more appropriately be applied.

Two all-computer competitive simulations will be examined in this section. The first model was developed by Kotler and explores the outcomes and the merits of several alternative competitive strategies that two competitors may employ against each other. The second competitive simulation was formulated by Tuason and features a competitive decision program for a coffee manufacturer which updates the company's marketing program from period to period.

Kotler Simulation

Kotler developed a simulation model for evaluating alternative competitive marketing strategies for guiding a new product through its life cycle (Kotler, 1965a). The market is a hypothetical one that bears highly plausible behavioral properties drawn wherever possible from empirical studies. The market exhibits S-shaped growth and seasonal variation, and it is responsive to the total amount and composition of competing companies' marketing efforts. A new product, an inexpensive portable tape recorder, is introduced simultaneously by two competitors at $t = 0$. Market acceptance is rapid at first and then tapers off; by the last month of the simulation, $t = 60$, the product is in the market saturation stage of the product life cycle.

Each of the duopolists has three marketing instruments available: price (P), advertising (A), and distribution (D). Firm i's marketing decisions at time t are given by $(P, A, D)_{i, t}$. *Marketing strategy* is defined by Kotler as a set of decision rules that would adjust $(P, A, D)_{i, t}$ to $(P, A, D)_{i, t+1}$ for all t. The following nine classes of marketing strategy are distinguished.

1. *Non-adaptive strategy:* Initial marketing mix is held constant throughout the product's life cycle.
2. *Time-dependent strategy:* Scheduled changes in the marketing mix take place through time.
3. *Competitively adaptive strategy:* Firm i adjusts its marketing mix to match marketing mix changes made by firm j in the previous period.
4. *Sales-responsive strategy:* Firm adjusts its marketing mix on the basis of its sales in the previous period(s).
5. *Profit-responsive strategy:* Firm adjusts its marketing mix on the basis of its profits in the previous period(s).
6. *Completely adaptive strategy:* Monthly changes are made in the

company's marketing mix in response to the passage of time, changes in company sales and profits, and changes in the competitor's marketing mix.

7. *Diagnostic strategy:* Changes in the marketing mix are made only after distinguishing between possible causes of current developments.

8. *Profit-maximizing strategy:* Plan that would maximize the firm's profits.

9. *Joint profit-maximizing strategy:* Plan that would maximize total industry profits under collusion.

Examples of three of these strategies are given in figure 10-8.

Thirteen specific variations of the first five classes of strategy were developed and tested. The simulation consisted of running different competitive confrontations between two duopolists for sixty months. For example, one of the runs consisted of the first duopolist using strategy 1 shown in figure 10-8 while the second duopolist used strategy 3. Since there were thirteen specific strategies, a total of seventy-eight different runs could be made, according to the combinatorial formula 13! / 2! 11!. The printout for each of the seventy-eight runs provided a month-by-month description of the marketing decisions of each duopolist, the level of total industry sales, each firm's sales and market share, and each firm's profits. In addition, the monthly profits of each duopolist were cumulated with compound interest (at 6 percent) and the total was posted at the end of sixty periods of play.

The profit and market share results of the seventy-eight runs were summarized and analyzed. Subject to a number of qualifications, the results made it possible to answer the following questions:

1. Which long-run strategy is the best to adopt if the firm wants to guarantee a minimum return regardless of what its competitor does? (The best maximum strategy was a completely nonadaptive one.)

2. Which long-run strategy subjects the firm to the greatest amount of risk? (One of the competitively adaptive strategies was ruinous. It led to the largest market share, but at substantial losses.)

3. Which strategy offers the chance of greatest profit? (The profit-responsive strategy was best.)

4. If the rival's strategy is known in advance with certainty, what is the firm's best strategy? (The best strategy varied with the firm's objectives.)

Comment. While the data and consequently the results of Kotler's simulation were artificial, they were nevertheless revealing of the power of simulation in evaluating marketing strategies available to a company facing a complex marketing environment. The strategies and environment were too intractable to analyze by classical methods, and so the power of computer simulation was harnessed to create the histories of various strategy confrontations.

Various improvements could be made in the model, as noted by Kotler in his article. If the approach were to be applied in a real situation, much data

Time-dependent strategy

$P_{i,t} = 5(0.95)^t + 15$

Price initially stands at $20 and falls exponentially and asymptotically to $ 15 as $t \to \infty$

$A_{i,t} = 1.01\ A_{i,t-1}$

$D_{i,t} = 1.01\ D_{i,t-1}$

Advertising and distribution expenditures are increased one per cent each period.

Competitively adaptive strategy

$P_{i,t} = 0.95\ P_{j,t-1}$

Firm i sets its price at 95 per cent of firm j's previous price.

$A_{i,t} = 1.02 A_{j,t-1}$

$D_{i,t} = 1.02\ D_{j,t-1}$

Firm i sets advertising and distribution expenditures at 102 per cent of firm j's expenditures in the previous period.

Sales-responsive strategy

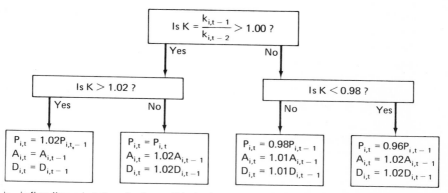

$k_{i,t}$ is firm i's market share in time t. K is the latest change in firm i's market share. If the firm's market share increased by more than two per cent, price is raised by two per cent and advertising and distribution expenditures are held constant. The other boxes are similarly interpreted.

FIGURE 10-8   Examples of some marketing strategies used in the simulation. Adapted from Kotler, "Competitive Strategies for New-Product Marketing over the Life Cycle."

gathering would be needed to develop the market model and the cost model. Furthermore, the assumption that the firms would use the same strategy for sixty periods without modification would be unrealistic. Myers has recently developed a simulation in which marketing managers "learn" and are able to change their strategies in response to disappointing results (Myers, 1968). Kotler's model may also be enhanced by incorporating stochastic features to allow for the typical uncertainty that the firm faces in the real world.

Kotler's conclusions on the relative performance of different classes of strategies are limited to this particular application. He did not test the sensitivity of the results to changes in the parameters. This would have been a worthwhile feature and would be mandatory if companies were actually going to base their actions on simulated tests of strategies.

Kotler's simulation pointed to the possibility of constructing large-scale dynamic models of the relationship between company competitive strategies and market response. The framework can be extended to include middlemen channels, multiple products, and multiple territories.

Tuason Simulation

Tuason carried out a simulation of the household coffee market in which he tested the competitive efficacy of a particular adaptive, diagnostic strategy (Tuason, 1965). His model consists of a coffee manufacturer who makes decisions on price, deals, and product blend each week. The decisions are made according to the strategy shown in Figure 10-9. Essentially the coffee manufacturer considers whether the last-period results satisfy its market share and profit goals. If not, the firm considers appropriate changes in its marketing program or product. These changes are guided by a diagnosis of any changes in the marketing program of competitors. The company evaluates the expected costs, sales, and profits from the contemplated marketing plan, and if these meet its goals, the plan is adopted.

Other programs generate the actions of competitors, retailers, and customers. For example, competition is represented by one other firm whose marketing program is generated stochastically each period. Retailers react to the marketing programs of the company and their competition by making adjustments in retail price and shelf space. Finally, different segments of the market respond differently to the competitive and retail developments in a manner specified by empirically derived equations from a study by Massy and Frank (1965).

Tuason's main purpose was to see how well the manufacturer's strategy shown in figure 10-9 would perform under different conditions, such as rapid versus slow market growth, high versus low market variability, high versus low marketing costs, high versus low weight given to the market share objective, and so forth. This meant designing a set of experiments involving rerunning the model under many different combinations of conditions. To make the matter more complicated, the model would have to be rerun many times within each experimental setting to measure the average effectiveness of the decision program in that circumstance. To keep down the computer cost and time involved in exploring every possible experimental setting, Tuason used a fractional factorial experimental design similar to that used by Bonini (1963). It permitted the testing of a more limited number of experimental settings while providing sufficient information for estimating the effect of different parametric settings on the outcome. Through this procedure, he was able to draw various conclusions about the effectiveness of the manufacturer's decision program under different conditions. He found, for example, that the decision program performed more effectively in an environment of high sales variability than in one of low sales variability.

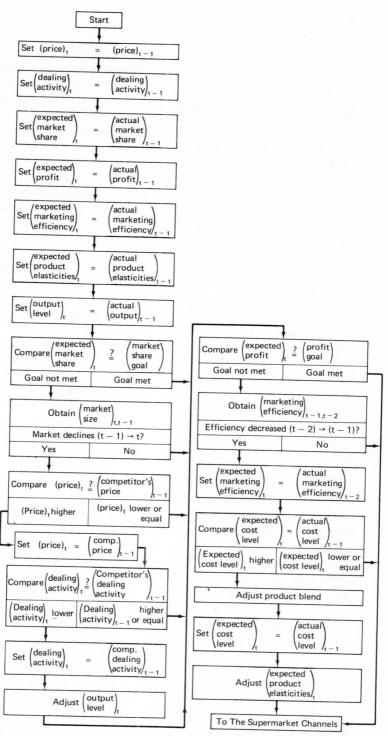

FIGURE 10-9   Decision-making processes for the marketing firm of Brand M.
Source: Tuason, "Experimental Simulation on a Predetermined
Marketing Mix Strategy," p. 88.

Comment. Tuason demonstrated that a computer program could be designed that would develop weekly decisions on pricing, dealing, and product blend that produced satisfactory profits and market share despite unpredictable changes in competitive and environmental conditions. To this extent, it follows the lead of Kotler's and Haines's (1961) simulations, but it goes beyond them in a number of respects. In addition to taking the form of a more elaborate decision program, it was put through a series of sensitivity tests to see how well the strategy performed under different environmental conditions. Through the use of factorial experimental design, Tuason managed to test effectively a number of hypotheses, demonstrating as Bonini had the importance of factorial experimental design in simulation studies. Tuason also used some empirically determined equations from a study by Massy and Frank as his model for determining sales, showing the manner in which positive models could be incorporated within the framework of an essentially normative model.

This work can be extended in several ways. Tuason generated competition's strategy stochastically although in reality it is developed in much the same way as the company's strategy. This suggests the possibility of including two elaborate strategy decision programs and simulating the results. Furthermore, the study was based on artificially generated data, and a logical next step would be to try to develop an explanation of an actual set of data describing competitive marketing activity in the coffee industry. Executives could be interviewed about their actual decision processes as a way of refining the computer program.

PRICING SIMULATIONS

We now turn to a series of simulations dealing with a more limited number of decision variables. We will start with pricing, which has historically received the most attention. In the last thirty years, several studies have appeared that have sought to examine the actual practices of price setting in American industry. For example, Kaplan, Dirlam, and Lanzillotti (1958) surveyed a number of business executives of major American corporations to learn what criteria they adopted in setting prices. They found a variety of pricing objectives and procedures, most of which were at variance with the normative model of economists. As a result of this and other studies, a prolonged and still unsettled debate has taken place between those who defend the classical pricing models of economists and those who defend the pricing models that are closer to what many businessmen say they do (Machlup, 1946; Lester, 1946).

It is only in the last decade that some of the behavioral-oriented models of pricing have been expressed and tested in a simulation context for their effectiveness in explaining pricing behavior. Two such models will be described here. The first was designed by Cyert, March, and Moore and involves the pricing behavior of a department store buyer of shirts. The second was designed by Morgenroth and attempts to describe the price reaction behavior of a pricing executive in the retail gasoline industry.

Cyert, March, and Moore
Department Store Pricing Model

Cyert, March, and Moore (1962) became interested in the types of business decisions that have a fairly routine character and the extent to which computer programs could be developed to represent the decisions of actual executives. As an example of a potentially programmable decision area, they singled out the job of the department store buyer who must periodically forecast demand, place orders, and price his merchandise. The researchers worked with a particular buyer in the shirt department of a Pittsburgh department store. The buyer was intensively interviewed about his procedures, and the investigators examined various data he used in his job. The researchers felt that if they could develop a decision program that would produce results that corresponded to the buyer's past decisions on pricing and order quantity, the program would essentially constitute an explanation of his behavior. We will restrict the discussion here to their pricing model.

The authors provide for three different pricing situations: normal, sales, and mark-down. Normal and sales pricing are planned for regular intervals; mark-down pricing is contingent upon failure or anticipated failure to reach organizational goals.

Normal pricing is used for new merchandise in the department. The basic pricing procedure is the application of a mark-up to the cost of merchandise. The researchers had little difficulty in determining a "normal" mark-up for each product group, since these mark-ups had remained relatively stable for many decades. They found the normal mark-up for standard items to be 40 percent and for import items to be 60 percent. They also found that the normal retail practice for these items was to round prices to the nearest ninety-five cents. The pricing decision rules for standard items, exclusive items (those items not made available to competition), and import items are summarized in part *A* in figure 10-10.

Regular sales pricing is used periodically throughout the year—for example, after the big Christmas, Easter, and "Back-to-School" promotions. The basic procedure is to use either a standard sales reduction from normal price or a standard mark-up on cost (like normal pricing) subject to several general policy constraints. Figure 10-10 describes these constraints in part *B*.

Mark-down pricing is used as an adaptive device when feedback indicates an unsatisfactory sales or inventory position. Defective merchandise accounts for only a small part of mark-down pricing. As in other pricing, the department uses a set of standard decision rules which provide for determination of the *timing* and *amount* of mark-downs. Part of the pricing decision rules for mark-downs is described in part *C* in figure 10-10.

The results of the pricing model showed remarkably high levels of prediction. To test the normal pricing situation, a random sample of 197 invoices with actual prices was drawn. The cost and classification data from this sample were used as inputs to the computer model; computer output was a predicted price. Thus, the researchers were able to compare the predicted price with the actual price. They found that 95 percent of the predictions were correct to the penny. Similar results were obtained for tests of sales pricing (96 percent accurate) and mark-downs (88 percent accurate).

Comment. Cyert, March, and Moore believe that the high levels of prediction support the validity of their model—and they do. But if pricing is regulated by store policy, then it is not surprising that their program based on this policy can reproduce actual results. In fact, it suggests that top management should try to leave these relatively routine decisions to computer programs and thus free the buyers to examine pricing and other situations not completely covered by store policy, as well as mark-downs, where competitive and environmental changes play a more important role.

The authors did not use the simulation to consider optimization issues in pricing, although this would have marked a worthwhile extension. Specifically, they could have examined the effect on sales of changing the pricing decision rules or decision parameters. This would have required the development of functions relating sales rates to prices on which management had some ideas. The simulation could then have provided a tool for exploring alternative mark-up and mark-down policies.

A. NORMAL PRICING

Standard items: divide each cost by 0.6 and move the result to the nearest $.95.

Exclusive items: calculate the standard price from the cost, then use the next highest price on the standard schedule.

Import items: divide the cost by 0.4 and move the result to the nearest standard price. If this necessitates a change of more than $ .50, create a new price at the nearest appropriate ending (i.e., $.95 or $.00).

B. REGULAR SALES PRICING: GENERAL CONSTRAINTS

If normal price falls at one of the following lines, the corresponding sale price will be used:

Normal price ($)	Sale price ($)
1.00	0.85
1.95	1.65
2.50	2.10
2.95	2.45
3.50	2.90
3.95	3.30
4.95	3.90
5.00	3.90

For all other merchandise, there must be a reduction of *at least* 15 per cent on items retailing regularly for $3.00 or less and *at least* 16 2/3 per cent on higher-priced items.

All sales prices must end with 0 or 5.

No sale prices are allowed to fall on any price lines normal for the product group concerned.

Whenever there is a choice between an ending of 0.85 and 0.90, the latter ending will prevail.

FIGURE 10-10 Pricing decision rule. Source: Cyert, March, and Moore, "A Model of Retail Ordering and Pricing by a Department Store," p. 518.

C. MARK-DOWN PRICING: FLOW CHART

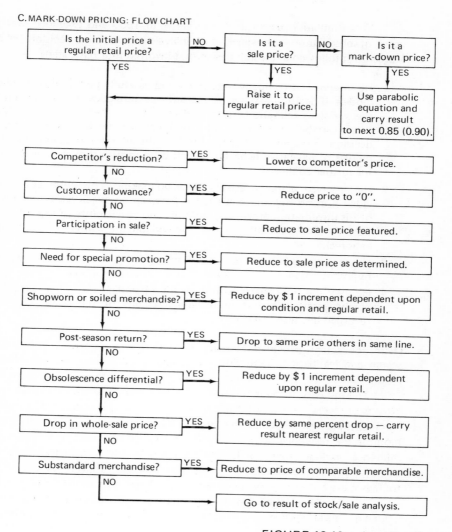

FIGURE 10-10   (continued).

Morgenroth Pricing Simulation

The Morgenroth simulation deals with a more dynamic pricing situation in which executives may face sudden unexpected changes in competitive prices and must make quick and decisive responses (Morgenroth, 1963; Howard and Morgenroth, 1968). Although Morgenroth did not identify the industrial situation, it is not unlike the price wars in the retail gasoline industry where a major oil producer periodically confronts a price cut by another producer and must quickly decide whether to match the reduction or maintain his price.

This simulation began as a case study in 1959 and extended over three years of observation, interviews, and analysis of executive decision behavior. Morgenroth followed a research design that involved four basic steps.

1. He interviewed company executives and carefully probed their implicit models of decision-making. The executives were asked to try to identify the variables relevant to pricing decisions and then to verbalize how these variables were mentally combined as the basis for a pricing decision. Flow charts were used to turn the interview data into working models of executive behavior, which were in turn shown to executives for further comment.

2. He observed actual executive decision-making. This observation provided a check for the consistency between an executive's description of his information processing and the actual availability of this information. At this stage interest was focused on the company's communication channels.

3. He analyzed company records of past pricing decisions and reviewed hypotheses based on these data with the executive group.

4. He used the flow charts to test decision paradigms of other executives at the same level but in different regions of the country against recorded decision behavior.

These four steps provided Morgenroth with a comprehensive procedure of search and feedback to represent pricing decision processes in a dynamic simulation.

The industry is made up of a few large *major firms* and several quite small *private branders.* A key element in the model is the concept of a *reference marketer,* the firm that has the largest share of a particular market. All firms monitor the reference marketer's pricing activity and react to it, although the reference marketer may not be the price leader if he is perceived as failing to interpret market conditions correctly. For example, a given company quickly follows a price decrease by the reference marketer, but follows a price increase more slowly. These timing dynamics together with differing pricing behavior among majors and private branders set off the Morgenroth simulation from other decision models.

Figure 10-11 depicts a sequential binary choice process triggered by an initial price change. It can be interpreted as follows. The company first monitors the reference marketer's wholesale price in each local market at time t (box 1). If the price is not the same as the company's price, the next step is to determine whether it has increased or decreased (box 3). The direction of the price change is very important because upward price moves are much simpler than downward price moves. The rationale for this asymmetry is economic: Since the market is such that total purchases are relatively unaffected by price changes, a mutual price rise is always profitable. Competitive reactions to price declines vary according to the company's strength in the market (as evidenced by its market

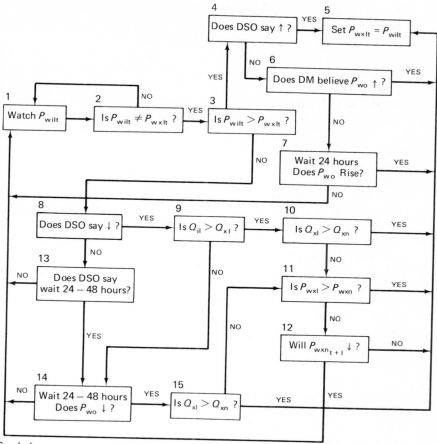

Symbols:

P = price
w = wholesale
x = our company
o = other major competitors in local market
i = initiator
t = time, at present
Q = quantity, i.e., sales volume in physical terms
l = local market wherein price change is being considered
n = nearby market with funnel influences
DSO = District Sales Office (District Sales Manager)
$\uparrow$ = raise price
$\downarrow$ = drop price
DM = decision maker

FIGURE 10-11. Model of the pricing decision process. Source: Howard and Morgenroth, "Information Processing Model of Executive Decisions," p. 419.

share) and location of the affected market. Thus, a price increase results in only three steps (boxes 4, 6, and 7) before a decision is made, whereas a price decrease can require up to eight steps (boxes 8-15).

The timing of price changes is also considered with a given company *delaying* more in following the reference marketer up than in following him down. While several reasons are put forth for this behavior, our concern here is with the general nature of the model, and therefore it is sufficient to account for only the general direction and timing mechanisms.

Morgenroth tested his model on two levels. A random sample of decisions was taken from company records of the division managed by the primary executive under study, and these sample data served as input to the simulation model. The output was a series of simulated pricing decisions. Only the *direction* of a price change was taken into account. By comparing the simulated direction indicators with the actual data, Morgenroth found perfect agreement and considered the model to be confirmed by the output test.

Further validation was accomplished by taking extensive company data and analyzing the model shown in figure 10-11 box by box. Here again Morgenroth found perfect agreement and considered the model to be confirmed by the process test.

A final step in the verification procedure was to review the model with executives of other company divisions. Here both the plausibility of the decision process and the actual decisions were again tested to assure interdivisional comparability.

Comment. Like the Cyert, March, and Moore pricing model, this model showed a high level of validation—perfect if directional changes are alone considered. It is also subject to the same kind of criticism. If pricing is regulated by company policy, it is not surprising that Morgenroth's model is able to produce results similar to the historical output. His success primarily indicates that company executives follow a well-defined pricing policy.

Models such as this one make a contribution in highlighting the specific decision rules and parameters that management is following.[3] The next logical step is to turn the model into an instrument for investigating alternative pricing strategies that may improve company performance. This was not done by Morgenroth but remains an unexploited major value of constructing such models.

ADVERTISING SIMULATIONS

The phenomenon of advertising with its great cost and highly uncertain effects has stimulated a large number of model-building efforts in recent years. The majority of models have been single-equation expressions of the relationship of sales to total advertising expenditures and a few other factors. For example,

[3]For another recent example of modeling management decision behavior, this time in the area of making forecasts, see G. C. Michael (1971).

Vidale and Wolfe (1957) expressed the rate of change in sales as a function of total advertising dollars, the percentage of the market that is saturated, a sales response coefficient, and a sales decay coefficient. Other interesting analytic formulations have been developed and tested successfully by Telser (1962) and Palda (1964). Recently Bass (1969) and Schultz (1971) developed and tested simultaneous-equation models of the sales-advertising relationship. Charnes *et al.* (1966) developed a model of advertising consisting of linked relations between intermediate variables connecting advertising and sales. The model, called DEMON (Decision Mapping Via Optimum GO-NO Networks), assumes that the advertising budget will be spent in a way that will achieve a certain percentage of market coverage and will produce a certain frequency of exposures per capita; and these will in turn produce a certain level of product awareness; a certain percentage of those aware of the product will try it, and a certain percentage of the triers will become regular users. By statistically determining the relationship between each pair of variables, the effect of a specific advertising budget on final sales can be estimated.

Some of the most useful simulation work in the advertising area has occurred with the problem of media selection. After an advertiser decides how much to spend on advertising, he must find the best media for carrying his message to the target population. He faces a vast number of media choices with different audience sizes and compositions, different visual and audio qualities, and different costs. He has to determine how large his ads should be and how often they should be seen by buyers in different market segments. It is not surprising that operations researchers have studied this problem and have proposed solutions in terms of linear programming, dynamic programming, or heuristic scheduling. These models, some of which are being used by advertising agencies, seek to find "best" media schedules.[4] In two cases, the proposed solution has taken the form of microbehavioral simulations. The simulation models do not seek to find the "best" schedule so much as they seek to measure the dynamic exposure characteristics over time of any proposed schedule. The first model was designed by the Simulmatics Corporation and involves a hypothetical population of viewers who individually watch media and are exposed on a probabilistic basis. The second model was designed by Dennis Gensch and involves testing a media schedule against an audience with known viewing habits.

Simulmatics Media Selection Model

Simulation models are used in media selection not to find optimal media plans but rather to test one or more proposed plans. The Simulmatics model uses a sample of 2,944 hypothetical persons supposedly representing a cross section

[4]For a review of previous media models and an excellently conceived new media model, see J. D. C. Little and L. M. Lodish (1969).

of the American population (1962a, 1962b). One hundred and forty different types of persons were defined by stratifying according to sex, age, type of community, employment status, and education. Several representatives of each type were located at ninety-eight sampling points in the continental United States.

This population was specifically created to test alternative media schedules and not buyer brand choice. An individual's media choices were determined probabilistically from his social-economic characteristics. Thus, the probability that a particular person would watch the "Tonight Show" during a given week was treated as a multivariate stochastic function of his age, education, and so forth. As an example, suppose the probability equation for watching the "Tonight Show" is:

$$P = .04 + .20 \, X_1 \, .44 \, X_2$$

where P = probability of watching the "Tonight Show"

X_1 = sex, with 1 for male and 0 for female

X_2 = education, with 1 for college and 0 for high school

The computer program now asks whether the first person in the population is watching this program (on which client advertising may be placed). Suppose the person is a high school educated male. Inserting $X_1 = 1$ and $X_2 = 0$ in the equation, the probability of his watching the "Tonight Show" is estimated at 0.24. Then a random two-digit decimal number is drawn, and if it is less than 0.24, this person is recorded as having seen the show; otherwise he missed it. The computer program then moves on to the next person and tests whether he has been exposed to this show. When this is done for all persons in the population, the computer program moves to the next media purchase in calendar time and repeats the procedure of generating and recording the number of exposures. The routine is summarized in figure 10-12. As the simulation of the year's schedule progresses, the computer tabulates the number and types of people being exposed. Summary graphs and tables are automatically prepared at the end of the hypothetical year's run, and they supply a multidimensional picture of the schedule's probable impact. The media analysts examine these tabulations and decide whether the audience profile and the reach and frequency characteristics of the proposed media schedule are satisfactory.

Comment. The fortuitousness of the media exposure process as it is currently understood makes Monte Carlo simulation an ideal tool for studying it. The Simulmatics model highlights the major issues faced by the designers of media models in seeking to achieve validity. The first issue common to all microbehavioral simulations involves the problem of constructing a repre-

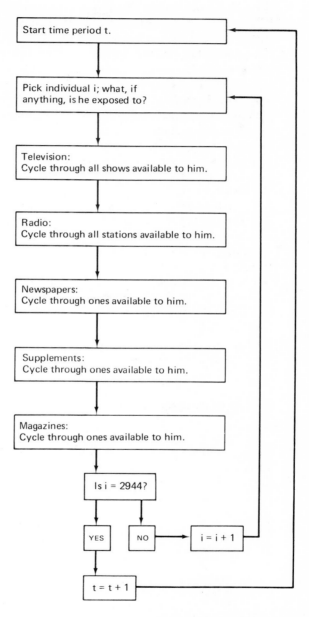

FIGURE 10-12   Simulmatics flow chart. Source: Simulmatics Corp., *Simulma-*
tics Media-Mix, p. 2.

sentative hypothetical population. The 2,944 consumers in the model should exhibit the distribution of age, sex, income, and education that exists in the general population or at least in the company's target population. If the sample is a biased representation of the actual population, the estimated number of total media exposures will likewise be biased. This point stresses the need for generously financed and well-designed surveys prior to building the simulation model.

The second issue concerns the representativeness of the hypothetical population's media habits aside from their demographics. Each media vehicle publishes a breakdown of its audience composition, but the figures are generally based on inadequate sampling procedures. The problem of the relationship between demographic characteristics and media preferences is a general one that must be faced by anyone building models in this area.

A third issue concerns the relationship between media exposure (a person is exposed to the media vehicle), advertising exposure (the person notes the ad), and purchase behavior (the person is motivated by the ad to buy the product). The Simulmatics model does not go beyond tabulating the number of probable media exposures, and the advertiser is therefore left without a standard for evaluating the probable sales results of a particular media schedule. This again is a general problem faced in all media model-building, not just simulation. Data are quite weak when it comes to adjusting media exposures into an estimate of ad exposures and finally into an estimate of sales.

The Simulmatics model contains no methods for finding improved or optimal media plans. Management must examine the dynamic media exposure output for its plan and then judgmentally formulate an alternative plan to be tested against the first one.

Gensch Media Selection Model

Gensch developed his computer simulation model, called AD-ME-SIM, to select the most effective national magazines and network television shows for a specified advertising message (Gensch, 1969). Unlike the Simulmatics model, Gensch's model employed the actual viewing pattern of a sample of people and recorded the number of exposures that took place with different proposed schedules. He also used a heuristic procedure to arrive at a better schedule. Finally, he made a specific application of the model to selecting media for a particular dog food product.

Gensch's simulation model was set up to satisfy the following criteria:

(1) the demographic, reading and viewing habits must all come from the same individual, (2) the data must come from real individuals, not hypothetical or imaginary individuals, (3) the sample must be large enough to be significant on a national level, and (4) the cost of gathering this data must be low enough to fall within the advertising agencies' budget.

He found that data supplied by Brand Rating Research Corporation met these criteria and only required extrapolation in order to predict the impact of *future* media schedules on *future* audiences.

An overview of the media simulation model is shown in figure 10-13. The input stage requires the following data: (1) the proposed media plan and schedule, (2) a set of weights for the effectiveness of different media, (3) a set of weights for the effectiveness of varying size and color ad forms, (4) a set of weights showing the value of different patterns of exposure frequency, (5) a list of the media costs and volume discounts, (6) a set of weights showing the value of an exposure to different types of persons in the target population, and (7) data from the Brand Rating Research Corporation showing the reading and viewing patterns over time of a real sample of individuals.

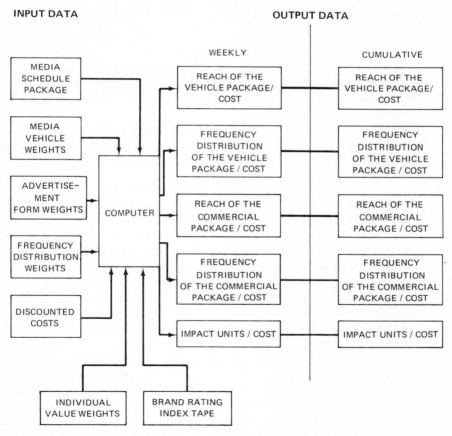

FIGURE 10-13   Gensch's media schedule evaluation model. Source: Gensch, "A Computer Simulation Model for Selecting Advertising Schedules," p. 205.

From these inputs, the computer program is able to generate weekly and cumulative output on several important variables bearing on the overall impact of the proposed media schedule: (1) the number and percentage of people in the target population reached by the proposed media schedule and its cost; (2) the number of persons reached zero, one, two, etc., times during the period by the proposed media schedule; and (3) an adjusted exposure number representing the overall impact of the proposed media schedule taking into account all the objective and subjective evaluations of the media and the value of exposures to different members of the target audience.

Gensch tested the applicability of his model to a real media selection problem facing the advertising agency handling the account of Ken-L-Ration canned dog food. The target population was defined to be female heads of households on a national basis. Information on past purchases was obtained from the Brand Rating Research Corporation. Subjective media weights were obtained from media planners in the company's advertising agency. Two alternate media plans had been drawn up by the agency before Gensch's model was applied. The media experts considered both plans to be good but held that one had an edge over the other. Gensch simulated the two plans, and the one that achieved higher overall impact agreed with the agency's judgment. Gensch also tested the sensitivity of the overall impact values to variations in the subjective weights.

The final step involved a heuristic that sought a "better" schedule. The result of this attempt was a schedule that the experts at J. Walter Thompson felt was superior to those they themselves had drawn up.

Comment. A major limitation of Gensch's model which he is careful to note is that it relies heavily on subjective judgment. This is almost unavoidable at this time in the state of the art because of inadequate media data, knowledge, and theory. The only way to validate the computer estimates of exposure value of two alternative media plans is to try both of them out in different but matched media areas and subsequently interview households to determine their actual exposure levels. This type of postanalysis of media campaigns is not only very expensive but also subject to many kinds of measurement errors. Short of this validation approach, the use of expert ranking procedures seems the next best alternative. Gensch's results would have been more convincing if the media planners had drawn up, say, five plans instead of two, along with a subjective ranking of them on expected overall impact. If Gensch's model had produced the same rankings, this would at least have indicated that the computer program had evaluated the plan in the same way that the media men had, although there was no evidence that the advertising men's judgment itself was sound. The computer could considerably shorten the amount of time required by the agency to arrive at an initially sound plan. Then, if Gensch's heuristics also worked well in finding better plans, the simulation model would provide a true advantage over completely judgmental approaches to media planning.

SALES FORCE SIMULATIONS

The problems of sales force management, such as recruiting, selecting, assigning, compensating, and motivating salesmen, would seem to be an area that would attract a number of simulation studies. But relatively few attempts have been made to study complex issues in sales force management through the medium of simulation. One type of effort has been directed to the problem of routing traveling salesmen. The problem is to find how to route a salesman through several cities that he must visit in a way that minimizes his total travel time, distance, cost, or some weighted combination. The problem is difficult to solve analytically, and it has spawned several types of solutions, including heuristic ones (Little *et al.,* 1963; Karg and Thompson, 1964). However, in actual sales management the problem of routing is far less important than the problem of which customers should be called on in the first place and how often during the year. Cloonan has recently examined this problem by using an interesting simulation model that combines both the sales call problem and the routing problem (Cloonan, 1966).

Cloonan Sales Call Simulation Model

Cloonan approached the sales call and the routing problems by designing a set of simulated territories, each territory containing a different number and spatial arrangement of customers. He then proposed and tested four alternative heuristic decision procedures for determining which customers to call on each period and which to bypass. Each call pattern was subsequently ranked in terms of the ratio of value to cost. Cloonan then compared these ratios for the four heuristics to the known optimum value/cost ratio to measure their relative performance as heuristics. The optimal value/cost ratio was determined through exhaustive enumeration and would not ordinarily be feasible in large-scale problems.

As an example, imagine a sales territory containing seven customers and a home office. Suppose the salesman selects four accounts to visit each period. In designing a hypothetical sales territory, the spatial arrangement of the accounts and the size of the acocunts must be considered. Figure 10-14(*A*) shows a circular and an irregular arrangement of account, respectively. The optimum routing is obvious for the circular pattern but not for the irregular pattern. Another feature of each territory is the size distribution of accounts. Companies frequently classify their customers into A, B, C, and D type accounts where A has the highest and D the lowest sales potential. Figure 10-14(*B*) illustrates a particular distribution and order of account sizes.

For his simulation, Cloonan designed four different spatial arrangements of accounts. The first two were actual configurations of the scatter of the major twenty-five cities in two different states (and were purposefully complex), and

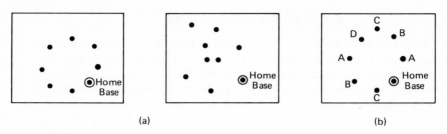

(a)   (b)

FIGURE 10-14   Simulated sales territories.

the second two were artificial, one resembling a circle and another a lattice-point design. Each territory had twenty-four cities plus a home base. Cloonan assumed six accounts of each kind (A,B,C, and D) and distributed them according to the following three main patterns:

1. H BCDA BCDA BCDA BCDA BCDA BCDA H
   No clustering of similar accounts
2. H AAA BBB DDD CCC AAA BBB H
   High level of clustering and the larger accounts are close to base
3. H CCC DDD BBB AAA BBB AAA CCC DDD H
   High level of clustering and the larger accounts are far from base

The four types of spatial arrangement and the three account size configurations provide twelve territory types. Each was used to test each of four heuristics for a total of forty-eight simulation runs. Furthermore, each simulation run consisted of a salesman making a sales call plan each week for twenty-four weeks.

Cloonan had his hypothetical salesman call on twelve customers each week, selecting them according to one of four heuristic decision programs:

1. The simplest program required the salesman to call on the first twelve adjacent accounts in the order of the optimal tour through the territory.
2. The second program required the salesman to make a call to six A's, three B's, two C's, and one D on each tour of twelve accounts.
3. The third program required the salesman to evaluate the value/cost ratio for every account in the territory. The ratio is calculated by considering the value and cost of a current call on that account. The cost of making the call is the cost of call time plus the difference between going directly between accounts $(i-1)$ and $(i+1)$ and going from $(i-1)$ to (i) to $(i+1)$. The program determines the twelve accounts with the highest value/cost ratio; they are the accounts called on in this tour. A subroutine is used to determine the optimum route through these twelve accounts.
4. The last program is the most complex. It selects a starting account and then considers whether to go to the next account in the optimal territory tour or skip it. If the account's value/cost ratio exceeds a

certain minimum value, it will be included. Otherwise the next account is considered for inclusion or bypassing. This reasoning is continued until twelve accounts are selected.

In a preliminary test, Cloonan found that the last three heuristic programs performed substantially better than the first one. This is not surprising since the first program completely neglected the different sizes of the accounts. But the last three heuristic programs did not show much difference among themselves. Although the second program comes closest to the procedures used by salesmen in practice and the fourth program represents a substantial sophistication of the choice procedure, the fourth program did not improve the results substantially. The last three in general were performing at over 90 percent of the optimum value for perfect selection. This may explain why sales management has shown little interest in sophisticating account selection procedures: They may suspect that the cost of additional information and analysis may exceed the gains in better account selection.

Comment. Cloonan's invention and testing of artificial sales territories and alternative heuristics for selection customers are a valuable contribution to the study of the sales call problem. Further refinements can be made in the measurement of the value of an account call (Cloonan made value a function of only two elements: account size and elapsed time since last call) and in the cost of skipping a call. It is also possible that experts can formulate a number of other heuristic decision programs to select accounts each period, and these should be tested against the present four.

NEW-PRODUCT DEVELOPMENT SIMULATIONS

A marketing function that is extremely complex and of increasing importance to the survival of the firm is that of new-product development. The life cycle of the average product or brand has been getting shorter, making it increasingly necessary for each company to plan successors to the current products in its line. The cost of launching new products is large and market acceptance remains, even under ideal conditions, uncertain. Many companies have been rationalizing their new-product planning function, but the number of new products that still fail in the marketplace, after considerable research and careful introduction, is substantial and points to the need for still better information and analysis.

All the data that are collected about the market prospects for a new product must be organized into some model for profit analysis. Companies have typically used quite simple profit models, such as breakeven models, early payout models, and so forth, and have done little systematically to handle risk, product interactions, marketing strategy alternatives, and explicit competitive and environmental assumptions (Kotler, 1968). Two models have recently appeared that attempt to handle these complexities through the medium of simulation. The first model was designed by Edgar Pessemier and is an adaptation of a technique known as risk analysis; the second model was designed by Glen L. Urban and uses simulation as a solution search technique.

Pessemier New-Product Profit Simulator

A major decision criterion in judging whether to launch a new product is the internal rate of return that the product is likely to yield on its investment. Management recognizes that any estimate of rate of return is subject to some uncertainty, and this had led analysts to recommend that management be provided with a probability distribution of the possible rates of return. A probability distribution will show the lowest and the highest rate of return that might be expected as well as the likelihood of different possible returns. A separate probability distribution would be estimated for each alternative marketing strategy and forecast environment. Strategies are to be preferred that yield a high rate of return and a low variance. The task facing the analysts is to find a way to estimate the probability distribution of possible rates of return for each alternative strategy.

The answer comes in the form of a technique known as risk analysis (Hertz, 1964). Management is asked for estimates of future prices, advertising, manufacturing costs, and investments in the form of probability distributions. These probability distributions are then sampled in Monte Carlo fashion to generate a time series of future sales, costs, profits, and investment from which an internal rate of return is calculated. By repeating this a sufficient number of times, the rates of return can be accumulated in a frequency distribution which then serves as the desired probability distribution.

Pessemier has developed the details of this technique in a model for evaluating new products (Pessemier, 1966). The entire model is represented in the flow diagram in figure 10-15. The following points should be noted:

1. The executives are asked to spell out a marketing strategy, a facilities strategy, and assumptions about the marketing environment.
2. They are then asked to estimate, on the basis of the given assumptions, sales, costs, and investments for each year of the planning horizon.
3. The executives are also asked to furnish pessimistic and optimistic estimates for each uncertain variable. These estimates allow a probability distribution to be constructed for each variable. A value is randomly drawn from each probability distribution for each year in the planning horizon. From these data, the cash flow is calculated.
4. The cash flows are discounted to find the rate of return (r) for the particular run. The simulation is repeated many times to derive a probability distribution of r's.
5. The model allows the calculation of the value of additional information.

FLOW DIAGRAM

Informed company executives spell out:
I. a specific marketing strategy, *s*
II. a specific marketing environment, *e*

The executives then estimate the annual expected values of the following variables over the *i* years in the new-product planning horizon:

1. unit sales (Q_i)
2. unit price (P_i)
3. unit cost (V_i)
4. investment (I_i)

implies a specific facilities strategy, *f*

In addition to the expected values of these variables, the executives estimate the respective 0.1 and 0.9 decile values. The decile values enable the analyst to treat each variable as subject to a lognormal probability distribution with a known mean and variance. (These will be used later.) Finally, the executives estimate annual expected values, but not quantile values, for the following three variables:

5. depreciation (g_i)
6. opportunity costs in dollars (h_i)
7. nonrecurring start-up costs (u_i)

The analyst now generates through Monte Carlo methods a sample value for each variable $(Q_i, P_i, V_i,$ and $I_i)$ for each year during the product-planning horizon. (A device is used to permit serial correlation of successively generated value of unit sales.)

FIGURE 10-15   Monte Carlo simulation model. Source: This flow diagram has been developed from Pessemier's discussion, *New Product Decisions: An Analytical Approach.*

These values enable a computation to be made of the positive cash flows (Z_i) each year. The formula is:

$$Z_i = T\left\{Q_i(P_i - V_i) - h_i - u_i\right\} + g_i$$

where T stands for one minus the tax rate. In other words, the profit contribution in year i is found by first taking the difference in year i between price (P_i) and unit cost (V_i) times unit sales (Q_i). This gives the unadjusted before-tax profit. Then opportunity costs (h_i) and nonrecurring start-up costs are subtracted (u_i). The after-tax profit is found by applying the profit retention rate. Depreciation (g_i) is then added back to the cash flow to cancel the effect of its being subtracted implicitly as part of V_i.

(E)

The investment cash-flow stream is discounted back to the present by the firm's opportunity cost of capital to yield the present value of the firm's investment flow (I).

(F)

Successive approximation is used to find the rate-of-return (r) that will discount the profit flow to an amount equal to the discounted investment (I).

(G)

Have enough $r's$ been generated yet?

(H) No

(I) Yes

Return to Step C

Order the $r's$ in a frequency distribution and show graphically. This is the approximate probability distribution of rates-of-return for marketing strategy s, marketing environment e, facilities strategy f, and various expected values and quantile estimates.

(J)

FIGURE 10-15 (continued)

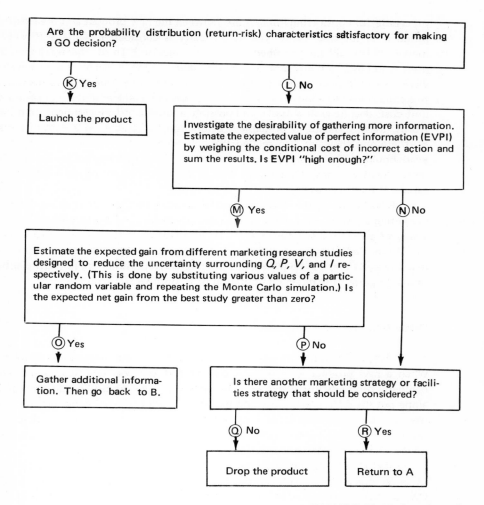

Are the probability distribution (return-risk) characteristics satisfactory for making a GO decision?

(K) Yes

(L) No

Launch the product

Investigate the desirability of gathering more information. Estimate the expected value of perfect information (EVPI) by weighing the conditional cost of incorrect action and sum the results. Is EVPI "high enough?"

(M) Yes

(N) No

Estimate the expected gain from different marketing research studies designed to reduce the uncertainty surrounding Q, P, V, and I respectively. (This is done by substituting various values of a particular random variable and repeating the Monte Carlo simulation.) Is the expected net gain from the best study greater than zero?

(O) Yes

(P) No

Gather additional information. Then go back to B.

Is there another marketing strategy or facilities strategy that should be considered?

(Q) No

(R) Yes

Drop the product

Return to A

FIGURE 10-15 (continued)

Thus, the model has many features that are desirable in a refined analysis of the new-product decision. Forms that executives may fill out and a computer program to perform the calculations have been developed. Further refinements of the model and field tests on new products are in progress.

Comment. At the same time, some weaknesses are apparent. First, Pessemier's procedure for determining the value of additional information requires considerable computation time (see box M). Second, except through the recognition of opportunity costs, the model does not contain an explicit procedure for considering interactions between the new product and the existing products in the line. (This type of difficulty is considered in the SPRINTER model, which will be examined in the next section.) Third, the treatment of price, unit costs, and investment as *independent* random variables is theoretically weak. It would seem that if a high price were drawn by Monte Carlo methods, this should lead to a lower sales level than expected, a higher unit cost (assuming unit costs decline with output), and possibly a lower annual investment. In other words, these variables in principle are functionally related on a current basis. Furthermore, they are also related through time, in that the value drawn this year should affect next year's expected value. These refinements are handled in more recent versions of the model, but at the cost of requiring many additional estimates from executives.

Finally, the model does not provide an algorithm for determining the best marketing strategy. Like many simulation approaches, the model basically offers a way of generating the dynamic results of a particular strategy. Comparison of the Monte Carlo results of different strategies often suggests potentially fruitful new strategies that can be tested, but functional relationships are not sufficiently simple to permit the direct computation of an optimal strategy.

Urban New-Product Profit Simulator

Most new-product models have neglected the product-interaction problem. When a new product is related on the demand and cost side to some current company products, the deciding factor should not be the new product's profits alone but the differential profits on the whole product line with and without the new product. The estimated profits for the new product must be adjusted downward if they reduce the profits on the company's current products and upward if they enhance the profits on the company's current products.

SPRINTER (Specification of PRofits with INteractions under Trial and Error Response) was developed by Urban and is oriented toward yielding a GO, ON, or NO decision (Urban, 1968). SPRINTER gives explicit consideration to the dynamic marketing program that is planned over the new product's life cycle for both the new product and any existing products in the line having profit interactions with the new product. In addition, variations in competitive response during the product life cycle are explicitly considered. Estimates of demand and costs are functionally derived from marketing plans, competitive actions, product trend, and cyclical and seasonal characteristics.

The two pivotal measures in the model are discounted differential profit and differential uncertainty. *Discounted differential profit* is the present value of the annual differences in the profits on the whole line with and without the new product. *Differential uncertainty* is the difference in total uncertainty (as measured by the variance of profit) with and without the new product. These two measures form the axes of the decision grid to decide between GO, ON, and NO.

The mathematical structure of the model proved to be too complex to solve analytically for the best marketing plan. Consequently, Urban resorted to numerical search procedures to approximate the best solution, which involved testing gross variations in the marketing program to identify near optimal neighborhoods and then searching finer intervals. This technique is efficient if the profit function is well behaved. The overall model is described in the diagram in figure 10-16.

Urban tried out his model on data supplied by a large chemical company that had just nationally introduced a new nylon compound that had significant demand and cost interactions with two current products in its line. His purpose was to determine whether the model would lead to the same decision and forecasts made by the executives on the basis of the data known to them at the time. Urban's model indicated that the company should have made a NO decision in view of its planned marketing program.[5] His model then proceeded to search for other configurations of the marketing program that might lead to an ON decision and possibly a GO decision. After much search, a marketing strategy and a facilities strategy were found that just managed to lead to a GO decision.

To summarize, the central features of the SPRINTER model are as follows:

1. Profits and uncertainty are estimated for the new-product line versus the old-product line.
2. A demand equation is formulated that incorporates life cycle, seasonal, industry marketing, competitive, and product interaction effects.
3. The model uses the GO, ON, NO framework, based on rate-of-return measures.

Comment. Urban originally set up his model for estimating new-product profits along analytical lines but found that the mathematical expressions were too complicated to allow finding the optimum marketing strategy through calculus procedures. Instead, he used the speed of the computer to try out all possible reasonable marketing mixes by varying price, advertising, and other decision variables in discrete steps. This enabled him to identify strategies with high local profit peaks which he then investigated closely by using finer discrete steps. Thus simulation in Urban's model means the use of numerical analysis as a method of studying the irregular response surface produced by a complex model.

Since Urban's model depends on a great number of subjective executive estimates, a desirable step is to test the sensitivity of his profit estimates to variations in these critical estimates. Because of the large number involved, this step would include some kind of factorial design procedure to hold computer time and cost to a reasonable level. It will be a desirable additional step when this technique is applied to a real product decision situation.

Since designing this model, Urban has created newer versions of

FLOW DIAGRAM

Informed executives supply estimates and data on:

1. A reference marketing program matrix (P_i, A_i, D_i) over the planning period for the new product and for any interacting products.
2. Estimate of probable time of competitive entry and marketing program of competitors. (Also high and low estimates.)
3. Estimates of market share of competitors in new-product market over planning period and effect of industry marketing effort on demand. (Also high and low estimates.)
4. Planned plant capacity for each year in the planning period.
5. Estimate of the reference life cylce demand for the new and interacting products on the basis of the marketing programs. (Also high and low estimates.)
6. Executive estimates of the demand response of the new and old products to systematic variations in (P_i, A_i, D_i), including cross-product response functions. Also needed is an estimate of the shift in the reference life cycle in response variations to the marketing program. (Also high and low estimates.)
7. New-product development costs and production for new and existing products.
8. Management constraints on advertising budget, plant output capacity, technical service, and price.
9. Management requirements regarding the minimum profit necessary for a GO decision (Z_g), the minimum profit for an ON decision (Z_o), the minimum probability necessary for a GO decision (Pr_g), and the minimum probability necessary for an ON decision (Pr_o).
10. A reference marketing program matrix for the old line (on the assumption that the new product is not introduced).

Calculate the total discounted profits for the new line and the old line under the new reference marketing program and the old reference marketing program. (All constraints must be satisfied.) Subtract old line discounted profit from new line discounted profit to find total discounted differential profit.

IGURE 10-16   Flow diagram of SPRINTER model. Source: This flow diagram has been adapted and modified from Urban's, "A New Product Analysis and Decision Model."

```
┌─────────────────────────────────────────────────────────────────────┐
│ Specify alternative values for price, advertising, and distribution   │
│ and for each combination find total discounted differential profit.   │
│ Select the marketing program that yields the maximum total discounted │
│ differential profit.                                                  │
└─────────────────────────────────────────────────────────────────────┘
                              (D)
┌─────────────────────────────────────────────────────────────────────┐
│ Calculate the differential uncertainty.                               │
└─────────────────────────────────────────────────────────────────────┘
                              (E)
┌─────────────────────────────────────────────────────────────────────┐
│ Plot the estimated maximum discounted differential profit and         │
│ differential uncertainty on the GO, ON, NO decision grid. Does the    │
│ estimate fall in the GO region?                                       │
└─────────────────────────────────────────────────────────────────────┘
     YES (F)                        (G) NO
┌────────────────────────────┐   ┌──────────────────────┐
│ Introduce the new product  │   │ Does the estimate fall│
│ with the prescribed        │   │ in the NO region?     │
│ marketing program.         │   │                       │
└────────────────────────────┘   └──────────────────────┘
                              (H)            (I)
                    ┌──────────────┐  ┌──────────────────────┐
                    │ Drop the     │  │ Determine the best   │
                    │ product.     │  │ marketing study      │
                    │              │  │ according to expected│
                    └──────────────┘  │ value of informa-    │
                                      │ tion/cost of         │
                                      │ information          │
                                      │ criterion. After the │
                                      │ information is       │
                                      │ gathered, return     │
                                      │ to A.                │
                                      └──────────────────────┘
```

FIGURE 10-16 (continued)

SPRINTER, resting somewhere between the highly analytic formulation of the original SPRINTER and the microbehavioral formulations of Amstutz. Called macrobehavioral process models, they require a reasonable amount of behavioral data without resorting to the modeling of individual buyers. They can be implemented, run, and analyzed on a limited research budget (Urban, 1969).

MARKETING RESEARCH SIMULATIONS

Marketing executives face the task of making decisions on price, advertising, sales force, and new-product development on the basis of imperfect information about their markets. The marketing executive's understanding is at the mercy of his information. Companies are aware of this problem and have been trying to improve the quantity and quality of their marketing information.

When the firm recognizes a need for a market survey, many questions arise about its design. Who should be sampled? How should they be selected? How should they be contacted? Different survey research plans imply different cost and accuracy levels. In this section, we will review a simulation model designed by Charles Mayer that aids in the planning of market surveys.

Mayer Market Survey Simulation

Mayer's research is focused on the task of managing the field interviewing process at the least cost (Mayer, 1964). He reasoned that by building an operating model of this process incorporating a dynamic environment and subject to experimental changes in the configuration of field procedures and sample designs, he could determine the cost implications of alternate field interviewing plans and arrive at a better solution to the market survey management problem.

The field interviewing process involves sending an interviewer into a field situation or environment to obtain completed interviews from a predetermined set of respondents. The interviewer makes decisions regarding ordering the interviews, allocating his time, arranging his travel routes, keeping records, and terminating the interviewing process. The last decision follows the rule of either limiting the number of calls per address or prescribing a minimum acceptable response rate, and the interviewer must plan his own activity and decide when he has accomplished his task.

Field interviewing takes place in the environment of the respondent, whether it is his home, office, or some other place. The "state" of this environment—for example, whether the respondent is available when the interviewer calls—enters into the outcome of the call.

The purpose of Mayer's simulation is to describe the interviewer's decision process in a changing and uncertain environment. A simplified flow chart of his field interviewing model is shown in figure 10-17. Selection of a respondent by the computer depends upon the geographic location of the interviewer and other "prevailing conditions," such as the number of unobtained interviews or the number of previous visits. After the computer calculates the travel time required to reach the selected respondent from the interviewer's present location, it determines the specific outcome of the call by generating a random number and matching it with a predetermined response probability for this respondent. Calls are continued in this manner over one or more simulated days until a predetermined response rate is achieved. The computer then prints out the relevant costs of the field interviewing procedure.

Mayer believes that the validity of his model depends "not on whether the

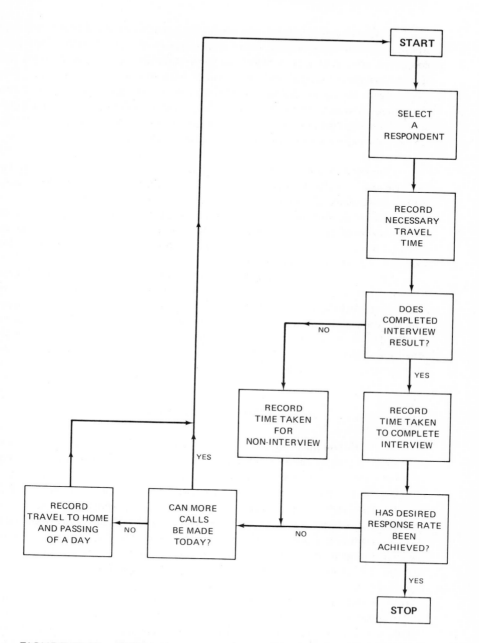

FIGURE 10-17 A flow chart of the field interviewing model. Source: Mayer, "Pretesting Field Interviewing Costs through Simulation," p. 49.

absolute values generated by the model compare favorably with reality, but whether the relative values seem appropriate." To check for validity, he compared simulated and actual cost differences among geographic areas and found that there was no significant departure from the reported experience of the Survey Research Center of the University of Michigan. Then Mayer attempted to validate individual parameter values by comparing them with other survey research data and the judgment of experienced field managers. Finally, an actual field experiment was replicated by the model with high correspondence.

Comment. Mayer is very convincing when he states that computer simulation can be a useful tool for estimating the probable costs of different field interviewing plans. Simulation is able to take into account the different decision rules used by interviewers in selecting their next call and the different probabilities of successfully completing a call. Without simulation, the survey designer has to rely on averages, and his estimates tend to be less accurate than the estimates yielded by computer simulation.

Mayer's work can be extended to include issues not discussed in his article. For example, the assignment of different interviewers to the different areas of the city can make a difference because call completion and information accuracy will be somewhat influenced by the interviewer's background. The computer simulation can introduce the interviewer as a variable. Also, Mayer did not apply sensitivity tests to see what difference would be made in costs if some of the uncertain input variables were varied.

MARKETING GAMES

We shall now turn from all-computer marketing simulations to some recent examples of man-machine simulations in the area of marketing. Gaming originally started out as a pedagogical tool to improve students' understanding and feeling for the situations they were studying. By taking on the roles of businessmen and making company decisions in a simulated competitive environment, students were expected to gain a number of insights not available through textbook study or case discussion. More recently, investigators have realized that gaming could be of value to research both in attempting to design realistic marketing games and in studying the organizational and decision behavior of the players during the game.

The early marketing games were very simple in design, consisting usually of one product, one territory, and only a few marketing instruments, such as pricing and advertising. The sales response functions were artificially developed. More recently, games have been designed around specific industries, and they have benefited from considerable research into industry behavior. Usually the first version of such a game will strike actual decision-makers in the industry as oversimplified and sometimes behaviorally incorrect. The game designers will revise it, and, after several cycles of criticism and revision, the final game will turn out to be quite a complicated and realistic model of the industry. It is at this point that some executives will begin to recognize the potential uses of the game for marketing planning purposes.

We shall review two outstanding games to show the variety of mechanisms and approaches that can be taken. The Carnegie Tech Marketing Game is modeled after the detergent market, and the M.I.T. Marketing Game is modeled after electric floor polishers for household use.

Carnegie Tech Marketing Game

One of the first and most fully developed business games is the Carnegie Tech Management Game, created by faculty members at Carnegie Tech's Graduate School of Industrial Administration (Cohen *et al.,* 1964). A special version called MATE (Marketing Analysis Training Exercise), developed by Kuehn and Weiss (1965), shows the marketing operation in much detail while simplifying the production and financial aspects of the original game. MATE revolves around three firms in the packaged detergent industry, operating in four geographical regions. Each firm may market from one to three brands of detergent in each of the four regions. Each month the firm may establish or alter price, advertising expenditures, sales-force size, call time allocation to the brands, and retail allowance. The firm may also purchase market survey reports containing estimates of total retail sales and market shares, retail distribution and stockouts, and competitive advertising expenditures. These estimates are subject to an amount of error varying with the funds appropriated for each survey and for chance factors. Furthermore, the firm can invest in product research to find new and better products or to imitate competitors' products, with the results also dependent on how much it invests and chance factors.

On the nonmarketing side, the firm owns one factory and a factory warehouse, and rents warehouse space in each marketing region. Products may be shipped directly from the factory to the regional warehouses or stored in the factory warehouse, the decision being influenced by the respective costs. Each month, the firm decides how much to produce, subject to having sufficient funds to pay for this level of production and subject to capacity limitations. The firm may seek additional funds to augment its working capital, and this must be approved by the game administrator. At the end of each month the decision-makers receive financial and accounting information on monthly sales, cost of goods sold, operating expenses and profit, and current balance-sheet figures.

The basic market model consists of an equation determining monthly total case demand for packaged detergent in each region as a function of a growth term, average income per capita, seasonality, and the company's price and sales-promotion expenditures relative to competitors. Total demand in a region is then allocated among the brands competing in the region in two steps. First, the holdover or habitual demand for each brand is determined. That is, a percentage of last period's sales is repeated for each brand, this percentage varying with its customer loyalty factor. The difference between total demand

and total holdover demand represents potential shifters' demand and is shared among the competing brands as a function of their relative product characteristics, retail price, retail distribution and availability, and advertising. Each brand consists of a particular mix of three product characteristics—washing power, gentleness, and sudsing—and the sales potency of the mix depends on their levels relative to the market's most desired levels and also the average ability of product users to discern differences in product characteristics.

The effects of the product characteristics and the other merchandising factors on market shares are worked out in interesting detail, including attempts to incorporate realistic interaction effects, which also applies to the modeling of the results of investments in product research and market survey studies.

Comment. The MATE game is a major contribution to marketing simulation in two different ways. First, it provides the players with a large number of marketing decision variables that carefully reflect the kinds of decisions facing actual executives in the detergent industry. Second, the game is distinguished for the theory underlying the modeling of the market response functions. Instead of simple response functions, the game is based on analytically elegant expressions to portray lagged effects, marketing mix interactions, stochastic determination of new product discoveries, and so on. Naturally, any game can be extended further (see the use of qualitative planning in the M.I.T. game), but MATE can stand on its present merits.

M.I.T. Marketing Game

The M.I.T. Marketing Game also models a complex and realistic marketing environment (King *et al.,* 1961). Developed around electric floor polishers for household use, this game requires the players to determine product quality, price, dealer margins, channels of distribution (including number and type of dealers), market area, advertising expenditures, advertising media and appeals, number and disposition of salesmen, and promotion within the retail store. Like the Carnegie Tech game, the M.I.T. game is designed as a training device, specifically for advanced marketing management courses. But in contrast with most other business games, the M.I.T. simulation requires the players to make some qualitative as well as quantitative decisions. For example, players of the M.I.T. game must develop advertising plans and copy for their product which are subsequently given a quantitative rating by a "control team."

Industry sales are determined by adjusting exogenously determined "normal sales" for the effects of price, promotion, and retail margins. Market share for each company is a function of competitive prices, distribution policies, and advertising. Company sales are determined by multiplying total industry sales by market share.

The control team ratings enter the model through certain "effectiveness functions" which adjust the sales for the quality of the respective company marketing programs. While these control team ratings are subjective, they provide an analogue of the subjective appraisal of real marketing programs by real customers.

Comment. The M.I.T. game possesses the same strong points found in the Carnegie Tech game. The modeling differences largely center on the differences that tend to exist between a consumer packaged goods market and a consumer durable market. The M.I.T. game also places greater responsibility in the hands of the game administrators, since they are required to analyze the marketing plans of the teams for their content as well as their budget levels. This overcomes a criticism that has been directed at most business games: that they stress only the dollar allocation dimensions of competition and not the bright ideas that could make for a brilliant advertising campaign or marketing strategy.

MAIN AREAS OF
FUTURE OPPORTUNITY FOR MARKETING SIMULATION

The preceding discussion demonstrates that marketing simulation, despite its short history, has already found many uses in the study of marketing problems. Simulation has been used to probe the behavior of complex marketing subsystems such as distribution, customer behavior, and competition as well as the management of important marketing decision areas such as pricing, advertising, personal selling, new product development, and marketing research. Simple algebraic models have also been designed in most of these problem areas, but some investigators have been dissatisfied with the degree of severe simplification and have turned to more complex modeling and simulation.

Although a rich variety of simulation techniques and solutions has already appeared in marketing, much work remains to be done. Many contributions might result from simply redoing simulations that have already been developed. At least two of these deserve to be imitated by applications to new situations. The first is Cohen's model on the shoe, leather, and hide industry. The vertical channels in several industries are quite complicated, and simulation would be a useful methodology for studying distribution dynamics. Another simulation that should be applied to other industries is Morgenroth's pricing model. Mapping the decision processes of executives in a number of pricing situations would lend empirical substance to traditional pricing explanations. It is very surprising that studies following the leads of Cohen and Morgenroth have not emerged during the several years that have passed since their appearance.

Beyond the value of studies following the lead of existing simulations, some areas of emergent work are extremely interesting. We will comment on several of these areas.

Marketing planning simulators. Corporate management is rapidly becoming aware of and excited by the possibility of using the computer to aid in the evaluation of alternative corporate plans. Already in existence are corporate simulators that produce a series of future company income statements and balance sheets resulting from different hypothetical investment and financial plans (Gershefski, 1969; Furst and Porter, 1968). These corporate simulators are usually oversimplified on the marketing side, and there is no reason why a computer model could not be developed that encompassed marketing decisions and their likely results. A game such as the Carnegie Tech Marketing Game (MATE) or the M.I.T. Marketing Game could provide a starting point for this type of *marketing planning simulator*. Instead of leaving the determination of competitive strategies to an assortment of teams, the company would express its own strategy and those of competitors in the form of computer inputs and run the model to create the outcomes. The major adjustment that has to be made in the game design is to particularize each competitor's resources and product offerings to represent those found in the industry. Market segments will have to be distinguished in the market model, and more regions and product lines may have to be introduced. With such adjustments, it is highly likely that the company will find a useful tool for testing alternative marketing strategies. A simulation run would reveal to the management some unanticipated problems and risks with the contemplated strategy and would indicate the kind of data that should be collected to increase their confidence. These potential gains would have to be measured against the substantial costs of designing, programming, and operating these complex simulations.

A small number of companies are reported to be constructing marketing planning simulators, although details of the construction are generally withheld. The Xerox Corporation has described the major components of its simulator (Zivan, 1967), which is designed in a modular fashion so that any component can be reworked without requiring much reprogramming of the other components. Among the components is one that forecasts the economy, another the industry sales, a third the market share, a fourth the costs, and a fifth the return on investment. The Minute Maid Company has developed a complex model for forecasting prices, output, and demand for Florida frozen orange juice (Buchin, 1968). The Arthur D. Little Company helped develop a market simulator of the fertilizer industry for forecasting long-run demand (Hegeman, 1965). The E. I. du Pont de Nemours and Company developed several simulations of specific markets and marketing plans under the general name of venture analysis (Andersen, 1961). Other companies active in developing complex computer models of specific markets are General Electric, Pillsbury, and Corning Glass. All these are pioneer models because no extensive precedents exist in this area and each company must create its concepts, data base, and means of validation. Marketing planning simulators will undoubtedly be important in the next decade, and they represent an excellent area for research for those interested in marketing simulation.

Test market simulators. Companies that develop new products, particularly in the consumer packaged goods area, often undertake to try out the product in a few selected cities to gain an initial reading on consumer interest and response. Putting products into a test market involves considerable company expense. In addition to the out-of-pocket costs of distributing free samples and arranging for promotion and market surveys, there is the potentially much higher opportunity cost of delaying the introduction of the product into the national market and thus allowing competitors a possible lead. Furthermore, test market results often fall in the middle range between indicating strong consumer interest in the product and very little interest. Even after expensive test marketing, the interpretation of the results and its projectability on a national basis are not always clear.

This had led some people to suggest that a computer model designed around one or more actual test market cities may aid as a tool in predicting the probable adoption rate of the new product or in helping interpret the weekly response data as they come in. For example, suppose that a new product is in its fifth week of test market and a market survey shows the percentages of the population in different stages of product awareness, interest, preference, trial, and usages. Assume that continuous communication takes place through specific selected commercial media and through definite patterns of word-of-mouth communications. Then, the task of the test market simulator would be to take these data as they develop and project future sales and profits. If competently done, the results may enable the company management to make an early determination of whether or not the product will be successful and may lead to an earlier termination of the test market phase and to considerable savings.

Thus far, there are no published examples of test market simulators or even conceptual expositions. Yet the conviction exists that something useful can be designed to help simulate the test market situation, and this idea represents an interesting area of research.

Marketing organization behavior simulator. When we look within the company at its marketing planning, decision, and control procedures, we are likely to find a maze of vaguely formulated and sometimes contradictory methods for getting things done. In addition to company standard operating procedures, various informal procedures and rules come into use for setting the various marketing budgets and allocating them to different products, territories, and so forth. This maze of practices is often difficult to comprehend, and there is no reason to believe that top management is fully aware of what is going on. According to Kotler:

> The typical picture is that of a marketing vice president requesting and receiving a budget and dividing it among different product managers, salesforce managers, and other executives. He develops broad policies that are interpreted and filled in at successively lower levels of the marketing organization and eventually emerge as micro-actions in the marketplace, such as a full-page ad in *Life* or three calls to a particular large account during the month of March. These marketing micro-actions were never specifically planned by the vice president of marketing. Yet they take

place and are the ultimate occurrences affecting the market's response to the firm's offerings. (Kotler, 1968b)

The effect of a proposed change in decision procedure in a local portion of the marketing planning system would be difficult to estimate, since it occurs in combination with so many other rules. This leads to the idea that there would be value in trying to simulate the marketing planning and control procedures in use by a particular company at a particular time. This would require identifying the major marketing jobs in the company (vice-president of marketing, brand managers, sales managers, etc.), determining the major decisions made by each type of executive, and determining the typical selection processes used in each type of decision. The totality of these decision programs could be set to operate in the calendar time of the planning and control cycle of the company. Once these decision and control procedures have been modeled and validated (in much the same manner as the Morgenroth pricing study), the analyst can experiment with specific changes in the rules (such as basing advertising budgets on last year's profits instead of last year's sales). This implies the inclusion of a model of market response designed in sufficient behavioral detail to respond to the decision detail of the firm. A preliminary version of this work is described by Kotler (1968b).

Standard marketing games. The potential of marketing games in aiding pedagogy and research in the marketing area has hardly been tapped as few industries have specifically been modeled in game form. There is a game designed around the rubber tire market (Basil, Cone, and Fleming, 1965), a retail store game (Schellenberger, 1965), a wholesale supply game (Kehl, 1961), and a few others. To our knowledge there is nothing designed around the defense market, chemicals, automobile dealers, and so forth. Yet the design of each new game requires coping with the facts of an actual industry and can lead to useful theoretical and practical findings.

Two marketing decision areas are particularly complex and, curiously, have not yet found full expression in gaming. Until recently there was no advertising agency game, although there are a number of sales management games (Day, 1968). The players in an advertising agency game would represent a team of account executives, creative and media planners, and marketing researchers. The teams would recommend an advertising budget to clients, develop creative and media plans, and initiate marketing research. The players would work with a large variety of reports found in the advertising world, such as Nielsen reports, Schwerin studies, Standard Rate and Data catalogs, media audience studies, and so on, and this would dramatize to the student the value and limitations of these information sources. An initial version of such a game was designed at Northwestern University, and an improved version was recently described by Tuason (1969).

The other marketing decision area that can usefully be studied in a gaming format is sales promotion. Companies in the consumer packaged goods industry spend millions of dollars every year on sales promotions such as cents-off, two-for-the-price-of-one sales, coupons, retail allowances, and so forth. Selecting the particular promotion and its timing is left to marketing managers who base their decisions on inventory levels, expected competitive timing of promotions, and a good deal of intuition. No marketing game has been reported that features a wide variety of sales promotion devices, and yet such a game would be a useful way to introduce students to the intricate issues involved in this area and possible models for improved decision-making.

In-basket simulations and scenarios. At least two variations of the standard business game have been exploited in other fields (particularly political science and international relations) but hardly touched upon in marketing. Both call for creating artificial situations modeled after real situations in which players participate and thereby hopefully gain insights into the real situation. Called all-man simulations, they have not generally been used experimentally to shed light on probable behavior in the corresponding real situations and therefore are not simulations as defined in this chapter. But common usage of the term *simulation* to cover these exercises justifies their brief consideration here.

The *in-basket simulation* presents a decision-maker with the kind of messages and memos that would come across his desk in the course of a day's work. Imagine a player assuming the role of a brand manager who starts his day by finding the monthly company sales report on his desk. The report shows sales by brand size and region, and he is to take whatever action, if any, he sees fit. He also finds a letter on his desk from the advertising agency proposing a specific change in media strategy. As these reports come into his basket, he peruses them and prepares memos in response. These memos go to a game umpire who uses certain rules for rating the responses on how well they reflect "good" management practice. The student benefits from this evaluation and also gains insight into the problems and situations that he would face as a brand manager. Although in-basket games can be designed around every type of marketing job, only one has been reported in the literature at the time of this writing. It describes the situations a marketing researcher would face in the consumer packaged food industry (Barksdale, 1963). An interesting in-basket simulation built around the job of personnel manager but modifiable into a marketing job position can be found in Plattner and Herron (1962). It is hoped that more in-basket simulations can be designed around marketing situations and eventually put to research purposes.

The other modified gaming device, the *scenario,* has found increasing use in the military, but so far it has been used very little in business. A scenario is a dramatic background, not unlike a simplified business case, for a historical or future situation which must be completed by participants who assume the roles. One recent scenario described the events and actions leading up to World War I but stopped short of the actual decisions made by the Great Powers. Another scenario described hypothetical events leading up to a possible nuclear war. In these and other cases, the participants assume identities and try to continue the action through time beyond the original situation. While there are grave questions surrounding the validity of the participants' responses as indicating how the real situation is likely to be resolved, there are several research values to the scenario as well as value to the players in giving them insight into the situation. Scenarios could be created around prominent marketing situations such as the competition among the auto giants, raising the questions of what American Motors could do to survive, Chrysler to improve its market share, and General Motors and Ford to keep the upstarts from moving up too much. Scenarios would provide a valuable tool for stimulating the students to imaginative marketing planning that would get around the limitations of both the conventional marketing game and the business case discussion method.

CONCLUSION

Although simulation is a relatively new approach in the study of social processes, it has already reached into many corners of marketing science and promises to make major contributions in the future. It attracts marketing scholars because the complexity of the marketing system seems to put simple analytic formulations on the defensive. On the other hand, the case study method often overrepresents the complexity and tends to discourage theory construction in marketing. Simulation steers a middle course in calling for a rich, but manageable, representation of the phenomena being studied. Its major value is in requiring that often vague ideas of how a marketing process or system works be reduced to operational terms. The need to specify and quantify relationships among marketing variables has proven a substantial stimulant to theory construction and fresh data collection in marketing. When simulations are run, they in turn produce output whose analysis can lead to interesting new findings. Altogether, we expect to find simulation assuming a permanent and major role in the battery of methodologies for studying marketing theory and practice.

At the same time, we would not want to understate the tremendous research effort required to construct, program, run, and analyze a simulation. The model builder will face the problem of finding appropriate and sufficient data to estimate the many parameters of the model; he will have to be prepared to update the model periodically if it is to be used for company decision purposes; and he will have to find ways to design and test the model without making computer running costs prohibitive. Someone recently said that although simulation methodology often produces new and useful perspectives, the analyst should always try to anticipate whether "the view will be worth the climb." We think the simulations reviewed in this chapter have made worthwhile contributions to marketing knowledge, and we expect this to continue to be true of future work of this kind.

REFERENCES

Alba, M. S., "Microanalysis of the Socio-Dynamics of Diffusion of Innovation." Ph.D. dissertation, Northwestern University, 1968.

Amstutz, A. E., *Computer Simulation of Competitive Market Response.* Cambridge, Mass.: The M.I.T. Press, 1967.

Andersen, S. L., "Venture Analysis: A Flexible Planning Tool," *Chemical Engineering Progress* (March 1961), 80-83.

Balderston, F. E., and A. C. Hoggatt, *Simulation of Market Processes.* Berkeley, Calif.: University of California, Institute of Business and Economic Research, 1962.

Barksdale, H. C., *Problems in Marketing Research: In-Basket Simulation.* New York: Holt, Rinehart & Winston, Inc., 1963.

Basil, D. C., P. R. Cone, and J. A. Fleming, *Executive Decision Making through Simulation: A Case Study and Simulation of Corporate Strategy in the Rubber Industry.* Columbus, Ohio: Charles E. Merrill Books, Inc., 1965.

Bass, F. M., "A Simultaneous Equation Regression Study of Advertising and Sales of Cigarettes," *Journal of Marketing Research,* 6 (August 1969), 291-300.

———, et al. (eds.), *Mathematical Models and Methods in Marketing.* Homewood, Ill.: Richard D. Irwin, Inc., 1961.

Bonini, C. P., *Simulation of Information and Decision Systems in the Firm,* especially chap. 7. Englewood Cliffs, N. J.: Prentice-Hall, Inc., 1963.

Buchin, S., "A Model of the Florida Orange Industry for Minute Maid Planning," Harvard Business School, AI-270, 1968.

Chamberlin, E. H., *The Theory of Monopolistic Competition.* Cambridge: Harvard University Press, 1933.

Charnes, A. W., W. W. Cooper, J. K. DeVoe, and D. B. Lerner, "DEMON: Decision Mapping via Optimum GO-NO Networks—A Model for Marketing New Products," *Management Science,* 12 (July 1966), 865-77.

Claycamp, H. J., and A. E. Amstutz, "Simulation Techniques in the Analysis of Marketing Strategy," in F. M. Bass, C. W. King, and E. A. Pessemier (eds.), *Applications of the Sciences in Marketing Management.* New York: John Wiley & Sons, Inc., 1968.

Cloonan, J. B., "A Heuristic Approach to Some Sales Territory Problems," in D. B. Hertz and J. Melese (eds.), *Proceedings of the Fourth International Conference on Operational Research,* pp. 284-92. New York: John Wiley & Sons, Inc., 1966.

Cohen, K. J., *Computer Models of the Shoe, Leather, Hide Sequence.* Englewood Cliffs, N.J.: Prentice-Hall, Inc., 1960.

———, et al., *The Carnegie Tech Management Game.* Homewood, Ill.: Richard D. Irwin, Inc., 1964.

Cox, R., *Distribution in a High Level Economy.* Englewood Cliffs, N.J.: Prentice-Hall, Inc., 1965.

Cyert, R. M., J. G. March, and C. G. Moore, "A Model of Retail Ordering and Pricing by a Department Store," in R. E. Frank, A. A. Kuehn, and W. F. Massy (eds.), *Quantitative Techniques in Marketing Analysis,* pp. 502-22. Homewood, Ill.: Richard D. Irwin, Inc., 1962. Also appears in R. M. Cyert and J. G. March, "A Specific Price and Output Model," *A Behavioral Theory of the Firm,* chap. 7, pp. 128-48. Englewood Cliffs, N.J.: Prentice-Hall, Inc., 1963.

Day, R. L., "Linear Programming in Media Selection," *Journal of Advertising Research,* 2 (June 1962), 40-44.

———, *Marketing in Action: A Decision Game.* Homewood, Ill.: Richard D. Irwin, Inc., 1968.

———, *Sales Management Simulation.* New York: Sales and Marketing Executives-International, 1968.

Forrester, J. W., "Advertising: A Problem in Industrial Dynamics," *Harvard Business Review,* 37 (March-April 1959), 100-10.

Frank, R. E., "Brand Choice as a Probability Process," *Journal of Business,* 35 (January 1962), 43-56.

Furst, J. K., and M. H. Porter, "Computerized Corporate Planning Model." A paper presented at the ORSA/TIMS Joint Meeting, San Francisco, California, May 1-3, 1968.

Gensch, D., "A Computer Simulation Model for Selecting Advertising Schedules," *Journal of Marketing Research,* 6 (May 1969), 203-14.

Gershefski, G. W., "Building a Corporate Financial Model," *Harvard Business Review* (July-August 1969), 61-72.

Haines, G., "The Rote Marketer," *Behavioral Science,* 10 (October 1961), 357-65.

Hegeman, G. B., "Dynamic Simulation for Market Planning," *Chemical and Engineering News* (January 4, 1965), 64-71.

Herniter, J., and V. Cook, "The NOMMAD Simulator: A Multidimensional Stochastic Model of Consumer Purchase Behavior." A paper presented at the 36th National Meeting, Operations Research Society of America, Miami Beach, Florida, November 10, 1969.

Hertz, D. B., "Risk Analysis in Capital Investment," *Harvard Business Review,* 42 (January-February 1964), 95-106.

Howard, J. A., *Marketing Theory,* chaps. 2, 4, 5. Boston: Allyn & Bacon, Inc., 1965.

———, and W. M. Morgenroth, "Information Processing Model of Executive Decisions," *Management Science,* 14 (March 1968), 416-28.

Howard, R. A., "Stochastic Process Models of Consumer Behavior," *Journal of Advertising Research,* 3 (September 1963), 35-42.

Kaplan, A. D. H., J. B. Dirlam, and R. F. Lanzillotti, *Pricing in Big Business.* Washington, D.C.: The Brookings Institution, 1958.

Karg, R. L., and G. L. Thompson, "A Heuristic Approach to Solving Travelling Salesman Problems," *Management Science,* 10 (January 1964), 225-48.

Kehl, W. B., "Techniques in Constructing a Market Simulator," in M. L. Bell (ed.), *Marketing: A Maturing Discipline*, pp. 75-84. Chicago: American Marketing Association, 1961.

King, P. S., *et al.*, "The M.I.T. Marketing Game," in M. L. Bell (ed.), *Marketing: A Maturing Discipline*, pp. 85-102. Chicago: American Marketing Association, 1961.

Kotler, P., "Competitive Strategies for New-Product Marketing over the Life Cycle," *Management Science*, 12 (December 1965a), 104-19.

———, "Computer Simulation in the Analysis of New-Product Decisions," in F. M. Bass, C. W. King, and E. A. Pessemier (eds.), *Applications of the Sciences in Marketing Management*, pp. 283-331. New York: John Wiley & Sons, Inc., 1968a.

———, "Decision Processes in the Marketing Organization," in D. M. Slate and R. Ferber (eds.), *Systems: Research and Applications for Marketing*, pp. 57-70. Urbana: University of Illinois, Bureau of Business and Economic Research, 1968b.

———, "Mathematical Models of Individual Buyer Behavior," *Behavioral Science*, 13 (July 1968c), 274-87.

———, "The Competitive Marketing Simulator—A New Management Tool," *California Management Review*, 7 (Spring 1965b), 49-60.

Kuehn, A. A., "Consumer Brand Choice—A Learning Process?" *Journal of Advertising Research*, 2 (December 1962), 10-17.

———, and M. J. Hamburger, "A Heuristic Program for Locating Warehouses," *Management Science*, 9 (July 1963), 643-66.

———, and D. L. Weiss, "Marketing Analysis Training Exercise," *Behavioral Science*, 10 (January 1965), 51-67.

Lester, R. A., "Shortcomings of Marginal Analysis for Wage-Employment Problems," *American Economic Review*, 35 (March 1946), 63-82.

Little, J. D. C., and L. M. Lodish, "A Media Planning Calculus," *Operations Research* (January-February 1969), 1-35.

———, *et al.*, "An Algorithm for the Traveling Salesman Problem," *Operations Research*, 11 (November-December 1963), 972-89.

Machlup, F., "Marginal Analysis and Empirical Research," *American Economic Review*, 36 (September 1946), 519-54.

Mack, R. P., *Consumption and Business Fluctuations: A Case Study of the Shoe, Leather, Hide Sequence*. New York: National Bureau of Economic Research, 1956.

Maffei, R. B., "Brand Preferences and Simple Markov Processes," *Operations Research*, 8 (March-April 1960), 210-18.

Massy, W. F., "Order and Homogeneity of Family Specific Brand-Switching Processes," *Journal of Marketing Research*, 3 (February 1966), 48-54.

———, and R. E. Frank, "Short Term Price and Dealing Effects in Selected Market Segments," *Journal of Marketing Research*, 2 (May 1965), 171-85.

Mayer, C. S., "Pretesting Field Interviewing Costs through Simulation," *Journal of Marketing*, 28 (April 1964), 47-50.

Michael, G. C., "A Computer Simulation Model for Forecasting Catalog Sales,"

Journal of Marketing Research, 8 (May 1971), 224-29.

Montgomery, D. B., "A Stochastic Response Model with Application to Brand Choice," *Management Science*, 15 (March 1969), 323-38.

Morgenroth, W. M., "Price Determinants in an Oligopoly," in W. S. Decker (ed.), *Emerging Concepts in Marketing*. Chicago: American Marketing Association, 1963.

Morrison, D. G., "Testing Brand Switching Models," *Journal of Marketing Research*, 3 (November 1966), 401-9.

Myers, J. G., "On-Line Simulation Experiments in Marketing: Problems and Prospects." A paper presented at the 1968 Fall Conference, American Marketing Association, Denver, Colorado, August 28-30, 1968.

Palda, K. S., *The Measurement of Cumulative Advertising Effects*. Englewood Cliffs, N.J.: Prentice-Hall, Inc., 1964.

Pessemier, E. A., *New Product Decisions: An Analytical Approach*. New York: McGraw-Hill Book Company, 1966.

Plattner, J. W., and L. W. Herron, *Simulation: Its Use in Employee Selection and Training*. New York: American Management Association, 1962.

Robertson, T. S., "An Analysis of Innovative Behavior and Its Determinants." Ph.D. dissertation, Northwestern University, 1967.

Schellenberger, R. E., *Development of a Computerized, Multipurpose Retail Management Game*, Research Paper Number 14. Chapel Hill: University of North Carolina, Graduate School of Business, 1965.

Schultz, R. L., "Market Measurement and Planning with a Simultaneous-Equation Model," *Journal of Marketing Research*, 8 (May 1971), 153-64.

Shycon, H. N., and R. B. Maffei, "Simulation—Tool for Better Distribution," *Harvard Business Review*, 6 (November-December 1960), 65-75.

Simulmatics Media-Mix I: General Description. New York: The Simulmatics Corporation, February 1962a.

Simulmatics Media-Mix: Technical Description. New York: The Simulmatics Corporation, October 1962b.

Telser, L. G., "Advertising and Cigarettes," *Journal of Political Economy*, 70 (October 1962), 471-99.

Tuason, R. V., "Experimental Simulation on a Pre-determined Marketing Mix Strategy." Ph.D. dissertation, Northwestern University, 1965.

———, "MARKAD: A Simulation Approach to Advertising Management," *Journal of Advertising Research*, 9 (March 1969), 53-58.

Turing, A. M., "Computing Machinery and Intelligence," *Mind*, 59 (1950), 433-60.

Urban, G. L., "A New Product Analysis and Decision Model," *Management Science*, 14 (April 1968), 490-517.

———, "SPRINTER Mod I: A Basic New Product Model," *Proceedings*, National Conference, American Marketing Association, 1969, 130-50.

Vidale, M. L., and H. B. Wolfe, "An Operations Research Study of Sales Response to Advertising," *Operations Research,* 5 (June 1957), 370-81.

Zivan, S. M., "Planning Using a Corporate Model." A paper presented at the TIMS Meeting, Boston, Massachusetts, April 5, 1957),

DEVELOPMENT, VALIDATION, AND IMPLEMENTATION OF COMPUTERIZED MICROANALYTIC SIMULATIONS OF MARKET BEHAVIOR

ARNOLD E. AMSTUTZ

Since 1959 I have been working with cooperating managements to develop, validate, and implement microanalytic behavioral simulations designed to aid formulation and evaluation of marketing policies and strategies.[1] This work has focused on a particular class of decision situation in which the simulation approach has shown unusual promise. These situations are characterized by two common elements. First, outcomes are largely determined by complex human behavior and second, management must influence actions by persuasion in order to achieve desired results—management is not able to exercise direct control.

The planning and implementation of marketing programs involve the coordination of many types of management activity designed to persuade the prospective customer to take actions or develop attitudes and beliefs favorable to the company's brands. Market-oriented simulation systems focus on the processes through which management attempts to influence market behavior. The models on which such simulations are based encompass microanalytic representations of retailer, distributor, salesman, and consumer and industrial-purchaser behavior in the environment external to the firm.

The objective of this article is to describe the process followed in developing, validating, and implementing a market-oriented microanalytic computer simulation.[2] This process is illustrated by examples drawn from three representative simulation-based management systems.

Reprinted with permission from D. B. Hertz and J. Malese (eds.), *Proceedings of the Fourth International Conference on Operational Research* (New York: Wiley-Interscience, 1966). Copyright©1966 by International Federation of Operational Research Societies. A portion of this research was done at Project MAC and the Computation Center at the Massachusetts Institute of Technology, Cambridge.

[1] A more detailed discussion of work with one representative company is provided in A. E. Amstutz and H. J. Claycamp, "Simulation Techniques in Analysis of Marketing Strategy," in *Applications of the Sciences in Marketing Management,* Purdue University, Lafayette, Inc., 1966.

[2] A thorough discussion of this process is provided in A. E. Amstutz, *Computer Simulation of Competitive Market Response,* M.I.T. Press, Cambridge, Mass., 1967.

THE SYSTEM DEVELOPMENT PROCESS

When developing a microanalytic computer simulation of market behavior, the firm and its competitors are viewed as input generators. The external market simulation is then designed to duplicate response characteristics of a comparable real-world market to inputs of the type generated by the firm and other market sectors.

The process followed when working with management to develop a microanalytic simulation of a company's markets is illustrated by reviewing the steps taken by one management using this technique.

Boundary Definition

The first step in simulation is to establish boundary definitions that will determine the detail and scope of the system to be developed. In most instances this preliminary "macro specification" is relatively crude. Management is encouraged to define a limited number of basic system elements. Relationships between these elements are then summarized graphically, as illustrated in Figure 1. This figure, developed in the course of discussions with a food-product manufacturer, indicates management's concern with interactions among manufacturers, distributors, retailers, and consumers in a perishable food-product market.

Although Figure 1 may appear overly simplified it is an important first step in the structuring process. Realistically complex representations evolve by gradual refinement of initially simple structures.

Objective Formulation

In the course of boundary definition management also specifies the objectives it hopes to achieve by the use of the system once it has been developed, validated and implemented. Objectives explicate criteria of relevancy that determine whether a particular aspect of the environment will be included in or excluded from the system. Objectives also indicate the level of system detail and accuracy required by management.

Description of Macro Behavior

Once the desired system scope and objectives to be achieved by system use have been specified, the structuring process continues. Each sequential step involves increasingly micro (detailed) description of behavior within the environment to be simulated.

Figure 2 illustrates a later stage in management progress toward system specification. Concepts illustrated in Figure 1 have been expanded by recognition of government and salesmen. The description of interactions has become

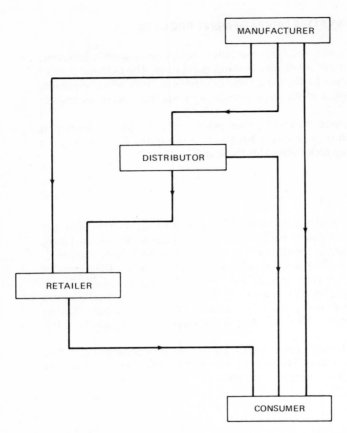

FIGURE 1 A first-stage macro description of market interactions.

more explicit. Information flow is now differentiated from product flow. (In later stages additional recognition was given to the flow of capital.) At this level of specification discussion is focused on bilateral channels relating the manufacturer and his competitors to distributors and retailers through their respective salesmen. Media promotion is represented by unilateral communication channels from the manufacturer to the consumer directly and through trade channels.

Once major interactions have been identified, attention is focused on processes associated with each interaction, taking account of backlogs, delays, and transfer points at which the rate of product, information, or value flow may be measured. Figure 3 illustrates this development for the channel of product flow represented by solid lines in Figure 2.

After interactions *between* sectors have been described, major decision points *within* each sector are identified. When examining the manufacturer

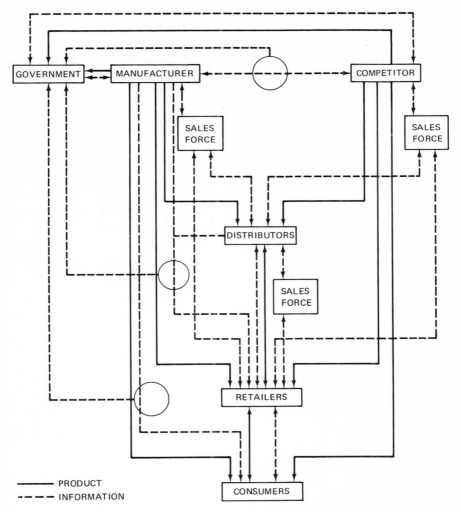

FIGURE 2   A second-stage macro description.

sector, the objective is to identify major decisions *affecting* the generation of inputs to the external environment.[3] Examination of other sectors is directed toward decisions *affected by* manufacturer generated inputs. Figure 4 illustrates one such description of decision and response factors within the manufacturer, retailer, and consumer sectors.[4]

[3]Input-associated decisions associated with a wide range of consumer and industrial products are discussed in A. E. Amstutz, "The Manufacturer-Marketing Decision Maker," *ibid.*, Chapter 7.

[4]This decision structure is considered extensively in A. E. Amstutz, "A Model of Consumer Behavior," *ibid.*, Chapter 8.

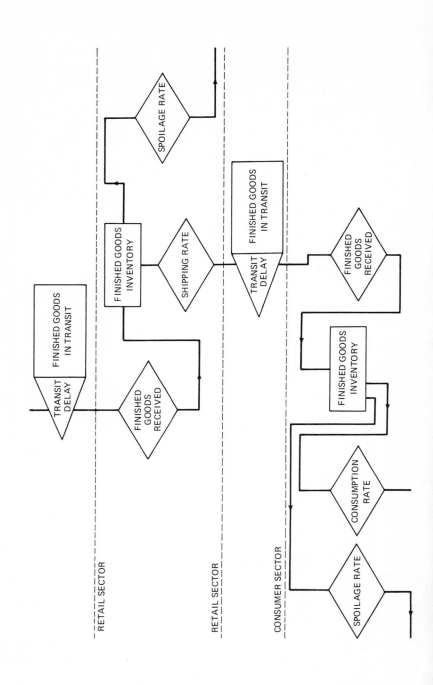

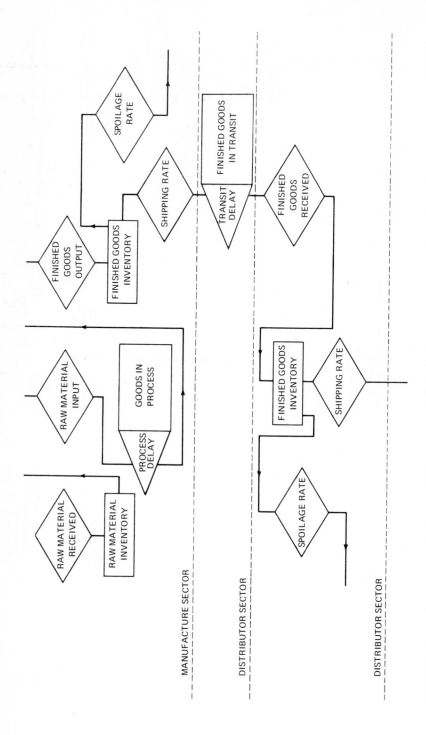

FIGURE 3 Macro description of product-flow based processes.

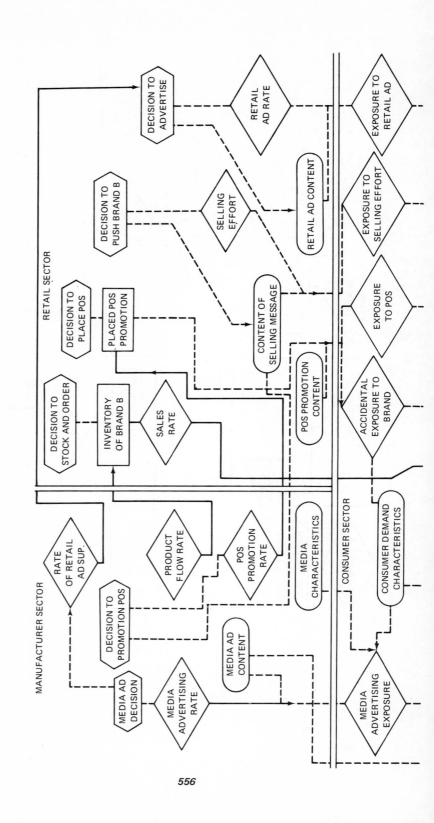

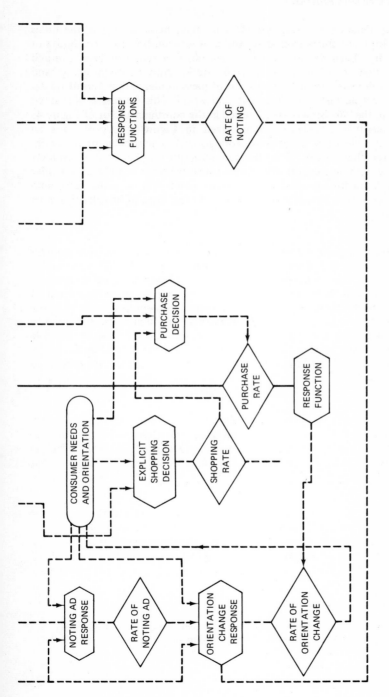

FIGURE 4 Decision and response function specification for three market sectors.

Description of Decision Processes

Once key decision and response elements have been identified, the focus of model development shifts from description of relationships to formulation of behavioral theory. Each decision point is described in terms of inputs to and outputs from that decision. Hypothesized relationships between inputs and observable behavior are formulated in terms of measurements that permit validation of the model against data from the real world. This process is illustrated with reference to the "decision to shop" noted in the consumer sector of Figure 4.

The conceptual framework summarized in Figure 4 hypothesizes an explicit consumer decision to go to a store to seek a particular brand or information about that brand. This decision structure implies that consumers entering a store with an explicit intention to investigate or acquire a particular brand exhibit behavior significantly different from that of consumers who accidentally encounter a brand or product line in the course of broader shopping (search) activity.

The perceived need concept—an example. Evaluation of management hypotheses regarding the decision to shop led to a qualitative concept of "perceived need." This concept, which was later quantified in terms of a measure of expressed intention to buy, might be viewed as an extension of utility theory. When formulating this model, management proposed that the consumer's motivation to take action to acquire a particular brand is related to his perceived need for that brand, which increases with (a) positive attitude toward the brand, (b) opportunity for brand use, and (c) time since purchase. This qualitative concept was later refined to the series of relationships illustrated in Figures 6, 7, and 8.

The effect of attitude. Using a modified Osgood scale consumer orientation (attitude) toward a brand is measured by asking a respondent to rate the brand on an eleven point scale from +5 (strongly favor) through 0 (indifferent) to −5 (strongly dislike).[5] The observed relationship between attitude (measured with the scale shown in Figure 5) and "perceived need" is illustrated in Figure 6.

Use opportunity. Use opportunity is measured in terms of the number of times that the consumer had an opportunity to use a brand within the product class being studied during the preceding quarter. This information is obtained by direct interview as well as diary maintenance. As illustrated in Figure 7, a linear association was established between the use-opportunity and perceived-need measures.

[5]The derivation and use of this and other intermediate measures of consumer disposition are discussed in A. E. Amstutz, "Quantification of Marketing Processes," *ibid.,* Chapter 5.

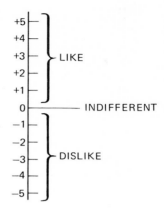

FIGURE 5   The attitude scale.

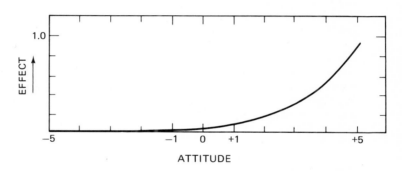

FIGURE 6   Effect of attitude on perceived need.

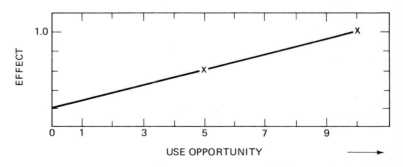

FIGURE 7   Effect of consumer-use opportunity on perceived need.

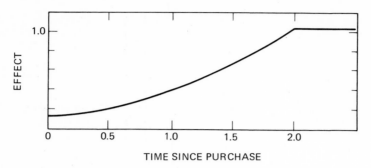

FIGURE 8   Effect of time since purchase on perceived need.

Time since purchase.   The time since purchase is measured, as the name suggests, by determining the time (in weeks or average product life) since the consumer last purchased a brand in the product class being studied. Figure 8 illustrates the general form of this relationship expressed in multiples of average product life for the current perishable food-product example.

Income stratification.   Initial attempts to validate the perceived-need construct produced evidence that the relationship between the three perceived-need measures and actual shopping behavior is income-dependent. Further investigation revealed that behavior could be differentiated by population subsegments established on the basis of income-level stratification, as illustrated in Figure 9.

Probability-of-shopping function. Combining the three elements of perceived need with income stratification produced a function of the type illustrated in Figure 10 relating the probability of shopping for the food product to perceived need and income. This figure specifies the perceived-need-based function for each of the income levels stratified in Figure 9.

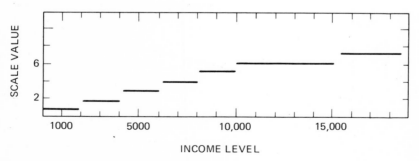

FIGURE 9   Sample income stratification.

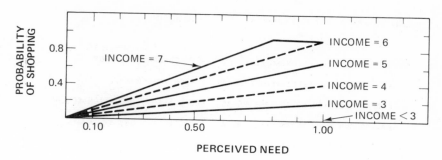

FIGURE 10   Probability of shopping as a function of perceived need and Income.

Additional function formulation.   In a similar manner each decision and response function defined by macro specifications is investigated. In some instances initial theoretical constructs are validated. In others empirical evidence suggesting alternative constructs is obtained and the process of formulation is repeated for revised structures.

Explicit Decision Representation

Decision and response functions are formulated and tested as probabilities, since referenced data are in the form of frequency distributions. Generation of explicit decision outputs for each cell within a simulate population requires conversion of the probabilistic statement into explicit yes/no decisions. A number drawn randomly from a rectangular distribution of range 0 to 1.0 is compared with the stated probability to determine the occurrence of the probabilistic event.[6]

BEHAVIOR OF AN ARTIFICIAL POPULATION

Once validated at the function level, decisions and response formulations of the type just described are combined in a simulation structure. Operating within the framework supplied by the simulation system, these functions determine the actions and responses of simulated population members.

This stage in the simulation process is illustrated by using output from an appliance-market simulation. This system of models was developed following procedures comparable to those just described for the food product. Similar concepts and measures as well as parallel model structures are evident.

---

[6]If the number drawn is less than or equal to the stated probability, a positive outcome is assumed.

SIMULATION APP-03 TEST RUN APRIL 4, 1965   1400 HOURS

-- CONSUMER 0109 NOW BEGINNING WEEK 117 --   FEBRUARY 19, 1962

   - REPORT MONITOR SPECIFIED. TO CANCEL PUSH INTERRUPT.

   - CHARAC - REGION NE SU, AGE 25-35, INCOME 8-10K, EDUCATION COLLEG

   - BRANDS OWN 3, 6 YEARS OLD. RETAILER PREFERENCE 05, 11, 03

   - MEDIA AVAILABLE 1 0 0 1 0 0 0 0 1 1 1 1 0 0 0 0 0 0 0 0 0 0 0 0

   - ATTITUDES . 1   2   3   4   5   6   7   8   9 10 11 12

     ....

     PROD CHAR .   0 +1 +1   0 -3 -1   0 +5   0 +3   0   0
     APPEALS   . -3   0 +1 +5   0 -3 +3   0   0   0 +5   0
     BRANDS   . +2 +1 +3 +2
     RETAILERS . +1 -5 +3 +1 +5 -5 -5 +1 -1 -3 +5 +1
               . -3 +1 -1 +3 +1 +1
     AWARENESS .   1   0   0   0

   - MEMORY DUMP FOLLOWS. BRANDS LISTED IN DESCENDING ORDER 1 TO 4

   PRODUCT CHARACTERISTIC MEMORY   APPEALS MEMORY

 . . . .
 1 2 3 4 5 6 7 8 9 10 11 12 . 1 2 3 4 5 6 7 8 9 10 11 12
 . . . .

 2 3 15 0 5 5 4 14 8 7 1 3 . 8 9 7 3 1 11 7 4 4 3 9 3
 8 0 6 4 9 5 4 13 0 3 6 7 . 6 8 0 7 0 9 2 4 3 10 3 1
 0 6 15 7 0 3 11 3 5 2 5 7 . 0 4 8 10 9 2 14 3 9 7 9 5
 7 9 3 7 3 2 7 2 6 12 14 2 . 0 9 7 8 13 9 11 6 0 2 5 9

- MEDIA EXPOSURE INITIATED

       - MEDIUM 003 APPEARS IN WEEK 117 -- NO EXPOSURES
       - MEDIUM 004 APPEARS IN WEEK 117
           - EXPOSURE TO AD 013, BRAND 3 -- NO NOTING
           - EXPOSURE TO AD 019, BRAND 4
               - AD 109, BRAND 4 NOTED. CONTENT FOLLOWS
               - PROD. C 11 P = 4, 4 P = 2,
               - APPEALS 5 P = 2, 7 P = 2, 12 P = 2,
       - MEDIUM 007 APPEARS IN WEEK 117 -- NO EXPOSURES
       - MEDIUM 012 APPEARS IN WEEK 117
           - EXPOSURE TO AD 007, BRAND 2
               - AD 007, BRAND 2 NOTED. CONTENT FOLLOWS
               - PROD. C 8 P = 3, 12 P = 1,
               - APPEALS 2 P = 1, 4 P = 1, 6 P = 1, 10 P = 1,
           - EXPOSURE TO AD 013, BRAND 3 -- NO NOTING
           - EXPOSURE TO AD 004, BRAND 1 -- NO NOTING
       - MEDIUM 016 APPEARS IN WEEK 117 -- NO EXPOSURES
       - MEDIUM 023 APPEARS IN WEEK 117 -- NO EXPOSURES

FIGURE 11

```
- WORD OF MOUTH EXPOSURE INITIATED

            - EXPOSURE TO CONSUMER 0093 -- NO NOTING
            - EXPOSURE TO CONSUMER 0104 -- NO NOTING
            - EXPOSURE TO CONSUMER 0117 -- NO NOTING

   - NO PRODUCT USE IN WEEK 117

   - DECISION TO SHOP POSITIVE -- BRAND 3 HIGH PERCEIVED NEED
                                -- RETAILER 05 CHOSEN

   - SHOPPING INITIATED

         - CONSUMER DECISION EXPLICIT FOR BRAND 3 -- NO SEARCH
         - PRODUCT EXPOSURE FOR BRAND 3
            - EXPOSURE TO POINT OF SALE 008 FOR BRAND 3
               - POS 008, BRAND 3 NOTED.  CONTENT FOLLOWS
               - PROD. C  3 P = 4,  6 P = 4,
               - APPEALS  5 P = 2,  7 P = 2,  10 P = 2,  11 P = 2,
         - NO SELLING EFFORT EXPOSURE IN RETAILER 05

   - DECISION TO PURCHASE POSITIVE -- BRAND 3, $ 38.50

         - DELIVERY IMEDAT
         - OWNERSHIP = 3, AWARENESS WAS 2, NOW 3

   - WORD OF MOUTH GENERATION INITIATED

         - CONTENT GENERATED, BRAND 3
            - PROD. C  3 P = +15,  8 P = +15,
            - APPEALS  4 P = +50,  11 P = +45

   - FORGETTING INITIATED -- NO FORGETTING

-- CONSUMER 0109 NOW CONCLUDING WEEK 117   --   FEBRUARY 25, 1962

-- CONSUMER 0110 NOW BEGINNING WEEK 117   --   FEBRUARY 19, 1962


 QUIT,
 R 11.633+4.750
                                      FIGURE 11   (Continued).
```

A Week in the Life of a Simulated Consumer

Figure 11 was obtained by monitoring the "thoughts and actions" of one member of a simulated appliance-market population during a simulated week in which the population experienced events comparable to those encountered by a real-world population during the week beginning February 19, 1962.

Identifying characteristics. The information provided, beginning with the third line of output in Figure 11, identifies characteristic attributes of consumer 109. He is a suburban (SU) resident of New England (NE) between 25 and 35 years of age, with an income between $8,000 and $10,000 a year, and has a college education. He now owns a brand 3 appliance purchased six years ago.

Consumer 109 presently favors retailers 5, 11, and 3, in that order. He subscribes to or otherwise has available media of types 1, 4, 9, 10, 11, and 12. Media of types 2, 3, 5, 6, 7, 8, and 13 through 24 are not available to him.

Consumer 109's attitudes are summarized in a matrix beginning on line 6 of Figure 11. This matrix indicates his orientation toward 12 product characteristics, 12 appeals, 4 brands, and 18 retailers. From these figures it may be established that the most important (highest attitude) product characteristic insofar as consumer 109 is concerned is characteristic 8, which he regards very highly (+5). Appeals 11 and 4 are similarly indicated as of primary importance to this artificial consumer. From the retailer attitude portion of this matrix his preference for retailers 11 and 5 (both +5 attitudes) and 3 or 16 (both +3 attitudes) may be established. The final entry in the orientation matrix indicates that consumer 109 is aware of brand 1.[7]

Consumer memory content. The line stating "MEMORY DUMP FOLLOWS. BRANDS LISTED IN DESCENDING ORDER 1 THROUGH 4" introduces the print-out of consumer 109's present simulated memory content. This memory dump is a record of noted communications retained by the consumer relating specific product characteristics and appeals to each of four brands. From this report it can be established, for example, that consumer 109 has retained 14 communication exposures associating product characteristic 8 with brand 1, 13 exposures relating product characteristic 8 with brand 2, and 14 exposures associating appeal 7 with brand 3.

Media exposure and response. The entry in the report following the memory dump indicates that the segment of the simulation representing media exposure processes has been entered. Six media appear (are published or broadcast) during week 117. Consumer 109 is not exposed to medium 3, for that medium is not available to him (see media availability indicator in the characteristic output). Medium 4 also appears in week 117, and because it is available to consumer 109 he may be exposed to relevant ads appearing in it. The output indicates that he is exposed to an advertisement for brand 3 but does not note that communication. On the other hand an advertisement for brand 4 also present in medium 4 during week 117 is noted as indicated by the line reading, ADVERTISEMENT 19, BRAND 4 NOTED. CONTENT FOLLOWS. The output message then indicates that advertisement 19 contains a high prominence (4) reference to produce characteristic 11 and a medium prominence (2) reference to characteristic 4.[8] Advertisement 19 also contains medium prominence references to appeals 5, 7, and 12.

---

[7]The awareness measure used in this system is indicative of the respondents' top-of-mind cognizance determined by eliciting the name of the first brand in a product class that "comes to mind," in A. E. Amstutz, "Quantification of Marketing Processes," *op. cit.,* Chapter 5.

[8]A five point (0-4) prominence scale is used to code content of all communications inputted to the model. Each communication is evaluated by using the following coding structure:

Level of Prominence	Evaluation Scale
Extremely prominent—impossible to miss	4
Very prominent—major emphasis given	3
Average prominence—normal identification	2
Present but not prominent—easily missed	1
Not present—impossible to determine	0

A. E. Amstutz, "The Manufacturer—Marketing Decision Maker," *op. cit.,* Chapter 7.

Consumer 109 does not see medium 7, although it appears in week 117. However, he is exposed to three advertisements in medium 12, which also appears during that week. The advertisement for brand 2 is noted, whereas those for brands 3 and 1 are not. Media 16 and 23 also appear in week 117 but are not seen by consumer 109.

Word-of-mouth exposure. Report entries following the media exposure section indicate that consumer 109 is exposed to word-of-mouth comment generated by consumers 93, 104, and 117, but fails to note communication from any of these individuals. Had noting occurred, a message-content report comparable to that generated for advertising would have specified the information noted.

Product experience. Consumer 109 did not have product experience during week 117. Had he made use of the product, a report of his response to product use indicating product characteristics or appeals, if any, emphasized by the use experience would have been printed.

Decision to shop. The next entry in the Figure 11 output indicates that consumer 109 has made an explicit decision to shop, that his highest perceived need is for brand 3 and that his first choice retailer is 5. Simulation models representing in-store experience have been loaded.

In-store experience. The first entry within the SHOPPING INITIATED section notes that the consumer is exhibiting behavior associated with the explicit-decision-to-shop option and is seeking brand 3 (there is therefore NO SEARCH activity—no opportunity for accidental exposure). Simulated retailer 5 is carrying brand 3; therefore consumer 109 finds the brand he is seeking (3).

Retailer 5 has placed point-of-sale display material for brand 3. The consumer is exposed and notes its content, which emphasizes appeals 3 and 6 and product characteristics 5, 7, 10, and 11 as attributes of brand 3. Retailer 5's simulated salesmen are either not pushing brand 3 or are busy with other customers. In any event, consumer 109 is not exposed to selling effort while shopping in retailer outlet 5.

Decision to purchase. The output statement DECISION TO PURCHASE POSITIVE—BRAND 03, $38.50, specifies that consumer 109 has made a decision to purchase brand 3 at a price of $38.50. The line following indicates that retailer 5 can make immediate delivery of brand 3.

Response to purchase. Since consumer 109 has now purchased brand 3, his awareness, which was favoring brand 2, is changed to favor brand 3.

Word-of-mouth generation. Since consumer 109 is now the proud owner ·of a brand 3 product, it is not surprising to find him initiating word-of-mouth comment regarding his new purchase. The content of his communication regarding brand 3 emphasizes product characteristics 2 and 8 and appeals 4 and

11—the appeals and product characteristics toward which he has the highest perceived brand image as indicated in the previous memory dump.

Forgetting. Consumer 109 did not lose any of his existing memory content during week 117.

Simulated Population Behavior

The behavior of population groups within each simulation sector is described by accumulating simulated individual behavior. Population behavior may be summarized in terms of the proportion of purchases allocated to each brand (brand shares), changes in population attitude distributions toward brands or changes in the perceived brand images held by significant population segments.

One Year in the Lives of Two Simulated Doctors

Once the reasonableness of simulated behavior of the type outlined in Figure 11 has been established, the system may be used to produce behavior over time. This aspect of simulation testing is illustrated by output obtained from a third system—a microanalytic simulation of the doctor population in the United States.[9] Figure 12 illustrates the cumulative prescriptions of 10 drugs written by two general practitioners in the simulated medical environment. These two doctors prescribed only one relevant drug during the first two weeks of simulated activity. As the simulated year progressed, however, they used other drugs in the set of 10, and by year end their cumulative prescription shares were 37.5, 28.3, 21.1, 5.5, and 5.0 for drugs 1-Y, 2-X, 2-0, 1-0 and 1-+, as illustrated at the end of the time plot.

Total Population Behavior

Output of the type illustrated in Figure 12 is used primarily to test system stability. Two simulated G.P.'s are no more representative than two real-world doctors. Meaningful tests of system response require examination of the behavior exhibited by major population segments. Figure 13 illustrates the weekly prescriptions of 10 drugs by 100 members of an artificial population segment during the simulated year 1961. These simulated drug-usage figures may be directly compared with data generated during a comparable period in a real-world test market.

SYSTEM VALIDATION

Once a system has been developed and tested to the point at which management is convinced of its viability, validation tests designed to determine the extent to which the simulation is an accurate representation of a real-world

---

[9]This system is discussed extensively in A. E. Amstutz and H. J. Claycamp, *loc. cit.*

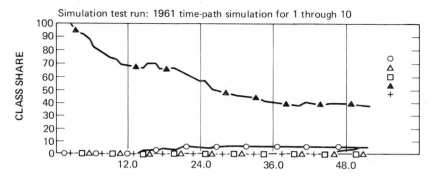

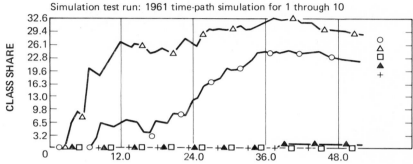

FIGURE 12   Sample output of two doctors.

environment must be undertaken. Validation testing generally proceeds sequentially from function analysis through cell and population validation.

Function-Level Validation

The first step in function validation is a sensitivity analysis that indicates the relative sensitivity of total system performance to various functions within the system structure. Sensitivity testing establishes priorities for functional validations, for it is most reasonable to expend effort in validating those functions on which system performance appears to be most dependent.

In validating a functional relationship the generally followed procedure is to test the null hypothesis that observed relationships are due to random variation by the application of a chi-square test. Once the null hypothesis is rejected, usually at the 1 per cent level, the degree of correspondence between real-world data and theoretical function form is established by using standard curve-fitting techniques.

Cell-Level Validation

Validation at the cell level is to establish that the behavior of an individual within the simulated population cannot be differentiated from that of a similar member of the real-world population. First-level testing is of the type suggested by Turing.[10] Later tests are designed to ensure that the distribution of relevant parameter values (e.g., frequency of brand purchases and changes in attitude and knowledge) exhibited by the simulated and real-world consumers under comparable conditions are statistically indistinguishable.

Population-Level Validation

Tests focusing on the simulated population are designed to establish the degree of correspondence between behavior exhibited by members of the simulated population and that exhibited by members of the real-world population measured in terms of variables relevant to management.

Reliability testing. In beginning population testing it is necessary to establish that model performance is relatively insensitive to different random-number seeds used on sequential system runs. As an example, in the simulation used to generate the output illustrated in Figure 13 terminal drug share deviation between runs is less than 1 per cent.

Performance Validation

The acid test of simulation validity is the ability to duplicate historical real-world population behavior under comparable input conditions. In conducting such tests, the population is initialized to duplicate the distribution of all relevant parameters as they existed at a specified point in time in the real-world environment. In the case of the Figure 13 run the artificial population had been initialized to correspond to conditions existing at the beginning of 1961.

Inputs to the simulation during performance tests described conditions existing in the real world during the referenced time period. In the test illustrated in Figure 13 conditions were those existing during 1961. Inputs specified the content and related media allocation for all journal, direct mail, and detail man promotion generated by competitors operating in the market during 1961. In addition, drug characteristics and distribution conditions were established to correspond to conditions in the real world during 1961.

Analytical procedures applied in performance testing may be summarized with reference to data plotted in Figure 13. The first test performed following this simulation established that the rank order of drug shares at the end of 1961 in the real and simulated worlds were equivalent. Actual simulated-data comparisons are presented in Figure 14.

---

[10]Turing has suggested that if a person knowledgeable in the area of simulated decision making cannot distinguish the modeled behavior from reality the model is realistic. See A. M. Turing, "Computing Machinery and Intelligence," *MIND*, 433-460 (October 1950).

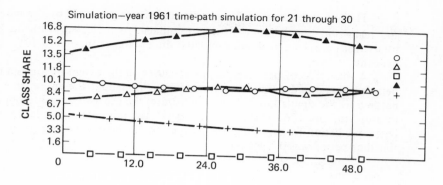

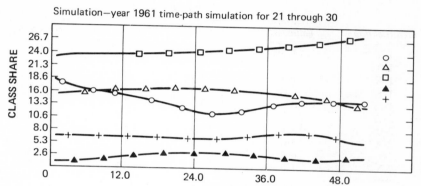

FIGURE 13 Sample output of 100 doctors.

Identification	Rank as Initialized	Year End Rank	
		Simulated	Actual
1-Y	4	2	2
1-0	5	6	6
1-X	6	5	5
1-+	8	8	8
1-□	10	10	10
2-□	1	1	1
2-0	2	3	3
2-X	3	4	4
2-+	7	7	7
2-Y	9	9	9

FIGURE 14 Drug rank order comparison.

The absolute values of therapeutic class shares generated by the simulation and real-world population were then examined. As indicated in Figure 15, the total error between actual and simulated drug shares at the end of 1961 was 5.1 per cent.

A final class of performance tests focuses on the extent of correspondence between actual and predicted market shares throughout the entire time period covered by the simulation. Figure 16 illustrates the procedure used to obtain this measure for the 1961 simulation test data. The minimum error in simulation-based prediction for any drug was 5.2 per cent, whereas the average error over this time period was 0.7 per cent.

Identification	Initialization Value (%)	Year End Value (%)		Difference (magnitude) (%)
		Simulated	Actual	
1-Y	13.7	15.0	16.1	−1.1
1-0	9.7	9.1	8.7	+0.4
1-X	7.3	9.3	9.0	+0.3
1-+	5.0	3.2	2.8	+0.4
1-□	0	0	0	−
2-□	23.2	27.6	28.8	−1.2
2-0	18.1	13.0	12.7	+0.3
2-X	15.6	13.9	14.4	−0.5
2-+	6.2	5.9	5.5	+0.4
2-Y	1.0	2.5	2.0	+0.5
Σ	99.8	99.5	100.0	5.1

FIGURE 15   Absolute share of therapeutic class comparison.

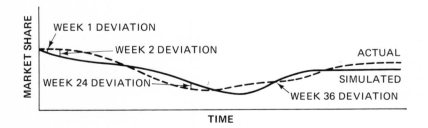

Total deviation over time for Drug d:

$$\sum_{t=1}^{52} |\text{actual } (d,t) - \text{simulated } (d,t)|.$$

Average error for 10 drugs:

$$\frac{\sum_{d=1}^{10} \text{total deviation}}{520}$$

FIGURE 16   Over-time market share deviation—measurement illustration.

MANAGEMENT USES OF SIMULATION

Given systems of the type described in this article, management must assess system performance in terms of intended applications. If, in their opinion, performance is sufficient to warrant use of the simulation as a representation of the real-world environment, applications of the type outlined below may be appropriate. If, in their opinion, however, the simulation fails to duplicate salient attributes of the real-world environment, further development leading to a more refined system must be undertaken or the use of the technique rejected.

Testing Implicit Models

One of the first benefits to accrue from the development of a simulation system is the systematic testing of management conceptions of the environment in which they operate. In reviewing alternative formulations and evaluating functions, cell-model behavior, and total population performance, management must make explicit the often implicit models on which their decision making is based.

The "What If?" Question

Given that management accepts simulation performance as indicative of real-world response under comparable conditions, the simulation becomes a test market without a memory in which management may examine with impunity the implications of alternative policies and strategies. Whether introducing new products or considering modifications of a marketing program, management may apply alternative strategies in the simulated environment and evaluate their implications under various assumed competitive conditions.

The effectiveness of such pretesting is dependent on management's ability to predict probable competitive responses to proposed actions as well as the accuracy of the simulation system. Management may find it profitable to examine the impact of best- and worst-case competitive response patterns. In most instances the best case assumes that competition will continue with programs developed before the initiation of company actions, whereas the worst case assumes full competitor knowledge of the proposed company program and combined action to thwart company efforts.

Performance References

The simulated environment provides the references against which the progress of operations in the real world may be measured. Given a simulation pretest, management can determine by monitoring appropriate variables whether or not a program is progressing as planned. If conditions producing satisfactory performance in the simulated environment are encountered in the real world, it is assumed that final results will be comparable. Differences between simulated and experienced results are viewed as potential bases for failure to achieve real-world performance comparable to simulation results.

SUMMARY

This article has examined procedures followed in developing, testing, and validating computerized microanalytic behavioral simulations. The process of boundary definition, macro and micro behavior description, and decision-function formulation has been discussed with reference to sample system structures. System performance characteristics and procedures for simulation validation at the function, cell, and population level were considered in context of output obtained from three representative microanalytic simulation based systems. It has been suggested that such systems have the potential to contribute significantly as vehicles for testing preconceptions regarding complex environments, evaluating the implications of alternative policies and strategies, and providing performance references against which the effectiveness of implemented plans may be assessed.

11

Public System Simulations

GUSTAVE J. RATH

INTRODUCTION

The term *public system* as employed in this chapter applies to a widely diversified field of activities. Public systems may have their boundaries at national, regional, or local levels. The principal activity of the system studied is usually service of some sort; hence, the transportation field (public and private) has been included, as well as health care, education, and government agencies— both executive and judicial. Each of these topics is treated in some depth in as many aspects of the field as possible.

No general functional classification of simulations has been made principally because of the fragmentation of current activities. Possible system simulation classification includes types of subsystem simulated; simulation by type of entity; simulation by type of purpose such as planning, control, analysis; method of development; and level of detail. This chapter will treat such areas of interest as transportation in general categories.

The simulations included here are classified individually. Each simulation and its model is reviewed as to objectives, structures, empirical bases, results, and validation. When no specific validation information is discussed, the reader should assume that such information was not available in the cited references.

Each topic area has characteristics that specifically determine the applicability and potential use of simulation. System definition is generally a problem for simulations, but the problem may be compounded by the nature of the simulation. For example, urban applications suffer from the difficulties involved in defining the urban phenomena in the midst of fantastic complexity. On the other hand, transportation does not typically have this problem because the objectives, functions, and phenomena are well defined. Passengers, destinations,

The preparation of this chapter was partially supported by PHS Grant LM 00098-02 and the National Center for Health Services Research and Development.

freight, modes of transportation such as airplanes, ships, trucks, railroads, and buses, and pathways such as roads, railroads, canals, rivers, and oceans establish the entities; measures of cost, safety, and speed can immediately be applied. Thus it is not surprising that transportation is the most successfully developed field.

Somewhere in the middle lies the field of urban growth and land use. It is clearly less structured than transportation; however, it has enough well-defined features (geography) to allow broad simulation activities. Similarly, for hospital care, education, and criminal justice, the applications are scattered.

These areas (excluding transportation) have only recently begun to employ the tools of simulation and they have therefore been difficult to evaluate. Each area has problems associated with identifying, classifying, and describing entities as well as problems involving the amount of detail and the type of organization incorporated within the model. Attempts to apply simulation on a global level, but only as applied to one specific problem, generate biased or skewed models that fail to reflect real problems except for immediate data concerning the subsystems for which they were originally developed.

Simulations in which entities exist at operational levels (Schroeder and Rath, 1965) seem most feasible. Many studies might best be accomplished on operational levels rather than on dramatic or global levels. Data obtained would then specifically be applicable and could ultimately be synthesized into a global systems model.

Another difficulty encountered is linking simulations at low levels to those at high levels. Is it possible to simulate a hospital ward and examine the effects of modifying this ward on the entire hospital without involving the rest of the hospital at the same level of detail? This question has yet to be answered.

Finally, lack of data could potentially destroy the validity of simulation studies. In many cases the writers do not present enough data to evaluate the success of the simulation. In this chapter, each subject will be analyzed individually. Sections have been arranged according to the amount and quality of simulation work done, as well as to the time involved, in each subject area.

Other public systems have used simulation techniques in such areas as water resources, forest fire fighting, and mining, and some of the recent work in these areas is worth mentioning.

The Harvard Water Program (see Hufschmidt and Fiering, 1966; Fiering, 1966; and Bower, Hufschmidt, and Reedy, 1962) has been a key contribution. Hufschmidt (1963) has presented a simulation of a reservoir system on the Lehigh River, and one of the most ambitious development simulations has been presented by Fiering (1966), who studied the Punjab region of West Pakistan.

Forests have been another area of interest (Howard, Gould, and O'Regan, 1966). A comprehensive simulation in the conservation area has been presented by Stade (1967) in a cost-effectiveness study of forest fire control, which was undertaken to evaluate various configurations of water bombers for fire control. Clough and Levine (1965) tested alternative subsidy policies for the Canadian uranium mining industry.

TRANSPORTATION SYSTEM SIMULATIONS

Administrative decisions in transportation systems require an under-standing of the system properties, environment, and interactions between the system and its environment.[1] Analytical mathematical methods have described and predicted certain specific, consistent, and standardized transportation processes. Physical models have replicated in miniature the physical environ-ments that engineering hardware might encounter. Simulation, however, is often the only useful method of evaluating the operation of a system consisting of several complex and dynamically interacting processes and subsystems.

Transportation simulations may be subdivided into two groups of the type of system studied. The first group, which we will call channelized systems, consists of railroad, highway, and inland waterway systems, in which the environment is continuously (or indirectly) considered to be a determinate of system contradictory performance. (The number of channels is usually limited, although it may be very large.) Entities within the channels compete and conflict en route as well as during times of arrival or departure. The second group, air and ocean systems, will be called spatial systems. Here arrival and departure are primary, with aggregate figures reflecting en route experiences and variations. In spatial systems the capacity between terminals is assumed to be infinite (oceans) and irrelevant to the analysis.

These groups may further be subclassified as flow oriented or demand oriented. In flow-oriented systems, levels of demand are normally considered to be fixed. The objective of the simulation is to evaluate flow with respect to the requirements that demand implies. In demand-oriented models, economic and flow characteristics are, hopefully, well understood; the objectives of this simulation would then be to make inferences about the effects of demand on the facilities or services involved.

Four simulations, one within each of the above categories, will be described. Each simulation represents the investigation of a large, complex, and more or less complete transportation system. Most of these simulations contain subprograms which in themselves simulate activities of some part of the total system. The high level of simulation activity may be seen by key studies listed in the references.

Flow-Oriented Channelized Transportation System

The most important simulation work produced for the flow-oriented channelized transportation is that done of the St. Louis-San Francisco Railway (Bellman, 1967). In this simulation, the industrial engineering department of the San Francisco Railway used the railroad network model originally developed by

---

[1] This section was prepared with the assistance of David Isleb and Peter Westphal, Northwestern University Transportation Center.

Allman at the National Bureau of Standards and at Northwestern University (Allman, 1967), which was designed to be used by policy makers in the operations department. The study examined a selected section of the St. Louis-San Francisco railroad network, which handles approximately 90 percent of the total traffic. The objective was to evaluate alternative operating policies involving scheduling, train length, preblocking, car sorting, and work distribution.

By using different sets of operation policies, the model predicts railroad operating performance. The objectives of the model originally developed by Allman were to investigate railroad freight train scheduling and car sorting, to determine car sorting and scheduling rules, to identify key total-system problems, to exploit the use of SIMSCRIPT and GPSS, and to demonstrate the applicability of simulation to a real-world railroad.

For the construction of the model, four entity types are involved: (1) sets of cars called cuts, (2) terminals or freight yards, (3) railroad lines or tracks, and (4) trains. Each cut has one of four attributes—it is an origination connection cut, an origination classification cut, a cut set-off but pregrouped through yard, or a cut set-off but *not* pregrouped through yard. The attributes of a freight yard are its grouping policies, priorities, and yard costs; the attributes of a train are its hauling costs, route, schedule, capacity, and train-take list. Sets of cars flow through a network of twenty-five nodes representing twenty-five freight yards connected by links representing railroad lines, all representing the San Francisco Railway network. These sets of cars are picked up by trains (there may be as many as fifty-one) in freight yards as cuts according to yard grouping policies. All yards have standard structures, and ordered sets of operations are performed on each cut. Pregrouped and connection cuts go through a single expediting operation, while classification cuts and non-pregrouped cuts endure as inbound yard operations and classification operations. Outbound cuts are then reserved for trains based on train cut-off lead times for each yard, according to train-take lists which are subject to train capacities and cut priorities. Each train takes all the cuts reserved for it.

Cut information based on historical data is read from an exogenous event tape. Although the study incorporates a two-week period, any specific number of days is applicable. The model is a status historical model.

Very little detail is included in this simulation because it is a large system, and this deficiency can be seen by comparing the real-world freight yards with the simulated freight yards. Real-world freight yards lack standardized structures, and the time and costs involved to perform an operation differ from yard to yard. This model assumes that resources such as motive power, operations facilities, and manpower are always available at each yard.

Allman used 1965 data from the New York Central System in developing his model, which consisted of twenty nodes and eighty-five trains and was written in GPSS to be run on an IBM 7094. Due to inefficiencies and difficulties in data handling on IBM 7094 class computers, a real-world railroad of this size could not be simulated using GPSS, but these problems were overcome by using SIMSCRIPT with its greater design freedom and greater storage capabilities. Allman was also able to simulate the Southern Railway System by using SIMSCRIPT.

The San Francisco Railway has made some major extensions since it began working with Allman's model. Different types or classes of traffic had not been recognized. Cuts are now taken in any specified sequence—instead of applying the same cut-off lead times for all trains, each train now has a specific cut-off lead time. Train capacities are now quoted in car length and train tons rather than number of cars and train tons.

The San Francisco Railway uses three SIMSCRIPT programs and one FORTRAN program on a CDC 6000 series computer. The SIMSCRIPT programs are (1) a preanalysis program to analyze input data for logical inconsistencies, (2) the network simulation, and (3) a postanalysis program. The FORTRAN program, also a postanalysis program, compares the time it takes real-world cuts to move through the network with the time it takes model cuts to move through the model network.

Allman was able to validate his model and his simulation of the Southern Railway by comparing the model's behavior with the behavior of the real-world railroad. Both were considered valid because the schedules maintained by the simulated trains in the model approximated those of the scheduled trains in the real world. In other words, the model behaved in the same manner as the real-world railroad. The San Francisco Railway has also been able to validate its simulation on the same basis and should be using it as a policy-making tool. However, to use it in this capacity, it has to alter the model to allow handling of unscheduled extra trains (needed for greater realism).

The major contribution by Allman's model and the San Francisco Railway is that it is the first successful simulation of an entire railroad network in which operating policies affect an entire system rather than a small segment.

It is also the single most important simulation of a flow-oriented channelized transportation system. Other simulations of flow-oriented channelized transportation systems include railroad networks such as the British Railways (Coates and Clark, 1967); the Netherlands Railways (de Cock, 1964); railroad freight yards (Crane, Brown, and Blanchard, 1955; Eberhardt, 1966; Guins, 1966; Nadel and Rovner, 1967; Nippert, 1966; Nippert and Guins, 1966); railroad junctions (Dodd, Hawkes, and Muddle, 1966; Niemond and Kostalos, 1966); single railroad lines (Coupal, Garves, and Smith, 1960; Soberman, 1967; Wilson and Lach, 1967); railroad train operations (Bontadelli and Hudson, 1966; Feiler, 1966; Hogan, 1958; *Data Processor,* 1968; Wilson and Lach, 1967); subway lines (Day, 1965); a transit transfer point (DiCesari and Strauss, 1967); a bus terminal (Jennings and Dickins, 1958); canal locks (Carroll, 1965); rail highway interchanges (Davenport, 1968; Vaswani, 1956); and highway-marine port interchanges (Feiler, 1966; O'Neill, 1957).

Demand-Oriented Channelized Transportation Systems

Simulation of the total transportation system of any given region developed by General Motors Research Laboratories (U.S. Department of Housing and Urban Development, 1964), is an example of a demand-oriented channelized transportation system. The objective was to forecast the distribution of demand for transportation among the choices available for a given transportation configuration and specified demand. Simulations of this type have been called traffic assignments. Because they include all transportation modes, they have greater significance.

In this type of simulation different modes and different routes of the same mode are the entities. Attributes are fare, travel time, walking time, frequency of service, and so forth. The usual convention of a network containing modes connected by links, or arches, is used for a model. The nodes are used as origins and destinations and/or transfer points. Each arch is a given mode, and there can be more than one arch between two nodes. Since each attribute has a price or disutility for each type of traveller, each mode also has a price or disutility for each type of traveller.

The mode and route utilized by each type of traveller are determined by finding those that are the least expensive. After the volume on each arch is determined, new weighted values are given to some of the attributes which are changed due to congestion on each arch. Using the new attributes, the volume on each arch is again determined. This is called an iteration. By comparing the new volumes with the previous volumes on each arch after each iteration and by using new weighted attributes for each new iteration, the volumes converge. The process is stopped when the difference between volumes from one iteration to the next reaches a point at which any more iterations would no longer be beneficial.

In this dynamic historical model, volumes can be determined for any time period—one hour, two hours, twenty-four hours, during the rush hour, and so forth. The model's structure explicitly describes the system being modeled. The development of this simulation was based on preceding work done in traffic assignment. For example, simulation has been used in most of the metropolitan area transportation studies made since the Chicago Area Transportation Study first used traffic assignment simulation in 1955. This form of simulation was made possible by the Moore minimum path algorithm, a method of determining the least cost or least time path between any nodes of a network. Using the experience gained from previous metropolitan area transportation studies, General Motors Research Laboratories have made innovations in demand determination and have developed an improved minimum path algorithm. The core of the simulation was written in machine language in order to handle large networks and the numerous iterations that are required when some attributes are made functions of arch volumes and demand is allowed to be a variable.

This simulation had not yet been validated as of early 1968. It was thought necessary to validate this simulation against a home interview tape and then against actual counts, since errors could arise in both the model and the sampling scaling size. Previous traffic assignment models have been validated in a similar manner.

This simulation was an improvement not only over previous traffic assignment work but also over other public systems simulations, due to the amount of work that had previously been done. One of the earlier simulations was the source of many of the ideas used by General Motors Research Laboratories, such as the Toronto Metropolitan Area (Hill and Dodd, 1962; Irwin and von Cube, 1962; Kates, 1963; Mertz, 1961; Ridley, 1968; Smock, 1960; Zettel and Carll, 1962) done by Traffic Research Corporation. Another major simulation involving traffic assignment is being developed as part of the Northeast Corridor Project (U.S. National Bureau of Standards, 1968).

Although the model produced by General Motors Research Laboratories does not consider land use as a demand determinate, it is constructed so that it can easily include land use as a demand determinate. As such, demand determination is more realistic because it is more microscopic. Different types of travelers are examined; induced demand can be considered. This model can also be used for larger networks, in more situations, and needs less time to operate than previous models because of programming improvements. It also has a faster method of determining minimum costs or time paths employed.

Flow-Oriented Spatial Transportation System

As an example of a flow-oriented spatial system we have chosen the MARAD Fleet Operations Simulation (U.S. Maritime Administration, 1964), which was initiated as a generalized continuation of an earlier project undertaken privately by the Matson Steamship Lines. In the Matson simulations, their California-Hawaii fleet operations had been abstracted and modeled for the analysis of various alternative operating policies on this one fixed route. The usefulness of the early Matson exercises stimulated the development of a more elaborate and flexible model for general merchant fleet operations. In January 1964, Arthur D. Little, Inc., submitted the MARAD simulation to the United States Maritime Administration, Department of Commerce.

The objective of the simulation was to provide a tool for the study and evaluation of a fleet of ships operating on one or more services. *Services* are patterns of operation defined by a sailing schedule and an itinerary. The simulation was designed to evaluate the influence of individual ship design characteristics, voyage itineraries, schedules, cargo availability, and handling characteristics on the economic performance of the fleets studied. Economic performance was measured in terms of cost-revenue and vessel utilization accounts. This simulation was equipped to handle situations where fleets were of mixed and varying composition. The competitive effects of the rival fleets were not handled in detail, since the author believed that this did not adversely affect the usefulness of the model because competition had a relatively small effect in the real system.

The general operation of the model involved following the progress of each vessel under consideration through several voyages. Overall time periods were from six months to a year. The measurement used to end the simulation was the completion of a given number of voyages. Because voyage length was affected by events within the simulation, the simulation was terminated after the completion of all required activity rather than after a given time period.

The system is described in terms of two general groups of entities: ships and ports. Cargo, or more meaningfully, cargo revenue, is a third entity but serves mainly as a unit of measurement and exchange between the ships and the ports. Cargo availability is also a major source of variation in performance within the model.

Ships are more specifically defined by several sets of attributes in terms of physical configuration. For example, capacity is defined in terms of *cargo space types* for each vessel: *reefer, 'tween decks, lower holds, deep tanks,* and *on deck.* Various cargo types are *reefer, vehicles, general cargo unitized, general cargo, liquid bulk, steel,* and *dry bulk.*

In addition, each ship belongs to a *class,* which quantifies several qualitative aspects of its identity and contains information on operating performance and costs. Each ship also belongs to a *fleet,* which is defined as a group of ships that are interchangeable in a given service. Finally, each ship is described by the *line* to which it belongs. This attribute determines the cargo-generating ability of the vessel and reflects competitive superiority, if any.

Ports are defined in terms of their location within the total system and in terms of proximity and relationship to other ports. These definitions are used not only to identify originally planned movements of vessels and cargoes but to provide logical aids for determining alternative methods of handling a movement—for example, off-loading at other than the original destination when the locations are such that inland final delivery is a realistic possibility. This mechanism eliminates many short local moves and thus helps to keep the problem manageable.

Because of the long time periods and the low time granularity involved, a dynamic clock is used which is advanced from event to event as required. As for this clock, the principal events are arrivals and sailings of vessels. En route performance, cargo availability and loading, port delays, and other endogenous events are computed and related, and the economic result is determined at the appropriate arrival or sailing time. After the computations, these events are treated in the aggregate.

Two major programs provide for the actual operations of the simulation. The first, the Schedule Selector, outlines the proposed strategy for fleet operations. Sailing schedules and itineraries are outlined with sufficient slack to allow for internal variation in the course of the simulated operations. General information about fleet characteristics is utilized to develop a more specific plan of operation.

The second basic program, the Roll Call of Ships, follows and determines the activity of each individual vessel as it proceeds on a given voyage through the system. The program is highlighted for the provisions it makes for intermediate decision-making within the rules of the system. Its normal functions provide for compliance with the proposed schedule and itinerary, but it also simulates alternate decisions that may be made in mid-voyage. After a vessel has taken on cargo (the amount and nature of which are generated by a subprogram) and has been given the general voyage plan, a search procedure is followed to determine what additional ports could be used "on the way" to provide additional revenue while incurring minor additional costs. This search may involve extra cargo loading for ports not originally scheduled or sailing to new ports to pick up extra cargo for the original destinations. The sequence of activities within this program is shown in figure 11-1. The output of the model is in the form of *voyage logs* which give relevant data on:

Ports– Visited in order of arrival. Voyage and berth times as well as information on the volume of cargo unloaded, loaded, and available but not loaded.

Delays—A listing by port of the hours early or late for arrival or departure, also loading and unloading overtime.

Cargo Carried—Inbound, outbound, and intermediate cargo handled by cargo type.

Revenue—From above cargoes.

The simulation has been run in various configurations. Maximum capacity is such that it can at present manage thirty ports, forty ships of four different types, six ship space types, ten cargo classes, and three fleets handling up to five different types of services.

These activities are complex enough for meaningful analysis, and for these problems validation has been considered complete.

Demand-Oriented Spatial Transportation Systems

A nonchanneled or spatial simulation differs from a channeled simulation in that only the significance, origins, and destinations are considered. As a result the model may be simplified to eliminate detailed investigations of the demand for facilities between these points. Such a simulation, the SST Worldwide Route Simulation (Operations Research, Inc., 1964) for the Department of Commerce, was used to analyze world demand for air travel for a twenty-five year period. With specific reference to supersonic transport, information on the competitive effects of airline equipment purchase policies was desired. The simulation is based on the operation of an Economic Demand Model. Inputs to the model are considered and related to provide estimates of the impact of new SST and subsonic aircraft on air travel demands and to determine the effects of change in the number and type of aircraft and in routes and services for all points that may be considered serviceable by SST aircraft.

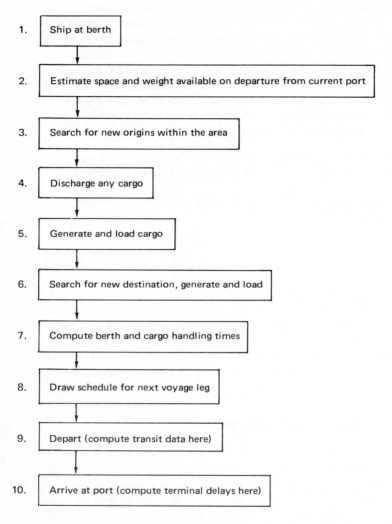

1. Ship at berth

2. Estimate space and weight available on departure from current port

3. Search for new origins within the area

4. Discharge any cargo

5. Generate and load cargo

6. Search for new destination, generate and load

7. Compute berth and cargo handling times

8. Draw schedule for next voyage leg

9. Depart (compute transit data here)

10. Arrive at port (compute terminal delays here)

FIGURE 11-1   Roll call of ships program. (Loop ends with arrival at final destination.) Source: U. S. Maritime Administration, 1964.

A basic assumption of the model is that the market for air travel is increasing and that new demand may be generated by increasing service or lowering rates. The amount of new traffic to be analyzed is calculated after evaluating technical and economic factors. In the initial simulation application, technology and policy, although considered in detail, are used as fixed input parameters for the demand model. After the output, under present policy, has been used to predict future demand, other alternative policies are utilized and the outcomes are compared with previous runs.

An especially significant feature of this simulation is the extensive comparison of competing carrier services and the assignment of the aggregate demand for a given service to each of the competing carriers. Provision is made for the alteration of carrier policy if a service is not well received.

In addition to the network description, the input to the model contains information on the number of aircraft assigned to each route by each carrier. Additional input provides details on the technical performance of the aircraft so as to provide a detailed source of information for calculating service times. Costs of numerous types are separately computed to measure the full effects of any scheduling or operating changes. The importance of connecting or alternate service by the same carrier is also being considered.

Another major group of input information involves fare and scheduling structures. The first set of input data defines those items that will affect the supply of available transport service, and the second set of input data outlines the determinates of the demand for these services.

In operation, the model considers the presented system configuration as input, and it evaluates the services offered in a complex manner. After the aggregate demand for a given service has been determined, the simulation takes the computed demand and reconsiders each of the competing carrier services to determine how the given demand can rationally be allocated. The costs and revenues are then computed for each carrier on each route segment and for each service or fare if more than one package is available to the public. These data are then collected and summarized to provide information on the effectiveness (measured in terms of profit and loss) of the use of aircraft types, schedules, fare structures, types of service, or any others.

The output then will resemble the operating reports of the carriers involved, with additional data available to provide cross-carrier analysis on the profitability of certain equipment and marketing policies.

The major entities considered are individual flights between city pairs. The city points studied were selected either as potential SST terminals or as tag-on points to be served as extensions of the major long-distance routes. The network used was worldwide, with sixty-seven cities involved. Routes to serve these cities were proposed, and the combination of these proposals led to the identification of 2,214 links between these cities. The model was designed to recognize thirty-one separate carriers as well as thirty types of aircraft. The variables for the evaluation of each aircraft include configuration, speed, fuel economy, crew requirements, carrier maintenance policies, and assigned route. Scheduling, pricing, and development policies are considered in a similar manner.

While the basic model is in itself a matter of totaling expected costs and revenues for the multiple operations involved, the number of individual items considered important is so large that the computer offers the only practical aid to analysis. The assignment models contained within this simulation also consider a dynamic competition between carriers which can best be handled by simulation methods.

This simulation has been used to examine existing route structures and marketing practices. Although the simulation includes only some portions of present carrier operations, the correlation between predicted and observed economic performance suggested validation of the model. Additional applications of the model have been undertaken as more detailed data on new SST aircraft become available.

Planning and renewal have been the focus of many studies. Urban studies by Hill (1965), Irwin (1963), Lathrop (1965), Lowry (1966), and Schlager (1964) have centered on the very broad subject of metropolitan growth. Much less ambitious objectives were set by Chapin (1965) in simulating Residential Development, and Robinson, Wolfe, and Barringer (1965) and Steger (1965) in studying urban renewal planning.

The following section examines simulations dealing with the problem of land use, the interaction problem in urban development, and a gaming approach that simulates this phase of urban policy. Finally, it describes simulations attempting to aid the urban planning process.

SIMULATION OF
URBAN GROWTH, RENEWAL, AND LAND USE

A logical broadening of transportation studies leads one to review the urban literature.[2] The first step would be to examine simulations of overall city planning and then simulations of specific functional areas of the city operation. Unfortunately, this logical approach fails because simulations devoted to overall city planning either have not adequately been developed or have been difficult to find. Researchers in the field have instead adopted an aggregative, modular approach, preferring to model and simulate specific problem areas. As a result, any state-of-the-art review of the application of simulation techniques to urban development problems must be done from a functional point of view.

Land Use Simulations

The origins of land use simulations in urban planning and development are found in transportation research. Transportation and land use forecasting are

[2]This section was prepared with the assistance of James M. Comer, Graduate School of Management, Northwestern University.

inseparably bound by both concept and methodology. An interaction exists between land use activity in an urban area and the network of streets that connects these activities. Transportation is a major force in directing the development of land uses in a growing and changing metropolitan region and this influence operates in many different ways (Harris, 1962). As a consequence, research in the area of land use forecasting has a road network as a primary entity. The first simulation described is a characteristic analytic solution to a land use forecasting problem, and the second exhibits the present state of the art.

EMPIRIC land use forecasting model.[3] The EMPIRIC Land Use Forecasting Model (Traffic Research Corporation, 1967) is a simulation that distributes future-year regional totals of urban activities such as population and employment. The model was developed by the Traffic Research Corporation for the Metropolitan Area Planning Council and the Massachusetts Department of Commerce and Development as part of the work of the Eastern Massachusetts Region. EMPIRIC was designed to produce land use forecasts to be used as input to traffic forecasting techniques being applied to the region and to be sensitive to public policy inputs in order to use the model as a regional planning tool to test alternative development plans, that is, to measure the impact on development patterns of alternative sets of public policies.

The EMPIRIC model is composed of a system of simultaneous linear equations which quantify relationships among the dependent output and independent causal variables. That is, more than one dependent variable is in each equation; the effects of these on the independent variables is linear and additive. The model is calibrated (i.e., variables are specified and coefficients are estimated) by hypothesizing relationships among activities and by applying statistical techniques to historical data. Solution of the set of equations comprising the calibrated model for each subregion yields future-year estimates of activity levels within each subregion. As presently utilized, the dependent variables are formulated as changes in subregional shares of activities, which are added to the base-year shares to obtain the shares of activities within each subregion at the end of the forecasting interval. These new shares are then multiplied by exogenously forecast regional control totals for each activity to obtain future-year activity levels. The model is operated recursively; after each forecast, the state of the region is updated, transportation service levels deriving from the forecast land use patterns are checked to ensure compatibility with the service levels estimated for the prediction of these land use patterns, and forecasts are made for the next time interval. The process continues until the final target year is reached.

Early attempts at developing land use forecasting models generally yielded

[3]The following summary was taken almost verbatim from the report "Development and Calibration of the EMPIRIC Land Use Forecasting Model for 626 Traffic Zones" by Traffic Research Corporation, February 1967.

separate models for each activity being forecast; for example, models of residential, commercial, industrial locations. Separate models, however, required judgment as to which activities to forecast or locate first. A simultaneous model like EMPIRIC eliminates this requirement.

Three types of data are input to the EMPIRIC model: (1) subregional values at the beginning of the forecast interval of various significant land use activities and characteristics, (2) estimated regional totals at the end of the forecast interval for those land use activities that it is to allocate among the subregions, and (3) subregional values of various pertinent policy variables, which reflect future characteristics of features of the environment that to some degree are directly controllable by government—transportation systems, public open space, zoning restrictions, and public utilities service. By making several forecasts using different sets of policy variables, the EMPIRIC model may be used as a regional planning tool to explore the implications of various sets of public policies and their effects on regional development. Planners can thereby determine the range of projects achievable by means within their control and can develop a set of policies calculated to produce a desirable regional development plan.

With minor variations, the EMPIRIC model is representative of land use models. Such models are static mathematical models; many employ some type of linear programming algorithm, producing analytic forecasts of land use based on fixed parameters and constraints. These models have one objective function which they are attempting to either maximize or minimize. They also usually do not either include a time dimension or provide an opportunity for sensitivity analysis.

A simulation model for land use forecasting. The EMPIRIC model utilized its land use forecasts as an imput to traffic forecasting techniques. The model developed by Fairburn (1967), on the other hand, studies the effect produced on land use when a change in road network is made. The simulation can be viewed as a realistic approach to relating the large range of parameters to be considered in forecasting land use.

The two basic entities in the simulation are land and street networks. Fairburn categorizes land by usage and arranges the categories in a hierarchy of permanence:

Municipal
Industrial
Commercial
Residential
Vacant

His forecasting model assumes that municipal and industrial land are fixed; therefore, the simulation deals only with commercial, residential and vacant land, and the primary consideration is the relative permanence. The street network consists of streets (arcs) and intersections (nodes). There are two types of nodes—schools and CBD (Central Business District). The CBD nodes were determined by inspection of the central business district of East Syracuse, New York, and the school nodes by reference to a tax map of the same area.

The basic data required by Fairburn's model are of four types:

1. Network data consisting of the street system topology and the location of both school and CBD nodes.
2. Probability density maps, one for each vacant, residential, and commercial occupancy.
3. Assembled data consisting of the existing land use, size, and assessments for all parcels.
4. Map making data consisting of character-coordinate to lot number data for printing out land use maps.

The procedure for producing forecasts utilizes a program to analyze the input data. For each lot, a resistance equation is defined which is a function of building, value of building, land value, difference in density on map, and coefficient of resistance. This equation then calculates the lot's resistance to change in present status. When this is done for all available land, the minimum path of change is determined, and then correlated into a forecast land use change for each lot. Finally, a land use map is printed.

In the validation process five forecasts were made and compared to existing land usage. The effects of these different conditions on the distribution of land use between the three flexible categories were displayed. Fairburn concluded that the forecasting procedure yields rational results, the probabilistic density map method of ranking land parcels is valid and useful, an observable coefficient of resistance to change exists, and the procedure is flexible enough to allow for estimated changes in economic conditions and variations in the road network and other phenomena.

So little work has been done in the area of land use forecasting employing simulation techniques that any initial effort seems noteworthy. One of its specific contributions is its departure from the strict mathematical optimizing techniques characteristic of the general literature of the area.

Gaming Simulation in Urban Development

Simulation can generally be classified as (1) all machine, (2) man-machine, and (3) all manual (Rath, 1968). The vast majority of simulations in this chapter are machine and man-machine. The simulation discussed below is an exception for it is of the third type—all manual.

The objective of the urban development simulation game called METROPOLIS is to examine the decision processes involved in the allocations of land resources (Duke, 1964). It was designed for the training of university students or young professionals in the basic decision processes involved in urban land use changes.

The program sequence is as follows. *The Citizens Gazette,* which contains appropriate national, state, and local news, is distributed. The public opinion poll, which requires decisions on three community issues from each group, is distributed. The amount of revenue available during the year is calculated. Decision forms are then made available to each team and consequences of the players' moves are calculated. End-of-cycle records are made, and every three simulated years an election is held. The process is then repeated. There are three entities: administrator, politician, and speculator.

The administrators, composed of a city manager and the department heads responsible to him, recommend the capital improvement program for future years. Presumably they are seeking ego satisfaction through proper performance of their job; however, they are influenced by the political situation and are responsive to population attitude on particular issues. The medium of exchange measuring the administrator's satisfaction level is in *utils* of satisfaction in the physical form of poker chips.

Politicians are assumed to be operating under the motivation of holding political power. This was operationally defined as a set of probabilities for reelection in each of the three major wards.

The real estate speculator is assumed to be primarily motivated by the idea of making a profit on his investment. These three roles are linked together mechanically through the device of the capital improvement program which bears some influence on, and in turn is influenced by, each of them. Figure 11-2 shows these relationships.

METROPOLIS is played as a series of cycles, each representing a year. For descriptive purposes each cycle is portrayed as a separate unit of activity; in actual play they become almost continuous except for an arbitrary interruption when the end-of-cycle calculations are announced.

The development, operation, and testing of METROPOLIS required eleven separate runs over a period of four months. Seventy players were utilized, including undergraduate students, graduates, young professionals, experienced professionals, allied administrative personnel (other than urban planners), lay planning commissioners, and faculty. Over seventy-five cycles were completed, requiring eighty-eight hours of elapsed time in play. Play was conducted on the basis of continuous periods, two periods interrupted by an evening, on a discontinuous play basis.

The validation of METROPOLIS as an instructional method was undertaken by the administration of an examination to a control group and to participants. Based on a final exam, both groups that played the game (I and II) showed a marked improvement in score—an average increase of 37.5 percent between them as opposed to a six percent increase for the control team.

METROPOLIS has received much more extensive use than the related Community Interaction Game (Yates, 1967). The urban planner should also

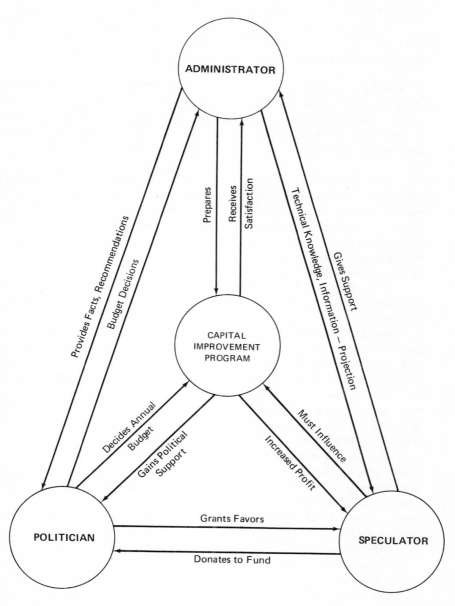

FIGURE 11-2   METROPOLIS static relationships (Form 50). Source: Duke, *Gaming Simulation in Urban Research*, p. 17.

review other attempts and commentary by Lee (1967), Meir and Duke (1966), Redwin (1967), and Schlager (1964).

Simulation in Planning

Simulation may be used for forecasting and planning where either empirical data or entities serve to describe the phenomena. Of the several simulations that have been developed with applications in urban planning, three will be discussed: a housing model, a land development model, and an urban resource allocation model.

A housing market model.   The San Francisco simulation model has been designed to generate conditional predictions suitable for impact analysis and thereby analyze for city planners the impact of including alternative public actions in recommended programs (Robinson, Wolfe, and Barringer, 1965). It incorporates the interactions and effects on residential housing of public policies, programs, and actions; investment behavior of the private market; and location decisions of households.

The model can be divided into three major elements: housing stock, space transitions, and space pressures. The entire housing stock is divided into (1) fourteen location or amenity-oriented categories involving topography, number of dwelling units under one roof, and rent range, (2) twenty-two land-use type categories describing the tenure of the residents, the number of rooms in the unit, and the structural type, and (3) four condition categories describing the condition of the dwelling unit. The city is divided into 106 neighborhoods in which homogenous types of location, land-use, and condition dwelling units are grouped together and counted as *fracts*. One fract consists of two acres which are close but not necessarily contiguous in a neighborhood. San Francisco was divided into 4,980 fracts.

The second element, space transitions, involves the change of locational, conditional, and land-use categories of a fract of land. Such transitions can take place through public or private actions. Public actions influencing spatial forms of a normal aging process could be described as a first-order Markov process. The operation of the model is described in part of the abbreviated flow chart (see figure 11-3).

After adjustment of some parameters, the simulation replicated total new construction investment figures satisfactorily. However, some failures were encountered in reproducing single-family dwelling unit construction investments, probably because of translation of single-family owner-occupied unit costs into rentals for the model. The treatment of owner-occupied units is one of the few serious defects in the model. The primary strength of the model lies in its ability to suggest the relative impact of public actions over a two-time period interval rather than the exact impact.

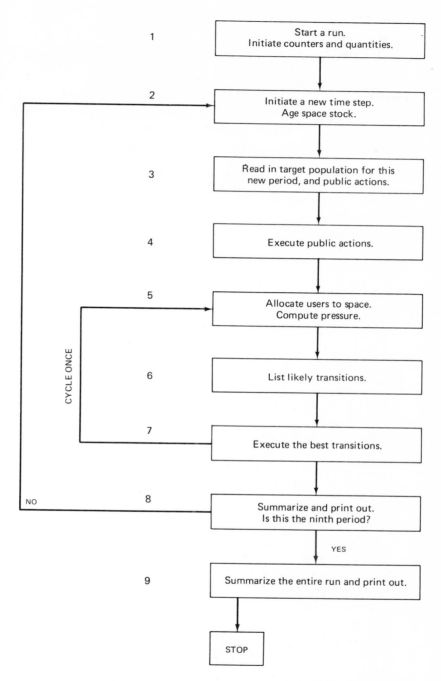

FIGURE 11-3   Abbreviated flow chart of housing market model. Source: Robinson, Wolfe, and Barringer, "A Simulation Model for Renewal Programming."

Proposed extensions to the model include reduction of the level of detail to reduce computational costs and the incorporation of output summarization beyond present levels. The author also proposes incorporation of endogenous location changes, population feedback loops, and extensions in the range of public actions. The expansion of the simulation to include industrial and commercial sectors, reallocation and population movement patterns, and regional housing markets is also under consideration.

The M.E.T.R.O. simulation. Ray (Ray and Duke, 1967) designed M.E.T.R.O., a model that simulates an urban government agency interacting with the forces affecting the agency's actions. M.E.T.R.O. is in the form of a game, where the player can assume the role of politician, planner, educator, or land developer. Any individual can exert influence on individuals belonging in one of the other three groups, or on a group itself. For example, the land developer makes zoning requests from planners, while the planner furnishes technical information and projections to the land developer. The land developer might also contribute to the politician's campaign fund, thus expecting favorable budget decisions by the politician. The educator, naturally, attempts to maximize his own share of the city budget. The model also considers how the actions of these players affect voting patterns and future decisions by households and firms.

A municipal resource allocation model. Crecine's model (1967a, 1967b), which consists of three submodels, attempts to simulate municipal resource allocation in large metropolitan communities. The department submodel depicts the formation of a budgetary request by department heads. The output of the department submodel provides the input to the mayor submodel. The mayor submodel illustrates the decision processes within the mayor's office in reviewing the budgetary requests, and its output, in turn, provides the input to the council submodel. The council submodel simulates the council's approval or alteration of the mayor's proposed budget. All transactions in the model are depicted in the form of dollar amount requests per account per department, each of which has between two and five salary or nonsalary accounts.

The mayor's budget letter along with the standard budget forms initiates the system, and this procedure allows for a uniform start in the simulated model. The department heads determine their requests by estimating how much they will need to carry out their programs, deciding what they believe will be acceptable by the mayor's office, and requesting a reasonable share of the total budget. These considerations are accounted for in the simulated model by comparing current appropriations with next year's expected appropriations, analyzing last year's expenditures, establishing increases in account requests, and testing for constraints.

In the actual system, the mayor obtains preliminary revenue estimates from people in the city government and from an outside source. In the simulated model, as well as in reality, revenue estimates are separate from expenditure estimates and provide the constraint function for the analysis of the expenditure requests.

The mayor's objective is to present a balanced budget to the council, maintain existing service levels, provide raises for city employees, and, if possible, avoid tax increases. The input to the mayor submodel enters one of four procedures. The procedure chosen depends on the "honest" and "realistic" budget requests by the departments, behavioristic characteristics accounted for in the model. This choice of procedures and parameter values was made on the basis of empirical tests using a regression model. The four procedures prepare a preliminary budget. Next, to achieve the main objective of a balanced budget, the data are analyzed in a surplus or deficit routine. On a hierarchical scale, nonsalary accounts are adjusted prior to salary accounts and tax-rate changes are determined. In reality, the complexity of the budget does not allow the council to consider account items by department on an individual basis. The simulated council submodel also only analyzes the data on the basis of total revenue and expenditure estimates. Adjustments are made in a surplus or deficit routine.

The formal model was tested by generating budget decisions for Pittsburgh, Detroit, and Cleveland, and simple models were constructed for the actual data from the three cities. The simulation models' results were then compared with the simple models for the cities' data. The author found a high degree of goodness of fit between the simple and the simulated models. A change in administration in one of the cities accounted for discrepancies in the goodness of fit. The simulated model was excellent for predicting changes resulting from continuation and elaboration of existing policy. Policy "shifts," such as changes in federal regulations, as well as creation of new departments, were not satisfactorily anticipated by the model.

Crecine's study concludes that environmental corrections appear to be more related to revenue changes than to expenditure changes. As an example, rapid growth is observed in areas that provide the city with an opportunity to expand because of revenue rather than in areas that have rapid changes in "needs."

Crecine admits to certain limitations in the model. For example, few behavioristic characteristics such as public pressure, lobbyists, and personal relationship between the mayor and the departments heads are considered. The simulated model does not utilize the approach of optimally balancing community resources, allocating funds among functions to achieve overall community goals, but instead analyzes strictly dollar totals, neglecting optimal allocation of manpower and material. It is a year-by-year process, neglecting provision for long-range planning. Crecine's model, however, is an excellent introductory attempt in a relatively unexplored field. It is reprinted as a case-example following this chapter.

Problems in urban areas have been attacked from two angles: (1) land use, its allocation and growth, and (2) planning and organization of metropolitan functions. This is analogous to the *stock* and *flow* concepts of economics.

The first approach is heavily indebted to transportation models of the

1960s for nomenclature and methodology. This debt, however, can become a burden, for far too often land use analyses have resorted to mathematical techniques that simplify behavioral relationships and thus force them into nonrealistic simplistic equational straightjackets.

The second, planning of metropolitan functions, is a process approach, which attempts to examine the flow of goods and services or the process of planning or decision-making in urban problem areas. Here both behavioral and computerized models have been utilized with some success.

It is important to recognize that both approaches must be integrated for coping with cities with complexly interacting social, psychological, and economic forces. A unidimensional approach of either type is of questionable validity. Because the area is so complex and available data are so scant and disreputable, only piecemeal attacks have been feasible. The mid-1960s marked the high point of these mathematical and simulation approaches. The outputs of these approaches, although valuable in their own right, were less than satisfactory to the modelers. The problem involved was not so much methodology as lack of available data on various sublevels of metropolitan planning and operating. Since that time, efforts have been directed not at methodology but at understanding the basic processes and interactions entailed in city operations. Thus it seems reasonable that this area will expand and mature with the addition of fresh outlooks.

Simulation of Hospitals

Hospital costs have skyrocketed as traditionally low labor costs reach more competitive levels, as more complex and expensive equipment becomes available, and as societal changes such as the growing importance of the nuclear family increase the variety of demands for services. Hospitals have had to turn to professional management and to more sophisticated management tools to achieve efficient utilization, and one of these tools has been simulation. Although simulation work has been limited, much has been accomplished to clarify the important role it can play in reducing an otherwise inevitable rise in hospital costs.

A number of hospital areas lend themselves to a simulation approach, such as professional services, which comprises pathology, radiology, physical therapy, and occupational therapy. Personnel in these professional and technical specialties provide services at the request of the attending physician. Resources in these areas are often relatively scarce, demand fluctuates considerably, and services are usually required for only a relatively short period to aid the physician in his diagnosis and the patient in his recovery. Simulation work on a model of clinical pathology has been completed by Alvarez (Rath, *et al.,* 1970). Another area that lends itself to the simulation approach is medical operations. Fetter and Thompson (1963, 1965) have modeled a maternity suite and have made specific recommendations. Occupancy rates for hospital beds have also been examined (Goldman, Knappenberger, and Eller, 1968; Handyside, 1967; Lapp, 1968; Robinson, Wing, and Davis, 1968).

While the modeling process for this topic has been relatively straightforward, difficulties have been encountered in assigning realistic costs. For example, costs are assigned to wards other than those to which they would ordinarily be assigned. Choice of criteria in hospital simulations has created major difficulties. Gauging the effects of various levels and types of medical care requires some assessment of saving lives or improving the quality of life. A hospital may from month to month have different rates of success with patients having extreme afflictions, due solely to differences in the physical conditions of incoming patients. Most investigators in the area of hospitals have therefore chosen problems in which quantitative criteria are possible; for example, the number of patients processed or the speed of processing.

One interesting exception to the use of this type of criterion, by Colley (Colley, Hallan, and Packer, 1967), concerns medical care in the immediate postnuclear attack period. In this instance, the number of survivors was considered a measure of the success of various priority rules for medical service. While this is in a sense quantitative, the lives of human beings do enter the equation. However, no attempt was made to assess the value of human life. Such an assessment might well be necessary to determine the amount of equipment, effort, and supplies devoted to emergency use.

The remainder of this section will describe four simulations in: (1) medical operating areas, (2) professional services, (3) hospital bed occupancy, and (4) doctor's office and waiting room scheduling.

Simulation in Medical Operating Areas

In a series of articles, Fetter and Thompson describe a model of a maternity suite at Grace-New Haven Community Hospital (Fetter, Thompson, 1963, 1965), which they chose to simulate mainly because the process is relatively self-contained with well-defined stages of care directly related to observable patient states. Due to the nature of the process, input queues in the normal sense are not allowable, except for Caesarean operations. The problem is characterized as providing an adequate number of process stations of each type to best meet given demands.

Fetter and Thompson's model consists of four basic types of process stations: labor rooms, delivery rooms, Caesarean rooms, and postpartum rooms. Patients enter labor rooms, give birth in either delivery or Caesarean rooms, and are then taken to postpartum rooms for the immediate postoperative period. Given a set of arrival data, the percentage occupancy of rooms of each process type is computed. At each process stage, the first room of each type receives 100 percent occupancy before another room is used.

This model was validated with Grace-New Haven Community Hospital data. Occupancy statistics found by the model using actual hospital decision rules corresponded to a high degree with actual occupancy statistics. Repeated sampling was used to establish reliability; about five thousand patients were considered necessary to stabilize parameters. Resulting data revealed that for this particular hospital, a 25 percent increase in load would cause a significant decrease in the service provided. In this respect, the model provides a means for measuring the effects of projected increases or decreases in demand. It was also discovered that a policy allowing more frequent admissions, but with fewer patients per admission, would equalize the bed requirements.

Simulation of Professional Services

The hematology section of the clinical pathology laboratory at Passavant Memorial Hospital, a hospital of about 350 beds, in Chicago, was simulated by Rath *et al.* (1970). It performs a variety of tests on blood samples at physician request. The simulation model divides personnel into teams of one or more, and each team performs specified tests. Requests are assumed to come in batches, although provision was made for alternative arrival schemes, and are sorted and given to the first free members of the appropriate team. If tests are to be performed in batches, a queue is formed until a sufficient batch size is obtained. This simulation was written in FORTRAN IV, utilizing the simulation routines available at Northwestern University in the SPURT package (Vogelback Computing Center, 1968). An eight-hour shift was simulated with processing time estimates obtained from an independent time and motion study and compared with time data for the laboratory. The characteristics of clinical laboratories are such that the model devised could be used for other sections of the laboratories simply by varying parameters.

Unfortunately, technical difficulties at present preclude the simultaneous operation of an entire clinical laboratory the size of that at Passavant Hospital. However, the simulation clearly and concisely represents the operation of this laboratory section, and work is continuing to broaden the model. Further validation of the present model would be useful with time data obtained from the specific laboratory modeled.

Simulation of Hospital Bed Occupancy

The problem of bed allocation faces the majority of hospitals. Whenever beds are specifically allocated, the random fluctuations in demand increase the likelihood that those beds will be idle, as opposed to general purpose beds being idle. Thus allocation of beds to specialties decreases overall utilization, although it permits utilization of specially trained personnel, greater availability of special equipment, and convenience for the physician. In emergency wards, bed availability may be crucial. Yet, most of the time many beds are idle.

Robinson, Wing, and Davis (1968) consider the problem of scheduling elective patients for a hospital, creating a three-part simulation. The first section, written in SIMSCRIPT, is called the Request Generator. Patients entering the hospital are assigned four attributes: (1) desired admission day, (2) alternative admission days, (3) probable length-of-stay information, and (4) demands for hospital services. The second and third items are not at present generally available to the hospital admissions office. The authors feel, however, that the second atrribute could be generated through cooperation of the admissions personnel and the attending physician. The third may be generated from historical information. Only elective patients are included. This failure to treat emergency patients in an explicit manner, however, is a major weakness of this simulation. Also, no allowance is made for no-shows. All beds are considered to be general purpose.

The second, or scheduling, stage, also written in SIMSCRIPT, contains three possible scheduling policies: fixed number of admissions each day; expected length of stay; and a rule assigning probability levels to a range of possible lengths of stay. The third stage computes costs and was written in FORTRAN IV. Costs were taken to be a linear function of the number of turnaways, empty beds, and overflows where these costs were set in a subjective fashion.

The authors found that if estimates of length of stay were unreliable, the fixed admission per day policy would be the best choice. With proper scheduling, cost savings did occur. The authors further suggested creating a third category of patients with a priority between that of emergency and elective patients to give greater flexibility to the handling of the patient stream. This, however, was not incorporated into the model. The simulation could have been improved if some measure of patient satisfaction had been considered in assigning cost. These authors are not alone, however, in disregarding this factor. The article is particularly noteworthy for its use of probabilistic estimates of length of stay.

Doctor's Office and Waiting Room Scheduling

Fetter and Thompson (1966) studied the problem of balancing the time patients must wait in a doctor's office with the time a doctor must idly wait for patients. The basic method now used by doctors in treating patients is a first-in/first-out (FIFO) process. Fetter and Thompson (1966) considered the effect of a number of variables on waiting time for patients and idle time for doctors: patient arrival time, doctor's arrival time, no-shows, walk-ins, service times, and rest times for the doctor. A simulation was made for a clinic with three physicians. Whenever a patient arrived, the first physician available would attend that patient. The program generated a schedule of appointments for each day. The model was quite simple, with a single arrival queue for patients of all types, although patients entered queues by appointment or by walking in.

It was found that if patients would arrive within five minutes of their appointed time, physicians would be idle only about 20 percent of the time. Keeping physicians idle for at most 10 percent of the time required a very large increase in the waiting time for patients. Walk-in patients had greatly increased waiting times (up to eight hours). In addition, the physician would be much more likely to be kept working late. It was also discovered that were the doctor to arrive late, waiting times would significantly be increased. Choosing an optimal policy for balancing idle time for the doctor and waiting time for the patient requires an assessment not only of the values of time for patients and doctor but also of the cost of implementing the desired policy. For instance, if patients are to arrive within five minutes of their appointed time, a significant amount of time and effort will be involved.

Conclusions

A great deal of work remains to be done in the field of hospital simulations as only a few areas have been modeled. Each simulation completed, however, stands by itself as a small island. Obviously what is needed is a model that interrelates the operation of different hospital areas into a functioning whole. Some work has been done in setting up criteria for such a model (Lapp, 1968). To keep the size of the model within reasonable bounds, however, the basic unit must exist at the hospital department levels, with sufficient attributes to represent each department skeletally and also to provide for interdepartmental interactions.

SIMULATION IN EDUCATION

Computers have been used for years to assign classes, make payrolls, design schools' construction schedules, allocate bussing routes, and design district boundaries (Rath, 1968). Given social and economic data, growth can fairly well be predicted and future resource demands can be scheduled. Many school districts, however, fail to allocate resources far enough into the future, some planning only a year or two ahead. Because schools are more dependent than most institutions on tax money and special tax levies, planning-for-aid campaigns must be also included for the future. There is an unmistakable need for tools to aid educational economic planning. Current work includes educational subsystems (Haussman and Rath, 1965) and university studies (Weathersby, 1967; Firmin *et al.,* 1967; Koenig, n.d.).

The following section will describe two simulations, each of which models an entire educational subsystem. The first simulation models a school system in terms of resources and policies; the second, a university in terms of resources, both human and monetary.

A School System Model

The first simulation, named S:D:TWO by its authors (Szekely, Stankard, and Sisson, 1968), attempts to model an entire school system. The model it uses is extremely general, since it makes few assumptions about the system simulated and both school policies and school resources are therefore inputs to the simulation and are entirely dependent on the system being modeled. Policies are described in terms of their strain on resources—for example, budgets, equipment, and manpower.

The objective of this system model is to aid in exploring the consequences of alternative resource allocation policies. The simulation meets this objective well. However, the objective of the education system being simulated is to change the potential behavior of the students. Originally, the authors proposed to measure the success of a school system in this light. Their representation of achievement is still under study and will soon be released, it is hoped.

In the systems inputs, there are several presimulation groupings into (1) areas within the district, (2) grouping of students, types, (3) grades, (4) equipment types, and (5) educational programs. Inputs are policies and plans for the district. They are constrained by operating budgets, capital budgets, limitations on the number of teachers and staff available, tenure, and desire for continuity of programs, for example. The inputs are transformed into outputs that take the form of operating expenditures, capital expenditures, programs actually implemented and changes in student achievement.

The administrator's desires are input by assigning priorities of 1, 2, or 3 to the various programs and by designating preferred areas as key or poverty areas. Priority programs are given full implementation, and others are implemented as resources will allow. The authors have set minimum levels so that trivial projects will not be funded. This is perhaps advisable when running a simulation, but a real district may have a number of small programs that produce quite significant results. The administrator must use the model as a tool, not as an excuse for not making decisions. After the inputs have been specified, the operation of the simulation proceeds in the following manner:

1. Programs are applied to students in each area to determine resource needs.
2. Space and equipment needs are projected into the future and compared with the future availability of buildings and equipment now existing or soon to be completed.
3. Prediction of student achievement is carried out, assuming the operating policy decisions have been made, and yearly operations are summarized.

This sequence is performed each year to be simulated.

One of the problems encountered is the designer's failure to realize that the policies that he is changing when he runs the simulation may have some effect upon the input functions he is using, such as when teacher hiring policy is changed. Changing policy may cause the supply of teachers to fall to a drastically low level, which invalidates any previous results of the simulation. The authors seem to realize this; in fact, their model can account for this type of phenomenon because it can be used to specify policies, to simulate one school year, and to present the outputs to a group of students, teachers, administrators, and parents, using their reactions as inputs for another run. This would almost be a "business game" method and might well be employed in running a model such as S:D:TWO. This model is therefore an example of a model that can be used either as a training device (management game) or as a planning tool for specific school districts.

The CAMPUS Model

The CAMPUS model simulates university operations and accepts descriptions of university structure with level of activities. The inputs allow the experimenter to compute the requirements for staff, space, and money. This model is becoming a key tool at the University of Toronto for planning, data collection, and analysis. A viewpoint of higher education and resource management has developed in which a single utility function describes total university performance. Key activities in the university are structured into categories of instruction, research, library, scholarly development, committee service, and others, with a great deal of detail under specific subject matter areas such as science and its subdivisons. University education functions and activities are dependent upon resources. Many uncontrollable variables in this study serve as inputs, such as cost of new facilities, competitive academic salaries, and price of library acquisitions.

The CAMPUS model stands as a comprehensive analytical model for planning in the university sphere. The administration summary report prepared by the simulation offers a breakdown of the departmental budget into academic and nonacademic salaries, miscellaneous expenses, academic staff information for each faculty member, physical facilities, lecture lab study, supporting staff, and data on student enrollment. Each given department has similar budgetary breakdowns specifically of this type. Class size data and output in terms of performance by academic year and student involvement are printed out, as well as further data on associated courses. Graphs analyzing the departmental budget as a function of time are broken down into staff salaries, assisted research that has been completed, academic salaries, and miscellaneous. Staff distributions that can also be broken down in a similar manner are demonstrated in another output by rank.

The CAMPUS model, limited to undergraduate activities at the University of Toronto within the faculty of arts and sciences, builds up instructional workloads for each department yearly and calculates the resources needed. The model entails four sections: enrollment formulation, resource loadings, space requirements, and budgetary calculations. This program was written in FORTRAN over approximately two man-years of work covering a period of fourteen man-months. A team reports a learning curve in the performance during that time. Application potentials have also increased; for example, a model for the medical school has been created. Other applications are sure to follow.

SIMULATION OF SELECT GOVERNMENT FIELDS—
CRIMINAL JUSTICE SYSTEM AND JUVENILE PROBATION

The majority of the simulations conducted in the criminal justice system are at present only in the working stage. Few published reports are available on projects in process or on those very recently completed. SIMBAD (McEachern, 1968), an acronym for "SIMulation as a Basis of social Agent Decisions" in the juvenile probation system, is a proposed simulation that will attempt to analyze the decision process and the probation process. Specifically, the objectives of the project are to develop an operating model in the form of a computer program of the probation process from the point of initial intake decisions, investigations, court decisions, disposition, and treatment and placement alternatives to subsequent re-referral.

Sager (1968) is currently developing a model of the United States Supreme Court which analyzes the decision processes of individual justices.[4] The model predicts the voting pattern of the lawmakers on particular cases by considering their opinions on previous cases.

The first simulation to be described (Taylor and Navarro, 1968) is COURTSIM. This model represents the United States District Court and simulates the passage of cases through the legal process. Of primary concern are the time delays experienced by individual cases because they represent delay in receiving the due process of the law.

The second simulation to be described (Rath and Braun, 1968) has analyzed the communications center of a police department. Improvement of the console efficiency was considered by introducing new concepts of manning and job assignments as well as assignment priorities for handling incoming phone calls. Different configurations of space and layout, along with changing zone structures for the city, were also considered, as was improvement of the allocation of police resources. Nilsson and Olsen (1969) have further developed this topic and are using simulation to evaluate both communications center responses and police response capability.

---

[4]For a more complete detailed examination of the project's antecedents and present status the reader should refer to the January-February 1968 issue of *American Behavioral Scientist.* The entire issue is devoted to a review of this project.

COURTSIM

COURTSIM (Taylor and Navarro, 1968) simulates the processing of felony defendants in the United States District Court in the District of Columbia.[5] The model examines the time delay problem between stages of the court proceedings, and it constructs the step-by-step process of a criminal case in the courts. The first step following arrest is presentation, which occurs before a judge of the Court of General Sessions or the commissioner. This presentment is screened and prepared by a United States assistant attorney. If the felony charge is reduced to a misdemeanor or if there is no probable cause, the case exits from the system. If the case is to be prosecuted or if preliminary hearing has been waived, the case is screened again in the office of the United States attorney, Grand Jury Unit. If the indictment is signed by the Grand Jury, the case proceeds to the district court; otherwise, the case exits from the system. In the district court, the defendant is arraigned, and if he pleads guilty, the case exits after sentencing without trial. For those who plead not guilty, an attorney is chosen or assigned and preparations are made for the trial.

Felony cases (11,550), many occurring in 1965, were chosen as input data to the model. The computer programs were written in GPSS. Among the observed cases, not a single defendant's case went from the arrest to the arraignment stage in less than six weeks unless the case was dismissed at the preliminary hearing stage or by the Grand Jury. Out of the total of six weeks, approximately five weeks were spent in the Grand Jury Unit. One defendant waited 463 days between being arraigned and entering a guilty plea. An average of thirteen weeks elapsed between arraignment and the beginning of the trial.

Experts in the field of law have observed that none of the actual court appearances, from the time of arrest to the commencement of the trial or the entrance of a guilty plea, takes more than thirty minutes. The average period from arrest to trial of felony cases is four months. The initial results concur closely with the actual results. By simulating the system with a second Grand Jury, sitting part of the time, the time spent in the Grand Jury Unit was reduced from five weeks to less than one week. The additional Grand Jury also reduced the average time from arrest to trial from four months to approximately three months.

At present 55 percent of the defendants plead guilty and only 30 percent of the cases result in a trial. The effect of a decreased number of defendants pleading guilty was simulated, but the results could not be validated because of the absence of any data regarding trial times. An amendment to a *rule* was considered in the simulation. The rule provides restrictions on time lapses between arraignment and filing motions. The adoption of the rule resulted in reducing the average time between arraignment and trial from thirteen weeks to

[5]The study was conducted at the Institute for Defense Analysis and is part of the Report of the Science and Technology Task Force of the President's Commission on Law Enforcement and Administration of Justice.

seven weeks. Additional statistical tests will be performed to certify the validity of the model.

Evaluation of COURTSIM shows that it is an effective model because it considers the procedural, organizational, and behavioristic aspects of the court system and demonstrates that the simulation of the court process is feasible. The extent of the existing data, with the exception of actual trial times, helps to illustrate the model's effectiveness in simulating the present process.

Process Simulations—Police Communications Centers

The growth rate of the urban areas in the United States has been outstripped only by the accelerated crime rate. The result has been an ever-increasing pressure on city police departments to acquire more personnel and more updated equipment and techniques. However, metropolitan governments have been unable to allocate the money required and police are therefore faced with a traditional economic problem—the allocation of scarce resources to maximize benefits. The Chicago Police Department is no exception.

In an attempt to improve both its data collection and its communications system, the Chicago Police Department has installed a communications center that handles all calls requesting police assistance. Each call is recorded on a radio dispatch card and is assigned a special number which is used as a reference number in its data-processing system. As one might expect, the department has a number of questions concerning both efficiency of the present communications center structure and proposed changes. As a result, a number of cooperative but unfunded research and development studies for the Chicago Police Department have been initiated by Northwestern University with the assistance of the Communications Division of Motorola.

The research and development studies undertaken at the police department utilize the systems analysis methodology. To prepare for the series of systems analyses, two activities have been initiated, the development of a SIMSCRIPT simulation for the communications center and the gathering of data on certain key operations where data are not typically available. This simulation will be reviewed in the following paragraphs.

The input-output black box model is the basic model from which the simulation is constructed. That is, a call (input) is made into the communications center, it is processed (black box), and direct action is taken (output).

The simulation has three parts: (1) the generation of calls and the three levels at which they may be answered—primary operator, overload operator, or supervisor, (2) the distribution and evaluation of calls, and (3) the assignment of resources to answer the calls.

There are thus three entities: calls, operators, and cars. Calls are created endogenously according to a normal distribution and are located in an hour according to a uniform distribution. The zone and beat origin of the call and its property are determined by random look-up tables. The length of the call is determined by a negative exponential distribution.

As mentioned, the operators are of three types: primary, overload, or supervisor. A call comes into the center and is switched to the appropriate console. If not busy, the call is answered. If busy, the call is held for twelve seconds with a check made on the queue every two seconds. If twelve seconds elapse, the call is shunted to the overload operator. There are seven overload operators, and if they are busy, the call is held in queue for thirty seconds, checking queue status every two seconds. If the thirty seconds elapse the call is shunted to the supervisor, who determines its priority and acts accordingly.

Other attributes of the entities that have been simulated are:

1. Time it takes console operator to fill out R/D card and hand to primary operator.
2. Time it takes overload or supervisor to fill out R/D card.
3. Time it takes overload or supervisor to go personally to primary operator.
4. Time it takes messenger to transfer to primary operator.
5. Time it takes for all nondispatch calls.
6. Time it takes to get response from patrol cars.
7. Field response time for beat car.
8. Field environment time for high priority.

The actual radio dispatch card call records of the communications center served as the data input to the simulation.

OTHER PUBLIC SYSTEMS

So far the simulations discussed have all been directed at specific entities—a transportation system, or a city—with respect to one factor. Recently other simulations have been developed that are macroscopic in nature; for example, a disaster environmental simulation which studies the allocation of resources during a disaster (Williams and Richie, 1969). It treats the problem of using medical, nonmedical, and transportation resources efficiently in a disaster area encompassing several population centers.

There have been simulations involving mass behavior that operate on a higher level, in terms of both geography and scope. An example of this type involves military conflict and political and economic stability (Rastogi, 1969). A model of a hypothetical underdeveloped country and war situation is formulated and attempts to view a prolonged military conflict in terms of the interrelated psychological and economic factors. The driving force is assumed to be military pressure. The simulation process has emerged as a powerful technique for studying such an interrelated system.

Carried to an even higher level, in a statement of personal beliefs, John McLeod (1969) calls for the development of a World Simulation. This model would include the factors of social, economic, political, and military forces as entities. The method of achieving such a goal would be to improve or develop where necessary the models for subsystems and combine them into a hierarchy of systems to develop ultimately a useful model of the total environment.

These models have a common direction. They all attempt to deal with situations that are made up of other systems. One model ties medical treatment to transportation; the higher-level models tie together social, economic, and political structures. This is the new direction of simulation, the coupling of systems.

CONCLUSIONS

Each area reviewed varies greatly in the extent to which it is backed by a well-developed body of data and theory. The success and validity of each application seem to be highly correlated with the ability to identify entities, attributes, and structures. Transportation is clearly the most concrete and successful field covered in this chapter. Objective function, objectives, and criteria become even worse problems as one moves progressively through urban growth, renewal and land use, natural resources, education, hospitals, and the criminal justice system.

The lack of data is the key to many problems. The lack of clear variables in an understood structure makes simulation a major intellectual challenge, a task made more difficult by the goal in many cases to aid real decision-making. For instance, Allman's (1966) contribution is probably more important in terms of formulation than of execution, even though several railroads are using his approach to making key decisions.

REFERENCES

Alkin, Marvin, "The Use of Quantitative Methods as an Aid to Decision Making in Educational Administration." Paper delivered at the American Educational Research Association Annual Meeting, Los Angeles, February 1969.

Allman, William P., Jr., "A Computer Simulation Model of Railroad Freight Transportation Systems," in *Bulletin I.R.C.A.: Cybernetics and Electronics on the Railways* (February 1967), 45-57.

———, "A Network-Simulation Approach to the Railroad Freight Train Scheduling and Car Sorting Problem." Ph.D. dissertation, Graduate School, Northwestern University, 1966, 234 pp.

Bellman, J. A., "Railroad Network Model," in the International Union of Railways and Canadian National Railways (eds.), *Second International*

Symposium on the Use of Cybernetics on the Railways, Montreal, October 1-6, 1967, pp. 148-54.

Bontadelli, James A., and Colin J. Hudson, "Simulation of Systems Motive Power," *Simulation of Railroad Operations.* Chicago: Railway Systems and Management Association, October 1966, pp. 109-22.

Bower, Blair, M. M. Hufschmidt, and W. W. Reedy, "Operating Procedures: Their Role in the Design of Water Resource Systems by Simulation Analysis," in Arthur Maass (ed.), *Design of Water Resource Systems.* Cambridge, Mass.: Harvard University Press, 1962.

Carroll, Joseph L., "Waterway Lock Simulation Model," *Sixth Meeting Transportation Research Forum,* 1965.

Chapin, F. Stuart, Jr., "A Model for Simulating Residential Development," *Journal of the American Institute of Planners,* 31, No. 2 (May 1965), 120-26.

Chicago, City of, *Basic Policies for the Comprehensive Plan of Chicago,* Department of Development and Planning, Chicago, Ill., 1964.

———, *The Comprehensive Plan of Chicago,* Department of Development and Planning, Chicago, Ill., 1966.

Clough, Donald J., and J. B. Levine, "A Simulation Model for Subsidy Policy Determination in the Canadian Uranium Mining Industry," *Journal of the Canadian Operational Research Society,* No. 3 (November 1965), 115-28.

———, and William P. McReynolds, "State Transition Model of an Educational System Incorporating a Constraint Theory of Supply and Demand," *Ontario Journal of Educational Research,* 9, No. 1 (Autumn 1966).

Coates, P. J., and P. W. Clark, "A Simulation Method Applicable to Railway Operation," in the International Union of Railways and Canadian National Railways (eds.), *Second International Symposium on the Use of Cybernetics on the Railways,* Montreal, October 1-6, 1967, pp. 171-75.

Colley, J. L., J. B. Hallan, and A. H. Packer, "A Model of a Saturated Medical System," in *Proceedings of the 18th Annual Institute Conference and Convention, AIIE,* May 1967.

Cook, Desmond L., "An Overview of Management Science in Educational Research," Symposium on Management Science in Educational Research, 15th International Meeting of the Institute of Management Science, September 1968, 25 pp.

Cope, Robert G., "Simulation Models Should Replace Formulas for State Budget Requests," *College and University Business* (March 1969).

Coupal, R. T., L. L. Garves, and W. R. Smith, "A Digital Computer Simulation of Single Track Railroad Operation," *Electrical Engineering,* No. 79 (August 1960), 648-53.

Crane, R., F. Brown, and R. Blanchard, "An Analysis of a Railroad Classification Yard," *Operations Research,* 3 (November 1955), 262-71.

Crecine, John P., "A Computer Simulation Model of Municipal Budgeting," *Management Science,* 13, No. 11 (July 1967a), 786-816.

———, *Governmental Problem Solving: A Computer Simulation of Municipal Budgeting.* Chicago: Rand McNally & Co., 1967b.

Data Processor, "Railroad without Trains," 11 (August 1968), 12-15.

Davenport, D. Harold, "Crane-Ramp Equivalents at Railroad TOFC Terminal: A Computer Simulation." Master's thesis, Graduate School, Northwestern University, January 12, 1968.

Day, D. J., "Simulation of Underground Railway Operation," *Railway Gazette* (U.K.), 121 (June 1965), 438-41.

de Cock, E., "Simulation-Specimen on Behalf of Wagonload Transport on Netherlands Railways," *Cybernetics and Electronics on the Railways*, 1 (April 1964), 125-30.

DiCesari, F., and J. C. Strauss, "Simulation of an Urban Transportation Transfer Point," *ACM Symposium on Application of Computers to the Problems of Urban Society, New York,* November 10, 1967.

Dodd, K. N., J. Hawkes, and R. P. Muddle, *Digital Simulation of Traffic Flow with Application to Railway Train Movements.* Farnborough, England: Royal Aircraft Establishment, January 1966, 56 pp.

Duke, Richard D., "Gaming Simulation in Urban Research," Institute for Community Development and Services, Michigan State University, 1964.

Eberhardt, J. S., "The Effects of Shorter, More Frequent Trains on Railroad Classification Yards: A Computer Simulation," *Cybernetics and Electronics on the Railways* (September-October, 1966).

Etherington, Edwin D., and Richard F. Vancil, "Systems and Simulations: New Technology Goes to Work on Decision-Making," *College and University Business* (March 1969).

Fairburn, Donald T., "A Simulation Model for Land Use Forecasting in an Urban Area." Ph.D. dissertation, Syracuse University, 1967, 128 pp.

Feiler, A. M., "TRANSIM—A General Purpose Transportation Simulation," *Simulation of Railroad Operations.* Chicago: Railway Systems and Management Association, October 1966, pp. 141-168.

Fetter, R. B., and J. D. Thompson, "Patient Waiting Time and Doctor's Idle Time in an Outpatient Setting," *Health Services Research Journal,* 1,1 (Summer 1966), 66-90.

———, "Predicting Requirements for Maternity Facilities," *Hospitals,* JAHA, 37 (February 16, 1963), 45 ff.

———, "The Simulation of Hospital Systems," *Operations Research,* 13,5 (September-October 1965), 689-711.

Fiering, Myron B., "Revitalizing a Fertile Plain: A Case Study in Simulation and Systems Analysis of Saline and Waterlogged Areas," *Water Resources Research* (U.S.), 1 (April 1966), 41-61.

Firmin, Peter A., *Methods and Statistical Needs for Educational Planning.* Paris: Organization for Economic Cooperation and Development (OECD), 1967.

———, *et al.,* "University Cost Structure and Behavior," Final Report Contract NGF-C451, 214 pp., Tulane University, 1967.

Garfinkel, David, "Simulation of Ecological Systems," in R. Stacy and B. Waxman (eds.), *Computers in Biomedical Research,* Vol. II. New York: Academic Press Inc., 1965.

———, and R. Sach, "Digital Computer Simulation of an Ecological System Based on a Modified Mass Action Law," *Ecology,* 45 (1964), 502-7.

Goldman, Jay, H. Allan Knappenberger, and J. C. Eller, "Evaluating Bed Allocation Policy with Computer Simulation," *Health Services Research,* 3,2 (Summer 1968), 119-29.

Guetzkow, Harold, "Simulation in International Relations," *Proceedings of the IBM Scientific Computing Symposium on Simulation Models and Gaming.* Yorktown Heights, N. Y.: Thomas J. Watson Research Center, December 7-9, 1964, pp. 249-78.

Guins, Sergei G., "Application of the RSRG Terminal Model to Analyze Yard Consolidation in the Toledo Area," *Simulation of Railroad Operations.* Chicago: Railway Systems and Management Association, October 1966, pp. 181-88.

Handyside, A. J., "Simulation of Emergency Bed Occupancy," *Health Services Research,* 2, No. 3 (Fall-Winter 1967), 287-97.

Harris, Britton, "Linear Programming and the Projection of Land Uses," Pennsylvania-New Jersey Transportation Study Paper No. 20, November 9, 1962, p. 1.

Haussman, R. D., and G. J. Rath, "Automatic Teacher Assignment—a GPSS Simulation," *Journal of Educational Data Processing,* II, No. 3 (Summer 1965).

Hill, Donald M., "A Growth Allocation Model for the Boston Region," *Journal of the American Institute of Planners,* 31, No. 2 (May 1965), 111-20.

———, and Norman Dodd, "Travel Mode Split in Assignment Programs, *Highway Research Board Bulletin,* No. 347 (1962), 290-301.

Hogan, J. E., "Train Performance and Tonnage Ratings Calculated by Digital Computer," *Electrical Engineering,* 77 (1958), 424-29.

Hoover, Thomas E., "A Systems Approach to Long Range Facilities Planning in Higher Education." Presented at the TIMS/ORSA Joint Meeting, San Francisco, 1968.

Howard, R. A., E. M. Gould, and W. G. O'Regan, "Simulation for Forest Planning," *Simulation,* 7 (July 1966), 44-52.

Hufschmidt, M. M., "Simulating the Behavior of a Multiunit, Multipurpose Water Resource System," in A. C. Hoggatt and F. E. Balderston (eds.), *Symposium on Simulation Models: Methodology and Applications to the Behavioral Sciences.* Cincinnati: South-Western Publishing Co., 1963.

———, and M. B. Fiering, *Simulation Techniques for Design of Water Resource Systems.* Cambridge, Mass.: Harvard University Press, 1966.

Irwin, N. A., "Review of Existing Land Use Forecasting Techniques." A technical report presented to the Boston Regional Planning Project prepared by Traffic Research Corporation, July 29, 1963.

———, and H. G. von Cube, "Capacity Restraint in Multitravel Assignment Programs," *Highway Research Board Bulletin,* No. 347 (1962), 258-89.

Jennings, Norman H., and Justin H. Dickins, "Computer Simulation of Peak Hour Operations in a Bus Terminal (New York Port Authority Bus Terminal)," *Management Science,* 5, No. 1 (October 1958), 106-20.

Judy, Richard W., and Jack B. Levine, "A New Tool for Educational Administrators," University of Toronto Press, 1965, 33 pp.

———, and Jack B. Levine, "Systems Analysis for Efficient Resource Allocation in Higher Education." Paper presented at the 24th National Conference on Higher Education, 1969.

———, Jack B. Levine, and Richard Wilson, "Systems Analysis of Alternative Designs of a Faculty." Presented to the Organization for Economic Cooperation and Development in Paris, France, 1968.

Kates, J., "Traffic Analysis, Forecast and Assignment by Means of Electronic Computers," *International Road Safety and Traffic Review* (U.K.), 11 (Winter 1963), 28-34.

Kershaw, J. A., and R. N. McKean, "Systems Analysis and Education," the Rand Corporation, RM-2473-FF, October 1959.

Koenig, Herman, "Systems Model for Management, Planning and Resource Allocation in Institutions of Higher Education." Manuscript, n.d.

Lapp, Robert, "A General Hospital Model." Working Paper, Northwestern University, Department of Industrial Engineering and Management Sciences, Evanston, Illinois, 1968.

Lathrop, George T., and John R. Hamburg, "An Opportunity Accessibility Model for Allocating Regional Growth," *Journal of the American Institute of Planners*, 31, No. 2 (May 1965), 95-103.

Lee, J. W., "The Dimensions of U.S. Metropolitan Change," *Looking Ahead*, National Planning Association, June 1967.

Lowry, Ira S., "A Model of METROPOLIS," The RAND Corporation, Santa Monica, California, RM-4035-RC, August 1964.

———, *Migration and Metropolitan Growth: Two Analytical Models.* San Francisco, Calif.: Chandler Publishing Company, 1966, 118 pp.

McEachern, A. W., *et al.* "Simulation for Research and Decision Making in the Juvenile Probation System," *American Behavioral Scientist*, XI (January-February 1968).

McLeod, John, "A Statement of Personal Beliefs and Intentions," *Simulation*, 13, No. 1 (July 1969), vii.

Meir, Richard L., and R. D. Duke, "Gaming Simulation for Urban Planning," *AIP Journal* (January 1966).

Mertz, W. L., "Review and Evaluation of Electronic Computer Traffic Assignment Programs," *Highway Research Board Bulletin*, No. 297 (1961), 94-105.

Nadel, R. H., and E. M. Rovner, "The Use of a Computer Simulation Model for Classification and Yard Design," in the International Union of Railways and Canadian National Railways (eds.), *Second International Symposium on the Use of Cybernetics on the Railways*, Montreal, October 1-6, 1967, pp. 158-62.

Nordell, Lawrence P., "A Dynamic Input-Output Model of the California Educational System," Technical Report No. 25, University of California, Center for Research in Management Sciences, Berkeley, August 1967.

Niemond, Kenneth S., and John Kostalos, "Simulation of a Commuter Railroad," in *Simulation of Railroad Operations.* Chicago: Railway Systems and Management Association, October 1966, pp. 23-52.

Nilsson, Ernst, and David Olsen, "Analysis of Police Response and Preventive

Sub-Systems," a paper presented at the 36th National Meeting, Operations Research Society of America, Miami Beach, Florida, November 10-12, 1969.

Nippert, David, "Simulation of Terminal Operations," in *Simulation of Railroad Operations.* Chicago: Railway Systems and Management Association, October 1966, pp. 169-79.

———, and Sergei G. Guins, "Demonstration of the RSRG Terminal Model," in *Simulation of Railroad Operations.* Chicago: Railway Systems and Management Association, October 1966, pp. 189-92.

O'Neill, R. R., "Simulation of Cargo Handling," in Malcolm (ed.), *Report of the Systems Simulation Symposium*, American Institute of Industrial Engineering, Columbus, Ohio, May 16-17, 1957. Baltimore: Waverly Press, 1957, 7 pp.

Operations Research, Inc., *SST World Route Simulator.* Silver Spring, Md.: Operations Research, Inc., 1964.

Rastogi, P. N., "Protracted Military Conflict and Politico-Economic Stability," *Simulation,* 11, No. 7 (January 1969), 23.

Rath, G. J., "Management Science in University Operations," *Management Science,* Application Series, 14, No. 6 (February 1968), 373-84.

———, Jose M. Balbas Alvarez, Takehiko Ikeda, and O. George Kennedy, "Simulating a Clinical Pathology Department," *Health Services Research Journal,* Spring, 1970.

———, and William Braun, "Systems Analysis of a Police Communication Center," *Proceedings of the Second National Symposium on Law Enforcement Science and Technology,* Chicago, April 18, 1968.

Ray, Paul H., and Richard D. Duke, "The Environment of Decision-Makers in Urban Gaming-Simulation," 1967, mimeographed.

Redwin, Lloyd, "The Promise and Failure of Urban Research," *Planning,* ASPO National Planning Conference, Houston, Texas, April 1967.

Ridley, Tony M., "General Methods of Calculating Traffic Distribution and Assignment," University of California, The Institute of Transportation and Traffic, Berkeley, February 1968, 30 pp.

Robinson, G. H., Paul Wing, and Louis E. Davis, "Computer Simulation of Hospital Patient Scheduling Systems," *Health Services Research,* 3, No. 2 (Summer 1968), 130-41.

Robinson, Ira M., Harry B. Wolfe, and Robert L. Barringer, "A Simulation Model for Renewal Programming," *Journal of the American Institute of Planners,* 31, No. 2 (May 1965), 126-34.

Sager, Alan M., "Decision Making in the U.S. Supreme Court—A Judicial Process Simulation." Ph.D. dissertation, Northwestern University, Department of Political Science, Evanston, Illinois, 1968.

Schenck, Hilbert, Jr., "Simulation of the Evolution of Drainage Basin Networks with a Digital Computer," *Journal of Geophysical Research,* 68 (1963), 5739-45.

Schlager, Kenneth J., "Simulation Models in Urban and Regional Planning,"

Southeastern Wisconsin Regional Planning Conference Technical Record.
2, No. 1 (1964). Waukesha, Wisc.: Southeastern Wisconsin Regional
Planning Commission, 1964.

Schroeder, R. G., and G. J. Rath, "The Role of Mathematical Models in
Educational Research," *Psychology in the Schools,* II, No. 4 (October
1965).

Smock, Robert, "An Iterative Assignment Approach to Capacity Restraint of
Arterial Networks," *Highway Research Board Bulletin,* No. 347 (1960),
60-66.

Soberman, Richard M., "A Railway Performance Model for Developing
Countries," in *Transport Research Program, Discussion Paper No. 45.*
Cambridge, Mass.: Harvard University Press, June 1967, 32 pp.

Stade, Marinus, "Cost Effectiveness of Water Bombers in Forest Fire Control,"
Journal of the Canadian Operational Research Society, 5 (March 1967),
1-8.

Steger, Wilbur A., "The Pittsburgh Urban Renewal Simulation Model," *Journal
of the American Institute of Planners,* 31, No. 2 (May 1965), 144-50.

Szekely, Miguel, Martin Stankard, and Roger Sisson, "Design of a Planning
Model for an Urban School District," in *Socio-Economic Planning Science,*
Vol. 1, pp. 231-42. Great Britain: Pergamon Press, 1968.

Taylor, Jean G., and Joseph A. Navarro, "Simulation of a Court System for the
Processing of Criminal Cases," *Simulation,* 10 (May 1968), 235-41.

Traffic Research Corporation, "Development and Calibration of the EMPIRIC
Land Use Forecasting Model for 626 Traffic Zones," February 1967.

U.S. Maritime Administration, *User Manual for the MARAD Fleet Operations
Simulation-TR8,* April 1964.

U.S. National Bureau of Standards, "Computer Model Simulates Northeast
Corridor: NBS Aids System Planners," *Technical News,* November 1968,
8 pp.

Vaswani, Ram, "Intracity Trailer Movement," *Operations Research Quarterly*
(U.K.), 7 (September 1956), 91-96.

Vogelback Computing Center, *A Description of SPURT, A Simulation Package
for University Research and Teaching,* Northwestern University, August
1968.

Weathersby, George, "The Development and Application of a University Cost
Simulation Model." Master's thesis, University of California, Department
of Business Administration, Berkeley, June 1967.

Williams, Lewis H., and William C. Richie, "A Disaster Environmental Simulator," *Simulation,* 13, No. 3 (September 1969), 121.

Wilson, P. B., and D. C. Lach, "Computer Simulation Developments in Canadian
Railways," in the International Union of Railways and Canadian National
Railways (eds.), *Second International Symposium on the Use of Cyber-
netics on the Railways,* Montreal, October 1-6, 1967, pp. 176-82.

Yates, Jules David, "Community Interaction Game." Cambridge, Mass.: The
Simulation Corporation, December 27, 1967.

Zettel, Richard R., and Richard R. Carll, "Survey of Study Methods," *Summary Review of Major Metropolitan Area Transportation Studies in the United States.* University of California, Institute of Transportation and Traffic Engineering, Berkeley, November 1962, pp. 44-65.

A COMPUTER SIMULATION MODEL
OF MUNICIPAL BUDGETING

JOHN P. CRECINE

An increasing awareness of the importance of urban governmental activities draws attention to municipal expenditures and to the municipal operating budget. Little effort has been directed toward the question of how municipalities allocate resources. The work reported here represents an attempt to develop a positive theory of municipal resource allocation for large metropolitan communities and to subject this theory to empirical tests. Cleveland, Detroit, and Pittsburgh are used as data points.

A POSITIVE, EMPIRICAL THEORY
OF MUNICIPAL BUDGETING

The model is stated in the form of a computer program. The nature of the budgetary decision process suggests such an approach. Even a superficial examination of the municipal resource allocation procedure indicates that it is the result of a *sequence* of decisions—departmental requests, mayor's executive budget, and final council appropriations. A computer program is really a collection of *instructions* executed in a *specific* sequence. If a computer program is to be an appropriate way to describe the budgetary process, the individual rules ought to be stable over time and executed in a specific sequence (be part of a "stable" structure). Fortunately, there are some compelling reasons why this should be so.

One obvious reason why programmed decisions tend to deal with repetitive problems, and *vice versa,* is that "if a particular problem recurs often enough, a routine procedure will usually be worked out for solving it." Certainly the municipal budget is a recurrent problem (yearly). Evidence is growing that recurrent, complex problems are solved by individuals by breaking the global problem into a series of less complex ones, and then solving the simplified problems sequentially.

Reprinted with permission from *Management Science,* 13 (July 1967), 786-815. Copyright©1967 by The Institute of Management Sciences. Based on a paper presented at the 1966 Midwest Political Science Association Meeting, Chicago, Illinois. The research reported here was undertaken while the author was a Ford Foundation Fellow, Richard D. Irwin Foundation Fellow (at Carnegie Institute of Technology), and staff member of the Institute of Public Administration, The University of Michigan. The author gratefully acknowledges this generous assistance.

In particular, the problem-solving behavior of individuals has been described using a computer program, for a trust investment officer by Clarkson (1962), a department store buyer by Cyert, March, and Moore (1963), laboratory subjects solving simple problems (Newell and Simon, 1963a and 1963b), and chess players by Newell, Shaw, and Simon (1963).

Computer Simulation of Municipal Budgeting

A simulation of the budgetary process must describe the behavior of many individuals—department budget officers, budget officials in the mayor's office, and the council. There is no reason to think that the hundred or so actors involved in the formal budgetary decision system will be any more difficult to "program," or simulate than *single individuals,* however. The difficulty, if any, arises from the number of decisions and decision makers in our model, and the quantity of data to be analyzed.[1]

Simulation as a research tool. "Simulation is a technique for building theories that reproduce part or all of the output of a behaving system" (Clarkson, 1962, p. 16). In addition, some simulation models have the goal of reproducing not only final results but intermediate outputs as well. This is the task to which the model addresses itself. The attempt to reproduce output and procedures is in the form of a computer program representing the structural form of the decision process (sequence of decisions), the functional form of the individual decision rules (individual equations representing actual decision rules), and the decision parameters (values of "constants" or empirically determined variables embedded in the structure and functional relations of the model).

OVERVIEW OF MUNICIPAL BUDGETING

The entire decision process can usefully be thought of as an organized means for the decision maker to deal with the potential complexity of the budgetary problem. The most prominent feature of the "original" problem in terms of its contribution to complexity is an externally imposed constraint of a

---

[1]The implications of the "magnitude of the problem" are many. First, it should be fairly obvious that each actor in our simulation model will be described in a simpler manner than the individual problem solvers in most of the works cited above. Secondly, assumptions will have to be made which will detract from the overall accuracy (i.e., completeness) of the model. For example, it will be necessary to assume that each department head in the system behaves according to the same decisional model, with only parameters changing. It is obviously not practical or reasonable to interview all department heads and all parties involved in the budgetary process. Behavioral rules attributed by others to our decision makers will have to be incorporated in the model without individual verifications. The reasonableness and "accuracy" of these necessary "short cuts" will, of course, be measured empirically when the model is tested.

balanced budget[2] —by requiring that, *at some level of generality,* all budget items be considered simultaneously.

Problem Perception

Before proceeding, we should note that the "problem" we are referring to is the budgetary problem as seen by the actual decision makers (department officials, mayor and mayor's staff, and council members). It is quite clear (from interviews) that decision makers *do not* see the problem as one of optimally balancing community resources, allocating funds among functions to achieve overall community goals, and the like. The problem is generally "seen" by department heads as one of submitting a budget request that 1) assures the department of funds to carry on existing programs as part of a continuing attack on existing problems, 2) is acceptable to the mayor's office, and 3) provides for a reasonable share of any increase in the city's total budget, enabling the department to attack new problems (if any). The mayor's problem is largely one of recommending a budget that 1) is balanced, 2) at least maintains existing service levels, 3) provides for increases in city employee wages if at all possible, and 4) avoids tax increases (especially property tax increases, in the belief that increased property taxes cause business and industry to move from the city, reducing the municipal tax base). If, after achieving some of the above objectives, the mayor has "extra" funds, they will be used to sponsor programs or projects the mayor has on his "agenda," or to grant a portion of departments' supplemental requests.

The "problem" for the council is to review the mayor's budget recommendations and check for "obvious" errors and omissions. Because of the complexity and detail in the mayor's budget and lack of council staff, the council's options are limited largely to approving the mayor's budget. The requirement of a balanced budget means that a change in one expenditure category, for instance, implies a balancing change in the other account categories, administrative units, or revenues—i.e., one change in the budget (by council) implies many changes which the council has neither the time nor staff to consider.

Partitioning the Problem into Manageable Subproblems

One of the ways municipal decision makers deal with the *potential* complexity of the municipal resource allocation problem is through their necessarily simplified perception of the problem as discussed above. Other simplifying heuristics observed were:

1. The operating budget is treated separately from the capital budget as a generally independent problem. The only behavioral connection

[2] Required in the city charter, articles of incorporation, or by the State Legislature.

between the operating and capital budgets is the "logical" elaboration of capital budgeting decisions in the operation budget.[3]

2. The budget is formulated within a system of administrative units (departments and bureaus) and account categories (salaries, supplies and expenses, equipment, etc.) that is extremely stable from year to year. This partial structuring of the problem "allows" most of the decision makers to treat the appropriation question for one account category in one administrative unit as a (sub-) problem, separate from the overall resource allocation problem. Thus, the overall problem is transformed into a series of smaller problems of determining appropriations for individual departments.

3. The revenue estimates are generally separate from expenditure estimates. That is, estimates of yields from a given tax are treated independently from expenditures. While, on occasion, tax rates may be adjusted somewhat on the basis of preliminary calculations of *total expenditure estimates,* in order to balance the budget, *tax yield estimates* are seldom manipulated to achieve a balance.

4. The structure of the decision process itself represents a division of labor between department heads, the mayor's office, and the council—reflecting not only the administrative hierarchy, but a set of simplifying heuristics for making a complex problem manageable.

5. Finally, an additional simplifying policy is found in all cities investigated. The presence of a uniform wage policy which maintains relative positions of employees within a city-wide civil service pay scale, eliminates the potentially complex problem of deciding wage rates on an individual basis while attempting to maintain "similar-pay-for-similar-jobs" standards.[4]

Governing by Precedent

Perhaps the overriding feature of the mayor's budgetary "problem" is the balanced budget requirement. *If* the mayor took even the majority of items in the budget under serious consideration, his task would be enormous. The requirement of a *balanced* budget *could* mean that not only would the mayor have to consider every budget item, but he would have to consider each item relative to all other items. Somehow the entire level of police expenditures would have to be justified in light of the implied preemption of health department services, public works, fire department expenditures, etc. Obviously the

---

[3]The "legitimate" claim on operating funds by the capital budget is reflected in the following, found in the Mayor's Message accompanying the 1965 Pittsburgh Budget:

"A big item in the Lands and Buildings request pertains to the opening and operation of the new Public Safety Center next Spring. . . . There is a non-recurring expenditure of $150,000 for new furniture . . . and $91,000 is sought for maintenance personnel."

[4]On occasion, uniformed policemen's and firemen's salaries are treated separately from the others.

mayor does not have either the staff, cognitive abilities, or time to undertake such a study—even if the necessary knowledge and information existed.

Instead, as we have seen above, the mayor perceives this year's budget problem as basically similar to last year's with a slight change in resources available (new revenue estimates) for dealing with a continuing set of municipal problems (police and fire protection, urban renewal, public works, transportation) augmented by a small number of newly emerging problems and a small number of partial solutions to old problems. In this context, a "logical" way to proceed in solving the complex budgeting problem is to take "last year's solution" (current appropriations) to the problem and modify it in light of *changes* in available resources and shifts in municipal problems to obtain "this year's solution." This, of course, means that the budget is a slowly changing thing, consisting of a series of "marginal changes" from previous budgets.[5] Only small portions of the budget are reconsidered from year to year, and consequently, once an item is in the budget, its mere existence becomes its "reason for being" in succeeding budgets.

This "government by precedent" is an integral part of most positive models of decision making in the literature. Cyert and March's *A Behavioral Theory of the Firm* describes the usage of previous solutions and solution procedures to solve new problems and is largely a model of *incremental* adaptations of economic organizations to their internal and external environment (Cyert and March, 1963, p. 104). Braybrooke and Lindblom argue that "precedent" is justified and defensible as a "rational" decision strategy (Braybrooke and Lindblom, pp. 225-245). Wildavsky emphasizes the role of "precedent" as an "aid to calculation" in the Federal budgetary process (Wildavsky, 1964, pp. 13-18, 58-59).

Openness of Public Decisions

A basic property of decision making in the public sector (vs. the private) is the realization that both decisions and decision procedures are always subject (at least potentially) to public scrutiny. Decisions in the public sector would tend to be more "defensible" than corresponding ones in the private sector, and each particular decision (budget item) in a decision system (entire budget) ought to be able to stand on its own "merits." In addition, decision *procedures* are also subject to public question. We would argue that the openness of public decisions reinforces the use of rather straightforward methods of partitioning the budgetary problem, the use of *precedent* as a defensible[6] decision strategy, and encourages the use of simpler, easier-to-understand decision procedures than might otherwise be found.

---

[5]These notions are very similar to those of "disjointed incrementalism." (Braybrooke and Lindblom, 1963; Lindblom, 1959).

[6]We would also argue that, in general, the need for "defensible" decisions leads to more conservative decisions in the public sector than in the private.

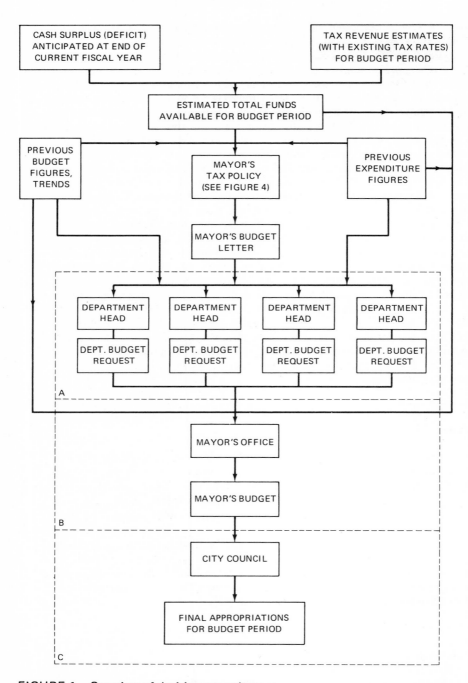

FIGURE 1   Overview of decision procedures.

FORMAL MODEL OF MUNICIPAL BUDGETING

In the context of the problem complexity and devices used to deal with that complexity, we now turn to an analysis of the model's behavioral characteristics. An overview of the model is found in Figure 1. Inasmuch as the model can be broken down into three reasonably independent submodels (the existence of these submodels illustrates the use of partitioning and division of labor in dealing with complexity), we will discuss each submodel separately.

Scope of Model

The formal computer model explicitly considers three decision processes—departmental requests as formulated by the various department heads in city government, mayor's budget for council consideration, and the final appropriations as approved by city council. These three processes are indicated by areas A, B, and C in Figure 1. The outputs of the departmental-request submodel are inputs to the mayor's budget submodel and outputs of the mayor's submodels are inputs to the council appropriations submodel.

The outputs of each submodel correspond quite closely in number and level of detail with the outputs (or decisions) found in the municipal budgetary process. In the model, each department included in the general fund or operating budget has requests for appropriations for each of 2 to 5 standard account categories—depending on the city involved. For example, the model produces at each of the 3 stages of the decision process, the following dollar estimates for the City Planning Department:

Cleveland	*Detroit*	*Pittsburgh*	
Personal Services $X Materials, Supplies, Expenses, Equipment, Repairs and Improvement } $Y	Administrative Salaries } Non-Adm. Salaries } $X Materials, Supplies and } Expenses } $Y Equipment and Repairs $Z	Administrative Salaries	$X
		Non-adm. Salaries	$Y
		Materials, Supplies and Expenses	$W
		Equipment	$U
		Maintenance	$Z

Forty-four to 64 departments and administrative units are involved in the cities examined, with each unit having estimates for 2-5 accounts. Between 128 and 220 decisions are produced at each of the three stages of the model, for each year in the study, in each of the three cities examined.

At this point, one might legitimately ask two important questions:

1. Why are accounts categorized in the manner indicated?
2. Why is "dollars" the unit of resource allocation rather than men, number of street lights, etc.?

Both questions are "crucial" ones for a normative theory of budgeting. In a positive theory, however, the answers are rather straightforward—and essentially identical. People interviewed in all three cities think and talk in terms of "dollars"; they differentiate (at least in interviews) expenditures in terms of the same categories used in their city's accounting system. Apparently, dollar amounts provide the relevant reference points for dealing with the conceptual framework provided by the city's accounting system[7] and provide the basis for the participants' cognitive maps of the process.

DEPT. Submodel

Role—The role of the Department head is similar to that of the agency or bureau chief in the Federal government as described in the Wildavsky study (1964, pp. 8-21). His objective is to obtain the largest possible amount of funds for his department and his purposes. Just as "'Washington is filled'. . . 'with dedicated men and women who feel that government funds should be spent for one purpose or another,'"[8] so are municipal governments. In general, department heads, through experience and the process of socialization into their positions, and by "learning" that their request is likely to be cut by the mayor's office or council, tend to ask for more than they "expect" to get. This "padding" of the budget is one part of a system of mutual expectations and roles. Department heads are expected to ask for more than they really "need"; the mayor's office is expected to cut these requests in order to balance the budget.

Context—The decisions we are speaking of set the limit on spending for the coming fiscal year. They are limits on manpower, supplies, material, and equipment. They are not program budgets in the sense that exact activity mixes are included in the municipal budget. In a sense, what we are talking about is an intermediate decision. This decision provides the constraints under which decisions about particular activities that a department will undertake must be made. The setting of *levels* of expenditures is just one part of the department head's continuing problems. Within a given expenditure ceiling, many different activity mixes can be utilized. "Low ceilings, in short, can still permit several rooms (Sorensen, p. 414)."

DEPT. model characteristics. The role of the mayor's budget letter and the budget forms sent to the department head is a clear one. Together with the time schedule for submission of the completed budget forms, these items have the effect of structuring the department head's problem for him. Budget forms are typically sent to department heads less than two months in advance of the presentation of the completed budget to council. The department head usually

[7]A legitimate question would be, "Why was a city's accounting system designed around a particular set of account categories?" This, however interesting and important, is beyond the scope of study.

[8]A quote of President John F. Kennedy (Sorensen, 1965, p. 414).

has about one month before his completed request forms are due in the mayor's office.

The importance of the time deadline should not be underestimated. In that there is no moratorium on the department head's problems, budget compilation represents an additional workload. In the context of a myriad of nonbudgetary problems and duties, most department heads are more than willing to accept the problem structure provided by the budget forms. To do otherwise would not only involve creating an alternative structure, but would place the "burden of proof" on the department head as far as justifying the alternative to the mayor's office.

Just how is the problem presented to the department head so as to pre-structure it for him?

Budget forms—Budget forms seem to be nearly one of the physical constants of the universe (Wildavsky, 1964, p. 59). They are laid out as follows:

	Expenditures Last Year	*Appropriations This Year*	*Next Year's Request*
Standard Account 1	$54321.00	$57400.00	?
Itemization of 1	—	—	
Standard Account 2	$43219.00	$45600.00	?
Itemization of 2			
⋮			
Standard Account N	$100.00	$120.00	?
Itemization of N			

By structuring the department head's problem, the forms "bias" the outcome or decision in two ways:

1. They provide a great deal of incentive for the department head to formulate his requests within the confines of the existing set of accounts.
2. They provide for an automatic comparison between "next year's" request and "this year's" appropriation—which automatically determines that "this year's" appropriation provides one criterion or reference point for "next year's" request.

The Mayor's budget letter always contains instructions which reinforce the structuring of the problem provided by budget forms—to provide a ". . . written explanation for any change in individual code accounts," "(e)xperience for the years 1962 and 1963 is shown . . . to assist you in estimating your needs for 1965," "(u)nder the heading 'Explanation of Increases and Decreases' must be explained the factors . . . which make up the increase or decrease over or under," "the current budget allowance is shown above on this form."

The level of detail in line items has its influence on the department head's decision process also. (In one city studied, one of the line items listed a $3.00 current appropriation for "Mothballs.") In general, each item broken out in the budget (each line item) "forces" one historical comparison and, hence, represents one more constraint the department request must satisfy. In the face of an increasing number of constraints (increasing as budget detail increases), it is not so surprising that the department head resorts to simpler decision rules to handle this potentially difficult problem. In addition, we would predict that the more detailed the budget (in terms of line items), because of the structure of the budget forms, the less change in requests (and appropriations) from year to year.

The need for effective budgetary control in the mayor's office, made more difficult by the presence of a small staff,[9] (small in relation to a similar organization in the private sector), is met by a large number of simple historical comparisons and has, in many instances, resulted in a burdensome amount of detail — responded to by busy department heads with little change in budget behavior from year to year.

The "tone" of the letter accompanying the budget forms has the effect of providing an arbitrary ceiling on the department's request (Figure 2, item 5). If the department total exceeds the "ceiling," the overage is generally submitted as a "supplemental" request (Figure 2, items 7 and 8). In addition, changes in salary *rates* through raises or promotions are submitted as a supplemental request (or not at all). Supplemental requests are accompanied with a detailed explanation and are treated separately by the mayor's office—and are always on the "agenda" when the department head meets with the mayor's office to discuss his requests.

So far, we have discussed only the constraints a department head must satisfy and the procedures he must follow. There is, obviously, some room for maneuvering. Many of the department head's "calculations" involve figuring "what will go" with the mayor's office.[10] This calculation involves using current appropriations as a base and adjusting this amount for recent appropriation trends, discrepancies between appropriations and corresponding expenditures, and the like (Figure 2, item 3). The results of this "calculation" are then tested to see if they satisfy the constraints discussed above. Preliminary decisions are then adjusted until constraints are satisfied, and the final request is entered on the standard budget forms and sent to the mayor's office for consideration.

Behavior not included in formal DEPT. model. A quick look at the DEPT. model would indicate that (at least according to our theory) department budgetary behavior varies from department to department only by the relative weights assigned to previous appropriations, trends, and expenditures by the various department heads (Figure 2, item 3). Furthermore, it is contended (by the model) that these relative weights are stable over time. Missing from the formal model are notions of non-regular innovation (or change) by department administrators and notions of the department as a mechanism for responding to particular kinds of complaints from the citizenry—in short, the department is conceived of as explicitly responding to only the mayor's pressure. Also missing are changes in the budget requests as logical elaborations of other policy

---

[9]For instance, in the City of Pittsburgh, no more than four people examine the entire budget in great detail. Of these four, at least one is faced with the purely physical task of putting the budget together, checking, and compiling city totals.

[10]Similar to Wildavsky's observations of department heads at the large end of the budgetary funnel (1964, pp. 25-31) and Sorensen's at the small end of the federal decision funnel (1965, p. 414).

1. Budget letter and Budget Forms received from mayor containing: a. current appropriations for all account categories in the department; b. current total appropriation; c. previous year's expenditures in various account categories; d. estimate of allowable increase over current appropriations implied from the "tone" of the mayor's budget letter.

2. Trend of departmental appropriations— direction and magnitude of recent changes in amounts of appropriations in departmental account categories.

3. Department, using information from 1. and 2., formulates a "reasonable-request" for funds in its existing account categories, using current appropriations as a "base" or reference point and adjusting this estimate according to whether there was an increase in appropriations last year (for some accounts, an increase for the current year means a decrease for next year—equipment—, for others, an increase for the current year indicates another increase next year), and the difference between last year's expenditures and appropriations.

4. Using "reasonable requests" calculated in 3., a preliminary department total request is calculated.

5. Is the total department request outside the guidelines set by the mayor's office (implied from the "tone" of the mayor's budget letter)?

   no   yes

6. Check to see if there are any increases in salary accounts over current appropriations

7. All department requests in all categories are adjusted so that any increase (proposed) over current appropriations is submitted as a supplemental request. Go to 6. to check for salary increases.

   no increase   increase

8. Make regular request equal current appropriations and put increase in as supplemental request.

9. Calculate total of regular departmental request.

10. Send regular requests and departmental total to mayor's office along with supplemental requests.

FIGURE 2   General DEPT. request decision process. For a more detailed flowchart of the DEPT. Submodel and a listing of the FORTRAN II computer program, see Crecine (1969).

commitments–implied increases in operating budget because of capital budgeting considerations, changes in intergovernmental support for services (the classic problem in this category involves the highly volatile state-local split of welfare payments), transfer of activities to (and from) other governing units (transfer of hospital system to State or country, etc.), and changes in activity level and scope because of funds obtained from sources (Urban Renewal planning and demonstration grants, the Federal Anti-Poverty Program, etc.) other than the general fund. Our model does not preclude innovative behavior, however. It merely states that innovation (if any) takes place within a regularly changing budget ceiling. It could be argued that a system of weights attached to current appropriations, trends, etc., that leads to relatively large, regular request increases represents a greater potential for innovation than do those leading to smaller increases (or decreases)–providing, of course, that a portion of the request is granted. On the other hand, it could be argued that the presence of a budget ceiling in the face of changing citizenry needs and pressures (precipitating a change in department goals and program needs) forces a department head to "innovate" to survive. Cyert and March, citing the work of Mansfield (1961), side with the former concept of innovation rather than the latter. They argue that the presence of "organizational slack" (evidenced by budgetary increases) ". . . provides a source of funds for innovations that would not be approved in the face of scarcity but that have strong subunit support." Major technological innovations, it is argued, are not problem-oriented innovations (Cyert and March, 1963, p. 279). At any rate, our model does not restrict certain kinds of innovation-producing behavior. The model is, however, unable to predict or recognize the acceptance of "major" innovations (major changes in expenditure and appropriations).

The other "charge" the model is open to is that it fails to deal with "outside" influences at all. This is particularly true if by departmental responses to pressure one assumes that total (for the department) external pressure and influence is a thing that varies a good deal from year to year and that mechanisms for responding to that pressure would lead to irregular budget decisions reflecting this variation. If, however, one assumes that each department has, over the years, "made its peace" not only with the mayor's office, but with the extra-governmental environment, then the pressure response mechanisms (i.e., constant responses to constant pressures) would also be reflected in the system of weights, above. The model does not exclude a pressure-response kind of budgetary behavior, but has a good deal to say about the nature and context of the response (and pressure). The reasonableness of our characterization of "innovation" and "pressure" is reflected in the model residuals.

MAYORS Budget Recommendation Model

Role–The function of the mayor's office relative to the budget is to fulfill the legal obligation of submitting a balanced budget to the city council for its consideration. The key word, of course, is "balanced." Most of the problem solving activity and behavior in the mayor's office revolves around attempts to eliminate a deficit or reduce a surplus. Like most other organizations, subunit requests (stated needs) almost always exceed available resources. So, *vis-à-vis* the departments, the mayor's office's role is that of an economizer, cutting departmental requests to the "bare minimum" in lean years and keeping the cost of government "under control" when revenues are more plentiful.

Characteristics of the MAYORS model. The decision process in the mayor's office can usefully be thought of as a search for a solution to the balanced-budget problem. In a sense, the mayor has guaranteed the existence of a solution through use of budget guidelines set up in his letter of instruction to department heads. Approximately four months before the final budget is due for council passage, the mayor obtains preliminary revenue estimates from people in city government and from an outside source. Armed with a rough estimate of money available for expenditures in the next budget period, current appropriations, and a knowledge of "required" and predetermined budgetary changes for the coming year, the mayor is able to make a rough guess of the total allowable increase or decrease over current appropriations. From this figure, an estimate of the "allowable" percent increase (or decrease) is made and transmitted to department heads *via* the budget letter. (Only the output from this part of the process is explicitly included in our model—"tone of mayor's letter.") In most instances, then, the "sum" of the budget requests reaching the mayor's office represents a "nearly" (within 10%) balanced budget.

The revenue estimate enters into the process at this point as an independent constraint to be satisfied. On very few occasions are revenue or tax *rates* changed to bring the budget into balance. In the municipalities investigated there was no evidence of any altering of tax *yields* to balance the budget.[11] Almost all tax rate increases are tied to general wage increases. Our formal model does not include the part of the decision process evoked when the revenue constraint becomes so restrictive (or loose) as to necessitate a change in tax rates. (See Figure 4.) Tax rate decisions are made prior to sending the budget letter to department heads and are considered as given from that point on.

Just as the budget forms and account categories structure the problem for the department head, they also structure it for the mayor's office (Figure 3, items 1 and 3). The legal requirement of a balanced budget also helps structure the problem for the mayor's office and partially determines its role behavior. Together the system of accounts and balanced budget requirement specify the cognitive map of the decision situation for mayor's office participants.

[11]One exception to the general rule that there is no alteration of the revenue yield estimates (revenue side) to achieve balance with expenditures, was found. For a couple of years in Detroit (1960-62), part of the cost of government operations was financed through "overly optimistic" revenue estimates which ultimately resulted in operating deficits. Those deficits (technically illegal) were then refinanced, with debt service charges for this refinancing showing up in subsequent operating budgets as deductions from revenue available for general fund expenditures. This brief "operating practice" was quickly discontinued by a new city administration. The magnitude of the effect of this practice on the planning process (budget formation) is unclear and is not incorporated in the formal model. The effects were reflected as larger deviations of model estimates from actual decisions during particular years in the City of Detroit.

General MAYORS Budget Recommendation Model*

1.   Department regular and supplemental budget requests received

2.   Latest Revenue Estimate

3.   Historical Data— Current appropriations, last year's expenditures, and appropriation trends

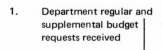

4.   Preliminary check of all departmental requests—if departmental request is less than current appropriations, it is tentatively accepted; otherwise a tentative "calculation" of the mayor's recommendation is made based on the department's regular and supplemental requests together with the change in appropriation from last year to the current year and the last available expenditure data.

5.   Preliminary calculation of total budget—sum of preliminary calculations

6.   Check of preliminary total against revenue estimate to determine if a surplus or a deficit is anticipated. If "surplus," a set of "surplus reduction" routines is evoked. If "deficit," "deficit elimination" routines are evoked.

surplus reduction procedures

deficit elimination procedures (Go to 15).

7.   Calculate magnitude of anticipated surplus or residual.

8.   Find total salaries and wages for the city (preliminary estimates).

9.   Is the anticipated surplus large enough to finance a minimum salary increase?

yes   no

10.   If so, increase salary levels for all departments and reduce calculated surplus

11.   Is there enough anticipated surplus left to distribute among departments?

FIGURE 3   General MAYORS budget recommendation model. For a more detailed flowchart of the MAYORS Submodel and a listing of the FORTRAN computer program, see Crecine (1969).

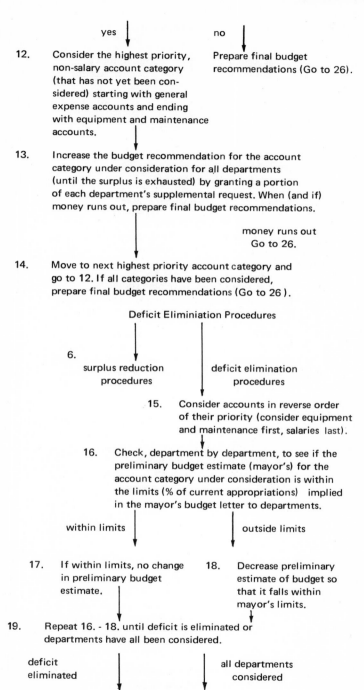

yes   no

12.   Consider the highest priority,   Prepare final budget
non-salary account category   recommendations (Go to 26).
(that has not yet been con-
sidered) starting with general
expense accounts and ending
with equipment and maintenance
accounts.

13.   Increase the budget recommendation for the account
category under consideration for all departments
(until the surplus is exhausted) by granting a portion
of each department's supplemental request. When (and if)
money runs out, prepare final budget recommendations.

money runs out
Go to 26.

14.   Move to next highest priority account category and
go to 12. If all categories have been considered,
prepare final budget recommendations (Go to 26).

Deficit Eliminiation Procedures

6.

surplus reduction   deficit elimination
procedures   procedures

15.   Consider accounts in reverse order
of their priority (consider equipment
and maintenance first, salaries last).

16.   Check, department by department, to see if the
preliminary budget estimate (mayor's) for the
account category under consideration is within
the limits (% of current appropriations) implied
in the mayor's budget letter to departments.

within limits   outside limits

17.   If within limits, no change   18.   Decrease preliminary
in preliminary budget   estimate of budget so
estimate.   that it falls within
mayor's limits.

19.   Repeat 16. - 18. until deficit is eliminated or
departments have all been considered.

deficit   all departments
eliminated   considered

FIGURE 3 (continued)

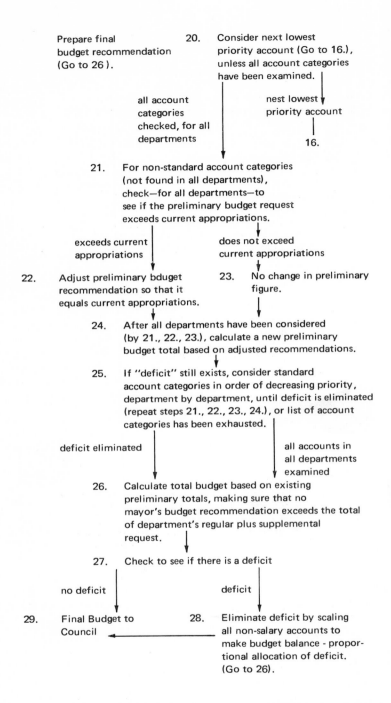

Prepare final
budget recommendation
(Go to 26).

20. Consider next lowest
priority account (Go to 16.),
unless all account categories
have been examined.

all account
categories
checked, for all
departments

nest lowest
priority account

16.

21. For non-standard account categories
(not found in all departments),
check—for all departments—to
see if the preliminary budget request
exceeds current appropriations.

exceeds current
appropriations

does not exceed
current appropriations

22. Adjust preliminary bduget
recommendation so that it
equals current appropriations.

23. No change in preliminary
figure.

24. After all departments have been considered
(by 21., 22., 23.), calculate a new preliminary
budget total based on adjusted recommendations.

25. If "deficit" still exists, consider standard
account categories in order of decreasing priority,
department by department, until deficit is eliminated
(repeat steps 21., 22., 23., 24.), or list of account
categories has been exhausted.

deficit eliminated

all accounts in
all departments
examined

26. Calculate total budget based on existing
preliminary totals, making sure that no
mayor's budget recommendation exceeds the total
of department's regular plus supplemental
request.

27. Check to see if there is a deficit

no deficit

deficit

29. Final Budget to
Council

28. Eliminate deficit by scaling
all non-salary accounts to
make budget balance - proportional allocation of deficit.
(Go to 26).

* For a more detailed flowchart of the MAYORS Submodel and a
listing of the FORTRAN II computer program, see [3].

Preliminary screening of requests. As budget requests are received from departments by the mayor's office, they are screened individually (Figure 3, item 4). The screening process reflects particular biases and relationships between the mayor's office and individual department heads (and departments). "Department heads are dealt with differently during the (budget) hearings. Some department heads can be depended on for an honest budget request. Others have a history of being less-than-realistic in their budgets."[12] Different perceptions of different departments are reflected in both model structure and model parameters (Figure 3, item 4). The interaction of perceptions and role (to cut requests) describes the preliminary screening process.

Basically, if the department request for a given account category is less than current appropriations, a preliminary, automatic acceptance of the request is made. If the request is larger than current appropriations, a request evaluation procedure is evoked that "calculates" or subjectively determines preliminary appropriation figures (Figure 3, item 4). A particular department can evoke one of four subjective evaluation procedures. The procedure evoked represents the cognitive map used by the mayor's office in dealing with that department.

The four basic procedures consist of two which arrive at a preliminary appropriation figure by making marginal adjustments in the department's request figures—representing departments that submit "honest" or "realistic" budget estimates—and two which make adjustments in current appropriations to arrive at preliminary recommendation figures—representing less "realistic" or "honest" departments. The choice of procedures and parameter values was made on the basis of empirical tests using regression models. The four models used were:

i) department head's request respected and adjusted by his supplemental request and current trends

ii) department head's request ignored, and current appropriations adjusted to reflect recent trends and over or underspending in the past

iii) department head's request used as a basis for calculation and changes in it are based on the magnitude of the requested change in appropriations, supplemental requests, and past change in appropriations

iv) department request ignored and change from current appropriations based on previous changes and magnitude of underspending or overspending in the past

The values of the estimated parameters represent the relative weights given to variables in the particular model by decision makers in the mayor's office.

From the preliminary screening of requests outlined above (Figure 3, item 4), a preliminary budget total is compiled (Figure 3, item 5).

The next step in the process is to balance the preliminary budget. The "directives" issued by the Mayor's Office in the budget letter to department heads may be viewed as devices for guaranteeing that the budget will be "nearly" balanced. All alterations in regular departmental requests are aimed at balancing the budget. "Balancing techniques" are:

[12]November, 1964 interview with chief budget officer in one of the three sample cities. Name withheld on request.

1. Raise tax rates or add a new tax to eliminate anticipated deficit.
2. Cut "lower priority" account categories (maintenance, equipment) to bring expenditures into line with revenues.
3. Grant some supplemental requests to reduce anticipated surplus.
4. Eliminate an "undesirable" tax or reduce tax rates to reduce anticipated surplus.

In general, strategies 1 and 4 are used when the anticipated discrepancy between revenues and expenditures is high, while techniques 2 and 3 are used if revenues and expenditures are reasonably close. The general tendency is to move toward a balance between revenues and expenditures by changing either revenue *or* expenditures, but not both. Only "techniques" 2 and 3 are a formal part of the model.

Surplus elimination procedures. If a surplus is anticipated, several standard spending alternatives are considered in order of their priority:

1. General salary increase (Figure 3, items 8 to 10)
2. Grant portion of supplemental requests (Figure 3, items 11 to 14)
   a. general expense accounts (Figure 3, item 12)
   b. equipment accounts
   c. maintenance accounts

Although the formal model includes only the above alternatives, others are clearly evoked. It can be said with reasonable assurance, though, that the first alternative considered is a general salary increase whenever a surplus is anticipated.

The model is also "incomplete" in the sense that some departmental priority list obviously exists in the granting of supplemental requests. Thus, the sequence in which departments are considered (the order of departments in Figure 3, items 11 and 13) is important under a revenue constraint. The model's assumption that departments are considered in the order of their account numbers is a poor one, but not enough department request data existed to establish any other reasonable priority list.[13] An analysis of the model residuals, however, failed to reveal any discernible pattern (or "list").[14] A priority list of account categories does exist, though, and is shared by departments, the mayor's offices, and council. The salience of wage and salary accounts is readily discernible through interviews.

Deficit elimination routines. If, instead of an anticipated surplus after preliminary screening of requests, a potential deficit appears (the usual case),

[13]It should be noted that a substantial portion of the department-priority phenomena is accounted for in the preliminary screening of requests.

[14]In deficit-elimination years, an underestimate would be expected for departments with low account numbers (in the computer-coded data) and an overestimate would be expected for departments with high account numbers. The opposite expectations would exist for surplus-elimination years. These phenomena were not observed.

routines are evoked to eliminate the deficit. One routine not evoked in the formal model, but one of the alternatives evoked in practice, is the routine that says "raise taxes."

The alternatives are evoked in the following order.

a) Check preliminary recommendations (lower priority accounts first) to see if they are within limits on increases[15]—bring all preliminary recommendations within limits.
b) Eliminate all recommended increases over current appropriations in non-salary items, considering low priority accounts first.
c) Uniform reduction of all non-salary accounts to eliminate deficit, if all else has failed.

The order in which alternatives are considered represents a priority list for the alternatives (in order of their decreasing desirability) and a search routine evoked by the problem of an anticipated deficit (Figure 3, item 15).

The order-of-account sanctity for the mayor's office is identical to that of the department. This shared preference ordering[16] is as follows:

1. administrative salaries,
2. non-administrative salaries and wages,
3. operating expenses, supplies, materials, etc.,
4. equipment,
5. maintenance,

with maintenance and equipment the first accounts to be cut (and the last to be considered for an increase in the surplus elimination routines) and salaries the last. This deficit elimination procedure is executed only as long as a "deficit" exists. The first acceptable alternative (balanced budget) found is adopted and search activity is halted.

One item that is never reduced from current appropriations in the usual sense is salaries and wages. The salary and wage accounts are different from other accounts in that they represent commitments to individuals currently employed. There are no mass layoffs, etc., rather, a freeze is placed on the filling of positions vacated by retirement, resignation, and death and scheduled step-raises and salary increments are deferred.

Finally, either by reducing the surplus or by eliminating a deficit, the mayor's office arrives at a balanced budget.

Behavior not included in formal MAYORS model. Perhaps the most prominent omission of problem-solving behavior is the lack of a priority list for departments. The model assumes that the priority list is ordered the same way as the account numbers. The overall importance of this faulty assumption is, of

---

[15]The limit is roughly equivalent to the limit indicated in the mayor's letter to department heads.

[16]Shared also with the council.

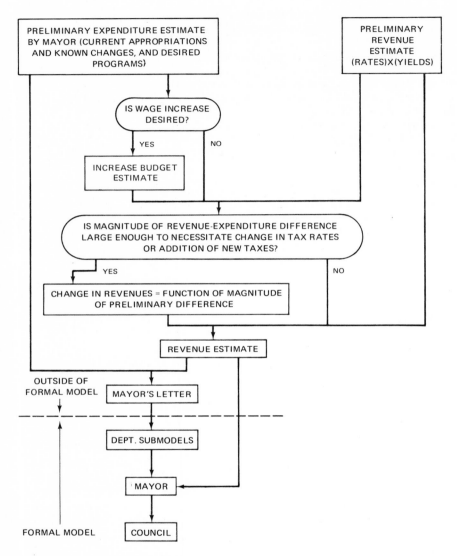

FIGURE 4   Mayor's Tax Decision Process.

course, an empirical question. An analysis of model residuals suggests this was not important or was reflected in estimated parameter values.

The entire budgetary process model we have constructed hypothesizes a stable decision structure between cities, and a stable decision structure over time within cities. Stability in decision structure between cities is "explainable" through problem similarity. Stability within cities reflects stable sets of relationships existing between positions and roles through processes of learning, reinforcement, and socialization. This assumption of stability and uniform socialization is predicated on the assumption that only a relatively few occupants of government positions change in a given period of time. The obvious exception to this situation occurs when an administration is defeated at the polls. This results in a complete reordering of position occupants and relationships. The gradual socialization, learning process will no longer hold. So, we expected and found the largest systematic model errors in those years immediately following the start of a new administration.

Another kind of behavior not included in the model is the kind reflecting the mayor's response to external (to the government) pressure and constraints. Again, as in the DEPT. models, the MAYORS model does not preclude a mayoralty response to requests for services from "powerful" interest groups or individuals. It only postulates that the response is *within* the budget constraint for the department involved. The model, as constructed, implies that either the "response-to-pressure" is systematic and regular over the years (implying a stable system of "pressure" or "influence" in the community) and is reflected in the model parameters, or it does not enter the part of the budgetary process represented by our model at all. The only case where external "influence" could be conceived of as imposing a decisional constraint is in the revenue estimate. Most systematic "pressure" from the business community concentrates on keeping tax rates constant, and not on particular expenditure items.

The importance of these conscious omissions is reflected in the empirical tests of the model.

Characteristics of the COUNCIL Model

The role of the city council is a limited one. The primary reason is more one of cognitive and informational constraints than of lack of interest. The city budget is a complex document when it reaches the council. The level of detail makes it virtually impossible to consider all or even a majority of items independently. An example of this complexity is the Mayor's budget for the Pittsburgh Department of Public Works, the Division of Incineration, Miscellaneous Services Accounts, found in Table 1. It illustrates the kind of document the council must deal with.

The council is asked to deal with the budget at this level of detail. The sheer volume of information to be processed limits the ability of a council, without its own budget staff, to consider the budget in a sophisticated or complex manner.

Perhaps a more important computational constraint is the balanced budget requirement. If there is no slack in the budget the mayor presents to council (Figure 5, item 8), then any increase the council makes in any account category must be balanced with a corresponding decrease in another account or with tax increase. So, in the presence of a revenue constraint, the council cannot consider elements of the budget independently as is done in Congress. Davis, Dempster, and Wildavsky found that Congressional budgetary behavior could be described extremely well using a series of linear decision rules (Davis, Dempster, and Wildavsky, 1966). Behavior of this nature would not be possible if it were

TABLE 1   Sample of Municipal Account Complexity

		Department of Public Works			
1963 Code		*Departmental Estimates 1964*	*Appropriation Year 1963*	*Expenditures Year 1962*	*Increase or Decrease '64 over '63*
Acct. No.	*Title of Account*				
Division of Incineration					
1687	Miscellaneous Services				
B-5	Recharge Fire Extinguishers	$50.00	$—	$89.26	$—
B-5	Extermination Service	200.00	—	—	—
B-8	Towel Rental	25.00	—	.26	—
B-9	Supper Money	100.00	—	—	—
B-18	Freight and Express Charges	89.00	—	—	—
B-17	Public Property and Property Damage Insurance	125.00	—	—	—
B-18	Water Cooler Rental	390.00	—	390.00	—
B-18	Power Shovel Rental	12,000.00	—	14,880.00	—
B-18	Truck Rental for Incinerator and Boil Farm	3,765.00	—	2,295.00	—
B-20	Waste Disposal Permits	50.00	—	50.00	—
B-20	Demurrage on Oxygen and Acetylene Tanks	170.00	—	200.40	—
B-20	Services, N.O.C.	275.00	—	—	—
B-21	Test Boring, Survey and Report, for Landfills	1,000.00	—	—	—
Totals		$19,199.00	$18,199.00	$17,904.92	$1,000.00

required that the sum of the changes in budgets made by Congress add to zero—*i.e.*, the budget must add up to an amount predetermined by the President. Congressmen and Congressional committees also have staffs, councilmen do not.

Another reason for the limited effect of the council on the budget reflects the nature of the "pressures" they face. All interest groups, neighborhood organizations, department heads, etc. feel that some department's budget should be increased. The pressures transmitted to council concerning the operating budget are of one kind—those advocating increases in the mayor's recommendations. The other side of the argument—curtailment of government activities—is seldom, if ever, presented to council. This countervailing influence enters the decision process not at the council level, but generally through the mayor's office and in particular, through the mayor's revenue estimate.

Given the above limitations, the council is "forced" to use the mayor's decisions as the reference points for their decisions. The constraints—"pressure," informational, and computational—coupled with a recommended budget with no slack to allocate (not enough difference between estimated revenues and

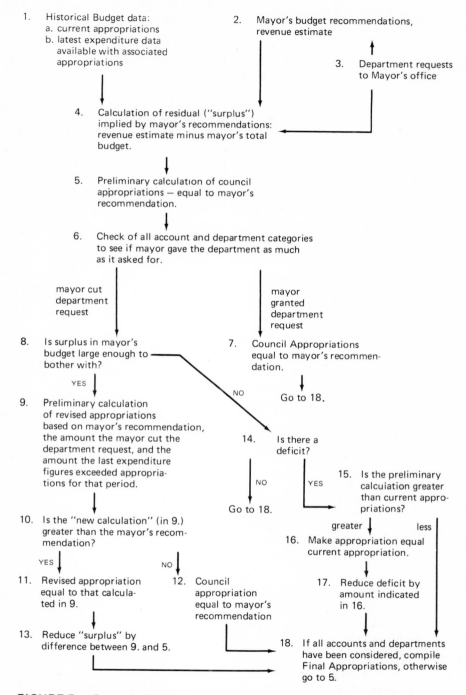

1. Historical Budget data:
   a. current appropriations
   b. latest expenditure data available with associated appropriations

2. Mayor's budget recommendations, revenue estimate

3. Department requests to Mayor's office

4. Calculation of residual ("surplus") implied by mayor's recommendations: revenue estimate minus mayor's total budget.

5. Preliminary calculation of council appropriations — equal to mayor's recommendation.

6. Check of all account and department categories to see if mayor gave the department as much as it asked for.

mayor cut department request

mayor granted department request

8. Is surplus in mayor's budget large enough to bother with?

YES

7. Council Appropriations equal to mayor's recommendation.

Go to 18.

NO

9. Preliminary calculation of revised appropriations based on mayor's recommendation, the amount the mayor cut the department request, and the amount the last expenditure figures exceeded appropriations for that period.

14. Is there a deficit?

NO

Go to 18.

YES

15. Is the preliminary calculation greater than current appropriations?

10. Is the "new calculation" (in 9.) greater than the mayor's recommendation?

greater

less

16. Make appropriation equal current appropriation.

YES

NO

11. Revised appropriation equal to that calculated in 9.

12. Council appropriation equal to mayor's recommendation

17. Reduce deficit by amount indicated in 16.

13. Reduce "surplus" by difference between 9. and 5.

18. If all accounts and departments have been considered, compile Final Appropriations, otherwise go to 5.

FIGURE 5 General COUNCIL appropriations model. For a more detailed flowchart of the COUNCIL Submodel and a listing of the FORTRAN II computer program, see Crecine (1969).

recommended expenditures "to be bothered with") makes it extremely difficult for council to veto[17] or change the mayor's budget significantly.

Overview. Generalizing, the entire model is one of a systematic, bureaucratic administrative decision process. The stability of the decision system is portray   as evolving from the restrictive revenue environment, an assumed continuity in the actors manning the system, and an implied stable or non-existent "community power network." The interaction of problem complexity and need for decision, combined with the lack of extra-governmental reference points or standards, produces a decision system which uses historical experience and precedent as its operating standards; a system which handles interest conflicts (high service rates, low taxes) by largely ignoring divergent viewpoints and using feasibility as the prime decision criterion; a system which handles complexity by fragmenting and simplifying the problem. By assuming (implicitly) that "this year's problem" is nearly identical to "last year's," "this year's solution" will be nearly identical to "last year's." It is a system that structures a complex problem, formulates alternatives and makes choices using simple decision rules.

MODEL TESTS

The formal model of the budgetary process was subjected to many forms of empirical tests. Basically the model was used to generate budget decisions for six years in Pittsburgh (1960-65), seven years in Detroit (1958-59 to 1964-65), and ten years in Cleveland (1956-65). Model results were then compared with the observed budgetary decisions in the cities.

Three primary goodness-of-fit indicators were used: "modified-r^2" statistics, a comparison of the relative predictability of the simulation model with three naive models, and a regression of observed budgetary decisions on model predictions.

We can view the linear regression, r^2-statistic as a measure of the relative precision of the linear hypothesis *vs.* the alternative hypothesis that the dependent variable is randomly distributed about its mean:

$$r^2 = 1.0 - \Sigma(\text{observed} - \text{regression prediction})^2 / \Sigma(\text{observed} - \text{mean of observed})^2$$

By substituting "model estimate" for "regression prediction" and more reasonable alternative hypotheses for "mean of observed," modified-r^2 statistics were constructed,[18] and calculated for each year in each city. The model performed satisfactorily on these measures.

---

[17]Occasionally the council will defeat a proposed new tax—income tax in Cleveland, tax for meat inspectors in Detroit—but seldom will defeat expenditure recommendations.

[18]The "more reasonable" alternative hypotheses used were: "This year's budget for a particular standard account in a given administrative unit equals
  a. Last year's budget for that item, or
  b. the average, over the study period, for that item."

Three alternative, naive models were also tested and compared with the simulation model:

1. Constant-increase model

$$B_{i,t} = (1.0 + a_i)\, B_{i,t-1}$$

2. Constant-share-of-the-budget-total model

$$B_{i,t} = \beta_i (\, \Sigma\, {}^{n}_{j=1} B_{j,t}\,)$$

3. Constant-share-of-the-budget-increase model

$$B_{i,t} - B_{i,t-1} = \delta_i [(\, \Sigma\, {}^{n}_{j=1} B_{j,t}\,) - (\, \Sigma\, {}^{n}_{j=1} B_{j,t-1}\,)]$$

where

$B_{i,t}$ = budget for account i, year t

n = total accounts in city general fund budget

a_i, β_i, δ_i = empirically estimated parameters (regression coefficients).

The simulation model results were then compared with the naive model predictions. Choices between our simulation model and each of the naive models were then made using two statistics suggested by Hunt.[19] In nearly every case, the simulation model "performed" better on both choice measures. It should be noted, however, that the constant-share-of-the-budget-total model also predicted quite well.

Finally, the following relationship was tested:

[Model Estimate of Appropriations] = a[Observed Appropriations] $+ b$

For an unbiased model that predicts perfectly, the expected value of "a" is 1.0 and the expected value of "b" is 0.0. The results for model tests, where inputs were updated at the beginning of each budget year, are found in Table 2.

The true test of goodness-of-fit, however, is the ability of the model to describe the actual budgetary decision process. From all indiciations, our model does this quite well.

We found a change in goodness-of-fit associated with a change in administration in Detroit, but not in Pittsburgh or Cleveland. This was to be expected, since the "change-overs" in Cleveland and Pittsburgh represented a change only in the person occupying the mayor's position and represented a kind of handpicked replacement by the incumbent party (the departing mayors moved on to higher political office). This indicates that, in Cleveland and Pittsburgh, the mayors underwent a process of socialization. No perceivable

[19]The statistics used (Hunt, 1965, pp. 40-44) were:
Min Σ [(estimated - observed)^2/(estimated)], and Min Σ [(estimated - observed)^2//(observed)].

TABLE 2   Regression of Observed Decisions on Model Predictions

City	a	Std. error of "a"	b	r^2	n
Cleveland....................	.977	.001	−$5282	.9980	999
Detroit (excluding Welfare Dept.)..	.984	.005	$7281	.9772	918
Pittsburgh...................	.991	.002	−$28	.9975	1002

differences in goodness-of-fit were associated with increasing or decreasing revenues, indicating our surplus and deficit elimination routines were equally valid.

ANALYSIS OF MODEL RESIDUALS

In general, there are two kinds of "budgetary" change.

1. Those changes resulting from the continuation and elaboration of existing policies, and

2. Those changes resulting from shifts in municipal policies.

Our model is clearly one describing changes of the first kind. It is a model of the standard procedures which result in particular forms of marginal adjustments in resource allocation from year to year.

The model does not describe changes of the second kind—significant shifts in municipal policies. The model, however, by filtering out (i.e., "predicting" or "explaining," in the statistical sense) incremental changes, draws attention to those items ("unexplained") in the budget that are not marginal adjustments or elaborations of previous policies.

By focusing on the "unexplained" changes in resource allocation, we can discover a great deal about the budgetary process as a change process. "Unexplained" changes include:

1. Incremental changes whose cumulative effect results in a "non-incremental" change.

2. Non-incremental policy shifts.[20]

3. Significant changes in policy, not reflected in the budget.

It should be noted that not all large changes are "changes resulting from shifts in municipal policies," and not all small changes are "changes resulting from the continuation and elaboration of existing policies." For example, a

---

[20]The use of the term "innovation" has been consciously avoided because of lack of a generally-agreed-upon, operational definition of the concept. Rather, "policy shift" will be our theoretical construct. An allocation decision represents a "policy shift" when either through cumulative effects of small changes or immediate effects, it brings about a "significant" reallocation of resources between account categories.

significant policy shift may result from the decision to handle the city's welfare load through the welfare department, rather than have the program administered by the county or the state for a fee. The total budget cost may be nearly the same, so this "significant" change may never be reflected in the operating budget. On the other hand, suppose the city decides to build an office building of its own to house a number of departments, rather than rent office space. Once the building has been completed, several years after the initial decision, a large change is noted in the budget—a change our formal model is not equipped to handle. This change, representing an increase in personnel and building maintenance expenses and large decreases in rental expenses for the departments affected, does not represent a significant shift in policy, however. It is merely an elaboration of a long-existing policy (resulting from the decision to build rather than rent). The original decision to build represents a significant "policy shift," however, and anticipated operating budget changes may or may not have been an important part of this capital decision. Our point is that for purposes of analyzing the 1966 operating budgetary process, the items resulting from previous capital decisions represent "automatic" changes in appropriations.

Model deviations[21] were classified by their perceived "cause." Four types of "causes" appear reasonable:

1. Change in External Environment
   a. Intergovernmental transactions
      i. State and Federal subsidies and regulations
      ii. Transfer of functions involving other governments
   b. Catastrophic event, emergency, crises, etc.—reaction to focus of public attention
2. Changes in Internal Environment
   a. New administration (new actors in system of interrelationships)
   b. Change in departments or functions
      i. Transfers of activities—change in organizational structure
      ii. Changes in programs, functions
3. Lack of Model Information
   a. Implications of capital budgeting decisions
   b. Additional revenue sources discovered
   c. Change in system of accounts
   d. Other
4. Unexplained, Miscellaneous, and Other
   a. Model coding errors and missing data

---

[21]In each city, for each year, the five deviations largest in magnitude and the five largest percentage deviations were examined. In each city, for each year, the five deviations largest in (absolute) magnitude and the five largest percentage deviations were examined. "Causes" were associated with individual deviations on the basis of published information. It should be noted that, in nearly every case, the "reason" for the deviation was found in the Mayor's Budget Message to the Council. In other words, the actual devision system identified nearly the same set of unusual decisions as did our model. This, perhaps, is a more significant indication of goodness of fit than the many statistical measures calculated (Crecine, 1969, Section 6).

b. "Improper" accounting procedures (Detroit only—capital items included in operating budget, 1958-59 to 1961-62)
c. Increased work load (or decreased)
d. Other, unexplained

Those "causes" that represent "policy shifts" would be:

1.a.ii. Transfers of functions involving other governments
1.b.   "Catastrophic event," emergency, etc.
2.a.   New administration
2.b.i.   Transfers of activities—organizational change
2.b.ii. New programs, functions
3.b.   Additional revenue sources discovered

"Policy elaborations" would correspond to:

1.a.i.   State and Federal subsidies and regulations
3.a.   Implications of capital decisions
3.c.   Change in system of accounts
3.d.   Other information not part of allocation (timing of elections) process
4.c.   Increased workload

Results

An analysis of model residuals revealed some consistent patterns of change in Cleveland, Detroit, and Pittsburgh. Two principal patterns were noted, only one of which could be described as a "policy shift." One class of revealed "unprogrammed" changes represented changes dictated by the external environment. These were largely due to changes in levels of "ear-marked" revenues (especially in Cleveland) and the terms of negotiated contracts (in both Cleveland and Pittsburgh). The other area of change, representing a kind of "policy shift," was observed in those problem areas and activities where Federal funding and involvement was greatest.

The presence of citizenry demands, needs, etc. does not appear to be related to "policy shifts" in any systematic way. This is probably due to the presence of "needs," demands, etc. for additional services in *all* areas of municipal activity, none of which can be fully "satisfied" given revenue conditions.

SUMMARY

Traditional studies of public finance and governmental decision making, by trying to couple economic, political, and population characteristics to municipal expenditure items, attempt to identify those forces that determine the direction of budgetary drift. Their (implicit) contention is that the "role" of governmental decision makers is that of a translator of environmental characteristics into expenditure items.

By emphasizing the "short-run," we have stressed internal characteristics of the "Government" decision process and the relationship between current and historical decisions. In Figure 6, our findings are that in the short-run, items 5 and 6 are the most significant. By studying the budgetary phenomena over time, others have emphasized items 1, 2, 3, and 4 almost to the exclusion of 5. The question now remains—do our short-run findings apply in the long run? Our model described a somewhat "drifting" budgetary process. Do long-run "pressures" determine the overall direction of that drift?

Causes of Model Drift

Model drift could be biased by external constraints. "Expenditures" would be "allowed" to drift "only so far" without being corrected. They would then be brought back into line with "national standards," party or pressure group demands, population needs or tastes, etc. If, in fact, this were the case, evidence of the use of correcting mechanisms should exist in our model deviations because of the lack of provisions for these mechanisms in our model.

"Observed" Environmental Corrections in Drift

In Detroit, corrections in drift (model deviations) seemed to consist of establishing new departments in the urban renewal area, adjusting appropriations to correspond with State and user revenue changes, and one adjustment in Police

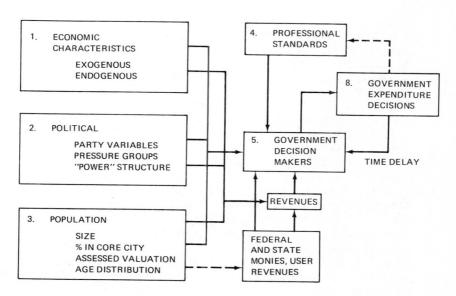

FIGURE 6   Environmental factors in municipal finance.

and Fire salaries resulting from a Public Administration Survey report, (that could be interpreted as a correction in drift to correspond to some national operating standard).

Cleveland's drift corrections consisted of new departments in the urban renewal area, adjustments in appropriations because of changes in State and user revenue contributions and costs of negotiated contracts, and special wage increases for Police and Fire (corresponding to "national" rates?).

Pittsburgh appears much the same, with changes in State revenue contributions, the terms of negotiated service contracts (street lighting), and the emergence of city activities in areas of Federal program involvement accounting for most of the environmental corrections in model drift.

It appears from an analysis of our residuals over time that environmental corrections:

1. are seldom (if ever) evoked directly to bring specific expenditure items "into line," *or*
2. are filtered through the revenue constraint (see Figure 6), blurring the "cause" of increased (decreased) revenues and blunting possible direct impact on specific budget items.

In any event, environmental corrections appear to be more related to revenue changes than to expenditure changes. Hence, their *impact* on expenditures appears to be a blurred one that is excised through the administrative allocation and decision process rather than through any direct "expenditure-correction" mechanism.

Some *direct* environmental corrections were observed, however. Negotiated contracts and changes in "ear-marked" revenue (State and user) provided some clear "corrections." The existence of Federal monies for municipal programs also appears to have "caused" a change in the "budgetary drift."

What seems to emerge from this study is an opportunity model of budgetary change. The broad "pattern of drift" is accelerated or depressed due to changes in general revenues. The "drift" in specific expenditures items changes in response to changes in "ear-marked" revenues or the terms of negotiated contracts. Rapid spurts of growth are observed in those areas where the city has the *opportunity* to expand activities because of the presence of revenues (Federal funds), rather than in areas having rapid changes or spurts in "needs." This also could be due to the fact that "needs" do not change in "spurts" either.

From a normative standpoint, the drifting *general fund* budget has some appeal. If, in fact, we were *able* to specify desired changes in municipal expenditures as a function of environmental changes, the "system" (if we have a Darwinian view of the world) would tend to place these expenditures *outside* of the general fund. The funding of activities where we *can* "logically" connect environmental changes (demands, ability to pay, etc.) to expenditure (or activity) changes, is common. The extreme case results in a "private" good where "supply equals demand" and the level of activity is determined by the price mechanism. Somewhere in between lie the public power and utility companies where price roughly equals cost of goods sold and supply (activity level) equals "demand." Public transportation companies, hospitals, community colleges, etc. all have a system of user taxes where the municipal government provides a partial subsidy. Generally, only activities where user-tax financing is not feasible or undesirable receive a full municipal subsidy and hence are "eligible" for inclusion into the general fund. *It should not be surprising, then, that in the absence of a system of standard costs or ways of determining activity levels* (characteristics of general fund activities) *the decision systems exhibit drifting, opportunistic characteristics.*

REFERENCES

Braybrooke, D., and Lindblom, C. E., *Strategy of Decision*, The Free Press of Glencoe: Glencoe, Illinois, 1963.

Clarkson, G. P. E., *Portfolio Selection: A Simulation of Trust Investment*, Prentice-Hall, Inc.: Englewood Cliffs, N. J., 1962.

Crecine, J. P., *Governmental Problem Solving: A Computer Simulation of Municipal Budgeting*, Rand McNally, Inc.: Chicago, Illinois, 1969.

Cyert, R. M., and March, J. G., *A Behavioral Theory of the Firm*, Prentice-Hall, Inc.: Englewood Cliffs, N. J., 1963.

Davis, Otto A., Dempster, M. A. H., and Wildavsky, A., "On the Process of Budgeting: An Empirical Study of Congressional Appropriations," in Tullock, G. (ed.), *Papers on Non-Market Decision Making*, Thomas Jefferson Center for Political Economy: Charlottesville, Virginia, 1966.

Hunt, E. B., "The Evaluation of Somewhat Parallel Models," in Massarik and Ratoosh, *Mathematical Explorations in Behavioral Science*, Irwin-Dorsey: Homewood, Illinois, 1965.

Lindblom, C. E., "The Science of Muddling Through," *Public Administration Review*, Spring, 1959.

Mansfield, E., "Technical Change and the Rate of Initiation," *Econometrics*, No. 29, 1961.

Newell, A., Shaw, J. C., and Simon, H. A., "Chess-Playing Programs and the Problem of Complexity," in Feigenbaum, E., and Feldman, J. (eds.), *Computers and Thought*, McGraw Hill, Inc.: New York, N. Y., 1963, b.

Newell, A., and Simon, H. A., "Computers in Psychology," in Bush, R. R., Galanter, E., and Luce, R. D. (eds.), *Handbook of Mathematical Psychology: Volume One*, John Wiley and Sons: New York, N. Y., 1963, a.

Newell, A., and Simon, H. A., "GPS, A program that Simulates Human Thought," in Feigenbaum, E. and Feldman, J. (eds.), *Computers and Thought*, McGraw Hill: New York, N. Y., 1963, b.

Sorensen, T. C., *Kennedy*, Harper and Row, Inc.: Scranton, Pennsylvania, 1965, p. 414.

Wildavsky, A., *The Politics of the Budgetary Process*, Little, Brown, and Co.: Boston, Massachusetts, 1964.

PART FOUR

Methodology and Theory

12

Methodological Considerations in Simulating Social and Administrative Systems

THOMAS H. NAYLOR

INTRODUCTION

The earlier chapters of this book have documented the development of simulation in social and administrative science. The purpose of this chapter is to discuss simulation methodology as it relates to modeling social and administrative systems. The scope of this chapter is limited to *computer* simulation. Although all-man and man-computer models are important forms of simulation work, the essential methodological considerations can be examined in the all-computer framework.

In his introduction to simulation and gaming, Barton discusses all-man and man-computer simulation techniques (Barton, 1970). Examples of the use of models with human participants can be found in Guetzkow (1962) and Guetzkow *et al.* (1963).

COMPUTER SIMULATION DEFINED

Although the reader has been exposed to a number of case studies involving the use of simulation, before turning to a discussion of the methodology for designing and implementing simulation experiments it is appropriate that we redefine the term *simulation* as it is used in the following discussion. We shall consider simulation as a *numerical technique* for conducting *experiments* with mathematical and logical models describing the behavior of social and administrative systems on a *digital computer* over extended periods of time.

Several key words merit special attention. First, numerical technique implies that simulation is a technique of "last resort" to be used only when analytical techniques are not available for obtaining solutions to a given model. Being a technique of last resort by no means implies that simulation will find only limited usefulness in the social sciences, for it is well known that only a small number of problems in the social sciences give rise to mathematical models for which standard analytical techniques exist for finding solutions. For most problems in the social and administrative sciences, simulation may be the only technique available to the analyst.

Second, a computer simulation is an *experiment.* With the advent of the high-speed digital computer, social scientists can now perform controlled, laboratorylike experiments in a manner similar to that employed by physicists and other physical scientists, only using a mathematical model programmed into a computer rather than some physical process such as a nuclear reactor. The only difference between a simulation experiment and a real world experiment is that with simulation the experiment is conducted with a model of the real system rather than with the real system. Since a simulation is an experiment, special consideration should be given to the problem of *experimental design*—a point that has been ignored all too often by social scientists.

Third, although a computer is not a necessary tool for carrying out a simulation experiment with a mathematical model of a social system, it certainly speeds up the process, eliminating computational drudgery and reducing the probability of error. For these reasons we shall concentrate only on computer simulation experiments. Although it is indeed possible to conduct simulation experiments with models of social systems on analogue computers, the programming flexibility that one gains by utilizing *digital computers* is sufficient to induce most social scientists to restrict themselves to digital computers. There are, of course, exceptions to this rule. For example, simulation experiments that are directly coupled to some physical process such as the human body or a machine in a factory can be performed most efficiently by using a hybrid computer system, that is, a combination analogue digital computer system. Hybrid simulations will not be discussed in this chapter.

Fourth, with computer simulation we can conduct experiments with our model at a particular point in time or we can conduct experiments over extended periods of time. The former is a *static,* or *cross-section,* simulation; the latter, a *dynamic,* or *time-series,* simulation. A static simulation is achieved by replicating a given simulation run, that is, by changing one or more of the conditions under which the simulation is being conducted. A dynamic simulation results when we simply extend the length of a given simulation run over time without changing any of the conditions.

Fifth, frequently simulation experiments with models of social and administrative systems are stochastic simulations as opposed to purely deterministic simulations. Models of social systems may include random variables over which decision-makers can exercise little or no control. By including these random or stochastic variables in the model, a simulation experiment can be used to make inferences about the overall behavior of the system of interest based on the probability distributions of these random variables. Deterministic simulations are characterized by the absence of random error, that is, all stochastic variables are suppressed.

METHODOLOGY

Having defined computer simulation we now turn to a brief summary of the methodology of computer simulation.[1] Experiments with models of social systems usually involve a procedure consisting of six steps:

1. Formulation of the problem
2. Formulation of a mathematical model
3. Formulation of a computer program
4. Validation
5. Experimental design
6. Data analysis

A description of each of these steps follows.

Formulation of the Problem

Not unlike other forms of scientific inquiry, computer simulation experiments should begin with the formulation of a problem or an explicit statement of the objectives of the experiment, since little benefit is to be derived from experiments that involve simulation for the sake of simulation. These objectives usually take the form of (1) questions to be answered, (2) hypotheses to be tested, and (3) effects to be estimated.

If the objective of a simulation experiment is to obtain answers to one or more specific questions, then obviously one must attempt to specify these questions with a high degree of detail at the outset of the experiment. Among the questions or decision problems that may be solved by computer simulation in the social sciences are the following: What effect will a particular drug have on an individual's personality? Will a change in federal import quotas adversely affect the profitability of a given business firm? What effect will a strong stand on civil rights have on the popularity of a gubernatorial candidate? It is not sufficient to specify the questions that are to be answered by a simulation experiment; we must also specify objective criteria for evaluating possible answers to these questions. For example, in international relations we must define exactly what we mean by an *optimum* foreign policy if we expect to recognize such a policy when we are confronted by one. Unless we specify

[1]For a comprehensive treatment of the methodology of computer simulation see Thomas H. Naylor *et al., Computer Simulation Techniques.* New York: John Wiley & Sons, Inc., 1966.

precisely what is meant by a *suitable* answer to a question that has been raised, we cannot hope to achieve meaningful results from computer simulation experiments.

On the other hand, the objective of a simulation experiment may be to test one or more hypotheses about the behavior of a complex social system. Does television affect voting behavior in presidential elections? Will a tax increase lead to a recession? Is there any significant difference in the effects of five alternative advertising policies on the sales of a firm? What effects does birth control have on economic growth in underdeveloped countries? In each case, the hypotheses to be tested must be stated explicitly, as well as the criteria for "accepting" or "rejecting" them.

Finally, the objective of a simulation may be to estimate the effects of certain changes in the controllable decision variables of a social system on the endogenous or dependent variables describing the behavior of the system. For example, we may wish to estimate the effects of alternative monetary and fiscal policies on GNP. Generally speaking, we would want to construct confidence intervals for parameter estimates of endogenous variables generated by simulation experiments, where these output variables represent the results of the use of alternative policies, decision rules, parameters, and so forth.

Formulation of a Mathematical Model

Having formulated our experimental objectives, the next step in the design of a simulation experiment is the formulation of a mathematical model relating the endogenous variables of the system to the controllable and exogenous variables. The exogenous variables are assumed to be determined by forces outside the system. Some of the exogenous variables may be random variables; others may be expressed in the form of time trends. As we shall soon observe, the inclusion of random, or stochastic, variables in a computer model gives rise to a number of unique methodological problems which do not exist with deterministic methods.

One of the first considerations that enter into the formulation of a mathematical model of a social system is the choice of variables to be included in the model. As a general rule, we encounter little or no difficulty with the endogenous variables of a model because these variables are usually determined at the outset of the experiment when we formulate the objectives of the study. For example, the endogenous variables of a socioeconomic model of a city might include per capita income, employment, level of education, crime rate, age distribution of population, and so forth. However, the real difficulty arises in the choice of the input (exogenous and controllable) variables affecting the output variables. Too few input variables may lead to invalid models, whereas too many exogenous and decision variables may render computer simulation impossible because of insufficient computer memory or may make computational procedures unnecessarily complicated.

A second major consideration in the formulation of mathematical models is the complexity of the model. It can be argued that social and administrative systems are indeed quite complicated and that mathematical models that claim to describe the behavior of social systems must also necessarily be complicated; however, we do not want to go to the extreme of constructing such complex models, regardless of how realistic they may be, because they require an unreasonable amount of computation time. In general we are interested in formulating mathematical models that yield reasonably accurate descriptions or predictions about the behavior of a social system while minimizing computational and programming time. The complete interdependence of these characteristics of mathematical models cannot be overemphasized. For example, the number of variables in a model and its complexity are directly related to programming time, computation time, and validity. By altering any one of the characteristics of a model we in turn alter all the other characteristics.

Computer programming time represents a third consideration in formulating mathematical models for computer simulation. The amount of time required to write a computer program for generating the time paths for the endogenous variables of a particular mathematical model depends in part on the number of variables used in the model and the complexity of the model. If some of the variables utilized in the model are stochastic, both programming time and computation time are likely to be increased significantly. The amount of effort one expends in attempting to reduce programming time must, of course, be balanced against the questions of validity and computational speed. If the costs in terms of realism are not too great, it may even pay the analyst to formulate his model in such a manner that it satisfies the requirements of one of the simulation languages such as SIMSCRIPT (Naylor, 1970), GPSS (Gordon, 1962; General Purpose Simulator II, n.d.), DYNAMO (Pugh, 1963), or SIMULATE (Holt *et al.,* 1964). The gains made in terms of reduced programming time may completely offset the loss in validity that may result from such a modification.

The fourth consideration in model building is the validity of the model, or the amount of "realism" built into it. That is, does the model adequately describe the behavior of the system being simulated in future time periods? Unless the answer to this question is yes, the value of our model is reduced considerably, and our simulation experiment becomes merely an exercise in deductive logic.

The fifth and final consideration in formulating a computer simulation model is its compatibility with the type of experiment that is going to be carried out with it. Since our primary objective in formulating mathematical models is to enable us to conduct simulation experiments, some thought must be given to the particular type of experimental design features that must be built into our models.

Having formulated a mathematical model describing the behavior of a social system, we must estimate the values of the parameters of the model and test the statistical significance of these estimates.

Once we have formulated a model describing the behavior of our system and have estimated the parameters of the operating characteristics on the basis of observations taken from the real world, we must then make an initial value judgment concerning the adequacy of our model. That is, we must test the model. Very little is to be gained by using an inadequate model to carry out simulation experiments on a computer because we would merely be "simulating our own ignorance."

Among the questions that we may wish to raise at this point in our procedure are the following:

1. Have we included any variables that are not pertinent in the sense that they contribute little to our ability to predict the behavior of the endogenous variables of the system?
2. Have we failed to include one or more exogenous variables that are likely to affect the behavior of the endogenous variables in our system?
3. Have we inaccurately formulated one or more of the functional relationships between the system's output and input variables?
4. Have the estimates of the parameters of the model's operating characteristics been properly estimated?
5. Are the estimates of the parameters in our model statistically significant?
6. On the basis of hand calculations (since we have not yet formulated a computer program), how do the theoretical values of the endogenous variables of our model compare with historical or actual values of the endogenous variables?

If, and only if, we can answer these questions satisfactorily should we proceed to step 3 and the formulation of a computer program. Otherwise, we should repeat steps 1 and 2 until such time as we can achieve satisfactory answers to all six questions.

Formulation of a Computer Program

The formulation of a computer program for conducting simulation experiments with a model of a complex system requires that special consideration be given to three activities: (1) computer program, (2) data input and starting conditions, and (3) data generation.

Computer program. The first step in writing a computer simulation program involves formulating a flow chart outlining the logical sequence of events to be carried out by the computer in generating the time paths of the endogenous variables of the model. The importance of flow charting in writing computer programs cannot be overemphasized. Next we must consider the matter of writing the actual computer code that will be used to run our experiments. In general, there are two **alternatives** available. We can write our program in a general purpose language such as FORTRAN, ALGOL, or PL/I or we can use a special purpose simulation language such as GPSS (Gordon, 1962), SIMSCRIPT (Markowitz, Hausner, and Karr, 1962), DYNAMO (Pugh, 1963), or SIMULATE (Holt *et al.,* 1964). The principal advantage of using a special purpose simulation language is that it requires less programming time than that needed by general purpose compilers. These languages have been written to facilitate the programming of certain types of systems. For example, SIMU-LATE was designed primarily for simulating large-scale economic systems that have been formulated as econometric models consisting of large sets of equations. On the other hand, GPSS and SIMSCRIPT are particularly well suited for queuing problems. Although we can reduce programming time by using a simulation language, we must usually pay a price for this benefit in terms of reduced flexibility in models and increased computer running times. Another important advantage of special purpose simulation languages is that they usually provide error-checking techniques far superior to those provided by FORTRAN, ALGOL, and the like. One final consideration in the development of a computer program for a simulation experiment is the kind of output reports needed to provide the required information about the behavior of the simulated system. If we use a general purpose language such as FORTRAN, there will be a minimum number of restrictions imposed on the format of our output reports. However, if we use a special purpose simulation language such as SIMSCRIPT, we must adhere to the output format requirements of the language.

Data input and starting conditions. Another aspect of the computer programming phase of the development of simulation experiments is the matter of input data and starting conditions for the simulation experiments. Since simulation experiments are by their very nature dynamic experiments, a question arises as to what values should be assigned to the model's variables and parameters at the point in time when we begin simulating the system. That is, we must break into the system at some particular point in time. When we do so, what assumptions should we make about the state of the system being simulated? This question is not easily answered for most systems, and the investigator must usually resort to trial-and-error methods for determining a set of initial conditions that will not lead to biased results in future time periods.

Data generation. Directly related to the problem of writing computer simulation programs is the development of numerical techniques (which can be programmed on a computer) for data generation. Data used in computer simulation experiments either can be read into the computer from external sources, such as punched cards and magnetic tapes, or can be generated internally by special subroutines. If one or more of the exogenous variables included in our model are stochastic variables with a known probability distribution, we are confronted with the problem of devising a process of random selection from the given probability distribution so that the results of the repetition of this process on a digital computer will give rise to a probability distribution of sampled values that corresponds to the probability distribution of the variable of interest.

In considering stochastic processes involving either continuous or discrete random variables, we define a function $F(x)$ called the cumulative distribution function of x, which denotes the probability that a random variable X takes on the value of x or less. If the random variable is discrete, then x takes on specific values and $F(x)$ is a step function. If $F(x)$ is continuous over the domain of x, it is possible to differentiate this function and define $f(x) = dF(x)/dx$. The derivative $f(x)$ is called a probability density function. Finally, the cumulative distribution function may be stated mathematically as

$$(1) \quad F(x) = P(X \le x) = \int_{-\infty}^{x} f(t)dt$$

where $F(x)$ is defined over the range $0 \le F(x) \le 1$, and $f(t)$ represents the value of the probability density function of the random variable X when $X = t$.

Uniformly distributed random variables play a major role in the generation of random variables drawn from *other* probability distributions. We will denote uniform variables by r, when $0 < r < 1$, and $F(r) = r$. Naylor *et al.* (1966, chap. 3) contains a survey of the theory and methods of generating uniformly distributed random variables on the interval (0,1). These numbers are called pseudorandom numbers because, although they are generated from a completely deterministic recursive formula by a computer, their statistical properties coincide with the statistical properties of numbers generated by an idealized chance device that selects numbers from the unit interval (0,1) independently and with all numbers equally likely. So long as these pseudorandom numbers can pass the set of statistical tests (frequency test, serial test, lagged product test, runs test, gap test, etc.) (Naylor *et al.*, 1966) implied by an idealized chance device, these pseudorandom numbers can be treated as "truly" random numbers even though they are not.

Since pseudorandom numbers generators (in the form of subroutines) are available for all computers, we shall not delve further into the topic. It will be assumed that the pseudorandom numbers used in the subroutines described in the following pages will be generated by a preprogrammed FORTRAN function denoted by $R = RAND (R)$.

If we wish to generate random variables, x_i's, from some particular statistical population whose distribution function is given by $f(x)$, we first obtain the cumulative distribution function $F(x)$. (See figure 12-1.) Since $F(x)$ is defined over the range 0 to 1 we can generate uniformly distributed random numbers and set $F(x) = r$. It is clear that x is uniquely determined by $r = F(x)$. It follows, therefore, that for any particular value of r, say r_0, which we generate, it is possible to find the value of x, in this case x_0, corresponding to r_0 by the inverse function of F if it is known,

$$(2) \quad x_0 = F^{-1}(r_0)$$

where $F^{-1}(r)$ is the inverse transformation of r on the unit interval into the domain of *x*. We may summarize this method mathematically by saying that if we generate uniform random numbers corresponding to a given *F(x),*

(3) $r = F(x) = \int_{-\infty}^{x} f(t)dt$

then

(4) $P(X \leq x) = F(x) = P(r \leq F(x)) = P(F^{-1}(r))(x)$

and consequnetly $F^{-1}(r)$ is a variable which has *f(x)* as its probability density function. This is equivalent to solving *(3)* for *x* in terms of *r*. This procedure is called the inverse transformation method.

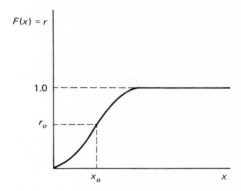

FIGURE 12-1   A cumulative distribution function.

Validation

The problem of validating simulation models (Naylor and Finger, 1967) is indeed a difficult one because it involves a host of practical, theoretical, statistical, and even philosophical complexities (Reichenbach, 1951). Validation of simulation experiments is merely part of a more general problem, namely, the validation of any kind of model or hypothesis. The basic questions are, What does it mean to validate a hypothesis? and What criteria should be used to establish the validity of a hypothesis?

In view of the complexity of most computer models of social and administrative systems, a multistage validation procedure seems appropriate.

The first stage of this procedure calls for the formulation of a set of postulates, or hypotheses, describing the behavior of the system of interest. To be sure these are not just any postulates, for what is required in stage one is a diligent search for what Kant has called "synthetic a priori" using all possible information at our disposal.

> Like the scientist, the scientific philosopher can do nothing but look for his best posits. But that is what he can do; and he is willing to do it with the perseverance, the self-criticism, and the readiness for new attempts which are indispensable for scientific work. If error is corrected whenever it is recognized as such, the path of error is the path of truth. (Reichenbach, 1951, p. 326.)

This set of postulates is formed from the researcher's already acquired "general knowledge" of the system to be simulated or from his knowledge of "similar" systems that have already been successfully simulated. The point we are striving to make is that the researcher cannot subject all possible postulates to formal empirical testing and must therefore select, on essentially a priori grounds, a limited number of them for further detailed study. He is, of course, at the same time rejecting an infinity of postulates on the same grounds. The selection of postulates is taken here to include the specification of components and the selection of variables as well as the formulation of functional relationships. But having arrived at a set of basic postulates on which to build our simulation model, we are not willing to assume that they require no further validation. Instead we merely submit these postulates as tentative hypotheses about the behavior of a system.

The second stage of our multistage validation procedure calls for an attempt on the part of the analyst to "verify" the postulates on which the model is based subject to the limitations of existing texts. Although we cannot solve the philosophical problem of What does it mean to verify a postulate?, we can apply the "best" available statistical tests to them.

But in the social sciences we often find that many of our postulates are either impossible to falsify by empirical evidence or extremely difficult to subject to empirical testing. We have two choices—we may either abandon the postulates entirely, arguing that they are scientifically meaningless because they cannot conceivably be falsified, or we may retain the postulates merely as "tentative" postulates. If we choose the first alternative we must continue searching for other postulates that can be subjected to empirical testing. However, we may elect to retain these "tentative" postulates that cannot be falsified empirically on the basis that there is no reason to assume that they are invalid just because they cannot be tested.

The third stage of this validation procedure consists of testing the model's ability to predict the behavior of the system under study. C. West Churchman asserts that the purpose of simulation is to predict, and he considers the point so obvious that he offers no defense of it before he incorporates it into his discussion of the concept of simulation (1963). This point does indeed seem obvious. Unless the construction of simulation models is viewed as a game with no purpose other than the formulation of a model, it is hard to escape the conclusion that the purpose of a simulation experiment is to predict some aspect of reality. To test the degree to which data generated by computer simulation models conform to observed data two alternatives are available—historical verification and verfication by forecasting. The essence of these procedures is prediction, for historical verification is concerned with retrospective predictions, whereas forecasting is concerned with prospective predictions.

If an experimenter uses a simulation model for descriptive analysis, he is interested in the behavior of the system being simulated and would therefore attempt to produce a model that would predict that behavior. The use of simulation models for prescriptive purposes involves predicting the behavior of the system being studied under different combinations of controllable conditions. He would then decide on the most desirable set of conditions to put into effect by choosing the one that produced the most desirable set of outcomes. When a simulation model is used for descriptive analysis, the actual historical record produced by the system being simulated can be used as a check on the accuracy of the predictions, and hence on the extent to which the model fulfilled its purpose. But prescriptive analysis involves choosing the one historical path along which the system will be directed. Therefore, only the historical record of the path actually traveled will be generated, and the historical records of alternative paths corresponding to alternative policies will not be available for comparison. Although, in this instance, the historical record cannot be used as a direct check on whether the model did actually point out the best policy to follow, the actual outcome of the policy chosen can be compared with the outcome predicted by the simulation model as an indirect test of the model. In either case, the predictions of the model are directly related to the purpose for which the model was formulated, whereas the assumptions that make up the model are only indirectly related to its purpose through their influence on the predictions. Hence the final decision concerning the validity of the model must be based on its predictions.

Thus far, we have concerned ourselves only with the philosophical aspects of the problem of verifying computer simulation models. What are some of the practical considerations that the social scientist faces in verifying computer models? Some criteria must be devised to indicate when the time paths generated by a computer simulation model agree sufficiently with the observed or historical time paths so that agreement cannot be attributed merely to chance. Specific *measures* and *techniques* must be considered for testing the goodness of fit of a simulation model, that is, the degree of conformity of simulated time series to observed data. Richard M. Cyert has suggested that the following measures may be appropriate (1966):

1. Number of turning points.
2. Timing of turning points.
3. Direction of turning points.
4. Amplitude of the fluctuations for corresponding time segments.
5. Average amplitude over the whole series.
6. Simultaneity of turning points for different variables.
7. Average values of variables.
8. Exact matching of values of variables.

To this list we would add the probability distribution and variation about the mean (variance, skewness, kurtosis) of variables.

Although a number of statistical techniques exist for testing the goodness of fit of simulation models, for some unknown reason social scientists have, more often than not, restricted themselves to purely graphical (as opposed to statistical) techniques of goodness of fit for validating computer models. The following statement by Cyert and March (1963) concerning the validity of their duopoly model is indicative of the lack of emphasis placed on goodness of fit by many practitioners in this field.

> In general, we feel that the fit of the behavioral model to data is surprisingly good, although we do not regard this fit as validating the approach [p. 97].

This statement was made on the basis of a graphical comparison of the simulated time series and actual data. Not unlike most other simulation studies described in the literature, Cyert and March did not pursue the question of validation beyond this point.

Within the confines of this chapter it is impossible to enumerate all the statistical techniques available for testing the goodness of fit of simulation models. However, we shall list some of the more important techniques and suggest a number of references which describe them in detail.

Analysis of variance. The analysis of variance is a collection of techniques for data analysis which can be used to test the hypothesis that the mean (or variance) of a series generated by a computer simulation experiment is equal to the mean (or variance) of the corresponding observed series. Three important assumptions underlie the use of this technique—normality, statistical independence, and common variance. The paper by Naylor, Wertz, and Wonnacott (1967) describes the use of the F-test, multiple comparisons, and multiple ranking procedures to analyze data generated by simulation experiments.

Chi-square test. The chi-square test is a classical statistical method which can be used for testing the hypothesis that the set of data generated by a simulation model has the same frequency distribution as a set of observed historical data. Although this test is relatively easy to apply, it has the problem of all tests using categorical type data, namely, the problem of selecting categories in a suitable and unbiased fashion. It has the further disadvantage that it is relatively sensitive to nonnormality.

Factor analysis. Cohen and Cyert have suggested the performance of a factor analysis on the set of time paths generated by a computer model, a second factor analysis on the set of observed time paths, and a test of whether the two groups of factor loadings are significantly different from each other (Cyert and March, 1963).

Nonparametric tests. The books by Siegel (1956) and Walsh (1962, 1965) describe a host of nonparametric tests that can be used for testing the goodness of fit of simulated data to real-world data.

Regression analysis. Cohen and Cyert (1963) have also suggested the possibility of regressing actual series on the generated series and testing whether the resulting regression equations have intercepts that are not significantly different from zero and slopes that are not significantly different from unity (Cyert and March, 1963).

Spectral analysis. Data generated by computer simulation experiments are usually highly autocorrelated. When autocorrelation is present in sample data, the use of classical statistical estimating techniques (which assume the absence of autocorrelation) will lead to underestimates of sampling variances (which are unduly large) and inefficient predictions. Spectral analysis considers data arranged in a series according to historical time. It is essentially the quantification and evaluation of autocorrelated data at which spectral analysis is aimed, after the data have been transformed into the frequency domain. For purposes of describing the behavior of a stochastic variate over time, the information content of spectral analysis is greater than that of sample means and variances. Spectral analysis provides a means of objectively comparing time series generated by a computer model with observed time series. By comparing the estimated spectra of simulated data and corresponding real-world data, one can infer how well the simulation resembles the system it was designed to emulate (Blackman and Tukey, 1958; Fishman and Kiviat, 1967; Naylor, Wertz, and Wonnacott, 1968).

Theil's inequality coefficient. A technique developed by Theil has been used by a number of economists to validate simulations with econometric models (Theil, 1961). Theil's inequality coefficient U provides an index that measures the degree to which a simulation model provides retrospective predictions of observed historical data. U varies between 0 and 1. If $U = 0$, we have perfect predictions. If $U = 1$, we have very bad predictions. There is no obvious reason why this technique cannot be used to validate other models of social systems, as well as econometric models.

Experimental Design

In a computer simulation experiment, as in any experiment, consideration should be given to the problem of experimental design. Although a number of researchers have considered the need to utilize experimental design techniques with simulation experiments and have noted the extensive literature on the subject of experimental design,[2] social scientists have virtually ignored experimental design considerations in carrying out simulation experiments with models of social and administrative systems. For the most part, the existing experimental design literature deals with the problems and techniques of designing real-world experiments, whereas computer simulation experiments are in effect experiments on a mathematical model.[3] The task of deciding which material in the experimental design literature is applicable to simulation experiments is extremely difficult. This situation is likely to be particularly acute to social scientists who (prior to the advent of computer simulation) have had only limited opportunity to perform experiments with social systems. Our objective in this and the following sections is to show the relationship between existing experimental design and data analysis techniques and the design of computer simulation experiments with models of social systems.

The two most important terms in the language of experimental design are *factor* and *response*. Both terms refer to variables. Whether a variable in a particular experiment is a factor or a response depends upon the role played by the variable in the experiment in question. To illustrate the difference between a factor and a response, suppose we have two variables, X and Y. If our experiment is designed to answer the question, How does a change in X affect Y?, then X is a factor and Y is a response. In an experiment with a computer model of a social system, a response must of necessity be an endogenous (output) variable, whereas a factor will normally be an exogenous or policy (input) variable.

For example, with a computer model of a firm, profit, output, or utility might be response variables. On the other hand, advertising expenditures, labor inputs, capital outlays, GNP, and per capita income might be treated as factors.

Many of the terms and concepts in the theory of experimental design result from the classification of the factors in the experiment by the following dichotomous questions:

1. Is the factor in question controlled or not?
2. Are the values (levels) of the factor observed or not?
3. Is the effect of the factor a subject for study or is the factor included merely to increase the precision of the experiment?
4. Are the levels of the factor quantitative or qualitative?
5. Is the factor fixed or random?

---

[2] Among the publications that have acknowledged the relevance and importance of experimental design problems in simulation experiments are Burdick and Naylor (1966), Conway (1963), Naylor, Burdick, and Sasser (1967), Naylor *et al.* (1966), and Tocher (1963). In addition, Bonini (1963), Cyert and March (1963), Fishman and Kiviat (1967), Hufschmidt (1966), Naylor, Wertz, and Wonnacott (1967, 1968, 1969), and Preston and Collins (1966) have considered specific experimental design problems.

[3] The "classical" experimental design literature includes the works of Cochran and Cox (1957) and Winer (1962).

A factor is *controlled* if its levels are purposefully selected by the experimenter. (In economic and political models, the experimenter is usually called a policy maker.) Campaign issues and advertising policies are subject to the control of political strategists. Wars, foreign competition, labor strikes, and national disasters are factors that may affect the outcome of an election but may not be controllable by political strategists.

A factor is *observed* if its levels are observed or measured and recorded as part of the data. More often than not the observed factors consist of just the controlled factors in a particular experiment, but there are frequent exceptions. It is unwise to control a factor without observing it, but an uncontrolled factor may often be observed. For a model of the political structure of the United States, wars and strikes, although uncontrolled, can be observed. Observations on uncontrolled factors are often called concomitant observations. In the analysis of data, concomitant observations should be treated differently from observations on controlled factors. The *analysis of covariance* is a technique of data analysis that utilizes concomitant observations. Although concomitant observations are useful, in the real world it is never possible to observe *all* the factors that may affect a given response.

The distinction between factors that are of basic interest and those that are included to increase precision is an important one because it demonstrates that for almost all experiments the factors of basic interest are not the only ones to significantly affect the outcome. In the literature, controlled factors that are included to increase precision are often called block factors, and their levels are called blocks. In simulation experiments we never have uncontrolled or unobserved factors. The role that uncontrolled and unobserved factors play in the real world is played in a computer simulation model by the random character of exogenous variables. The effects or variations in response that these factors cause in the real world have been incorporated in the computer simulation model in the form of experimental errors or random deviations. Once we have a model, the factors are determined, and it is not possible in an experiment on the model to identify additional factors as sources of variation.

A factor is *quantitative* if its levels are numbers that are expected to have a meaningful relationship with the response. Otherwise, a factor is *qualitative.* The amount of money spent on advertising and the size of the campaign staff might be among the quantitative factors for a model of political processes. If part of the input to a simulation model consists of a decision rule or policy and if several policies are under consideration, the policy can be a qualitative factor.

When an experimenter is investigating the effect of a factor on a response, he will be interested in drawing inferences with respect to a certain range of population of levels for the factors. If all the levels of interest of a particular factor are included in the experiment, that factor is said to be *fixed.* If, however, the levels of a factor that are actually included in the experiment constitute a random (or representative) sample from the population of levels in which the experimenter is interested, the factor is said to be *random.* The notion of random factors permits probabilistic inferences to be made about factor levels that do not actually appear in the experiment.

Next we describe four problems that arise in the design of simulation experiments and identify some of the techniques that have been developed to solve them. The four experimental design problems are (1) the problem of stochastic convergence, (2) the problem of size, (3) the problem of motive, and (4) the multiple-response problem.

The problem of stochastic convergence. Most simulation experiments are intended to yield information about population quantities or averages, such as average level of education in sociological simulations. As estimates of population averages, the sample averages we compute from several runs on a computer will be subject to random fluctuations and will not exactly be equal to the population averages. However, the larger the sample (i.e., the more runs we observe), the greater the probability that the sample averages will be very close to the population averages. The convergence of sample averages for increasing sample size is called stochastic convergence.

The problem of stochastic convergence is that it is slow. A measure of the amount of random fluctuation inherent in a chance quantity is its standard deviation. If σ is the standard deviation of a single observation, the standard deviation of the average of n observations is a $\sigma / \sqrt{n}$. Thus, to halve the random error one must quadruple the sample size n; to decrease the random error by a factor ten, one must increase the sample size by a factor of one hundred. It can easily happen that a reasonably small random error requires an unreasonably large sample size.

Because of the slowness of stochastic convergence, we are led to seek methods other than increasing sample size to reduce random error. In real-world experiments, error reduction techniques commonly involve including factors such as blocks or concomitant variables that are not of basic interest to the experimenter. If some of these factors, instead of being uncontrolled and unobserved, can be controlled or observed, then their effects will no longer contribute to the random error, and the standard deviation σ of a single observation will be reduced.

In a computer simulation experiment on a given model, it is not possible to include more factors for error reduction purposes, as inclusion of more factors requires a change in the model. There are, however, error reduction techniques that are suitable for computer simulation experiments. They are called Monte Carlo techniques (Clark, 1961; Ehrenfield and Ben-Tuvia, 1962; Hammersley and Handscomb, 1964). The underlying principle of Monte Carlo techniques is the utilization of knowledge about the structure of the model, properties of the probability distributions of the exogenous inputs, and properties of the observed variates actually used for inputs to increase the precision (i.e., reduce random error) in the measurement of averages for the response variables.

Hammersley and Handscomb (1964) have written an excellent book on the subject of Monte Carlo techniques. They discuss importance sampling, control variates, correlation (i.e., regression methods and antithetic variate methods), and conditional Monte Carlo. The book also contains an extensive bibliography.

The problem of size. What we have called the problem of size arises in both real-world and simulation experiments. It could just as easily be called the problem of too many factors. In a factorial design for several factors, the number of cells required is the product of the number of levels for each of the factors in the experiment. Thus, in a four-factor experiment with a model of a firm, if we have six different employment policies, five alternative marketing plans, five possible inventory policies, and ten different equipment replacement policies, then a total of $6 \times 5 \times 5 \times 10 = 1,500$ cells (or factor combinations) would be required for a full factorial design. If we had a ten-factor experiment and if we only used two levels for each of these factors, the full factorial experiment would require $2^{10} = 1,024$ cells. It is evident that the full design can require an unmanageably large number of cells if more than a few factors are to be investigated.

If we require a complete investigation of the factors in the experiment, including main effects and interactions of all orders, there is no solution to the problem of size. If, however, we are willing to settle for a less than complete investigation, perhaps including main effects and two-factor interactions, there are designs that will accomplish our purpose which require fewer cells than the full factorial. Fractional factorial designs, including Latin square and Greco-Latin square designs, are examples of designs that require only a fraction of the cells required by the full factorial design.

In any design that utilizes fewer cells than the full factorial there will be some *confounding* of effects. A main effect, for example, may be confounded with an interaction effect, which means that the statistic that measures the main effect is exactly the same statistic that measures the interaction effect.

Thus, the statistic in question can tell us that some effect is present, but it cannot tell us whether the main effect, the interaction effect, or some combination of the two is present. Only if the interaction effect can be assumed to be zero (or at least negligibly small) are we justified in stating that the observed effect is in fact a main effect.

Experimenters are usually most interested in main effects. It is important therefore that main effects not be confounded with other main effects. In almost all the commonly used fractional factorial designs, main effects are confounded with interactions (preferably high-order interactions) and not with other main effects. If an experimenter uses one of these designs to measure main effects, he must be willing to assume, at least tentatively, that the interactions with which the main effects are confounded are zero. Few experimenters are deterred from the use of fractional factorial designs by the necessity of such assumptions. Although the assumption that a high-order interaction is zero is frequently justifiable, we suspect that in many instances the difficulty in interpreting a high-order interaction influences the experimenter's willingness to assume it to be zero.

The problems that arise in obtaining fractional factorial designs by confounding main effects with interactions have proved appealing to mathematical statisticians. As a result, much has been written in this area both in books and in articles in the professional journals. Tables of designs can be found in Cochran and Cox (1957) and in the Applied Mathematics Series of the National Bureau of Standards (1961). Bonini (1963) has an example of a fractional factorial design employed in a computer simulation experiment.

So far the problem of size reduction has been discussed in an analysis of variance framework. This collection of techniques for data analysis (i.e., the analysis of variance) is appropriate when the factors are qualitative. However, if the factors $X_1, X_2, \ldots, X_k$ are quantitative, and the response Y is related to the factors by some mathematical function f, then regression analysis rather than the analysis of variance may be an appropriate method of data analysis. The functional relationship $Y = f(X_1, \ldots, X_k)$ between the response and the quantitative factors is called the response surface (Box, 1954; Box and Wilson, 1951; Hill and Hunter, 1966; Hufschmidt, 1966). Least squares regression analysis is a method for fitting a response surface to observed data in such a way as to minimize the sum of squared deviations of the observed responses from the value predicted from the fitted response surface.

For an experiment that utilizes regression analysis to explore a response surface, a factorial design or a fractional factorial design may not be optimal. Several authors, primarily George Box (1954; Box and Wilson, 1951), have developed designs called response surface designs which are appropriate when response surface exploration via regression analysis is the aim of the experiment. An important advantage of the response surface designs in comparison with factorial designs is the reduction in the required size of the experiment without a corresponding reduction in the amount of information obtained.

Response surface designs have not been given the attention they deserve in most of the books on experimental design. An exception is chapter 8A in the second edition of the book by Cochran and Cox (1957). The recent paper by Hill and Hunter (1966) contains a survey of response surface designs and a complete bibliography. Response surface designs were used by Hufschmidt (1966) to design a computer simulation experiment with a model of a water-resource system. Austin C. Hoggatt has used response surface designs with simulation experiments with a computer model of a market (Preston and Collins, 1966).

The problem of motive. The experimenter should specify his objectives as precisely as possible to facilitate the choice of a design that will best satisfy his objectives. Two important types of experimental objectives can be identified: (1) the experimenter wishes to find the combination of factor levels at which the response variable is maximized (or minimized) in order to optimize some process; (2) the experimenter wishes to make a rather general investigation of the relationship of the response to the factors in order to determine the underlying mechanisms governing the process under study. The distinction between these two aims is less important when the factors are qualitative rather than quantitative. Unless certain interactions can be assumed to be zero, the only way to find the combination of levels of qualitative factors that will produce an optimum response is to measure the response at all combinations of factor levels (i.e., the full factorial design). Even if interactions are assumed to be negligible in an experiment with qualitative factors, the design is likely to be the same whether the aim is to optimize or to explore.

In an experiment with quantitative factors the picture is quite different. Hence, the continuity of the response surface can generally be used to guide us quickly and efficiently to a determination of the optimum combination of factor levels. There are two commonly used sampling methods for finding the optimum of the response surface: systematic sampling and random sampling. Systematic sampling methods include (1) the uniform-grid or factorial method, (2) the single-factor method, (3) the marginal analysis method, and (4) the steepest ascent method. The article by Hufschmidt (1966) contains a case study involving the use of both systematic and random sampling methods for the design of a simulation experiment. A detailed description of several of these methods can be found in Cochran and Cox (1957).

When general exploration of a response surface is the aim, it is difficult to identify a "best" experimental design because general exploration is usually a less precisely specified goal than optimization. However, we can state a guiding principle: When the aim of an experiment is to further general knowledge and understanding, it is important to give careful and precise consideration to the existing state of knowledge and to the questions and uncertainties that we desire the experimental data to elucidate.

The multiple-response problem. This problem arises when we wish to observe many different response variables in a given experiment. The multiple-response problem occurs frequently in computer simulation experiments with social systems. For example, salary, security, status, power, prestige, social service, and professional excellence, to mention only a few, might all be treated as response variables in a simulation experiment with a model of an organization.

It is often possible to bypass the multiple-response problem by treating an experiment with many responses as many experiments, each with a single response. Or several responses could be combined (e.g., by addition) and treated as a single response. However, it is not always possible to bypass the multiple-response problem; often multiple responses are inherent to the situation under study. Unfortunately, experimental design techniques for multiple-response experiments are virtually nonexistent.

Any attempt to solve the multiple-response program is likely to require the use of utility theory. Gary Fromm (1966) has taken an initial step in this direction by using utility theory to evaluate the results of policy simulation experiments with the Brookings model. The specific problem that confronted Fromm was how to choose among alternative economic policies that affect a large number of different response variables in many different ways. He treated utility as a response variable and developed a discounted utility function over time which depends on the values of the endogenous variables of the model, as well as the mean, variance, skewness, and kurtosis of these variables.

Output Analysis

In a well-designed experiment consideration must be given to methods of analyzing the output (or output data) once it is obtained. Most of the classical experimental design techniques described in the literature are used in the expectation that the output data will be analyzed by one or both of the following methods: analysis of variance and regression analysis. The analysis of variance is a collection of techniques for data analysis which is appropriate when qualitative factors are present, although quantitative factors are not excluded. Regression analysis is a collection of techniques for data analysis which utilizes the numerical properties of the levels of quantitative factors. From a mathematical point of view the distinction between regression and the analysis of variance is somewhat artificial. For example, an analysis of variance can be performed as a regression analysis by using dummy variables that can assume only the values zero or one. An excellent treatise on the application of regression analysis has been written by Draper and Smith (1966). Since the great bulk of experimental design techniques described in the literature have the analysis of variance as the intended method of data analysis, we shall investigate several special cases of analysis of variance. These techniques include the F-test, multiple comparisons, multiple rankings, spectral analysis, sequential sampling, and nonparametric methods. We shall investigate the application of each of these techniques to the analysis of output data generated by simulation experiments with models of social systems.

F-test. Suppose that we are interested in testing the null hypothesis that the expected payoffs associated with each of five political strategies are equal. The F-test is a straightforward procedure for testing hypotheses of this type. If the null hypothesis is accepted in our example experiment, then one tentatively concludes that the sample differences between strategies are attributable to random fluctuations rather than to actual differences in population values (expected payoffs). On the other hand, if the null hypothesis is rejected, then further analysis, such as multiple comparisons and multiple rankings, is recommended. The F-test rests on three important assumptions: (1) *normality,* (2) *equality of variance,* and (3) *statistical independence.* The papers by Naylor, Wertz, and Wonnacott (1967, 1968) contain two applications of the use of the F-test to analyze output data generated by simulation experiments.

Multiple comparisons. Typically, social scientists are interested not only in whether alternatives differ but also in *how* they differ. Multiple-comparison and multiple-ranking procedures often become tools relevant to meeting the latter issue, for they have been designed specifically to attack questions of how means of many populations differ.

In contrast with the analysis of variance, multiple-comparison methods (Dunnett, 1955) emphasize the use of confidence intervals rather than the testing of hypotheses. For example, if one is interested in comparing the means of different populations, then a number of (100- a) percent confidence intervals for the differences between population means may be constructed. Gupta (1965) and Winer (1962) have written comprehensive surveys of multiple-comparison procedures. Naylor, Wertz, and Wonnacott (1967, 1968) have applied multiple comparisons to the analysis of output data from simulation experiments.

Multiple rankings. Frequently, the objective of computer simulation experiments with models of social and administrative systems is to find the "best," "second best," "third best," and so forth. Although multiple-comparison methods of estimating the sizes of differences between policies (as measured by population means) are often used as a way of attempting, indirectly, to achieve goals of this type, multiple-ranking methods represent a more direct approach to a solution of the ranking problem.

A good estimate of the rank of a set of alternatives is simply the ranking of the sample means associated with the given alternatives. Because of random error, however, sample rankings may yield incorrect results. With what probability can we say that a ranking of sample means represents the true ranking of the population means? It is basically this question that multiple-ranking procedures attempt to answer.

Bechhofer, Dunnett, and Sobel (1954) have developed a procedure for selecting a single population and guaranteeing with probability P that the selected population is the "best" provided some other condition on the parameters is satisfied. This procedure assumes normality, statistical independence, and a common *unknown variance*. It has been used by Naylor, Wertz, and Wonnacott (1967) in simulation experiments with a model of a multiprocess firm to evaluate the profitability of alternative managerial plans and strategies.

Spectral analysis. Spectral analysis is a statistical technique frequently employed in the physical sciences and more recently applied by economists to analyze the behavior of economic time series (Blackman and Tukey, 1958; Fishman, 1967; Fishman and Kiviat, 1967; Granger and Hatanaka, 1964; Jenkins, 1961; Parzen, 1961; Tukey, 1961). There are at least four reasons for considering spectral analysis a possible technique for analyzing output data generated by simulation experiments with a model of a social system.

First, output generated by computer simulation experiments is usually highly autocorrelated—for example, births in period t are likely to be highly correlated with births in period $t - k$. As we have stated previously, it is well known that when autocorrelation is present in sample data the use of classical statistical estimating techniques that assume the absence of autocorrelation will lead to underestimates of sampling variances (which are unduly large) and inefficient predictions. Several methods are available for treating this problem:

1. Ignore autocorrelation and computer sample means and variances over time, thereby incurring the aforementioned statistical problems.
2. Divide the sample record length into intervals that are longer than the intervals of major autocorrelation and work with the observations on these supposedly independent intervals (Fishman and Kiviat, 1967). This method suffers from the fact that "the choices of sample record length and sampling interval seem to have neither enough prior nor posterior justification in most cases to make this choice much more than arbitrary" (Fishman and Kiviat, 1967).
3. Replicate the simulation experiment and compute sample means and variances across the ensemble rather than over time. This method may lead to excessive computer running time and fail to yield the type of information desired about a particular time series.
4. Use a technique such as spectral analysis, which is based on a model in which the probabilities of component outcomes in a time series depend on previous outcomes in the series. With spectral analysis the problems associated with methods (1) and (2) can successfully be avoided without replicating the experiment.

Second, "When one studies a stochastic process, he is interested in the average level of activity, deviations from this level, and how long these deviations last, once they occur" (Fishman and Kiviat, 1967). Spectral analysis provides this kind of information.

Third, with spectral analysis it is relatively easy to construct confidence bands and to test hypotheses for comparing two or more alternative simulation runs. Frequently, it is impossible to detect differences in time series generated by simulation experiments when one restricts himself to simple graphical analysis. Spectral analysis provides a means of objectively comparing time series generated with a computer model.

Fourth, spectral analysis can also be used as a technique for validating a model of a social system. By comparing the estimated spectra of simulated output data and corresponding real-world data one can infer how well the model resembles the system it was designed to emulate (Fishman and Kiviat, 1967).

Fishman and Kiviat (1967) have written a pathbreaking article on the use of spectral analysis in analyzing data generated by computer simulation models. The books by Blackman and Tukey (1958) and Granger and Hatanaka (1964) and the papers by Jenkins (1961) and Parzen (1961) are recommended for obtaining the basic elements of spectral analysis. Tukey (1961) has written a paper in which spectral analysis and the analysis of variance are compared in detail. Naylor, Wertz, and Wonnacott (1969) have applied spectral analysis to the analysis of simulation experiments with econometric models. Spectral analysis has also been used to compare output generated by computer simulation experiments on a model of the textile industry with corresponding real-world data as a technique of verification (Naylor, Wallace, and Sasser, 1967).

Sequential sampling. Since computer time is not a free gift of nature, output generated by computer simulation experiments (observations) is costly. The cost of experimentation may greatly be reduced if at each stage of the simulation experiment the analyst balances the cost of additional observations (generated by the computer) against the expected gain in information. With computer simulation experiments, the objective of sequential sampling is to minimize the number of observations (sample size) for obtaining the information that is required from the experiment. Instead of setting in advance the number of observations to be generated, the sample size n is considered a random variable dependent on the outcome of the first $n - 1$ observations. In terms of computer time, the cost of a simulation run is minimized by generating only enough observations to achieve the required results with predetermined accuracy.

For example, a sequential test on a computer model of a firm could be designed to determine if the profits obtained by using a certain investment policy in combination with various production policies differed significantly. The sequential method sets a procedure for deciding at the ith observation whether to accept a given hypothesis, reject the hypothesis, or continue sampling by taking the $(i+1)$th observation. Such a procedure must specify for the ith observation a division of the i-dimensional space of all possible observations into three mutually exclusive and exhaustive sets: an area of preference A_i for accepting the hypothesis, an area of preference B_i for rejecting it, and an area of indifference C_i where no statement can be made about the hypothesis and further observations are necessary. The fundamental problem in the theory of sequential sampling is that of a proper choice of the sets—A_i, B_i and C_i.

Wald's *Sequential Analysis* (1947) is the best-known reference on sequential procedures. The article by Chernoff (1959) is also worthy of consideration.

Nonparametric methods. In addition to the aforementioned techniques, numerous nonparametric techniques of data analysis are available. See the books by Siegel (1956) and Walsh (1962).

REFERENCES

Barton, Richard F., *A Primer on Simulation and Gaming.* Englewood Cliffs, N.J.: Prentice-Hall, Inc., 1970.

Bechhofer, Robert E., C.W. Dunnett, and M. Sobel, "A Two-Sample Multiple Decision Procedure for Ranking Means of Normal Populations with a Common Unknown Variance," *Biometrika*, XLI (1954), 170-76.

Blackman, R.B., and J.W. Tukey, *The Measurement of Power Spectra.* New York: Dover Publications, Inc., 1958.

Bonini, Charles P., *Simulation of Information and Decision Systems in the Firm.* Englewood Cliffs, N.J.: Prentice-Hall, Inc., 1963.

Box, G.E.P., "The Exploration and Exploitation of Response Surfaces: Some

General Considerations and Examples," *Biometrics*, X (1954), 16-60.

———, and K.B. Wilson, "On the Experimental Attainment of Optimum Conditions," *Journal of the Royal Statistical Society B*, XIII (1951), 1-45.

Burdick, Donald S., and Thomas H. Naylor, "Design of Computer Simulation Experiments for Industrial Systems," *Communications of the ACM*, IX (May 1966), 329-39.

Chernoff, Herman, "Sequential Design of Experiments," *Annals of Mathematical Statistics*, XXX (September 1959), 755-70.

Churchman, C. West, "An Analysis of the Concept of Simulation," in Austin C. Hoggatt and Frederick E. Balderston (eds.), *Symposium on Simulation Models*. Cincinnati: South-Western Publishing Co., 1963.

Clark, C.E., "Importance Sampling in Monte Carlo Analyses," *Operations Research*, IX (1961), 603-20.

Clarkson, G.P.E., and H.A. Simon, "Simulation of Individual and Group Behavior," *American Economic Review*, L, 5 (December 1960), 920-32.

Cochran, W.G., and G.M. Cox, *Experimental Designs*. New York: John Wiley & Sons, Inc., 1957.

Cohen, Kalman J., *Computer Models of the Shoe, Leather, Hide Sequence*. Englewood Cliffs, N.J.: Prentice-Hall, Inc., 1960.

———, and R. M. Cyert, "Computer Models in Dynamic Economics," in R. M. Cyert and James G. March, *A Behavioral Theory of the Firm*. Englewood Cliffs, N.J.: Prentice-Hall, Inc., 1963.

Conway, R.W., "Some Tactical Problems in Digital Simulation," *Management Science*, X (October 1963), 47-61.

Cyert, Richard M., "A Description and Evaluation of Some Firm Simulations," *Proceedings of the IBM Scientific Computing Symposium on Simulation Models and Gaming*. White Plains, N.Y.: IBM, 1966.

———, and James G. March, *A Behavioral Theory of the Firm*. Englewood Cliffs, N.J.: Prentice-Hall, Inc., 1963.

Draper, N.R., and H. Smith, *Applied Regression Analysis*. New York: John Wiley & Sons, Inc., 1966.

Dunnett, C.W., "A Multiple Comparison Procedure for Comparing Several Treatments with a Control," *Journal of the American Statistical Association*, L (1955), 1096-1121.

Ehrenfield, S., and S. Ben-Tuvia, "The Efficiency of Statistical Simulation Procedures," *Technometrics*, IV (May 1962), 257-75.

Fishman, George S., "Problems in the Statistical Analysis of Simulation Experiments: The Comparison of Means and the Length of Sample Records," *Communications of the ACM*, X (February 1967), 94-99.

Fishman, G.S., and Philip J. Kiviat, "The Analysis of Simulation-Generated Time Series," *Management Science*, XIII (March 1967), 525-57.

"Fractional Factorial Designs for Factors at Two and Three Levels," U.S. Department of Commerce, National Bureau of Standards, Applied Mathematics Series 58, U.S. Government Printing Office, Washington, D.C. (September 1, 1961).

Fromm, Gary, "An Evaluation of Monetary Policy Instruments." Paper presented at the annual meeting of the Econometric Society, San Francisco, December 1966.

Gafarian, A.V., and C.J. Ancker, "Mean Value Estimation from Digital Computer Simulation," *Operations Research* (January-February 1966).

General Purpose Simulator II, Program Library, Reference 7090-CS-13X, International Business Machines Corporation.

Gordon, G., "A General Purpose Systems Simulator," *IBM Systems Journal,* I (1962).

Granger, C.W.J., and M. Hatanaka, *Spectral Analysis of Economic Time Series.* Princeton, N.J.: Princeton University Press, 1964.

Guetzkow, Harold (ed.), *Simulation in Social Science: Readings.* Englewood Cliffs, N.J.: Prentice-Hall, Inc., 1962.

———, *et al., Simulation in International Relations: Developments for Research and Teaching.* Englewood Cliffs, N.J.: Prentice-Hall, Inc., 1963.

Gupta, S.S., "On Some Multiple Decision (Selection and Ranking) Rules," *Technometrics,* VII (May 1965), 225-46.

Hammersley, J.M., and D.C. Handscomb, *Monte Carlo Methods.* New York: John Wiley & Sons, Inc., 1964.

Hill, William J., and William G. Hunter, "A Review of Response Surface Methodology: A Literature Survey," *Technometrics,* VIII (November 1966), 571-90.

Holt, Charles C., Robert W. Shirley, Donald V. Steward, Joseph L. Midler, and Arthur Stroud, "Program SIMULATE, a User's and Programmer's Manual," Social Systems Research Institute, University of Wisconsin, May 1964, mimeographed.

Hufschmidt, M.M., "Analysis of Simulation: Examination of Response Surface," in Arthur Maass *et al.* (eds.), *Design of Water-Resource Systems,* Cambridge: Harvard University Press, 1966.

Jenkins, G.M., "General Considerations in the Analysis of Spectra," *Technometrics,* III (May 1961), 133-66.

Markowitz, H.M., Bernard Hausner, and H.W. Karr, *SIMSCRIPT: A Simulation Programming Language,* the RAND Corporation, RM-3310 (November 1962).

Naylor, Thomas H., *Computer Simulation Experiments.* New York: John Wiley & Sons, Inc., 1970.

———, (ed.), *The Design of Computer Simulation Experiments.* Durham, N.C.: Duke University Press, 1969.

———, Joseph L. Balintfy, Donald S. Burdick, and Kong Chu, *Computer Simulation Techniques.* New York: John Wiley & Sons, Inc., 1966.

———, Donald S. Burdick, and W. Earl Sasser, "Computer Simulation Experiments with Economic Systems: The Problem of Experimental Design," *Journal of the American Statistical Association,* LXII (December 1967), 1315-37.

———, and J.M. Finger, "Verification of Computer Simulation Models," *Management Science,* XIV (October 1967), 92-101.

———, William H. Wallace, and W. Earl Sasser, "A Computer Simulation Model of the Textile Industry," *Journal of the American Statistical Association,* LXII (December 1967), 1338-64.

———, Kenneth Wertz, and Thomas Wonnacott, "Methods for Analyzing Data from Computer Simulation Experiments," *Communications of the ACM;* X (November 1967), 703-10.

———, "Some Methods for Evaluating the Effects of Economic Policies Using Simulation Experiments," *Review of the International Statistical Institute,* XXXVI (1968), 184-200.

———, "Spectral Analysis of Data Generated by Simulation Experiments with Econometric Models," *Econometrica,* XXXVII (April 1969), 333-52.

Parzen, Emanuel, "Mathematical Considerations in the Estimation of Spectra," *Technometrics,* III (May 1961), 167-90.

Preston, Lee E., and Norman R. Collins, *Studies in a Simulated Market*, Research Program in Marketing, Graduate School of Business Administration, University of California, Berkeley, 1966.

Pugh, Alexander L., *DYNAMO User's Manual.* Cambridge, Mass.: The M.I.T. Press, 1963.

Reichenbach, Hans, *The Rise of Scientific Philosophy.* Berkeley: University of California Press, 1951.

Siegel, Sidney, *Nonparametric Statistics.* New York: McGraw-Hill Book Company, 1956.

Teichroew, Daniel, and John F. Lubin, "Computer Simulation: Discussion of Techniques and Comparison of Languages," *Communications of the ACM,* IX (October 1966), 723-41.

Theil, H., *Economic Forecasts and Policy.* Amsterdam: North-Holland Publishing Co., 1961.

Tocher, K.D., *The Art of Simulation.* Princeton, N.J.: D. Van Nostrand Co., Inc., 1963.

Tukey, John W., "Discussion Emphasizing the Connection between Analysis of Variance and Spectral Analysis," *Technometrics,* III (May 1961), 191-220.

Wald, A., *Sequential Analysis.* New York: John Wiley & Sons, Inc., 1947.

Walsh, John E., *Handbook of Nonparametric Statistics,* I & II. Princeton, N.J.: D. Van Nostrand Co., Inc., 1962, 1965.

Winer, B.J., *Statistical Principles in Experimental Design.* New York: McGraw-Hill Book Company, 1962.

13

Theory Construction and Comparison
through Simulation

This chapter consists of two readings, both written by Harold Guetzkow. They represent his original thinking and research on theory construction and comparison through simulation. Although they are both written in the setting of international relations, their significance spans the disciplines covered in this volume. We have included the articles in their entirety, believing that they serve both as "case-examples" of this important use of simulation and as capstones for the development of simulation that we have presented in this work.—The Editors

SIMULATIONS IN THE CONSOLIDATION
AND UTILIZATION OF KNOWLEDGE
ABOUT INTERNATIONAL RELATIONS

HAROLD GUETZKOW

In the last decade and a half, important gains have been made in the number and quality of research studies in the field of international relations, evidenced by such empirical pieces as those assembled by J. David Singer (1967) and by Dean G. Pruitt and Richard C. Snyder (1969). If the trend continues, there will be an increasing need for ordering and integrating the knowledge generated in such studies—and an opportunity for the application of the findings, in on-going decisions by policy-makers of the world. During this same period of time, a capability to simulate complex international processes was created by the development of a variety of simulation formats (Guetzkow, 1966a) and by the invention of simulation languages (Naylor, *et al.*, 1966, Chapter 7).

This essay explores the potentiality of using simulations as devices for ordering theories and for integrating empirical findings. Buttressed by verbal deliberations on the one hand and by mathematical formulations on the other, can simulations implement the ability of decision-influencers to use such theories and findings in international policy-making? At the end of this presentation, a proposal is developed to illustrate one way in which work with simulations might be organized for the consolidation and utilization of knowledge about international relations.

SIMULATIONS AS A FORMAT FOR THEORY

Cver the past centuries it has been customary to express political, economic, social, and psychological theory in words—in the vernacular of the

Reprinted with permission from Dean G. Pruitt and Richard C. Snyder (eds.), *Theory and Research on the Causes of War* (Englewood Cliffs, N. J.: Prentice-Hall, Inc., 1969). Copyright©1967 by Northwestern University. This paper was prepared as part of the activities of the Simulated International Processes project (Advanced Research Projects Agency, SD260) conducted within the International Relations Program at Northwestern University.

times, after the demise of Latin as the *lingua franca*. With the development of mathematics, scholars possessed a vehicle by which they might express their loose verbal formulations with more explicitness, separating their assumptions from derivations which follow as consequences of their analyses (Alker, 1965). But both of these formats for the development of theories have shortcomings. The serial nature of verbal exposition, with one thought following another on the written page, imposes serious limitations. Likewise, the intractability of many mathematical systems, once non-linear formulations are involved, seriously handicaps the investigators. Building upon both verbal and mathematical expositions, contemporary scholars are exploring the usefulness of simulations as devices for handling complex materials, both theoretical and empirical (Dawson, 1962; Naylor, *et al.*, 1966; Evans, *et al.*,1967, Chapter 1, pp. 1-15). In the social sciences, a simulation may be conceived as "an operating representation in reduced and/or simplified form of relations among social units [i.e., entities] by means of symbolic and/or replicate component parts" (Guetzkow, 1959, p. 184).

Simulations in international relations attempt to represent the on-going international system or components thereof, such as world alliances, international organizations, regional trade processes, etc. Clark Abt and Morton Gorden (1968) and their associates have represented such processes as perception, homeostasis, and bargaining, through a digital computer simulation called TEMPER, a *T*echnological, *E*conomic, *M*ilitary, *P*olitical *E*valuation *R*outine. Harold Guetzkow and his associates (Guetzkow, Alger, Brody, Noel, and Snyder, 1963) have developed man-machine constructions (sometimes called "games") in which the decision-making processes are handled by human participants serving as surrogates for the international actors, whilst national processes are formulated through some thirty equations, the computation of which serves to represent the capabilities and consequences of the decision-making. Lincoln P. Bloomfield and his associates (Bloomfield and Whaley, 1965), following earlier developments by Hans Speier and others at the RAND Corporation (Goldhamer and Speier, 1959), used the "political exercise" in which crisis gaming among area experts is monitored by a "control team" which serves to umpire moves developed in response to an on-going scenario. This "all-manual" (as distinguished from the "all-computer") format is now used intensively at high levels within some parts of the United States government (Giffin, 1965). Because the control team operates principally in terms of intuitive verbal theory as it directs the progress of the game, allowing and disallowing particular international behaviors and imposing consequences on the various countries' teams, this simulation style in the long run may not prove as useful a format for the consolidation of knowledge as will man-computer and all-computer simulations. Further, it seems that, as the state of the computer arts becomes more adequate and our knowledge about international affairs grows more explicit and is grounded on a better data-base, man-machine constructions will be replaced by all-computer simulations. Already this has occurred, for example, in the development of formulations about legislatures, which moved from James S. Coleman's all-manual game (Coleman, 1963) to the all-computer simulation of voting in the Eighty-eighth Congress by Cleo H. Cherryholmes (1966) and Michael J. Shapiro (1966).

Consolidating Knowledge about International Relations through Simulations

As we move into the latter third of the Twentieth Century, it seems feasible to catalyze the consolidation of our knowledge about international affairs through the use of simulations. Verbal efforts to present holistic integrations of extant knowledge are found in the textbooks of international relations. Yet, their contents are theoretically vague and their data bases are largely anecdotal as Denis G. Sullivan points out (Sullivan, 1963; especially his "Conclusions," pp. 305-313). Mathematical formulations, such as those by Lewis F. Richardson (1960), are more partial in scope, even though they are explicit in structure and systematic in their grounding in data. When an attempt is made to be comprehensive, as occurred in the work of Rudolph J. Rummel (1966), the mathematical theory tends to be at a metalevel, more statistically theoretical than substantively explicit.

How can simulations be used as vehicles for accumulating and integrating our knowledge, both in its theoretical and its empirical aspects, building upon the contributions of those who work in ordinary language as well as of those who use the language of mathematics?

Simulations may serve in three ways as formats through which intellectuals may consolidate and use knowledge about international relations: (1) Simulations may be used as techniques for increasing the coherence within and among models, enabling scholars to assess gaps and closures in our theories; (2) Simulations may be used as constructions in terms of which empirical research may be organized, so that the validity of our assertions may be appraised; (3) Simulations may be used by members of the decision-making community in the development of policy, both as devices for making systematic critiques, through "box-scoring" its failures and successes, and as formats for the exploration of alternative plans for action.

(1) Simulations in the differentiation and amalgamation of theories in international relations. Simulations of the international system are frameworks into which both verbal and mathematical formulations may be incorporated, therein combining something of the rigor of a mathematical model (which an all-computer simulation is) (Guetzkow, 1965) with the comprehensiveness of a verbal inventory (Snyder and Robinson, 1961 [*sic*]). Quincy Wright (1955) noted years ago in his *Study of International Relations* that our knowledge of international affairs develops in fragments. What is examined piecemeal, however, must eventually be reassembled, especially if the knowledge is to be used in policy work, where problems come as wholes. Once differentiated, our findings must be amalgamated.

The appearance of "handbooks" in the social sciences, consisting of chapters which attempt integrative summaries of bodies of literature, such as the one developed by Herbert C. Kelman (1965) on *International Behavior,* dramatizes how knowledge tends to be developed segmentally, composing "islands of theory" (Guetzkow, 1950, pp. 426, 435, 438, *et passim*). The contributors to a handbook single out a componential process within international affairs, such as "Bargaining and Negotiation" (Sawyer and Guetzkow, 1965), foci which are sometimes differentiated in much detail elsewhere, as in this case in the exciting verbal treatments by Fred Charles Iklé in *How Nations Negotiate* (1964) and by Arthur Lall in *Modern International Negotiation* (1966). In parallel, a body of quasimathematical work may develop, as in this same instance is found in the Theory of Games (Shubik, 1964). Within a simulation, aspects of both streams of theory may then be consolidated as a nodule or a modular—as each subroutine of a simulation is sometimes designated—as was done by Otomar Bartos in developing a negotiating routine for international trade in which he used the rubics of J. F. Nash's mathematically formulated "solution" (Sherman, 1963).

The "reader," exemplified in the influential compilation of *International Politics and Foreign Policy* (Rosenau, 1961), uses juxtaposition as a tool for the integration of knowledge. But, a more closely articulated and systematic integration of components is now possible through the use of simulation. As Paul Smoker and John MacRae demonstrate in the reconstruction of the Inter-Nation Simulation to explore the Vietnam situation, it is possible to take an already existing model of international affairs and incorporate additional and revised components within the existing framework (MacRae and Smoker, 1967, see Appendix, pp. 11-23). For example, in this Canadian/English simulation of the Vietnam situation, the collaborators were able to use Smoker's earlier work (Smoker, 1965) with the Richardson model in a rigorous development of polarization as dependent upon both trade and defense (MacRae and Smoker, 1967, pp. 16-17, *cf.*, "The Computer Model" columns) in defining "National Security"—an important feature heretofore absent from the Inter-Nation Simulation.

Simulations, especially those of the all-computer variety, demand a clarity that is unusual in theory building (Guetzkow, 1965, pp. 25-39) in the specification of the entities involved, in the exact involvement of variables used to describe the entities, and in the explicit formulation of the relations among both entities and variables. Once these components of theory have been assembled into a simulation, gaps within the framework become more readily apparent. One reason Walter C. Clemens (1968) elucidated the shortcomings of TEMPER with ease is found in its high level of explicitness (Guetzkow, 1966a). As more and more effort is put into amalgamating part-theories, there will be an increasing need for a standard language within which to construct each such "island of theory," so that they may be readily incorporated into large, more encompassing constructions.

Despite the difficulties involved, as one of the central architects of TEMPER knowingly testifies (Gorden, 1967), with improvements in simulation languages (Naylor, *et al.*, 1966) it will be possible to articulate one modular with another more easily, if they are all built originally in a common computer language. Then there may be a division of labor among scholars, in which each may work on his components with a thoroughness worthy of his specialization. Then, when his "islands of theory" are placed within a simulation, the researcher may become aware of the broader issues that are relevant to his area of focus.

The complexities of theory, which are impossibly cumbersome when the ideas are formulated verbally in textbooks (Scott, 1967) and intractable when the ideas are structured as models (Orcutt, 1964, pp. 190-191), may become more amenable when simulations are used to organize the division of labor more coherently among the scholars working within international affairs, providing for a differentiation of effort as well as for an amalgamation of findings.

(2) Simulations as vehicles in the validation of theories in international relations. Simulations of the international system are devices through which empirical findings may be organized, so that the validity of their theoretical contents may be assessed. With the coming increase in the number of "data-making" studies (Singer, 1965, pp. 68-70) of the "real world," as reference materials are sometimes designated, there is need for consolidation of these empirical findings, as well as integration of our theories. Theory of all kinds—be it verbal, verbal-mathematical, or simulation—needs to be validated. As Charles F. Hermann has pointed out, it is important for many purposes to determine the degree of correspondence between the simulation model and the reference system (Hermann, C. F., 1967, p. 220), whether interested in the variables and parameters of the model (*ibid.*, p. 222), in the similarity or dissimilarity of the array of events produced in both simulation and the world (*ibid.*, p. 222-23), or in determining whether the same hypotheses hold in both model and reference systems (*ibid.*, pp. 223-24). When policy work is based on explicitly formulated theory, it is possible to judge the adequacy of policy alternatives more adequately if the extent of its validation is known.

Man-machine and all-computer simulations, especially, provide a systematic, somewhat rigorous technique for the appraisal of the validity of theory. Richard W. Chadwick (1966) has shown how correspondences between hypotheses embodied as assumptions about the functioning of national political systems may be checked out against empirical data gathered from the reference system of years centering on 1955. For example, he found that, although the likelihood of a decision-maker to continue in office is assumed in the simulation to be a function both of the latitude the decision-maker has in constructing his policies and the extent to which his supporters are satisfied with the consequences of his policies, the hypothesis holds in the international reference system of 1955 only with respect to the latter (Chadwick, 1966, p. 11).

Guetzkow (1967) was able to examine over twenty studies in which one or more operations in simulations of international processes were each paralleled by an empirical finding. These ranged from correspondences in the form of anecdotes about events [such as the fact that a conference, called by Lord Grey of England in the prelude to World War I and never assembled, proved to be the vehicle by which the issue was resolved by the participants in the Hermanns' adaptation of the Inter-Nation Simulation representing European developments in the summer of 1914 (Hermann and Hermann, 1967, pp. 407-8)] to correspondences between relationships among variables [such as the linear function between national consumption standards and the satisfaction of the groups which validate the officeholders (Elder and Pendley, 1966, p. 31)]. In his summary of particular comparisons of simulation outputs with data from the reference system, Guetzkow found there was "Some" or "Much" congruence in about two-thirds of the fifty-five instances available from the twenty-three studies, providing a kind of "box-score" on the simulations. These findings, taken in conjunction with achievements being realized through simulation in other parts of the social sciences (Guetzkow, 1962a), foreshadow the fruitfulness of cumulating findings on the validity of theory as it has been integrated within an operating simulation model.

Simulations are not only apt vehicles for making studies of the systematic, rigorous validations of theory. When used for such purposes, they also heuristically spin off ideas for revising theory about international processes. For example, in the man-computer format Dina A. Zinnes demonstrated an inadvertent error in the construction of the Inter-Nation Simulation: by omitting the buffer role of embassies between the home nation's foreign office and the foreign offices of other nations, a "small groups" effect was elicited, in which a cycle of even more hostility leading to less communication leading further to even more hostility exacerbated itself (Zinnes, 1966, pp. 496, 498-99). This effect was not found in the relations among the European capitals in the summer of 1914. Now it is possible to reconstruct the simulation so as to avoid this so-called "autistic hostility" phenomenon (Newcomb, 1947). Another example of the way in which validation study of simulation theory aids in its revision is found in Robert E. Pendley and Charles D. Elder's re-definition of the meaning of "officeholding" in the programmed components of the Inter-Nation Simulation (INS) in terms of contemporary verbal theory and data on the stability of regimes and governments. After comparing ways in which the simulation and the reference system behave, they conclude that "INS theory is a fairly good predictor of stability, but that it is the stability of the political system rather than stability of particular officeholders that the theory explains" (Pendley and Elder, 1966, p. 25).

Thus, simulations are useful devices through which efforts in the validation of their theoretical soundness may be organized when outputs of simulations are compared with corresponding characterizations in the reference system. Further, in the very process of making the comparisons one has a heuristic tool through which verbal and mathematical speculation can be grounded empirically to provide a base for the revision of simulation theory. Unless simulation theory is validated, it would seem unwise to use it for "decision-making" in the policy-making community.

(3) Simulations in the utilization of knowledge for policy-making in international affairs. Were a body of consolidated knowledge about international affairs available, it would seem that simulations might aid in the utilization of that knowledge—for monitoring on-going events as well as for the construction of "alternative futures." The myriad of actors within the international system—be they members of planning units in foreign ministries, entrepreneurs in business operating overseas, or officials within governmental and non-governmental international organizations—base their decisions for actions upon their assumptions of the ways in which this system functions, combined with their assessments of its present state. Simulations may increase the adequacy with which knowledge about international affairs is utilized in the conduct of foreign affairs, by providing explicit theories as to how the system operates, as well as by providing a continuously up-dated data-base. A somewhat comprehensive list of "Some Areas of Knowledge Needed for Undergirding Peace Strategies," presented elsewhere (Guetzkow, 1962b, pp. 90-91), runs the gamut of such topics as "initiative and coordination within national security decision machinery" and "international communications." Simulations, geared to the policy problems confronting the public and private decision-makers of the world, can serve in two ways as aids in the utilization of this knowledge: (a) in being a framework within which the antecedents and consequences of on-going policy decisions can be examined, and (b) in being a way of considering alterna-

tive futures of the international system, either as contingencies or as ends whose paths-to-achievement may be plotted.

(a) *"Box-Scoring" Policy Decisions.* In the hurly-burly of organizational life, there is seldom time for explicit analyses of the effectiveness of "hits" and "misses." Seldom does the decision-maker systematically sort out the way the antecedents in his decision situations eventuate in their consequences. Yet, today there are increasingly adequate techniques of both verbal and mathematical varieties available, which help in structuring knowledge so that it may be used more effectively in decision processes. For example, the verbal analyses involved in program-planning and budgeting procedures now being urged throughout government (Chartrand and Brezina, 1967) are becoming more and more sophisticated in their specification of the means-ends chains involved in cost-benefit analyses (Grosse, 1967). With the mathematization of "optimizing techniques" within operations research (Carr and Howe, 1964), miniature theories are being constructed about the assessment of the influence of factors upon outcomes. These verbal and mathematical sources of explication are making it feasible to construct on-going simulations of policy processes of decision-making, as illustrated in extant all-computer models of budgetary processes of municipalities (Crecine, 1965). In the arena of international relations, it also would seem possible to use simulations as vehicles for the explication of decision-making processes, thereby "box-scoring" policy-making.

Suppose a policy-planning group in a country's disarmament and arms control bureau were interested in making an analysis of the relations among antecedents and consequences of their policy decisions with respect to the nation's postures in a multilateral "standing group" operating in Geneva. It might then erect a simulation of the international system and run it parallel with the policy deliberations. For example, it might combine a negotiation exercise (Bonham, 1967) within the context of a man-computer simulation of international processes (Smoker, 1967), tailored to fit conditions of the moment. Were there disagreement within the staff, two or more alternative simulations could be explored, changing the parametric weighting given to particular variables, and even substituting one module for another, allowing them to "compete" as to adequacy. Through such a double-nested simulation, it would be possible to examine developments at two levels—in terms of (1) the on-going conference situation in Geneva, and (2) the changes in the overall international political scene itself. In examining the immediate situation, they could simulate the action of the committee of principals within the foreign policy machinery of the government itself, along with responses of opposite numbers, of allies, and of non-aligned nations—once policy proposals were activated in the international arena. In examining the context of the work of the multilateral group, one would simulate the arms race within regions as well as globally, including the impact of the failure to achieve a non-proliferation treaty as well as the consequences of already agreed-upon treaties, such as the test-ban.

One operation of the simulation might be molded to be strictly congruent with on-going policies. Then the developments—both antecedents (in structure and process) and consequences (in outcomes and feedbacks)—could be "box-scored," as the events of the international system unfolded week-by-week, month-by-month. In fact, were alternative simulations operating simultaneously a few weeks or months ahead of the decisions of this committee of principals, their outputs might be used for policy development within the arms control and disarmament agency, were they satisfied with the extent of the validity of these simulation results.

Such a "box-scoring" procedure would demand an explication of the theories (which now are often being used without clear formulation by the policy-makers of the assumptions involved), so that an appropriate simulation (with its competing variations) might be adapted from extant models in forms useful for policy problem-solving. Further, in requiring the tallying of successes and failures, the procedure would eventuate in a careful validation of the simulation. The modulars composing the antecedent processes as postulated by the policy-makers would be assessed as to whether they yield their predicted consequences in the "real world." Note that these two steps are the same two procedures involved in the consolidation of knowledge by basic researchers concerned with international affairs, as outlined in the previous sections of this essay: (1) the amalgamation of verbal and mathematical theory in a simulation's constructions, and (2) its validation through empirical confrontations.

To use simulation as an instrument for policy development would be to have a powerful tool by which the ever increasing richness of theory and data might be brought to bear upon decision-making in international affairs by the policy-influencers of the world throughout the remainder of this century. But such "box-scoring" procedures would not only be of aid to the policy-makers; the applied work would have important feedbacks into research. It would provide a vehicle by which the thinking of outstanding political leaders might be fed back into the academic community, so that its work might benefit from the creativity of the policy community. Further, on-going comparisons between expectations and realities, made week-by-week over the years, would highlight congruences and incongruences between the simulation model—in its many variations—and central aspects of the reference system.

(b) *Exploring Alternative Policy Futures.* Simulations are an important heuristic in their potential for representing alternative, future state of affairs which to date have been non-existent (Boguslaw, 1965). As knowledge is consolidated through simulation, enough confidence may be gained eventually to use such constructions for the systematic exploration of alternative futures (de Jouvenel, 1963, 1965). Although, certainly, theory should be data-based, simulations must not be "data-bound" (Guetzkow, 1966b, pp. 189-91).

Efforts to use simulations for the exploration of possible futures are in their infancy. An example in miniature of such pioneering in an all-computer format is found in the U.S. Department of State's analysis by computer of the consequences of various voting arrangements within the United Nations. The simulation was applied to "178 key votes that took place in the General Assembly between 1954 and 1961," the weighting being based on population and contributions to the UN budget. Richard N. Gardner reports that while the weightings "would have somewhat reduced the number of resolutions passed over U.S. opposition, they would have reduced much more the number of resolutions supported by the United States and passed over Communist opposition. The same conclusion was reached in projecting these formulas to 1970, having regard to further increases in membership" (Gardner, 1965, p. 238). Using a man-computer format, Richard A. Brody experimented in the summer of 1960 with the effects of the proliferation of nuclear capabilities upon alliances (Brody, 1963). Working from an inventory of some thirty-six verbal propositions in the literature about the "Nth country problem," Brody designed a variation of the Inter-Nation Simulation so that consequences of an antecedent spread of nuclear weapons technology among nations might be studied. Brody found a "step-level change in the 'cold war system'" after the spread of nuclear capability: Threats external to each bloc were reduced, and threats internal to each block were increased, accompanied by a decrease in bloc cohesiveness; the original bipolarity of the system was fragmented (Brody, 1963, p. 745). A final example, employing the manual technique of the political-military exercise, is found in the recent exploration by **Bloomfield** and his colleagues of the "possible future employment of United Nations military forces under conditions of increasing disarmament" in the context of the U.S. proposals of April 18, 1962, on General and Complete Disarmament (GCD) (Bloomfield and Whaley, 1965). After investigating four hypothetical crises—indirect agression and subversion in Southeast Asia, a colonial-racial civil war in a newly independent African nation, a classic small-power war in the Near East, and a Castro-type revolution in Latin America—Bloomfield draws a set of policy inferences, including the notion that "disarmament planning might well consider whether an appropriate plateau for the GCD process can be found somewhere" (*ibid.*, p. 864).

The use of simulation for sketching alternatives may prove in the long run to be a useful implement in the reconstruction of our international system, should we ever devote enough resources toward the generation and consolidation of knowledge so as to give us the validity to make such constructions viable. In the creation of unprecedented alternative futures (Huntington, *et al.*, 1965), can we manage with verbal speculation alone—or with mathematical formulations only? Perhaps, as we gain experience in amalgamating modulars which have been grounded in empirical findings, we will, someday, have the ability to construct futures which are more than visionary.

These are the goals, then: to develop simulation theory, in the context of verbal speculation and mathematical constructions about the structures and processes involved in international affairs; to apply a variety of criteria in the validation of such theory, depending upon the purposes for which it is intended; to use for decision-making the knowledge which has been consolidated for purposes of policy development, both in terms of short run "box-scoring" and in the long run for the creation of alternative futures. If such is the potential, the query becomes, "How can we accelerate our rate of accomplishment in achieving these goals?"

ACCELERATING THE CONSOLIDATION AND UTILIZATION OF KNOWLEDGE ABOUT INTERNATIONAL RELATIONS THROUGH STANDING COLLOQUIA

Let this essay conclude with a proposal for acceleration in the study of international relations through the establishment of *colloquia* centering on simulations, so that there may be a continuous dialogue between theory builders, empirical researchers, and policy developers. Perhaps the time is now appropriate for a more integrative, long range effort than is possible alone through doctoral dissertations, textbooks, collections of juxtaposed readings, handbooks of summary pieces, substantive inventories developed for special occasions, and *ad hoc* conferences and committee reports. Given the pace of the explosion of knowledge within the international relations area (Platig, 1966, pp. 3–11), which the foregoing efforts are yielding, it seems imperative that a technique of potential efficacy be explored to hasten a tighter, more cumulative articulation of this knowledge.

There now are some fifteen to twenty sites throughout the world at which simulations of international affairs are being conducted. These operations vary widely in their magnitude and quality, as well as in their styles. Only a few of these will nurture their activities into full-fledged centers worthy of adequate, continuous support.

One or more such units probably will operate within a university setting, with its efforts undergirded by the relevant disciplines and its output available to all throughout the world. Such university-related centers would provide training in simulation for scholars and professionals interested in foreign affairs. With the recent emergence of the autonomous research organization, it is difficult to believe that such "think-tanks" (Reeves, 1967)—be they of a "for profit" or a "non-profit" variety—will not give serious attention to simulation work. Perhaps some of the international companies will develop their considerable knowledge, obtained in commercial operations overseas, through special corporate staffs concerned with the simulation of the international system in which they operate, with some focus on the role of the non-governmental organizations. In the decades ahead, at least a dozen or so units concerned with simulation may be established within agencies of different governments in the world. In developing their foreign policies toward each other, there may emerge a common core of data and theories, even though each foreign office—just like each international company—will probably develop for its own exclusive use a simulation base of secret contents. Little wonder, then, that the international organizations—most appropriately perhaps the United Nations Institute for Training and Research—will need to lead the development of simulation models, so that all countries, regardless of their resources in the social sciences, may have access to a universal model for the exploration of the antecedents and political consequences of their policies. Because of the exemplary work the United Nations Secretariat has done on its statistical services in decades past, a universal data-base for simulation work is already well advanced.

Within the last five years there has been a considerable growth in bodies of empirical data which have been and are being generated through elections, interview surveys, and the like (Bisco, 1966). Perhaps now is the time to develop somewhat analogous *standing colloquia*, so that those developing theory—be it verbal or mathematical in style—may consolidate their work. Were one such colloquium staffed with a secretariat, perhaps it could provide a means by which theorists could relate with more intimacy and rigor to the data-gatherers and data-makers. In addition, such a colloquium would provide a forum in which policy-influencers might have their formulations dialogued. It may turn out that just as a number of data consortia are developing throughout the world, there may be more than one colloquium—different modes of simulation and different purposes may demand different kinds of colloquia for varying styles of collaboration. For example, already it seems that those interested in the uses of simulation for education and training are centering their efforts in ways different from those using simulations as vehicles for theory construction and validation (Coleman, Boocock, and Schild, 1966).

How might a colloquium implement collaboration among four to five simulation centers located in different parts of the world? Through a working director—who would probably need to be a young, flexible theorist of some distinction—the secretariat of the colloquium might develop an integrated model of the international system, as was suggested above (*cf., supra,* (1), pp. 4-7). In the process of constructing the model over a period of some five to ten years, periodic sessions with theorists—regardless of the mode in which they work— would be convened, so that a consolidation of fruitful and adequately verified components of theory might be incorporated into the colloquium's simulation. The staff of the colloquium would compare the assumptions underlying extant simulations and design experiments for assessing the importance of the differences. To obviate the need to operate its own simulation in its early years, it might invite three or four simulations already in operation—perhaps on subcontract—to address the same inquiry, so that systematic comparison among the various alternative models might be made, in the style pioneered by Hayward R. Alker, Jr. and Ronald D. Brunner (1967).

Without attempting to operate its own data consortium, how might the staff of the colloquium manage to ground its model in the findings from such research? It might review the empirical literature to assess the extent to which components of its model are being validated. It might then develop recommendations, in the mode of inventories, as to where further empirical work was needed, coordinating the execution of such research so that there would be close matching between simulation and reference materials (*cf., supra,* (2), pp. 7-9). With an ever increasing volume of research being generated through Programs of Area Study as well as in the more traditional Centers of International Studies (Snyder, 1968), the staff of the colloquium might usefully provide a liaison service, so that the developing simulations throughout the world would be more closely articulated with outpourings in empirical research, both of a qualitative and quantitative variety. In fact, just as a member of the colloquium's secretariat might be designated to work with the verbal theory of the more speculative scholars in order to incorporate their ideas into the simulation, so might a special staff member be assigned the task of developing the theory which emerges from empirical studies into a form that could be phrased as modular sub-routines for use in the colloquium's simulation. In fact, it is easy to understand how the colloquium might pay special attention to achievements in simulation in other parts of the social sciences, too, so that full advantage could be taken of developments in the field of artificial intelligence (for the development of foreign policy decision-making models) and in the field of organizational simulation (for the development of inter-nation system models), for example.

In all the activities of the colloquium, policy-related professionals would be involved intimately so that their decision-makers—be such located in foreign offices, in international corporations, or in international organizations—might develop operations in tandem with those evolving in the colloquium. Were some consensus to emerge through the good offices of the colloquium, various simulation groups throughout the two hemispheres might exchange modules, as well as use each other's data-bases—on perhaps multilateral as well as bilateral bases. Which centers will effect a collaboration so that their competitive efforts will become cooperative, too—as in the fashion of the SSRC/Brookings Economic Quarterly Model of the United States (Duesenberry, *et al.,* 1965)? It is exciting to imagine the officers of the inter-parliamentary unions of the regions of the world contracting with colloquia for systematic exploration of some items on their agenda, so that their deliberations might be grounded in the fruits of social sciences, as such are represented in the consolidation of knowledge about international affairs through simulations.

Collaboration among simulation centers will be accelerated mightily with the coming of world-wide computer systems, which might be shared by centers comprising a core group which has proven its ability to work together in an integrative way. Through the leadership of the Western Behavioral Sciences Institute, John Raser and his overseas colleagues—from Japan, Mexico, and Norway—already are gaining experience in the practicality of cooperation in cross-cultural research involving man-computer simulations (Solomon, Crow, and Raser, 1965). Should the work of the colloquium be successful, such centers would operate in a common computer language, making possible the integration of their work. Should the efforts of the colloquium be achieving its goals, the same group of centers would be sharing data-bases using common variables, all commensurable with each other. Eventually the colloquium's function would be merely that of coordinating the operation of a communication-by-satellite system of validated simulations. Then, the policy-makers of the world might all join freely such an international net, building our world futures through cooperative endeavor.

As Hans J. Morgenthau asserts, "What is decisive for the success or failure of a theory is the contribution it makes to our knowledge and understanding of phenomena which are worth knowing and understanding. It is by its results that a theory must be judged . . ." (Morgenthau, 1967, from "PREFACE to the Fourth Edition). Will the use of simulation as a vehicle in the consolidation and utilization of knowledge in international relations enable us to develop theory whose results will give us a better world than the one we've lived in for the last quarter century, whose policy roots have been dominated by the babel of theory posed in ordinary tongues?

SUMMARY

What is simulation's potential for the consolidation and utilization of knowledge about international affairs? Although all-computer and man-computer simulations, as well as all-manual political exercises, may be employed as devices for training participants, simulations may also be used as a way of positing theory and deriving its consequences. Simulations may be used as a tool for the integration of widely used verbal theory, as well as for theory which is developed in mathematical language. Simulations encourage explicitness in formulation and permit the coherent amalgamation of sub-theories into inter-active, holistic constructions of great complexity. Further, using simulation as the format for formulating theory enables systematic and rigorous work to be achieved in the validation of its interrelated parts, feeding back heuristically into reformulations of aspects of the model which are less than congruent with the empirical materials. Finally, data-grounded simulations may aid in the development of policy, both in terms of its evaluation as well as in terms of its creation. By monitoring on-going events with simulations operated in parallel, "box-scores" can be derived for appraising the adequacy of unfolding policies. By using these same simulations as devices for the examination of alternative futures, modification in short term policies can be made and long-run forecasts can be mounted.

In conclusion, a proposal is made for acceleration of the consolidation and utilization of knowledge about international affairs through the establishment of standing colloquia. Analogous to the growing consortia being developed throughout the world for amassing and retrieving political data, it is proposed that special standing colloquia be developed among scholarly and governmental centers, so that competing simulations, along with their verbal theories and mathematical formulations, might be used integratively as ways of coordinating the development and use of knowledge about international relations.

REFERENCES

Abt, Clark, and Morton Gorden. "Report on Project TEMPER." In Dean G. Pruitt and Richard C. Snyder (Editors), *Theory and Research on the Causes of War*. Englewood Cliffs, New Jersey: Prentice-Hall, Inc., 1969.

Alker, Hayward R., Jr. *Mathematics and Politics*. New York: The Macmillan Company, 1965.

Alker, Hayward R., Jr., and Ronald D. Brunner. "Simulating International Conflict: A Comparison of Three Approaches." Mimeo. New Haven, Connecticut: Yale University, July, 1967.

Bisco, Ralph L. "Social Science Data Archives: A Review of Developments." *American Political Science Review*, 40, 1 (March, 1966), 93-109.

Bloomfield, Lincoln P., and Barton Whaley. "The Political-Military Exercise: A Progress Report." *Orbis*, 8, 4 (Winter, 1965), 854-870.

Boguslaw, Robert. *The New Utopians: A Study of System Design and Social Change*. Englewood Cliffs, New Jersey: Prentice-Hall, Inc., 1965.

Bonham, G. Matthew. "Aspects of the Validity of Two Simulations of Phenomena in International Relations." Ph.D. Dissertation. Cambridge, Massachusetts: Department of Political Science, Massachusetts Institute of Technology, 1967.

Brody, Richard A. "Some systemic effects of the spread of nuclear-weapons technology: a study through simulation of a multi-nuclear future." *The Journal of Conflict Resolution*, 7, 4 (December, 1963), 663-753.

Carr, Charles R., and Charles W. Howe. *Quantitative Decision Procedures in Management and Economics*. New York: McGraw-Hill Book Company, 1965.

Chadwick, Richard W. "An Empirical Test of Five Assumptions in an Inter-Nation Simulation, about National Political Systems." Evanston, Illinois: Simulated International Processes project, Northwestern University, August, 1966.

Chartrand, Robert L., and Dennis W. Brezina. "The Planning-Programming-Budgeting System: An Annotated Bibliography." Washington, D. C.: The Library of Congress Legislative Reference Service, April 11, 1967.

Cherryholmes, Cleo H. "The House of Representatives and Foreign Affairs: A Computer Simulation of Roll Call Voting." Ph.D. Dissertation. Evanston, Illinois: Department of Political Science, Northwestern University, August, 1966.

Clemens, Walter C. "TEMPER and International Relations Theory: A Propositional Inventory." In William D. Coplin (Editor), *Simulation Models of the Decision-Maker's Environment.* Chicago: Markham Publishing Co., 1968. Presented at Wayne State University Symposium, Detroit, Michigan, May 10-13, 1967.

Coleman, James S. "The Great Game of Legislature." *The Johns Hopkins Magazine* (October, 1963), 17-20.

Coleman, James S., Sarane S. Boocock, and E. O. Schild (Editors). *In Defense of Games. American Behavioral Scientist*, Part I, 10 (October, 1966); *Simulation Games and Learning Behavior. American Behavioral Scientist*, Part II, 10 (November, 1966).

Crecine, John P. "A Computer Simulation Model of Municipal Resource Allocation," Ph. D. Dissertation. Pittsburgh, Pennsylvania: Carnegie Institute of Technology, 1965.

Dawson, Richard D. "Simulation in the Social Sciences." In Harold Guetzkow (Editor), *Simulation in the Social Sciences: Readings.* Englewood Cliffs, New Jersey: Prentice-Hall, Inc., 1962, 1-15.

De Jouvenel, Bertrand (Editor). *Futuribles: Studies in Conjecture.* Geneva, Switzerland: Droz, 1963 and 1965.

Duesenberry, J. S., G. Fromm, L. R. Klein, and E. Kuh (Editors). *The Brookings Quarterly Economic Model of the United States.* Chicago: Rand, McNally & Co., 1965.

Elder, Charles D., and Robert E. Pendley, "An Analysis of Consumption Standards and Validation Satisfactions in the Inter-Nation Simulation in Terms of Contemporary Economic Theory and Data." Evanston, Illinois: Department of Political Science, Northwestern University, November, 1966.

Evans, George W., II, Graham F. Wallace, and Georgia L. Sutherland. *Simulation Using Digital Computers.* Englewood Cliffs, New Jersey: Prentice-Hall, Inc., 1967.

Gardner, Richard N. "United Nations Procedures and Power Realities: The International Apportionment Problem." *Proceedings of the American Society of International Law,* 59th Meeting (April, 1965), 232-245.

Giffin, Sidney F. *The Crisis Game: Simulating International Conflict.* Garden City, New York: Doubleday & Company, Inc., 1965.

Goldhamer, Herbert, and Hans Speier. "Some Observations on Political Gaming." *World Politics,* 12, 1 (October, 1959), 71-83.

Gorden, Morton. "Burdens for the Designer of a Computer Simulation of International Relations: The Case of TEMPER." In Davis B. Bobrow (Editor), *Proceedings of the Computers and The Policy-Making*

Community Institute. Englewood Cliffs, New Jersey: Prentice-Hall, Inc., 1967. Presented at the Institute held at Lawrence Radiation Laboratory, University of California, Livermore, California, on April 4-15, 1966.

Grosse, Robert N. "The Application of Analytic Tools to Government Policy: The Formulation of Health Policy." In William D. Coplin (Editor), *Simulation Models of the Decision-Maker's Environment.* Chicago: Markham Publishing Co., 1968. Presented at Wayne State University Symposium, Detroit, Michigan, May 10-13, 1967.

Guetzkow, Harold. "Long Range Research in International Relations." *The American Prespective,* 4, 4 (Fall, 1950), 421-440.

Guetzkow, Harold. "A Use of Simulation in the Study of Inter-Nation Relations." *Behavioral Science,* 4 (1959), 183-191.

Guetzkow, Harold (Editor). *Simulation in the Social Sciences: Readings.* Englewood Cliffs, New Jersey: Prentice-Hall, Inc., 1962a.

Guetzkow, Harold. "Undergirding Peace Strategies through Research in Social Science." In Gerhard S. Nielsen (editor), *Psychology and International Affairs: Can We Contribute? Proceedings of the XIV International Congress of Applied Psychology,* Volume I. Copenhagen, Denmark: Munksgaard, 1962b, 88-96.

Guetzkow, Harold. "Some Uses of Mathematics in Simulations of International Relations." In John M. Claunch (Editor), *Mathematical Applications in Political Science.* Dallas, Texas: The Arnold Foundation, Southern Methodist University, 1965, 21-40.

Guetzkow, Harold. "Simulation in International Relations." In *Proceedings of the IBM Scientific Computing Symposium on Simulation Models and Gaming.* York, Pennsylvania: Maple Press, 1966a, 249-278.

Guetzkow, Harold. "Transcending Data-Bound Methods in the Study of Politics." In James C. Charlesworth (Editor), Monograph 6, *A Design for Political Science: Scope, Objectives, and Methods.* Philadelphia: The American Academy of Political and Social Science, December, 1966b, 185-191.

Guetzkow, Harold. "Some Correspondences Between Simulations and 'Realities' in International Relations." Evanston, Illinois: Northwestern University, 1967. In Morton Kaplan (Editor), *New Approaches to International Relations.* New York: St. Martin's Press, 1967.

Guetzkow, Harold, Chadwick F. Alger, Richard A. Brody, Robert C. Noel, and Richard C. Snyder. *Simulation in International Relations: Developments for Research and Teaching.* Englewood Cliffs, New Jersey: Prentice-Hall, Inc., 1963.

Hermann, Charles F. "Validation Problems in Games and Simulations with Special Reference to Models of International Politics." *Behavioral Science,* 12, 3 (May, 1967), 216-231.

Hermann, Charles F., and Margaret G. Hermann. "An Attempt to Simulate the Outbreak of World War I." *American Political Science Review,* 61, 2 (June, 1967), 400-416.

Huntington, Samuel P., Ithiel De Sola Pool, Eugene Rostow, and Albert O.

Hirschman. "The International System." In *Working Papers of the Commission on the Year 2000 of The American Academy of Arts and Sciences,* Volume V. Boston: The American Academy of Arts and Sciences, *circa* 1965.

Iklé, Fred Charles. *How Nations Negotiate.* New York: Harper and Row, 1966.

Kelman, Herbert C. (Editor). *International Behavior: A Social-Psychological Analysis.* New York: Holt Rinehart and Winston, 1965.

Lall, Arthur. *Modern International Negotiation: Principles and Practice.* New York: Columbia University Press, 1966.

MacRae, John, and Paul Smoker. "A Vietnam Simulation: A Report on the Canadian/English Joint Project." *Journal of Peace Research,* 1 (1967), 1-25.

Morgenthau, Hans J. *Politics Among Nations: The Struggle for Power and Peace.* New York: Alfred A. Knopf, 1967 (Fourth Edition).

Naylor, Thomas H., Joseph L. Balintfy, Donald S. Burdick, and Kong Chu. "Introduction to Computer Simulation." In their *Computer Simulation Techniques.* New York: John Wiley and Sons, Inc., 1966, 1-22.

Newcomb, Theodore M. "Autistic Hostility and Social Reality." *Human Relations,* 1 (1947), 69-86.

Orcutt, Guy H. "Simulation of Economic Systems: Model Description and Solution." *Proceedings of the Business and Economic Statistics Section, American Statistical Association*, 1964, 186-93.

Pendley, Robert E., and Charles D. Elder. "An Analysis of Office-Holding in the Inter-Nation Simulation in Terms of Contemporary Political Theory and Data on the Stability of Regimes and Governments." Evanston, Illinois: Department of Political Science, Northwestern University, November, 1966.

Platig, Raymond E. *International Relations Research: Problems of Evaluation and Advancement.* New York: Carnegie Endowment for International Peace, 1966.

Pruitt, Dean G., and Richard C. Snyder (Editors). *Theory and Research on the Causes of War.* Englewood Cliffs, New Jersey: Prentice-Hall, Inc., 1969.

Reeves, Frank. "U. S. Think-Tanks: The New Centers for Research and Thought and Their Growing Impact on American Life." A series of five articles in the *New York Times,* June 12-16, 1967.

Richardson, Lewis F. *Arms and Insecurity.* London: Stevens and Sons, Ltd., 1960.

Rosenau, James (Editor). *International Politics and Foreign Policy: A Reader in Research and Theory.* New York: The Free Press, 1961.

Rummel, Rudolph J. "A Social Field Theory of Foreign Conflict Behavior." Prepared for Cracow Conference, 1965. *Peace Research Society (International) Papers*, 4 (1966), 131-150.

Sawyer, Jack, and Harold Guetzkow. "Bargaining and Negotiation in International Relations." In Herbert C. Kelman (Editor), *International Behavior: A Social-Psychological Analysis.* New York: Holt, Rinehart and Winston, 1965.

Scott, Andrew M. *The Functioning of the International Political System*. New York: The Macmillan Company, 1967.

Shapiro, Michael J. "The House and the Federal Role: A Computer Simulation of Roll Call Voting." Ph. D. Dissertation. Evanston, Illinois: Department of Political Science, Northwestern University, August, 1966.

Sherman, Allen William. "The Social Psychology of Bilateral Negotiations." M.A. Thesis. Evanston, Illinois: Department of Sociology, Northwestern University, 1963.

Shubik, Martin (Editor). *Game Theory and Related Approaches to Social Behavior*. New York: John Wiley and Sons, Inc., 1964.

Singer, J. David. "Data-Making in International Relations." *Behavioral Science*, 10, 1 (January, 1965), 68-80.

Singer, J. David (Editor). *Quantitative International Poltiics. International Yearbook of Political Behavior Research*, Volume VI. New York: The Free Press (Macmillan), 1967.

Smoker, Paul. "Trade, Defense, and the Richardson Theory of Arms Races: A Seven Nation Study." *Journal of Peace Research*, II (1965), 161-176.

Smoker, Paul. "International Processes Simulation." Evanston, Illinois: Simulated International Processes project, Northwestern University, 1967.

Snyder, Richard C. "Education and World Affairs Report." New York: 1968.

Snyder, Richard C., and James A. Robinson. "The Interrelations of Decision Theory and Research and the Problem of War and Peace." *National and International Decision-Making*. New York: Institute for International Order, 1961 (*sic*), 16-25.

Solomon, Lawrence N., Wayman J. Crow, and John R. Raser. "A Proposal: Cross-Cultural Simulation Research in International Decision-Making." La Jolla, California: Western Behavioral Sciences Institute, June, 1965.

Sullivan, Denis G. "Towards An Inventory of Major Propositions Contained in Contemporary Textbooks in International Relations." Ph. D. Dissertation. Evanston, Illinois: Department of Political Science, Northwestern University, 1963.

Wright, Quincy. *The Study of International Relations*. New York: Appleton-Century-Crofts, 1955.

Zinnes, Dina A. "A Comparison of Hostile Behavior of Decision-Makers in Simulate and Historical Data." *World Politics*, 18, 3 (April, 1966), 474-502.

SOME CORRESPONDENCES
BETWEEN SIMULATIONS AND "REALITIES"
IN INTERNATIONAL RELATIONS

HAROLD GUETZKOW

When the student of politics is a poet, his simulations of international relations are works of art, constructions that fulfill aesthetic needs. When the student of politics is a social scientist, his simulations of international processes are theories that need verification like other claims to knowledge. When the scholar is a policy-influencer, he seeks to make application of simulations so that he may guide the affairs of states and international organizations in directions he values and wishes to achieve. The aesthetician may work without constraints from reality; both social scientist and policy-influencer, in their attempts to understand and to shape the processes and outcomes of international politics, accept the challenge that their simulations must be anchored empirically albeit not bound by such "realities."

To gain confidence in his simulations, the social scientist may check them against scholarly work in general. Further, he should compare his constructions with "realities"—empirical descriptions of the world of nation-states and international organizations. Comparisons of simulation theory with traditional verbal theory have been made elsewhere.[1] This essay will compare some aspects of simulations of international processes with corresponding empirical materials obtained from political, economic, and military studies of international affairs.

[1] For example, Admiral Clyde J. Van Arsdall, former Chief of the Joint War Games Agency of the U.S. Department of Defense's Office of the Joint Chiefs of Staff, chaired a panel at the 1965 meeting of the American Political Science Association on *"Embedding Games and Simulations* in the Literature of International Relations." One of the papers from this panel has been published, namely that by William D. Coplin (1966).

Reprinted with permission from Morton Kaplan (ed:), *New Approaches to International Relations* (New York: St. Martin's Press, 1968). Copyright © 1967 by Harold Guetzkow. Preparation of this paper was part of the author's activities as Gordon Scott Fulcher Professor Decision-Making at Northwestern University. An overview of its contents was presented to the Norman Wait Harris Conference in celebration of the Seventy-fifth Anniversary of the University of Chicago in June, 1966. The author's own work has benefited greatly from research supported by the JWGA/ARPA/NU project on Simulated International Processes (Advanced Research Projects Agency, SD 260) conducted within the International Relations Program at Northwestern University. Special thanks are due the Carnegie Corporation of New York for providing the opportunity for work during 1966-67.

USE OF SIMULATION FOR THEORY-BUILDING
IN INTERNATIONAL RELATIONS

Throughout the centuries students of politics have used words as a vehicle for the building of theory; the viability of verbal theory, well done, is exemplified for us in the works of Aristotle. More recently, in the nineteenth and twentieth centuries, theoretical tools have involved a greater use of formal logics and mathematical devices. In the footnotes of his *A Preface to Democratic Theory,* Robert Dahl (1956) candidly explained the clarification he obtained from his use of formal symbolic logic in the development of ideas for his Walgreen lectures. Using simple algebra, Duncan Black (1958) developed a number of political hypotheses about committee behavior. In quite another style, working with a formalized "theory of data" (Coombs, 1964), a bevy of scholars (Alker, 1966; Gregg and Banks, 1965; Rummel, 1966b; Russett, 1968, and Tanter, 1967) have used statistical and mathematical techniques for quasi-inductive work. And now in the latter part of the twentieth century, the verbal and mathematical tools for theory building are being complemented by the use of simulations (Guetzkow, 1962).

Think of a simulation as a theoretical construction, consisting not only of words, not only of words and mathematical symbols, but of words, mathematical symbols, *and* surrogate or replicate components, all set in operation over time to represent the phenomena being studied. As in the case of other models, a simulation represents its object in but partial or simplified form. Richard C. Snyder pointed out that simulations are "scaled-down" in the "proportion of characteristics actually employed to the total sample space" and through "the use of substitute mechanisms or surrogate functions" (Snyder, 1963, pp. 4-5). In international relations, especially, it is disburbing to many scholars to have decision-makers simulated by students, even when they are multinational in background, as is the case in some constructions which use humans along with computers (Noel, 1963b, pp. 88-94). It seems inappropriate to some that Nash solutions are used to compute trade among nations (Sherman, 1963, pp. 89-92). Yet, through the use of simulations one gains a broadness of repertoire of materials and operations through which more adequate theory may possibly be constructed in the years ahead. To an extent, simulations reduce the looseness of traditional verbal theory and provide some relief from the intractability of multivariate formulations in classical mathematics. When the constructions are realized in action, they are an operating vehicle through which many implicit consequences of theory may be exposed, although not tested, in the sense of providing a situation which permits verification of facets of the theory they constitute.

Investigators of international politics at the RAND Corporation (Goldhamer and Speier, 1959) and at the Massachusetts Institute of Technology (Bloomfield and Padelford, 1959) have developed *political-military exercises* using words and men; the experienced professionals they employed as "black boxes" were usually all citizens of the United States. In creating *all-computer simulations* involving words and binary mathematics, Oliver Benson (1961) built upon the verbal theory of Morton Kaplan, Thomas Schelling, and Quincy Wright; while Clark Abt (1964) and his programmers at Raytheon built TEMPER (a *T*echnological, *E*conomic, *M*ilitary, *P*olitical *E*valuation *R*outine) from a pool of ideas contributed by designers from many "disciplines outside the recognized international relations arena" (Gorden, 1965, p. 2). Gaining benefits from both the all-man and all-computer formulations, Harold Guetzkow and his colleagues constructed a *man-machine simulation,* the Inter-Nation Simulation (INS), containing many "free" or unprogrammed activities, especially those representing behavior in the arena of international politics per se (Guetzkow, 1959). Therein, humans serving as surrogates interact with the computed programs. Both men and machines are embedded within a loose verbal framework derived from the contemporary literature about international politics including textbooks (Sullivan, 1963). As Chadwick F. Alger (1966a) aptly put it, a simulation of the man-machine variety "is a construction of surrogates in programmed theory."

The particular mix of computers and men used in a simulation of international affairs is a matter of expediency, except the extent to which the operations are explicitly programmed rather than being allowed to occur in a "free," *ad hoc* manner. In fact, in its early stages of development the Inter-Nation Simulation's programs were hand-calculated to avoid the premature expense of a digital computer. In principle even as complex a simulation as TEMPER could be handled by pencil-and-paper calculations, were there enough funds to assemble a large staff of clerks for an adequate period of time. Thus, the mix of simulation formats may be viewed as being of little consequence, except to the extent that the formats imply an explicitness in the constructions. In the "all-computer" format, all components must be explicitly programmed; in the "all-man" format, once the scenario has been written and roles assigned, the operation may be quite intuitive for both participants and umpires. The central matter of concern is not whether computers per se are used; the central issue is the extent to which the simulation is explicitly programmed, as contrasted with the extent to which its operations are hidden in "black boxes," whether the latter be complex computer languages or human surrogates (Gorden, 1968).

Although simulations are expensive, given the primitive technologies of today, some limited exploration of their potential values in the study of international relations is under way, as Sidney Verba (1964) has reviewed. Three possible values of their use by the social scientist are:

1. Simulations *enable scholars to build syntheses beyond the capability of the individual,* as is now being realized in the Brookings–SSRC Quarterly Econometric Model of the United States (Fromm and Klein, 1965).

2. Simulations *demand an explication and then an articulation of theory usually not required by the vernacular,* as Robert Abelson and Alex Bernstein (1963) illustrate vividly in their computerized modeling of the local politics of fluoridation.

3. Simulations, once created, *are vehicles for experimental work, providing devices for both replication and variation,* as Charles Bonini (1963) has dramatized in his prize-winning sensitivity analysis on the operation of a computerized model of a business firm.

However, a simulated construction is but theory. It provides no short-cut or magical route to the "proof" of the validity of the verbal and mathematical components it contains. Thus, there is a need for a systematic examination of the extent of the congruences between empirical analyses of world processes and simulations of international relations. This essay attempts such an examination.

As Vincent McRae (1963) reported, free-wheeling use has been made of simulations in the study of political-military affairs. When the focus is upon the participants, either as benefactors because of the training values involved or as subjects to be studied within a complex simulated environment, such uses of the man-machine constructions often are designated as "gaming." There has been an expansion of social psychological work in which persons are studied in laboratory situations, as in the mixed-motive game (McClintock, Harrison, Strand, and Gallo, 1963). It is beyond the scope of this essay to attempt to encompass the work in experiments focusing upon the behavior of individuals in simulated environments. There is an important link, however, between gaming and man-computer simulation. Humans are thought to be effective surrogates because they function in an operating environment which is designed to be homomorphic to the environment in which foreign policy decision-makers operate within the nations of the world.

It is still convenient to employ the definition[2] of simulation of behavioral processes written some years ago: "An operating representation, in reduced and/or simplified form, of relations among social units (or entities) by means of symbolic and/or replicate component parts" (Guetzkow, 1959, p. 184). As Richard F. Chadwick explicated for the Inter-Nation Simulation:

> It is this total system of action—not just the basic conditions and assumptions formally defined in INS theory—that constitutes a "simulation." Thus INS theory, in extended form, is not simply a set of definitions of conditions—political, military, and economic—linked by assumed interrelationships; it is presumed to encompass the participants (as surrogates for decision-making processes) and the patterns of behavior emerging from simulation as well. (Chadwick, 1967, p. 178.)

Within the perspective that simulation is operating theory, let us proceed with our central task: To what extent are simulations of international processes being verified? Probing reveals that the problems involved in verifying simulation theory are not basically different from those involved in establishing validity for symbolic theory in the languages of the historian or the mathematician (Hochberg, 1965; Kress, 1966). The formulations developed a decade ago by the Symposia of the American Association for the Advancement of Science on "The Validation of Scientific Theories" seem still to hold true (Frank, 1954, especially Chapter I, pp. 3-36). Charles F. Hermann (1967b) has recently investigated the sweep of validation problems in games and simulations. This present essay, however, addresses itself to but a segment of the gamut. Omitted herein, for example, is an examination of the validities involved when simulations are used for such purposes as prediction and exploration of policy alternatives and nonexistent universes, as well as a discussion of the use of simulation in education (Alger, 1963; Cherryholmes, 1966b; Coleman, Boocock, and Schild, 1966a, 1966b; and Robinson, Anderson, Hermann, and Snyder, 1966). However, as Hermann pointed out, "for the most part the various purposes for conducting games and simulations do not negate the need for

---

[2]As is often the case when new ideas are introduced, there is much difference of opinion among social scientists as to what terminological uses are fruitful, as Richard Dawson (1962, pp. 1-15) and Martin Shubik (1964, pp. 70-74) have documented.

criteria we can use to estimate the degree of fidelity with which one system (the operating model) reproduces aspects of another (the reference system)" (C. F. Hermann, 1967b, p. 220).

VALIDITY AS HOMOMORPHISM
OF THEORY AND EMPIRICAL ANALYSES

When policy influencers and scholars handle the contents of international politics, seldom is there explicit reference to the evidential base from which assertions are constructed, whether these assertions are based on flashes from cocktail conversation or on the propositions of a formalized verbal theory. "Facts on file" as reported in news media are but interpreted reportings of journalists (Gamson and Modigliani, 1965, pp. 47-78). Nor are the "inside dopester's" anecdotes to be taken at face value, for often such a participant observer of a decision group is only an intimate member with vested interests. As William D. Coplin pointed out in his comparison of verbal theories with simulation theory, "the lack of congruence between the assumptions of the simulation and the assumptions of the verbal theories does not necessarily indicate that the simulation model lacks validity since the verbal theorist has no monopoly on valid hypotheses" (Coplin, 1966, p. 562). So, in seeking to establish simulation theory, it is imperative to make checks against empirical observations, not merely against another set of theories.

It would seem impossible at the outset to achieve a strict and complete isomorphism between the "realities" of world processes (the reference system) and simulation materials, in compliance with May Brodbeck's stipulation that "there must be a one-to-one correspondence between the elements of the thing of which it is the model. . ." (Brodbeck, 1959, p. 374). In the early stages in the construction of simulations there is at best a homomorphism, in which many entities in one system may be represented by fewer entities in the other system, with but some correspondence in the relationships among the coalesced entities (Beer, 1965, pp. 223-231). The homomorphic "many-to-one" relationships may eventually be differentiated to the extent of forming isomorphic "one-to-one" relationships, in which the materials "resemble one another as systems . . . in ways which do not depend on the particular elements of which each consists . . ." (A. Kaplan, 1964, p. 263). But to have a complete isomorphism the model might be as complex and intractable as the reality. It is ironic that sometimes simulations are so large that they defy intuitive understanding. Yet, is this not exactly one reason why simulations hold promise—that they may permit us to reach beyond the limitations of verbal and mathematical formulations (*supra*, p. 692)?

Some homomorphy may exist among outputs as well as between the very processes which result in such outputs. As we analyze the correspondences between simulations and "realities" sometimes an internal process, like the representation of the decision-making within foreign offices, helps produce an outcome of some validity, such as the constellation of internation alliances. At other times—less often because of lack of appropriate research—an internal process will be judged to be of some validity because the very process itself has some congruence with corresponding processes in the reference data.

In verifying theory the simulations' constructions are compared with reference materials of the "real world" (A. Kaplan, 1964, Chapters 23 & 36). This testing of hypotheses may be done with different degrees of accuracy, ranging from rough-and-ready, intuitive appraisals of the behavior of the individuals and groups involved in world politics to rigorous, systematic comparisons with content analyses of diplomatic documents (North, Holsti, Zaninovich, and Zinnes, 1963) and with careful coding of interactions among international representatives (Alger, 1966b; 1968). The lack of correspondences between the simulations and the empirical materials indicates the extent of nonverification.

Reference materials used in assessing correspondences may be abstract and prototypic in nature, embodying processes which are supposed to hold more generally as they have been derived from a gamut of studies made in the field and laboratory, as Herbert C. Kelman and his associates have developed in their *International Behavior* (1965). Or, reference data may be used which are more concrete and systematically gathered from operations within the international system, such as Wright presented in *A Study of War* (1942; 1965). The extent to which processes within the international system differ from those found in noninternational materials is largely unknown today. To what extent does the simplification achieved in the conventional laboratory experiment of the psychologist make the findings a valid reference for the outputs generated in the simulation exercise (Barber, 1966, pp. 8-13)? May the phenomena found occurring in the field of noninternational affairs be used as a legitimate probe of the validity of an Inter-Nation Simulation?

In posing the verification problem in the study of international relations, my distinctions are overdrawn. As is widely recognized in the methodology of science, work is not done analytically with empirical materials without constructing at least implicit theories about the *denotata*—as events are selected from the entire population of such items and then a few reported events characterized. Sense impressions are "validated" by perceiving and then re-interpreting them in terms of individual theories of history, as Snyder and Glenn Paige believe happened in Truman's conceptualization of the cables from Korea on June 24, 1950 (Snyder and Paige, 1958, p. 359). Practitioner and scholar alike work with materials at varying levels of abstraction (Guetzkow, 1966a, pp. 264-267), sometimes "correcting" data in terms of theories and at other times "editing" such theories because of the data. "Our longing is for data that prove and certify theory, but such is not to be our lot," according to contemporary methodologists (Webb, Campbell, Schwartz, and Sechrest, 1966, p. 10). However, by using some systematic rigor in making comparisons between simulations and "realities," by taking reference data largely from extant international systems rather than from laboratory and field research about noninternational phenomena, and by finding in simulations internal processes and outputs which correspond to reference processes as well as reference outcomes, a convergence of evidence is gained which increases the credibility of the theoretical constructions of simulations.

ASSESSMENTS OF CORRESPONDENCES: SIMULATIONS AND EMPIRICAL FINDINGS

Let us now examine assessments of the congruence which simulations of international relations have with analyses made of empirical materials derived from world processes. Inasmuch as but one work (Bonham, 1967) on the validation of either the RAND/MIT political exercise or TEMPER has been recorded so far, the studies to be surveyed in this essay center on the Inter-Nation Simulation in its many variations.[3] This limitation is great; therefore, our findings may *not* be extrapolated broadly.

No attempt is made to rehearse materials presented elsewhere. *Simulation in Social Science: Readings* (Guetzkow, 1962) provides general background. The book my colleagues and I assembled, entitled *Simulation in International Relations: Developments for Research and Teaching* (Guetzkow, Alger, Brody, Noel, and Snyder, 1963) describes the Inter-Nation Simulation (INS) in considerable detail. A summary presentation is given in an earlier essay, "A Use of Simulation in the Study of Inter-Nation Relations" (Guetzkow, 1959), which is reprinted as Chapter Two of our "Developments" volume (Guetzkow, 1963a); this essay was updated for IBM's 1964 Scientific Computing Symposium on Simulation Models and Gaming (Guetzkow, 1966a). The newcomer may find Verba's (1964) review of our work useful in providing further orientation to simulation in international relations. Were the details of each of the cited studies to be repeated in this essay, its length would be intolerable.

Findings on the correspondences between simulations and "realities" will be discussed from a substantive point of view, as outlined in Figure 1 below, following the rubrics outlined in our chapter, "Structured Programs and Their Relation to Free Activity Within the Inter-Nation Simulation" (Guetzkow, 1963b). The studies summarized will cover first the "Decision-Makers and Their Nations" and then "Relations Among Nations." The assessment of correspondences between each unit of simulation and reference materials is reported on a five-point scale after the relevant research is discussed, as a *Correspondence Rating*.[4] A summary of these assessments of congruence is contained at the end of each of three substantive sections of this essay in Tables 5, 8, and 17, with a final summation in Table 18 (pp. 711, 719, 734, and 736).

While viewing the substantive findings, it is important to realize not only that the correspondences are based on very limited analyses of "realities"—the reference data—but that severe methodological difficulties are encountered in studying the comparisons, as has been found in making comparisons among

---

[3]In utilizing one work done on the Princeton/Rutgers "Tactical Game" (Streufert, Clardy, Driver, Karlins, Schroder, and Suedfield, 1965) at one point in this essay, no claim is made that this simulation has the scope of complexity of the other simulations described herein.

[4]A *Correspondence Rating* is given in parentheses following the discussion of each of the fifty-five comparisons between simulations and "realities" in international relations, and summaries of these ratings are contained in Tables 5, 8, 17, and 18. The rating of correspondence is either *"Much"* (M), *"Some"* (S), *"Little"* (L), *"None"* (N), or *"Incongruent"* (I). *"Much"* and *"Some"* were used to describe the close correspondences where the findings obtained in simulation and reference materials were judged not to be different from each other at levels of significance of 5 per cent or less, or between 5 per cent and 10 per cent, respectively. A rating of *"Little"* was assigned when the sets of materials indicated relationships between variables to be in the same direction but clearly of different magnitudes. If such relationships were in opposite directions, a rating of *"Incongruent"* was given. When the findings from reference materials were absent in the simulation materials, or vice versa, the designation of *"None"* was made. The reader who is acquainted with the twenty-four source materials used herein may wish to make his own ratings; the author used intuition to arrive at his assessments.

nations (Eckstein and Apter, 1963, Part II, pp. 34-94) or between societies (Naroll and Cohen, 1968). Two difficulties of special prominence indicate that the entire effort of this essay to survey the correspondences at this time should be viewed with much circumspection: (1) In all instances the matching of variables in the reference and simulation materials has been a most intuitive process, usually based upon phenotypic judgments of the author; (2) in most instances in neither the reference nor simulation materials has attention been given to the problem of the reliability of the data and outputs. Yet, unless both sets of materials can be replicated, a lack of correspondence may be due to **variations coming from unreliability rather than to invalidities of the simulation,** per se. Now add to these two sources of error the coarseness of our measuring instruments themselves (Guetzkow, 1965, pp. 25-31).

"DECISION-MAKERS AND THEIR NATIONS"   "RELATIONS AMONG NATIONS"

Organizational Relations
Among Decision-Makers   Communications

Office-Holding

Inequalities Among Nations

Decision Latitude

Orderly and Disorderly
Transference of Power   Armaments: Defensive and Aggressive
Use of Force

National Capabilities

Consequences of Pressured Changes   Interaction Patterns:
in Decision Latitude   Conflict and Cooperation
in War and Peace

Occurrence and Consequences
of Revolution

Inter-Nation Organizations

Nationalism

FIGURE 1   STRUCTURED PROGRAMS AND THEIR RELATION TO FREE ACTIVITY WITHIN THE INTER-NATION SIMULATION. (Outlined in Chapter 5 in Harold Guetzkow, Chadwick F. Alger, Richard A. Brody, Robert C. Noel, and Richard C. Snyder, *Simulation in International Relations: Developments for Research and Teaching,* Prentice-Hall, Inc., 1963, pp. 103-149.

We then can readily agree with C. F. Hermann's position that "multiple validity criteria are needed because of the error in measurement and because of the recognition that criteria can be only assertions about 'reality'" (C. F. Hermann, 1967b, p. 225). It is hoped that the reader will not find the demands on attention involved in pursuit of the fifty-five comparisons which follow, too great to bear. When patience yields, skip to the four summary tables (p. 711, p. 719, p. 734, p. 736) and then proceed to the Conclusions and The Findings in Broader Perspective—and Some Implications (pp. 732-741).

"DECISION-MAKERS AND THEIR NATIONS"

In considering "Decision-Makers and Their Nations" it is convenient first to discuss individuals serving as surrogates for the decision-makers of the world. Attention will be given next to these humans assembled in groups, along with the political, economic, and military programs which function together as representations of the nation-state, following the topical arrangement presented in the outline contained in Figure 1.

Humans as Surrogates

Because of the difficulties involved in programming the decision-making within a nation, given the present development of work in "artificial intelligence" (Reigenbaum and Feldman, 1963), humans are used in the Inter-Nation Simulation to handle these activities. These simulations are *not* being used as a "synthetic environment" in which to explore the psychological characteristics of individuals as subjects in an experiment (Kennedy, 1962, pp. 27-29). Given the perspective of simulation as theory, the participants may be considered as surrogates, focusing upon their outputs as inputs into other components of the simulation taken as a whole. It was realized early that humans "bring both their own personal characteristics into the model and their own implicit theories of the way in which nations should behave" (Guetzkow, 1959, p. 188). It was noted also that "eventually it will be necessary to appraise these personal styles of decision-making and organizational presuppositions, so that their influence on the evolving inter-nation interaction may be studied" (Guetzkow, 1959, p. 188). There is now evidence on the ways in which the decision-making occurs within simulation, and these findings may be compared with results obtained from analyses of empirical materials. The following discussion will consider three aspects of the behavior of the surrogates as such that derive (1) from their personal characteristics, (2) from their education, and (3) from their ethnicity.

Personal characteristics. Perhaps the most focused evidence available on the impact of personal style on outputs from the humans who constitute the decision-making units within the Inter-Nation Simulation is presented by Michael J. Driver (1965) in his essay, "A Structure Analysis of Aggression, Stress, and Personality in an Inter-Nation Simulation." With Richard A. Brody (1963) Driver selected 336 participants for their sixteen runs of the simulation on the basis of each individual's cognitive simplicity/complexity. Driver found that outputs of the high school seniors and graduates who served as his decision-makers conform to the findings obtained in many other situations (Harvey, Hunt, and Schroeder, 1961; Schroeder, Driver, and Streufert, 1967). Driver noted how these surrogates with simpler conceptual structures, as determined on a pretest, tended to involve their nations in more aggressive behavior than did those with more complex, abstract conceptual structures, as is illustrated in Table 1 (*Correspondence Rating: "Much"*) (*infra*, Footnote 4, p. 000).

TABLE 1   Driver's Content Analysis of the Incidence of Aggression by Large Powers as Related to Cognitive Structures in the Brody-Driver Realizations of an Inter-Nation Simulation

Individual Cognitive Structures and Incidence of Aggression by Large Powers					
		Kind of Aggression			
Large Powers	*Wars*	*Unprovoked arms increases*	*Provoked arms increases*	*War plans*	*No aggression*
Nations with decision-makers with simple cognitive structures	4	$7 + 2^§$	$2 + 1^§$	2	1
Nations with decision-makers with complex cognitive structures	1	3	3	$0 + 1^§$	9

†Adapted from Michael J. Driver, "A Structure Analysis of Agression, Stress, and Personality in an Inter-Nation Simulation." Institute Paper No. 97. Lafayette, Indiana: Institute for Research in the Behavioral, Economic, and Management Sciences, Herman C. Krannert Graduate School of Industrial Administration, Purdue University, January, 1965, Table 5, pp. 30–31.
‡There were sixteen large nations manned by decision-makers with *simple* cognitive structures; there were sixteen large nations manned by decision-makers with *complex* structures. Ninety-six high school juniors and seniors served in these thirty-two nations as decision-makers. A chi-square test is statistically significant ($\chi^2 = 7.1^{**}$; $p < .005$)(Driver, 1965, Table 6, p. 33; p. 34.).
§As indicated by the numbers appearing after the "plus" signs, in some runs a given nation was involved in more than one act of aggression. Thus, among the sixteen powers manned by decision-makers with *simple* cognitive structures, of four going to war, two in addition initiated unprovoked increases in arms and one also responded similarly to provocation on other occasions within its run. Likewise, among the sixteen powers manned by decision-makers with *complex* structures, of the three provoked into arms increases, one also developed war plans.

The operation of personal characteristics of surrogates within simulations may be pinpointed, too,—both in terms of a particular set of personality traits, namely "self-esteem" and "defensiveness," and in terms of a particular situation, namely "crisis." Personal characteristics are related intimately to the way in which individuals handle crises (Basowitz, Persky, Korchin, and Grinker, 1955; Funkenstein, King, and Drolette, 1957; Janis, 1958; Lazarus, 1964; and Selye, 1956). Dr. Margaret F. Hermann (1965) has reproduced aspects of these phenomena (Lazarus and Baker, 1956) concerned with self-esteem and defensiveness in her observations of 163 U. S. Navy petty officers (average age, 32 1/2 years) who conducted decision-making in eleven replications of a crisis-permeated simulation of foreign policy-making (C. F. Hermann, 1965). Along with many other outcomes she found that as the simulated crisis produced more negative affect, the decision-makers *high* in self-esteem and *high* in defensiveness ("avoiders") decreased their attempts to seek aid from other nations and they decreased their search for information about the threat. Conversely, those *low* in self-esteem but *high* in defensiveness ("affiliators") increased their attempts to affiliate and increased their search for information (M. G. Hermann, 1965, p. 73). The affiliation attempts by these two types of participants (along with four other types, reflecting the intricacies involved in personality analyses) are presented in Figure 2; the results are statistically significant (F = 4.37**[5]). Further details may be obtained in Mrs. Hermann's doctoral dissertation (1965, pp. 48-70) (*Correspondence Rating: "Much"*).

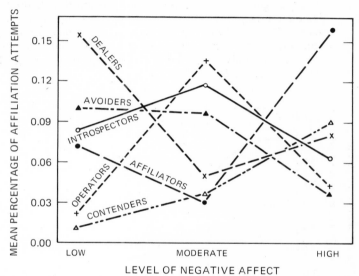

FIGURE 2   *Margaret G. Hermann's Study of the Relation of Rated Negative Affect to Affiliation Attempts in an Inter-Nation Simulation.* (Adapted from Fig. 1, p. 56 in M. G. Hermann. "Stress, Self-Esteem, and Defensiveness in an Inter-Nation Simulation," Ph.D. Dissertation. Department of Psychology, Northwestern University, 1965. The levels of negative affect were obtained from ratings of three phases [e.g., "injurious to my nation"] made in response to each of seven threatening international events, by 163 U.S. Navy petty officers serving as decision-makers for the six nations in eleven simulations.)

[5]There is controversy on the contemporary scene as to how useful levels of significance are in utilizing statistical tests. In this essay whenever levels of significance were developed by the scholars whose work is being discussed, they are cited. The use of two asterisks (**) in this essay indicates that the test is significant at the .01 level or better; one asterisk (*) indicates that the test is significant somewhere between the .01 and .05 levels. That is to say, the chances of obtaining such a result by pure chance are 1 per cent or less for results marked with ** and between 1 per cent and 5 per cent for those marked with *, were the assumptions involved in deriving the statistic applicable. When no levels of significance are mentioned in the text, tables, or figures, none were reported in the original source.

Hermann-like findings, corresponding to field and laboratory work reported by Harold M. Schroder, Driver, and Siegfried Streufert (1967), have been obtained in a "tactical game situation" less rich than the Inter-Nation Simulation, in which crisis was created by increasing information loads. Using three measures of information handling—delegated information searches; self-initiated information searches, as presented in Figure 3; and integrated utilization of sought-for-information in subsequent decision-making—Streufert, Peter Suedfeld, and Driver (1965) obtained statistically significant impact of levels of information load upon information handling by 185 college students, assembled into fourteen teams, serving as decision-makers. And using the same personality measures employed by Driver in his operation with Brody of the Inter-Nation Simulation (*supra*, p. 699), these three researchers obtained dramatic—as well as statistically significant—differences in the effects of crisis upon information handling for those surrogates with structurally *complex* styles as contrasted with those structurally *simple* styles (*cf.*, broken versus solid lines in Figure 3) (*Correspondence Rating: "Much"*).

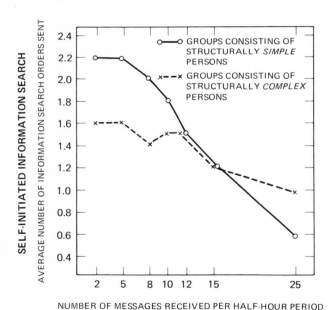

FIGURE 3 *Findings from "A Tactical Game" on the Relation of Levels of Information Load upon Information Handling by "Complex" and "Simple" Individuals under "Crisis" and "Non-crisis" Conditions.* (From Siegfried Streufert, Peter Suedfeld, and Michael J. Driver, "Conceptual Structure, Information Search, and Information Utilization" *Journal of Personality and Social Psychology.* 2, 5 [1965]. Fig. 2, p. 739. One Hundred eighty-five college students, assembled into fourteen teams, served as decision-makers in this game.)

Thus, both in ordinary and in crisis situations within two simulations, the surrogate decision-makers behaved in ways similar to which other individuals act in field and laboratory studies. In the end, however, it may be expeditious to use surrogates who match particular international actors for work within the Inter-Nation Simulation. An attempt to encompass the entire personality of the participants was made by Mrs. Hermann (Hermann and Hermann, 1967) in her use of a semantic differential instrument and the California Psychological Inventory (CPI), which yielded a profile of some thirteen traits, in a disguised simulation of the activities within and among the countries which became involved in World War I during the summer of 1914. Working as a clinical psychologist, Mrs. Hermann prepared personality profiles for each of ten actors who played significant roles in the 1914 crisis, on the basis of personal letters, autobiographical materials, and biographies. Then she matched these profiles to those of potential participants, choosing ten from an available population of 101 high school graduates for use as surrogates. The findings from this pilot study are suggestive: one realization with matched participants (M-Run) came closer to producing an output similar to the unfolding of historical events, as they are described by the historian Luigi Albertini (1953), than did those in a second realization (A-Run) in which another set of ten surrogates were less well-matched (*Correspondence Rating: "Some"*). More definitive validation study covering key personal characteristics of relevance to policy-makers acting in the international scene awaits the production of comparable research in reference materials.

Inspection of John R. Raser's matrix of comparisons among Western and non-Western political decision-makers, who were subgrouped into bureaucratic, military, and nonmilitary categories, suggests that there is a tendency for such leaders to be located at the high end of each continuum describing personal characteristics of ordinary populations. For example in characterizing political decision-makers of the twentieth century, Raser said, "They are typically dominant, articulate, flexible, opportunistic, able to think *abstractly* as well as concretely, occasionally low in *self-esteem* but more often very self-confident, able to tolerate ambiguity, and willing to take at least moderate risks" (Raser, 1965, p. 25; italics mine). If this is the case, then, surrogates who have too wide a range of variation in their personal characteristics should not be selected (*Correspondence Rating: "Little"*). Driver, however, argues that such is not the case, inasmuch as extreme differences occur on occasion, as in such pairs of leaders as different as Stalin and Khrushchev, with many in the middle ranges as represented in men as similar as Adenauer and Erhardt. Driver fears contrarily that "our surrogates may not be widely distributed enough. What high school boy can match de Gaulle's rigid grandiosity" (Driver, 1966)?

To this point our review of the operation of personal characteristics in the Inter-Nation Simulation has been directed to their impact upon the outputs of the surrogates. It is also of interest to know to what extent the very processes which produce the consequences are homomorphic themselves to those which create outcomes in the reference system—are the "right" outputs being produced for the "wrong" reasons? There are three researches of relevance which examine how processes of perception mediate the impact of personal characteristics upon outcomes in the simulation.

1. Studies of President Woodrow Wilson and Secretary of State John

Foster Dulles made by other researchers suggested to Michael J. Shapiro that their personal styles exemplified the frequently verified relations between cognitive rigidity and tendencies to perceive conflicts in moral rather than instrumental terms, as well as being relatively unreceptive to change (Shapiro, 1966b). Using Driver's measures of cognitive style, as derived from the Adorno/California F. Scale and the Schroeder/Streufert Situational Interpretation Test (SIT), Shapiro checked whether the same processes held within these Brody-Driver simulations. He found that cognitive rigidity correlated significantly ($r = .51**$ for the F Scale; $r = .69**$ for SIT) (Shapiro, 1966b, Table 1, p. 10) with the extent to which 336 high school students, serving as participants, evaluated environmental stimuli in moral categories, as revealed in coding the messages generated in the course of the simulation. But he found that neither rigidity measure was correlated with fixity of beliefs and attitudes about decision-makers in other nations (the r's being $- .01$ and $.12$ respectively) (Shapiro, 1966b, Table II, p. 12) (*Correspondence Rating: "Some"*).

2. Using a model developed by Ledyard Tucker and Samuel L. Messick (1963), Driver (1963) was able to measure changes in the dimensionality of the perceptions of the nations' decision-makers as they moved from ordinary to tense to dangerous situations within the 112 simulated nations operated in his research with Brody (1963). Corresponding to the findings in laboratory and field situations, including those analyzed in studies of natural disasters (such as panics, during fires and floods), Driver noted that the dimensionality of the perceptions of the 336 high school students used as surrogates changed curvilinearly, from simple to more complex to less complex, as the inter-nation situation moved from run-of-the-day interaction through conflict into war (Schroder, Driver, Streufert, 1967, pp. 68-81). Driver found that even the content of the framework, in terms of which the other nations were perceived as being similar or different from one another, varied as the distinctions among the nations were made in terms of two to three, and then three to five, and then reduced to two or three dimensions again. For example, Driver noted that "economic power dimensions are first transformed into military power, and finally replaced altogether by alliance concerns as the clouds of war gather" (Driver, 1962, p. 243) (*Correspondence Rating: "Much"*).

3. In quite a different way, C. F. Hermann obtained findings which converge with those of Driver and noted that in crisis as compared with non-crisis there was a tendency—slight, albeit statistically significant—for his petty officers to perceive "events as involving a number of different alternatives or only one or two alternatives," as displayed in Table 2, even though content coding revealed no such differences in frequency in alternatives found in messages and conference statements exchanged in the course of the simulation (C. F. Hermann, 1965, p. 224). In his illustrations from the literature of international crisis behavior of political decision-makers, C. F. Hermann listed observations (1965, pp. 220-222) that also are congruent with Driver's findings (*Correspondence Rating: "Some"*).

In all three of these sets of findings there is somewhat convergent evidence as to how processes of perception operate within the surrogate decision-makers in realizations of the Inter-Nation Simulation. Two processes are displayed: namely, the correlation of cognitive rigidity with the extent to which participants evaluate environmental stimuli in moral categories and the tendency toward reduction in the perceived richness of the situation in crisis. The reference data used in making the comparisons with the simulations consisted of case materials, along with anecdotal observations. The evidence samples but limited aspects of perceptual phenomena, even though both ordinary and political decision-makers were compared with the surrogates. Within these confines, it seems the outputs deriving from the personal characteristics are being produced for at least "some" of the "right" reasons.

TABLE 2   C. F. Hermann's Questionnaire Findings on Perceptions of Alternative Solutions Identified in Non-Crisis and Crisis Situations in an Inter-Nation Simulation†

	Participants' Perception of Alternative Solutions‡	
Alternative solutions:	*In non-crisis situations*	*In crisis situations*
Number of participants who perceived few alternatives	87	103
Number of participants who perceived many alternatives	34	18

† Adapted from Charles F. Hermann, "Crises in Foreign Policy Making: A Simulation of International Politics." Ph. D. Dissertation. Evanston, Illinois: Department of Political Science, Northwestern University, 1965, Table 29, p. 224.

‡ A total of 121 U. S. Navy petty officers served as decision-makers for six nations operated by them in eleven simulations. Fewer alternative solutions were perceived in crisis than in non-crisis. A chi-square test was statistically significant ($x^2 = 6.2; p < .05$).

Impact of training/professionalism/background. Just as the Joint War Games Agency of the government of the United States from time to time has involved very high level civilian and military officials in its political-military exercises, eventually as part of each year's opening activities at the United Nations the central decision-makers of the countries of the world may want to explore their problems through simulation. Thus far, achievements in securing surrogates who actually participate in international affairs have been modest. Dr. Dorothy L. Meier (1965) has completed two runs of the Inter-Nation Simulation involving some forty-one members of the diplomatic corps at the consul and secretarial level drawn from twenty-six foreign embassies in Washington, D.C., and from the U.S. Department of State. These runs, set to play the contemporary situation for a year or two forward from 1964, have been replicated using both high school and college students, seventy-two of each (Nardin and Cutler, 1967). Impressionistic observation suggests there were not important differences in at least many of the ways in which the simulations were operated by the three different types of surrogates (Meier and Stickgold, 1965) (*Correspondence Rating: "Some"*). This observation is congruent with one of the "tentative" findings of Lincoln P. Bloomfield and Barton Whaley which "shows an amazing uniformity of 'U.S.' responses to dangerous situations" in replication of the "same games using different levels and types of personnel" (Bloomfield and Whaley, 1965, p. 869) (*Correspondence Rating: "Some"*).

In pilot work in a set of exercises within a socioeconomic model of American society enabling "simulated interest groups . . . to make decisions on given public policy issues during specific future time periods," Robert Boguslaw, Robert H. Davis, and Edward B. Glick (1966) found that their seven "professional and administrative staff" participants secured the adoption of more policies than did seven graduate students in political science and business administration. "Conference transcripts show that more experienced subjects focused their attention on fewer issues which they regarded as critical." The authors also remarked that "it is, of course, possible that some variable other than experience, e.g., personality, may account for these observed differences . . ." (Boguslaw, Davis, and Glick, 1966, p. 54) (*Correspondence Rating: "Incongruent"*). More adequate analyses of the impact of professionalism on the behavior of surrogates is imperative, so that continued reliance upon impressionistic reports will be unnecessary.

Certainly the ideology and orientation of the decision-makers toward international affairs would seem to be an important component in the background of international decision-makers. Yet, Philip F. Beach and Guetzkow found no relationship between the content of policy decision exhibited within the simulations and the participants' orientation toward international affairs as assessed before their participation in the simulations (Guetzkow, Brody, Driver, and Beach, 1960) (*Correspondence Rating: "None"*). However, this finding may be merely an indication of the inadequacy of our measuring instruments, inasmuch as Shapiro (*supra*, p. 704) demonstrated a statistically significant tendency in the simulations for participants with rigid cognitive structures to attach moral rather than instrumental values to the actions used in executing foreign policy, parallelling the tendency of Wilson and Dulles—both men of intense religious orientation—to bring moral orientations to affairs of state. In addition, Wayman J. Crow and Robert C. Noel (1965) were able to show that at least one component of the surrogates' world view, namely militarism, was related to their escalation responses.

This latter work is now considered in more detail. In a complex environment specially designed to mirror features of the TEMPER model, Crow and Noel secured written responses on a pretest from 384 U.S. Navy enlisted personnel (in training), which ranged from "Disengagement with Insurgents" (Level 1) through "Declaration of War, Occupation of Winding River Basin, Establishment of Military control there, and establishment of defensive and/or offensive positions in mountains separating the Basin from southeastern Utro" (Level 11) (Crow and Noel, 1965, p. B-17). Using Gerald H. Shure's questionnaire (Shure, Meeker, and Hansford, 1965) on "aggressive militarism versus nonbelligerence," Crow and Noel found that those surrogates who indicated on the pretest they would rely on the "use of force, threats, and power in dealing with other foreign powers, versus avoidance of belligerent means" (Crow and Noel, 1965, p. 8), when confronted with the necessity for military response tended to levels invoking larger and more offensively oriented forces in making their verbal responses about escalation during the simulations, as is indicated in Table 3 (Crow and Noel, 1965, p. D-3) (*Correspondence Rating: "Much"*).

TABLE 3   Crow and Noel's Findings on the Relation of Militarism to the Level of Escalation Response in the East Algonian Exercise†

Relation of Participants' Militarism to Level of Escalation Response	
Militarism *Participants' score on protest*	*Mean response level chosen* ‡ *Pre-group individual judgments*
High	5.83
Low	4.83

† Adapted from Wayman J. Crow and Robert C. Noel, "The Valid Use of Simulation Results." La Jolla, California: Western Behavioral Sciences Institute, June, 1965, Table I-C, p. D-3.

‡ Higher response levels involve use of greater and more offensive force. One hundred seventy-six Navy recruits served as decision-makers for four nations in eleven replications. No statistical tests of significance are reported.

Impact of cultures/nationalism. A simulation of international affairs is different from simulations of intra-national social and economic processes, in that it needs to represent the multicultural, multinational characteristics of the reference system. On the face of it, this factor seems to be of obvious importance. Yet, the anecdotal report of a day's run in 1959 (INS4-1) with eighteen foreign college students (Noel, 1963b, pp. 89-94) indicated no radical "difference of kind" obtained in using surrogates from different cultures, even though the representation from nation to nation consisted of Africans, Asians, Europeans, Latin Americans, and North Americans (*Correspondence Rating: "Some"*).

Usually simulations are operated with participants with similar cultural backgrounds, all of whom are of the same nationality. Raser and Crow have realized two runs of the Inter-Nation Simulation with Mexican college students at the National University of Mexico. They state, "Looking at the over-all pattern of the results it can be concluded in general that the observed regularities . . . obtained with U.S. participants . . . held under the changed condition of the cultural background . . ." (Raser and Crow, 1964, p. 10) (*Correspondence Rating: "Some"*). There were differences between the U.S. and Mexican participants "that seemed to be related to a culturally-based predisposition for the Mexican participants to respond to stress and frustration with a passive rather than an active response" (Raser and Crow, 1964, p. 11). Consider the nature of the differences: The Mexicans exchanged more messages and they placed more emphasis on the formalities and phrasing of their communications, and in the course of the simulation they focused more on international issues and less on matters of internal economic growth. Could these differences also have been the result of differences in literary achievement (the Mexican participants had about two more years education than the participants from the United States) and policy orientation (the forty-six Mexican participants were students in liberal arts, while the 240 U.S. participants were Navy recruits)?

In fact, it may be possible to produce many of the differences mentioned between the cross-cultural populations through preselection of surrogates, as Mrs. Hermann did with respect to personal characteristics (*supra*, pp. 700-703) Crow and Noel, too, demonstrated this capability later in their East Algonian Exercise (*supra*, pp. 706-707) with respect to aggressive militarism. Based on such evidence as that assembled by H. C. J. Duijker and N. H. Frijda (1960, Chapter V, "Register of National Character Studies"), Otto Klineberg concluded that "we may find differences in 'mode' or average or central tendency, but there will always be overlapping" (Klineberg, 1964, p. 142). Thus, it would seem possible to represent at least some important cross-cultural differences through the judicious selection of participants from within a single culture.

Suppose such capability in the representation of cross-cultural differences were realized as part of the initial conditions of a simulation designed to incorporate multinational properties; would it lack a representation of the nationalism which plays such a dominant role in contemporary international affairs? In terms of the work of Crow and Noel, it seems not—as they have shown that nationalistic predispositions, as measured through presimulation attitudes on authoritarian nationalism versus equalitarian internationalism, carry over into the operation of the simulation itself. In their East Algonian Exercise (Table 3), the more nationalistic participants tended toward responses involving more escalation (Crow and Noel, 1965, p. 20) (*Correspondence Rating: "Much"*).

More dramatic perhaps is the congruence which Daniel Druckman (1968) obtained with findings from the empirical literature about ethnocentrism, when he obtained special data from Raser and Crow's (1964) exploration of the effects of capacity to delay response in nuclear war in their "WINSAFE II" research. In the trial-by-trial development of the eleven Raser and Crow simulations, Druckman found an enhancement of the ingroup bias for their own nations of the 176 U.S. Navy recruits serving as participants (*Correspondence Rating: "Much"*). Thus, just as the perceptual processes were shown to have important impact upon the personal outputs of the surrogates (*supra*, pp. 704-705), so it would seem that the surrogates' identifications with their own nations in these simulations produced evidence of predispositions which operated in a fashion homomorphic to the ethnocentrism of cultures and nationalities.

Using an all-man simulation technique, G. Matthew Bonham (1967) has made checks of the congruences involved in his eleven efforts to replicate the East-West disarmament negotiations of the mid-nineteen-fifties. In the style of the political-military exercise, his work employed a conventional role playing of the concrete historical situation through a somewhat abstract representation, as was done for World War I by the Hermanns (1967). Although Bonham felt the "background conditions of cultural traditions and relations within parties to the East-West negotiations are not duplicated very well" (Bonham, 1967, p. 20), "the simulation was to some degree successful in representing the attitudes between the nations in the historical negotiations" (Bonham, 1967, p. 25), as exhibited in Table 4 (*Correspondence Rating: "Some"*). These findings seem similar to those obtained by Druckman (1968), in the sense that there is friend-liness among ingroups and unfriendliness toward outgroups.

TABLE 4   Bonham's Findings on the Attitudes of Friendliness and Unfriendliness and Unfriendliness between Nations in a Simulation of Disarmament Negotiations†

Attitudinal Relationships between the Nations ‡

For all runs: attitudes of:	Nations being rated Mean ratings		
	"United States" *("USA")*	*"Soviet Union"* *("USSR")*	*"United Kingdom"* *("UK")*
"USA" toward	——	−2.5	+3.3
"USSR" toward	−2.8	——	−1.3
"UK" toward	+3.3	−1.3	——

† Adapted from G. Matthew Bonham, "Aspects of the Validity of Two Simulations of Phenomena in International Relations." Ph. D. Dissertation. Cambridge, Massachusetts: Department of Political Science, Massachusetts Institute of Technology, 1967, Table 2, p. 26.

‡ A total of sixty-six college students, half undergraduate and half graduate, served in eleven simulation runs. The two participants on each team were asked to rate their attitudes toward the other two nations in the negotiations, on a scale of +5 (Friendliness) to −5 (Unfriendliness). No statistical tests of significance are reported.

Thus, it would seem that some approximation of the functioning of cultural and national characteristics can be realized within the man-machine simulation of international affairs. In his examination of the relevancies of "national character" for international relations, Bernard C. Hennessy concluded that foreign "policies are made largely by cosmopolitan elite groups who appear to be on the whole little affected by national character or modal personality traits" (Hennessy, 1962, p. 46). Even though the importance of personal characteristics in decision-making is well accepted—be they linked to leadership or grounded in nationality and culture—could it be that the degree of congruence now capable of achievement between simulations and "realities" in international relations suffices, for the moment?

On the other hand, when an all-man or a man-computer format is used for simulations, why not attempt to take advantage of the ethnic characteristics and nationality of the surrogates by employing participants from many cultures? Instead of operating simulations with all participants of a single cultural background, it would seem useful to operate multinational simulations in which the nations are each composed of surrogates with different nationalities, although the participants within each of the nations might have the same nationality. Perhaps now that academic groups on two other continents are beginning research on the simulation of international activities using man-computer vehicles, such designs will become feasible. "Face validity," as C. F. Hermann put the matter (1967b, pp. 221-222), then will be gained and the correspondences of the surrogates in the simulations to the decision-makers of the world may become a bit less problematic.

Summary: humans as surrogates. A compilation of the author's individual judgments rating the correspondences between simulations of international processes with "General Materials from Field and Laboratory" and "Materials from International Relations" is given in Table 5. It is not surprising that the surrogates in simulations behave remarkably like humans—although the subtotals in Table 5 suggest that they resemble the decision-makers of the international scene (*"Much"* = 2; *"Some"* = 6) less than they do more ordinary mortals observed in field situations and within the laboratory (*"Much"* = 5; *"Some"* = 2). In general, the homomorphies hold for both crisis and non-crisis situations. Correspondences are found for the outputs, as in terms of the inter-entity production of violence (Driver, 1965) and alliance (Bonham, 1967); correspondences are found, also, for the intervening processes, as for perception (Driver, 1962) and nationalism (Druckman, 1968). However, only a few of the multitude of components comprising the varying personalities, diverse backgrounds, and multinationalities have been assessed. Through the use of pre-selection of the participants, it may be possible to represent in the human surrogates in simulations an important sample of the attributes of decision-makers of the world—be they decision-makers for nations, for companies overseas, or for governmental and nongovernmental international organizations.

Surrogate Groups and Programs as "Nations"

In making an assessment of "Decision-Makers and Their Nations," it is useful not only to consider the decision-makers per se, but also to explore how the surrogates function when assembled as decision-making groups, as well as how the consequences of their decisions are programmed as the outputs of their nations. "Individual and group components of the Inter-Nation Simulation are meshed into an operating model through both structured and free, self-developing interactive processes. In general, programmed assumptions are used for setting the foundations of the simulation, serving to provide operating rules for the decision-makers whereby they may handle the political, economic, and military aspects of their nations" (Guetzkow, 1963b, p. 148). Let us now examine aspects of the validity of the processes within these "nations" in the simulation. In addition to giving attention to the extent of the congruences which occur between simulations and "realities" with respect to the roles, group structures, and international communication patterns within the decision-making organizations involved, there will be an evaluation of the national programs.

Organizational characteristics: decision-making groups and roles therein. In their East Algonian Exercise, Crow and Noel demonstrated the effects of an organizational context upon their decision-makers, at least in one experiment with respect to one output; namely, the level of military response used to control a simulated military insurrection (Crow and Noel, 1965, pp. 19-20). As Table 6 indicates, those with high-risk preferences tended to respond throughout at a higher level than those with low-risk preferences. But in both instances, as the individual moved from private decision-making to a situation in which he needed to come to consensus with three other "top-level leaders of

TABLE 5   Assessments of Correspondences† between Simulations and "Realities" in International Relations Regarding "Decision-Makers and Their Nations"

	Humans as Surrogates		
Decision-makers:	*Correspondence with "General Materials from Field and Laboratory"*	*Correspondence with "Materials from International Relations"*	*Ratings:*
Personal characteristics	(M) Complexity (Driver, 1965) (M) Self-esteem, defensiveness, affect and information search in crisis (M. G. Hermann, 1965) (M) Complexity and information handling (Streufert, Suedfeld, & Driver, 1965) (M) Dimensionality of perceptions in crisis (Driver, 1962)	(S) Matched characteristics (Hermann & Hermann, 1967) (L) Skewness of characteristics‡ (Raser, 1965) (S) Rigidity and moral orientation‡ (Shapiro, 1966b) (S) Decision alternatives in crisis‡ (C. F. Hermann, 1965)	(M) 4 (S) 3 (L) 1 (N) 0 (I) 0 # = 8
Impact of training/ professionalism/ background	(I) Professionals vs. students (Boguslaw, Davis, & Glick, 1966) (N) IR beliefs (Guetzkow, Brody, Driver, & Beach, 1960)	(S) Diplomats vs. students‡ (Meier & Stickgold, 1965) (S) U. S. officials vs. students (Bloomfield & Whaley, 1965) (M) Militarism (Crow & Noel, 1965)	(M) 1 (S) 2 (L) 0 (N) 1 (I) 1 # = 5
Impact of culture/ nationalism	(S) Multi-cultures (African, Asian, European, Latin American, North American (Noel, 1963b) (S) Cross-cultures (Mexican/U. S.) (Raser & Crow, 1964) (M) Ethnocentrism (Druckman, 1968)	(M) Nationalism (Crow & Noel, 1965) (S) Ally/adversary (Bonham, 1967)	(M) 2 (S) 3 (L) 0 (N) 0 (I) 0 # = 5
Correspondence ratings:	(M) *"Much"* 5 (S) *"Some"* 2 (L) *"Little"* 0 (N) *"None"* 1 (I) *"Incongruent"* 1 # = 9	(M) *"Much"* 2 (S) *"Some"* 6 (L) *"Little"* 1 (N) *"None"* 0 (I) *"Incongruent"* 0 # = 9	(M) 7 (S) 8 (L) 1 (N) 1 (I) 1 # = 18

† See Footnote 4. p. 697.
‡ Materials involving anecdotes and case illustrations.

Algo, equal in authority and responsibility" (Crow and Noel, 1965, p. 3), there was a reduction in the level of response. Some writers about politics (e.g., Acheson, 1960; Neustadt, 1960) believe that a committee system tends "to inhibit innovation, boldness, and creativity, resulting usually in consensus or compromise based on the lowest common denominator of agreement" (Crow and Noel, 1965, p. 10). As Henry A. Kissinger speculates, ". . . the system stresses avoidance of risk rather than boldness of conception" (Kissinger, 1962, p. 356).

TABLE 6   Crow and Noel's Findings on the Impact of Group Consensus on Changes in Individual Judgments about Level of Escalation Response in the East Algonian Exercise†

	Individual Judgments and Group Consensus Levels of Escalation Response	
	Probability of winning war *Mean response level chosen*‡	
Participants' risk-taking preference *Score on pretest:*	*Individual judgment*	*Group consensus*
High	6.2	5.1
Low	4.7	3.9

†Adapted from Wayman J. Crow and Robert C. Noel, "The Valid Use of Simulation Results." La Jolla, California: Western Behavioral Sciences Institute, June, 1965, Table II-C, p. 6.

‡Higher response levels involve use of greater and more offensive force. One hundred-twelve Navy recruits served as decision-makers, operating in groups of "four top-level leaders of Algo, equal in in authority and responsibility" (Crow and Noel, 1965, p. 3). "Mean response levels showed that those preferring high risk chose a higher response level as compared to those preferring low risk; while, . . . group consensus levels were lower than group means for individual response levels" (Crow and Noel, 1965, pp. 19-20). No statistical tests of significance are reported.

In two of Crow and Noel's other experiments in the East Algonian Exercise, in which military response levels made in the course of rendering individual "pre-group" judgments were compared with the outcomes of group consensus, there were no clear effects of organizational context shown, despite the similarity of these experiments to the one mentioned earlier. In one, there was an interaction effect between the simulated situation and the organizational context, but in a contrary direction. When the opponent was presented as highly aggressive, then the decision-makers shifted to a significantly (F-test = 11.1**) higher level of military response as a result of group decision—in this experiment from a level of 7.4 to 9.2 (Crow and Noel, 1965, Table III A, p. D-7; and Table III C, p. D-9)—a result contrary to current verbal speculation among students of politics. Yet, such findings are in keeping with results from social psychological experiments by D. J. Bem, M. A. Wallach, and N. Kogan (1965) in which group discussions permit shifts to accept greater risks, "because the individual can feel less than proportionally to blame for the possible failure of a risky decision" (Crow and Noel, 1955, p. 11) (*Correspondence Rating: "Incongruent"*). Both results may be valid—although there is dissatisfaction with the limitations of both criteria: the unsystematic nature of field observation and the lack of "richness" of the laboratory.

In his simulation study of "Crises in Foreign Policy Making," C. F. Hermann (1965) probed the development of consensus within sixty-six decision-making groups comprising eleven runs of an Inter-Nation Simulation with U.S. Navy petty officers as participants (*supra*, pp. 704 & 705). In an "Event and Decision Form," Mr. Hermann queried his participants a number of times as to whether a crisis which they "recently or are now experiencing" had made the nation's goals "easier/harder to attain," covering such goals as "office-holding," "alliance development," and an ability to "preserve nation as separate unit" (C. F. Hermann, 1965, pp. 295-296). Although the experimenter demanded no actual group decision after focused discussion on the matter, as was the case in the East Algonian Exercise conducted by Crow and Noel (1965), crisis induced considerably more consensus, as measured by the agreement among three or four office-holders within each nation that "one or more goals had been made more difficult to attain" (C. F. Hermann, 1965, p. 214). In a set of forty-eight paired samples of crisis vs. non-crisis events, consensus existed for two-thirds of the non-crisis situations; the consensus increased significantly ($x^2 = 7.2, p < .005$) to 100 percent in the crisis situations (C. F. Hermann, 1965, Table 26a, p. 216). In discussing his hypothesis that "in crisis as compared to non-crisis, the frequency of consensus among decision-makers as to the national goals effected by the situation is increased," Hermannn indicated (1965, pp. 210-211) that such a tendency toward increased consensus is documented by the general literature on conflict (Mack and Snyder, 1957, p. 234) and on disaster (Thompson and Hawkes, 1962, p. 278), and by the specific case studies of U.S. decision-making within the Korean (Snyder and Paige, 1958, p. 375) and the Cuban (Larson, 1963, p. 225) crises (*Correspondence Rating: "Some"*).

As part of the starting conditions within the Inter-Nation Simulation, roles are designated within each group responsible for the nation's decision-making—a procedure which contrasts with the usual RAND/MIT practice of having each "team" work without assigned activities for any participant in their political-military exercises. In this way, an attempt is made within INS to induce a "division of labor" among the participants so that each position gains its perspective, as commonly occurs in roles found in bureaucracies (Katz and Kahn, 1966, chapter 7, pp. 171-198). Then, the group as a surrogate tends to function less as a small "face-to-face" group and instead perhaps takes on some characteristics of an organization (Guetzkow and Bowes, 1957).

Druckman has gathered data in his study of ethnocentrism (Druckman, 1968) which indicated that tendencies toward "bias" as found in laboratory and field studies (Rosenblatt, 1964) occur among the roles within the simulation. For example, those in low status roles within the simulated executive decision-making groups in WINSAFE II (*supra,* p. 708) (Raser and Crow, 1964), especially the marginal decision-maker who was aspiring to office, were found to rate their "outgroups, ally and enemy, least favorably and rate their ingroups most favorably" (Druckman, (1967, p. 117) 1968). Likewise, following observations made by Gordon Allport, Leonard Berkowitz, Robert Hamblin, and George Homans, Druckman noted "the more equal status contacts with outgroups, . . . the less ethnocentric the raters" (Hamblin, 1962, p. 106). Druckman then found, "The foreign minister or external decision-maker was the least ethnocentric role" (Druckman, (1967, pp. 124-125) 1968). The external decision-maker rated his own group least favorably and the outgroup's allies and enemies most favorably (*Correspondence Rating: "Much"*). Thus, role differentiations in the Inter-Nation Simulation may be homomorphic to those which occur in government offices handling decision-making for countries within the international system; there is no direct evidence on this matter from a study (Argyris, 1967) made within the U.S. Department of State.

Organizational characteristics: organizational structure and internal communication processes. In seeking to represent a structured government rather than a decision-by-committee process, the Hermanns used five men within each of their "nations" in the simulations. In this way these experimenters were able not only to specify roles for their surrogates, but also to develop "special-task subgroups" (by establishing "two policy groups in each government"), to provide indirect "mediated communication," and to introduce an "authority hierarchy" with "formalized rules" and "defined subordination" (C. F. Hermann, 1965, pp. 53-64) (*supra*, pp. 700, 704-5, 713). These elements were chosen so as to conform to the prescriptions developed by Morris Zelditch and Terrence Hopkins (1961, pp. 472-473) for establishing an organization in a laboratory setting.

An analysis of the communications within the "nations" (C. F. Hermann, 1965, pp. 58-60), as shown in Table 7, indicates that the Hermanns succeeded in

TABLE 7   C.F. Hermann's Tabulation of Two-Way Communication Channels Based on Number of Messages Written by Decision-Makers in an Inter-Nation Simulation†

Messages written via two-way channels: §	Communication Analysis‡									
	Within government						Between government members and aspiring decision-makers			
	CDM & IDM	CDM & FDM	CDM & EDM	EDM & IDM	EDM & FDM	IDM & FDM	ADM & CDM	ADM & FDM	ADM & IDM	ADM & EDM
Number of two-way channels	43	44	45	38	19	2	34	7	4	14
Total messages in two-way channels	524	461	411	300	86	9	209	33	16	46
Average number of messages in two-way channels	12.2	10.5	9.1	7.9	4.5	4.5	6.2	4.7	4.0	3.3

†Adapted from Charles F. Hermann, "Crises in Foreign Policy-Making: A Simulation of International Politics." Ph. D. Dissertation. Evanston, Illinois: Department of Political Science, Northwestern University, *1965*, p. 59.

‡CDM = Central Decision-Maker, IDM = Internal Decision-Maker, FDM = Force Decision-Maker, EDM = External Decision-Maker, and ADM = Aspiring Decision-Maker.

§ The messages were generated by 325 U.S. Navy petty officers serving as decision-makers for the six nations operated in eleven simulations. Between each combination of roles, the highest possible number of two-way communication channels was sixty-six. No statistical tests of differences are given.

inducing the development of organizational structures (*Correspondence Rating: "Some"*). Further, the communication patterns responded to the impingement of crisis, as shown to have been the case in the international system during the summer of 1914 (Holsti, 1965), as well as during October and November, 1962, during the Cuban crisis (Runge, 1963). Hermann discovered, "In crisis as compared to non-crisis, the volume of communication within a nation is significantly greater in crisis," with a probability of this result occurring by chance less than .01 (one-tail test) (C. F. Hermann, 1965, pp. 198-199) (*Correspondence Rating: "Some"*).

The correspondences between the literature of organizational behavior and the operation of the simulated nations are not as clear with respect to C. F. Hermann's hypothesis on the contraction of authority during crisis. He asserts that "in crisis as compared to non-crisis, the number of decision-makers assuming a major role in the decision process will be reduced; that is, there will be a contraction in the number of individuals exercising authority" (C. F. Hermann, 1965, p. 225). Although Hermann determined "how many individuals were required to account for seventy-five percent of the total influence attempts made by all the members of a government in response to an induced situation" (C. F. Hermann, 1965, p. 228), he found no significant differences between crisis and non-crisis behaviors. However, the "participants more frequently perceived fewer decision-makers involved in crisis decisions" (C. F. Hermann, 1965, Table 30, p. 229).

It is interesting to note that in Dean Pruitt's systematic research on twenty-eight case studies within the "office of XYZ Affairs" of the U.S. Department of State, no significant relationship was found between crisis and the number of either officers or agencies involved in a decision, although a number of his seventeen respondents volunteered the information "that fewer people are consulted as time pressures go up" (Pruitt, 1964-1965, p. 25). The evidence C. F. Hermann presented from his reference materials in supporting his authority-contraction hypothesis (C. F. Hermann, 1965, pp. 225-227), however, is largely illustrative—citing the perceptions of men who have occupied the position of Assistant Secretary of State, such as Roger Hilsman (1959, p. 372) and Harlan Cleveland (1963, p. 638) describe. Hermann's simulation seems to be homomorphic to both these sets of reference data which showed no contraction at the behavior level, yet participants from both the simulation and the reference systems stated they perceived such a contraction of authority (*Correspondence Rating: "Some"*).

Evaluation of variables comprising national programs.[6] To this point in this essay, our comparisons of outcomes in simulations to those in the reference

---

[6]Some readers may find this next section on the "Evaluation of Variables Comprising National Programs" more technical than most other parts of this essay, inasmuch as "through simulation we are able to force ourselves to develop more explicit theory" (Guetzkow, 1966a, p. 250). Throughout this essay an effort is made to handle the research being reported without need for the reader to consult the original sources. However, before reading the following sections on national programs involving political, economic, and military variables in INS, some readers may want to acquaint themselves with further substantive details by referring to my essay, "Structured Programs and Their Relation to Free Activity Within the Inter-Nation Simulation" (Guetzkow, 1963b), where the contents of the Equations and Programmed Assumptions are explicated at greater length. Others may find it useful to gain a methodological orientation to the rigorous style involved by reading

systems have centered upon the functioning of the participants, as individuals or in groups. It also is possible, however, to examine aspects of the validity of the programmed segments separately from their man-machine outputs. This research is analogous to the findings presented earlier (*supra*, pp. 704-705) on the extent to which personal outputs are mediated by intervening perceptual mechanisms which have some correspondence to those occurring in the reference materials— are the "right" results being produced for the "right" reasons? The following discussion is devoted to results from studies undertaken at Northwestern University. To date, validities of some twelve of the twenty-nine "Programmed Assumptions" of the Inter-Nation Simulation (INS) have been assessed by the work of Chadwick (1966a-d; 1967). He operationalized seventeen variables (1967, Table 1, p. 179) from INS theory in data derived from Rudolph Rummel's Dimensionality of Nations (DON) project (Rummel, 1966a)— comparing correlations among indices of INS variables (1966d, Table 3, p. 15) represented in an Inter-Nation Simulation conducted by Brody (1963) and Driver (1962) with those among empirical indices of these variables in a reference system of some sixty-four nations in the mid-nineteen-fifties (1967, Table 2, p. 180). Charles Elder and Robert Pendley, using another approach (1966a, pp. 10-11) in which relations hypothesized in INS theory were developed as differential equations, have contributed assessments of aspects of the political (Pendley and Elder, 1966b) and of the economic (Elder and Pendley, 1966b) programs. The core variables used in INS theory are analogous to the use of human beings as surrogates, in that "whole sets of variables in the complex of national and international life are represented by simplified, generic factors supposedly the prototypes of more elaborate realities." Then, a programmed relation among such prototypic variables hopefully "provides a condensed version of a gamut of real-life activities, similar to the way in which probability distributions are used by simulators to represent elaborate, underlying mechanisms that are too complicated to detail" (Guetzkow, 1963b, p. 105).

National Programs Involving Political Variables. Pendley and Elder (1966) have developed an assessment of Programmed Assumption No. 1 (Guetzkow, 1963b, Equation (1), p. 111). This hypothesis asserts a relationship between the probability of the decision-makers' holding of office (pOH) and the way their chances for remaining in office depend upon how satisfied their validators (be such masses and/or elites) are with their performance (VS_m). This programmed assumption indicates that the closeness of relationship of pOH and VS_m depends upon the decision latitude (DL) which their political structures allow; governments with wide latitude find their office-holding (pOH) is less directly dependent upon the extent to which they please those who validate their office-holding (VS_m). Pendley and Elder challenged the definition of office-holding given in INS theory, wondering whether it did not pertain to the stability of the government rather than to the regime, per the formulations of

"Some Uses of Mathematics in Simulation of International Relations" (Guetzkow, 1965) or essays by my younger colleagues, "Simulation as Theory Building in the Study of International Relations" (Elder and Pendley, 1966a) and "Relating Inter-Nation Simulation Theory with Verbal Theory in International Relations at Three Levels of Analysis" (Chadwick, 1966b).

David Easton (1965, pp. 153-219). By using measures for some sixty-two nations in the mid-nineteen-fifties, as derived from various data sources, Pendley and Elder discovered that the equation underlying Programmed Assumption No. 1 accounted for only 5 per cent of the variance in a pOH-type measure of regime stability, but accounted for some 25 per cent of the corresponding measure of system stability (Pendley and Elder, 1966, p. 24) (*Correspondence Rating: "Some"*).

In contrast to Pendley and Elder's approach, Chadwick for purposes of analysis simplified INS theory into a series of two-variable, monotonic functions (Chadwick, 1966a).[7] Components of Programmed Assumption No. 1 were analyzed one by one with similar results. In a reference system of some sixty-four nations in the mid-nineteen-fifties, Chadwick (1967, pp. 181-184) found that the relation between pOH, indexed as regime (rather than as system) stability, and VS_m was not statistically significant ($r = .06$); there was a statistically significant relationship ($r = .30*$) between pOH and DL. However, when the simulation was operated, both of these relationships were realized—the relation of pOH to VS_m, was $.99**$, an artifact due to a regrettable error by the calculators in omitting the mediating effects of DL; the relation of pOH to DL was $.56**$ (Chadwick, 1966d, p. 15). (*Correspondence Rating: "Little"*). Programmed Assumption No. 2 relates validator satisfaction (VS_m) to its two components: satisfaction with consumption standards (VS_{cs}) and satisfaction with the nation's national security (VS_{ns}) (Guetzkow, 1963b, p. 114). In Chadwick's research VS_{cs} correlated somewhat significantly with VS_m ($r = .19*$) in the reference system of the mid-nineteen-fifties (Chadwick, 1967, pp. 184-186); it correlated significantly ($r - .51**$) within the simulation. National security satisfaction (VS_{ns}) did not correlate with VS_m in this same reference system ($r = -.15$); it correlated significantly but weakly ($r = .23*$) in the simulation (Chadwick, 1966d, pp. 14-15) (*Correspondence Rating: "Little"*).

The behaviors of the variables included in Programmed Assumptions No. 8 and No. 21, concerned with revolution (Guetzkow, 1963b, p. 119: p. 130), were not found by Chadwick to be congruent, with one exception: The relationship specified between the probability of a revolution occurring (pR) and validator satisfaction (VS_m) proved to be significantly negative ($r = -.41**$) in the reference materials (Chadwick, 1967, pp. 186-188); although of lesser magnitude, it also was significantly negative in the simulation ($r = 0.29**$) (Chadwick, 1966d, p. 15) (*Correspondence Rating: "Little"*). Chadwick's research also indicated that although Programmed Assumptions No. 17 through No. 20 concerned with "Consequences of Pressure Changes in Decision Latitude" (Guetzkow, 1963b, pp. 127-130) all were realized in the simulation, the relationships postulated between DL and VS_m in Programmed Assumption No. 17, between DL and basic capability (BC) in Programmed Assumption No. 19, and in the feedbacks from these three in Programmed Assumption No. 20 did not

[7]In later work in developing his proposals for modifications in the Inter-Nation Simulation, Chadwick (1966c) used factor analysis and partial correlation techniques to elicit the complex interdependencies existing in the reference material; in this way he wished to assess non-additive explanatory relations through confluence analysis (Alker, 1966, pp. 639-653). Because of the overall lack of relations among the seventeen variables in the simulation, Chadwick was unable to apply these more adequate techniques to the outputs of the simulation itself, as he had hoped to do. Hence, they are not discussed in this essay.

prove out in the reference materials (Chadwick, 1966d, pp. 14-15) (*Correspondence Rating: "None"*).

A tabulation of the findings involving the political variables is included in Table 8. Overall, there was not much congruence between the reference materials from international relations and the outputs generated in the simulation.

National programs involving economic variables. In keeping with the challenge issued by Noel (1963a) in his theoretical work on economics within the Inter-Nation Simulation, Elder and Pendley (1966b) examined portions of the INS model in terms of both economic theory and data. Focusing on Programmed Assumptions No. 11, No. 12, and No. 13 (Guetzkow, 1963b, pp. 123-124), the interrelations among a nation's basic capability (*BC*), consumption standards (*CS*), and validator satisfaction (VS_m) were probed in considerable detail. In validating the relations posited in Equations (4) and (7) (Guetzkow, 1963b, pp. 124-125) of the INS theory between the minimum consumption standards (CS_{min}) and the maximum possible (CS_{max}), given the nation's capability, Elder and Pendley found reference data for some thirty-eight nations in the mid-nineteen-fifties related curvilinearly rather than linearly (Elder and Pendley, 1966b, Figure 4, p. 25) (*Correspondence Rating: "Incongruent"*). In checking the validity of the relation of ongoing consumption (*CS*) (as a ratio of CS_{min}) to an index validator satisfaction with consumption (VS_{cs}), they found a correlation of .75**, calculated over forty-one nations (Elder and Pendley, 1966b, p. 27) (*Correspondence Rating: "Much"*).

Chadwick's component-by-component analysis revealed less congruent results. The correlation between basic capability (*BS*) and consumption standards (*CS*) in some sixty-four nations in the mid-nineteen-fifties was .87** in the reference system (Chadwick, 1967, p. 186); an extremely high relationship of .95** occurred in the simulation (Chadwick, 1966d, p. 15). *Vis-a-vis* Equation (7) (Guetzkow, 1963b, p. 125), the correlation between VS_{cs} and *CS* was .60** in the same reference system: however, the programmed relation failed to be realized ($r = .15$) in the simulation (Chadwick, 1966d, pp. 14-15) (*Correspondence Rating: "Some"*). These discordant results between reference system and simulation, with the simulation producing either too strong a relation or too weak a relation between the variables, indicate that adjustments are needed in the model in order to increase its value.

National programs involving military variables. In developing the INS model which distinguishes nuclear from conventional force capabilities, Brody (1963, p. 698) secured counsel in 1959-1960 from military experts in the specification of both his parameters and his equations. Further work in the style of Elder and Pendley, however, has not yet been done in evaluating his programs. Chadwick's findings on military variables are few and negative. In researching the components of Programmed Assumption No. 16 [Equation (8)] (Guetzkow, 1963b, pp. 126-127), Chadwick (1967, pp. 188-190) found that neither the reference data on sixty-four nations in the mid-nineteen-fifties nor the simulation materials produced any clear relationship between validator

TABLE 8   Assessments of Correspondences† between Simulations and "Realities" in International Relations Regarding "Decision-Makers and Their Nations"

Surrogate Groups and Programs as "Nations"

"Nations": Correspondence with "Materials from International Relations" Ratings:

Organizational characteristics	(I)	Risk-taking in groups (Crow & Noel, 1965) Consensus in groups in crisis‡	(S)	Internal communications volume increases in crisis (C.F. Hermann, 1965) Contraction of authority	(M) 1 § (S) 4 § (L) 0 (N) 0 (I) 1
	(S)	(C.F. Hermann, 1965)	(S)	(C.F. Hermann, 1965)	#

In politics:

National programs	(S)	Office-holding as system stability (PA#1) (Pendley & Elder, 1966)	(L)	Validator satisfaction (PA#2) (Chadwick, 1966d)	
	(L)	Office-holding as regime continuance (PA#1) (Chadwick, 1966d; 1967)	(L)	Revolution (PA#s 8 & 21) (Chadwick, 1966d; 1967)	
			(N)	Decision-latitude (PA#s 17-20) (Chadwick, 1966d)	(M) 1 (S) 2 (L) 3

In economics:

	(I)	Consumption standards (Eq #s 4 & 7) (Elder & Pendley, 1966b)	(S)	Consumption satisfaction (PA #s 11-13 and Eq #7) (Chadwick, 1966d)	(N) 2 (I) 1 # = 9
	(M)	Consumption satisfaction (PA#s 11-13) (Elder & Pendley, 1966b)			

In security:

(N) Security Satisfaction (PA #16) (Chadwick, 1966d)

Correspondence ratings:	(M)	*"Much"*	1	(M)	2 §
	(S)	*"Some"*	5	(S)	6 §
	(L)	*"Little"*	3	(L)	3
	(N)	*"None"*	2	(N)	2
	(I)	*"Incongruent"*	2	(I)	2
			# = 13		# = 15 §

---

†See Footnote 5, p. 701.
‡Materials involving anecdotes and case illustrations.
§Includes *Correspondence Ratings* of the following "General Materials from Field and Laboratory": (M) Bias in internal/external roles (Druckman, 1968)
            (S) Organizational structure (C.F. Hermann, 1965)
    PA = Programmed assumption, and Eq = Equation; both as numbered in Guetzkow, 1963b.

satisfaction with respect to national security (VS_{ns}) and the nation's present force capability or its war potential (as represented in its *BC*) (Chadwick, 1966d, pp. 14-15) (*Correspondence Rating: "None"*). Chadwick challenged INS theory, too, developing a somewhat different index of national security; he then found "a general relationship between this (reformulated) VS_{ns} index and (as a set) force capability, threats, accusations, and protests" (Chadwick, 1967, p. 189).

Summary: surrogate groups and programs as "nations." Table 8 is the record of my assessments of correspondences between simulations of inter-national relations and "realities" with respect to the operations of the surrogate groups and the programs which are meshed into "nations." Only two of the fifteen assessments compared simulation materials with more "General Materials from Field and Laboratory"; thus, almost all the comparisons were made with materials referring to the international system. Some of these were quite rigorous in conception, particularly the work on the operation of the pro-grammed political, economic, and military variables, taken as prototypes. It is interesting to note that, as the comparisons became more rigorous, the con-gruence seems to become less. It is recognized that only a third of the programmed assumptions of the Inter-Nation Simulation model have been assessed; undoubtedly, the number of free variables which were examined is small, contrasted with the number of those sampled within the simulation which might have been examined. It is disturbing to have two incongruences occurring simultaneously with the eleven homomorphies, even though the latter vary widely in their levels of correspondence.

"RELATIONS AMONG NATIONS"

In our examination of "Decision-Makers and Their Nations," surrogates and programs were composed so as to constitute nation-actors within the inter-nation complex. Let us now examine the validities obtained within the simulations when the relations among these "nations" are contrasted with "realities" of international relations—the reference system.

In generating "Relations Among Nations" the Inter-Nation Simulation exhibits characteristics of a "self-organizing system" (Guetzkow, 1962, pp. 89-90). How can these aspects be included in an assessment of the corre-spondences between the simulations and the reference systems? Terry Nardin utilized notions of Brody and Rummel in creating variables which characterized the relations among the nations, taken two at a time (Brody, 1963, p. 714; Rummel, 1963, p. 22). He measured "distance" with respect to economic prowess between the U.S.S.R. and India in the reference system, for example, and measured the difference between Yora and Zena in the Inter-Nation Simulation, using the differences for the (1962-1963) gross national product and (for the third through sixty periods) for basic capability (*BC*), respectively. With fifteen countries in the reference system, data may be compiled for 105 pairs; with nine nations in the simulation, for each run thirty-six pairs may be con-stituted. Nardin, with the help of Neal E. Cutler, utilized materials produced in two simulations, an original simulation (INS-16) consisting of six runs (Meier,

1965; Meier and Stickgold, 1965) and its replication (INS-19) consisting of four runs. A population of 360 pairs was available from the ten simulations (Nardin and Cutler, 1967). In addition to characterizing each pair of nations in terms of its dissimilarity in national wealth and development, Nardin and Cutler composed differences among the nations with respect to type of regime, as characterized by decision latitude (*DL*). The alliance memberships of each pair were characterized, also, as to how they differed in preceptions of the likelihood of war occurring between them. Further, Nardin and Cutler used measures of interaction among the pairs, including their communications, their exports, and the number of treaties they signed with each other. The twenty-one possible relations among these measures were checked for their reliability by making comparisons of the Nardin replication (INS-19) to the Meier-Stickgold simulation (INS-16), with the finding that only two relations were importantly discrepant (Nardin and Cutler, 1967, p. 10).

Nardin and his associates, including Harry Targ (Targ and Nardin, 1966), devoted themselves exclusively to the use of variables constituted from pairs of nations. Others have probed "Relations Among Nations" using variables which characterized the external behaviors of the nation, even though the variable itself was an attribute to the nation, in the style of the fifty-seven "raw characteristics" of Arthur S. Banks and Robert B. Textor (1963, pp. 18-20). For example, Chadwick included for each nation its exports and imports, as well as the number of agreements into which it entered which were of a military and/or of an economic/cultural nature—all variables which reflect its relations with other members of the international system. In these contrasting ways, researchers have captured some of the "variations in the unprogrammed activities, which emerge as the nations relate to each other within the developing overall system" (Guetzkow, 1963b, p. 104).

Following the order listed in the right hand tree of the outline in Figure 1 (*supra*, p. 698), communication structures and processes used for handling the "Relations Among Nations" will be examined, first. Then, characteristics of alliances and their involvement in the use of force will be surveyed briefly. Finally, aspects of the "interaction patterns," including those concerned with cooperation and conflict, will be assessed. It should be noted that in the Inter-Nation Simulation, "with the exception of the rules for the conduct of war, there are no programs prescribing the relations among nations. The basic strategy used in the construction of the simulation has been to allow free development of the inter-nation relations" (Guetzkow, 1963b, pp. 148-49). Inasmuch as no research on the homologies between simulations of international relations and "realities" has as yet been reported on war as embedded in an international simulation (McRae, 1963)—nor in such computerized models of relations among nations as those developed by Benson (1962) and Abt and his associates (1964), for that matter—it is not possible at this time to include an evaluation of programmed work in this area.

Communications

At first glance there seems to be "face validity" in the communications which develop among the nations of a simulation, with the bilateral interchanges often being complemented by a one-to-many issuance of messages and conferences among the external decision-makers (EDMs). Credibility is given to this impression by such an occurrence as when the participant positioned to represent the English Foreign Minister (Lord Grey), in the Hermanns' effort to simulate (M-Run) the events leading into World War I, assembled an international conference of the principal actors. During this meeting Austria-Hungary was pressured into withdrawing her claims against Serbia (Hermann and Hermann, 1967, p. 407). Historically, England did call for such an international conference which subsequently was rejected. The Hermanns concluded, "Thus, an alternative actually considered and subsequently excluded by the historical figures provided the avenue which the simulation participants followed for the resolution of the imposed situation" (Hermann and Hermann, 1967, p. 407-408) (*Correspondence Rating: "Some"*). It is interesting to note in the less well-matched run (A-Run) that the surrogate nations—Austria, England, France, and Russia—then pursued paths which led to dampening effects, perhaps similar to those which were obtained in the 1908-1909 Bosnian crisis (Hermann and Hermann, 1967, Table 2 and accompanying text, p. 410).

When evidence was examined in a less anecdotal fashion, the validity of communication processes of the simulation was not found to be well substantiated. C. F. Hermann (1965, pp. 192-203) found but little congruence between simulation and reference materials when he checked his hypothesis that "in crisis as compared with non-crisis, the volume of communication between a nation's decision-makers and other international actors, external to the nation, will increase" (C. F. Hermann, 1965, p. 193). The relation of crisis/non-crisis to the frequency of external communication proved to be significant at the .09 level for a one-tail Mann-Whitney U-test (C. F. Hermann, 1965, Table 22, p. 200). However, Robert C. North and his colleagues found that in 1914 "as the crisis developed, decision-makers in the various capitals received rapidly increasing volumes of messages from various parts of Europe" (North, Holsti, Zaninovich, and Zinnes, 1963, p. 164); this phenomenon appeared again in the course of the Cuban missile crisis in 1962. Hermann noted, "For example, the American Secretary of State related that in addition to the adversary and the United Nations, the United States communicated with more than seventy-five governments in the Cuban crisis (Larson, 1963, p. 268)" (C. F. Hermann, 1965, p. 196) (*Correspondence Rating: "Little"*). This lack of correspondence seems not related to instability of simulation output, inasmuch as Hermann reported an "absence of significant variation between one run and another" on external communication, as well as on three additional variables (out of five) for which a reliability analysis was made (C. F. Hermann, 1967a, Footnote 12, p. 31).

A somewhat more rigorous comparison of communication processes between simulation and reference materials was presented by Nardin and Cutler (1967) (*supra*, p. 721) when they related a communication variable ("SALIEN") to six other structural characteristics and behavioral processes. In three instances the relationships proved null in both systems; in three other instances, however, the simulation failed to generate the relationships which did appear in the reference system of the early nineteen-sixties. Although in this reference system, diplomatic exchanges were related to the magnitude of the difference in the gross national product for the two nations, to the nonmilitary exports of one of the nations of the pair to the other, and to the number of treaties they signed, such relationships were not obtained for homologous variables in the simulation (*Correspondence Rating: "Little"*).

Even the modicum of congruence disappeared in examining in some further detail the structure of the communication within the simulations. Dina A. Zinnes carefully noted the differences between the representative role of the ambassador in the pre-World War I situation and the policy role of the external decision-maker (EDM) in the Inter-Nation Simulation, making *a priori* analysis that would produce important differences in communication effects (Zinnes, 1964, pp. 2-6). The fact that the EDMs in the Brody-Driver runs had much direct interaction with each other, without "intermediary or buffer," perhaps induced the nations to "behave in a manner comparable to findings in 'small-group' studies; namely, that the hostility between two individuals results in lessened contact between them and that there is a preference for interaction with one's alliance or in-group . . . relationships [which] do not hold in the 1914 data" (Zinnes, 1966, p. 496). The three hypotheses making alternative tests of the relationship in the two sets of materials are presented in Table 9 (*Correspondence Rating: "Incongruent"*).

Bonham's (1967) experimental simulation of face-to-face negotiations on the post-World War II disarmament negotiations produced the overall effects described by Zinnes. When he intervened by inducing strong disagreement between parties about the relative importance of the inspection versus the reduction-in-arms issues, Bonham obtained more hostility, as measured by attacks made in the course of the simulated negotiations (Bonham, 1967, Table 3, p. 110). There were also reductions in the mean number of messages exchanged per move period (Bonham, 1967, Table 2, p. 107). Bonham interpreted the matter: "Attempts to reduce the differences of opinion about the salience of the issues may be unsuccessful. Misunderstanding may result and lead to increased hostility. Consequently, the negotiators may tend to withdraw and communicate less with each other" (Bonham, 1967, p. 106). It is interesting to note that in the detailed analyses completed on the 1955 Subcommittee negotiations using an identical coding scheme, Bonham also found high amounts of expression of hostility (Bonham, 1967, p. 35) associated with divergence between the U.S.S.R. and the U.S. on the importance of reduction-in-arms vs. inspection. It is to be regretted that Bonham could not make comparative analyses of the volume of communications in the reference and simulation materials (*Correspondence Rating: "Much"*).

These complementary findings by Zinnes and Bonham with respect to interaction among nations, as reflected in their communication processes, are reminiscent of the earlier finding that decision-makers behave in simulations "remarkably like humans." Our negotiators, be they "ambassadors" (as in the Zinnes' work) or "representatives" (as in the Bonham research), behave as though they were members of "face-to-face" decision-making groups (Collins and Guetzkow, 1964). In making reconstructions of the Inter-Nation Simulation, Zinnes' insight must be taken with seriousness.

Alliances and Involvement of Force

Because of the important part alliances play in national security within the

TABLE 9 Zinnes' Findings on the Relationship between Communication Interaction and the Perception and Expression of Hostility in the Brody-Driver Realizations of an Inter-Nation Simulation and 1914 Historical Data†

Communication Interaction and the Perceptions and Expressions of Hostility as Represented by Spearman Rank Order Correlation Coefficients‡

Hypotheses:	Simulations § (42 pairs of nations possible)			1914 Historical data (30 pairs of nations possible)			
			Sum of three	Number of messages		Word message volume:	
	Pre- nuclear	Post- nuclear	hostility themes	Total	In crisis	Total	In crisis
Hypothesis 11. There is a negative relationship between the *perception* of threat and the frequency of interaction, i.e., the greater the perceived threat, the less the interaction.	−.36 *p*<.05	−.20	...	−.04	−.03	.04	.01
Hypothesis 12. There is a negative relationship between the *perception* of unfriendliness and the frequency of interaction.	−.56 *p*<.001	−.41 *p*<.001	...	−.02	−.03	.04	.11
Hypothesis 13. There is a negative relationship between *x's hostility* to *y* and *y's* frequency of interaction with *x*.	−.71 p<.001	−.77 p<.001	.07	−.00	−.11	.03	−.04

†Adapted from Dina A. Zinnes, "A Comparison of Hostile Behavior of Decision-Makers in Simulate and Historical Data." *World Politics,* 18, 3 (April, 1966), Table IX, p.491.

‡The level of significance of the Spearman Rank Order Correlation Coefficients above is given only when it was less than either .001 ($p<.001$), .01 ($p<.01$), or .05 ($p<.05$).

§ Three hundred thirty-six high school juniors and seniors served as decision-makers in the seven nations constituted for each of sixteen runs of the simulation, from which the pairs of nations were drawn.

Inter-Nation Simulation (Guetzkow, 1963b, pp. 140-142), Druckman's ethnocentric evidence is of much relevance. He found that the ingroup bias for one's own nation (*supra.* p. 708) also held to an extent for those nations within their alliances—and that simultaneously there was depreciation of "out-nations," as is illustrated in Table 10 (Druckman, 1968). When alliances shifted in the course of the WINSAFE II simulation (Raser and Crow, 1964), in which Druckman's results were gathered, the bias shifted correspondingly. These effects are documented within the general literature of ethnocentrism by Paul C. Rosenblatt (1964) (*Correspondence Rating: "Much"*).

TABLE 10   Druckman's Analysis Indicating Ethnocentrism at the End of Ten Realizations by Raser and Crow of an Inter-Nation Simulation†

Evaluation by Participants of Members of Own Nation, Allies, and Enemies‡				
Ratings by all decision-makers:	*Mean ratings of traits*			
	Overall evaluation	*"Liking"*	*"Respect"*	*"Strong personality"*
—of members of own nation	−.44	−.50	−.32	−.21
—of allies	+.01	−.02	+.04	−.01
—of enemies	+.22	+.28	+.13	+.10

†Adapted from Daniel Druckman, "Ethnocentrism in the Inter-Nation Simulation." *The Journal of Conflict Resolution,* 12,1 (March, *1968*), Table 1.

‡One hundred sixty Navy recruits served as decision-makers, operating four nations in each of ten runs for Raser and Crow. Data on ratings of all participants as persons were collected at the end of each run. The more negative the mean rating, the more favorable the evaluation. All differences are significant at the .001 level, as tested through analyses of variance.

Such ingroup vs. outgroup differences are reflected also in the alliance behaviors generated in the Brody-Driver runs (Brody, 1963). Although it seems that the patterning of communication, as discussed above, does not correspond closely to "realities" of the reference system, the communication net *vis-a-vis* the bipolar alliances seemed to have simulated somewhat satisfactorily the patterns among those countries existing in the mid-nineteen-fifties. After demonstrating that "there is more interaction within the blocs than between the blocs" (Brody, 1963, p. 725), using the terminology of Morton Kaplan (1957), Brody then stated that in their sixteen simulations "there is a tendency for non-nuclear bloc members to communicate with the leader of their bloc rather than with the external nuclear power—the system prior to the spread of nuclear capability is not only bipolar, it is tightly bipolar" (Brody, 1963, p. 731). Presaging the late nineteen-sixties, these simulations (operated during the summer of 1960) then developed a fragmentation of the alliance systems, with the coming of nuclear proliferation. The schematics presented in Figure 4 (Brody, 1963, Figures 4.3 and 4.4, pp. 743-744) highlight the dramatic change which accompanied the experimental introduction of widespread nuclear capability, mirroring in a general way the fragmentation which seems to be taking place during the nineteen-sixties in the alliances of both the East and the West (*Correspondence Rating: "Some"*).

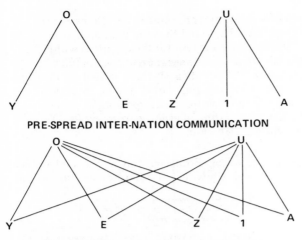

PRE-SPREAD INTER-NATION COMMUNICATION

POST-SPREAD INTER-NATION COMMUNICATION

FIGURE 4   *Richard A. Brody's Conceptualization of Pre-Spread and Post-Spread Inter-Nation Communication.* (Adapted from Richard A. Brody, "Some Systemic Effects of the Spread of Nuclear-Weapons Technology: A Study Through Simulation of a Multi-Nuclear Future." *The Journal of Conflict Resolution*, 7, 4 (Dec., 1963), Figs. 4.3 and 4.4, pp. 743-744. Letters represent different nations in the simulation; lines represent modal communication links.)

The relations of the alignments of nations to six other structural and behavioral aspects of the nations in the simulations of international relations and in the "realities" of the reference system were explored with some rigor by Nardin and Cutler (1967). There was a close correspondence between alignment and five of the six variables thought to have been homologous, ranging in parallel fashion from strong through weak relationships for such pairs of variables as alignment *vis-à-vis* openness of the regime (strong) and alignment *vis-à-vis* communication (weak), the latter being in contradiction to the finding by Brody presented immediately preceding (*Correspondence Rating: "Much"*). Yet, the relationships between the number of treaties signed and the variables used to characterize the system were simulated with adequacy in only three of the six possible relationships; the three non-congruences were of the kind in which the simulation failed to generate a relationship strong enough to match the operation of the variables in the reference system (*Correspondence Rating: "Some"*).

Turning from alliances to consideration of potential uses of violence within the international system, the findings of Nardin and Cutler (1967) on the involvement of force in the simulations are parallel to those found in the reference material they examined. Once again the correspondences ranged over strong through weak relationships, for such pairs of variables as expectation of war *vis-à-vis* openness of the regime (strong) and the likelihood of war *vis-à-vis* communication (low) (*Correspondence Rating: "Much"*). In fact, the "case" materials generated in the effort to reproduce the events of the summer of 1914

(Hermann and Hermann, 1967) provide ground for speculation that perhaps the simulation developed by the Hermanns could be "turned" to produce war. "In the researchers' opinion, if the M-Run simulation [the better-Matched run, in which an attempt was made to develop homology between the personality characteristics of the historical decision-makers and their counterparts in the simulation] had been continued for another 50-to-100 minutes (one or two more simulated days), war would have been declared along lines similar to the historical situation. This position is confirmed by ten of the fifteen M-Run participants and messengers (66.7 per cent) in their debriefing questionnaires" (Hermann and Hermann, 1967, p. 407) (*Correspondence Rating: "Little"*).

Less anecdotal is the statistical finding with respect to the involvement of force by Crow and Noel (1965) in their East Algonian Exercise. While there was wide variation in the level of escalation response chosen, almost half of the 384 individuals and more than half of the 96 groups chose the level the authors "judged to represent the historical decision" (Crow and Noel, 1965, p. 20).[8] They noted that to obtain such congruences in 45 and 57 per cent of the cases, respectively, "the validity coefficients would have to be .67 and .75 respectively" (Crow and Noel, 1965, p. 21) (*Correspondence Rating: "Much"*).

Interaction Patterns: Cooperation and Conflict

In illustrative data from the "realities" of the reference system, a cooperative phenomenon has been found which seems to correspond in a measure to cooperation patterns in simulations. C. F. Hermann reported that among his decision-makers in the simulation a "search for support was much more frequent in crisis" (C. F. Hermann, 1965, p. 207); he also noted that the parallel "lesson" was pointed out "a year after the Cuban crisis [by] an Assistant Secretary of State, . . . 'Even when the decision to employ power is essentially our own, we find it highly desirable to widen the community of the concerned—to obtain sanction for the necessary "next step" from the broadest practicable segment of the international community' (Cleveland, 1963, p. 645)" (C. F. Hermann, 1965, p. 209) (*Correspondence Rating: "Some"*).

Bonham's analysis of the timing and contents of concession-making by the United States and the Soviet Union in disarmament negotiations checked validities of simulation materials against the codings presented by Lloyd Jensen (1963). "The United States tended to make concessions earlier in the round than the Soviet Union" (Bonham, 1967, p. 39). In the simulation the "United States made over 70% of its concessions during the *first* half of the negotiations, while the Soviet Union made 58% of its concessions during the *second* half," a difference statistically significant at the .01 level (Bonham, 1967, p. 41). But, there was little correspondence between the types of concessions reached. Further, Jensen found that the United States and the Soviet Union tended to reciprocate concessions during the twenty-one rounds of the post-World War II disarmament negotiations which he coded. "No relationship was found between American and Soviet [reciprocation in] concession-making ($R_s = -.01$) for the

---

[8]The authors explain, "Identification of the historical situation would impair its usefulness for further research. It will be identified for qualified investigators" upon inquiry (Crow and Noel, 1965), p. 2).

data from all the simulation runs. However, when the control runs were examined separately, the Spearman correlation (R_s = .55) was positive and significant" (Bonham, 1967, p. 47) (*Correspondence Rating: "Some"*).

Alliances may be used for peaceful purposes, too, as in the development of economic and cultural exchanges. In fact, Chadwick found both in reference and in simulation materials that there were tendencies for economic/cultural and military cooperation to occur simultaneously, the correlations between such agreements being .50** and .26* in the two systems, respectively (Chadwick, 1966d, pp. 14-15) (*Correspondence Rating: "Some"*). When trade per se was examined, Chadwick found that although there was a very high correlation between imports and exports in the reference system of the mid-nineteen-fifties (r = .93**), in the Brody-Driver simulation there was no linkage (r = −.02) (Chadwick, 1966d, p. 14-15) (*Correspondence Rating: "None"*). As the upper portion of Table 11 indicates, in Chadwick's analysis of the economic/cultural treaties in relation to trade within the simulation and reference systems, there was no correspondence. In the analyses presented by Nardin and Cutler, the trade variable ("EXPORT") behaved with little congruence, even though both reference and simulation materials were different from those used by Chadwick. For example, the correlation of .68** found between trade ("EXPORT") and

TABLE 11   Chadwick's Correlations of Economic/Cultural Agreements with Trade and Conflict Behaviors in the Brody/Driver Simulation and in the (1955-1957) Reference Materials†

	Correlation of Economic/Cultural Agreements Variable (Aec)	
With *Inter-Nation Simulation* *behavior variables below:*	*Within* *Brody-Driver* *Simulation‡*	*Within* *reference materials§* *(1955-1957)*
Trade:		
Import (Ti)	.04	.79**
Export (Te)	−.09	.70**
Conflict behaviors:		
Threats (T)	−.10	.29*
Accusations (A)	−.08	.18
Protests (P)	.00	.40**

†Adapted from Richard W. Chadwick, "Theory Development Through Simulation: A Comparison and Analysis of Associations Among Variables in an International System and an Inter-Nation Simulation." Evanston, Illinois: Simulated International Processes project, Northwestern University, September, 1966d, Tables 2 and 3, pp. 14-15. Correlations are product-moment coefficients, and for both the simulation and reference systems the tests of significance are two-tailed.

‡The three hundred thirty-six participants who served as decision-makers were high school juniors and seniors. The correlations are based on data from 112 "nations," with no missing figures. See Richard A. Brody, "Some Systemic Effects of the Spread of Nuclear-Weapons Technology: A Study Through Simulation of a Multi-Nuclear Future." *The Journal of Conflict Resolution*, 7, 4 (December, 1963), pp. 663-753.

§In the reference material of the mid-nineteen-fifties, the correlations are based on sixty-four countries, with a maximum of five cases of missing data per variable. See Rudolph J. Rummel, "Divensions of Conflict Behavior Within and Between Nations." *General Systems: Yearbook of the Society for General Systems Research*, 8, 8 (1963), pp. 1-50.

treaties ("INTAGR") in the reference system was only .02 in the simulation (Nardin and Cutler, 1967, Appendix II) (*Correspondence Rating: "None"*).

Correspondences between simulations and "realities" may be examined also with respect to conflict among entities comprising the international system. Chadwick's (1966d) analyses revealed almost no congruence among the measures of conflict, as adapted from the work of Rummel (1963) and Raymond Tanter (1966) for the last half of the nineteen-fifties, exhibited in Table 12 following. Nor do the measures of threat and protest yield even low correlations in the simulation when they are related to economic/cultural treaties which occurred in the reference data, as shown in the lower half of Table 11 preceding (*Correspondence Rating: "None"*).

TABLE 12   Chadwick, Rummel, and Tanter: Correlations among Coded Measures of Conflict Behaviors Exhibited by Nations in an Inter-Nation Simulation and in 1955—1960 Reference Materials

	Correlations between Conflict Behavior Variables		
Conflict behaviors:	*Within Inter-Nation Simulation†* *(112 nations)*	*Within 1955—1960 reference systems†*	
		1955-1957 (77 nations)	*1958-1960 (83 nations)*
Threats/Accusations	.21*	.81	.62
Threats/Protests	.03	.66	.59
Accusations/Protests	.03	.69	.64

†Adapted from Richard W. Chadwick, "Theory Development Through Simulation: A Comparison and Analysis of Associations Among Variables in an International System and an Inter-Nation Simulation." Evanston, Illinois: Simulated International Process project, Northwestern University, September, *1966d,* Table 3, p. 15. Correlations are product-moment coefficients, with no missing figures. Three hundred thirty-six participants who served as decision-makers were high school juniors and seniors.

†Adapted from Raymond Tanter, "Dimensions of Conflict Behavior Within and Between Nations, 1958—60." *The Journal of Conflict Resolution,* 10, 1 (March, 1966), Table 1, p. 46, in which the 1955—1957 correlations were taken from Rudolph J. Rummel, "Dimensions of Conflict Behavior Within and Between Nations." GENERAL SYSTEMS: *Yearbook of the Society for General Systems Research,* 8, 8 (*1963*), pp. 1—50. No significance tests are reported.

However, there was considerable homomorphism in the examination of 1914 reference data in two instances. The Hermanns (1967) made a micro-analysis of messages, using content analysis categories very similar to those used by the Stanford Conflict and Integration Project, as displayed in Table 13, finding considerable congruence. When a more refined analysis was made by them, using differences between the numbers of hostility and capability statements, the "resulting rank-order correlation (.90) between the 1914 data and that of the M-Run [well-matched] was statistically significant. The correlation for the A-Run [less well-matched] was not significant, however" (Hermann and Hermann, 1967, p. 411) (*Correspondence Rating: "Much"*). Thus, the Hermanns realized experimentally—in a very limited case study—the Stanford hypothesis: "If a state's perception of injury (or frustration, dissatis-

TABLE 13   The Hermanns' Comparison of the Number of Hostility and Capability
Statements for Nations in an Inter-Nation Simulation and in 1914[†]

	Hostility and Capability Statements					
	In an Inter-Nation Simulation[§]				*In 1914*[§] *reference materials*	
By nations below	*M-Run*[1]		*A-Run*[1]		*1914*	
	Hostility	*Capability*	*Hostility*	*Capability*	*Hostility*	*Capability*
Austria-Hungary	91	38	63	28	179	26
England	17	13	24	7	32	16
France	41	36	21	5	26	18
Germany	53	35	107	26	138	34
Russia	30	16	69	20	50	28

[†]Adapted from Charles F. Hermann and Margaret G. Hermann, "An Attempt to Simulate the Outbreak of World War I." *American Political Science Review,* 61, 2 (June, 1967), Table 3, p. 411.

[‡]The ten decision-makers in each of the two runs were twenty students who recently had graduated from high school. "Perceptions of hostility exceed those of capability for every nation" in the simulations and in 1914 and are statistically significant by a sign test (Hermann and Hermann, 1967, p. 411).

[§]The 1914 data used by the Hermanns were taken from Dina A. Zinnes, Robert C. North, Howard E. Koch, "Capabilities, Threat, and the Outbreak of War." In James A. Rosenau (Editor), *International Politics and Foreign Policy.* New York: Free Press of Glencoe, *1961,* p. 476; they were based upon the coding of "1,165 communications exchanged between the European States in the weeks prior to the beginning of the First World War" (Hermann and Hermann, 1967, p. 411).

[1]In the M—Run an attempt was made to develop homology between the personality characteristics of the historical decision-makers and their counterparts in the simulation; in the A-Run these characteristics were less well-matched.

faction, hostility, or threat) to itself is 'sufficiently' great, this perception will offset perceptions of insufficient capability, making the perception of capability much less important a factor in a decision to go to war" (Zinnes, North, and Koch, 1961, p. 470)—a hypothesis which has been confirmed recently by Chihiro Hosoya (1967) in his analysis of the outbreak of World War II with reference to the Japanese decision-makers in 1941.

Statements made by negotiators from the U.S. and the U.S.S.R. during the post-World War II disarmament meetings were coded in comparable ways by Jensen and Bonham for "(a) insecurity about the stability of disarmament, (b) propaganda, and (c) hostility or attacks on the motives of the other side" (Bonham, 1967, p. 32). The frequencies presented in Table 14 following are rankings "assigned as a result of statistical comparisons of the frequencies by Mann-Whitney U-tests" (Bonham, 1967, p. 33). Two (insecurity and propaganda) of the three negotiation processes "were replicated successfully by the American teams in the simulation runs," but "the Soviet teams, on the other hand, did not replicate successfully these processes" (Bonham, 1967, p. 32). As Bonham remarked: "Apparently the Soviet participants played their roles like Americans" (Bonham, 1967, p. 36) (*Correspondence Rating: "Some"*).

In a considerably more elaborate analysis by Zinnes (1966), which was also based on a content analysis of messages in both simulation and reference

TABLE 14   Bonham's Findings on the Rankings and Frequencies of Expressions of Insecurity, Propaganda, and Hostility for Two Nations in Historical Materials and in Simulations of Disarmament Negotiations†

Expressions of Insecurity, Propaganda, and Hostility ‡

	In simulations § of disarmament negotiations		*In historical reference material: 1955 disarmament subcommittee*	
Expressions of:	*By "United States"*	*By "Soviet Union"*	*By "United States"*	*By "Soviet Union"*
Insecurity	High (41)	High (53)	High (65)	Low (27)
Propaganda	Low (18)	Low (21)	Low (25)	High (92)
Hostility	Low (16)	Low (28)	High (60)	High (83)

† Adapted from G. Matthew Bonham, "Aspects of the Validity of Two Simulations of Phenomena in International Relations." Ph. D. Dissertation. Cambridge, Massachusetts: Department of Political Science, Massachusetts Institute of Technology, 1967, Table 3, p.35.

‡ The rankings of High and Low were assigned on the basis of statistical comparisons of the frequencies by Mann-Whitney U-Tests; frequencies are given in parentheses following the rankings.

§ Sixty-six college students, half undergraduate and half graduate, served as participants in eleven runs.

systems, again much congruence was found in the effects of alliance systems on perceptions of threats and unfriendliness as well as in hostile behavior, as is presented in Table 15. Further, Zinnes found considerable congruence in the correlations between perceptions of hostility and hostile behavior, as is summarized in Table 16 (*Correspondence Rating: "Much"*). These results are confirmed in the Jensen/Bonham comparisons for insecurity and hostility (but not for propaganda) in the disarmament negotiations. For example, "the correlation between American and Soviet hostility for Jensen's data was high, R_s = .61** ... the hostility relationship was high in the simulation, R_s = .65**" (Bonham, 1967, p. 38) (*Correspondence Rating: "Much"*).

Summary: "Relations Among Nations"

The assessment of correspondences between simulations of international relations and "realities" with respect to "Relations Among Nations" as they are generated by the decision-makers and their nations, has been summarized in Table 17. In only one of the twenty-two ratings was comparison made with general materials derived from work in the field and laboratory; in all other instances the reference sources involved observations on world affairs, ranging from apt anecdotes through rigorous, systematic analyses. It seems that simulations are not realizing a communication structure which is adequately homologous to reference materials, nor do simulation trade activities parallel those

TABLE 15   Zinnes' Findings on the Effects of Alliances in the Brody-Driver Realizations of an Inter-Nation Simulation and in 1914 Historical Data†

Alliance Effects on Nations' Perceptions of Threat and Unfriendliness and on Nations' Subsequent Expressions of Hostility‡

Hypotheses:	Simulations§ (z scores) (24 Inter-alliance pairs) (18 Intra-alliance pairs)	1914 Historical Data (U-Test scores) (16 Inter-alliance pairs) (15 Intra-alliance pairs)
Hypothesis 1. Nations outside the bloc will be seen as more *threatening* than nations within the bloc.	5.20	22.5
Hypothesis 2. Nations outside the bloc will be seen as more *unfriendly* than nations within the bloc.	5.40	18.0
Hypothesis 3. There will be more *hostility* transmitted between blocs than within blocs.	3.81	40.0

†Adapted from Dina A. Zinnes, "A Comparison of Hostile Behavior of Decision-Makers in Simulate and Historical Data." *World Politics,* 18, 3 (April, 1966), Table IV, p. 486.

‡The larger the z score in the Simulations above, the greater its significance. In the 1914 Historical Data, the smaller the measure of unlikeness between the two groups, as seen on the Mann-Whitney U-Test, the greater its significance. The .05 level of significance was used as the criteron for rejection or acceptance of a hypothesis. All tests above were statistically significant at the .001 level.

§Three hundred thirty-six high school juniors and seniors served as decision-makers in the seven nations constituted for each of sixteen runs of the simulation from which the pairs were drawn.

found in the international arena. With respect to some activities involving cooperation and conflict, including alliances and the use of force, the correspondences overshadowed the non-correspondences, whether the interaction was in internation negotiation or at more macro-levels. However, the cited research examined only a few of the myriad of relations among nations which have been mentioned as being of potential importance in the study of world affairs.

CONCLUSIONS

The fragmentary nature of the evidence available for assessing validities in the simulation of international processes indicates that no firm conclusions

TABLE 16 Zinnes' Findings on Hostility in the Brody-Driver Realization of an Inter-Nation Simulation and in 1914 Historical Data†

Relationships Between Perceptions of Hostility and Hostile Behaviors as Represented by Spearman Rank Order Correlation Coefficients

Hypotheses:	Simulations ‡ (42 pairs)	1914 Historical Data § (31 pairs)
Hypothesis 4. There is a positive relationship between perceptions of *threat* and perceptions of *unfriendliness.*	.88	.91
Hypothesis 5. There is a positive relationship between *x*'s expression of *hostility* to *y* and *y*'s perception of *threat.*	.43	.70
Hypothesis 6. There is a positive relationship between *x*'s expression of *hostility* to *y* and *y*'s perception of *unfriendliness.*	.49	.73
Hypothesis 7. There is a positive relationship between the perception of *threat* and the expression of *hostility.*	.34	.74
Hypothesis 8. There is a positive relationship between the perception of *unfriendliness* and the expression of *hostility.*	.49	.73

† Adapted from Dina A. Zinnes, "A Comparison of Hostile Behavior of Decision-Makers in Simulate and Historical Data." *World Politics,* 18.3 (April, 1966), Table V, p.487.

‡ Three hundred thirty-six high school juniors and seniors served as decision-makers in the seven nations constituted for each of sixteen runs of the simulation from which the pairs were drawn. The simulation results above were statistically significant at the .01 level.

§ Correlations in the historical data were significant at the .001 level.

about the correspondences between simulations and world "realities" may be drawn at this time. The findings indicate that further work in simulation development should be guided closely by concurrent validity checks. The results do not signal that work in the simulation of international relations be abandoned.

When one returns to the broad rubrics outlined in our chapter "Structured Programs and Their Relation to Free Activity Within the Inter-Nation Simulation" (Guetzkow, 1963b, p. 103), which were used as guidelines in organizing

TABLE 17   Assessments of Correspondences† between Simulations and "Realities" in International Relations Regarding "Relations Among Nations"

"Relations Among Nations":	Correspondence with *"Materials from International Relations"*				Ratings:	
Communications	(S)	Conferences‡ (Hermann & Hermann, 1967)	(I)	Structure (Zinnes, 1966)	(M)	1
					(S)	1
			(M)	Salience and hostility (Bonham, 1967)	(L)	2
	(L)	Volume increases in crisis (C. F. Hermann, 1965)			(N)	0
					(I)	1
	(L)	Salience (Nardin & Cutler, 1967)			# =	5
Alliances and involvement of force	(S)	Fragmentation‡ (Brody, 1963)	(M)	Expectation of war (Nardin & Cutler, 1967)	(M)	4 §
					(S)	2
	(M)	Alignment (Nardin & Cutler, 1967)	(L)	Armed confrontation‡ (Hermann & Hermann, 1967)	(L)	1
					(N)	0
	(S)	Treaties signed in alliances (Nardin & Cutler, 1967)			(I)	0
			(M)	Escalation (Crow & Noel, 1965)	# =	7 §
Cooperation and conflict	(S)	Support in crisis‡ (C. F. Hermann, 1965)	(N)	Threats, accusations, protests (Chadwick, 1966d)		
	(S)	Concession-making (Bonham, 1967)	(M)	Hostility and capability in crisis (Hermann & Hermann, 1967)	(M)	3
	(S)	Treaties (Chadwick, 1966d)			(S)	4
					(L)	0
	(N)	Trade & treaties (Chadwick, 1966d)	(S)	Insecurity, propaganda, and hostility · (Bonham, 1967)	(N)	3
	(N)	Trade (Nardin & Cutler, 1967)			(I)	0
					# =	10
			(M)	Hostility (Zinnes, 1966)		
			(M)	Hostility (Bonham, 1967)		
Correspondence ratings:	(M)	*"Much"*	7		(M)	8 §
	(S)	*"Some"*	7		(S)	7
	(L)	*"Little"*	3		(L)	3
	(N)	*"None"*	3		(N)	3
	(I)	*"Incongruent"*	1		(I)	1
			# = 21		T =	22 §

†See Footnote 4, p. 697.
‡Materials involving anecdotes and case illustrations.
§Includes *Correspondence Rating* of the following "General Materials from Field and Laboratory":
(M) Ingroup vs. Outgroup‡
    (Druckman, 1968)

the findings for this essay (*supra,* Figure 1, p. 698), it is gratifying to note that correspondences, even though they may have been only incidental, have been examined within many of the topic areas listed. Researchers straightforwardly interested in the validity problem center their work upon intra-national processes rather than upon international processes within the simulation of world affairs; there are almost three entries in Tables 5 and 8 combined, to two entries in Table 17. But at least a beginning has been made in checking the homomorphic relationships of simulations to the world processes to which they refer.

Fifty-five entries derived from some twenty-four different studies, as asterisked in the bibliography, are made in Tables 5, 8, and 17. A series of overall summaries of these is given in Table 18. Almost two-thirds of the entries consist of assessments in which the findings from the simulations correspond with *"Some"* or *"Much"* similarity to the findings in the reference materials.

To what extent does the nature of the reference material affect the extent of congruence? Note that about three-quarters of the comparisons were made with materials derived from the "realities" of international relations, rather than from more general materials found in laboratory and field work. As indicated in Table 18, part (B), there is less correspondence when materials from the international system are used as the reference data. Is this finding an artifact, inasmuch as work with international materials permits closer scrutiny of the fit of simulation and reference materials than is the case in analyses of intra-national materials, which center largely on human components in a "general" comparison?

Over half of the comparisons were made in an articulated fashion, with systematic materials being presented for both simulation and reference systems. Although one has less confidence in anecdotal, illustrative materials, it is interesting to note that in the comparisons made to materials from (only) international relations, as exhibited in Table 18 (C), such materials present no more correspondence than do the systematic materials. Is this finding merely a reflection of the author's deprecatory attitude toward unsystematic material (pp. 695-96), inasmuch as he never allowed an assessment of *"Much"* on the anecdotal, illustrative variety?

There is greater congruence among the materials assembled in assessing the validation related to "Decision-Makers and Their Nations" than among entries concerned with "Relations Among Nations," as shown above in Table 18 (D). Yet, we are not in a position, now that "some elements of a model have been supported by validity operations," even to *"cautiously* infer a degree of validity to those related elements which have not," as C. F. Hermann (1967b, p. 225) hoped. The findings to date do not indicate that validity in one part of the model induces validity elsewhere—or vice versa. The international reference system seems to contain many interdependencies, and a simulation is more like an "empty world"; for therein, as Herbert A. Simon says, "most things are only weakly connected with most other things . . ." (Simon, 1965, p. 73). This state of affairs is reflected in the many times a relationship observed in the reference system failed to be generated in the simulation. Chadwick's hope that "if one were in possession of simulation and international system materials by which comparisons in activity patterns could be made, one could estimate the areas in which simulation theory was relatively insufficient" (Chadwick, 1967, p. 178), seems to be optimistic, given the findings reported above.

TABLE 18   Summaries of Assessments of Correspondences† between Simulations and "Realities" in International Relations

Assessments of correspondences between simulations and "realities"	Correspondence ratings					
	"Much"	"Some"	"Little"	"None"	"Incon-gruent"	Totals:
(A) Overall summary of assessments of cor-pondences from Tables 5, 8, & 17	17	21	7	6	4	55
(B) Comparison of cor-respondences in "Materials from Field and Laboratory"	7	3	0	1	1	12
In "Materials from International Rela-tions"	10	18	7	5	3	43
					T =	55
(C) Comparison of Cor-respondences in "Materials from (only)‡ International Relations"						
—of anecdotal, illustra-tive variety	0	8	2	0	0	10
—of articulated, sys-tematic variety	10	10	5	5	3	33.
					T =	43
(D) Comparison of cor-respondences in ma-terials representing "Decision-Makers and Their Nations" from Tables 5 & 8	9	14	4	3	3	33
"Relations Among Nations" from Table 17	8	7	3	3	1	22
					T =	55

†See Footnote 4, p. 697.
‡"Materials from Field and Laboratory" are *not* included in (C).

Except for the work of Bonham (1967), Chadwick (1966a-d; 1967), and Nardin and Cutler (1967), however, each entry represents a somewhat isolated relation among two (or, at most, three) variables. This represents a severe limitation on the assessments of the correspondences, given the posture that simulations gain leverage in modeling complex phenomena just because of the way in which they portray complex nets in interrelationships with many interactions among the variables. Could it be that in analyzing the simulation and reference materials we inadvertently imposed an "empty world"? Such hardly seems the case, inasmuch as quite similar techniques were employed in both analyses—and yet only outputs from the simulations were found so "weakly connected."

Although many of the topics within an encompassing theory of international relations have been touched upon—perhaps to the tedium of the reader—only the most superficial coverage of attributes was obtained, as has been pointed out again and again in the summaries (*supra*, pp. 710, 720, and 731-32.

The relative absences of incongruent findings in Table 18 (A) suggests that the simulations embody a gamut of "possibilisms" (Sprout and Sprout, 1965, Chapter 5 & p. 192), without generating much nonsense. Even in Chadwick's systematic examination of the 136 relations stemming from his seventeen variables, he rarely obtained an output from the simulation which inverted a finding from the reference system (Chadwick, 1966d, "Zone of Reversals," Figure 1, p. 17). On reflection, is it for this reason that Noel (1963b) and Guetzkow (1959, especially pp. 188-190; & 1963a) were able to embellish their essays with illustrations of simulation behaviors which seem so credible when a gross, anecdotal comparison is adduced between simulations and verbal theory? Does the fact that simulations seem in the main to generate states of affairs which seem to be potentially realizable, bolster my speculation that perhaps simulations can aid us as a heuristic in "Constructing Alternative Futures," thereby "Loosening the Bounds of the Past" (Guetzkow, 1966b, p. 186)? One "plays" with ideas in the vernacular; it may be that the manipulation potential of simulations is the characteristic which will enable theorists in international relations to reconstruct history and explore alternative futures (de Jouvenel, 1963; 1965) in ways beyond our present comprehension.

The conclusions about homomorphies of the simulations with the world, summarized above, are subject not only to much basic error, but to differences in intuitive interpretations of the assessments of the correspondences. Recall the tremendous liberties taken in composing the judgments summarized in Tables 5, 8, and 17. As was mentioned in the introductory passages of this essay (*supra*, p. 698), little attention was given by the researchers to the problem of reliability of the materials in either the simulations or the world reference materials. Nardin (Nardin and Cutler, 1967) and C. F. Hermann (1967b) provide the two exceptions (*supra*, pp. 720-21 & 722) with respect to simulations. Zinnes herself complained that the "1914 data . . . represents only one historical case, and possibly an atypical case since it was a crisis that ended in war" (Zinnes, 1966, Footnote 23, p. 494). Can one expect valid results if the analyses themselves are based on possibly unreliable materials? Further, the problem of matching variables proved thorny indeed. Chadwick rightly complains that "the variables defined in INS theory were not provided with empirical indices by INS theorists . . ." (Chadwick, 1966a, p. 48). And at times Chadwick quarrels even with himself as to which alternative measures should be employed, as in relating indices of office-holding to those of revolution (Chadwick, 1966a, Footnote 29, p. 66). Finally, the researches reviewed are most limited in being cross-sectional by and large; despite the fact that simulations, being models in operation across time, are amenable to longitudinal analysis.

When the reader examines each of the studies used in assembling this essay, he will note how rife the sources are with potential incomparabilities. As Snyder so rightly mused some years ago: "The scaling-down process—the sampling of attributes and the use of surrogate processes—involves the peril of inadvertent changes in kind when the counterpart of what is 'out there' is constructed 'in here'" (Snyder, 1963, p. 5). Why can Chadwick and Zinnes both use different reference materials, the one the world of the mid-fifties and the other the pre-1914 world, as checks against the same simulation material, the pre-proliferation period of the Brody-Driver runs? Is it not startling to find that the same simulation which yielded a patterned analysis for Brody (1963) produced an almost randomlike result for Chadwick (1966a)?

Perhaps the most serious limitation in the assessments of correspondences between simulations and "realities" in international relations in this essay is found in its almost total—albeit reluctant—reliance upon Inter-Nation Simulations. It is to be regretted that those developing political-military exercises and all-computer simulations have not produced more assessments of the validity of their contributions, so that the homomorphic relations could be checked out for more than the INS simulations.

It is within the framework of these shortcomings that one realizes the tentativeness of the overall findings of this review: a preponderance of correspondence, with proclivity toward errors of omission rather than commission.

THE FINDINGS IN BROADER PERSPECTIVE —AND SOME IMPLICATIONS

This critique of validities within the simulation of international processes is stringent, culminating in such closely developed analyses of the correspondences between simulations and "realities" as those by Bonham (1967), Chadwick (1966a-d, 1967); and Zinnes (1966). As an essay by Coplin indicates, there have been few comparable efforts to assess empirically the validities of segments of verbal theory (Coplin, 1966). This appraisal must be put into context, given the state of the art (Naylor, Balintfy, Burdick, and Chu, 1966; Abt Associates Inc., 1965; and Coplin, 1968). There are growing numbers of instances in which simulations in other areas in social science are proving their validities. The homomorphies of artificial learning processes with those obtained in the laboratory are many (Hunt, Marin, and Stone, 1966). The efforts of John and Jeanne Gullahorn in producing group phenomena in their computer simulations are yielding validity coefficients in the .80's and .90's (Gullahorn and Gullahorn, 1965). Cleo H. Cherryholmes (1966a) and Shapiro (1966a) modeled a legislative system as a whole, so as to simulate its voting with an accuracy of some 85 per cent. It seems there are no intrinsic reasons why simulations of international processes differ in kind from simulations being made of other psychological, social, and political processes.

However, it may be argued with reasonableness that the attempts to assess the validities of aspects of extant simulations of international processes are premature, given the inadequate state of both verbal theory and empirical work within international relations as a discipline (Platig, 1966). However, a "central reason for using simulation in international relations is [just] this lack of development" (Guetzkow, 1966a, p. 249). The present dissatisfaction with the Inter-Nation Simulation has grown out of both intuitive appraisal of its invalidities and partial studies demonstrating its inadequacies, as cited above. By taking time out now to check correspondences of constructions against the "realities," we may perhaps avoid the speculative extreme to which the work of Abt and his associates (Abt, 1964) at Raytheon went, in weaving assumption upon assumption into the gigantic TEMPER formulation, as critiqued by George Draper (1966). Likewise, by insisting upon a simulation format which permits checks of the outcomes to be made against reference materials, we would avoid an unending fabrication of case studies, as seems to be the proclivity of the scenarists (Bloomfield and Whaley, 1965; Kahn, 1965). Simultaneously, a challenge is presented to the "data-makers," as J. David Singer (1965) designates those developing empirical studies: simulations are "data hungry" monsters, as the analyses of Elder and Pendley (1966a, 1966b; Pendley and Elder, 1966) illustrate. Perhaps data work will become more coherent, as the components of our simulations are tested empirically. In the long run, the use of simulation as a theory-building device may provide an effective vehicle within a very complex area of behavior, allowing us to elaborate on a holistic, all-encompassing model with ever-increasing differentiation of its components. There are wide differences in the extent to which the simulations force explication of assumptions and development of interrelations among variables. In the political-military exercise, for example, the country-teams bring much implicit material to the simulation, as they role-play their scenarios; "Control" implements its conceptions of international processes as it umpires the "moves" by the teams. In all-computer simulations, each assumption and every relation among the variables are made explicit when they are programmed—even though such knowledge at times is available only to the programmers. As Singer and Hirohide Hinomoto (1965) found in building their "modest computer simulation" of the inspection of weapons production, the very construction of a model induces the explication of theory in more detail than is required by ordinary language. Because they are written in FORTRAN, one of the languages for computer programming, however, it is difficult to grasp the theory, even when the thousand-odd program cards are "explained" in a manual (TEMPER, Part III, 1966). Because of the complexities involved, the dynamics of the simulation are revealed only later when sensitivity analyses are completed on the operating simulation. In the mixed form of simulation, involving both men and machines, the use of prototypic variables and humans as surrogates to represent whole complexes allows the theory builders to dodge many of the "tough ones," at least temporarily, by employing such variables and surrogates as "black boxes" which are in no need of explication for the moment.

It is important that there not be premature closure upon a single formulation in the simulation of international processes. The potential existence of a gamut of simulations, as described above (*supra,* p. 692), seems of heuristic value. Even within a given formulation, such as the Inter-Nation Simulation, there has been much tendency to vary its contents. Some changes have involved mainly only revision of the parameters of the INS, as in the new weightings given components of validator satisfaction (*VS*) by Donald D. Skinner and Robert N. Wells, Jr. (1965, p. 15). Over the years the Western Behavioral Sciences Institute has made creative, incremental modifications of the Inter-Nation Simulation

(Crow, 1963; Crow and Raser, 1964; Raser and Crow, 1964; Crow and Noel, 1965; and Solomon, 1965). The changes and additions introduced by the Hermanns have been most innovative (Hermann and Hermann, 1967; M. G. Hermann, (1966). Coplin's venturesome transformation of the INS into a "World Politics Simulation" for the Center for International Systems Analysis of the U.S. Department of State (Coplin, 1967) indicates how features within an extant simulation may be reworked in a cumulative way. The group at The Ohio State University (Robinson, Burgess, and Fedder, 1966) are making important modification of the INS, as they proceed with the simulation of processes involved in international coalitions. Recently, the Peace Research Centre of Lancaster, England, and the Canadian Peace Research Institute have constructed an extensive elaboration of the INS, so that it serves as a model of the inter-nation environment within which the Viet Nam struggle may be explored (Laulicht, 1966; MacRae and Smoker, 1967). No one to date seems to have made drastic structural changes in the INS in which the very relationships themselves have been so modified as to make a quite different model, as Paul Smoker proposes now to do by constructing an "international" system, replete with international corporations, to compare with a "nation-state" system (Smoker, 1967, pp. 62-65).

In moving onward with the reconstruction and elaboration of simulations for the study of international relations, it would seem possible to borrow components from simulations constructed by others. Such soon may be possible within each segment of the Inter-Nation Simulation, as the following examples illustrate. To simulate "Decision-Makers and Their Nations," perhaps humans might be replaced to an extent by computer programs, such as the General Problem Solver developed by Alan Newell and Simon (1963). Eventually, even aspects of personality might be incorporated into such surrogates for the "decision-makers," as Kenneth Colby (1965) and others have demonstrated (Tomkins and Messick, 1963). Aspects of computer models both of artificial intelligence and of affect have been incorporated, for example, in the Crisiscom simulation (Pool and Kessler, 1965). The "nations" might be operated by programmed bureaucracies, as presaged in the work of the Gullahorns and of Beatrice and Sydney Rome (1966), including even budgetary processes (Crecine, 1965). The economists have made important advances in simulating the entire economy of a national entity, as embodied in *The Brookings Quarterly Economic Model of the United States* (Duesenberry, Fromm, Klein, and Kuh, 1965) and as projected in "A Programme for Growth" by members of Cambridge University's Department of Applied Economics (Stone, 1966). Does Noel's pioneer work (Noel, 1963a) toward enrichment of the economic part of the Inter-Nation Simulation provide guidelines for incorporation of features of these efforts? Perhaps some of the models built for intrasocietal processes could be extrapolated to the inter-nation area, so that the "Relations Among Nations" might be simulated with more adequacy. The work of Abelson and Bernstein (1963) on community referenda controversies included a communication process, as did the computer simulations of William N. McPhee (1963) and Ithiel de Sola Pool (Pool, Abelson, and Popkin, 1964), of Shapiro (1966a) and Cherryholmes (1966a). Hans Thorelli's international business game ("INTOP") (Thorelli and Graves, 1964) might be used wholesale in adding companies operating overseas to a simulation of the international system, as Smoker has suggested (Smoker, 1968). Is now the time to begin experimentation with ways by which component programs might be incorporated within an overall framework of simulated international processes?

Improvements and extensions of extant simulations, however, depend upon how well our colleagues assume the burdens of validating their componential work. To date, emphasis among simulation builders has been more upon the venture of model construction, with the scholar working as an artist, rather than upon the disciplines' involvement in checking correspondences between their simulations and their respective reference systems, as is incumbent when the scholar works as a social scientist or policy-influencer. This obligation, however, does not belong solely to the simulator. As Morton Gorden (1967) pointed out, the verbal theorist also shares in these same obligations. When simulator and verbal theorist ground their work in empirical materials, then they may fruitfully join hands with mathematical and simulation theorists in constructing homomorphic models of the international system which will have fidelity. Then, their constructions will represent the world with more adequacy—as it is now and as it may evolve—with simulated alternative futures unfolding the "realities" of the decades ahead.

REFERENCES

References marked with an asterisk (*) constitute the twenty-four studies from which the fifty-five *Correspondence Ratings* were made.

Abelson, Robert P., and Alex Bernstein. "A Computer Simulation Model of Community Referendum Controversies." *Public Opinion Quarterly*, 27 (1963), 93-122.

Abt, Clark C. "War Gaming." *International Science and Technology*, 32 (August, 1964), 29-37.

Abt Associates Inc. *Report of a Survey of the State of the Art: Social, Political, and Economic Models and Simulations.* For the National Commission on Technology, Automation, and Economic Progress, Washington, D.C. Cambridge, Massachusetts: Abt Associates, Inc., November, 1965.

Acheson, Dean G. "The President and the Secretary of State." In D. K. Price (Editor). *The Secretary of State.* Englewood Cliffs, New Jersey: Prentice-Hall, Inc., 1960, 27-50.

Albertini, Luigi. *The Origins of the War of 1914.* Edited and Translated by Isabella M. Massy. London, England, Oxford University Press, 1953.

Alger, Chadwick F. "Use of the Inter-Nation Simulation in Undergraduate Teaching." Chapter Six in Harold Guetzkow, Chadwick F. Alger, Richard A. Brody, Robert C. Noel, and Richard C. Snyder,. *Simulation in International Relations: Developments for Research and Teaching.* Englewood Cliffs, New Jersey: Prentice-Hall, Inc., 1963, 150-189.

Alger, Chadwick F. Personal Communication, Geneva, Switzerland, November, 1966a.

Alger, Chadwick F. "Interaction in a Committee of the United Nations General Assembly." Abridged version in *Midwest Journal of Political Science*, 10, 4

(November, 1966b), 411-447. In J. David Singer (Editor, Volume VI)—Heinz Eulau (General Editor). *Quantitative International Politics: Insights and Evidence, International Yearbook of Political Behavior Research.* New York: The Free Press (Macmillan), 1968.

Alker, Hayward R., Jr. "The Long Road to International Relations Theory: Problems of Statistical Non-Additivity." *World Politics,* 18, 4 (July, 1966), 623-655.

Argyris, Chris. *Some Causes of Organizational Ineffectiveness Within the Department of State.* For Center for International Systems Research, Washington, D. C. Department of State Publication No. 8180, U.S. Government Printing Office, January, 1967.

Banks, Arthur S., and Robert B. Textor. *A Cross-Polity Survey.* Cambridge, Massachusetts: The MIT Press, 1963.

Barber, James David. *Power in Committees: An Experiment in the Government Processes.* Chicago: Rand McNally & Co., 1966.

Basowitz, H., H. Persky, S. Korchin, and R. Grinker. *Anxiety and Stress.* New York: McGraw-Hill Book Co., 1955.

Beer, Stafford. "The World, the Flesh, and the Metal." *Nature.* 205, 4968 (January 16, 1965), 223-231.

Bem, D. J., M. A. Wallach, and N. Kogan. "Group Decision-Making Under Risk of Aversive Consequences." *Journal of Personality and Social Psychology,* 1 (1965), 453-460.

Benson, Oliver. "A Simple Diplomatic Game." In J. N. Rosenau (Editor). *International Politics and Foreign Policy.* New York: Free Press of Glencoe, 1961, 504-511.

Black, Duncan. *The Theory of Committees and Elections.* London and New York: Cambridge University Press, 1958.

Bloomfield, Lincoln P., and Norman Padelford. "Three Experiments in Political Gaming." *American Political Science Review,* 53 (1959), 1105-1115.

*Bloomfield, Lincoln P., and Barton Whaley. "The Political-Military Exercise: A Progress Report." *ORBIS,* 8, 4 (Winter, 1965), 854-870.

*Boguslaw, Robert, Robert H. Davis, and Edward B. Glick. "A Simulation Vehicle for Studying National Policy Formation in a Less Armed World." *Behavioral Science,* 2, 1 (January, 1966), 43-61.

*Bonham, G. Matthew. "Aspects of the Validity of Two Simulations of Phenomena in International Relations." Ph.D. Dissertation. Cambridge, Massachusetts: Department of Political Science, Massachusetts Institute of Technology, 1967.

Bonini, Charles P. *Simulation of Information and Decision Systems in the Firm.* Englewood Cliffs, New Jersey: Prentice-Hall, Inc., 1963.

Brodbeck, May. "Models, Meaning, and Theories." In Llewllyn Gross (Editor). *Symposium on Sociological Theory.* New York: Harper and Row, 1959, 373-402.

*Brody, Richard A. "Some Systemic Effects of the Spread of Nuclear-Weapons Technology: A Study Through Simulation of a Multi-Nuclear Future." *The Journal of Conflict Resolution,* 7, 4 (December, 1963), 663-753.

Burgess, Philip. "Nations in Alliance: A Simulation of International Coalition Process." Columbus, Ohio: Mershon Seminar, The Ohio State University, February, 1966.

Chadwick, Richard W. "Developments in a Partial Theory of International Behavior: A Test and Extension of Inter-Nation Simulation Theory." Ph.D. Dissertation. Evanston, Illinois: Department of Political Science, Northwestern University, June, 1966a.

Chadwick, Richard W. "Relating Inter-Nation Simulation Theory with Verbal Theory in International Relations at Three Levels of Analysis." Evanston, Illinois: Simulated International Processes project, Northwestern University, July, 1966b. (Revision of 1966a, Chapter I.)

Chadwick, Richard W. "Extending Inter-Nation Simulation Theory: An Analysis of Intra- and International Behavior." Evanston, Illinois: Simulated International Processes project, Northwestern University, August, 1966c. (Revision of 1966a. Chapter III).

*Chadwick, Richard W. "Theory Development Through Simulation: A Comparison and Analysis of Associations Among Variables in an Inter-national System and an Inter-Nation Simulation." Evanston, Illinois: Simulated International Processes project, Northwestern University, September, 1966d. (Revision of 1966a, Chapter II, Section C.)

*Chadwick, Richard W. "An Empirical Test of Five Assumptions in an Inter-Nation Simulation, About National Political Systems," *GENERAL SYSTEMS: Yearbook for the Society of General Systems Research,* 12 (1967), 177-192. (Revision of 1966a, Chapter II, Section B.)

Cherryholmes, Cleo H. "The House of Representatives and Foreign Affairs: A Computer Simulation of Roll Call Voting." Ph.D. Dissertation. Evanston, Illinois: Department of Political Science, Northwestern University, August, 1966a.

Cherryholmes, Cleo H. "Some Current Research on Effectiveness of Educational Simulations: Implications for Alternative Strategies." *American Behavioral Scientist,* 10, 2 (October, 1966b), 4-7.

Cleveland, Harlan. "Crisis Diplomacy." *Foreign Affairs,* 41 (1963), 638-649.

Colby, Kenneth M. "Computer Simulation of a Neurotic Process." In S. S. Tompkins and S. Messick (Editors). *Computer Simulation of Personality.* New York: John Wiley and Sons, Inc., 1963, 165-179.

Coleman, James S., Sarane Boocock, and E. O. Schild (Editors). Part I. "In Defense of Games." *American Behavioral Scientist,* 10, 2 (October, 1966a).

Coleman, James S., Sarane Boocock, and E. O. Schild (Editors). Part II. "Simulation Games and Learning Behavior." *American Behavioral Scientist,* 10, 3 (November, 1966b).

Collins, Barry E., and Harold Guetzkow. *A Social Psychology of Group Processes for Decision-Making.* New York: John Wiley and Sons, Inc., 1964.

Coombs, Clyde H. *A Theory of Data.* New York: John Wiley and Sons, Inc., 1964.

Coplin, William D. "Inter-Nation Simulation and Contemporary Theories of International Relations." *American Political Science Review,* 60, 3 (September, 1966b), 562-578.

Coplin, William D. "Toward an All-Computer Simulation of the Foreign Policy-Making Forces at the Level of General Policy Trends." Research Memorandum No. 3 for Center for International Systems Analysis of the U.S. Department of State, Detroit, Michigan: Comparative Foreign Policy Systems Program, Wayne State University, 1967.

Coplin, William D. (Editor). *Simulation in The Study of Politics.* Chicago: Markham Publishing Company, 1968.

Crecine, John P. "A Computer Simulation Model of Municipal Resource Allocation." Ph.D. Dissertation. Pittsburgh, Pennsylvania: Carnegie Institute of Technology, 1965.

Crow, Wayman J. "A Study of Strategic Doctrines Using the Inter-Nation Simulation." *The Journal of Conflict Resolution,* 7, 3 (September, 1963), 580-589.

*Crow, Wayman J., and Robert C. Noel. "The Valid Use of Simulation Results." La Jolla, California: Western Behavioral Sciences Institute, June 19, 1965.

Crow, Wayman J., and John R. Raser. "A Cross Cultural Simulation Study." La Jolla, California: Western Behavioral Sciences Institute, November, 1964.

Dahl, Robert A. *A Preface to Democratic Theory.* Chicago: University of Chicago Press, 1956.

Dawson, Richard E. "Simulation in the Social Sciences." Chapter 1. in Harold Guetzkow (Editor). *Simulation in Social Science: Readings.* Englewood Cliffs, New Jersey: Prentice-Hall, Inc., 1962. 1-15.

De Jouvenel, Bertrand (Editor). *Futuribles: Studies in Conjecture.* Geneva, Switzerland: Droz, 1963 and 1965.

Draper, George. "Technological, Economic, Military, and Political Evaluation Routine (TEMPER)—An Evaluation." Washington, D.C.: NMCSSC, July, 1966.

*Driver, Michael J. *Conceptual Structure and Group Processes in an Inter-Nation Simulation.* Part One: "The Perception of Simulated Nations: A Multidimensional Analysis of Social Perception as Affected by Situational Stress and Characteristic Levels of Cognitive Complexity in Perceivers." Ph.D. Dissertation. Princeton, New Jersey: Department of Psychology, Princeton University, and Educational Testing Service, April, 1962.

*Driver, Michael J. "A Structure Analysis of Aggression, Stress, and Personality in an Inter-Nation Simulation." Institute Paper No. 97, Lafayette, Indiana: Institute for Research in the Behavioral, Economic, and Management Sciences, Herman C. Krannert Graduate School of Industrial Administration, Purdue University, January, 1965.

Driver, Michael J. Personal Communication. November 22, 1966.

*Druckman, Daniel. "Ethnocentrism in the Inter-Nation Simulation." Evanston, Illinois: Simulated International Processes project, Northwestern University, 1967; *The Journal of Conflict Resolution,* 12, 1 (March, 1968).

Duesenberry, J. S., G. Fromm, L. R. Klein, and E. Kuh (Editors). *The Brookings*

Quarterly Economic Model of the United States. Chicago: Rand McNally and Co., 1965; Amsterdam, Netherlands: North-Holland Publishing Co., 1965.

Duijker, H. C. J., and N. H. Frijda. *National Character and National Stereotypes.* Amsterdam, Netherlands: North-Holland Publishing Co., 1960.

Easton, David. *A Systems Analysis of Political Life.* New York: John Wiley and Sons, Inc., 1965.

Eckstein, Harry, and David E. Apter (Editors). *Comparative Politics: A Reader.* London: Collier-Macmillan, Limited; New York: The Free Press of Glencoe, 1963.

Elder, Charles D., and Robert E. Pendley. "Simulation as Theory Building in the Study of International Relations." Evanston, Illinois: Simulated International Processes project, Northwestern University, July, 1966a.

*Elder, Charles D., and Robert E. Pendley. "An Analysis of Consumption Standards and Validation Satisfaction in the Inter-Nation Simulation in Terms of Contemporary Economic Theory and Data." Evanston, Illinois: Simulated International Processes project, Northwestern University, November, 1966b.

Feigenbaum, E. A., and J. Feldman (Editors). *Computers and Thought.* New York: McGraw-Hill Book Co., 1963.

Fromm, Gary, and Lawrence R. Klein. "The Brookings-S.S.R.C. Quarterly Econometric Model of the United States: Model Properties." *American Economic Review,* 55, 2 (1965), 348-361.

Frank, Philip G. (Editor). *The Validation of Scientific Theories.* Boston: The Beacon Press, 1954.

Funkenstein, D. H., S. H. King, and Margaret E. Drolette. *Mastery of Stress.* Cambridge, Massachusetts; Harvard University Press, 1957.

Gamson, William A., and Andre Modigliani. "The Carrot and/or the Stick: Soviet Responses to Western Foreign Policy. 1946-53." *Peace Research Society (International) Papers,* 3 (1965), 47-78.

Goldhamer, Herbert, and Hans Speier. "Some Observations on Political Gaming." *World Politics,* 12, 1 (October, 1959), 71-83.

Gorden, Morton. "International Relations Theory in the TEMPER Simulation." Cambridge, Massachusetts: Abt Associates Inc., September, 1965.

Gorden, Morton. "International Relations Theory in the TEMPER Simulation." Evanston, Illinois: Simulated International Relations project, Northwestern University, 1967.

Gorden, Morton. "Burdens for the Designer of a Computer Simulation of International Relations: The Case of TEMPER." In Davis B. Bobrow (Editor). *Proceedings of the Computers and the Policy-Making Community Institute.* Presented at the Institute held at Lawrence Radiation Laboratory, University of California, Livermore, California, on April 4-15, 1966. Englewood Cliffs, New Jersey: Prentice-Hall, Inc., 1968.

Gregg, Philip M., and Arthur S. Banks. "Dimensions of Political Systems: Factor Analysis of a Cross-Polity Survey." *American Political Science Review,* 59 (September, 1965), 602-614.

Guetzkow, Harold. "A Use of Simulation in the Study of Inter-Nation Relations." *Behavioral Science,* 4, 3 (1959), 183-191.

Reprinted as Reprint No. PS-112 in the *Bobbs-Merrill Reprint Series in the Social Sciences.* Indianapolis, Indiana: The Bobbs-Merrill Company, Inc.

Reprinted as Chapter Two, "A Use of Simulation in the Study of Inter-Nation Relations," in Harold Guetzkow, Chadwick F. Alger, Richard A. Brody, Robert C. Noel, and Richard C. Snyder. *Simulation in International Relations: Developments for Research and Teaching.* Englewood Cliffs, New Jersey: Prentice-Hall, Inc., 1963a, 24-42.

Updated for IBM's 1964 Scientific Computing Symposium on Simulation Models and Gaming: Harold Guetzkow. "Simulation in International Relations." In *Proceedings of the IBM Scientific Computing Symposium on Simulation Models and Gaming.* York, Pennsylvania: Maple Press, 1966a, 249-278.

Guetzkow, Harold (Editor). *Simulation in Social Science: Readings.* Englewood Cliffs, New Jersey: Prentice-Hall, Inc.,1962.

Guetzkow, Harold. "Structured Programs and Their Relation to Free Activity Within the Inter-Nation Simulation." Chapter Five in Harold Guetzkow, Chadwick F. Alger, Richard A. Brody, Robert C. Noel, and Richard C. Snyder, *Simulation in International Relations: Developments for Research and Teaching.* Englewood Cliffs, New Jersey: Prentice-Hall, Inc., 1963b, 103-149.

Guetzkow, Harold. "Some Uses of Mathematics in Simulation of International Relations." In John N. Claunch (Editor). *Mathematical Applications in Political Science.* Dallas, Texas: The Arnold Foundation, Southern Methodist University, 1965, 21-40.

Guetzkow, Harold. "Transcending Data-Bound Methods in the Study of Politics." "Commment on Professor Deutsch's Paper." In James C. Charlesworth (Editor). *A Design for Political Science: Scope, Objectives, and Methods. Monograph 6.* Philadelphia: The American Academy of Political and Social Science, December, 1966b, 185-191.

Guetzkow, Harold, Chadwick F. Alger, Richard A. Brody, Robert C. Noel, and Richard C. Snyder. *Simulation in International Relations: Developments for Research and Teaching.* Englewood Cliffs, New Jersey: Prentice-Hall, Inc., 1963.

Guetzkow, Harold, and Anne E. Bowes. "The Development of Organizations in a Laboratory." *Management Science,* 3 (1957), 380-402.

*Guetzkow, Harold, Richard A. Brody, Michael J. Driver, and Philip F. Beach. "An Experiment on the N-Country Problem Through Inter-Nation Simulation." St. Louis, Missouri: Washington University, 1960.

Gullahorn, John R., and Jeanne E. Gullahorn. "Some Computer Applications in Social Science." *American Sociological Review,* 30, 3 (1965), 353-365.

Hamblin, R. L. "The Dynamics of Racial Discrimination." *Social Problems,* 10 (1962), 103-121.

Harvey, O. J., D. E. Hunt, and H. M. Schroder. *Conceptual Systems and Personality Organization.* New York: John Wiley and Sons, 1961.

Hennessy, Bernard C. "Psycho-Cultural Studies of National Character: Relevances for International Relations." *BACKGROUND: Journal of the International Studies Association,* 6, 1-3 (Fall, 1962), 27-49.

*Hermann, Charles F. "Crises in Foreign Policy-Making: A Simulation of International Politics." Ph.D. Dissertation. Evanston, Illinois: Department of Political Science, Northwestern University, 1965.

Hermann, Charles F. "Threat, Time and Surprise: A Simulation of International Crisis." Princeton, New Jersey: Princeton University, April, 1967a.

Hermann, Charles F. "Validation Problems in Games and Simulations with Special Reference to Models of International Politics." *Behavioral Science,* 12, 3 (May, 1967b), 216-231.

*Hermann, Charles F., and Margaret G. Hermann. "An Attempt to Simulate the Outbreak of World War I." *American Political Science Review,* 61, 2 (June, 1967), 400-416.

*Hermann, Margaret G. "Stress, Self-Esteem, and Defensiveness in an Inter-Nation Simulation." Ph.D. Dissertation. Evanston, Illinois: Department of Psychology, Northwestern University, 1965.

Hermann, Margaret G. "Testing a Model of Psychological Stress." *Journal of Personality* 34, 3 (September, 1966), 381-396.

Hilsman, Roger. "The Foreign Policy Consensus: An Interim Research Report." *The Journal of Conflict Resolution,* 3 (1959), 361-382.

Hochberg, Herbert. "Simulation, Models, and Theories." Paper developed for the Seminar on Simulation of Human Organizational Systems 1961-63, sponsored by the Office of Naval Research, U.S. Department of Defense, Grant No. 1228(22), 1965.

Holsti, Ole R. "The 1914 Case." *American Political Science Review,* 59, 2 (June, 1965), 365-378.

Hosoya, Chihiro. "Japan's Decision for War in 1941." *Hitotsubashi Journal of Law and Politics,* 5 (April, 1967), 10-19.

Hunt, Earl B., Janet Marin, and Philip J. Stone. *Experiments in Induction.* New York: Academic Press, 1966.

Janis, I. L. *Psychological Stress.* New York: John Wiley and Sons, Inc., 1958.

Jensen, Lloyd. "Soviet-American Bargaining Behavior in the Post-War Disarmament Negotiations." *The Journal of Conflict Resolution,* 7, 3 (1963), 522-541.

Kahn, Herman. *On Escalation: Metaphors and Scenarios.* New York: Frederick A. Praeger, 1965.

Kaplan, Abraham. *The Conduct of Inquiry: Methodology for Behavioral Science.* San Francisco: Chandler Publishing Co., 1964.

Kaplan, Morton A. *System and Process in International Politics.* New York: John Wiley and Sons, 1957.

Katz, Daniel, and Robert L. Kahn. *The Social Psychology of Organizations.* New York: John Wiley and Sons, Inc., 1966.

Kelman, Herbert C. (Editor). *International Behavior: A Social-Psychological Analysis.* New York: Holt, Rinehart and Winston, Inc., 1965.

Kennedy, J. L. "The Systems Approach: Organizational Development." *Human Factors,* 4, 1 (1962), 25-52.

Kissinger, Henry A. *The Necessity for Choice: Prospects of American Foreign Policy.* New York: Doubleday, 1962.

Klineberg, Otto. *The Human Dimension in International Relations.* New York: Holt, Rinehart and Winston, Inc., 1964.

Kress, Paul. "On Validating Simulation: With Special Attention to the Simulation of International Politics." Evanston, Illinois: Northwestern University, 1966.

Larson, D. L. *The "Cuban Crisis" of 1962.* Boston: Houghton-Mifflin Co., 1963.

Laulicht, Jerome. "The Vietnam Peace Game: A Simulation Study of Conflict Resolution." Clarkson, Ontario: Canadian Peace Research Institute, 1966.

Lazarus, R. S. "A Laboratory Approach to the Dynamics of Psychological Stress." *American Psychologist,* 19 (1964), 400-411.

Lazarus, R. S., and R. W. Baker. "Personality and Psychological Stress: A Theoretical and Methodological Framework." *Psychology Newsletter,* 8 (1956), 21-32.

Mack, R. W., and Richard C. Snyder. "The Analysis of Social Conflict—Toward an Overview and Synthesis." *The Journal of Conflict Resolution,* 1 (1957), 212-248.

MacRae, John, and Paul Smoker. "A Vietnam Simulation: A Report on the Canadian/English Joint Project." *Journal of Peace Research.* Groningen, Netherlands: International Peace Research Association, 4, 1 (1967), 1-25.

McClintock, Charles G., Albert A. Harrison, Susan Strand, and Phillip Gallo. "Internationalism-Isolationism, Strategy of the Other Player, and Two-Person Game Behavior." *Journal of Abnormal Psychology,* 67 (1963), 631-636.

McPhee, William N. "Note on a Campaign Simulator." Chapter Four in *Formal Theories of Mass Behavior.* London, Collier-Macmillan, Ltd.; New York: The Free Press of Glencoe, 1963, 169-183.

McRae, Vincent V. "Gaming as a Military Research Procedure." In Ithiel de Sola Pool (Editor). *Social Science Research and National Security.* Washington, D.C.: Smithsonian Institute, March 5, 1963, 188-224.

Meier, Dorothy L. "Progress Report: Event Simulation Project." Evanston, Illinois: Simulated International Processes project, Northwestern University, 1965.

*Meier, Dorothy L., and Arthur Stickgold. "Progress Report: Analysis Procedures." St. Louis, Missouri: Event Simulation Project, Washington University, 1965.

*Nardin, Terry, and Neal E. Cutler. "A Seven Variable Study of the Reliability and Validation of Some Patterns of International Interaction in the Inter-Nation Simulation." Evanston, Illinois: Simulated International Processes project, Northwestern University, December, 1967.

Naylor, Thomas H., Joseph L. Balintfy, Donald S. Burdick, and Kong Chu. *Computer Simulation Techniques.* New York: John Wiley and Sons, Inc., 1966.

Naroll, Raoul, and Ronald Cohen (Editors). *Handbook of Methodology in Cultural Anthropology.* New York: Natural History Press, 1968.

Neustadt, R. *Presidential Power.* New York: John Wiley and Sons, Inc., 1960.

Newell, Alan, and Herbert A. Simon. "GPS, A Program That Simulates Human Thought." in E. A. Feigenbaum and J. Feldman (Editors). *Computers and Thought.* New York: McGraw-Hill Book Co., 1963, 279-293.

Noel, Robert C. "A Simplified Political-Economic System Simulation." Ph.D. Dissertation. Evanston, Illinois: Department of Political Science, Northwestern University, 1963a.

*Noel, Robert C. "Evaluation of the Inter-Nation Simulation." Chapter Four in Harold Guetzkow, Chadwick F. Alger, Richard A. Brody, Robert C. Noel, and Richard C. Snyder. *Simulation in International Relations: Developments for Research and Teaching.* Englewood Cliffs, New Jersey: Prentice-Hall, Inc., 1963b, 69-102.

North, Robert C., Ole R. Holsti, M. George Zaninovich, and Dina A. Zinnes. *Content Analysis: A Handbook with Application for the Study of International Crisis.* Evanston, Illinois: Northwestern University Press, 1963.

*Pendley, Robert E., and Charles D. Elder. "An Analysis of Office-Holding in the Inter-Nation Simulation in Terms of Contemporary Political Theory and Data on the Stability of Regimes and Governments." Evanston, Illinois: Simulated International Processes project, Northwestern University, November, 1966.

Platig, E. Raymond, *International Relations Research: Problems of Evaluation and Advancement.* New York: Carnegie Endowment for International Peace, 1966.

Pool, Ithiel de Sola, and Allan Kessler. "The Kaiser, the Tsar and the Computer: Information Processing in a Crisis." *American Behavioral Scientist,* 8 (1965), 31-38.

Pool, Ithiel de Sola, Robert P. Abelson, and Samuel L. Popkin. *Candidates, Issues, and Strategies.* Cambridge, Massachusetts: The MIT Press, 1964.

Pruitt, Dean G. *Problem Solving in the Department of State, Monograph 2,* Denver, Colorado: The Social Science Foundation and Department of International Relations, University of Denver, 1964-65.

*Raser, John R. "Personal Characteristics of Political Decision-Makers: A Literature Review." La Jolla, California: Western Behavioral Sciences Institute, September 15, 1965.

*Raser, John R., and Wayman J. Crow. *WINSAFE II: An Inter-Nation Simulation Study of Deterrence Postures Embodying Capacity to Delay Response.* La Jolla, California: Western Behavioral Sciences Institute, 1964.

Robinson, James A., Lee F. Anderson, Margaret G. Hermann, and Richard C. Snyder. "Teaching with Inter-Nation Simulation and Case Studies." *American Political Science Review,* 60 (March, 1966), 53-65.

Rome, Beatrice K., and Sydney C. Rome. "Leviathan: An Experimental Study of Large Organizations with the Aid of Computers." In Raymond V. Bowers (Editor). *Studies on Behavior in Organizations: A Research Symposium.* Athens, Georgia: University of Georgia Press, 1966, 257-311.

Rosenblatt, Paul C. "Origins and Effects of Group Ethnocentrism and Nationalism." *The Journal of Conflict Resolution,* 8, 2 (1964), 131-146.

Rummel, Rudolph J. "Dimensions of Conflict Behavior Within and Between Nations." *GENERAL SYSTEMS: Yearbook of the Society for General Systems Research,* 8 (1963), 1-50.

Rummel, Rudolph J. "The Dimensionality of Nations Project." In Richard L. Merritt and Stein Rokkan (Editors). *Comparing Nations: The Use of Quantitative Data in Cross-National Research.* New Haven, Connecticut: Yale University Press, 1966a. 109-129.

Rummel, Rudolph J. "A Social Field Theory of Foreign Conflict Behavior." Prepared for Cracow Conference, 1965. *Peace Research Society (International) Papers,* 4 (1966b), 131-150.

Runge, W. A. *Analysis of the Department of State Communications Traffic During a Politico-Military Crisis.* Menlo Park, California: Stanford Research Institute, 1963.

Russett, Bruce M. "Delineating International Regions." In David Singer (Editor, Volume VI)—Heinz Eulau (General Editor). *Quantitative International Politics: Insights and Indicators, International Yearbook of Political Behavior Research.* New York: The Free Press (Macmillan), 1968.

Schroder, Harold M., Michael J. Driver, and Siegfried Streufert. *Human Information Processing.* New York: Holt, Rinehart and Winston, Inc., 1967.

Selye, Hans. *The Stress of Life.* New York: McGraw-Hill Book Co., 1956.

Shapiro, Michael J. "The House and the Federal Role: A Computer Simulation of Roll Call Voting." Ph.D. Dissertation. Evanston, Illinois: Department of Political Science, Northwestern University, 1966a.

*Shapiro, Michael J. "Cognitive Rigidity and Moral Judgments in an Inter-Nation Simulation." Evanston, Illinois: Simulated International Processes project, Northwestern University, 1966b.

Sherman, Allen William. "The Social Psychology of Bilateral Negotiations." Master of Arts Thesis. Evanston, Illinois: Department of Sociology, Northwestern University, 1963.

Shubik, Martin (Editor). *Game Theory and Related Approaches to Social Behavior.* New York: John Wiley and Sons, Inc., 1964.

Shure, Gerald H., R. J. Meeker, and E. A. Hansford. "The Effectiveness of Pacifist Strategies in Bargaining Games." *The Journal of Conflict Resolution,* 9, 1 (1965), 106-117.

Simon, Herbert A. "The Architecture of Complexity." *GENERAL SYSTEMS: Yearbook of the Society for General Systems Research,* 10 (1965), 63-76.

Singer, J. David. "Data-Making in International Relations." *Behavioral Science,* 10, 1 (January, 1965), 68-80.

Singer, J. David, and Hirohide Hinomoto. "Inspecting for Weapons Production: A Modest Computer Simulation." *Journal of Peace Research.* Groningen, Netherlands: International Peace Research Association, 2, 1 (1965), 18-38.

Skinner, Donald D., and Robert N. Wells, Jr. "Participant's Manual: Michigan Inter-Nation Simulation." Ann Arbor, Michigan: University of Michigan, 1965.

Smoker, Paul. "Nation-State Escalation and International Integration." *Journal of Peace Research.* Groningen, Netherlands: International Peace Research Association, 4, 1 (1967), 60-75.

Smoker, Paul. "An International Processes Simulation: Theory and Description." Evanston, Illinois: Simulated International Processes project, Northwestern University, 1968.

Snyder, Richard C. "Some Perspectives on the Use of Experimental Techniques in the Study of International Relations." Chapter One in Harold Guetzkow, Chadwick F. Alger, Richard A. Brody, Robert C. Noel, and Richard C. Snyder. *Simulation in International Relations: Developments for Research and Teaching.* Englewood Cliffs, New Jersey: Prentice-Hall, Inc., 1963, 1-23.

Snyder, Richard C., and Glenn D. Paige. "The United States Decision to Resist Aggression in Korea. The Application of an Analytical Scheme." *Administrative Science Quarterly,* 3 (1958), 341-378.

Solomon, Lawrence N. "Simulation Research in International Decision-Making." In G. Sperrazzo (Editor). *Psychology and International Relations.* Washington, D.C.: Georgetown University Press, 1965, 37-52.

Sprout, Harold, and Margaret Sprout. *The Ecological Perspective on Human Affairs: With Special Reference to International Politics.* Princeton, New Jersey: Princeton University Press, 1965.

*Streufert, Siegfried, Peter Suedfeld, and Michael J. Driver. "Conceptual Structure, Information Search, and Information Utilization." *Journal of Personality and Social Psychology,* 2, 5 (1965), 736-740.

Streufert, Siegfried, M. A. Clardy, Michael J. Driver, Marvin Karlins, Harold M. Schroder, and Peter Suedfeld. "A Tactical Game for the Analysis of Complex Decision-Making in Individuals and Groups." *Psychological Reports,* 17 (1965), 723-729.

Stone, Richard. "British Economic Balances in 1970." Chapter 17 *in Mathematics in the Social Sciences and Other Essays.* Cambridge, Massachusetts: The MIT Press, 1966, 249-282.

Sullivan, Denis G. "Towards an Inventory of Major Propositions Contained in Contemporary Textbooks in International Relations." Ph.D. Dissertation, Evanston, Illinois: Department of Political Science, Northwestern University, 1963.

Tanter, Raymond. "Dimensions of Conflict Behavior Within and Between Nations, 1958-60." *The Journal of Conflict Resolution,* 10, 1 (March, 1966), 41-64.

Tanter, Raymond. "Toward a Theory of Political Development." *Midwest Journal of Political Science,* 11, 2 (1967), 145-172.

Targ, Harry, and Terry Nardin. "The Inter-Nation Simulation as a Predictor of Contemporary Events." Evanston, Illinois: Northwestern University, 1966.

TEMPER: Technological Economic, Military, and Political Evaluation Routine. Bedford, Massachusetts: Raytheon Company, Volumes 1-7, 1965-1966.

Thompson, J. D., and R. W. Hawkes. "Disaster, Community Organization, and Administrative Process." In G. W. Baker and D. W. Chapman (Editors). *Man and Society in Disaster.* New York: Basic Books, 1962, 268-300.

Thorelli, Hans B., and Robert L. Graves. *International Operations Simulation with Comments on Design and Use of Management Games.* New York: The Free Press (Macmillan), 1964.

Tomkins, S. S., and Samuel L. Messick (Editors). *Computer Simulation of Personality.* New York: John Wiley and Sons, Inc., 1963.

Tucker, Ledyard R., and Samuel L. Messick. "An Individual Differences Model for Multi-Dimensional Scaling." *Psychometrika,* 28, 4 (1963), 333-367.

Verba, Sidney. "Simulation, Reality, and Theory in International Relations." *World Politics,* 16, 3 (April, 1964), 490-519.

Webb, Eugene J., Donald T. Campbell, Richard D. Schwartz, and Lee Sechrest. *Unobtrusive Measures: Nonreactive Research in the Social Sciences.* Chicago: Rand McNally and Co., 1966.

Wright, Quincy. *A Study of War.* Chicago: University of Chicago Press, 1942; revised edition, 1965.

Zelditch, Morris, Jr., and Terrence K. Hopkins. "Laboratory Experiments with Organizations." In Amitae Etzioni (Editor). *Complex Organizations: A Sociological Reader.* New York: Holt, Rinehart and Winston, Inc., 1961, 464-478.

Zinnes, Dina A. "Support of Simulation Studies Through a Parallel Historical Study." Memorandum. January, 1964.

*Zinnes, Dina A. "A Comparison of Hostile Behavior of Decision-Makers in Simulate and Historical Data." *World Politics,* 18, 3 (April, 1966), 474-502.

Zinnes, Dina A., Robert C. North, and Howard E. Koch. "Capabilities, Threat, and Outbreak of War." In James A. Rosenau (Editor). *International Politics and Foreign Policy.* New York: Free Press of Glencoe, 1961, 469-482.

Index

Abelson
    computer simulations in social psychology, 13, 21-23, 121-23
    use of experiments in testing model validity, 127-28
Abelson-Bernstein
    electoral simulation model, 174-77, 247, 251-56, 261-62
        strengths and weaknesses of, 255-56
        treatment of validation problems in, 176-77, 182
    interaction process, 246
Abt
    TEMPER, 304, 675, 692
Ackoff's classification of models, 6, 8, 18, 297
Action selection, 125
Adaptive decision process, 152
Adelman's classification of simulations, 16
AD-ME-SIM, 520
Adoption, purposeful delay in, 166
Adorno/California F. Scale, 704
Adult dispositions toward political parties, 198-99
Adult socialization in urban setting, simulation of, 166-67
Advertising, media selection in, 517-20
Advertising agency game, 542
Advertising simulations, 516-22
    Gensch Media Selection Model, 520-22
    Simulmatics Media Selection Model, 517-20
Aggressive interaction, Coe's simulation of, 126
Air travel simulation. *See* S.S.T. Worldwide Route Simulation
Alba simulation. *See* Customer simulation, Alba's model
Aldous, 124-27
ALGOL, 18-19
    in Colby's model, 123
Alker's comparison of TEMPER, PME, and INS, 314

All-computer simulations, 24, 325, 675, 678, 692-93
Allman's railroad network model. *See* Railroad system, simulations of, Allman's model
"All-man" format, explicitness in construction of, 693
All-man simulation techniques, 647, 693
"All-manual" simulations, 675
Allport, 117
ALTAIR, 78
American economy, microanalysis of, 152
American Voter, The, (Campbell *et al.*), 250
Amstutz simulation. *See* Customer simulation, Amstutz's model
Analogies, hydraulic, 213-14
Analogs, computation of, 136; Fig. 4, 139
Analogue models, 10-18
Analogues, electrical, 216
Analytic simulations, 492
Analytical models, disadvantages of, in research, 24-25
Andlinger's manual business game, 455
Anger responses, 179
Anthropologists, 156
Anthropometric traits, systematic classification of, 145
Artificial intelligence, purpose of studies of, 51-52
Artificial subject, 103
Asimov's business firm simulation, 434-35
Assembly line balancing, computer approaches to, 449-50
Assertion match, 175
Association units in perceptron, 56
Attitudinal characteristics of potential adopter, 159
Attitudinal predisposition, 171
Attitude structure
    Abelson's model of, 122-23
    in Aldous, 125
    represented in computer program, 121
Aural mode, 111

Neighborhood effect, 369
  in adoption of innovations, 158
  probabilities, computation of, 158-59
Neisser, 54-55
Net node, 104-5
Network detection, 145
Neural network and pattern recognition, 54
Neurotic personality
  computer simulations of
    Colby's model, 123-24
    limitations of, 138-40
Neurotic processes
  computer simulation of, 132-42
  methodological problems in simulating, 140-41
New Deal, 206, 264
New-product development simulations
  Pessemier New-Product Profit Simulator, 526-30
  Urban New-Product Profit Simulator, 530-33
    central features of, 531
Noncentral place activities, 378
Noncritical path, 426-27
Nonlinear discriminant functions, use of, in pattern recognition research, 58
Nonlinearities, 10
Nonmigratory movements of population, 385
Nonparametric tests, 659
Nonprogrammed operation, 11
Nonsense syllables, 110-14
Nonverbal output, 135
NORC panel survey, 158
"Normal" science versus "revolutionary", 52-53
Normative orientations of simulated groups, 156
Northeast Corridor Project, 579
NUCLEUS for a POOL, 138
Nystuen, 385

Object scheme, 72, 75
Objective reality, 195, 201
  cross-checking, with social reality, 207
  of external events, 204
Objective stimuli, variations of strength in, 206
Occupational groups, conflicts among, 156
Oligopoly, 504
"On arrival constants and variables", 166
Operational gaming, 4, 453
Operations research, use of models in, 27
Operator, 73
Operators
  derived
    in pattern-recognition program, 59-61
    in GPS, 69
  preprogrammed
    in pattern-recognition program, 59-61
Opinion formation, 156
Oppenheim, political games, 313
Opportunity structure of city, 166
Optimum values, 419, 422
Orcutt *et al.*, 220
Organismic analogies
  in functionalist theories of social systems, 144
  in social thought, 144

Organization of Behavior, The (Hebb), 54
Orientation to source, 177
Oscillating stimuli, 205
Oscillation in verbal learning, 103
Osgood's graduated independent tensions-reduction initiatives, 316
Output norm, 181
Overt behavior, social influence on, 171

Packer, 444
Paired-associate learning, 83, 107, 112
  effects of familiarization on, 114
  of nonsense syllables, 107-9
  stages in, 111
Paired-associate method in verbal learning, 103
Paired-associate paradigm, 104
Paired-associate task in EPAM-III, 105
Palda, 517
Pandemonium, 59
Panic behavior, simulations of, 161
Paradigm
  defined, 53
  development of new, in science, 53
Parameters
  in Aldous, 125
  representation of personality traits by, 121
Pareto's rational mechanics, 144
Park, 301
Participant-type simulations, 246
Partial modeling, overcoming problem of, 301-2
Party affiliation, group influences on, 170
Party dispositions, 203
Party loyalty, 202
  growth of, Figure 2, 197
  of youths, 198
Patrilateral marriage, 147, 149-51
Pattern recognition, 52, 85
  attributes, 54-55
  character recognition devices, 57
  and concept formation, 53
  defined, 54
  procedure, 58
  use of decision tree in, 59
  use of feature identification in, 59
Pelowski
  multinational corporate futures simulation, 300
  Taiwan Straits crisis simulation, 299
PERCEP, 304
Perception, 68
Perceptrons, "multiple layer", 56-57
Performance measurement, 459-60
Performance subroutines in verbal learning, 104
Performance validation, 498
Permanent memory, 125
Personal communication, 156
Personal predispositions in individual voter simulations, 167
Personal preference values, 180
Personality, 117
  computer models of, 126-29
  definitions of, 117-18
  processes, 121
  research, strategies in, 119-20
  simulations, 117-29

Pseudorandom variate generators, 19
Pseudorandom variates, 11
PSW simulation, 300
Psychic profits through social behavior, 181
Psychiatric research, lag in use of computer in, 142
Psychological resistance, 162
Psychological states of simulated group, 156
Psychology, simulation in, 36, 86
Psychometrics in personality research, 119, 127
Public system, 573
Pueblo social organization, 151-52
Pure competition, 504

Quantitative geographers
  influence of, by Karlsson's syntheses, 161
  use of Monte Carlo diffusion simulations by, 158
Question-answering systems, 77-80

Radcliffe-Brown homeostatic mechanisms, 144
Railroad system, simulations of, 575
  Allman's model, 576-77
RAND Corporation, 692
RAND/MIT political exercise, validation of, 697
Random nets, 61
Random variables, use of, in simulation, 10
Raser, 14, 23, 298, 300, 685
Rate of presentation, effect of, on verbal learning, 103
Rational mechanics, 144
"Raw characteristics," 721
Raytheon, 304
Reaction potential, 171-72
REASON MATRIX, 136
"Recent past behavior variables," 166
Receptivity to source, conditions affecting, 175
Recoding behavior of Ss, 109
Recognition in Aldous, 125
Recurrence-relation models, 16
Reference group processes, simulations of, 170-72
Reference groups, influence of, on communication probabilities, 161
REGNANT, 137
REGNANT BELIEF, 136
Regression analysis, 659, 664, 666
Reinforcement theory, 204
Relations among nations simulations
  alliances and involvement of force, 723-27
  communications in crisis and non-crisis, 722
  correspondences of, with "realities," Table 17, 734
  interaction patterns, 727-31
  validity of communications in, 721-23
  validity within, 720-38
Relative deprivation effects, 171-72
Republican Party, 264
Resistance to adopting an innovation, 163-65

Resistance considerations in diffusion simulations, 159
Resistance probabilities in adopting innovations, 160
Resistance threshold, 166
Response familiarization, 113-15
Response generalization, 177
Response similarity, 108
Response surface designs, 664
Response syllable, 105, 109
Response units in perception, 56
Restimulation process, 172
Retail pricing, 475
Retail store game, 542
Retrieved attitudes in Aldous, 125
Reward, value of, in Homans' social exchange theory, 177
Riker's game theory, 246
Risk analysis, 526-30
  of capital investments. *See* Business problems, risk analysis of capital investments
  in research and development planning, 30
Role conflict simulations, 180
Role definitions, complex, 182
Role dilemmas, decision-process in resolving, 180
Roll-call voting in House of Representatives
  communication effect on, 293-94
  communication phase, 270-79; Figure 7, 280; Figure 8, 281
  communication process in, 265
  constituency effect on, 289-90
  memory effect on, 292-93
  party effect on, 286-89
  predisposition phase of, 265-68
  predisposition-communication linkage, 268-69
  region effect on, 290-91
Romes's Leviathan model of hierarchical organization, 173
"Rootless situation," 166
Rosenblatt's "perceptron" approach to visual pattern recognition, 56
Rubber tire market game, 542
Rural society, 156-57

Sales-advertising relationship, simultaneous-equation models of, 517
Sales call problems, 523-25
Sales Force Simulations, 523-25
Sales management games, 542
Sales promotion game, 542
Sales variance, unfavorable, Figure 6, 472
Salesmen, routing of, 523
SALIEN, 722
SANTEX, 155
Saturation, 113
  hypothesis of, 111-12
Saunders, 121
Schelling, 692
Schema, 82
Schrieber's Top Management Decision Game, 455
Schroeder/Streufert Situational Interpretation Test (SIT), 704
Schulz, 103
Schwartz's crisis decision-making simulation, 298

765